Norwich's Maps of Africa

Detail from the frontispiece to *Atlas Novus* by J.B. Homann, Nuremberg: ca. 1707

Norwich's Maps of Africa

An illustrated and annotated carto-bibliography

OSCAR I. NORWICH

Bibliographical descriptions by

PAM KOLBE

Second edition

Revised and edited by

JEFFREY C. STONE
University of Aberdeen

TERRA NOVA PRESS

TERRA NOVA PRESS
G.B. MANASEK, INC.
NORWICH VERMONT
05055-1204 USA

First edition © 1983 I. Norwich
Second edition © 1997 G.B. Manasek, Inc.

First published 1983 as:
Maps of Africa
Second edition published 1997 as:
Norwich's Maps of Africa

ISBN 0-9649000-4-1

The publisher expresses appreciation for the kindness and gracious assistance offered by the Norwich family, especially Mrs. Rose Norwich, without whose enthusiasm and kind cooperation this project could not have been done. Our debt to Ms. Pam Kolbe is acknowledged. After a period of more than a decade she once again became involved with the Norwich Collection and helped make the second edition possible.

Dr. Norwich's dedication in the first edition:

To my wife

CONTENTS

Foreward to the First Edition

Dr I. Norwich of Johannesburg has collected maps, both printed and manuscript, for more than thirty-five years. He has some five hundred and fifty maps in all, embracing Africa, the Holy Land, Europe and the world; of these some four hundred are of Africa only.

No collector seems to be able to obtain all the maps he would like, but Dr Norwich has succeeded in getting most of any importance in his field of collecting, and thereby greatly added to the usefulness of this catalogue. The Norwich collection not only contains maps of the land masses, but also sea-charts. The latter category is on occasion very decorative and of particular appeal to the collector of today. This collection is made up of loose maps and some from a limited number of the author's rare atlases. A few were issued in books of travel or elucidated individual voyages.

Dr Norwich has not only collected maps, he has made a serious study of his own and other people's maps by visiting important public map collections in many countries in Europe and America, as well as making contact with numerous map dealers overseas. He is not only on the mailing lists for map catalogues, but as the dealers know him to be an active collector, he is often offered items before they appear in the sale catalogues.

Dr Norwich has spent the last few years in compiling this illustrated catalogue of his own African map holdings. To his description and notes on the background information relating to each map, he has added an illustration – sometimes in colour. This method will be of great help to other collectors, as words alone are not always enough to distinguish one map from another very similar one. An illustration makes the difference clear at once and answers numerous queries that other collectors inevitably must have.

Although much work on printed maps of the last five centuries has been done during the previous hundred years, there are still many unsolved problems to which the average collector would like to find answers, and Dr Norwich's catalogue will undoubtedly help. To make it easier to locate a particular map in this catalogue Dr Norwich has supplied an alphabetic index which includes all the personal names appearing on each map.

Maps can often be so attractive, that the collector is inclined to forget that they were primarily intended for use in most cases. They are based on the general geographic knowledge of the time and consequently can be of great interest as they reflect the state of knowledge at a specific date.

Dr Norwich's catalogue is a welcome addition to books on maps, and it is hoped that it will prove of interest to other collectors both of maps and cartographic bibliographies.

Anna Smith, 1983

Preface to the First Edition

After some thirty-six years' serious collecting of old maps, particularly of Africa and its parts, I was stimulated to publish this volume for a number of reasons: to provide an illustrated guide for other map collectors, and to produce a book which, by virtue of its contents' historical interest as well as beauty, would give pleasure and instruction to absolutely every reader, whether their interest lie in works of art in general, in Africa's history and geography, or in early exploration and discoveries.

It was clear from the start that to do justice to the maps – to highlight their true historical worth as well as their more obvious esthetic qualities – they should be thoroughly researched. It is often extremely difficult to establish a map's authenticity and authorship, its date, engraver and publisher, to say nothing of the provenance of the information on which it was based. To achieve this, many standard works on cartography, old atlases (whether original or in more recent facsimile copies), learned articles from cartographic studies and other relevant travel books and journals were essential. Detailed descriptive catalogues from reputable map dealers were also found to provide much invaluable information. In my early period of research, my wife and I travelled extensively, especially in the United Kingdom, Europe and the United States, where personal contacts made with libraries, museums and map dealers proved to be even more advantageous.

It has been stated that maps provide a pictorial history with the greatest economy of space, and I soon became aware of the vast amount of information they could yield on the history of the continent of Africa, and on the progress of discovery and development of the southern African subcontinent and of the Cape of Good Hope in particular. This pictorial history proved also to be of great ethnological interest, as many maps include illustrations or comments on the indigenous population together with details of their appearance, dress and customs.

Map collecting has a long history. Although the majority of early maps and charts were designed for practical use, it is also known that the wealthy merchants of Holland and the Low Countries, where map publishing excelled in the eighteenth century, were avid collectors of highly decorative maps, which they used to decorate their homes and offices, and kept, like other works of art, as an investment.

Both black and white and originally coloured maps stand out as fine examples of the art of some of the greatest masters in the techniques of wood engraving, the intaglio process of metal and copper engraving and the use of lithography. The schools of Albrecht Dürer and Hans Holbein, together with innumerable other artists and engravers, are recorded to have given their assistance in creating the colourful cartouches, figures, sailing vessels and land and sea-animals so often seen on old maps. In conjunction with these artists, a school of skilled professional colourists soon revealed their prowess in producing hand-coloured maps. The Italian schools of engraving, originally initiated by the publisher Giacomo Gastaldi, followed by Forlani, Camocio, Duchetti and others, are particularly renowned for their characteristic beautiful copperplate line engravings and their particularly fine calligraphy, which has not been excelled.

The maps in this volume are all individually illustrated, each with its own cartobibliography. In the bibliographies parentheses have been used to indicate information which does not appear on the map itself. The name of the person to whom the map is attributed is followed by the map's title, its translation (in brackets), its engraver, place of publication, publisher and date. The map's measurements follow, together with its other identifying features. The maps have been subdivided arbitrarily into those of the continent of Africa; of southern Africa; sea charts; the Cape of Good Hope; North, East and West Africa; islands, ports and town plans. The earliest map included is the incunabulum of Ptolemaeus-Ulm (in its second edition of 1486), and the latest is a surveyor's plan (regarded as a map) of the early town of Johannesburg in December 1886, only three months after its proclamation as the centre of the gold fields discovered in September that year.

Included in the text is a description of the land of Ophir, that mythical kingdom described in James Bruce's map of East Africa, taken from his five-volume account of his *Travels to discover the source of the Nile from 1768 to 1773*. In view of the repeated appearance of 'Monomotapa' (as a kingdom and as a city) in early maps of South and Central Africa, some research on this has also been included. (See Maps 307 & 310.)

Because of the proliferation of names – engravers, designers, printers and publishers, etc. – on maps, I have long looked forward to the opportunity of compiling an index cross-referencing names and the maps on which they appear. Complications of terminology can further obscure the identification of maps: an engraver can be denoted by the word (sometimes abbreviated) – *sculpsit* or *fecit* or, more rarely, *caelavit* or *incedit*. For references to printers or publishers the following words or their abbreviations are also used: *excudit, apud, formis, sumptibus* and *ex officiana*. For the name of an author or cartographer, use is made of such words as a*uctore, delineavit, descripsit* and *inventit*. If this name is not included in the title or the cartouche, it is often placed in the lower left border (where the engraver often appears), but there are no hard and fast rules. Hopefully the abovementioned index will facilitate the identification of many problem maps.

ACKNOWLEDGEMENTS

It is fitting that I should make special mention of Dr Anna Smith, formerly Director of the Africana Museum and Librarian of the Johannesburg Public Library. Since the early 1950s she has been a friend and my early mentor and adviser on old maps of Africa. She is, of course, well known in cartographical spheres, both in South Africa and internationally, for her descriptive catalogue *Exhibition of Decorative Maps of Africa up to 1800* (1952), a book that is eagerly sought after and used by collectors and map dealers. Dr Smith has always been available to offer sound advice and assistance with unlimited references to cartographic riddles, especially of southern Africa and the Cape of Good Hope. To her I am indebted for her advice and guidance, which she always offers with her usual modesty.

I have been fortunate to obtain the services of an experienced, well-qualified librarian, Pam Kolbe, who, although a co-author of this volume, must be thanked for her expertise and assistance with the cartobibliography of each and every map, carried out intelligently and diligently. She herself will unhesitatingly state that she has learnt a great deal about old maps of Africa. In times of certain difficulties Pam was equally proficient in assisting with the much-needed typing of the text.

I must acknowledge the optimistic advice of my good friend Professor Reuben Musiker, Professor of Librarianship at the University of the Witwatersrand, who succeeded in acting as a catalyst in the genesis of this book. His technical assistance in the layout of its contents has been of inestimable value to me.

I am indebted to many institutions and libraries, both in South Africa and in many overseas countries, especially the United Kingdom and the United States of America, for the information they so willingly provided in researching the provenance of maps. I refer particularly to the staff of the Africana (Strange) Library and to the Librarian of the Johannesburg Public Library for permission to photograph those maps in the early travel volumes not in my collection. I wish to thank particularly the staff of the Africana Department of the Library of the University of the Witwatersrand for similar kind assistance. I am grateful to the John Carter Brown Library, Brown University, for permission to reproduce the Wilczek-Brown map. I am greatly indebted to Dr Helen Wallis, Director of the Map Library of the British Library, and to her professional staff, for their kind co-operation and assistance in tracing cartographical and historical information on the lesser-known maps, and especially for bringing the Rotz atlas to my notice. I am grateful for permission to reproduce Rotz's maps in this book.

For the photographic illustrations of each map I am sincerely grateful to Nat Cowan, F.R.P.S., formerly Custodian of the Bensusan Photographic Museum, and a doyen of photographic spheres in Johannesburg.

To Miss A.A. Holloway I am indeed grateful for her remarkable proficiency in shorthand and typing and for her assistance in what to her was at first a truly somewhat varied English language.

Wendy van Schalkwyk's skilful typesetting made proof-reading a real pleasure and Frances Perryer, Ad. Donker's editor, was a most valuable asset in the final editing of the manuscript – I am truly grateful for her conscientious and wise assistance. To my recently acquired friends, Mr and Mrs Ad. Donker, who whole-heartedly and with enthusiasm entered into this project, I am indebted for their publication of this book.

Lastly, I am grateful to my wife, Rose, to whom this book is dedicated, for her affectionate companionship as well as her understanding and appreciation of the acquisitive syndrome of antique map collecting.

Oscar I. Norwich, July 1983.

SOURCES CONSULTED

Africana Notes and News. Johannesburg, 17(4) Dec. 1966 ; 22(2) Jun. 1976 ,22(5) Mar. 1977.

Bagrow, L. *History of Cartography*. London: Watts, 1964.

Beach, D.N. *The Shona and Zimbabwe, 900-1850*. London: Heinemann, 1980.

Brown, A. *The Story of Maps*. London: Cresset Press, 1951.

Cartwright, J.F. *A Bibliography of Maps of Africa and Southern Africa in Printed Books, 1750-1856*. Cape Town: University of Cape Town Libraries, 1976.

Cartwright, M.F. *A Bibliography of Maps of Africa and Southern Africa in Printed Books, 1550-1750*. Cape Town: University of Cape Town Libraries, 1976.

Map Collectors' Series. London: Map Collectors' Circle.

- No. 6 (1963). *Early Maps and Views of the Cape of Good Hope*, by R.V. Tooley.

- No. 17 (1965). *The Cape of Good Hope. 1782-1842*, by D. Schrire

- Nos. 29-30 (1966). *Printed Maps of Africa, 1500-1600*, by R.V. Tooley. 2 vols.

- Nos. 47-48 (1968). *Maps of Africa: a Selection of Printed Maps from the Sixteenth to Nineteenth Centuries.* 2 vols.

- No. 61 (1970). *Printed Map of Southern Africa and its Parts*, by R.V. Tooley.

- No.82 (1972). *A Sequence of Maps of Africa*, by R.V. Tooley.

Monumenta Ethnographia, - Frähe zölkerkundliche Bilddokumenta - volume 1: Black Africa. Akademische Druk u. Verlag Sanstalt, Graz/Austria, 1962-67.

Nordenskiöld, A.E. *Facsimile Atlas*. Stockholm, 1889. Reprint New York: Kraus, 1961.

Rotz, Jean, *The Maps and Text of the Boke of Idrography, presented by J. Rotz to Henry VIII;* ed. by H. Wallis. Oxford: Roxburghe Club, 1981.

Sanuto, L. *Geografia dell 'Africa*. Venice, 1588.

Skelton, R.A. *Decorative Printed Maps of the 15th to 18th Centuries*, London: Staples Press, 1952

Smith, A.H. *Exhibition of Decorative Maps of Africa up to 1800: Descripive Catalogue*. Johannesburg: Public Library, 1952.

Theatrum Orbis Terrarum: Series of Atlases in Facsimile. Amsterdam: N. Israel.

- No. 1 (1963) *Cosmographia* by C. Ptolemaeus. Ulm, 1482.

- No. 2 (1968) *Cosmographei* by S. Munster. 1550.

- No. 3 (1969) *Libro Dei Globi* by V.M. Coronelli. 1701.

- No. 4 (1965) *Civitates Orbis Terrarum* by G. Braun and F. Hogenberg. 1572-1618.

- No.5 (1964) *Theatrum Orbis Terrarum* by A. Ortelius. 1570.

- No. 6 (1964) *The Light of Navigation* by W.J. Blaeu. 1612.

- No. 7 (1965) *Thresoor der Zeevaert* by L.J. Waghenaar. 1592

- No. 8 (1973) *The English Pilot*, Part V by J. Seller and C. Price. 1701.

Tooley, R.V. *Collectors' Guide to the Maps of the African Continent and Southern Africa*. London: Carta Press, 1969.

Tooley, R.V. *Dictionary of Map Makers*. London: Map Collectors' Publications, 1979.

Tooley, R.V. and Bricker, C. *History of Cartography*. London: Thames and Hudson, 1969.

The World Encompassed: An Exhibition of the History of Maps, October 7th to November 23rd, 1952. Baltimore: Trustees of the Walters Art Gallery.

Preface to the Second Edition

The late Dr Oscar I Norwich (1910-1994) compiled what is probably the finest private collection of early maps of Africa, including many rare items. What he published in 1983 was, with very few exceptions, a permanent record of that very extensive collection, together with the results of his own literature and library searches into the provenance of the maps in his collection. This second edition continues to be the record of a superb private collection. It is the sheer extent of that collection, together with the associated bibliographical descriptions and the inclusion of almost 400 illustrations, which elevates the record of a private collection into a work of reference.

The book is sub-titled a "carto-bibliography" and a valuable feature of the first edition is the very professional bibliography of every illustrated map. However, it was never the author's intention to compile a complete carto-bibliography of early maps of Africa. Dr Norwich's collection is extensive but no private collection and indeed no public collection can hope to approach completion. A more comprehensive carto-bibliography of early maps of Africa would necessitate an examination of the records of all of the major map repositories the world over. But a bibliographical exercise of that sort is not only beyond the scope of this work, it pre-supposes a quite different objective.

In revising the first edition, I was concerned to retain the integrity of the collection, and specifically to respect the insights of the compiler into his collection, particularly in relation to southern Africa, where Dr Norwich's personal knowledge is so clearly apparent. Hence, few new maps have been added in this edition to the three hundred and forty-five which were originally selected for individual commentary by Dr Norwich. Nevertheless, new research has been published since Dr Norwich completed his "Historical Survey of Maps of Africa" for the first edition. His introductory chapter has therefore been replaced by a new summary over-view of the history of African maps so that the collection is now presented in the context of the most recent knowledge of maps of Africa. There are new illustrations in this section, relevant to the expanded subject matter and taken from my personal library. A substantial bibliography of African cartography has been added, to supplement the short list of sources which Dr Norwich originally consulted. A list of linen-backed folding maps of Africa has also been appended. These maps are important in the more recent mapping of Africa and are but little-known. Many of the individual descriptions of the maps in the collection have not been altered substantially. They continue to reflect the author's own personal interests, knowledge and pleasure in his collection. With few exception, notably map number 345, the descriptions have been revised only where errors were noted, whether typographical errors, errors of omission or more seriously, errors of fact. In this context, careful note has been taken of the comments of reviewers of the first edition.

I am grateful for the assistance of Mr Francis Herbert, Professor R.C. Bridges, Mr Thomas Kloeti, Ms Alice Hudson, Mr Rudolph J. H. Lietz, Mr Gerry Levin, Mr Francis J. Manasek and Mrs Valerie G. Scott.

JCS

MAPS OF AFRICA: A SUMMARY HISTORY

In looking for the oldest maps of Africa, we should turn to the peoples of the continent itself. After all, Africa is associated with the origins of the human species and the earliest known maps of any part of Africa are archaeological artefacts, in the form of Neolithic rock paintings in southern Algeria. However, our knowledge of early map making within indigenous African cultures is rudimentary. Evidence of the compilation of maps or plans in Pharonic Egypt is often quoted but surprisingly modest. There seems to be no evidence of the use of maps, plans or other cartographic devices among the later Bantu-speaking societies of central, east and southern Africa, or among the sub-Saharan states of west Africa, although schematic diagrams are recorded from eighteenth-century Ethiopia. Perhaps the maps of non-literate societies in Africa were so ephemeral that none

have survived. A sketch map drawn in the sand may have been the characteristic means of communicating mental images of the relative location of places, but it would not last very long. Alternatively, perhaps we have not yet recognised amongst the surviving artefacts of past societies in Africa what are, in the broadest sense, maps. Certainly, the evidence of recorded maps and plans in libraries, museums and private collections implies that the history of African cartography is a history of the interaction of external and internal cultures and societies, with the actual production of the documents taking place outside the continent. The written history of the cartography of Africa is a peculiarly distorted history. It is the outsider's view of Africa, primarily a Eurocentric point of view with Islamic inputs, in which the internal African contribution is not always adequately identified.

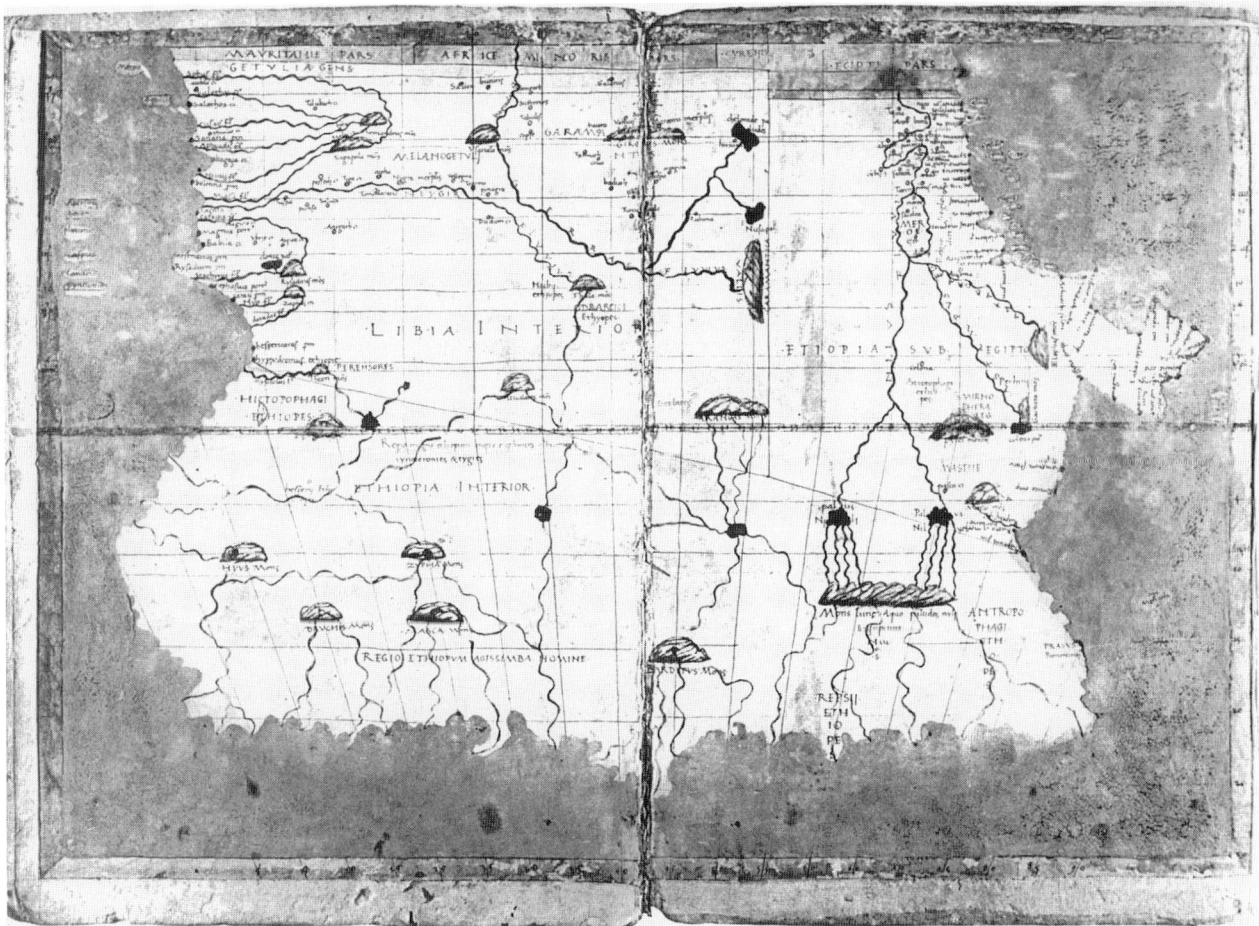

Fig. I Ptolemy-Wilczek Brown (c. 1450). Original in John Carter Brown Library, Brown University, Providence, USA.

Correction of the distortion presents a challenge to historians of cartography, but in the meantime we can only summarise what is currently known.

EARLIEST ORIGINS

The earliest printed map (Fig. VI) of the whole of Africa is a woodcut map without title which is found at the first opening of a rare book entitled *Itinerarium Portugallensium e Lusitania in Indiam*, published in Milan in 1508. The map is in the Ptolemaic mould, that is, the outline of the continent and the depiction of the interior accord with the late medieval reconstructions of the work of Claudius Ptolemy, who lived on the north shores of the African continent during the second century AD, at a time when Greek theory and Roman practice came together to produce remarkable advances in cartography, including the depiction of Africa.

Ptolemy was the heir to centuries of Greco-Roman discourse about Africa, going back to about 440 BC and to Herodotus, who is a prime source in the controversy over the supposed circumnavigation of Africa by Phoenicians as early as 600 BC. The significance of any such voyage is, of course, that it would have demonstrated that Africa is a peninsula, delimiting but not separating the eastern and western parts of the great southern ocean,

an important realisation in describing the shape of the continent. This was the view of Africa carried forward successively, with additional information by Eratosthenes, Strabo, Mela and Pliny, although it may have been Ptolemy who disseminated a different depiction of southern Africa, in the form of an eastward extension of the southern-most part of the continent, almost enclosing the Indian Ocean. Alternatively, this may derive from later Arab sources.

Ptolemy's more enduring legacy, which is apparent in the printed map of 1508 mentioned earlier, is his depiction of the headwaters of the Nile. There are several schools of thought in Greek and Roman writing about the source of the Nile, and Ptolemy may have chosen to rely on the account by Marinus of Tyre or the Greek trader Diogenes, who is said to have landed on the coast of East Africa in about 50 AD, where he learned that the source of the Nile lay in the lakes of East Africa. Ptolemy depicted the Nile as rising south of the Equator, from two lakes that are fed in turn by the streams from the Mountains of the Moon. This has given rise to further debate about whether the Mountains in fact represent the mountains of Ethiopia, substantially misplaced, but in reality, the source of the Blue Nile.

Herodotus seemed to favour an alternative source of the Nile, far to the west in sub-Saharan West Africa, and his interpretation is the origin of a second inland feature which was carried forward from Greek scholarship,

Fig. II Martellus (1489). Original in British Library.

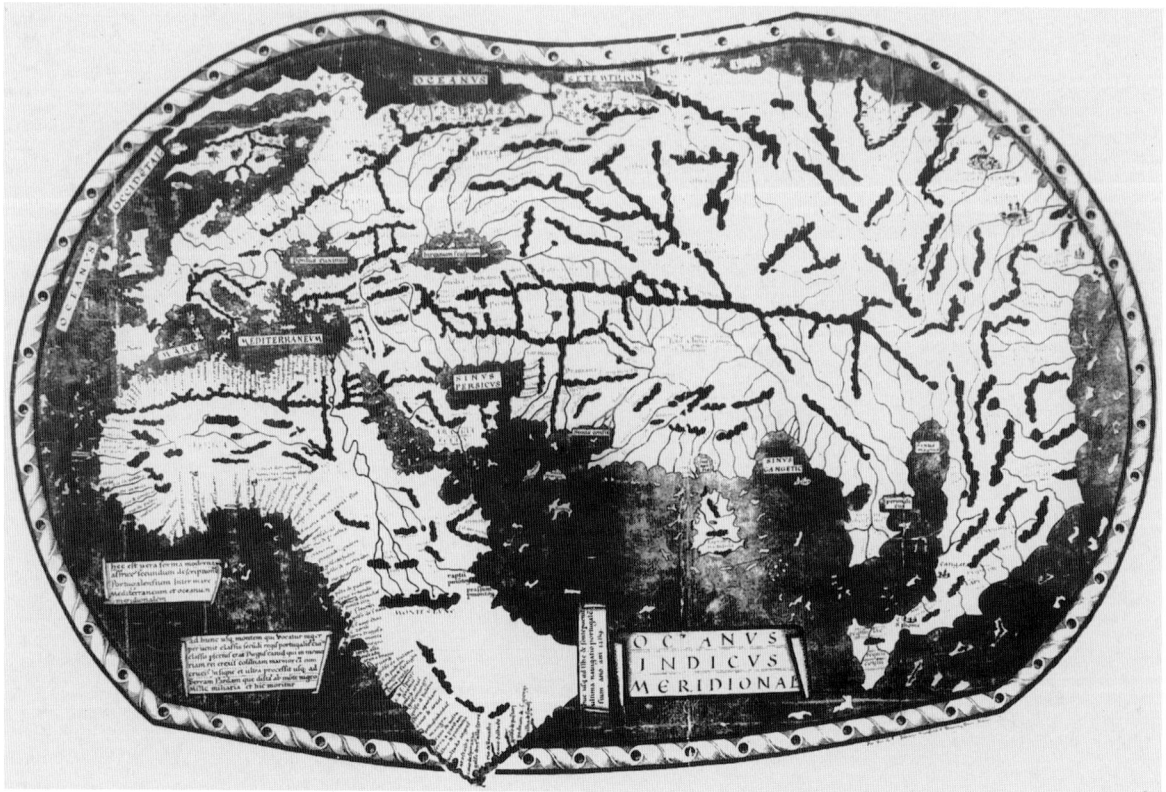

xiv

namely, a major river aligned latitudinally across West Africa. However, Ptolemy's source of the Nile allowed the retention of this river, but he reversed its direction and showed it as flowing westward into the Atlantic Ocean. This was an error that is apparent on the printed map of 1508 and which persisted into the nineteenth century.

It is surprising that the diffusion of Islam and therefore of literacy throughout much of the western Sudan by the end of the thirteenth century seemingly did not result in a legacy of Arabic cartography in Africa. However, in sub-Saharan Africa, Islam was grafted on to pre-existing societies, rather than subsuming them. Arab geographies were mainly concerned with North Africa, but the east coast of Africa was known and recorded by the Arabs. It is shown in a sixteenth-century Ottoman maritime atlas. However, Arab navigational techniques did not require charts based on rhumb lines or grids of latitude and longitude, so that there were no pre-Portuguese navigational charts of the Indian Ocean that might have been expected to provide source material for

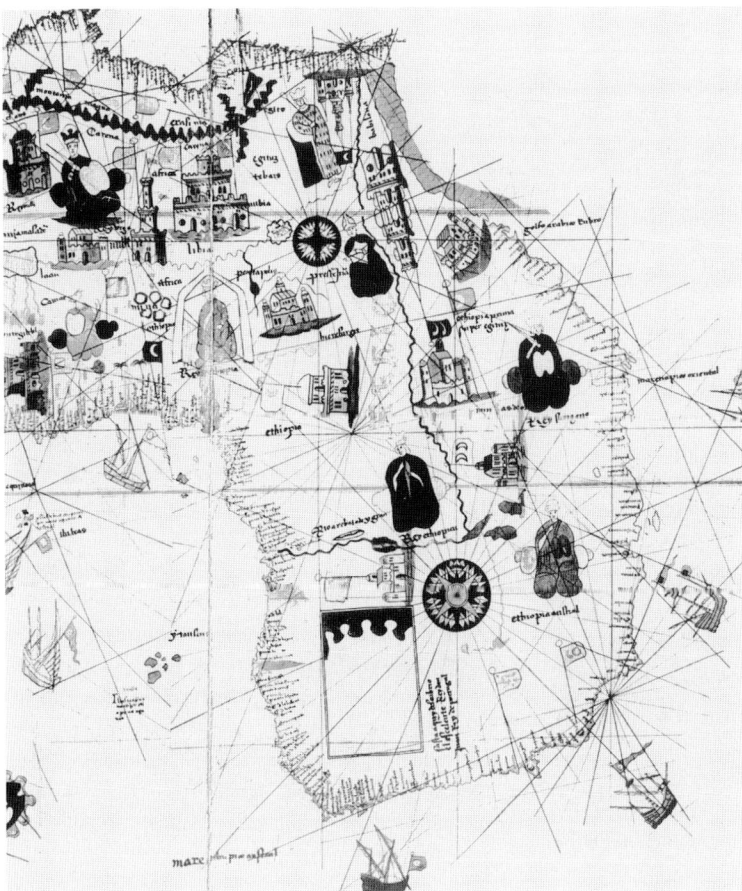

Fig. III Juan de la Cosa (1500). Original in Naval Museum, Madrid.

improvements in the depiction of the coastline in early maps of East Africa. Nevertheless, Arab travellers compiled their own pilots and recorded coastal settlements as far south as the Limpopo River. The eleventh-century scholar, al-Biruni corrected the convention of an eastward extension of Africa south of the Equator. Also by the eleventh century, Idrisi had mapped the East African coast. Inland, travellers such as Ibn Battuta seem to have had little use for maps, but from the twelfth to the fourteenth centuries, written records were compiled by Arab travellers into the interior. These records were translated much later, mostly into French, eventually becoming rather unreliable sources for nineteenth-century European cartographers.

By the end of the fifteenth century, knowledge of Africa had reached China and was incorporated into their concept of a Sinocentric world, but no major changes to the map of Africa are credited to the Chinese, who may have utilised Arab sources in part.

AFRICA AND RENAISSANCE EUROPE

It was the charting of the African coastline from the fourteenth century, firstly by the Majorcans of the Catalan school and then by the Portuguese, which delineated the coastal outline of Africa in portolan charts, an outline which was to be copied in the seventeenth century by more northerly European schools of map making. Hence, the west coast of Africa began to take shape in the Catalan world map at Modena of c.1448. In the fifteenth century, the Portuguese first opened up trade with the Guinea coast. Then, as a distinctly separate exercise, for which the exploration of the West African coast paved the way, they initiated direct maritime trade with India. The world map of 1489 by Henricus Martellus Germanus demonstrates how the rounding of the Cape of Good Hope by Bartolomeu Dias had an immediate impact on the map of Southern Africa.

The Cantino planisphere of 1502 is a particularly important example of Portuguese cartography, because it incorporates surveys of the African coasts extending from the Straits of Gibraltar to the Red Sea. The voyages of Dias and Vasco da Gama reformed the Ptolemaic shape of Africa. Over a much longer period, from the time of Prince Henry of Portugal to the later seventeenth-century work of the Teixeira family, charts of the African coastline were an increasingly accurate and

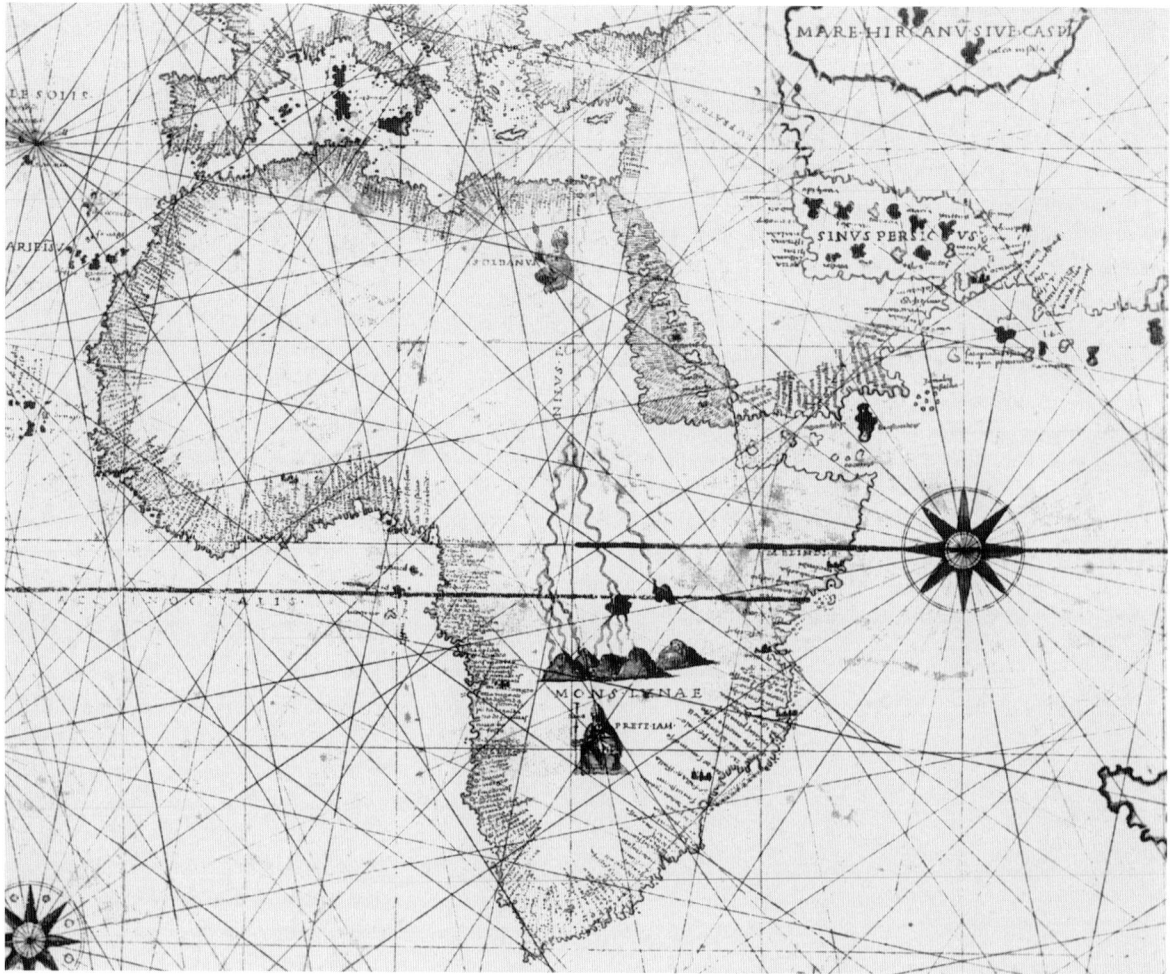

Fig. IV King-Hamy map (ca. 1504). Original in Henry E. Huntington Library, San Martino, California, USA.

detailed source for late seventeenth-century map makers in Holland, France and England.

Maps of Africa from the sixteenth to the eighteenth centuries were principally the work of Italian, Dutch, French, English and German publishers. However, separate schools of thought on the depiction of Africa did not coincide with the nationality of the publishers. Once published a map could be copied, and certainly in the latter part of the seventeenth century, maps of Africa were heavily influenced by the work of earlier map makers from different schools. It is essential to be aware of these trans-national influences in order to understand the shape and content of the late-seventeenth-century map of Africa.

In the sixteenth century, it was the Venetians who began to make significant changes to the way that the continent was depicted. Ramusio's *Delle Navigationi et Viaggi ... Africa*, published in 1550, was initially a great compendium of text on Africa which included *Descrittione dell'Africa* written by the Moroccan traveller al-Hasan (Leo Africanus) in 1526, whilst in captivity in Rome. Later

editions included a well-known map of the African continent with south at the top. Detail inland is widespread but sparse, but in 1564, Giacomo Gastaldi published his influential eight–sheet map of the continent, with its improved coastal outline and increased content inland. Then at an unknown date prior to 1576, Livio Sanuto drafted the map of Africa which was to be published in his *Geografia ... dell'Africa* in 1588. Sanuto's Africa is no mere copy of the much larger scale map by Gastaldi. Like Gastaldi, he also tackled the difficult task of translating imprecise descriptive statements in words to locations on maps. The interior of Sanuto's Africa differs from that of Gastaldi, particularly in the rivers of West Africa. Hence, the Venetians were independent of each other to some degree in their interpretation of the evidence, but they became the authorities on the interior of the continent in the mid-sixteenth century. Apart from the location of Madagascar, the transformation which the Venetians initiated was inland where the names of peoples and places began to appear. This was a break with the Greco-Roman tradition, represented for example in maps

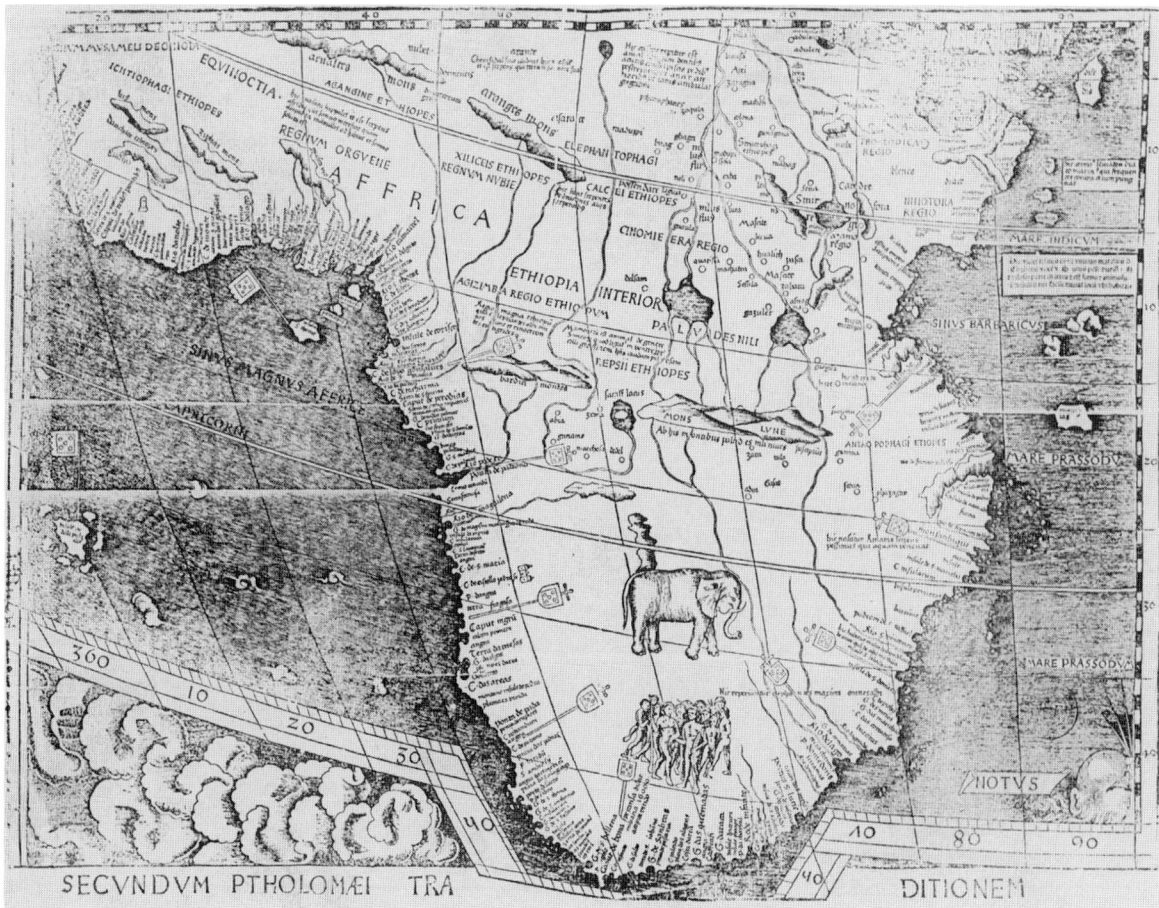

Fig. V Waldseemüller (1507). Original in Library of Wolfegg Castle, Germany.

of Africa by Martin Waldseemüller (fig. V) in 1507, Gaius Julius Solinius in 1538, and Sebastian Münster in 1540, in which the sub-Saharan interior is almost blank, apart from the Nile.

Cartographers who exerted an influence on the way inland areas were shown on late-sixteenth-century maps of Africa are widely held to include Gastaldi, but other influential forerunners include Abraham Ortelius's *Africae tabula nova* of 1570, Petrus Plancius's *Delineatio orarum Manicongo, Angolae, Monomotapae...* of 1592, Gerhard Mercator's *Africa* of 1595, and Henricus Hondius's *Africae nova tabula* of 1606. Gastaldi's *Africa* on eight sheets is rare today and was probably not widely circulated. However, the inland place names on the map by Ortelius are nevertheless closely derived from Gastaldi, so that the much wider circulation of Ortelius's work ensured that the Gastaldi image of inland Africa was similarly widely circulated.

Ortelius is sometimes seen as having established a standard pattern of rivers and lakes for inland Africa. Based on Gastaldi, this pattern was influential into the seventeenth century. However, regional maps of Africa by Ortelius do in fact contain significant differences (e.g. in the source of the 'Manicongo' River). Whereas the Gastaldi pattern reinforces Ptolemaic authority for sources of the Nile in twin lakes to the south of the Equator, named 'Zaflan' and 'Zaire' or 'Zembre,' it adds southeasterly flowing streams such as the 'Zuama' and the 'Spirito Sancto.' It also has the Niger flowing north from an equatorial lake before turning west into a Lake 'Bornu' (approximating to Lake Chad) and continuing towards Cape Verde and the Atlantic Ocean. Rather than conforming totally to the Gastaldi pattern, Ortelius's drainage pattern is a variant. It is one of a number of contemporary interpretations.

Gerhard Mercator's depiction of the rivers of Africa was another such variant and also differs from that of Ortelius. His Manicongo River is connected to the streams flowing southeast to the Indian Ocean, whilst his headwaters of the Nile are more widely distributed and the orientation of his Niger much more variable. Plancius loosened the connection between the headwaters of the Nile and the southeasterly flowing rivers and emphasised their easterly orientation, a form acceptable to Hondius, who otherwise followed the Gastaldi pattern for the Nile, Niger and Zaire, but with the addition of a major drainage line in the central Sahara. The late sixteenth century therefore contained a wider range of interpretations of

Fig. VI Montalboddo Fracan (1508). Collection of Henry C. Taylor.

the available evidence for inland areas than it did for the external shape of the continent, despite some differences in the degree of pointedness of the southernmost part of the continental outline.

Recent work has shown that the late sixteenth century also saw a much more radical reinterpretation of the interior, first published cartographically in Filippo Pigafetta's *Tavola generale dell' Africa* in 1591. In this map, an equally hypothetical but very different view of the source of the Nile derives from the account of Duarte Lopes, a Portuguese traveller who had spent five years in the Congo Basin. Later cartographers were reluctant to adopt the Lopes-Pigafetta model, in which two great lakes in the interior are placed, not on the same latitude in accordance with Ptolemaic tradition, but on the same meridian.

By the mid-seventeenth century, something approaching a concensus was emerging. The seventeenth century was, of course, dominated by the Dutch school of map making; cartographers associated with that school who chose Africa among their subject areas in the second half of the seventeenth century include Willem Blaeu, Jacob van Meurs, Frederick de Wit, Gerard Valk, Caroli Allard, Nicolas Visscher, Justus Danckerts and Joan de Ram. However, from a glance at their single-sheet maps

of the entire continent, it is apparent that continental interiors are now closely comparable. This was perhaps a consequence of plagiarization, but if that were the case, then the reason may have been a lack of new sources of information about the interior, by comparison with the ferment of the century or so earlier when Venice had become something of a clearinghouse for geographical knowledge. The standard format for the continental-scale map of Africa is exemplified by Blaeu's *Africae nova descriptio*, dating from about 1620 and published in 1630, but reprinted on at least fourteen occasions up to 1667. This meant that it was widely available in atlases with texts in several languages, making it a likely subject for the copyist. There has been little detailed scrutiny of the maps of Africa of this period and the pedigree of Blaeu's *Africae* is not established, but it is possibly comparable with his wall map of 1608 entitled *Nova et Accurata totius Africae Tabula*, which in turn resembles Hondius's *Africae Nova Tabula* of two years earlier.

A second cartographic model for seventeenth-century Dutch maps showing the whole of Africa has, however, been identified. The differences are the more pointed Cape in the south, the latitudinal orientation of the Congo River and the removal of islands from Lakes Zaire-Zembre and Zaflan. This pattern has been traced back

xviii

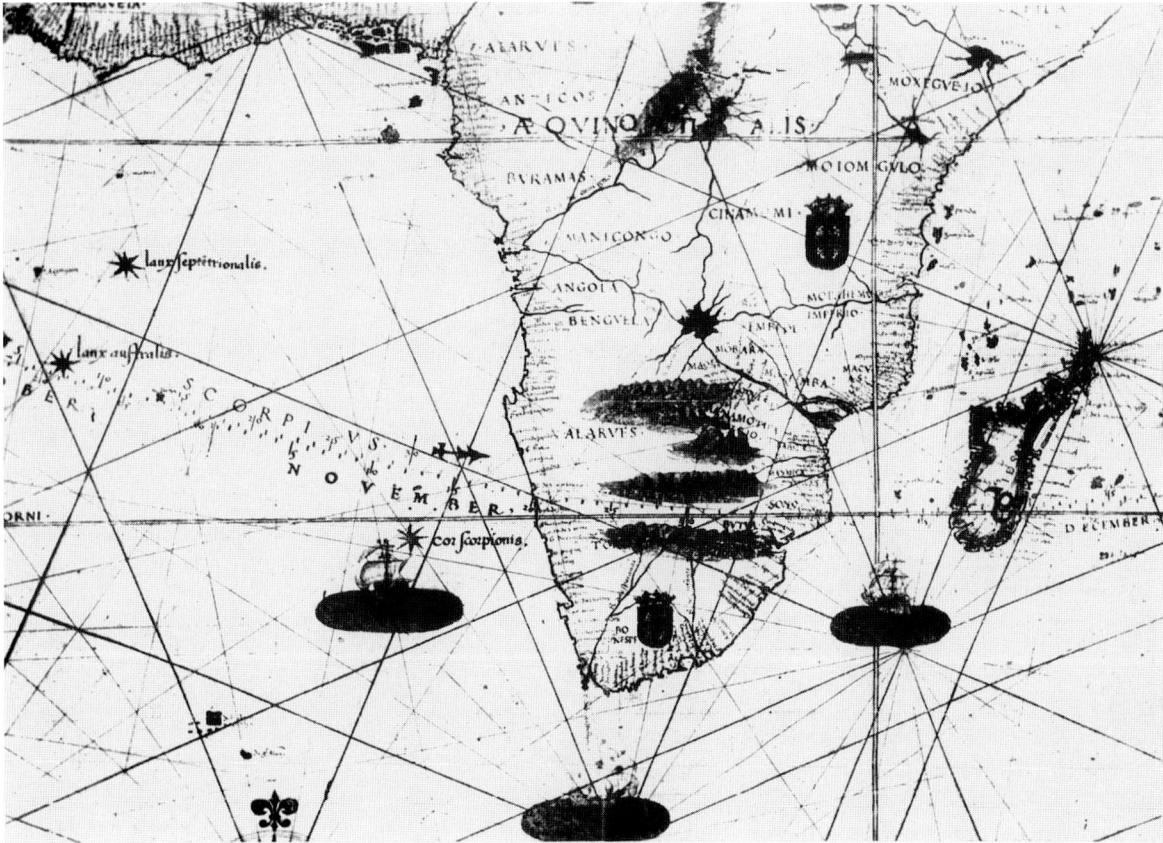

Fig. VII Bartholemeu Velho (1561). Original in Naval Museum, Spezia, Italy.

to Petrus Kaerius's wall map of 1614 entitled *Africae nova descr.*, and then through Visscher (*Africae nova descr.*, 1636), Joan Jansson (*Africae accurate tabula ex officina*, 1662) and de Wit (*Nova Africae descriptio*, 1660). The characteristics of these maps, in fact, have an even earlier origin in the second of two different maps produced by Hondius in 1606. However, given the small magnitude of the differences between the two cartographic models and given their common origins, perhaps they should be thought of as variants on a common theme, rather than two separate cartographic models of Africa. Apart from these two closely related strands in the depiction of Africa at the continental scale, there is remarkable similarity in the late-seventeenth-century Dutch maps of Africa.

Turning to larger scale seventeenth-century regional maps of parts of the African continent, adherence to a standard pattern is no longer apparent. For small scale maps of the continent as a whole, it seems that it was easier to copy existing maps. The decision to compile regional maps necessitated the search for new information, and it is indeed at the larger scales that a modest amount of change occurred in the map of Africa during the same period.

The Dutch publishing house of Blaeu was predominant in seventeenth-century regional mapping of Africa. Blaeu included four regional maps among the five maps of Africa in his *Novus Atlas* of 1635. However, at least two of these four were derived from Hondius, including *Fezzae et Marocchi, Aethiopia Superior* and also the inland areas but not the coastline of *Aethiopia Inferior*. *Aethiopia Inferior* of 1635 was something of a standard depiction of southern Africa during the seventeenth century and frequently copied, although the inland content of this larger scale regional map derives from the smaller scale coverage of the area in Hondius's map of all Africa. Seven regional maps were added in volume ten of Blaeu's *Atlas Major* in 1662, although again not all of their content was new. For example, for the area of the Nile delta, where most of the detail is to be found in his *Egypti*, Blaeu relies heavily on Hondius.

Although the map of Africa at the regional scale did evolve, not all regional maps contained significantly more information than continental-scale maps, or indeed than their predecessors at the regional scale. Change was desultory. Three of Blaeu's seven new maps of 1662, *Nigritaru, Regio* and *Regna Congo et Angola* included more information in their areas than previously available. In

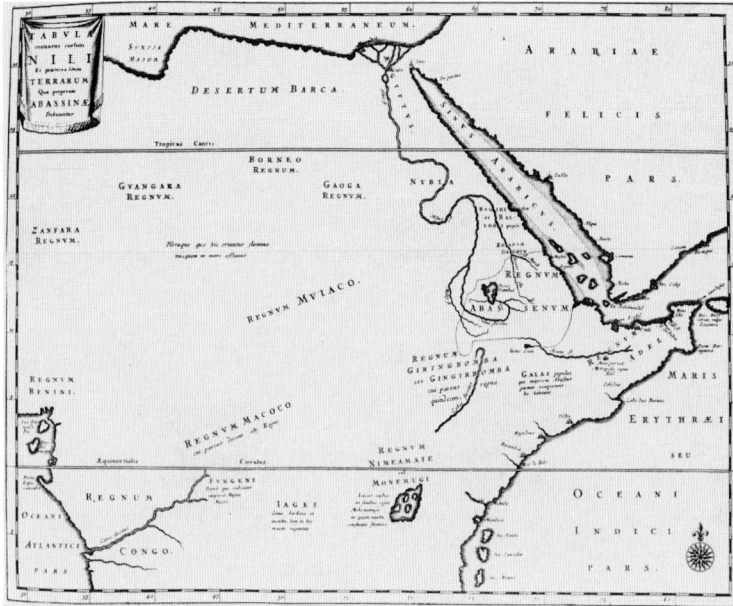

Fig. VIII Isaac Vossius (1666), From *Monumenta Cartographica: Africa et Egypti* by Youssouf Kamal, Cairo, 1951.

addition, *Barbaria* and *Guinea* contained major revision by comparison with Hondius. However, these alterations were to names of places and to detail. Only in the Congo area is the basic hydrological pattern reformulated in regional maps by comparison with the standard small scale pattern of the long-standing *Africae nova descriptio*.

Later seventeenth-century Dutch cartography relied to a large extent on Blaeu. For example, Nicolas Visscher II copied Blaeu in c.1689, in compiling his regional maps of *Barbaria* and *Nigritarum Regnum*, although on the other hand, Allard substantially reduced the extent of the upper Nile catchment and improved the alignment of the Zambezi in his *Novissima et Perfectissima Africae Descriptio*, a map at the continental scale included in Visscher's *Atlas Minor* of 1689-1698. The establishment of a settlement at the Cape of Good Hope by the Dutch in 1652 made little impact on the map of southern Africa since penetration inland was delayed, but with Dutch map makers so productive, it did eventually result in a few large scale maps of the extreme southwest of the continent, for example by Joachim Ottens, thus adding to the range of scales in the seventeenth-century mapping of Africa.

Turning aside from the Dutch school, French, English and Italian map makers also depicted Africa in the second half of the seventeenth century. However, the contents of their maps were often not significantly different from the standard Dutch format at the continental scale, and their regional maps were often also derivative. Occasionally they had access to original information about the interior of Africa, but otherwise they relied on the same textual sources as the Dutch, or more likely, on

published Dutch maps. They can be recognised as different schools on the basis of design but not on the basis of map content. A notable exception was Nicolas Sanson and his successors, who compiled both continental and regional maps. His regional maps of North Africa, Guinea and Madagascar differ substantially in coverage, outline and detail from his Dutch predecessors. The map of *'Afrique...corrigée et changée en plusieurs endroits...'* of 1669 by 'Sanson le Fils' also incorporates original detail.

The English map makers of this period to map Africa included Henry Seile, Philip Chetwind, Henry Overton, Richard Blome, William Berry and John Ogilby, but none made significant original contributions. Italian map makers included Joanne Nicolosia and more notably Vicenzo Coronelli, globe maker as well as map maker, whose best known map of Africa was printed in 1688 without title, in the form of six large gores intended for a globe, with later single and two-sheet editions. By comparison with his predecessors, he achieved an improved orientation of the Zambezi River. A German contribution is to be found in Hiob Ludolf's *Habessinia seu Abassia, Presbyteri Johannis regio* of 1683, which did provide a greatly improved depiction of Ethiopia from original sources.

A persistent characteristic of the cartographic evolution of Africa is the disparity between the accuracy and detail of the coast (with the notable exception of the southeast, where the orientation of the coast remained incorrect until the eighteenth century) and the paucity of precisely locatable information inland. From the European viewpoint, Africa was no more than a barrier on the route to Asia, the reason why the coast was charted but the interior remained relatively unknown. A significant proportion of the published maps of the African continent in the sixteenth and seventeenth centuries were coastal charts. Hence, the largest surviving seventeenth-century map collection, put together by the diarist Samuel Pepys and now in Magdalene College, Cambridge, England, includes eight maps of Africa, four of them coastal charts. From the fifteenth century, the Portuguese had been moving down the coast of West Africa to Guinea, the Congo and around the Cape of Good Hope to Malindi, and then across the Indian Ocean to Calicut. At least as far as West Africa, they were followed by French, English, Dutch, Swedish and Danish mariners, and hence the relatively accurate outline of Africa culminating in the better of the two outlines by Hondius in 1606.

By the seventeenth century, more than a century after Vasco da Gama, Portuguese mariners were no longer pre-eminent on the coasts of Africa but Portuguese chart

Fig. IX Mercator (1569), Original in Bibliothèque Nationale, Paris, France.

atlases of Africa of this type. The atlas thus influenced a number of well-known cartographers of Africa. It demonstrates the way in which unique cartographic information was a marketable commodity in Europe at this period.

The large proportion of the seventeenth century printed charts of the African coast are Dutch, partly in consequence of the predominance of their publishing houses. The landing by Jan van Riebeeck at the Cape in 1652 illustrates Dutch involvement in the developing maritime traffic around Africa. Hence, names associated with charts of Africa in the second half of the seventeenth century include Frederik de Wit, Johannes van Keulen, Pieter Goos and Arnold Colom. The English hydrographer, John Seller, also included Africa in his work, and the great sea atlas *Dell' Arcano del Mare* of 1646-7 by Sir Robert Dudley published in Florence contained some notable African charts.

THE ENLIGHTENMENT IN EUROPE

The extraordinary achievement of Portuguese soldiers and priests in the sixteenth century, in penetrating the Congo and Zambezi basins and also Ethiopia, was not to be emulated in the seventeenth century. In consequence, little new information was coming to hand for the European cartographers of the early eighteenth century, and in some maps of Africa the long-standing delineation of the interior, based in part on those early Portuguese journeys, began to be removed, with little or nothing inserted in its place. Whether change of this sort amounted to improvement or deterioration is debatable, and in any case it was not universally implemented by European cartographers.

The changes that were taking place in some depictions of Africa at the end of the seventeenth century have been seen as characteristic of the Enlightenment. Supposedly, what happened was the removal of the fictitious from the map of Africa, leaving what little remained as correct. An alternative interpretation of events is to suggest that an increasingly sceptical attitude towards the traditional Hondius/Blaeu model of the African interior (an amended Ptolemaic geography) was unfortunately not matched by sufficient new hard data to replace the traditional delineation, so that the logical outcome was a largely blank map of inland Africa. In any event, many cartographers were reluctant to pursue the new attitude to its logical conclusion. A blank map of Africa, even if justifiable, would presumably attract little interest. Instead,

making continued to flourish and had an enormous influence on the published charts of Africa. A supposed Portuguese policy of secrecy from the fourteenth to sixteenth centuries is sometimes seen as preventing the dissemination of maritime knowledge acquired by its countrymen. Whether or not there was such a policy and whether it could have been successfully implemented in the presence of so many foreigners in Portugal is the subject of debate. Certainly Portuguese charts of Africa from the early sixteenth century survive and if any such policy was ever formulated, it must eventually have been relaxed. Circumstances of strict secrecy are difficult to reconcile with the outstanding manuscript *Atlas of Africa* (1665) by João Teixeira Albernaz II in the Archives National, Paris, which contains twenty-nine coloured charts. The better known *Suite du Neptune François* published by Pierre Mortier in 1700, was based on and in part simply copied from the manuscripts of 1665, which were commissioned by a Frenchman, Jean Frémont d'Ablancourt. He lived in Portugal for five years up to 1664, before returning to France. Copies of the charts were made in France in the seventeenth century, and in all probability, João Teixeira Albernaz II made several descriptive textual entries of an ethnic nature, which did not require exact geographical location, were inserted inland. More information about the main physical features of the interior, particularly the rivers, was not to be forthcoming and did not appear on maps to any great extent until the journeys of European explorers in the nineteenth century.

The beginnings of change to the map of Africa in the so-called Age of Reason, when interest in cultural matters was greatly increased and when the emphasis in science was on the experimental method, is often associated with Guillaume Delisle. His first map of Africa, *L'Afrique Dressée...* of 1700 has been described as one of the landmarks in the history of mapping of Africa. This is because of the accurate longitudinal dimensions of the continent and the changes to the sources of the Nile and to southern Africa. Text on the map itself indicates a tentative attitude towards changes to the main drainage lines, however. In particular, the long-standing connection between the Nile and the Niger is questioned. Progress on resolving the problems of West African drainage can be followed in Delisle's *Carte de la Barbarie de la Nigritie et de la Guinée* of 1707, where the Nile-Niger connection is severed, albeit with a written qualification, but the Gambia River is now firmly separated from the Senegal River in an area where recent travels by a fellow Frenchman, André Brué, provided authority. The Niger and the Senegal are nevertheless still thought to be reaches of the same great river. A prime characteristic of both of these maps, particularly the larger scale map of West Africa, is the extensive use of text across the face of the map, providing not only the names of kingdoms and districts but also descriptions of peoples and places, with great numbers of conventional symbols for habitations and perspective drawings for hills. Much of the content of the map is clearly speculative. Delisle's *Carte du Congo et du Pays des Cafres* of 1708 is similarly replete with ethnological text, referring to the use of lions in fighting or to cannibalism. This map was copied by Henry Abraham Chatelain in 1709 and by John Senex in 1720. It was even reproduced by Carl Peters in 1902, in support of his theory for the location of the biblical gold mines of Ophir.

Other cartographers introduced similar changes in their delineation of the African rivers, as well as including increasingly frequent and speculative perspective drawings of mountain ranges, combined with extensive use of settlement symbols in association with names of peoples, kingdoms and written descriptions of ethnological traits. An example is Nicolas Visscher's *Carte de l'Afrique Meridionale...* which was published in a de Wit atlas of c.1705. In this instance, there is also authentic additional information on the Cape of Good Hope, which is portrayed in two large scale inset maps. However, the tentative nature of the delineation in the interior remains, with rivers which have no outlet to the sea but only inscriptions such as *On ne sait pas encore en quel endriot la Riviere de Cuneni tombe dans la Mer* (We do not yet know at which place the River Cuneni falls into the sea). This sort of compilation is not confined to French cartographers. The English cartographer John Senex followed Delisle and Visscher in both style and content in his maps of 1711 and 1714, significantly entitled *Africa corrected from the Observations of the Royal Society of London and Paris and dedicated to Sir Isaac Newton*. Charles Price published a similar map in 1721. However, Senex was to revert to the seventeenth-century format and hydrological content in 1721 in his *New Map of Africa from the latest observations*, a misnomer indeed. His reversion in content emphasises the uncertainty and tentativeness of the cartographic reformulation of the interior of Africa in the early eighteenth century.

Delisle made further changes in the African interior. His *Carte d'Afrique* of 1722 includes a Niger River which is separated from the Senegal River, and can be interpreted as flowing east, not west, into an inland drainage basin in the vicinity of Lake Chad but called 'Lac de Bournou,' Also a north-south aligned Lake 'Maravi' appears, north of the Zambezi. Later editions of Delisle's *L'Afrique* indicated awareness that these changes were not based on certain knowledge, because the map of 1730 adopts the hydrological pattern of his map of 1700, without the amendments of 1721. Nevertheless, others were willing to pursue the new format, for example Johann Hase, who published a map of Africa in 1737 which included the great bend of the Niger. At the same time, there were some European cartographers who did not adopt the new format. In the first three decades of the eighteenth century, the seventeenth-century delineation of Africa's lakes and rivers in the Hondius-Blaeu mould and with modest amounts of ethnological information, was still being published, for example by Pieter Schenk, Herman Moll, Johann Baptiste Homann, Gerard Valk, Henry Overton, Alexis Hubert Jaillot, Matthäus Seutter and Johann Wolfgang Heydt. The earlier depiction of Africa co-existed with the changes.

Whilst the tentative realignment of Africa's drainage and the addition of extensive ethnological notes was a commonly adopted format, for example in Emanuel Bowen's *New and Correct Map of Africa* of 1748, a map by Jean d'Anville dated 1749, entitled *Afrique publiée sous les auspices de Monseigneur le Duc d'Orleans...*, took a more purist approach. Large areas of the interior were left blank. D'Anville is seen as the major figure in the interpretation of blank spaces in the map of Africa as evidence of new standards of accuracy befitting the Age of Reason, whereby the fictitious would be removed in favour of only that which was correct. There are problems with this interpretation, however. Extensive blank spaces frequently occur on much earlier maps of Africa, e.g. Vossius (1666). Furthermore, eighteenth century cartographers seemingly did not all respond to the philosophical environment of the Age of Reason in the same way. Some eighteenth-century cartographers such as Johann Matthias Hase (1737), Pieter van der Aa (1713) or even Delisle (1700) were seemingly reluctant to leave great areas blank and only willing to leave blank spaces of modest proportions. Homann's map of *Gvinea propria...* (1743) has great contrasts in amounts of information in West Africa between better known areas along the coast and up the Senegal and Gambia rivers on the one hand, and extensive nearly blank areas elsewhere inland, although the quality of his information in coastal West Africa is not necessarily assured. Cartographers such as Vicenzo Maria Coronelli (1691) and Nicolas du Trallage Tillemont (1742) placed cartouches in otherwise blank areas. Moreover, much of the ethnological information which filled the otherwise blank spaces on the maps of Delisle and Senex was of uncertain authenticity. The reason for their inclusion may well have been the increasing number of accounts published in the first half of the eighteenth century containing just such information, particularly about West Africa. Many maps were still illustrative of texts, after all, and it was not until the later part of the century that interest in the artefacts of ethnography gave way to an interest in natural curiosities. Hence the impact of the Age of Reason on the map of Africa is perhaps not as straightforward as is sometimes suggested.

The evolutionary trend characterised by d'Anville, i.e. the acceptability of blank spaces on a map of Africa, can be distinguished from the other trend associated with Delisle, who deployed descriptive text on his maps in the early decades of the eighteenth century, not solely by the appearance of blank spaces, but by the manner in which the blank spaces came about. They were created firstly by the elimination of much of the ethnographic information which was so extensive in the Delisle format, information which may occasionally have been verifiable, but which by its nature could rarely be located precisely. People are mobile, so that ethnic traits cannot usually be delimited accurately by lines on a map. Secondly, physical features which perhaps lacked any authentication in recent accounts were sometimes removed. In consequence, it does not follow that all that remained was factually correct; merely that there was conceivably some justification for its inclusion. D'Anville's 1749 map of the continent still shows the Nile rising from two small circular lakes but now removed to the north of the Equator. Also, the drainage patterns in northern Angola and southeast Africa, where the density of detail is in contrast to extensive blank areas, are fraught with errors. With the benefit of hindsight, it is apparent that d'Anville's presumed attempts to verify or eliminate did not result in maps without major error or speculation. What they did do was to result in extensive blank areas by the elimination of mainly ethnographic information, which is the greatest difference with the Delisle format. An element of continuity with that format remained nevertheless. The use of extensive textual entries continued, though in small print. Also, the subject matter of the entries now tended to be physical features such as the unknown source of the Zambezi, which Delisle also noted. In these respects, there is a change discernible in d'Anville's attitude, between his compilation of *Carte de l'Ethiopia Orientale...* in 1727 and his map of all of Africa in 1749.

There are therefore three interrelated strands or models in the European cartographic depiction of Africa in the first half of the eighteenth century. There is the perseverance of the seventeenth-century Hondius-Blaeu delineation. Then there is the ethnographically influenced depiction associated with Delisle, among others. Lastly, there is the emergence of a more purist attitude which necessitated the great blank spaces by which it is characterised, and which was brought to a head latterly by d'Anville. These three strands are evident in the evolution of the map at the continental scale, but it is also instructive to look at the larger scale mapping which was produced contemporaneously.

During the first half of the eighteenth century, not many larger scale regional maps of Africa were published and they tended to fall into one or the other of only two of the contemporary strands of evolution. Taking the case of southern Africa by way of example, maps by Reinier and Josua Ottens (1700) and van der Aa (1720) exemplify the tendency for the seventeenth-century model to be carried forward, but maps in the Delisle mould, for example by Chatelain (1719) and Emanuel Bowen (1747), were more numerous. The third model, characterised by blank spaces, tends not to be represented among maps at scales appropriate to regional maps, nor indeed among maps at very much larger scales. However, large scale maps increased in numbers at this time, in response to the beginnings of European penetration of the interior, for example, d'Anville's *Cours de la Riviere de Senaga ou Senegal* at 1:250,000. Similarly, a survey of the Gambia River by an Englishman, John Leach, was published in 1748 at 1:1,000,000. Settlement at the Cape of Good Hope provided information for large scale maps and views by Dutch and French cartographers such as Johan Nieuhoff (1682 and 1703), Visscher (1701), de Fer (1705), François Valentyn (1726), Delisle (1740), Jacques Nicolas Bellin (1764) and van Keulen (1780). The Nile valley was also the subject of detailed mapping by European travellers.

During the second half of the eighteenth century, further evolution was modest. Change to the inland map of Africa along the lines associated with d'Anville had to await further travels into Africa by Europeans, and these were not many in the latter part of the eighteenth century. Furthermore, although the travels of James Bruce into Abyssinia began in 1769, the results were not published until 1790. Similarly, Anders Sparrman's travels in the south of the continent were not published until 1785. Consequently, all three cartographic models of Africa persisted, with the d'Anville model unable to evolve further or to eclipse the other two models, as it eventually would in the nineteenth century.

The d'Anville model is represented, for example, in a map of Africa by Isaak Tirion (1763), an Amsterdam publisher, at a time when French cartographers were more prolific in their portrayal of Africa. The French Royal Hydrographer, Rigobert Bonne exemplifies the same school of thought, despite a title reading *L'Afrique divisée en ses principaux états* (1778), which might suggest more ethnological subject matter. Then there were publishers such as Robert Sayer (1787), who largely copied d'Anville. All of these cartographers perpetuated the characteristic of extensive blank spaces. However, they also perpetuated the residual errors that were to remain for some time to come within this format. For example, the twin lakes at the fount of the White Nile, which are an identification mark of the d'Anville pedigree, were carried forward on these and other maps well into the nineteenth century, until finally disproved by the travels of Speke and Grant from 1858 to 1863. The two other cartographic models also persisted into the second half of the eighteenth century. Thomas Kitchen's map of 1770 divides all of Africa among its kingdoms and peoples, while Johann Michael Probst's *Africa* of 1778 contains the hydrology of Hondius and Blaeu, not for the last time.

There was little to add at the sub-continental scale in the second half of the eighteenth century, but there was an occasional interesting addition to maps of smaller areas. For example, Nicolas Louis de la Caille's *Carte du Cap de Bonne Esperance et de ses Environs* of 1752 locates triangulation north from Table Bay. A map of Egypt by le Sr. Robert de Vaugondy (1753) drew together the observations of several travellers in the Nile valley, but large scale maps of small parts of Africa were still sometimes compiled by copying and enlarging a detailed extract from a smaller scale map. For example, Bellin's map of *Congo Angola et Benguela* in volume 3 of *Le petit Atlas Maritime* of 1764, bears a striking resemblance to Delisle's map of Africa south of the equator of 1708. This particular form of plagiarism occurs throughout the eighteenth century, when change in map content may have more to do with the attitude of the cartographer than the availability of new data.

To summarise the changes that occurred in the Age of Reason, any suggestion that the map of Africa became factual and accurate requires qualification. Certainly, the content of some of the maps published during the eighteenth century changed substantially in consequence of what was reputedly the more scientific attitudes of the Enlightenment, but error was not necessarily eliminated. For example, geographical information deriving from the Portuguese penetration of the Zambezi basin in the seventeenth century was included by d'Anville in his maps of 1727 and 1749, but he knew too little of their provenance to avoid corruption or to avoid outdated portrayals. Recent work has also shown that the conventional concept of a scientific phase of map making associated with the Enlightenment is greatly exaggerated in the eighteenth-century depictions of West Africa. On the other hand, images of the interior of Africa which had been first published in the seventeenth century persisted alongside more recent images. Maps were also used for conveying ethnographic information by inserting text

Fig. X Part of: Detailed Map of the Revd. Dr. Livingstone's Route across Africa; ... by J. Arrowsmith, 1857.

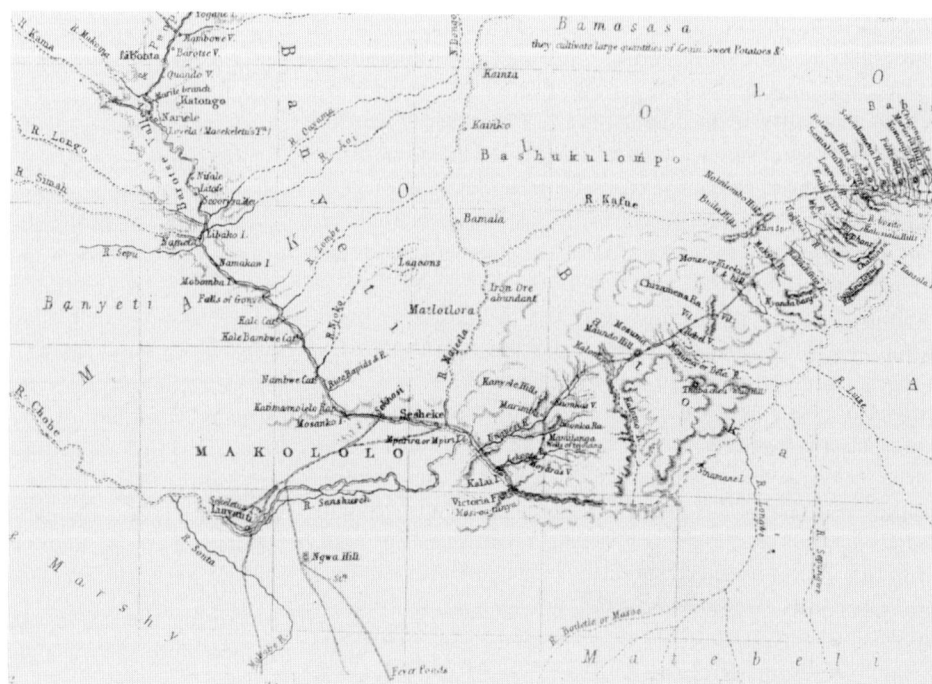

xxiv

extensively, although maps are better suited to portraying static features which can be precisely located. Indeed, it may have been this somewhat inappropriate use of maps combined with a shift of interest away from ethnographic to natural curiosities that prompted the emergence of the third type of maps of Africa, maps which have been seen as characteristic of the influence of the Enlightenment, and whose sole content consists of a small number of relatively permanent features.

NINETEENTH-CENTURY IMPERIALISM

Continuity with the past was also a feature of the changing map of Africa in the nineteenth century, when the accounts of classical authors such as Herodotus sparked off much controversy, for example, about the sources of the Nile. However, sixteenth- and seventeenth-century Portuguese sources such as João de Barros and João dos Santos were recognised as severely dated by the nineteenth century, when a new and expanding source of information came gradually to hand. It is the increasing availability of new data deriving from European explorers, plus military sources in the extreme south and northeast of the continent, that differentiates the nineteenth century as a separately identifiable period.

The build-up of information gradually gained momentum throughout the century but its quality was variable. It was not necessarily the case that every traveller's record rendered earlier sources obsolete. For example, the nineteenth-century armchair-geographer, W.D. Cooley exposed the discoveries of J.B. Douville in Angola in 1832 as fraudulent. Even where deceit was not intended, travellers could not necessarily identify geographical features that they saw, nor could they necessarily locate their position. Hence, cross-checking and verification from all possible sources continued to be necessary, particularly in the earlier part of the century. It was a century of cartographic transition rather than transformation.

In southern Africa, a succession of travellers, explorers, hunters, missionaries and frontiersmen penetrated northwards from the Cape, including John Barrow, John Campbell, Robert Moffat and David Livingstone, whose trans-Africa journey was particularly influential in the subsequent cartography of the Zambezi basin. The rivers and lakes of East Africa were investigated by many well-known explorers, including Lewis Krapf, Richard Burton, John Speke, James Grant, Samuel Baker, John Petherick and David Livingstone, culminating in perhaps the most significant contributor for the depiction of the hydrology of East Africa, Henry Moreton Stanley. In West Africa, it was the hydrology of the Niger and Chad basins that were eventually mapped by Mungo Park, Walter Oudney, Hugh Clapperton, Dixon Denham, Richard and John Lander, Gustav

Nachtigal and the remarkable Heinrich Barth, whose travels epitomise the best of nineteenth-century scholarly exploration in Africa.

The maps that were compiled by nineteenth century explorers and travellers found their way into print in different ways. Many of them first appeared appended to the explorers' own accounts, for example, in a *Detailed Map of the Revd. Dr. Livingstone's Route across Africa...*(1857) constructed by John Arrowsmith, or the route maps compiled by Auguste Petermann in Barth's Travels and Discoveries (1857). Another major outlet for the accounts of explorers were the journals of the national geographical societies of Europe, which reached the heights of their prosperity and influence in the nineteenth century, to the extent that they organised and financed major expeditions to Africa. Between 1830 and 1880, the Royal Geographical Society of London published thirty-five original maps of Africa. Then, atlas maps were compiled from these original sources, such as the two maps of West Africa in the atlas of *Maps of the Society for the Diffusion of Useful Knowledge* (1844), which include the routes of more than twenty explorers. Over a short period of five or six years in the 1860s, A.K. Johnston engraved no less than three different compilations of East Africa for Blackwood's *Royal Atlas of Modern Geography*, to take account of new reports from the region.

The detail that explorers selected for inclusion in their maps and the way in which they made their observations was the product of the political and economic environment that fostered their travels, what Cain and Hopkins (1993) have described as the 'gentlemanly capitalism' of the nineteenth century, at a time when European governments were not at all disposed to assume the financial burden of imposing and maintaining any form of direct colonial rule in Africa. It was an attitude of mind to be found amongst the commercial middle class of society, including businessmen, servicemen and missionary leaders, that is, among a wide alliance of unofficial interests who variously felt called upon to respond to what they saw as African inadequacy. Intervention in the lives of Africans was thought of as a laudable civilizing process to be fostered by the deployment of European organisational and technological skills. In this environment, the compilation of maps of the highest possible scientific calibre helped to legitimise European penetration and even interference in Africa.

The technical quality of explorers' maps in the nineteenth century was not entirely consistent, of course. Much depended on the skills of the traveller and how meticulous he was in observing and recording throughout his journey. Burton, for example, seems to have begun his journey to Lake Tanganyika in 1857 by taking a compass bearing at every stage of the walk and by recording what he saw on that stage, but eventually he lapsed into a more impressionistic written record. The Royal Geographical Society assisted travellers by providing instruction in surveying. It also published a manual on survey practice and other relevant subjects, entitled *Hints to Travellers*, as well as making expensive instruments such as sextants available to travellers. However, some explorers were more skilled and conscientious observers than others. David Livingstone was particularly conscientious and skilled, whereas Burton was inclined to leave the technicalities to Speke. However, the final form of the maps was usually the result of computation and compilation by an office-bound cartographer. The explorer's calculations were checked and adjustments made on the basis of observations from other sources. New subject matter might even be introduced from other sources. Hence, the technical quality and content of explorers' maps of similar dates may differ to some extent, but they exhibit common characteristics, nevertheless.

There are two recurring features which clearly relate to the circumstances surrounding late-nineteenth-century explorers' maps. It was common practice to insert altitudes at the locations where these had been measured, usually along the explorer's route. Measurements were often made using a spirit lamp to boil water, so that the steam passed over the bulb of a thermometer. These measurements must have been laborious, but observations of height above sea level obtained in this way were evidence of the scientific rigour with which the map had been compiled. A second and possibly surprising feature of many explorers' maps is the occurrence of descriptive phrases, on the face of it retaining a much earlier practice in African map-making. These are usually references to the resources of the country or to such subjects as river navigability, evidence of the commercial motivation contained within the unofficial mind of nineteenth-century imperialism.

An example of map making of this type is the remarkable *Survey of the Livingstone Falls (32 in number) of*

Fig. XI North-West Africa. From The World Wide Atlas of Modern Geography, W&A.K. Johnston, Edinburgh, 1901. Map shows five categories of "European Possessions."

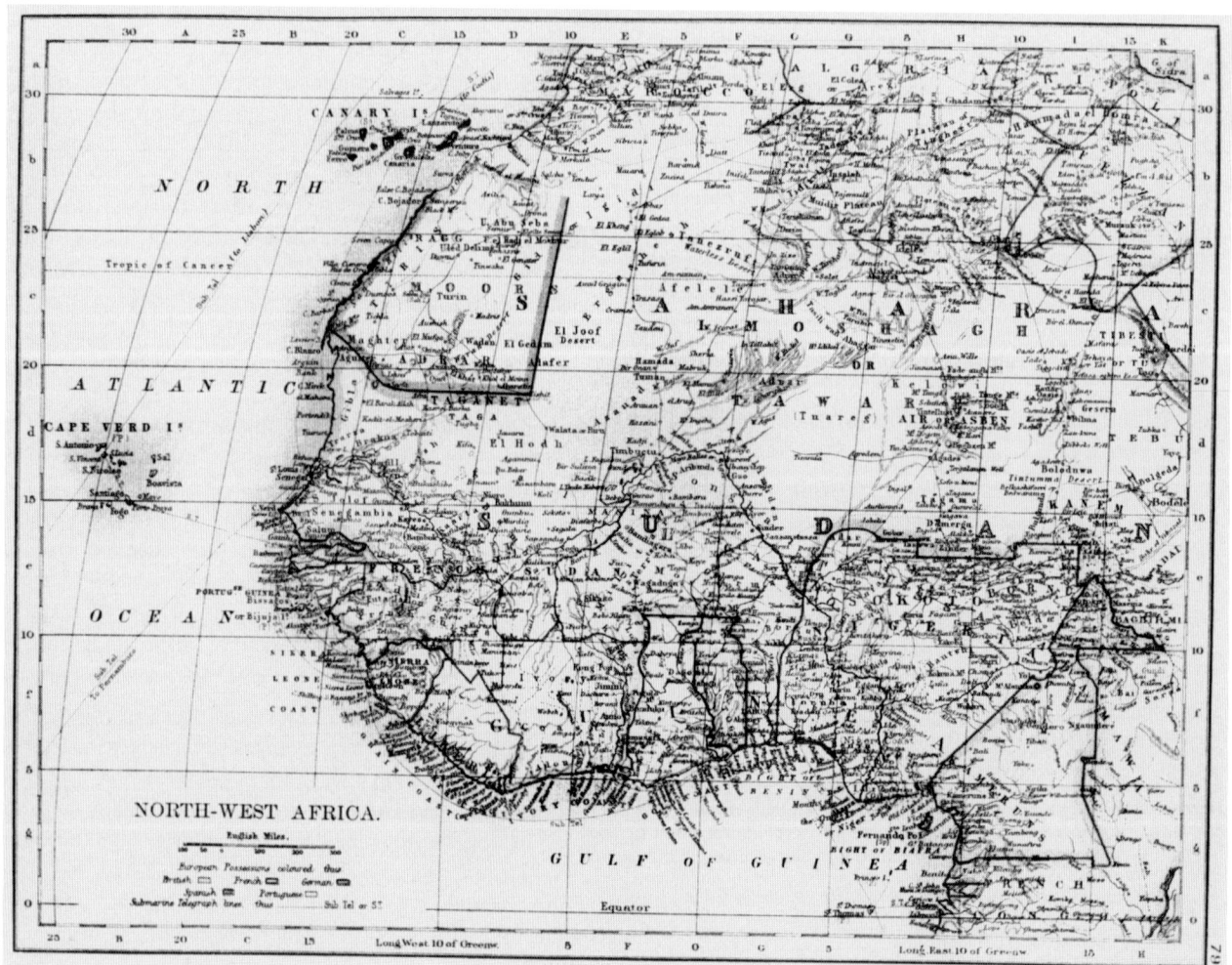

the *Lower Livingstone River from Stanley Pool to Boma* (1878) by H.M. Stanley, which includes precise measurements of the river. A good example of the explorer's traverse is to be found in the work of the artist, Thomas Baines, who travelled with James Chapman from Walvis Bay to Lake Ngami and on to the Victoria Falls in 1861, at a time when there was speculation about the possibility of opening up the Zambezi River for navigation and trade. Baines's observations are recorded, not only in his published *Map of the Route* (1864) but also in his manuscripts which are held in the Royal Geographical Society collections and provide further evidence of the amount of data that Baines collected. These were maps which both generated scientific data and informed about the potential for commerce.

The changes to the nineteenth-century map of Africa are synopsised in the continental-scale maps of Africa appearing in the many contemporary atlases of the world. At the beginning of the century, the picture of Africa which atlas publishers were presenting was not consistent. For example, Louis de la Rochette's *Africa* of 1803 incorporates recent exploration in West Africa, but south of the Equator, it continues to rely heavily on the long-standing Portuguese sources for the lower Congo and Zambezi basin. Elsewhere there are both blank areas and textual descriptions of an ethnographic nature or of potential routes. On the other hand, James Cundee's *New Royal Atlas* includes a map of Africa *Engraved for the Revd. Mr. Evans's New Geographical Grammar* in 1809, which is much more concerned with ethnic names distributed evenly across the entire face of the continent, albeit with the recent knowledge of the drainage of West Africa as a

part of the background information. It is as though maps were being compiled to match differing interests among readers.

During the first half of the nineteenth century, there was an increasing tendency to show large blank areas or to use the word 'Unexplored' extensively. Blank spaces do not necessarily imply rigorous removal of all that is uncertain. Speculative entries are still discernible, sometimes shown tentatively, whilst ethnic entries continue to be deployed extensively. For example, in *Chambers's Atlas for the People* of 1846, the map of Africa now has the outlet of the Niger correctly located on the Gulf of Guinea. In Central Africa, the phrase 'UNEXPLORED COUNTRIES' is used, the area fringed by the tentative depiction of rivers and lakes located by dashed lines. By comparison, a mountain range named 'Donga Mts.' is more confidently located to the south of a tentative White Nile, whilst names such as 'Monomotapa,' 'Tribes of Tonariks' and 'Tribes of Tibbous' are boldly inscribed in positions which, with hindsight, are at best approximate. The quality of the information is very variable, as the compiler must have known.

Nevertheless, the new information in explorers' maps was increasingly used by map publishers. For example, the results of Denham, Clapperton and Oudney's trans-Saharan journey, which were published in 1826, quickly made their mark and appeared in the *Map of Africa* by Gall and Inglis in their *Edinburgh Imperial Atlas* (n.d.), thereby helping to date the map. Their route remains distinctive on many subsequent maps. It appears on the

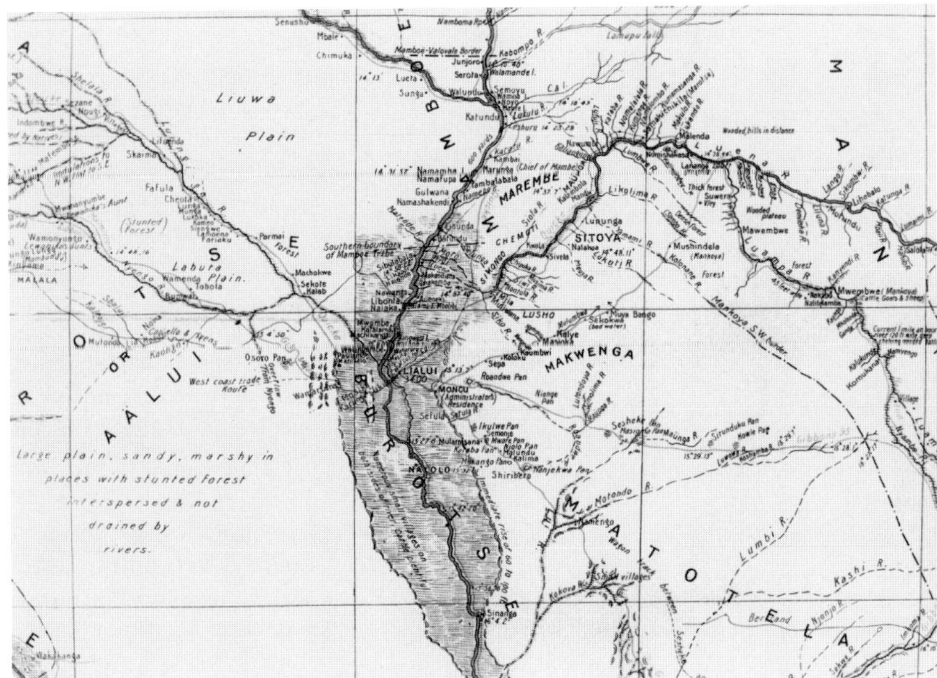

Fig. XII
Part of: "A Map Of The Upper Zambezi Basin, Showing the distribution of Tribes under the rule of the Paramount Chief of The Marotse, Principally from Surveys and Exploration by Major A. St. H. Gibbons. 1895-96 and 1898-1900."

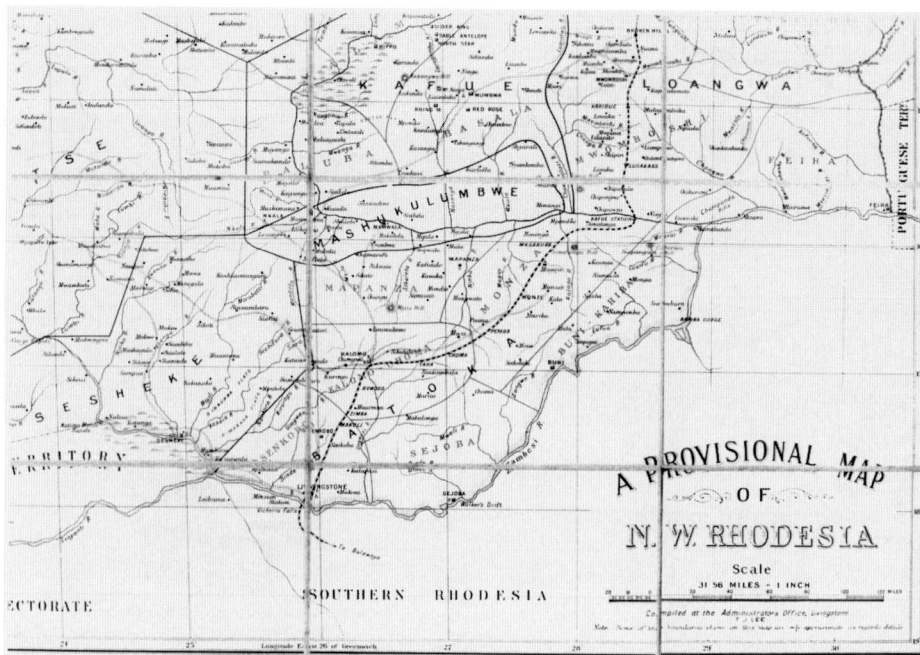

Fig. XIII
Part of: "A Provisional Map
OF N.W. Rhodesia
Compiled at the
Administrators Office,
Livingston. T.J. Lee."
c. 1908.

the railway lines to c.1905. There is a key at the foot, a feature that was not previously customary, listing the subject matter of this new map, namely a division into nine territorial authorities: British, French, German, Portuguese, Italian, Spanish, Belgian, Turkish and African States. This map is the product of a phase of change in the cartographic evolution of Africa, which transformed the content of maps. However, it required the use of very much larger scales than previously utilised, so that its impact is not easily seen on atlas maps of entire continents. Nevertheless, the change in the nature of the relationship between Europe and Africa, from imperial to colonial, that is from the exertion of external influence to the imposition of external rule, broke five centuries in the continuity of that relationship. Maps that served the colonial purpose were very different from those which had evolved continuously over the previous five centuries. Colonial rule was to transform the mapping of Africa.

Historians of Africa currently recognise three or four phases in the evolution of colonial government in Africa. The first phase was devoted to securing a hold on the colonies. During the second phase, colonial governments were more passive. They sought to do little more than maintain colonial rule in whatever form they had chosen to adopt, although in the later part of this phase, some colonial powers in Africa began to experiment in social and economic development, conceivably a separate phase in itself. Lastly, there was a phase of increasingly active involvement by the colonial state in the economic and social well-being of the territory. All of these phases resulted in distinctive cartography, much of it at increasingly large scales and formats.

The cartographic history of Zambia, formerly Northern Rhodesia, exemplifies the three typical phases of colonial government. In the second half of the nineteenth century, that part of southern Africa which was to become Northern Rhodesia was crossed and recrossed by explorers motivated in the pre-colonial imperial mould, who made maps accordingly. Amongst

map of Africa in *The Illustrated Atlas and Modern History of the World* published by John Tallis in 1851. Tallis also sought out other recent sources of African information; in East Africa, although the Nile was still shown as rising from prominent 'Mountains of the Moon,' he also showed 'M. Kilimandjaro' and 'M. Kenia.' Trans-Saharan routes also appear in A. & C. Black's *General Atlas of the World* dated 1854, which has new detail in the upper Zambezi basin and in East Africa, where one of the great dramas of African exploration was being played out. By 1882, *The Comprehensive Atlas and Geography of the World*, published by Blackie & Son, depicts the interior of Africa, with only a very small proportion of the total area left blank, although still with frequent dashed or dotted lines to indicate tentative alignments. The balance of content here swings heavily in favour of physical geography, with only a small number of ethnic references such as 'Independent Pagan Tribes.' The remaining textual entries have been displaced in Bartholomew's *Library Reference Atlas of the World* dated 1890, in which the map of the African continent still has blank spaces and in which the larger scale map of 'Central Africa' still has a great many broken lines for rivers. On this map, an entirely new category of information appears, including names such as 'British East Africa,' 'German East Africa,' 'Congo State' and 'French Congo Tery.' Here are the beginnings of a significant change, not just to the legends of maps of Africa but to the purpose and practice of African map making.

THE MAPPING OF COLONIAL AFRICA

Bartholomew's undated *Citizen's Atlas of the World* has a map of the African continent, which can be dated by

xxviii

these, David Livingstone is the best known, but quite characteristically, many lesser-known travellers continued to add to geographic knowledge. Also typical was the fact that the scientifically elegant maps that these explorers produced continued to be constructed in the early years of the implementation of colonial rule. The motives of the pre-colonial missionary, hunter or explorer were not instantly eclipsed by the declaration of authority over the territory by the colonial power. Indeed, very similar maps were compiled by a new breed of traveller, the emissaries of the European powers who sought treaties with tribal authorities. Hence, there is a short transition period in the first few years of colonial rule, when maps continued to be constructed in the pre-colonial mould, alongside very different maps which were the direct consequence of the coming of colonial rule.

Maps which match the older pattern but were constructed at the time of transition to colonial rule include the map of Lake Bangweulu by Joseph Thomson (1893) and the detailed maps of the Upper Zambezi basin compiled by Captain Alfred St. Hill Gibbons in 1898 and 1904. These maps were contemporaneous with a new type of map which was representative of the first phase of colonial government in that they were primarily compiled to record the location of the people who were the new subjects of colonial governments. Furthermore, they were compiled without considering accuracy to be a virtue for its own sake, or in the cause of the advancement of science. Many of these maps were not published and can be found only in manuscript form, in archives or private collections, although some have been reproduced. Examples include the pioneering map of the Hook of

the Kafue by Valdemar Gielgud (1901), and the maps of serving government officers such as C. de V. Hunt's map of Mapanza Sub-District (1905) and J. H. Venning's map of Balovale District (1907). Smaller scale compilations with some of the characteristics of the first phase of colonial rule were published and are therefore a little more widely available, despite small print runs. These include T. J. Lee's *Provisional Map of N.W. Rhodesia* (c.1908), maps of *Rhodesia: British South Africa Company's Territories* published in six sheets by Edward Stanford in several editions from 1895 to 1906, their map of *Central Africa* (c.1910) published in their Library Maps Series, as well as War Office publications under 'TSGS' (Topographical Section General Staff) designations.

The many manuscript maps compiled on the ground in the first phase of colonial rule were in part the subject matter of those of the second phase. In Northern Rhodesia, the second phase is well represented by a series of topographic maps at 1:250,000 published between 1917 and 1928. Map making did not have high priority at a time when colonial government sought to do little more than maintain the rule of law and when the limited official resources that were deployed were largely utilised in cadastral mapping, necessary for legal purposes to define the boundaries of farms or township plots. Professional surveyors could not be afforded for topographic mapping, so that the series was compiled from a wide variety of sources of very variable quality and with some areas remaining blank. Further compilation and reduction resulted in the publication of an historically interesting 4-sheet map of the entire

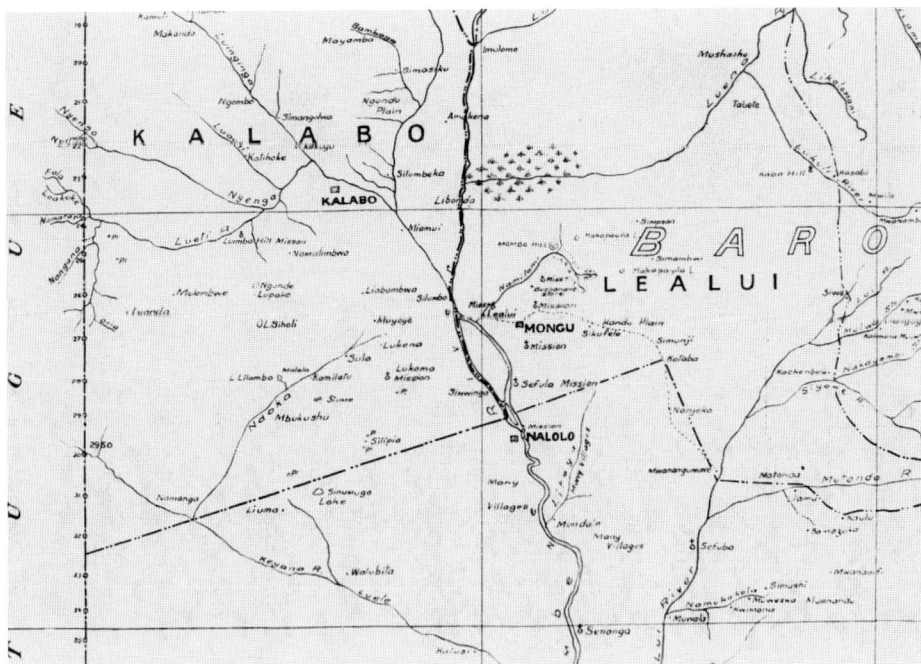

Fig. XIV
Part of "Sheet No. 3: Northern Rhodesia, 1:1,000,000, Compiled in the Survey Department. Northern Rhodesia. June 1921."

territory at 1:1,000,000 in several editions from 1921, as well as an attractive little single-sheet map at 1:4,000,000 from 1924.

In the 1930s, the government of Northern Rhodesia formulated what has been described as various impulses towards reform, in the context of wider debate over the role of colonial governments. The thinking of this decade also had cartographic consequence in the form of first one and then another new topographic series at 1:250,000, to be based in part on aerial survey, but both programmes were overtaken by events and only a very small number of sheets were published. These did include a sheet showing the Victoria Falls and its gorges, noteworthy because it was printed in five colours.

The last twenty years of colonial government in Northern Rhodesia, prior to independence in 1964, saw a transformation in topographic mapping. This third phase of colonialism was the developmental phase, when priority was accorded to the production of high-quality large scale maps utilising increasingly sophisticated technology. The choice of the parts of the country to be covered first was influenced by the location of development projects, such as the construction of the Kariba dam, and cover was far from complete at independence, but some remarkable maps were produced, such as another map of the Victoria Falls in 1947, this time at 1:25,000.

Colonial rule transformed the cartography of Zambia. From the outset, it set in motion the production of new forms of topographic mapping and of cadastral mapping. The initial momentum which was in part amateur was not sustained, however. Throughout the 1920s, only desultory progress was made by compilation, drawing partly on the previous phase. Then, in the 1930s, a modest increase in the demand for topographic mapping resulted in largely abortive attempts to provide improved cover. After the Second World War, the rate of production gathered momentum very rapidly and that momentum was sustained well into independence. Only recently has the sheer magnitude of political and economic problems in Zambia put a brake on that momentum. This sequence of events is discernible elsewhere in Africa, certainly in

Fig. XV Colonias Portuguesas Na Africa Meridional. Angola E Moçambique *Atlas Colonial Português, Ministerio des Colonias, 1914*

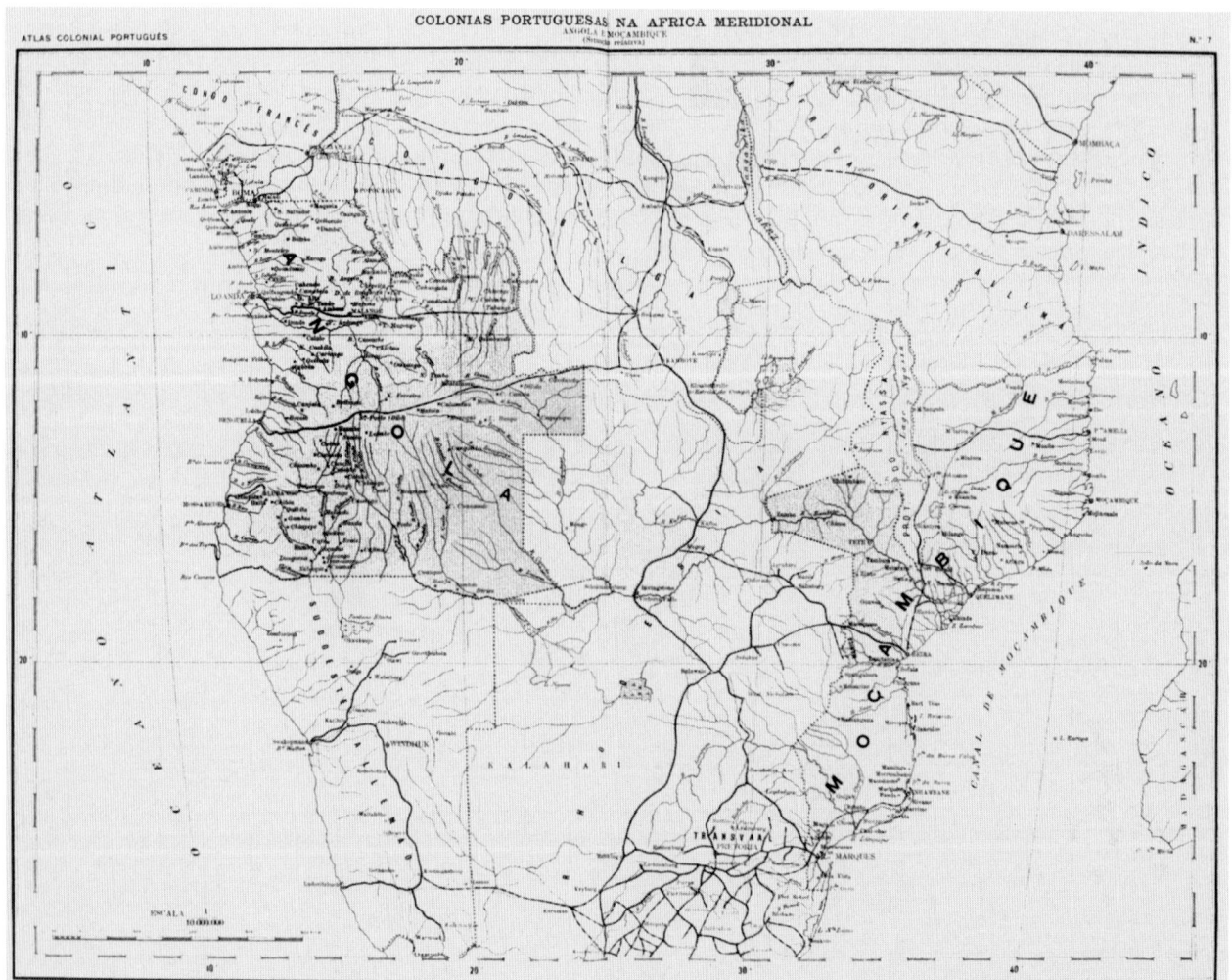

former British Africa, but to a greater or lesser degree, in countries which were the former colonial territories of the other six European powers. Furthermore, the cartographic transformation which was initiated at the level of individual territories during the colonial period, was quickly fed into mapping at the regional and continental scale, as is apparent from the twentieth-century maps and atlases of international firms such as John Bartholomew and McGraw-Hill.

CONCLUSION

'So Geographers in Afric-maps, with savage-pictures fill their gaps,' is part of a well-known quatrain from Jonathan Swift's *On poetry* which is often quoted by writers on maps of Africa, but it is far from apposite. Early map makers used such sources as were available to them, to match the perceived interests of their readers, thereby incorporating more contemporary knowledge than Swift implies. Thereafter, the Age of Reason did not sweep away Swift's rampaging elephants, as is sometimes suggested. Its impact was not so immediate or substantial. What followed represented less of a transformation and more of a trend through the eighteenth and into the nineteenth century, as African cartography evolved in response to new data of variable quality and as the interests of readers changed. The texts and the maps of nineteenth-century exploration in Africa can only be understood in terms of their several purposes, including commercial exploitation, missionary zeal and scientific enquiry.

With the failure of the Berlin West Africa Conference of 1884-85 to ensure the continuation of the traditional system of free-trading by European powers on the coasts of the continent, and the replacement of indirect imperial influence with direct colonial rule, the purposes of map making changed. This is the most significant turning point in the history of European map making in Africa. The previous five centuries of indirect imperial influence were relatively consistent and uniform in a functional sense, by comparison with the new requirement to govern Africa directly. The colonial administrator made maps to enable him to function and they were substantially different to those which previously existed. The maps compiled during the brief but influential period of colonial rule amounted to the nearest thing to a transformation in Africa's cartographic past.

BIBLIOGRAPHY
selected references for further reading

AFRICANUS, Leo. 1896. *The History and Description of Africa*. Trans. by John Pory (1600), The Hakluyt Society, first series, 92-94.

AMBROSE, D.P. 1976. Maps of Lesotho and surrounding areas, 1779-1828. *Mohlomi, Journal of Southern African Historical Studies* 1, 3-14.

ARDEN-CLOSE, C. 1950. The state of the surveys of British Africa in 1905-06. *Empire Survey Review* 10, 255-260.

AXELSON, E. 1960. *Portuguese in South-East Africa 1600-1700*. Johannesburg.

BAESJOU, R. 1988. The historical evidence in old maps and charts of Africa with special reference to West Africa. *History in Africa* 15, 1-83.

BAINES, T. 1864. *Explorations in South-West Africa*. London.

BAKER, S.W. 1866. *The Albert N'yanza Great Basin of the Nile and Explorations of the Nile Sources*. London.

BARBOUR, K.M. 1961. Maps of Africa before surveys. The location of Omodias' Map of Sudan. *Geographical Review* 51, 71-86.

BARROW, J. 1801. *An account of travels into the interior of Southern Africa*, vol.1. London.

BARTHOLOMEW, J.G. 1890. The Mapping of the World. Part II - Africa. *Scottish Geographical Magazine* 6, 575-597.

BASSETT, T.J. & PORTER, P.W. 1991. 'From the best authorities': the mountains of Kong in the cartography of West Africa. *Journal African History* 32, 367-413.

BAUD, M. 1989. La représentation de l'espace en Egypte ancienne: cartographie d'un itinéraire d'expédition. *Mappe Monde* 89/3, 9-12.

BLAKEMORE, M. 1981. From way-finding to map-making: the spatial information fields of aboriginal peoples. *Progress in Human Geography* 5, 1-24.

BOSMAN, W. 1705. *A New and Accurate Description of the Coast of Guinea*. London.

BRIDGES, R.C. 1982. The historical role of British explorers in East Africa. *Terrae Incognitae* 14, 1-21.

BRIDGES, R.C. 1994. Maps of East Africa in the nineteenth century. In: *Maps and Africa*, J.C.Stone (ed.), Aberdeen, 12-31.

BROEKEMA, C. 1971. Maps of the Canary Islands published before 1850. *Map Collectors' Circle* 74.

BRUCE, J. 1790. *Travels to discover the source of the Nile, In the Years 1768, 1769, 1770, 1771, 1772 and 1773*. London.

BUNBURY, E.H. 1959. *A History of Ancient Geography*, 2nd ed. New York.

BURTON, R.F. 1859. The Lake Regions of Central Equatorial Africa, with Notices of the Lunar Mountains and the Sources of the White Nile; being the results of an expedition undertaken under the patronage of Her Majesty's Government and the Royal Geographical Society of London, in the years 1857-1859. *Journal of the Royal Geographical Society* 29, 1-454.

CAIN, P.J. & HOPKINS, A.G. 1993. *British Imperialism: Innovation and Expansion 1688-1914*. Harlow.

CAMPBELL, J. 1835. *A Journey to Lattakoo, in South Africa*. London.

CARTWRIGHT, J.F. 1976. *Maps of Southern Africa in Printed Books 1750-1856*. Cape Town.

CARTWRIGHT, M. 1992. *Maps of the South Western Cape Of Good Hope*. Cape Town.

CARTWRIGHT, M. F. 1976. *Maps of Africa and Southern Africa in Printed Books 1550-1750*. Cape Town.

CHAPMAN, E.F. 1896. The Mapping of Africa. *Report of the Sixth International Geographical Congress held in London, 1895*. 571-578.

COLONIAL SURVEY COMMITTEE, 1928. *Special Report on the Triangulations of Eastern and Central Africa*, HMSO Colonial No. 33. London.

COMISSAO DE CARTOGRAFIA, 1914. *Atlas Colonial Português*. Lisbon.

COOLEY, W.D. 1852. *Inner Africa laid open in an attempt to trace the chief lines of communication across that continent....* London.

CORTESAO, A. & MOTA, T.A. da 1960. *Portugalia Monumenta Cartographica*, 6 vols. Lisbon.

CORTESAO, A. 1969. *History of Portuguese Cartography*. Coimbra.

CRAWFORD, O.G.S. 1949. Some Medieval Theories about the Nile. *Geographical Journal* 64, 6-28.

CUNNINGHAM, I.C. 1987. David Livingstone and his maps. In *Maps and Mapping of Africa*, P.M. Larby, ed., London, 14-20.

CURNOW, I.J. 1925. Topographic Mapping in Africa. *Journal of the Manchester Geographical Society* 41-42, 32-37.

DAMES, M L. 1918. *The Book of Duarte Barbosa I*. Hakluyt Society, London.

DAPPER, O. 1679. *Naukeurige beschrijvinge der Afriksensche gewestern van Egypten, Barbaryen, Lybien, Biledulgerid ...* 2 vols. Amsterdam.

DAPPER, O. 1689. *Description de l'Afrique ... Avec des cartes ...* Amsterdam.

DCS 1947. *Directorate of Colonial Surveys Annual Report 1946-47.* London.

DE BARROS, J. 1552. *Décadas da India*, pt. 1. Lisbon.

DE KOCK, W.J. 1957. Counting the pillars. A note on the earliest maps and coastal surveys of the South African coast. *Historia 2*, 66.

DE SMIDT, A. 1896. A brief history of the Surveys and the Cartography of the Colony of the Cape of Good Hope. In *Report of the Sixth International Geographical Congress held in London 1895*, London, 321-340.

DENHAM, D. & CLAPPERTON, H. 1826. *Narrative of Travels and Discoveries in Northern and Central Africa, in the years 1822, 1823 and 1824...* London.

DENUCÉ, J.B.F. 1938. Les sources de la carte murale d'Afrique de Blaeu, de 1644 (Amsterdam). *Cong. Géogr. Int. Comptes rendus*, 172-4.

DILKE, O.A.W. 1987. The Culmination of Greek Cartography in Ptolemy. In *The History of Cartography*, Vol. 1, J.B. Harley & D. Woodward, eds., 177-200. Chicago.

DOS 1969. *Directorate of Overseas Surveys Annual Report for the year ended 31 March 1969.* London.

DUYVENDAK, J.J.L. 1949. *China's Discovery of Africa.* London.

ELKISS, T.H. 1981. *The Quest for an African Eldorado: Sofala, Southern Zambezia, and the Portuguese, 1500-1865.* Waltham, Massachusetts.

ETHEL, L. and GRAY, J., 1948. *A History of the Witwatersrand Goldfields.* Johannesburg.

EVERSOLE, R. 1983. An Ethiopian Mountain in Maps and Literature. *Mapline 29*, 1-3.

FAGE, J.D. 1978. *A History of Africa.* New York.

FALL, Y.K. 1982. *L'Afrique a la Naissance de la Cartographie Moderne.* Paris.

FERNANDEZ-ARMESTO, F. 1991. *The Times Atlas of World Exploration.* London.

FISHER, A.G.B. & FISHER, H.J. 1987. *Sahara and Sudan, Volume III: The Chad Basin and Bagirmi by Gustav Nachtigal.* London.

FISHER, R. 1984. Land Surveyors and Land Tenure at the Cape 1657-1812. In *History of Surveying and Land Tenure in South Africa*, C.G.C. Martin and K.J. Friedlaender, eds., Vol. 1, Rondebosch, 55-88.

FORBES, V.S. 1965. Some early maps of South Africa 1595-1795. *Tydskrif vir Aardrykskunde* II, 6, 9-21.

FRACAN (Montalboddo). 1508. *Itinerarium Portugallensium e Lusitania in Indiam.* Milan.

GAMBLE, D.P. 1968. Published charts, maps and town plans of the Gambia. *Sierra Leone Studies 23*, 66-70.

GIBBONS, A.St.H. 1898. *Exploration and Hunting in Central Africa 1895-96.* London.

GIBBONS, A.St.H. 1901. Explorations in Marotseland and neighbouring regions (with) A map of Marotseland. *Geographical Journal 17*, 106-134 & 224.

GIBBONS, A.St.H. 1904. *Africa from South to North through Marotseland*, 2 vols. London.

GILL, D. 1896. On the Geodetic Survey of South Africa. In *Report of the Sixth International Geographical Congress held in London 1895*, London, 341-360.

GODLEWSKA, A. 1988. The Napoleonic Survey of Egypt. *Cartographica Monograph 38-39.*

GOODRICH, Th.D. 1986. The earliest Ottoman maritime atlas – the Walters Deniz Atlasi. *Archivum Ottomanicum 11*, 25-55.

GREY, C. 1919. *Some Notes on Mapping and Prospecting in Central Africa.* London.

GUGGISBERG, F.G. 1911. *Handbook of the Southern Nigeria Survey and Text Book of Topographical Surveying in Tropical Africa.* Edinburgh.

HAILEY, Lord 1956. *An African Survey.* London.

HARGREAVES, J.D. 1984. The Berlin West Africa Conference. A Timely Centenary. *History Today 34*, Nov., 16-22.

HARGREAVES, J.D. 1988. *Decolonization in Africa.* Harlow.

HASSENSTEIN, B. 1887. Die Portugiesische expedition quer durch Südafrica, 1884 & 1885. *Petermanns Geographische Mitteilungen 33*, Tafel 3.

HEAWOOD, E. 1916. Early knowledge of Central Africa. *Geographical Journal 47*, 304-306.

HERTSLET, E. 1909. *The Map of Africa by Treaty*, 3rd ed., 3 vols. and a collection of maps. London.

HINKS, A.R. 1921. Notes on the construction of a general map of Africa. *Geographical Journal 52*, 4, 218-237.

HMSO 1928. *Colonial Survey Committee Special Report on the Triangulations of Eastern and Central Africa.* Cd. No. 33, London.

HOLDITCH, T.H. 1901. How are we to get maps of Africa? *Geographical Journal 18*, 590-601.

HOPPEN, S. 1975. Fifty small and miniature maps of Africa. *Map Collectors' Series 108.*

HUNT, C. 1994. Some Notes on Indigenous Map-Making in Africa. In *Maps and Africa*, J.C. Stone, ed., Aberdeen, 32-35.

ICENOGLE, D. 1989. The Geographic and Cartographic Work of the American Military Mission to Egypt, 1870-1879, *The Map Collector 46*, 26-33

JAMES, W. 1988. The naming of places on African maps. *JASO (Oxford) 19*, 2, 181-187.

JEWITT, A.C. 1992. *Maps for Empire.* London.

JOHNSON, M. 1974. News from nowhere: Duncan and "Adofoodia." *History in Africa 1*, 55-66.

JOHNSTON, H.H. 1916. The Portuguese and their early knowledge of Central African geography. *Geographical Journal 47*, 210-212.

KAMAL, Y. 1951. *Monumenta Cartographica Africae et Aegypti*, 16 vols.. Cairo.

KAMMERER, A. 1952. *La Mer Rouge, L'Abyssinie et L'Arabie aux XVIe et XVIIe Siècles.* Tome III, Cairo.

KEANE, A.H. 1907. *Stanford's Compendium of Geography and Travel (New Issue), Africa*, Vol. 1. London.

KLEMP, E. 1968. *Africa on Maps Dating from the Twelfth to the Eighteenth Century.* Leipzig.

KOEMAN, C. 1952. *Tabvlae Geographicae qvibvs Colonia Bonae Spei Antiqua Depingitvr.* Amsterdam.

KRAPF, J.L. 1860. *Travels, Researches and Missionary Labours during an eighteen years' residence in Eastern Africa.* London.

KRETSCHMER, I. 1988 Österreichs Beitrag zur kartographischen Erschliessung Ostafrikas bis zum Ersten Weltkrieg. (In) *Abenteuer Ostafrika. Der Anteil Österreich-Ungarns an der Erforschung Ostafrikas. Ausstellung in Schloss Halbturn,* Eisenstadt, 129-160.

KUEI-SHENG CHANG, 1970. Africa and the Indian Ocean in Chinese Maps of the Fourteenth and Fifteenth Centuries. *Imago Mundi* 24, 21-30. See also Gao Jinyuan, 1984. China and Africa: the Development of Relations over Many Centuries. *African Affairs* 83, 241-50.

KUMM, H.K.W. 1925. The Arab Geographers and Africa. *Scottish Geographical Magazine* 41, 284-289.

LANDER, R. & J. 1832. *Journal of an Expedition to Explore the Course and Termination of the Niger: with a narrative of a voyage down that river to its termination.* London.

LANE-POOLE, E.H. 1937. Old Maps of Africa. In *Handbook of the David Livingstone Memorial Museum,* W.V. Brelsford, ed., Livingstone, 141-162.

LANE-POOLE, E.H. 1950. The Discovery of Africa. A History of the Exploration of Africa as Reflected in the Maps in the Collection of the Rhodes-Livingstone Museum. *Occasional Papers Rhodes-Livingstone Museum* 7.

LANGHANS, P. 1897. *Deutscher Kolonial-Atlas.* Gotha.

LANGLANDS, B.W. 1961. Maps of Africa 1540-1850. *Uganda Museum Occasional Paper* 5.

LANGLANDS, B.W. 1962. Concepts of the Nile. *Uganda Journal* 26, 1-22.

LARBY, P.M. 1987. *Maps and Mapping of Africa.* SCOLMA, London.

LIEBENBERG, E.C. 1973. *Die Topografiese Kartering van Suid-Afrika, 1879-1972.* MA Thesis, Potchefstroom University.

LINKE, M. 1970. Ägypten im europäischen Kartenbild vom 17. bis in die Mitte des 19. Jahrhunderts. *Petermanns Geographische Mitteilungen* 114, 3, 229-237.

LINKE, M. 1981. Die 'Carte topographique de l'Egypte' der Französischen Expedition 1798-1801. *Wiss. Abh. d. Geogr. ges d. DDR* 17, 267-280.

LIVINGSTONE, D. & C. 1865. *Narrative of an Expedition to the Zambezi and its Tributaries; and of the discovery of the Lakes Shirwa and Nyassa. 1858-1864.* London.

LIVINGSTONE, D. & OSWELL, W.C. 1852. Latest explorations into Central Africa beyond Lake Ngami. *Journal of the Royal Geographical Society* 22, 163-174.

LIVINGSTONE, D. 1857. *Missionary Travels and Researches in South Africa.* London.

MACKINDER, H.J. 1900. A journey to the summit of Mount Kenya, British East Africa. *Geographical Journal* 15, 453-485.

MACQUEEN, J. 1850. Notes on the present state of the geography of some parts of Africa. *Journal of the Royal Geographical Society* 20, 235-252.

MARTIN, C.G.C. & FRIEDLAENDER, K.J. 1984. *History of Surveying and Land Tenure in South Africa Collected Papers I. Surveying and Land Tenure in the Cape 1652-1812.* Cape Town

MARTIN, C.G.C. 1980. *Maps and Surveys of Malawi.* Rotterdam.

MASSON, J.R. 1986. Geographical Knowledge of Maps of Southern Africa Before 1500 A.D. *Terrae Incognitae* 18, 1-20.

MASSON, J.R. 1989. The first map of Swaziland, and matters incidental thereto. *Geographical Journal* 155, 3, 335-341.

McADAM, J. 1974. The flying mapmakers: some notes on early development of air survey in central and southern Africa. *Rhodesiana,* 30 June, 44-64.

McCAW, G.T. 1932. The African arc of meridian. *South African Survey Journal* 4, 53-57, 93-97 & 182-187.

McGRATH, G. 1976. The Surveying and Mapping of British East Africa 1890 to 1946. *Cartographica Monograph* 18.

McGRATH, G. 1983. Mapping for Development. The Contributions of the Directorate of Overseas Surveys. *Cartographica Monograph* 29-30.

McGRATH, G. 1983. The Directorate of Overseas Surveys: Colonial Development and Overseas Aid. *Cartographic Journal* 19, 2, 91-94.

MENDELSSOHN, S. 1973. Cartography of South Africa. In *Mendelssohn's South African Bibliography,* Vol. II. London, 1095-1113.

MERRETT, C.E. 1979. *A Selected Bibliography of Natal Maps, 1800-1977.* Boston.

MITCHELL, P.K. 1968. Eighteenth and nineteenth century printed maps of Sierra Leone: a preliminary listing. *Sierra Leone Studies* 22, 60-83.

MOFFAT, R. 1842. *Missionary Labours and Scenes in Southern Africa.* London.

MOTA, A.T. da 1965. *A cartografia antiga da Africa Central e a travessia entre Angola e Moçambique 1500-1800.* Lourenço Marques.

MOTA, A.T. da 1977. Africa in the Anonymous Portuguese "Cantino" Planisphere. In *Land und Seekarten im Mittelalter und in der Frühen Neuzeit,* C. Koeman, ed., Munich, 123-135.

MURRAY, J. 1981. *Cultural Atlas of Africa.* Oxford.

NGO, V.V. 1986. *Survey and Mapping of Cameroon 1884-1984.* Ph.D. diss., University of London, 2 vols.

NGO, V.V. 1987. Cartobibliography of the Cameroon. *O'Dell Memorial Monograph* 20.

NORWICH, O.I. 1966. Early Decorative Maps of Africa. *Africana Notes and News* 4, 179-187.

NORWICH, O.I. 1977. The Dudley maps of southern Africa from his atlas *Arcano del Mare*. *African Notes and News* 22, 5, 195-200.

NORWICH, O.I. 1979. Fat-tailed sheep on maps of Africa. *The Map Collector* 7, 31-34.

NORWICH, O.I. 1983. *Maps of Africa. An illustrated and annotated carto-bibliography.* Johannesburg.

NORWICH, O.I. 1989. Chinese maps of Africa in the fourteenth and fifteenth centuries. In *Maps in Africa*, South African Library General Series 14, Cape Town, 39-53.

NORWICH, O.I. 1989. Maps of Africa since 1486. In *Maps in Africa*, South African Library General Series 14, Cape Town, 13-26.

NORWICH, O.I. 1990. An unusual eighteenth century map of Southern Africa. *Quarterly Bulletin of the South African Library* 45, 1, 16-20.

O'BRIEN, C.I.M. 1985. African Surveys 1890-1940. In *Mapping the Commonwealth*, Commonwealth Institute, London.

ODA 1978. *Directorate of Overseas Surveys. Annual Report for the year ended 31 March 1978.* HMSO, London.

OGILBY, J. 1670. *Africa*. London.

OGUNSHEYE, F.A. 1964. Maps of Africa 1500-1800: a bibliographic survey. *Nigerian Geographical Journal* 7, 1, 34-46.

OLIVER, R. & FAGE, J.D. 1962. *A Short History of Africa.* Harmondsworth.

OUWINGA, M.T. 1975. *The Dutch Contribution to the European Knowledge of Africa in the Seventeenth Century: 1595-1725.* Ph.D. diss., Indiana University.

PANKHURST, A. 1989. An early Ethiopian manuscript of Tegre. *Proceedings of Eighth International Conference of Ethiopian Studies* 2, 73-78.

PARK, M. 1800. *Travels in the Interior Districts of Africa: performed under the direction and patronage of the African Association in the years 1795, 1796 and 1797.* London.

PENFOLD, P.A. 1979. *Maps and Plans in the Public Record Office 3. Africa.* London.

PENN, N.G. 1994. Mapping the Cape: John Barrow and the first occupation of the Colony, 1795-1803. (In) *Maps and Africa*, J.C. Stone (ed.), Aberdeen, 108-127.

PETHERICK, J. 1861. *Egypt, the Soudan and Central Africa with explorations from Khartoum on the White Nile to the regions of the Equator being sketches from sixteen years' travel.* Edinburgh.

PETHERICK, J. 1865. Land Journey Westward of the White Nile from Abu Kuka to Gondokoro. *Journal of the Royal Geographical Society* 35, 289-300.

PETHERICK, J. 1869. *Travels in Central Africa, and Explorations of the Western Nile Tributaries.* London.

PHIPPS, M. & SDIRI, A. 1994, Al-Hasan ibn Muhammad al-Wazzan az-Zayyati *alias* Leo Africanus. *Geographers Biobibliographical Studies* 15, 1-9.

PINTO, S. 1879. Major Serpa Pinto's Journey across Africa. *Proceedings of the Royal Geographical Society* 1, 481-489.

POGNON, E. 1976. Cartes et Plans. In *Sources de l'Histoire de l'Afrique au Sud du Sahara dans les Archives et Bibliothèques françaises.* London (UNESCO), 135-136.

PULLAN, R.A. 1977. The history and use of aerial photography in Zambia. *Zambia Geographical Journal* 31/32, 33-52.

PULLAN, R.A. 1978. A first check list of the published maps of Northern Rhodesia, 1890-1949. *Zambia Geographical Association Bibliographies* 3.

RAFFLE, J.A. 1984. Mapping the Kalahari Desert. *Botswana Notes and Records* 16, 107-116.

RAMUSIO, G.B. 1550. *Delle Navigationi et Viaggi nel qual si Contiene les Descrittione dele Africa.* Venice.

RANDLES, W.G.L. 1958. South East Africa and the Empire of Monomotapa as shown on selected printed maps of the 16th century. *Studia* 2, 103-163.

RANDLES, W.G.L. 1965. South-east Africa as Shown on Selected Printed Maps of the Sixteenth Century. *Imago Mundi* 13, 69-88.

RAVENSTEIN, E.G. 1891. The Lake Region of Central Africa: a Contribution to the History of African Cartography. *Scottish Geographical Magazine* 7, 299-310.

RELAÑO, F. 1995 Against Ptolemy: The Significance of the Lopes-Pigafetta Map of Africa. *Imago Mundi* 47, 49-65.

RENNELL, Major (J.) 1798. *Proceedings of the Association for Promoting the discovery of the interior parts of Africa: containing an abstract of Mr. Park's account of his Travels and Discoveries, abridged from his own minutes by Bryan Edward Esq. Also Geographical Illustrations of Mr. Park's journey, and of North Africa at Large by Major Rennell.* London.

RISTOW, W.W. 1967. Seventeenth-century wall maps of America and Africa. *Quarterly Journal Library of Congress* 24, 2-17.

ROBINSON, A.M.L. 1975. Lady Anne Barnard's Map, 1798. *Quarterly Bulletin South African Library* 30, 1, 14-19.

SANTOS, J. Dos. 1609. *Ethiopia Oriental.*

SANUTO, L. 1588. *Geografia ... dell' Africa.* Venice.

SAWYER, R. 1980. Collecting Maps of the African Continent. *Antiques in South Africa*, 77-78.

SCHILLING, O. 1968. Das Reich Monomotapa, sein erstes Bekanntwerden, sein Name und Seine Darstellung auf den Karten des 16. bis 19. Jahrhunderts. *Acta Cartographica* 2, 449-510.

SCHLICHTER, H. 1891. Ptolemy's Topography of Eastern Equatorial Africa. *Proceedings of the Royal Geographical Society* 13, 513-553.

SCHRIRE, D. 1965. The Cape of Good Hope 1782-1842 from De la Rochette to Arrowsmith. *Map Collectors' Circle* 17.

SDUK, 1844. *Maps of the Society for the Diffusion of Useful Knowledge*. London.

SHORE, A.F. 1987. Egyptian Cartography. In *The History of Cartography*, Vol. 1, J.B. Harley & D. Woodward, eds., Chicago, 117-129.

SIDDIQI, A.H. 1992 Ibn Battuta. *Geographers Biobibliographical Studies* 14, 1-12.

SKELTON, R.A. 1965. Bibliographical Note. In *Livio Sanuto Geografia dell' Africa Vencie 1588,* Theatrum Orbis Terrarum, Second Series, Vol. 1, Amsterdam.

SMITH, A.H. 1952. *Exhibition of Decorative Maps of Africa up to 1800: Descriptive Catalogue.* Johannesburg.

SMITH, A.H. 1960. Decorative Maps of Africa. *Samab* 7, 6, 117-121.

SMITH, A.H. 1960. Decorative Maps of Southern Africa. *South African Panorama* 5, 6, 38-41.

SMITH, A.H. 1973. African Decorative Maps. In *Africana Curiosities*, A.H. Smith, ed., Johannesburg, 11-28.

SMITH, C.D. 1987. Cartography in the Prehistoric Period in the Old World: Europe, the Middle East and North Africa. In *The History of Cartography*, Vol. 1, J.B. Harley & D. Woodward, eds., Chicago, 54-101.

SPARRMAN, A. 1785. *A Voyage to the Cape of Good Hope ... from the Year 1772 to 1776.* London.

SPEKE, J.H. 1863. *Journal of the Discovery of the Source of the Nile.* Edinburgh.

SPEKE, J.H. 1864. *What Led to the Discovery of the Source of the Nile.* London.

STANLEY, H.M. 1872. *How I Found Livingstone. Travels and Adventures and Discoveries in Central Africa; including four months' residence with Dr. Livingstone.* London.

STANLEY, H.M. 1878. *Through the Dark Continent.* London.

STIGAND, A.G. 1923. Ngamiland. *Geographical Journal* 62, 6, 401-419.

STONE, J.C. 1976. Pioneer geodesy: the arc of the 30th meridian in former Northern Rhodesia. *Cartographic Journal* 13, 2, 122-128.

STONE, J.C. 1980. Imported technology for alien purposes: the case of the Land Surveyor in North-Western Rhodesia. In *Experts in Africa*, J.C. Stone, ed., Aberdeen, 29-42.

STONE, J.C. 1982. The district map: an episode in British colonial cartography in Africa, with particular reference to Northern Rhodesia. *Cartographic Journal* 19, 2, 104-112.

STONE, J.C. 1983. An early map of the Hook of Kafue. In *An African Miscellany for John Hargreaves*, R. Bridges, ed., Aberdeen, 93-95.

STONE, J.C. 1984. The compilation map: a technique for topographic mapping by British colonial surveys. *Cartographic Journal* 22, 121-128.

STONE, J.C. 1988. Imperialism, colonialism and cartography. *Transactions of the Institute of British Geographers* NS 13, 57-64.

STONE, J.C. 1993. *A Colonial Surveyor at Work: the Field Diary of W.G. Fairweather, Assistant Surveyor, Northern Rhodesia Government, 1913-1914.* Aberdeen.

STONE, J.C. (ed.) 1994. *Maps and Africa. Proceedings of a colloquium at the University of Aberdeen April 1993.* Aberdeen.

STONE, J.C. 1995. *A Short History of the Cartography of Africa.* Lewiston.

STONE, J.C. 1995. The cartography of colonialism and decolonisation: the case of Swaziland. In *Geography and Imperialism 1820-1940*, M. Bell, R. Butlin & M.Heffernan, eds., Manchester, 298-324.

STONE, J.C. 1996. The early years of official topographic mapping in Northern Rhodesia. *South African Journal of Surveying and Mapping* 23, 4, 217-231.

STORRAR, C.D. 1989. Fourcade's map of Devil's Peak. In *Maps in Africa*, South African Library General Series 14, Cape Town, 29-38.

THOMAS, H.B. & SPENCER, A.E. 1938. *A History of Uganda Land and Surveys and of the Uganda Land and Survey Department.* Entebbe.

THOMAS, P.W. 1982. The topographical 1:50,000 map series of South Africa. *South African Journal of Photogrammetry, Remote Sensing and Cartography* 13, 77-88 & 171-184.

THOMSON, J. 1893. To Lake Bangweolo and the unexplored region of British Central Africa. *Geographical Journal* 1, 97-121.

THROWER, N.J.W. 1996. *Maps and Civilization.* Chicago.

TIBBETTS, G.R. 1971. Arab navigation in the Indian Ocean before the coming of the Portuguese. *Royal Asiatic Society Oriental Translation Fund*, New Series, 42.

TIBBETTS, G.R. 1992 The role of charts in Islamic navigation in the Indian Ocean. *In The History of Cartography*, Vol. 2, book 1, J.B. Harley & D. Woodward, eds., Chicago, 256-262.

TOLMACHEVA, M. 1991. Ptolemaic Influence on Medieval Arab Geography: The Case Study of East Africa. In *Discovering New Worlds*, S.D. Westrem, ed., 125-141., New York. See also: Ptolemy's East Africa in Early Medieval Arab Geography. *Journal for the History of Arabic Science*, 9 (1991), 31-43.

TOOLEY, R.V. 1963. Early maps and views of the Cape of Good Hope. *Map Collectors' Series* 6.

TOOLEY, R.V. 1966. Printed maps of the continent of Africa and regional maps south of the Tropic of Cancer. Parts I & II. *Map Collectors' Circle* 29 & 30.

TOOLEY, R.V. 1968. Maps of Africa. A selection of printed maps from the sixteenth to the nineteenth century. Parts I & II. *Map Collectors' Circle* 47 & 48.

TOOLEY, R.V. 1969. *Collectors' Guide to Maps of the African Continent and Southern Africa.* London.

TOOLEY, R.V. 1970. Printed maps of southern Africa and its parts. *Map Collectors' Circle* 61.

TOOLEY, R.V. 1972. A Sequence of Maps of Africa. *Map Collectors' Circle* 82.

TOOLEY, R.V. 1979. The Great Lakes of Africa. *The Map Collector* 7, 13-16.

TOOLEY, R.V., BRICKER, C. & CRONE, G.R. 1969. *A History of Cartography*, London. Republished (1976) as *Landmarks of Mapmaking*, Oxford.

TRIMINGHAM, J. S. 1975. The Arab Geographers and the East African Coast. In *East Africa and the Orient*, H.N. Chittick & R.I. Rotberg, eds., London, 115-146.

VALETTE, J. 1962. Cartographie ancienne de Madagascar. *Bulletin du Madagascar* 192, 619-622.

VENNING, J.H. 1955. Early days in Balovale. *Northern Rhodesia Journal* 2, 4, 53-57.

WALLIS, H. & MIDDLETON, D. 1987. The mapping and exploration of Africa and Joseph Banks. In *Maps and Mapping of Africa*, P.M. Larby, ed., London, 47-54.

WALLIS, H. 1986. So geographers in Afric maps. *The Map Collector* 35, 30-34.

WILKS, I. 1992. On mentally mapping Greater Asante: a study of time and motion. *Journal of African History* 33, 2, 175-190.

WILLIAMS, J.H. 1931. *Historical Outline and Analysis of the Work of the Survey Department Kenya Colony 1st April, 1903 - 30th September, 1929*. Nairobi.

WILLIAMS, S.G. 1960. Early maps of Nyasaland. *Society of Malawi Journal* 13, 2, 13-15.

WINTERBOTHAM, H.S.L. & McCAW, G.T. 1928. The triangulations of Africa. *Geographical Journal* 71, 1, 16-36.

WINTERBOTHAM, H.S.L. 1936. Mapping of the Colonial Empire. *British Association for the Advancement of Science. Report of the Annual Meeting, 1936*, 101-116.

WOLDAN, E. 1981. Die Altesten Gedruckten Modernen Karten Afrikas. *Anzeiger der Österreichische Akad. der Wiss. Philosophisch-Historische Klasse (Austria)* 118(1-10), 252-257.

WOOD, J.C. 1943. Sir David Gill and the geodetic survey of South Africa and the arc of the 30th meridian. *Empire Survey Review* 7, 50-56.

WORTHINGTON, E.B. 1938. *Science in Africa*. London.

YOUNG, A. 1968. Mapping Africa's natural resources. *Geographical Journal* 134, 2, 236-241.

THE CONTINENT OF AFRICA

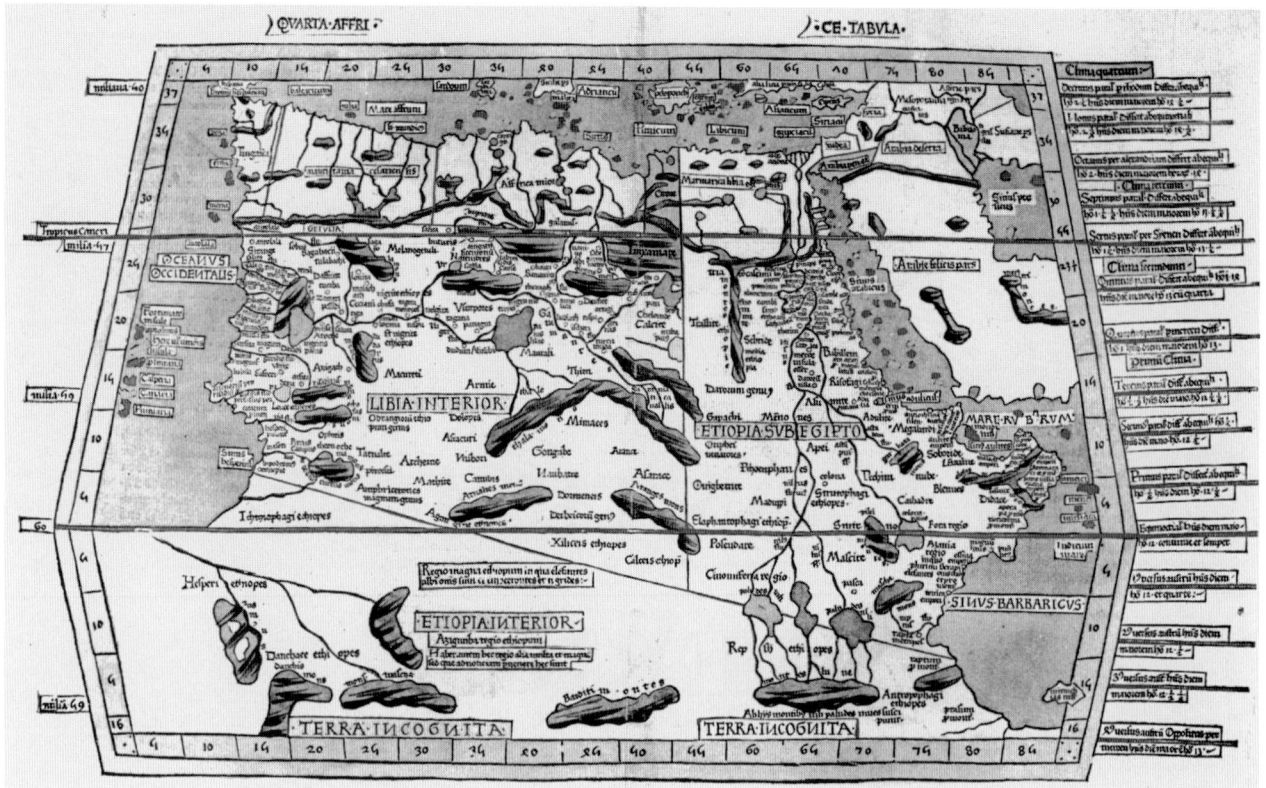

Map 1

ULM (Nicolaus Germanus) (c. 1470-1490)
Quarta Affrice Tabula. (Map of Africa.) (Ulm, Johann Roger, 2nd ed., for Justus de Alba
and Venette, 1486.)
Map, 31 x 47 cm, coloured.

An incunabulum woodcut map of Africa of 1486, taken from the second edition of an atlas based on Ptolemy's *Cosmographia*, which first came from the press of Leinhart Holle at Ulm in July 1482. The previous editions of Ptolemy's text were those printed at Vicenza, 1475 (without maps), Bologna, 1477 and Rome, 1478. The maps had been engraved on copperplate for the Bologna and Rome editions and to accompany the Italian version in *terza rima* by Francesco Berlinghieri, printed at Florence in 1482.

Leinhart Holle's Ulm edition of the *Cosmographia* was the first to appear outside Italy and the first to have woodcut maps. It is also the earliest edition in which the design and execution of the maps are ascribed to a named cartographer. This was the German Benedictine known as Donnus (Dominus) Nicolaus Germanus, who is named in the colophon of the Ulm edition and whose dedication to Pope Paul II is prefixed to it. The manuscript of the *Cosmographia* which provided copy for the printer of the two Ulm editions of 1482 and 1486 was identified by Joseph Fisher (1902) as a codex by Nicolaus preserved in Schloss Wolfegg, Wurtemburg.

Unlike the earlier edition of the *Cosmographia*, the descriptive text relating to the regional maps was printed at the back of the map in the 1486 edition. A curious feature of the work of Germanus is the engraving of the capital roman N in reverse. In the Ulm edition, as in the manuscript, the twenty-six regional maps are numbered Africa I - IV. Most of the extant copies of the 1482 and 1486 editions have maps, initials and borders coloured. The most usual conventions are blue for sea and rivers; olive green or bronze for mountain-chains; green (foliage) and yellow (stems) for tree symbols, denoting forest; red or brown lines along boundaries; red or yellow for the circular town symbols and bright colours for islands.

This handsome 1482 edition of the *Cosmographia* was unprofitable and by 1484 Leinhart Holle was out of business and all his stock, blocks and printed sheets passed into the hands of Johann Roger, who lost no time in bringing the *Cosmographia* back on the market. In the map section of his 1486 edition the descriptive texts were reset with only minor corrections but without borders (as existed in the 1482

2

edition). The maps were reprinted from Holle's blocks with no other change than the addition of woodcut headings in which the capital N is still reversed. The normal order of the map is as in the 1482 edition and the colour convention is vividly demonstrated. The map itself is printed on the trapezoid projection. The complete south is labelled *terra incognita* with 'Libia' and 'Etiopia' featuring prominently east and west. The basin of the Nile is Ptolemaic in concept and the Montes Lune are included.

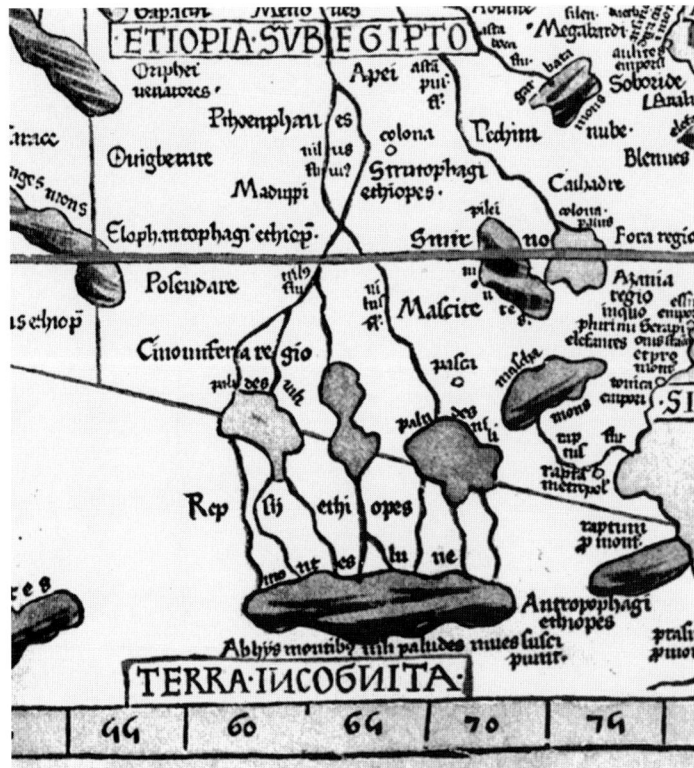

· Africa/ Libya/ Morland/ mit allen künigreichen so zü vnsern zeiten darin gefunden werden.

Map 2

MÜNSTER, Sebastian (1489-1552)
Africa, Libya, Morland, mit allen Kunigreichen so zu unsern zeiten darin gefunden werden. (Africa, Libya, Morland with their kingdoms to date.)
(Basle, S. Münster, 1544-1545.)
Map, 26 x 34 cm, uncoloured woodcut.

This black and white German edition of Münster's map of Africa is from an early edition of his *Cosmographia*. The title across the top border is in German and at the lower left there is a different German title within a woodblock surround. This is probably the 1544-1545 German edition, popularly referred to as the second edition of the original Münster map.

Münster was born in Hesse in 1489 and died of the plague in Basle in 1552. He studied in Heidelberg and became Professor of Hebrew at Basle, where he also taught mathematics and cosmography and, later in life, devoted himself almost exclusively to geography. He introduced the idea of having a separate map for each continent, and was responsible for editions of Solinus and Pomponius Mela, as well as some separately printed maps. His first printed maps were his new edition, in 1525, of Etzlaub's map of Germany; of the world, in 1532, and of Europe, in 1536. In 1530 he married the widow of Adam Petri, a printer and map publisher of Basle. Her son, Heinrich, named his son after Münster – Sebastian – and Henry Sebastian assisted in the subsequent editions of Münster's *Cosmographia*.

Münster's first maps of Africa appeared as woodcuts in 1540 in the *Geographia Universalis...* assisted by Sebastian Petri. Later editions of these maps appeared in Latin from 1541 to 1552. The first edition of Münster's famous *Cosmographia Universalis* was published in 1544 and contained 24 double-page maps. This important work was subjected to continual revision: certain maps were replaced by others and new maps were added in the text or on separate sheets. The greatly expanded edition of 1550 contained only fourteen maps, unaltered, from the earlier editions and of the fifty-two maps in the text some were based on the old ones and some were entirely new.

After Münster's death in 1552 successive editions were enriched by the addition of new maps by other hands and the 1578 edition is the last to contain his original maps. The work was so popular that it went through twenty-seven German editions, eight Latin, three French, three Italian, four English and one Czech. In Münster's lifetime six German editions, three Latin and one French had appeared. The latest edition seems to be that published at Basle in 1650, so that the work had a useful life of over a century. In this century, therefore, successive generations were schooled in two cosmographies; that with an astronomical and mathematical bias and that of Münster's more regional geography.

In 1968 a faithful facsimile copy of the 1550 edition was published in the Netherlands as one of the *Theatrum Orbis Terrarum* series, but no library or institution has anything like a complete series of the various editions of Münster. There are some means of identification, especially by means of the language, the lettering and the legends. Münster employed the practice – common at that time with publishers of wood engravings – of inserting movable type into the wood block so that the lettering for place names could be changed if necessary, in different editions. In some instances lettering appears upside down, with some place names incorrectly placed, presumably because those working with the type were not skilled typesetters, or were illiterate.

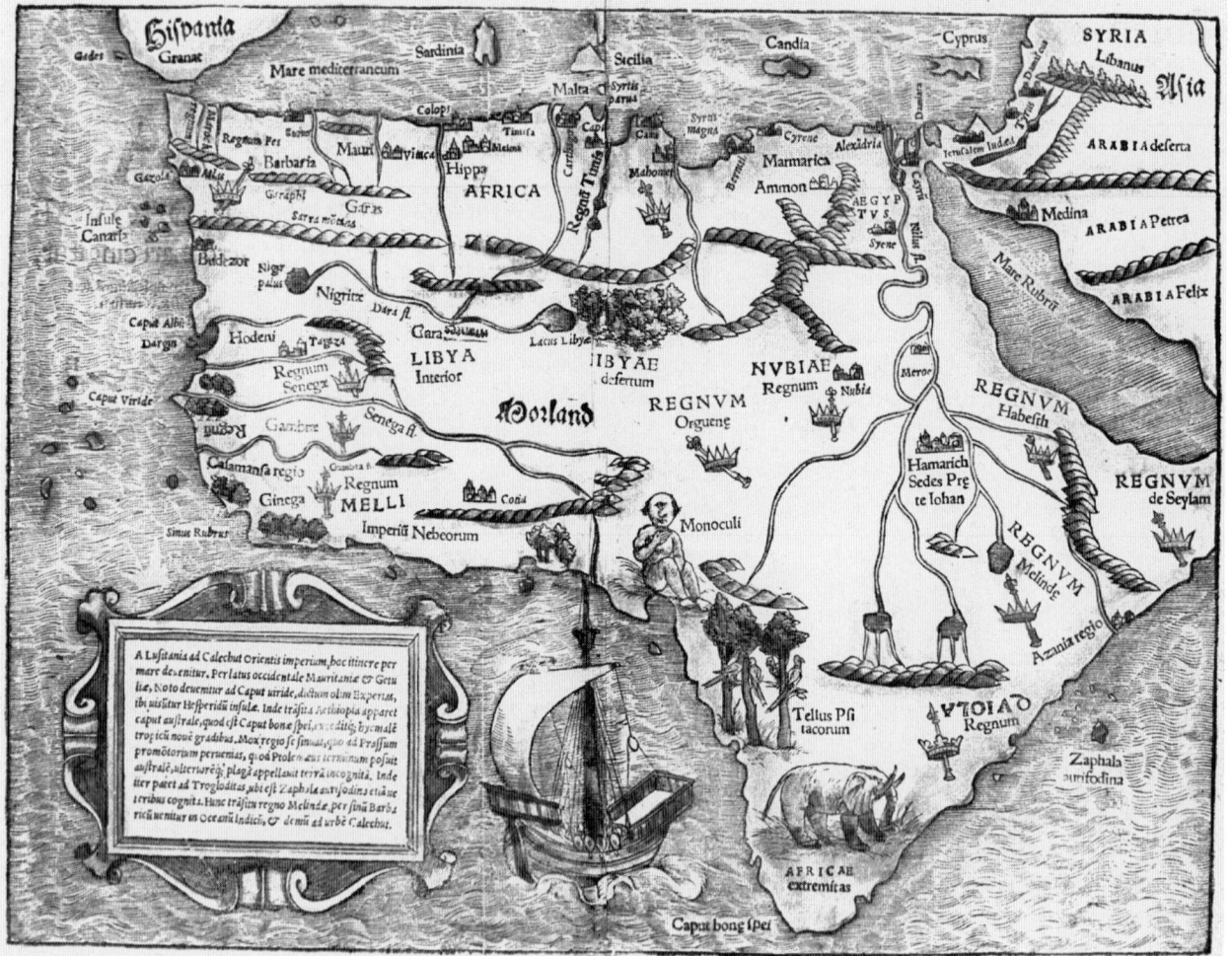

Map 3

MÜNSTER, Sebastian (1489-1552)
Totius Africae tabula et descriptio universalis etiam ultra Ptolemaei limites extensa. (Map of the whole of Africa based on Ptolemaic sources.)
(Basle, S. Münster, 1542.)
Map, 26 x 34 cm, coloured woodcut.

Another copy (coloured) of the Münster map, in a Latin edition of his *Geographia Universalis*, issued between 1540 and 1552. It shows the usual features of the early Münster woodcuts, together with the scroll-like cartouche within which the descriptive text in Latin is placed. This particular map has the interesting feature of showing three place names printed upside down: Quiloa, on the eastern coastline adjacent to the word Regnum; Mantes, being a portion of the word Caramantes which appears in the upper left part above Libya Interior; and Regnu[m], on the west coast. As previously mentioned, this is most probably due to the letter press of these early woodcuts having been inserted separately, so that either the letters were originally placed wrongly, or they could have fallen out prior to being printed and the typesetter, because of his presumed inability to read, did not replace them properly.

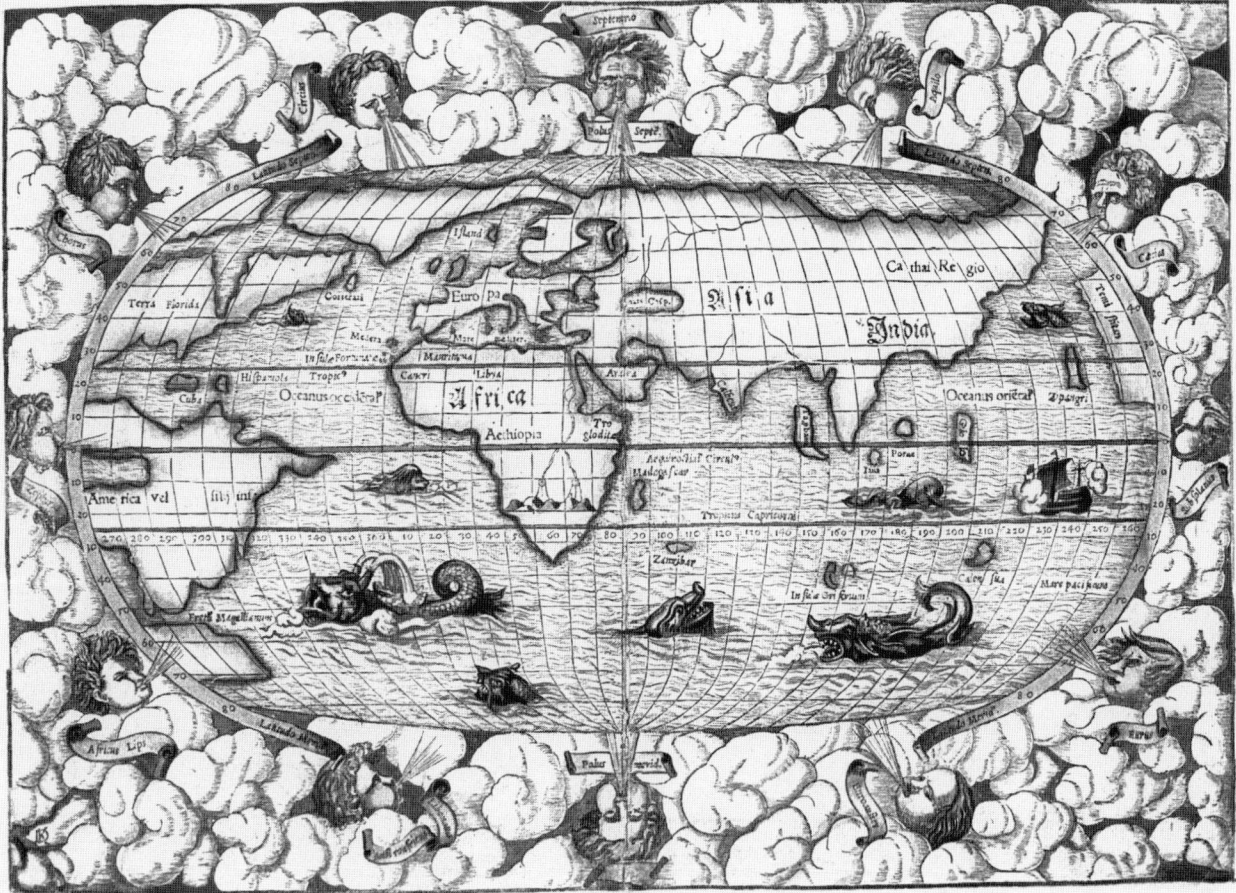

FIGVRA DEL MONDO VNIVERSALE.

Map 4

MÜNSTER, Sebastian (1489-1552)
Figura del Mondo Universale. (World map.) Engraved by David Kandel.
(Basle, S. Münster, 1550.)
Map, 25 x 37 cm, coloured.

Münster's *Cosmographia*, first issued in 1544, was greatly enlarged in the 1557 edition, with fourteen double-page woodcut maps and views and plans of cities, and numerous woodcut illustrations in the text, of costumes, real and mythical animals, trees, flowers and portraits. A double-page woodcut of the world exists which is almost identical to this map except for slight variations and the non-appearance of a DK monogram in the lower left corner. This monogram identifies David Kandel (1524-1596), a wood engraver and artist for the 1550 and 1567 editions of Münster's *Cosmographia*. The Latin title across the top border is further proof that this map appeared in one of the early Latin editions of Münster's *Cosmographia*.

This world map, an extremely finely engraved woodcut, features Africa, with the origin of the Nile prominent; America, Asia, India and Europe. The various continents are given a rounded appearance and are surrounded by numerous wind gods, whose heads appear through the billowy masses of clouds, blowing their winds onto the whole earth. The seas are filled with many unusual monsters and there is a sailing vessel in the east.

Map 5

MÜNSTER, Sebastian (1489-1552)
La description d'Affricque selon les divers pais, animaux et monstres horribles. (A description of Africa showing the different countries, animals and monsters.)
(Basle, S. Münster, 1550.)
Map, 12,4 x 15,5 cm, coloured.
Top right: 139.
At the top: *Universelle, livre VI Le Sixiesme Livre de la Cosmographie, recueilly par Sebastian Münstere des bons autheurs tant anciens que moderuer, & assemblé.*

La description d'Affricque selon ses divers
pais, animaux & monstres horribles.

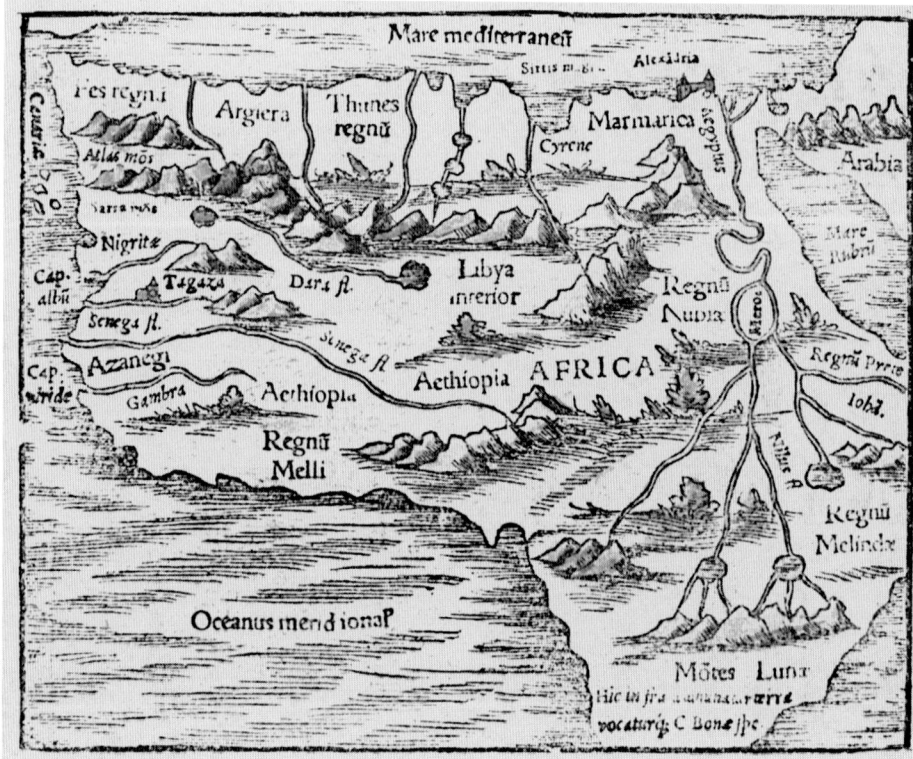

Most editions of Münster's *Cosmographia* contain a small woodcut map of Africa as a text illustration. This copy appeared in one of the French editions. It excludes the Horn of Africa and the Cape of Good Hope, but the Mons Lunae origin of the Nile appears, with many kingdoms of Central and North Africa. Additional text appears on the reverse.

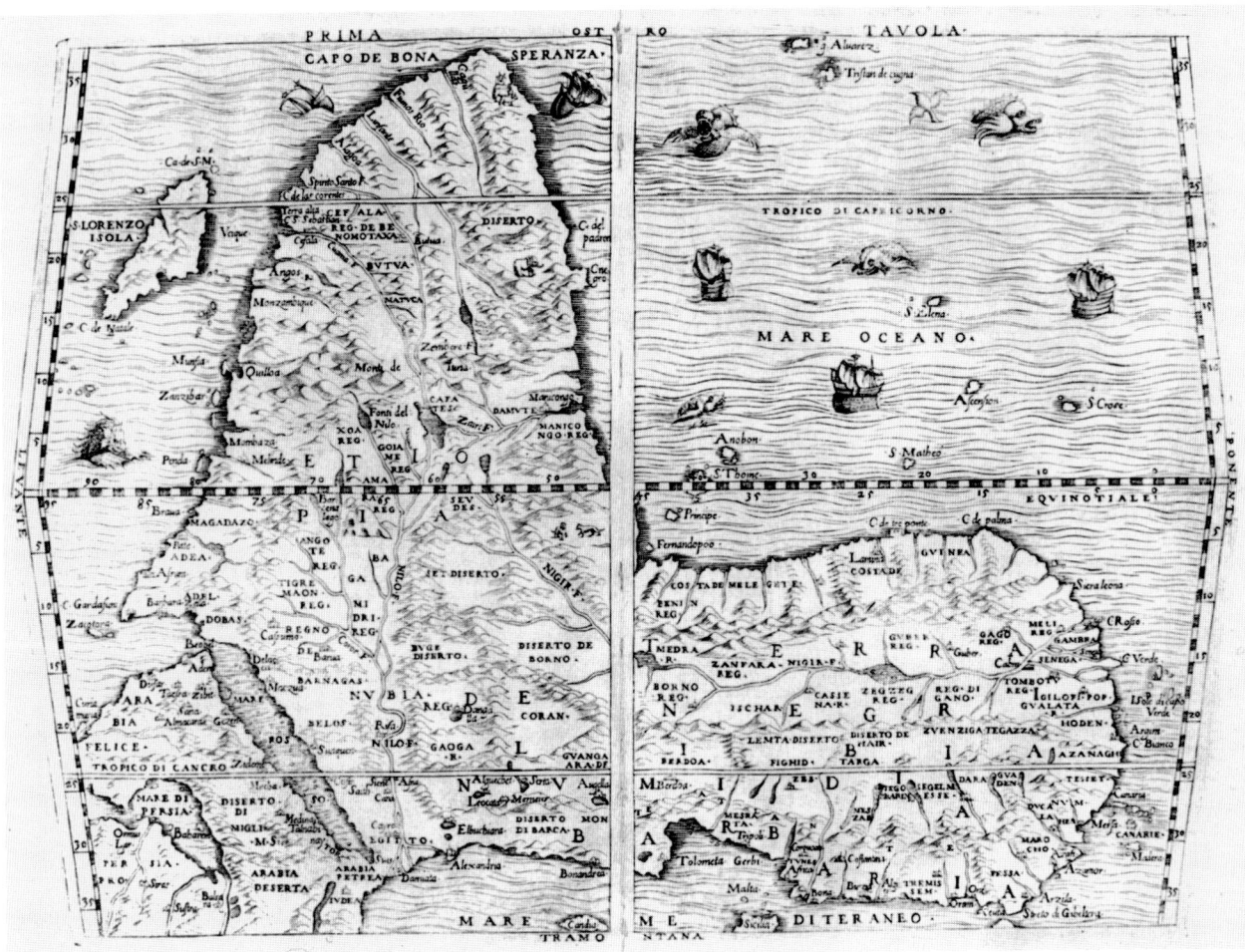

Map 6

RAMUSIO, Giovanni-Battista (1485-1557)
(Africa) Prima Tavola. (Plate 1.) (Venice, G.B. Ramusio, 1550.)
Map, 27 x 38 cm, uncoloured woodcut.
South at top of map – hence 'the upside-down map.'

This map of Africa appeared for the first time in 1550, in the second edition of Ramusio's travels, *Delle Navigatione et Viaggi nel qual si contiene les descrittione dele Africa.* This particular edition was re-engraved on a copperplate, the original having been executed as a wood-block. The engraving was completed for Ramusio by the celebrated Italian cartographer Giacomo Gastaldi. It was printed with south at the top of the map, and the sea is engraved in typical horizontal wavy lines decorated with numerous ships and sea monsters. This copy has no text on the reverse. Very little detail is provided in the southern part of the continent except for fictitious and inaccurate mountains and rivers. The Nile once again appears as arising from the two-lake system, in the Ptolemaic tradition. The information conveyed on the map is based on knowledge obtained from two Arab geographers (Edrisi, born in Ceuta in 1099, and Leo Africanus) and the Portuguese discoveries. A similar map appears in the early edition of Leo Africanus with minor differences. (See Map 7.)

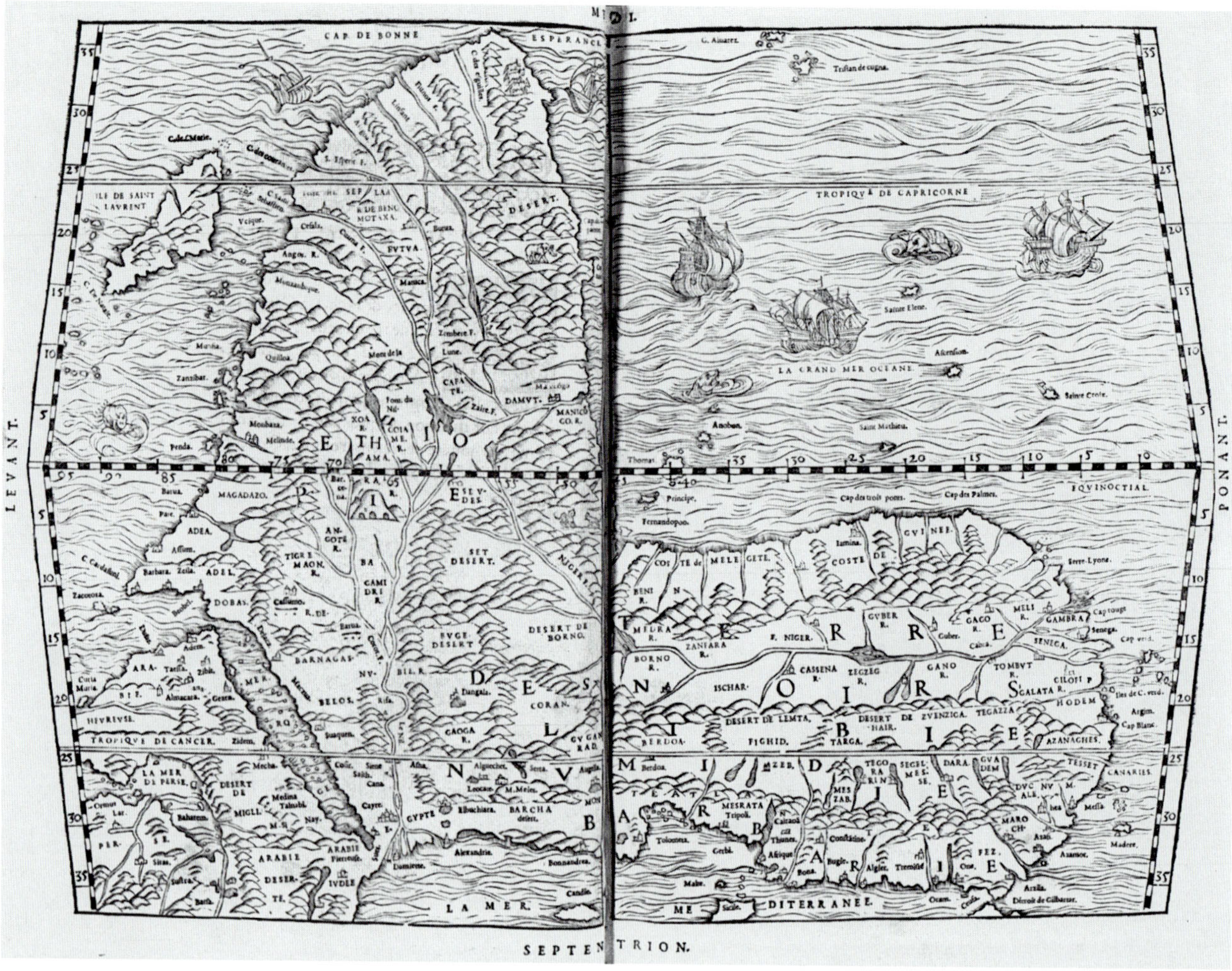

Map 7

AFRICANUS, Leo (1483-1552)
(*Africa.*) (Lyons, 1556.)
Map, 27,5 x 37,5 cm, uncoloured.

An Italian copperplate engraved map of Africa by the Arab geographer Leo Africanus, which appears in both volumes of the 1556 edition of his work *De l'Afrique*, published in Lyons. It is strongly reminiscent of the map appearing in Ramusio's travel book (see Map 6). This map has no title but the north, south, east and west appear on the centre of each map border as Septentrion, Midi, Levant and Ponant respectively. There are decorative ships and sea monsters scattered in the Indian and Atlantic Oceans, with what appear to be leopards and an elephant on the mainland. As with other maps of this period the south is shown to the top of the map, and large rivers, lakes and mountains are noted in the southern extremity. The origin of the Nile is shown here, too, to arise from the two lakes side by side in Central Africa.

NOT IN COLLECTION

10

Map 8 GASTALDI, Giacomo (c. 1500-1565).
Il disegno della geografia moderna de tutta la parte dell'Africa ... Engraved by
Fabricius Licinus. (Venice, G. Gastaldi, 1564.)
Map, eight sheets in two mounts, 53 x 142 cm each, uncoloured. Scale in Italian miles.

This rare separate map of Africa was printed on eight copperplates and is the finest and most important sixteenth-century large-scale map of Africa. It was published by the Italian geographer Gastaldi, who is known to have worked on it as early as 1545.

Giacomo Gastaldi (also spelt Castaldo and Castaldi) was born at Villa Franca in Piedmont and became the greatest of the sixteenth-century Italian geographers. After settling in Venice he became Geographer to the Venetian Republic. Because of Italy's favourable geographic position (before the discovery of the sea routes to the East and West), its centuries of experience in the construction of sea charts, its wealthy merchants, and the large number of Italian copperplate engravers, Italy became the map industry's centre of production and distribution at that time. Gastaldi's work had a seminal influence on many other European countries and his maps were in great demand. No less than 109 separate

map publications by him are known, the principal being the large map of the world, 1546; of America (three sheets), 1559; Africa (eight sheets), 1564; Europe (four sheets); Italy, 1561 and Lombardy, 1570. He also issued an edition of Ptolemy, for which he engraved a series of maps, in 1548.

The title of this map is placed within a simple rectangular cartouche; the sea is stippled, as is typical of the Italian copperplate technique, and illustrates ships and sea monsters, of which the greater number and most varied types appear in the left lower sheets. The rest of the map is not decorated. The interior is everywhere covered with mountain ranges, each shaded on the western side to indicate perpetual sunshine; capes and bays are named all round the coast from Portuguese and Arab sources and towns are shown in the interior. Simbaoe – which may refer to Zimbabwe, famous for its archaeological ruins – is featured and Buro Mina de Oro in Cefala undoubtedly refers to ancient gold mines,

11

whose king is referred to as Benomotopo. The Ptolemaic concept of the two-great-lake origin of the Nile is illustrated by one on the left, which is named Zaire in the north and Zembere in the south, while the right-hand lake is lettered Zafran. The Spirito Sancto and Cuomo rivers (Zambezi and Limpopo) join together to enter Lake Zembere. At the southern part of the Cape, Table and Saldanha bays are accurately illustrated although not named as such – a rather unusual feature. C de Bona Speranza is prominently labelled.

Gastaldi produced and published this large map of Africa in 1564. It was his finest and largest copperplate line engraving of this continent, measuring 143 centimetres east to west and 106 centimetres north to south. Beneath the title appears the name Fabricius Licinus – Gastaldi's engraver. On either side of the title appear three columns comparing modern and ancient place names. Many characteristic features of this map are recognisable in subsequent products of the sixteenth century, especially in those of the master cartographers and publishers such as Ortelius, de Jode and Magini as well as Mercator, Sanuto and Pigafetta. Tooley, in his article in *Imago Mundi* (1939), 'Maps in Italian Atlases of the 16th Century', records only six known copies of this map in different museums and libraries of the world, including the copy illustrated here. However, he later records one further copy, the seventh, owned by the Greenwich Museum Map Room. In an article in *Africana Notes and News* (Johannesburg, June 1976), an additional, eighth copy has been reported to have been found in a Lafreri-type atlas in the Marucelliana Library in Florence in 1975 (by the writer.) The Gastaldi Africa map described here is the original Lloyd Triestino copy, which Almagia described as having no date. This was printed in the *Imago Mundi* article on sixteenth-century Italian maps, but in fact the map, which is included in the author's collection, has 1564 very clearly marked on it.

Map 9

FORLANI, Paola de (fl. 1576).
Africa a veteribus (Venice, P. de Forlani, 1566.)
Map, 43 x 63 cm, uncoloured.
Dedicated to Nicolaus Stopius.

An example of a finely engraved Italian copperplate map of the late sixteenth century. These early Italian maps were issued separately and not in volumes, with no uniformity of size or number. Smaller maps had margins or were mounted to a uniform size so that the whole collection could be bound in one folio volume. Such collections are the so-called Lafreri atlases. The contents were either selected by the customer or made up by the publisher from his stock. They mostly preceded the standard Flemish atlas of Ortelius and no two copies were identical. An engraved title was occasionally added. The name Lafreri is partly a misnomer as Bertelli, Camocio, Duchetti and Forlani also issued similarly bound collections. The map featured here is an example of Forlani's work of 1566.

Forlani was a Veronese engraver and publisher, working in Venice, whose work was very popular around 1576. His prolific output included maps of the world and of Africa, Greece, Tuscany, France, Poland and America. He was one of that group of Italian publishers who excelled in the fine art of copperplate engraving, somewhat severe in style and with ornamentation in the oceans. In this map Forlani appears to have copied some of his large rivers and tribal names from those used by his compatriot, the Italian publisher Ramusio.

Above the simple cartouche surrounded by decorative scrollwork is the dedication to Nicolaus Stopius, the sixteenth-century cartographer. Within the cartouche are eleven lines in Latin describing Africa and below this the imprint, all in italics. There are sea monsters and ships in the Atlantic and Indian Oceans and, well to the north of the island of Madagascar, Noah's Ark is depicted, showing the dove returning with an olive branch in its beak. A building similar to the Doge's palace appears in Central Africa, possibly indicating Prester John's home. Geographically, Africa appears much wider than usual from east to west, just north of the equator. There are many mountain ranges with their western aspect shaded, a few lakes and rather sparse place-names on the mainland and along the coastlines. The Nile is shown as having its origin in two lakes between 'Manicongo' and 'Monzambique.'

13

Map 10

ORTELIUS, Abraham (1527-1598)
Africae Tabula Nova. (New map of Africa.) (Antwerp, A. Ortelius, 1570.)
Map, 37 x 50 cm, coloured.

Abraham Ortelius was born and died in Antwerp. He studied mathematics and the classics and commenced his career as a map colourist at the age of twenty. After his father died he supported his mother and family by trading in maps, especially at the big fairs at Frankfurt and other cities. His trips to the larger cities enabled him to view and purchase the best foreign maps, and decided him to publish a uniform atlas. After ten years of conscientious labour he produced his famous *Theatrum Orbis Terrarum*, the first atlas, in 1570, printed at the press of Christopher Plantin. This publication was a landmark in the history of cartography and had immediate success, requiring four printings in the first year. Ortelius was paid tribute by the famous geographer and cartographer Mercator and many geographers were so impressed with the work that they sent copies of their own maps for inclusion in the atlas, thus increasing the number

from 53 to 153. Almost fifty different editions appeared from 1570 to 1612 with texts in Latin, Dutch, French, German and Italian. The only English edition was published in 1606. Ortelius was thus responsible for the first systematic collection of uniform-sized maps of all parts of the world, based on knowledge dating back to Ptolemy. He was the first to separate the old from the new geography, and also to show their relationships by indicating the change from old to new nomenclature. Ortelius did what few other early map makers did – he gave the sources of his information. The map of Africa was, apparently, one of the few for which he himself was responsible, as he gives no source, but it has been pointed out that there is some similarity between this map and the copperplate engraved maps of Forlani and Gastaldi.

14

The *Theatrum Orbis Terrarum* added one name, Carneca, to Gastaldi's Africa, and made the Cape more pointed. Zanzibar is transferred from the east to the west coast. 'Zanzibar' was a designation often given to the East African coastlands north of Cape Delgado, but a further oddity is that Ortelius marks the island of Zanzibar in the correct place. The Nile is based on the Ptolemaic concept. The cartouche is classical in design, with two armless statues of women and four lions' heads incorporated in the supporters, and is completed with decorative scrollwork. Two swordfish and a whale decorate the Atlantic, while in the right-hand lower corner of the Indian Ocean a decorative group of three ships is enveloped in a cloud of smoke. Large lakes appear in various parts of the mainland and there are numerous rivers and cities, but no animal life is indicated. Despite a few mistakes, this beautiful map is one of the cornerstones of map making of the continent of Africa, and remained the standard map for the rest of the sixteenth century.

Map 10a

ORTELIUS, Abraham (1527-1598)
Africae tabula nova (New map of Africa.) (Antwerp, A. Ortelius, 1589.)
Map, 7,7 x 10,9 cm, uncoloured.

Ortelius published his maps in miniature version, in a series of small atlases. In the atlas, each map appeared as a full page illustration on the right side, and its corresponding text on the left, printed on the verso of the preceding map. These attractive copperplate images were very well engraved and contained much cartographic detail as well as embellishments such as ships and sea-creatures. The map of Africa shown above has been reproduced greatly enlarged in order to reveal these details.

Map 11

ORTELIUS, Abraham (1527-1598)
Presbiteri Johannis, Sive Abissinorum Imperii Descriptio. (The land of Prester John.) (Antwerp, Abraham Ortelius, 1573.)
Map, 37 x 43 cm, uncoloured. Dedicated to David.

This Ortelius map – 'Abyssinia, the Land of Prester John' – first appeared in his *Additamentum* and then in all subsequent editions of the *Theatrum Orbis Terrarum*. It depicts Africa from the Mediterranean to the Mountains of the Moon, which are placed below Mozambique. It is often referred to as the Prester John map. The dedication to biblical David is at the upper left, surrounded by the typical floral strap-like ornamentation. Above the dedication is the coat of arms of Prester John, and a long genealogical record traces his ancestors to King David. In the right lower corner is a similar flower-bordered rectangular cartouche with the map's title. An Arab dhow curiously appears off the west coast. Two dolphins are shown off the east coast and on land

four elephants. The two-lake system is represented, once again, with its tributaries providing the origin of the Nile. The east coast is shown in greater detail than in the 1570 map of Africa and the Arab settlements of Melinde, Mombaza, Quiloa and Mozambique are probably taken from the travels of Ibn Batuta, a renowned Arab explorer and writer.

Because Prester John's name and his kingdom appear so often on early maps of Africa it is highly appropriate that a descriptive history of this mythical figure should be included.

'About 1150 A.D. a rumour spread through Europe that somewhere in Asia there was a powerful Christian emperor

16

named Presbyter Johannes (with the court title of 'Gurkhan'), who had founded the kingdom of Kara Khitai. He had broken the power of the Musselman in his own domain after a fierce and bloody fight. The mysterious Priest-King became a symbol of hope in the Christian world beset by Mongol hordes. Pope Alexander III resolved to make contact with Prester John, and his first step was to address a letter to him (dated 27 September, 1177). The Pope's physician was dispatched to deliver the letter in person. He never returned. Pope Innocent IV was even more determined than his predecessor, and decided to convert the barbarians instead of trying to conquer them. Dominican and Franciscan missionaries as well as civil ambassadors of peace plodded back and forth between the Pope, the King of France and the Mogul Khan. These travellers soon learned that His Highness Presbyter Johannes and the Christian kingdom in deepest Asia were probably myths. But the popular fancy

was not easily dispelled, and instead of allowing their bubble to be punctured, the people merely transferred the kingdom of Prester John to Africa – specifically Abyssinia. No one knew very much about Abyssinia. A few die-hards like John de Plano Carpini and Marco Polo persisted in the belief that Prester John still reigned in all his splendour deep in the heart of the Orient. On the larger map in Higden's *Polychronicon* the empire of Prester John was located in lower Scythia within the limits of Europe, but on the map of Marino Sanuto it was placed in further India. It was moved again to Central Asia and ended up in Abyssinia. The legend persisted, however, and four hundred years after Pope Alexander III wrote his letter to Presbyter Johannes, Abraham Ortelius, a Dutch map publisher, issued a separate map entitled: *Presbiteri Iohannis Sive Abissinorvm Imperii Descriptio.* (L.A. Brown, *The Story of Maps*, pp. 98–99.)

Map 11a

ORTELIUS, Abraham (1527-1598)
Presbiteri Johannis, sive Abissinorum Imperii descriptio. (The land of Prester John.) (Antwerp, Abraham Ortelius, 1589.)
Map, 8 x 11 cm, uncoloured.

Ortelius pioneered the 'miniature atlas.' These charming little volumes contained essentially the same maps as did the larger folio atlases, but sold at a fraction of the cost of the large books. The map shown above is the miniature counterpart of Map 11. It is reproduced *slightly enlarged.* The

geography is essentially unchanged from that of the larger map, but the text panel in the upper left was removed. Two substantial elephants are in its place. The dhow has shifted to the Indian Ocean.

Map 12

THEVET, André (1502-1590)
Table d'Afrique. (Map of Africa.) (A. Thevet, Paris, 1575.)
Map, 35 x 44 cm, uncoloured.
Scale in Italian, French and sea leagues.

André Thevet, a Franciscan monk and Cosmographer to Henri III, published his *Cosmographie Universelle* in Paris in 1575 with maps of the four continents, of which Africa is reproduced above. He published other maps: of France; of the world, in the form of a lily (1583); of Switzerland and of Spain.

This map of Africa and the surrounding sea is engraved in very great detail. The wavy, billowy technique of depicting the sea rather overshadows the outline of the mainland. The north, south, east and west are marked as Septentrion, Midi, Orient and Occident, encircled by strap-like designs. The east and the west are similar in design, each showing a bunch of fishes being hauled out of the sea. The north and south each have their own design. The seas are teeming with numerous small sailing vessels of different shapes and sizes together with a number of kinds of sea monsters. In the western Atlantic Ocean the monster appears to have one horn – a narwhal, perhaps. The shape of the continent is made to bulge on the lower west coast. The map itself is full of place names, rivers, lakes and mountain ranges. In the south the Cape is marked 'Chef de bonne Esperance', Vigiti Magna, the Infanta River, Manica and Cafala are noted. Many areas are described as deserts, with unidentifiable names. The Ptolemaic concept of the origin of the Nile is once again evident, with a large number of tributaries flowing into it. The cartouche in the upper righthand corner consists of a podium supporting an oval stand, on top of which is placed a partially rolled sea chart, a compass and an hour-glass. Sitting on the outside of the oval on either side is a parrot and in its centre is the scale in Italian, French and sea leagues. At the bottom is the title, 'Table d'Afrique.'

18

Map 13

DUCHETO, Claudio (1554-1597)
Il disegno della Geografia moderna de tutta la parte' dell' Africa ...
(A modern outline of Africa...) Engraved by Hendrik Hondius. (Rome, C. Ducheto, 1579.)
Map, 43 x 59 cm, uncoloured.

This map of Africa was published in Rome by Ducheto, nephew and part successor to Lafreri, who is known for his maps of Europe, Africa, Naples and the world. There is a simple cartouche with decorative scrollwork at the lower left. The inscription within is in Italian, describing Africa and the surrounding islands, followed by the imprint of the cartographer and engraver, all in italics. Apart from the decorative calligraphy, the windrose in the Indian Ocean is the only decoration of the seas. The mainland is very similar to that of Forlani (see Map 9), including even Prester John's palace. The stippled ocean is characteristic of the early Italian copper engraved maps, as opposed to the wavy design of the Dutch maps of the seventeenth century. The map differs from the Forlani in that the Nile arises from the 'two side by side' lake system. It would appear that this particular copy was engraved by Hendrik Hondius from the Italian plate – a typical example of a slight change in the cartouche indicating another publisher's imprint, while the basic features remain the same.

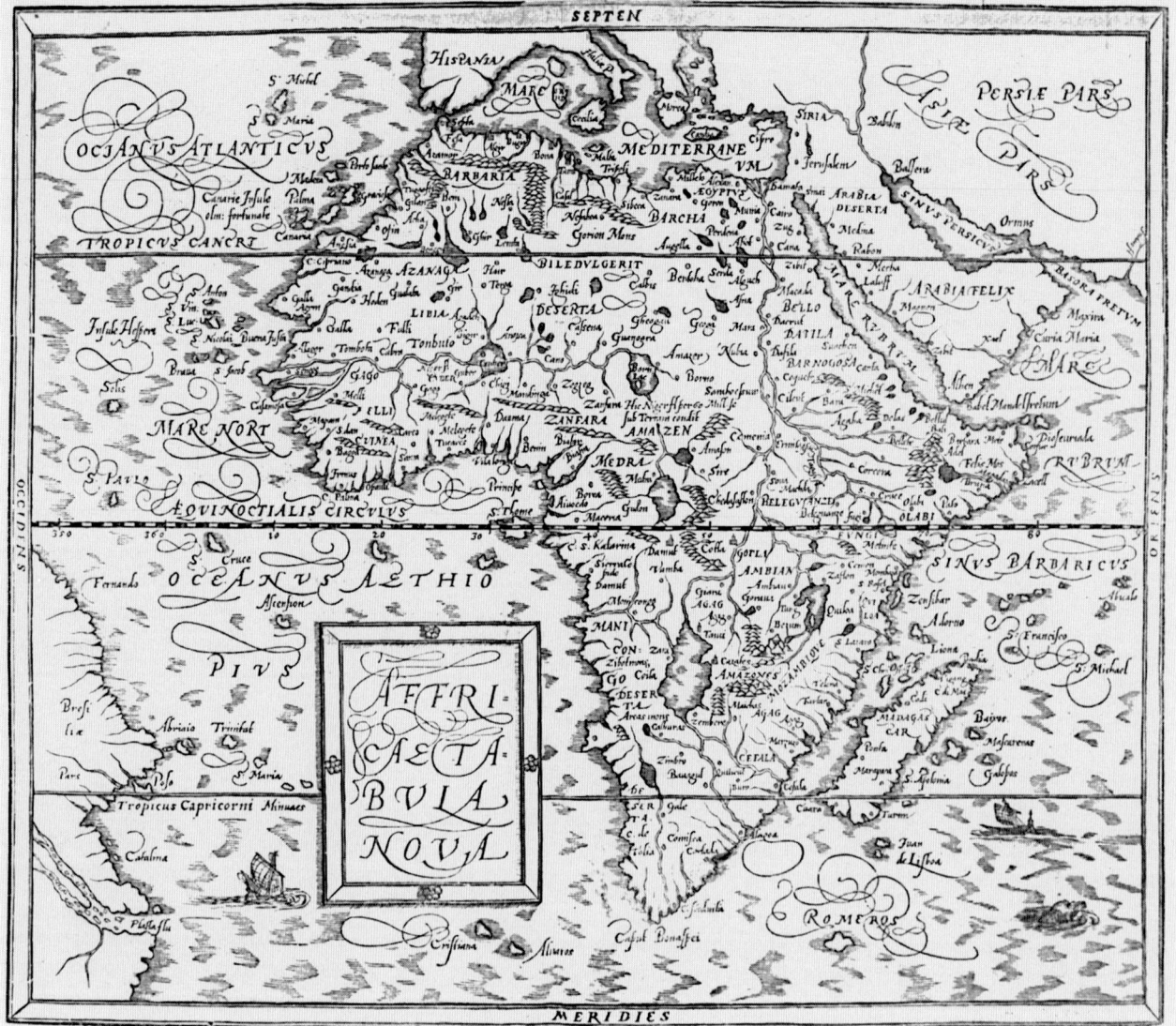

Map 14

MÜNSTER, Sebastian (1489-1552)
Affricae Tabula Nova. (New map of Africa.) (Basle, S. Münster, 1580.)
Map, 31 x 36 cm, coloured.
At the top: *Africa, Libya, Dozenlandt mit allen königreichen...*

Münster's *Cosmographia* continued to be issued from 1580 onwards, but with an entirely new set of maps when Sebastian Petri succeeded him. (Münster's original maps last appeared in the 1578 edition.) This new set of maps was based on those of Ortelius. This map could be regarded as coming from the second edition of Münster: the addition of a portion of the coast of Portugal makes it recognisable as such. The technique of the drawing is different from the Münster woodcuts of Africa. The coastline is jagged, with prominent bays, there are many large lakes and rivers and the geography is somewhat fictitious except for certain well-known places and rivers. There are two small sailing vessels and one sea monster. The left lower cartouche is a simple rectangular, slightly ornamented box containing the title in a florid calligraphy which is also used for the place names.

20

Map 15

SANUTO, Livio (1520-1576)
Africae Tabula XII. Engraved by brother, Giulio Sanuto. (Venice, L. Sanuto, 1588.)
Map, 39,5 x 51,5 cm, irregular shape, uncoloured. Scale in millaria.
In: Sanuto, Livio. *Geographia.* Venice: L. Sanuto, 1588.

The great achievement of German descriptive geography was Munster's *Cosmographia,* a massive work interspersed with woodcut maps and views of towns and many other vignettes and pictures. In Italy, where making manuscript maps and charts was still a flourishing industry, the popular form for printed maps was the separate sheet rather than the book illustration. Most maps printed in Italy before the middle of the sixteenth century were woodcuts. The first Venice edition of Ptolemy in 1511 had woodcut maps with legends printed from metal type. The craft of engraving on copper and the print-seller's trade had, however, long been established independently of the book industry, whose purposes were better served by wood engraving, and copper line engraving was exploited to greater advantage by the cartographer. The Ptolemy edition produced after 1511 in Venice was printed with maps engraved by Giacomo Gastaldi, and marked the turn of the tide in favour of copper.

The staple of the Italian map trade was the line-engraved map on copper, with which it captured the European market. The centres of this trade were in Rome and Venice (see note on Gastaldi, Map 8) and the map workshops of these centres were controlled by masters who combined the activities of cartographer, engraver, printer, publisher and map-seller. Lafreri, Tramezini and Duchetti (sometimes called Ducheto) established themselves as masters of the copperplate engraved map trade in Rome, and Camoccio, Forlani, Bertelli and Sanuto confined their activities to Venice.

Livio Sanuto was the son of François Sanuto, senator of the republic of Venice, an intellectual and a good orator, who gave his son a first-rate education. After Livio had been instructed in the arts and in music, he was sent to Germany to finish his studies, and made considerable progress in mathematics and cosmography. The latter science – which

21

was, at that time of maritime discovery, a great source of glory and riches – became the sole object of Sanuto's efforts; he used up all the resources at his disposal to accelerate his progress in this field. He invented instruments which gave greater precision in astronomical observations; he read the works of the historians and the seafarers; he tore out relevant information from those journals of navigators which he was able to procure. By combining all this information, he realised, he would be able to draw more exact maps than any previously known and to give, as he said, a new face to the world; in short, he proposed the publication of a complete and methodical description of the world, to be divided into three parts, conforming to the three great continents which were known to him:

1. le 'Pholémaigne,' the ancient world of the geographers of that time, i.e. Europe, Asia and Africa,

2. l' 'Atlantique' or the new world, i.e. the two Americas, and

3. l' 'Australie,' which was the name Sanuto gave to the third great continent of the globe.

Each part of Sanuto's work was to be divided into several books. The first book of the first part is dedicated to an explanation of the means of observation, to intellectual discussion of the declination of the magnetic needle and to the necessary corrections in his book. In the second book he establishes the main divisions of his work, determines the projection of his maps and assigns the extent and limits of each region, rectifying many mistakes which were common in geography at that time. The ten books of Sanuto's work which follow the first two contain his description of Africa – the first known atlas of the continent – accompanied by twelve maps drawn by the author and engraved very carefully by his brother Jules. Sanuto had hardly finished this portion of his vast undertaking when he died, at the age of 56. His work was printed just as he had left it, without even filling in figures or names which were in rough in the manuscript. It was his friend Saraceni who added tables of contents to the work and a preface on the life of the author.

His work appeared in Venice in one folio volume in 1588, under the title: *Geography of Livio Sanuto, divided into twelve books, in which besides the enlightening on several of the places of Ptolemy, on the compass, the magnetic needle, the provinces are made known, as well as the peoples, kingdoms, towns, ports, mountains, rivers, lakes and customs of Africa, with twelve maps engraved on copperplate, to which was added three indices composed by Jean-Charles Saraceni.*

The title page; the Africae Tabula VIIII (Abyssinia) (Map 304); Tabula X (Southern Africa) (Map 152) and Tabula XII (Continent of Africa), are reproduced in this book and represent sixteenth-century Italian line work on copper in its maturity. The seas are represented with the typical stipple design, the outlines are decisive and the diagrams and lettering appear in meticulous detail. The seas are occupied by sailing vessels of the time and the sea monsters are similar to those used to illustrate other maps of that period.

Map 16

PIGAFETTA, Filippo (1533-1603)
(Depicts Africa from the Mediterranean to The Cape, excluding West Africa.) Engraved by Bonifacio Natalis. (Rome, F. Pigafetta, 1591.)
Map, 62 x 43 cm, uncoloured. Scale in hundred leagues.
Dedicated to Antonio Miglioni Vescovo S. Marco.
In: Pigafetta, Filippo. *Relatione del reame di Congo et delle circonvicine contrade.* Rome: F. Pigafetta, 1591.

23

Filippo Pigafetta was an Italian historian and traveller and Chamberlain to Pope Sextus. He published his two-sheet map of Africa in his *Relatione del reame di Congo et delle circonvicine contrade* in Rome, 1591.

This was based on information drawn by Pigafetta from the writings of Duarte Lopes, a Portuguese explorer who collected a mass of authentic geographic material on the Congo and its basin to which he added speculations on the origin of the Nile. Pigafetta gave his view of the hydrography of Central Africa and the origin of the Nile, showing two lakes, one below the other, contrary to the Ptolemaic version. The Nile arises from a lake on latitude 23° South surrounded by very lofty mountains, flows for approximately 400 miles due north and enters another large lake, longer than the first, which lies under the equinoctial line. From this second lake the river continues to the island of Meroe, a distance of 200 miles. A comparison of the Pigafetta map and the portolano chart in the Bibliothèque Nationale, Paris, attributed to Pedro de Lemos (c. 1590), reveals that Pigafetta may have taken information from this chart. There are, however, even earlier manuscript charts with the same distinguishing features.

Pigafetta's concept of the origin of the Nile is much closer to the real relationship between Lake Victoria (the White Nile's true source) and its neighbour Lake Tanganyika further south, although there is no evidence to suggest that Pigafetta knew of the existence of these lakes. Other geographers, however, continued to copy Ptolemy. The map is thus highly important as well as rare, an important landmark in the historical cartography of the Nile water basin, completely changing the traditional Ptolemaic concept.

The map itself depicts Africa from the Mediterranean to the Cape but excludes West Africa. At the top right in the text is a scroll surmounted by papal arms, signed Filippo Pigafetta. At the lower right appears a dedication cartouche to Antonio Miglioni Vescovo di S. Marco, dated Roma 2d April MDXCX.

The writings of Filippo Pigafetta are to be found in an English translation of his *Relatione del Reame di Congo... A Report of the Kingdom of Congo... drawn out of the writings and discourse of the Portuguese Duarte Lopez in 1591*, by Margarite Hutchinson (London, John Murray, 1881).

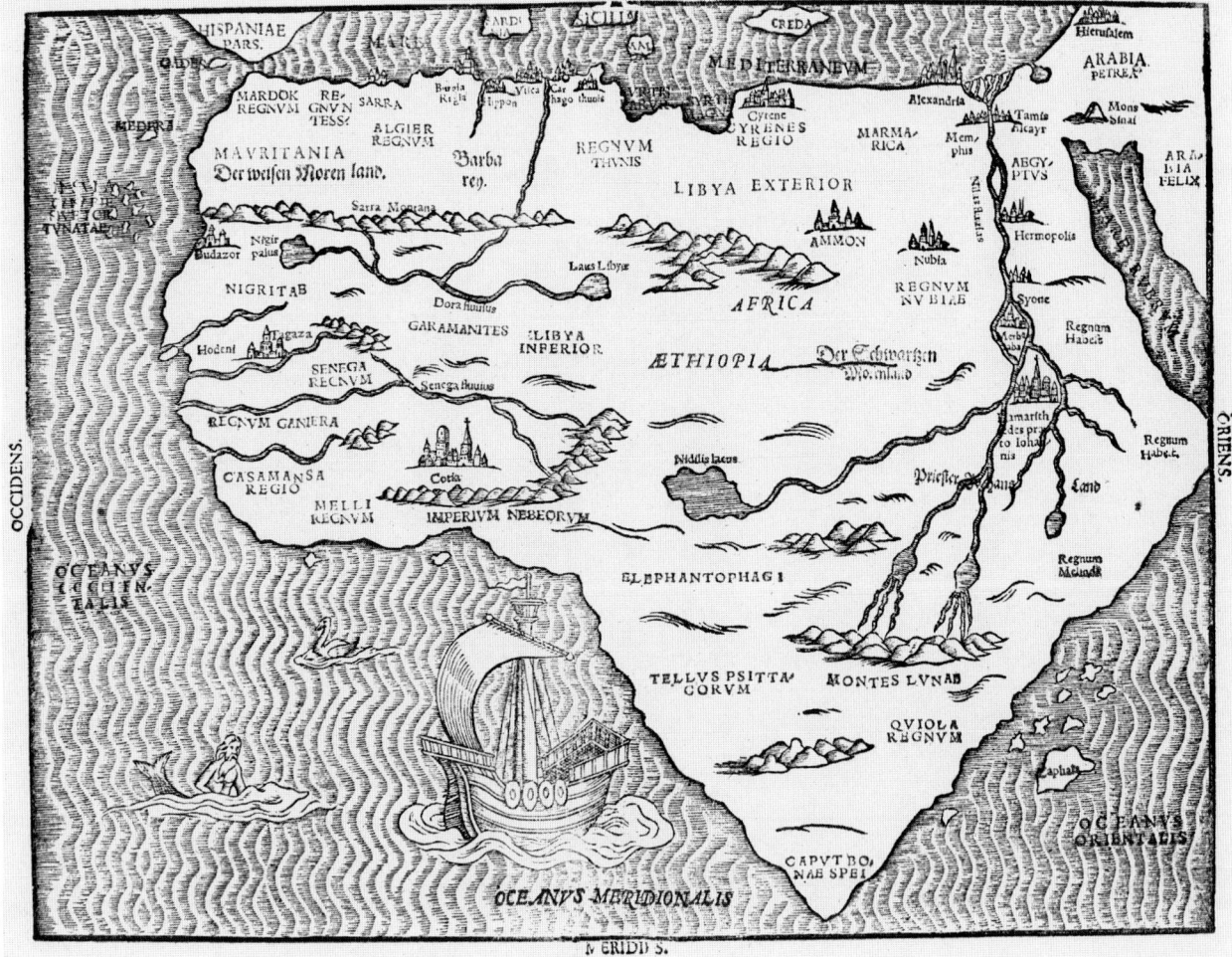

HISPANIAE PARS.   QVEN   MARE   SICILIA   CRETA   Hierusalem   ARABIA PETREA

MEDERI   MARDOK REGNVM GNVN TESS?   SARRA   ALGIER REGNVM   Bugia Regii   Vtica   Carthago thunis   MEDITERRANEVM   Alexandria   Tamis Alcayr   Mons Sinai

MAVRITANIA   Der weisen Morenland.   Barbarey.   REGNVM THVNIS   CYRENES REGIO   MARMARICA   Memphis   ABGYPTVS   ARABIA FELIX

INSVLAE FORTVNATAE   Sarra Montana   LIBYA EXTERIOR   AMMON   Nubia   Hermopolis

Nigir palus   Budazor   NIGRITAB   Laus Libyæ   AFRICA   REGNVM NVBIAE   Syene   Regnum Hauch

Hodeni   Tagaza   GARAMANITES   LIBYA INFERIOR   Dora fluuius   ÆTHIOPIA   Der Schwartzen Morenland.

SENEGA RECNVM   Senega fluuius   Regnum Habet.

REGNVM GANERA   Niddis lacus   Priester Johan Land   Regnum Melindi

CASAMANSA REGIO   Cotia   MELLI REGNVM   IMPERIVM NEBEORVM

OCEANVS OCCIDENTALIS   ELEPHANTOPHAGI

TELLVS PSITTACORVM   MONTES LVNAE   QVILOA REGNVM

OCEANVS ORIENTALIS

CAPVT BONAE SPEI

OCCIDENS   ORIENS

OCEANVS MERIDIONALIS

MERIDIES.

Map 17

BÜNTING, Heinrich (1545-1606)
Africa Tertia Pars Terrae. (Africa the third continent.) (Hanover, H. Bünting, 1592.)
Map, 26 x 34 cm, woodcut. Top left: 20. Top right: 21.

Heinrich Bünting (Buenting), a professor of Theology in Hanover, issued the first edition of *Itinerarum Saccrae Scripturae* in 1581. This atlas included symbolic maps: the world as a clover leaf, Europe as a lady or virgin queen, and Asia in the form of Pegasus (the winged horse). The text of this cosmography appeared in Danish, Swedish, Dutch and English. This woodcut map of Africa has an unusual shape, tapering to a narrow angle in the south. Numerous rivers appear in West Africa and as tributaries to the Nile, and mountain chains and various kingdoms are mentioned. The sea is engraved in narrow wavy lines, with a galleon, a merman and above him a sea-bird. It is interesting to note the different type of lettering used in this map, indicating the use of inset type for the woodcut technique.

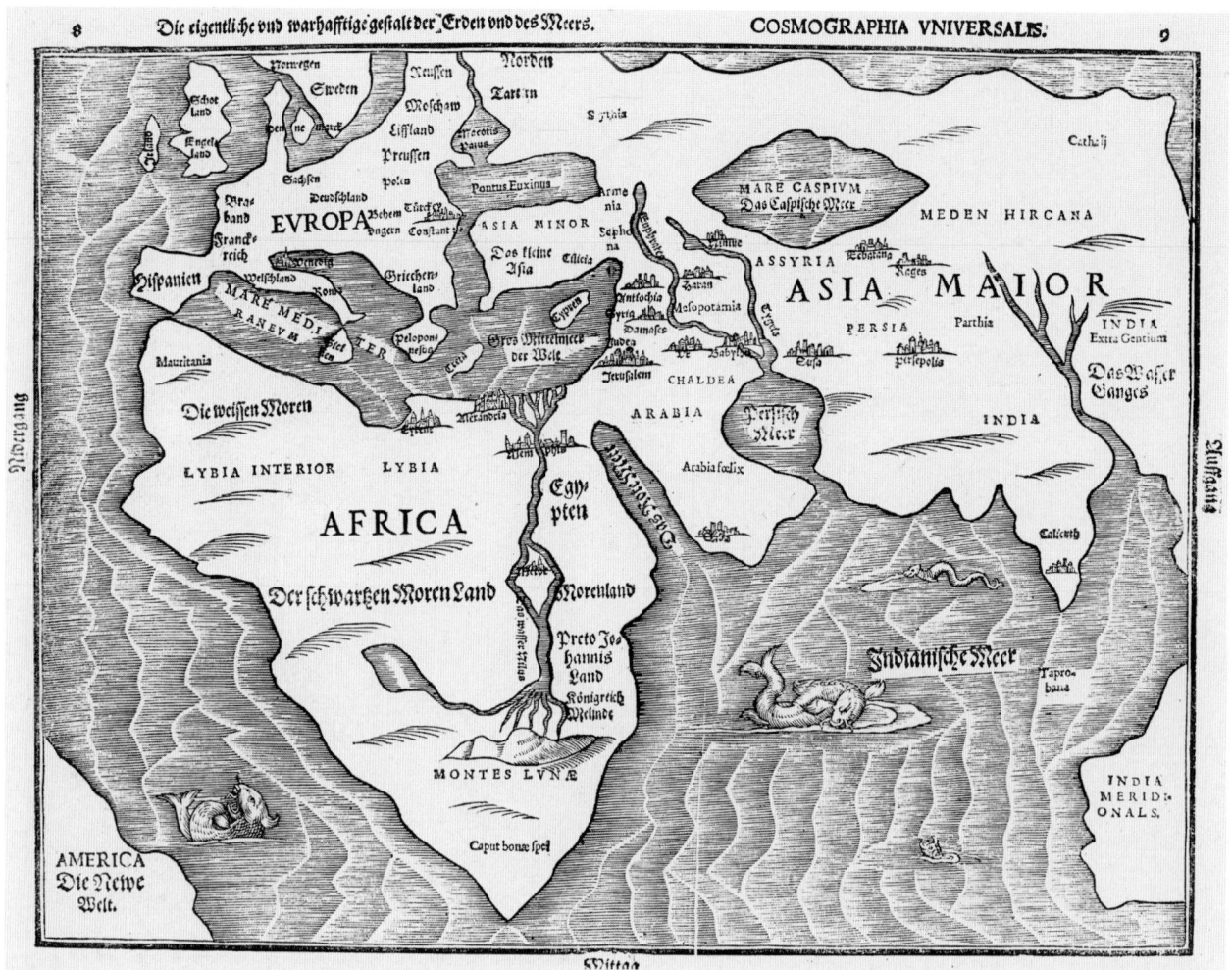

Map 18

BÜNTING, Heinrich (1545–1606)
*Die eigentliche und warhafftige gestalt der Erden und des Meers .
Cosmographica Universalis.* (The true shape of the world and the oceans.)
(Hanover, H. Bünting, 1592.)
Map, 27 x 36 cm, woodcut. Top left: 8. Top right: 9.

This is another Bünting woodcut map from his *Itinerarum Saccrae Scripturae*, featuring Africa prominently with the rest of the world. It is a curious map, including – at the east end of the Indian Ocean – a portion of land with an outline coincidentally much like that of Western Australia, named India Meridionals. In the south-west corner Bünting places a portion of America, **Die Newe Welt**. Not all the editions of the atlas had this map. Jerusalem is sited in the centre of the map because it was regarded then as the centre of the world. The Nile and its origin is prominent, arising from one large lake in the centre, with additional tributaries from the Montes Lunae. The sea is occupied by four sea monsters – two small and two large. This map differs entirely from Bünting's cloverleaf map of the world.

AFRICAE VERA FOR.MA, ET SITVS.

Map 19

JODE, Gerard de (1509-1591)
Africae Vera Forma, et Situs. (The true shape and situation of Africa.)
(Antwerp, G. de Jode ,1593.)
Map, 32 x 45 cm, uncoloured.

This map of Africa appeared in the second edition of Gerard de Jode's *Speculum Orbis Terrarum*, issued by his son, Cornelius, in Antwerp. The title is in large lettering across the top of the map, with jewelled bands on either side. A plaque of descriptive text at the top right terminates in a circular medallion with the words **Formes Haeredium Gerardi de Jode.** Six sea monsters and four sailing ships decorate the sea and small vignettes of natives, some with bow and arrow, others with spears, are placed on and around the map, which shows the Kingdom of Monomotapa extending down to the Cape, the Cuamo, the Rio de Spirite and the Rio Infanta. On the verso is the text in Latin.

Gerard de Jode and his son, Cornelius (1568-1600), were engravers, printers and publishers of Antwerp. Gerard published Gastaldi's world map of 1555; maps of Europe by Musinus in 1560, of Portugal by Seco in 1563, Ortelius's world map of eight sheets in 1544 and others.

He started work on his atlas, *Speculum Orbis Terrarum*, in 1570 and it was published in 1578, with a map of Africa by Jacopo Gastaldi. A revised edition was issued by Cornelius in 1593 with his own map of Africa.

Cornelius's first works, a world map (1589) and *Gallion Occidentale* (1592) were published by his father. His only other publication seems to have been the map featured here, which replaced the Gastaldi in the 1593 edition of the *Speculum*.

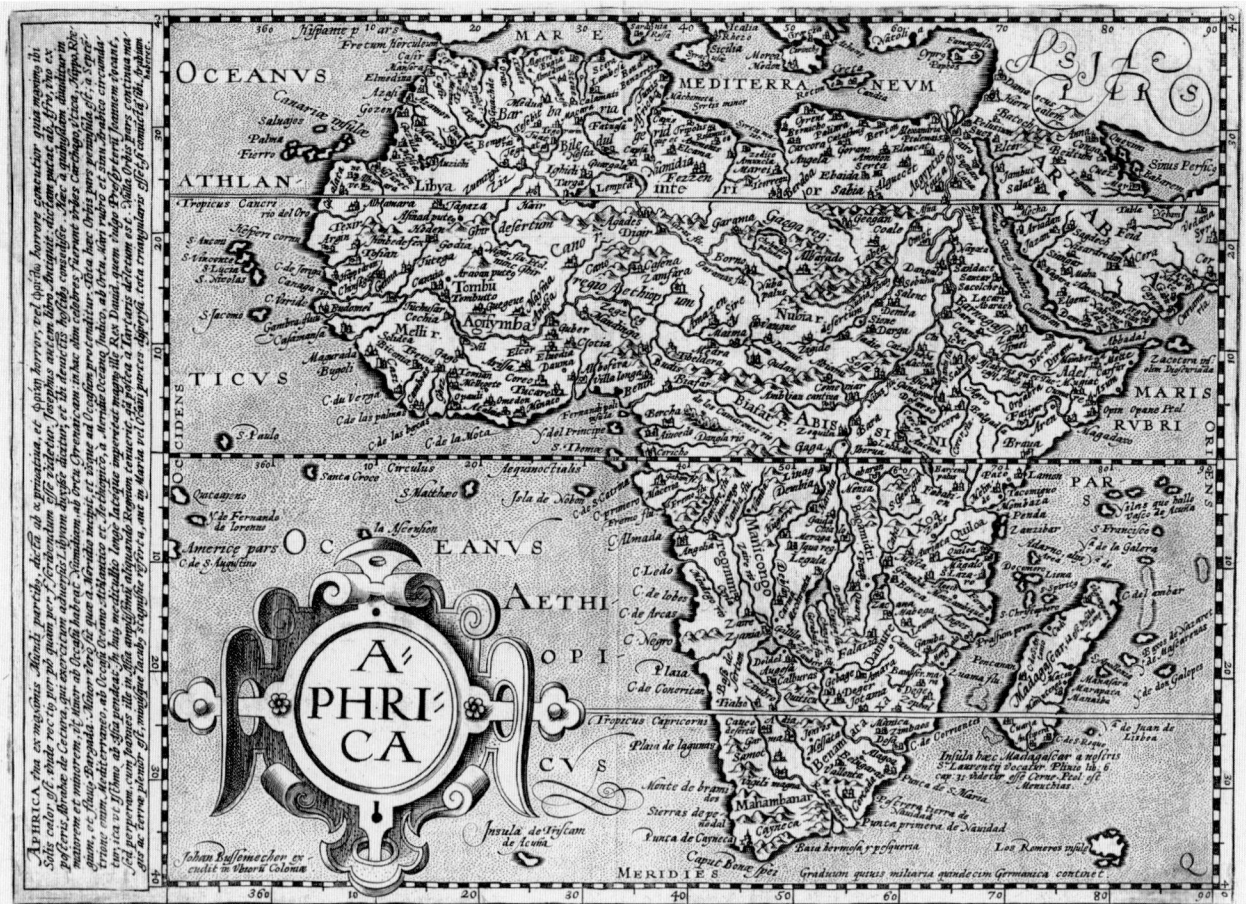

Map 20

BUSSEMACHER, Johan (fl. 1580-1613)
Aphrica. (Cologne, J. Bussemacher, 1600.)
Map, 21 x 29 cm, coloured. German text on verso.

This small German map has a simple scrollwork design cartouche. **Aphrica** is all that appears on the cartouche proper and adjacent to this just outside the map is a frame containing a vertically disposed Latin description. This map is from the German work by Matthias Quad issued in Cologne in 1600 under the title *Geographisch Hand Buch* and naming the author of the maps. According to the Library of Congress Catalogue the maps are mostly derived from the *Theatrum Orbis Terrarum* of Ortelius.

Bussemecher was an engineer printer of Cologne who collaborated with Quad (1557-1613), a geographer, humanist and printer, also of Cologne, who issued a number of works. In this map of Africa there are rivers, lakes and cities all over the continent, but in the south Vigiti Magna and the Rio de Infante are depicted with the surprising absence of Monomotapa. On the back of this copy there is a long description in German based partly on the work of Leo Africanus. Da Gama's voyage is mentioned and the fact that the inhabitants are black

Map 21

MERCATOR, Gerard (c. 1565-1656)
Africa. Ex magna orbis terre discriptione Gerardis Mercatoris desumpta.
(Africa. Taken from the world atlas of. . .) (Duisburg, G. Mercator, 1595.)
Map, 38 x 46 cm, slightly coloured. Scale in German miles.
Latin text on back of map.

This map was taken by Gerard Mercator from the eighteen-sheet world map in *Atlas; sive, Cosmographicae meditationes de fabrica mundi et fabricati figura*, first published in 1569 by his grandfather, Gerhard Mercator. Editions in many languages were published between 1585 and 1642, and for the 1595 edition Gerard extracted Africa from the original world map and updated it.

Gerhard Mercator senior was born Gerhard Kremer. Called 'the Ptolemy of our age' by his friend Ortelius, he coined the word 'atlas' for a collection of maps. At the university of Louvain, under the famous Gemma Frisius, he studied science and the art of constructing globes, and became famous as an engraver when he printed maps on separate leaves. He had three sons (who helped him in his business) and three daughters, moved with his family to Duisburg in 1552 and taught geography in a newly established university. Here he produced his most important works, including the eighteen-sheet world map. He worked on his atlas from 1585 onwards and it was completed in 1595, after his death, by his son Rumold. This map of Africa, compiled by his grandson, Gerard, was included.

The cartouche is circular, surrounded by decorative scrolls and surmounted by two satyrs. The beautiful calligraphy is a special feature of Mercator maps. The 'projection' associated with Mercator was not used in this particular map. Basically, much the same information is supplied in this map as in the Ortelius maps (10 and 11).

29

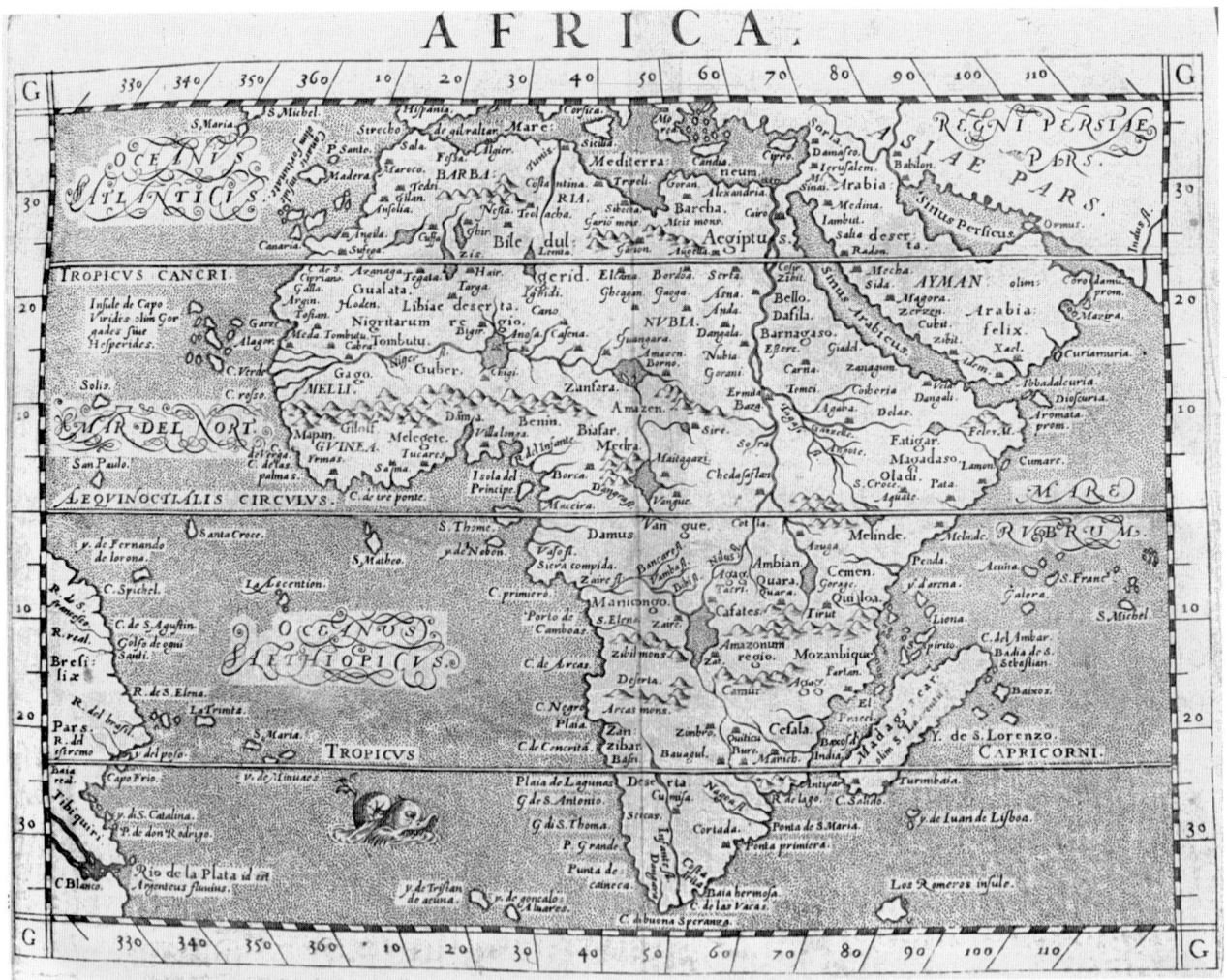

Map 22

BOTERO, Giovanni (1540-1617)
Africa. (Venice, G. Botero, 1596.)
Map, 18 x 24 cm, uncoloured.

Botero was an Italian priest and geographer, who was born in Piedmont and died in Turin. This map first appeared in *Relationi Universali*, published in Rome in 1591, and reissued in Rome, Bergamo and Venice in 1594, 1595, 1596 and 1605.

The shape of this map is fairly accurate considering that the Cape had been rounded only 100 years previously. The coastal names appearing around southern Africa are as originally laid down by the exploring Portuguese. The only decoration is a ferocious-looking sea monster in the Atlantic, and the calligraphy is florid and decorative. Cities are marked all over the continent, though neither Monomotapa nor Vigiti Magna are noted, and the river Infante is marked in a more easterly position than illustrated by later map makers. The

Nile appears to arise from the two-lake system although the western lake is lower in Central Africa, in the Congo. Zanzibar appears on the west coast – an error copied by even later cartographers. Natal is not described, which is strange, because it does appear in many maps of this period and even earlier.

This map is identical in almost every respect to the map put out by Girolamo Porro, an Italian engraver and publisher, working in Venice. (See Map 24.) Porro was a competent engraver and issued a new set of maps for the 1596 edition of Ptolemy's *Geographia* which superseded those of Gastaldi. His small maps of Africa were unsigned and undated. He also engraved maps for Porcacchi's *Isole* of 1572.

Map 23

PIGAFETTA, Filippo (1533-1603)
Africa–English Edition. Engraved by William Rogers. (London, John Wolfe, 1597.)
Map, 62 x 43 cm, uncoloured. Scale in leagues.

This English edition of the Pigafetta map appeared, in two parts, six years after its original publication by Théodore and Israel de Bry in 1591. The key giving a 'description of Aegypt from Cair downeward' at the top right is surrounded by the papal arms and a decorated scale appears below the equator. The sea is stippled and adorned with ships and sea monsters. This map is copied from the original Italian edition of 1591 (see Map 16), even to the decoration,

except that John Wolfe's armorial imprint replaces the legend index to the right of Madagascar, and a large sea serpent replaces the original Italian dedication in the lower right corner. The map also appears in the English edition of Linschoten's *His Discours of voyages to the East and West Indies*, printed in London by John Wolfe in 1598 and engraved by William Rogers, the English engraver.

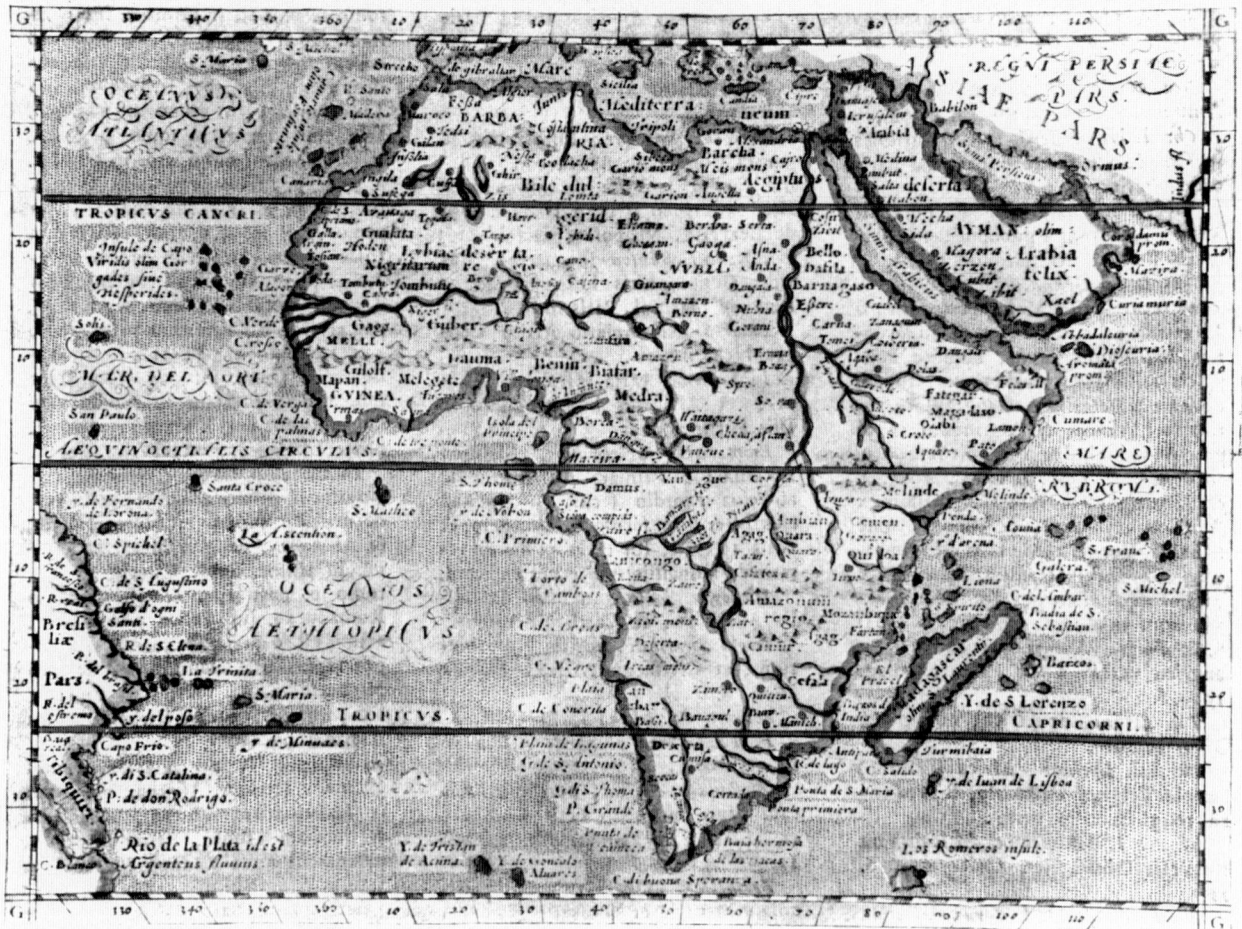

Map 24

PORRO, Girolamo (fl. latter 16th C.)
Africa. (Venice, G. Porro, 1598.)
Map, 12,5 x 17 cm, coloured.

A miniature map of Africa and Arabia showing part of the bulge of eastern Brazil down to the mouth of the Rio de la Plata. It was issued by Girolamo Porro, an engraver of considerable ability who was born in Padua and worked in Venice. He engraved a new set of maps for the 1596 edition of Ptolemy's *Geographia* which superseded those of Gastaldi and remained in use for about 24 years. This small map appears to be based on Ortelius in shape, with limited geographic content, but with decorative calligraphy. The rivers of Central and North Africa feature prominently, with the Nile appearing to have direct communication, via a lake, with the lower east coast of southern Africa. There were various editions from 1596 to 1620, differing only in the title of the map and the numbering of the text pages. A map identical to this one appears in the 1596 Venice edition of G. Botero's *Relationi Universali*, though the Botero map (measuring 18 x 24 cm) is larger than the Porro described here.

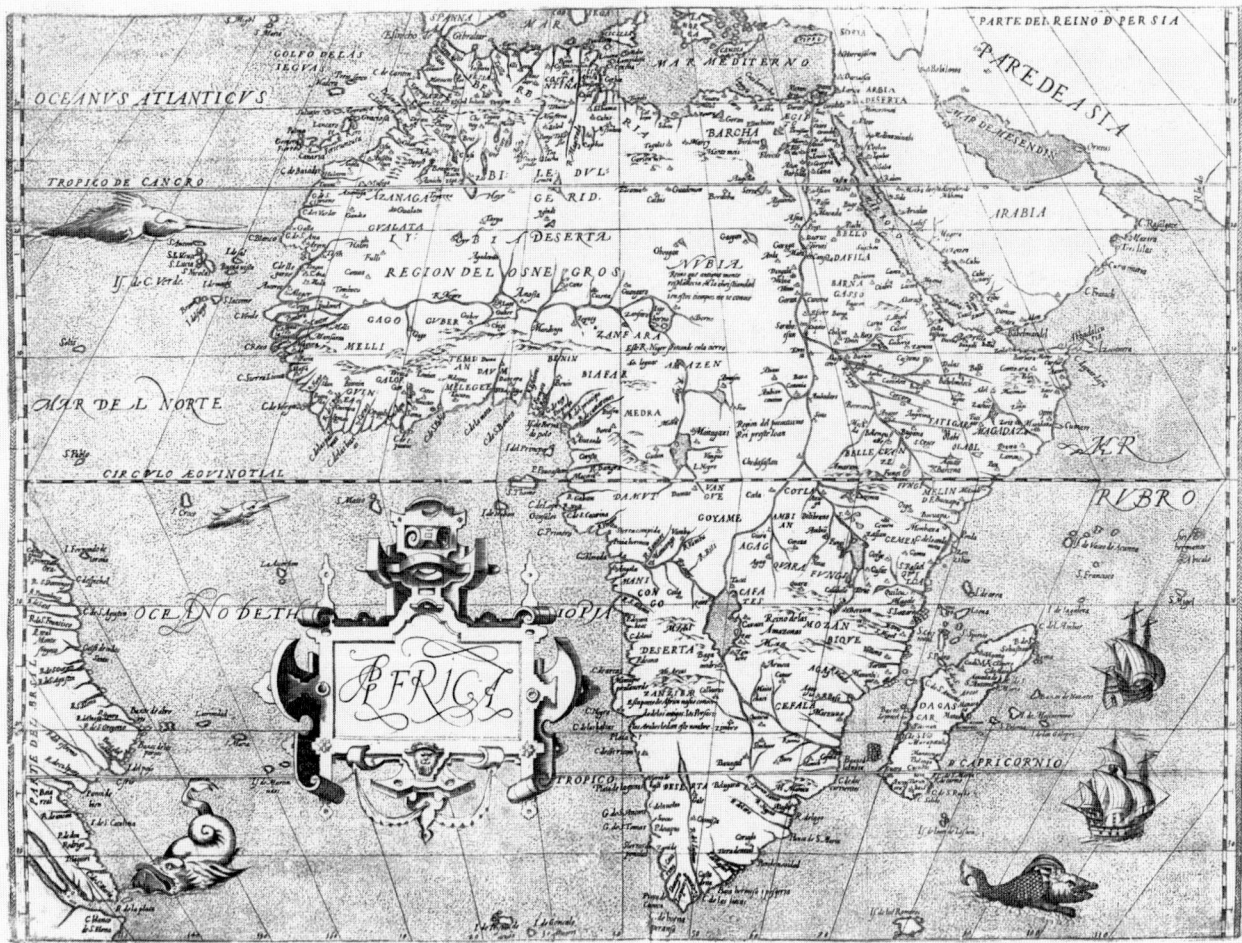

Map 25

SOLIS, Hermando de (fl. latter 16th C.)
Africa. (Spain, 1598.)

Map, 36,5 x 50 cm, uncoloured.

A finely engraved black and white map with its title within an attractive classical-type cartouche, scroll-like in character, with a lion's head appearing on the lower border. No names or dates appear on the map. The shape of this map is very similar to Abraham Ortelius's map of Africa, with the eastern bulge of Brazil and the same sea monster appearing on the left. Other swordfish-like sea monsters similar to Ortelius's appear in the Atlantic (Ethiopian) Ocean, one appearing to be jumping out of the sea below the bulge of the west coast and another larger one higher up. Madagascar and the surrounding islands are also illustrated. The naval engagement which appears in the bottom right corner of Ortelius's map is absent but two other sailing vessels and a different type of sea monster take its place. The seas have the stippled character commonly seen on Italian copperplate engraved maps. The lettering throughout is florid but not quite as flourishing as that in the Ortelius map, and Spanish words are used for descriptive purposes. Within the map itself the geography of the interior and the coastlines are almost identical to those of Ortelius. The two-lake origin of the Nile appears once more with Lake Zaire (on the left) lower than Lake Zaflan. Mer Rubro (Mer Rubrum in Ortelius) is also depicted below the Horn of Africa.

The dealer from whom this map was obtained provided the only further information concerning its provenance. The publisher is Hermando de Solis, a Spanish cartographer who published only four known maps: of Africa, America, Europe and the world. The only map on which his name is mentioned is that of the world, dated 1598. Further research revealed a reference in Tooley's *Dictionary of Map Makers* to one Hermando de Solis who produced a map of America in 1598, the same date as appears on the world map. Tooley adds that Giovanni Botero, the Italian priest and geographer, used this map of America by Solis in 1599, in one of his atlases, *Relationi Universali*.

Map 26

ARNOLDI, Fiamengo Aroldo di (d. 1602)
Africa. (Siena, F.A. di Arnoldi, 1600.)
Map, 37 x 48 cm, uncoloured.

In the latter half of the sixteenth century maps devoted exclusively to Africa became more numerous. Among the cartographer-publishers were a number of Italian authors using the copperplate line engraving technique. From Roberto Almagia's study *Il Planspero de Arnoldi (Monumenta Cartographica Vaticana,* vol. 2, Rome, 1934), we learn that Arnoldi was a Belgian who worked in Bologna from about 1555 to 1600, for the astronomer and geographer Magini, who edited the 1596 Ptolemy atlas. In about 1600 Arnoldi moved to Siena, where he executed the large world map which is the subject of the aforementioned monograph by Almagia. The earliest reference to Arnoldi to appear in English is recorded in *Imago Mundi 2* (1937), in which it is stated that Mr G.H. Beans of Jenkintown, Pennsylvania, well-known for his Tall Tree Library, had acquired four maps of the four continents, one of which (that of Europe) bears

the name Aroldo di Arnoldi. All seem to be by him, and to form a set, which in 1956/57 Mr Beans donated, together with a large number of other maps and geographically important writings, to the John Carter Brown Library in Providence, Rhode Island, USA. The map of Europe is labelled 'Aroldo di Arnoldi' and dedicated to Scipio Bargagli, the Genoese patron of scholars. The other maps all lack the name but have the same coat of arms. It must be inferred that all these large maps were engraved by Arnoldi in the last years of his life, in 1601 or 1602.

This map of Africa appears to have the shape of that of Ortelius. The cartouche appears at the lower left border adjacent to a portion of the eastern tip of Brazil. It is most decorative – oval-shaped with the title 'Africa' in its own oval box surmounted by the coat of arms. The outline of this cartouche is scrollwork with flower ornamentation at the lower border. Within it appears an Italian legend which refers presumably to the origin of the name of Africa in biblical times and also describes its physical features, particularly the very hot sun and scarcity of water. The sea, engraved in the characteristic Italian stipple technique, bears three ships in full sail and two rather unusual monstrous-looking sea animals not unlike those seen in Gastaldi's Africa. The lettering is all attractively florid. The detailed geographical content of the map itself is of varying validity and like so many of this period and before, the Nile is shown to arise from two separate lakes placed side by side. The Niger (Negri) is shown arising from Lake Niger, winding its way northward, passing through two lakes on its way to the Atlantic. The island of Madagascar is labelled as such although the legend below which describes the island refers to 'San Lorenzo,' adding that it is also called the 'Island of the Moon' and that Ptolemy, in his Book Six, Chapter 31, names it 'Sarne' and 'Ptolemeo Menuthias.' The legend goes on to describe Madagascar as being 300 miles in circumference and rich in gold, silver, wheat, sugar, etc., with forests of sandalwood and many elephants.

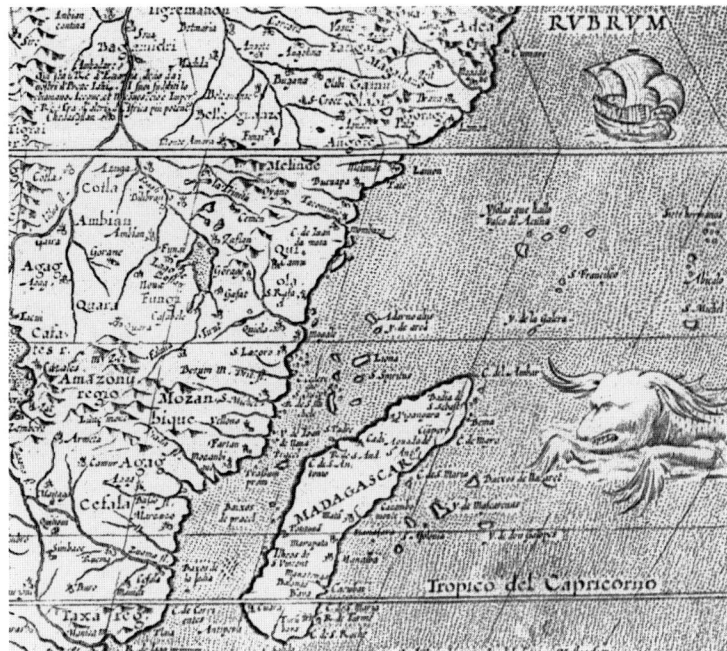

Map 27

HONDIUS, Jodocus (1563-1612)
Nova Africae Tabula. (New map of Africa.) (Amsterdam, J. Hondius, 1606.)
Map, 37 x 50 cm, uncoloured.

The title of this map is at the lower left within a large handsome cartouche of interlacing strap design, with a monkey's head at the top. Four ships, native craft, three sea monsters and two animals inland are expertly engraved, with particularly decorative calligraphy off the central west coast of the continent. Although this map was based on that of Mercator, the coastline, particularly that of Madagascar, is considerably altered. Hondius improved the Cape area by inserting Saldanha Bay and Cape False (Falso) and discarding Cayneca. The pleasure the ships from India experienced in obtaining fresh fruit and vegetables is noted in a Latin legend alongside St. Helena.

The head of the Hondius family, Jodocus Henricus, was born in 1563 at Walkene in Flanders but he and his family soon moved to Ghent. Here he taught himself to engrave on copper and ivory at the early age of eight years. Subsequently he became the most celebrated craftsman of his time and, in addition to his skill as an engraver, he became a good mathematician, an intelligent cosmographer and an able type-founder.

During the religious troubles in Flanders, Jodocus emigrated to London where he set up as a type-founder, mathematical instrument-maker and an engraver of maps and charts. Here he met many famous geographers and scientists, and married the daughter of a compatriot, Coletta van der Keere, by whom he had two sons, Jodocus and Henricus, and a daughter, Elizabeth, who later married Joan Jansson.

In 1593 Jodocus Hondius returned to Amsterdam to settle there and set up as a publisher and engraver of maps and charts. In 1606 he acquired the copperplates of the *Mercator Atlas* and re-published those maps, adding thirty-seven of his own composition, of which the above map is one. He added to this the further publication of an *Atlas Minor* and several more maps showing the latest English and Dutch voyages.

Map 27a

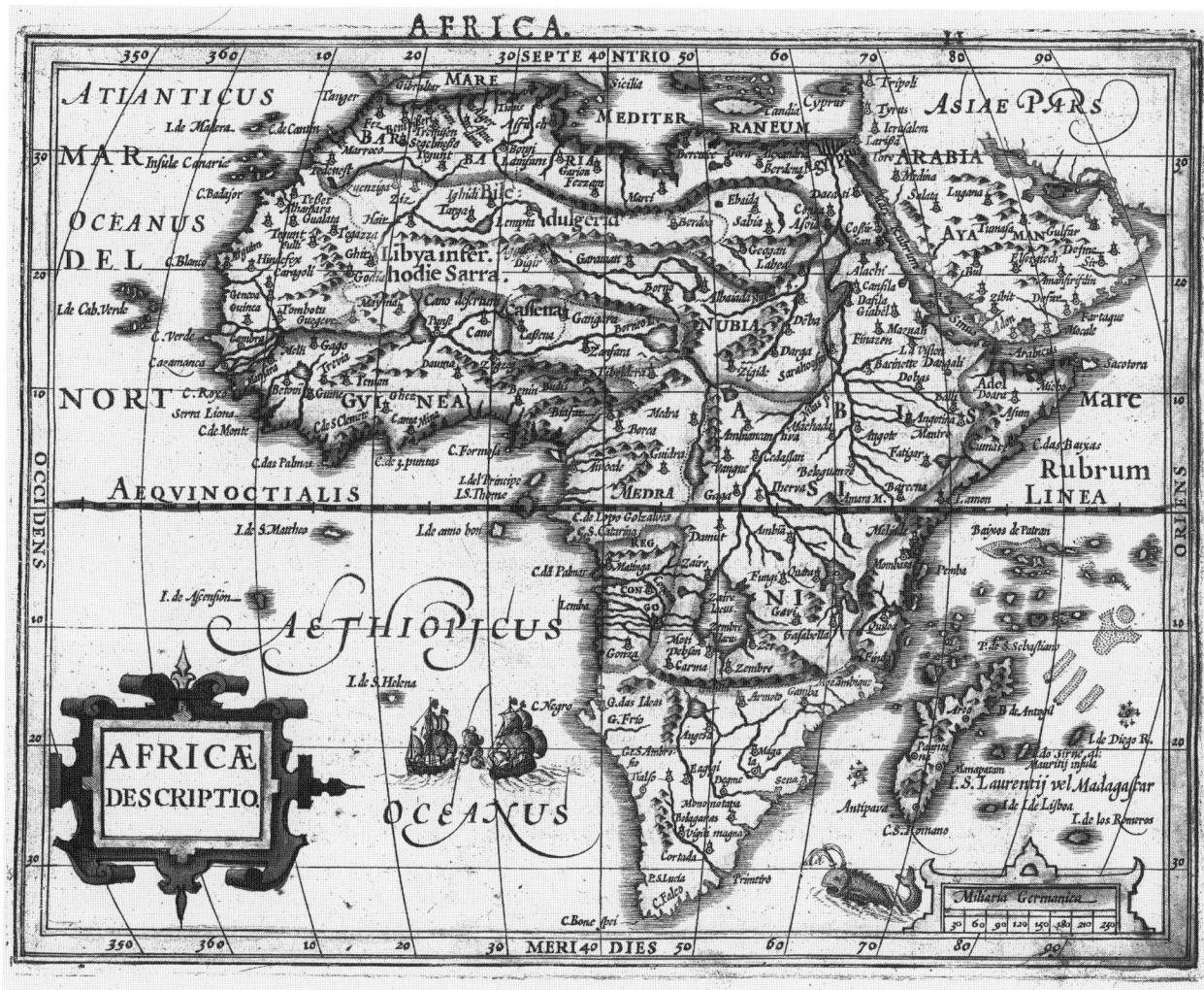

HONDIUS, Jodocus (1563-1612)
Africæ Descriptio. (Amsterdam, J. Hondius, 1621.)
Map, 15 x 20 cm, fully coloured.
In: *Atlas Minor*

Hondius acquired Mercator's copperplates and reissued the maps. As noted in the text for Map 27, he also issued an *Atlas Minor*. This was a compact, miniature atlas that was a great commercial success. Well engraved, the maps showed much detail despite their size.

Map 28

PONTANUS, Johannes Isaken (1571-1639)
Tabula Geograph in qua Europe, Africae, Asia et circu.... (Map of Europe, Africa and Asia.) (Amsterdam, 1611.)
Map, 27 x 42 cm, uncoloured.
Inset: figures in national dress.

A map of Africa extending eastwards to include the Middle East, India, and China. This map appears in both parts of the two-volume work *Rerum et Urbis Amstelodamensium Historia...* It is identical in volume 1 and is in a different state in the second volume. Johannes Isaken Pontanus came from Helsinki and was professor at Harderwyk, Holland. He published this 'Tabula Geograph' in 1611, with Jodocus Hondius as the engraver.

Along the top border is an inset illustrating the traditional dress of the natives of Ethiopia, Persia, Turkey, India, Sumatra and China. The details within the map reflect what was known in the seventeenth century. In the south there is not very much detail except for the usual Portuguese names known at that time. Once again the Nile arises from two lakes side by side. A portion of Spain is illustrated at the top left, with southern England and the North Sea.

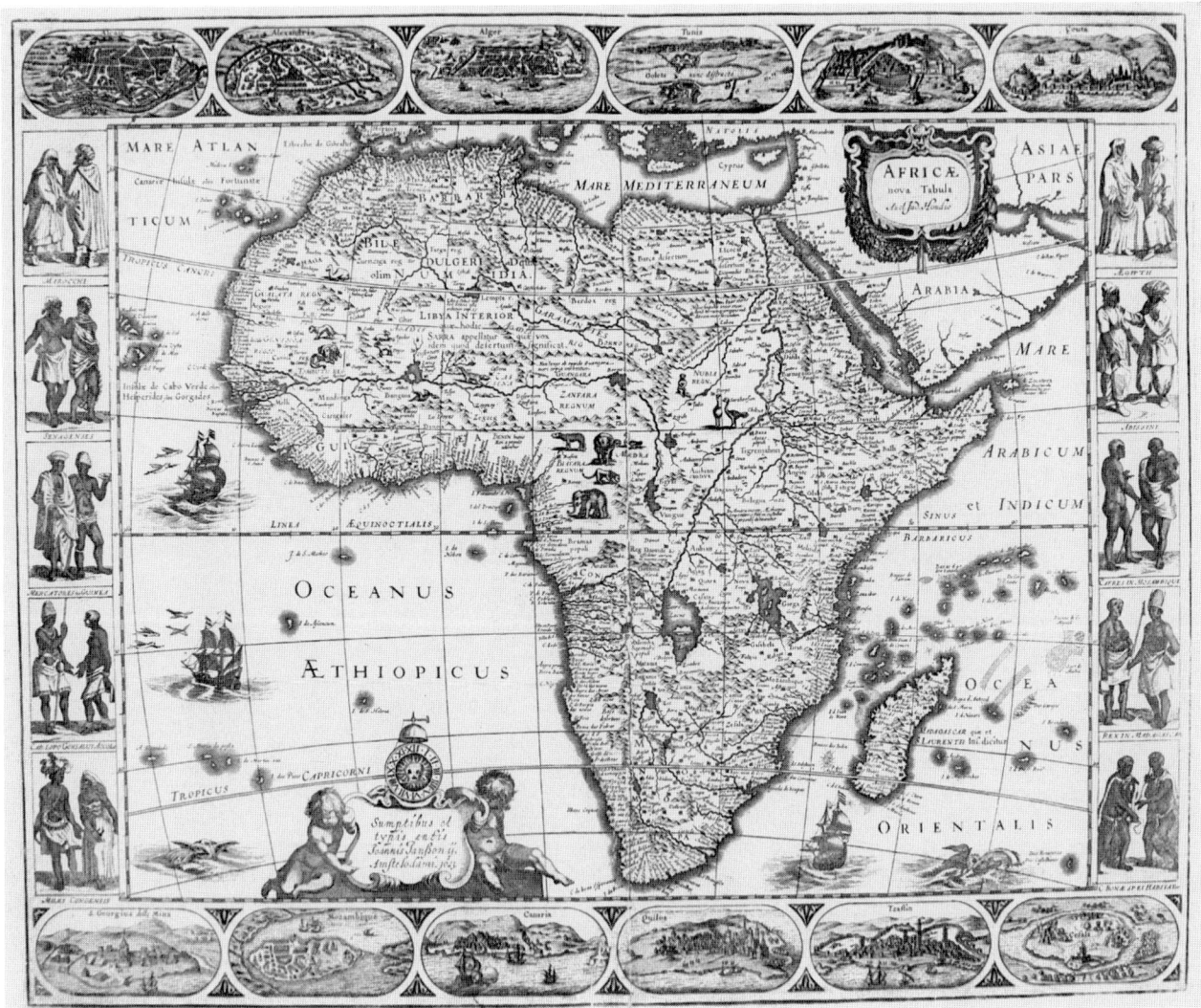

Map 29

HONDIUS, Jodocus (1563-1612)
Africae nova Tabula. (New map of Africa.) (Amsterdam, J. Hondius, 1623.)
Map, 45,5 x 56,5 cm, coloured.
Map surrounded by pictures of African towns and peoples.
Bottom left: Jansson imprint added; re-issue.

An attractive Hondius map, published in Amsterdam in 1623. The top and lower border illustrate twelve bird's-eye views of towns and the side borders each have figures in the traditional dress of their domain. Two Dutch ships are shown in the Atlantic Ocean, with flying fish and sea monsters. One ship and a sea monster are noted in the Indian Ocean. The geography is typical of the early seventeenth century with many fictitious names, rivers and lakes. The cartouche sited in the upper right corner is a simple rectangular scroll enclosing the title and surrounded by a decorative floral design on each lower border. Another cartouche appears at the lower left border and shows two cherubs supporting the sides of a rectangular design of scroll work surmounted by a clock with Roman numerals and a bell on top. The face of the clock is a skull surrounded by the rays of the sun.

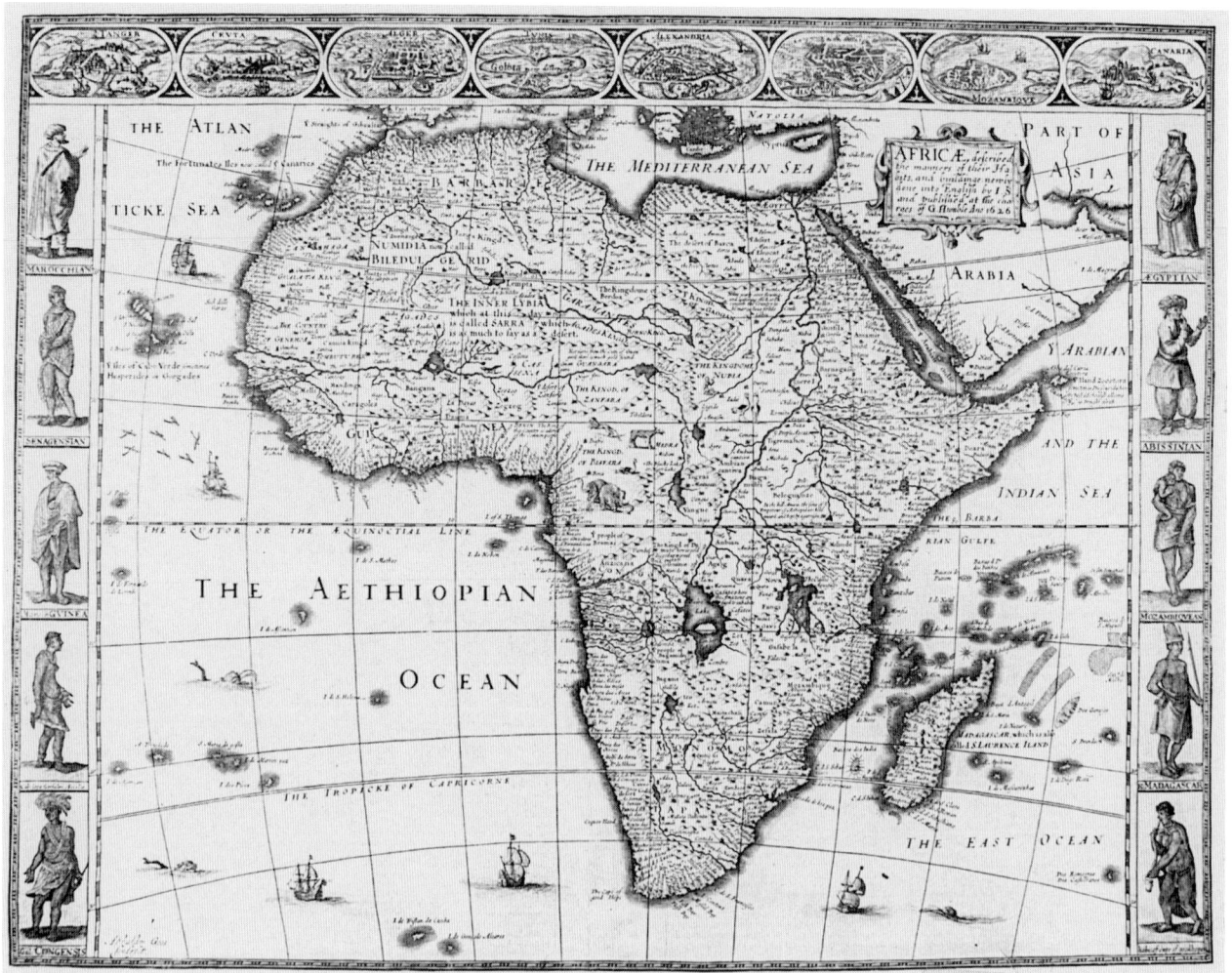

Map 30

SPEED, John (1552-1629)

Africae, described, the manners of their Habits, and buildinge: ... Engraved by Abraham Goos. (London, G. Humble, 1626.)
Map, 38 x 50 cm, slightly coloured.
Decorative frieze of cities and peoples. Description of Africa on verso.

The title appears at the top right in a cartouche of rectangular design with a decorative border. The side borders each contain five figures labelled according to their respective parts of Africa and dressed in their tribal costumes. The bottom right figure, labelled 'Habi of Cape of Good Hope,' is – as is often seen in travel books – a native eating entrails. Along the top border appear eight views of towns, mainly of North Africa, six of which are labelled at the top and two at the bottom. The Atlantic Ocean is decorated with four ships and another appears below Madagascar. Flying fish and spouting whales complete the decoration of the seas. Various types of animals such as elephants, lions, camels, monkeys and ostriches are to be found on the mainland. The whole interior of Africa is dotted with rivers, mountains, lakes and names, many of which are fictitious. Once again the origin of the Nile conforms to the Ptolomaic tradition of the two lakes side by side. The kingdom of Monomotapa occupies a large area of southern Africa.

On the back of the map there is a description of the various countries of Africa, which includes the kingdom of Prester John and Aethiopia Superior. One of the divisions of Aethiopia Inferior is named 'Manicongo,' whose inhabitants are said to be Christian in some parts but in others Anthropophagi, man eaters, who kill their own new born children to avoid the trouble of rearing them but, in order to preserve their race, steal babies from neighbouring countries.

John Speed was an English historian and cartographer who popularised the work of his predecessor Saxton and in 1627 issued *Prospect of the World*, the first printed general atlas to be compiled by an Englishman. This atlas was issued in several editions and, incidentally, contains the first map of Africa by an Englishman.

40

Map 31

SPEED, John (1552-1629)

Africa. Engraved by Pieter van den Keere. (London, J. Speed, 1627.)
Map, 8,5 x 12,5 cm, slightly coloured.

A small map of Africa engraved by Pieter van den Keere (1571–1646), a cartographer and book-seller of Amsterdam. He worked in London with his brother-in-law Hondius on maps and atlases of Europe, the English counties, the world, America and Germany.

This map is probably from *A Prospect of the Most Famous Parts of The World* (1646) – an atlas for William Humble, which is attributed to John Speed. Tooley described it as a small edition of Speed of 1627, reissued in several editions to 1676 (after Petrus Kaerius). The map has the usual features of this period with rivers and lakes in Central Africa, with the Nile arising from the two-lake system. It includes Arabia, Jerusalem, Samaria and part of Asia and is probably the first miniature map published by the Englishman, Speed.

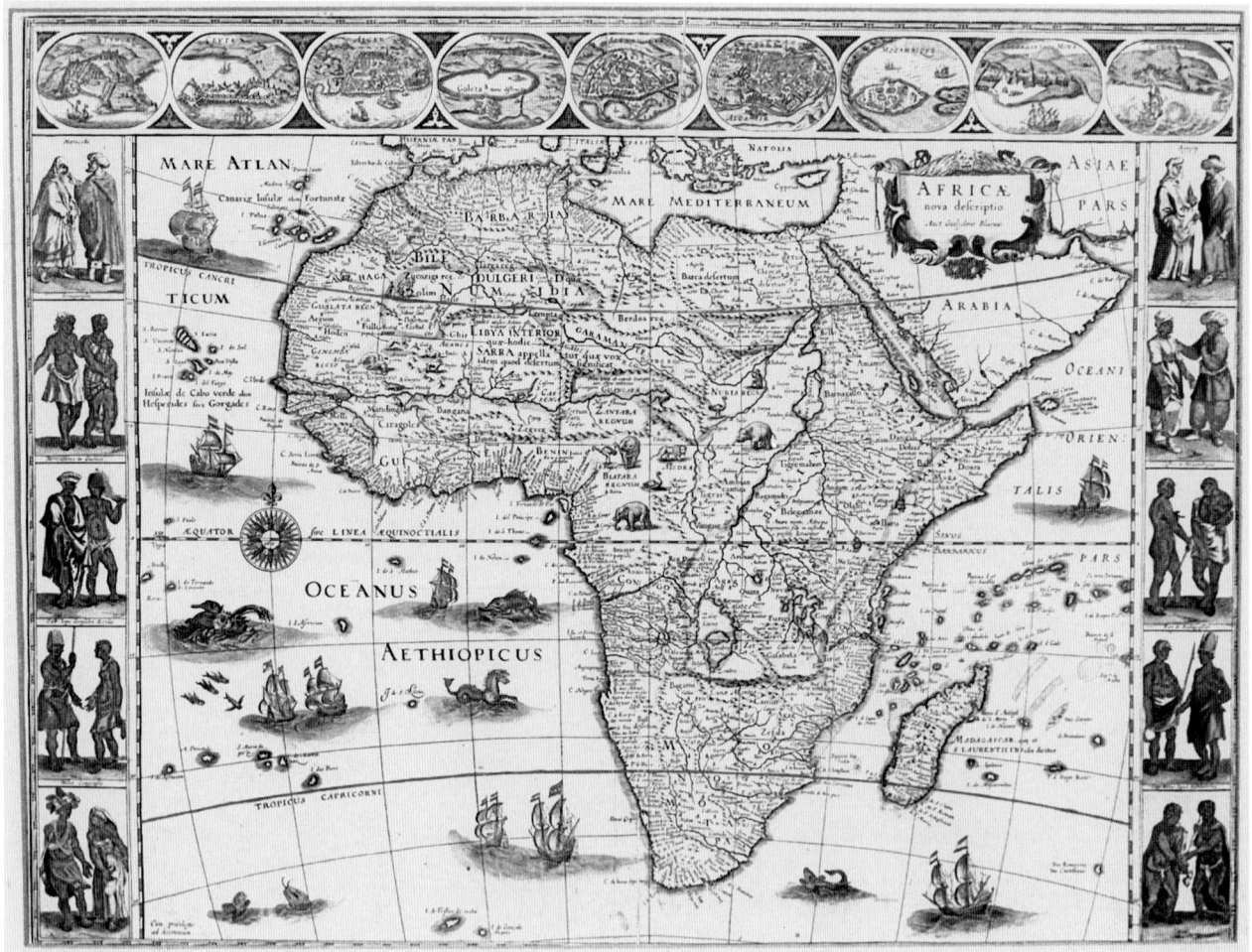

Map 32

BLAEU, Willem Janszoon (1571-1638)
Africae nova descriptio. (New map of Africa.) (Amsterdam, W.J. Blaeu, 1630.)
Map, 40 x 55 cm, coloured outline.
Decorative border of views of cities and peoples of Africa.

This is one of the most decorative and popular of all the early maps of Africa. The cartouche is not very elaborate but it is surmounted by what appears to be a lion with a flowing mane. The map itself is decorated with ships – seven in the Atlantic and two in the Indian Ocean – all flying Dutch flags; flying fish, whales and monsters are used to fill in the spaces in the oceans while elephants, monkeys and lions, ostriches and camels are depicted wandering all over the continent. Each side border is filled with decorative vignettes of costumed couples from various parts of Africa. The top border features oval vignettes of nine principal cities. The Nile is shown according to Ptolemy with its sources arising in Lakes Zaire and Zaflan; the fictitious Lake Sachaf of

Laurent Fries appears, and the R. de Spirito Santo. The only names shown at the Cape are coastal ones, engraved inwards to give an appearance of fullness to the map, leaving the contours clear and sharp.

There is a close resemblance, in both decoration and geography, between this map and that of John Speed (Map 30). This map, because of its extreme popularity, was reprinted many times from 1631 to 1667 with Latin, French, German, Dutch and Spanish texts on the reverse, and appears in volume X of Blaeu's elegant *Grand Atlas*. Blaeu published a number of fine large wall maps, some of which are in the Johannesburg Public Library's Africana Map Collection.

Map 33

MERIAN, Matthäus (1593-1650)
Nova descriptio Africae. (New map of Africa.) (Frankfurt, M. Merian, 1637.)
Map, 26 x 36 cm, uncoloured.

This map of Africa is copied from Blaeu's map but is smaller in size and lacks decorative borders. The cartouche in the right upper corner differs slightly in shape. (See Map 32 for further details.)

Matthäus Merian was born in Basle and was a prolific engraver and publisher who executed a great number of engraved copperplates. He studied in Zürich and became friendly with other well-known engravers, especially Jacques Callot and Théodore de Bry, whom he met in Frankfurt and whose daughter he married. Merian completed the maps of the last of the fourteen routes of the de Bry *Voyages*. His maps of Africa were issued in his *Topographia* of 1642-1688 and in Gottfied's *Neuwe Archontologia Cosmica* in 1638 and 1694. His work was continued by his son, Matthäus the younger, who also assisted his father with the *Topographia*.

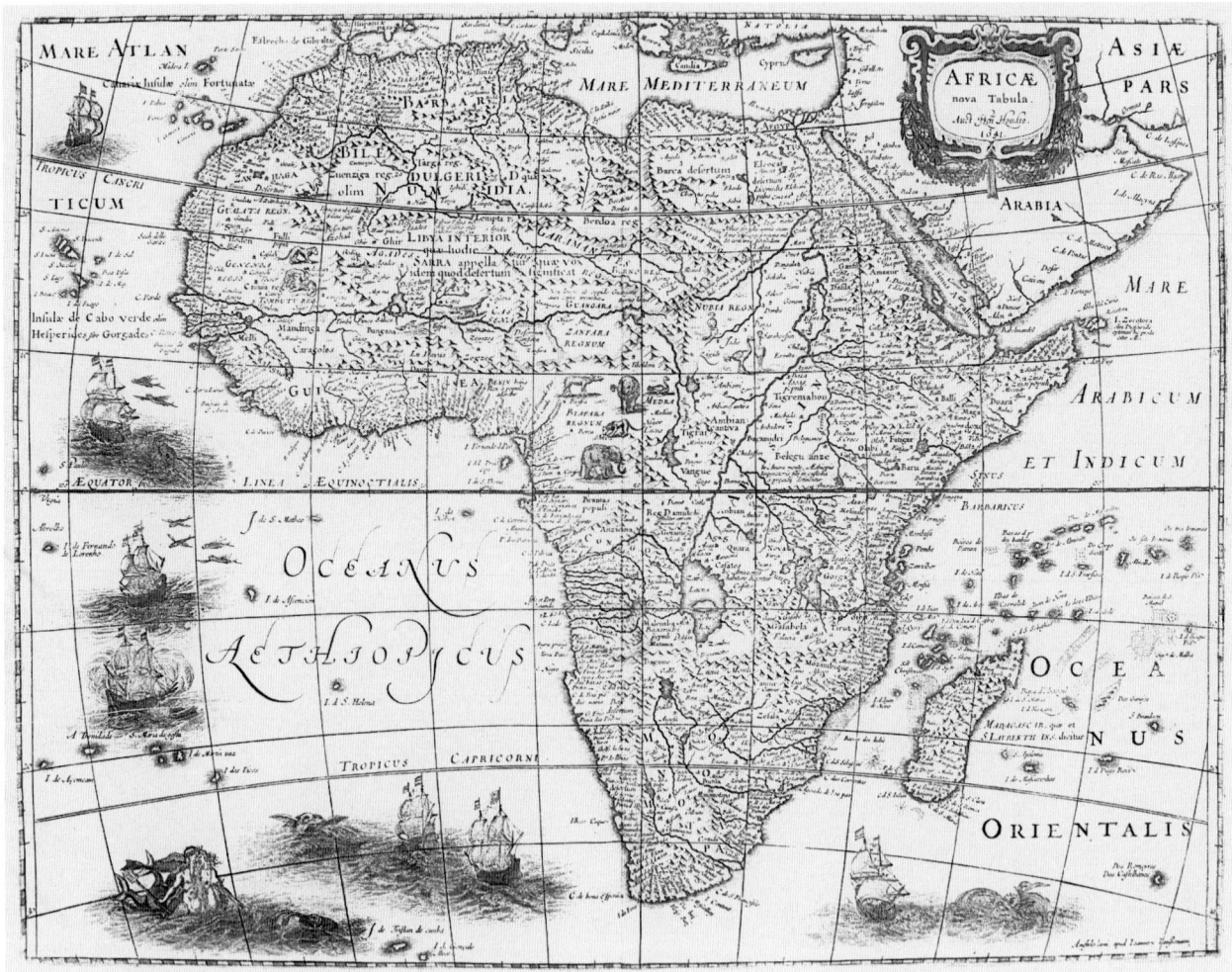

Map 34

HONDIUS, Henricus (c. 1597-1651)
Africæ nova tabula. (New map of Africa.) (Amsterdam, Joannem Jansonium, 1641.)
Map, 38 x 50 cm, coloured.

This map of Africa, a coloured copy with text in Latin on the verso, was executed by Henricus Hondius, and is a fine example of engraving of both the map and the surrounding oceans. It is a re-issue of the original Jodocus map, without the decorative borders, but with the publisher's imprint, Jansson, added at the lower right border. The cartouche has a wreath with an animal head at the top. There are six ships in full sail on the Atlantic and one to the south of Madagascar. Flying fish and various monsters decorate the sea and in the lower left corner Neptune and a mermaid are riding the waves. The usual elephants, lions, monkeys, ostriches, goats and other animals decorate the mainland. The very full geographical description of Africa is typical of the seventeenth century, and is similar to that of Speed and Blaeu.

After the death of Jodocus Hondius in 1612, his widow continued his business, assisted by her sons, Jodocus II and Henricus. The two sons often worked together in later life, but Henricus was the more active.

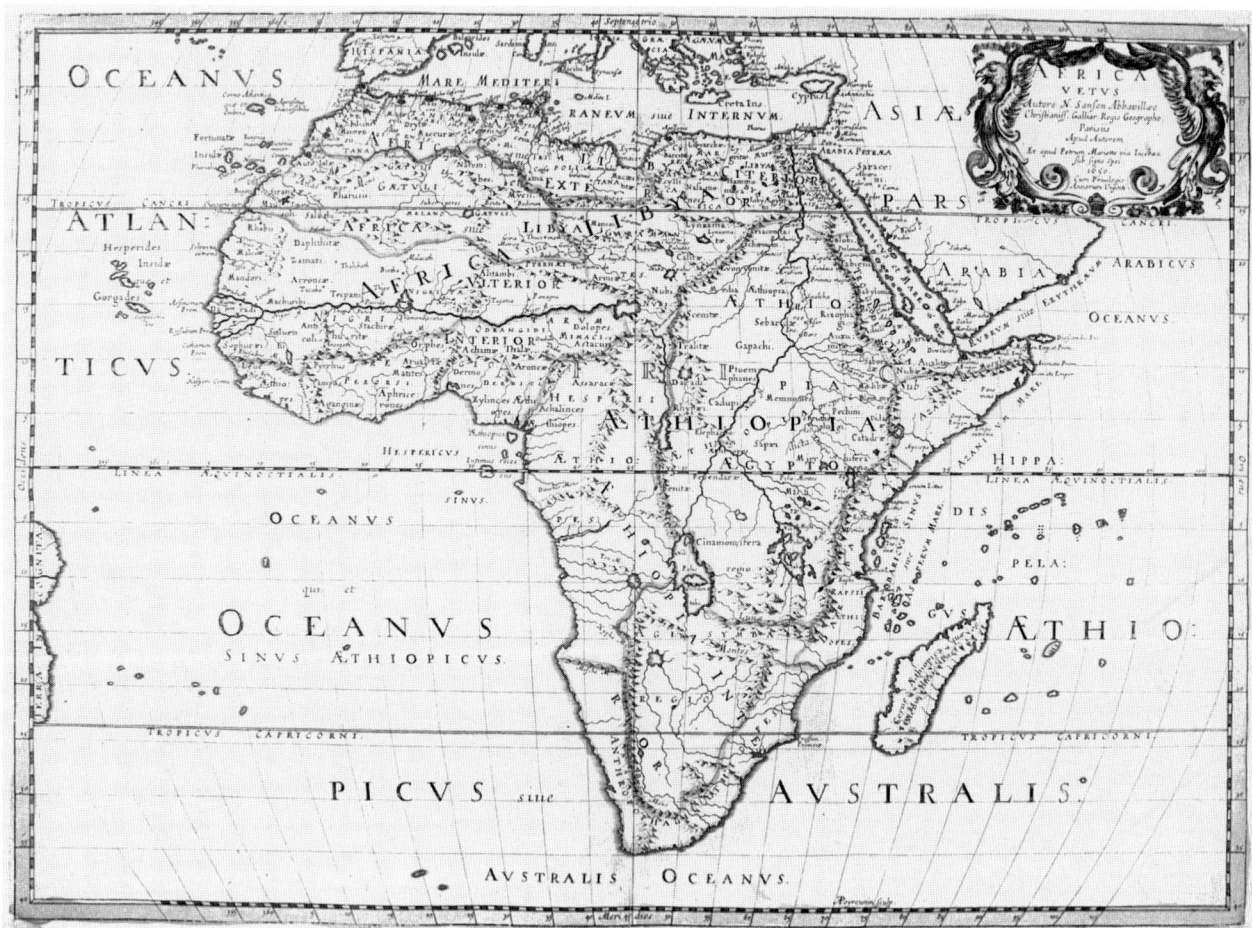

Map 35

SANSON, Nicolas (1600-1667)
Africa vetus. (Africa in antiquity...) Engraved by A. Peyrounin. (Paris, N. Sanson, 1650.)
Map, 39 x 55 cm, coloured outline.

This is an early Sanson map. The cartouche is made up of two heraldic eagles whose wings, together with a festoon of leaves, form the upper half. The lower part consists of festooned leaves and fruit. An odd feature is the name, Barditus Mons, given to a long chain of mountains running continuously from somewhere in Angola to Algoa Bay. No place names appear even on the coast of the southern part of the continent, which is labelled 'Agi Symba' (a name usually referring to the northern African region), and whose inhabitants are 'Anthropophagi Aethiopes' or cannibals.

Nicolaus Sanson was the first of three generations of this French family of cartographers. His work, though much less ornate than that of the Dutch school, was copied by many cartographers and publishers and his maps appear in many atlases and books of travel. (For further information on Sanson, see Map 41.)

Map 36

SEILE, Henry (fl. 1650s)
Africae Descriptio Nova Impensis. (Latest description of Africa.) Engraved by Will Trevethen. (London, H. Seile, 1652.)
Map, 34 x 43 cm, slightly coloured.

Henry Seile, the author of this map, was a publisher over St. Dunstan's Church in Fleet Street. His widow Anne (Anna) continued the business and on the title page of the 1666 edition of Heylin's four-volume *Cosmographie* appears 'Printed for Anne Seile and are to be sold at her shop in St. Dunstan's Church in Fleet Street.' This *Cosmographie* contains four maps and the one of Africa (see Map 39) is identical to the Seile map featured here except that the title name is Chetwind and the date 1666. The Seile map is dated 1652 and has in the left lower corner, 'Will Trevethen Sculp,' which also appears in the 1666 edition of the maps in the Heylin *Cosmographie*.

This map of Africa has a simple cartouche at the top right corner with a boxed title. There are four ships in the Atlantic Ocean and one in the Indian Ocean as well as flying fish and sea monsters, and on the mainland the usual seventeenth-century decorations – elephants, lions, ostriches and a giraffe appear. No tribal names appear in the south but many of our present-day names of bays can be recognised and Vigiti Magna and Monomotapa are both well marked. The Nile is shown as having its origin somewhere in the vicinity of the Congo and conforms to the Ptolemaic concept of a river from Lake Zaire joining with another river from Lake Zaflan. The course of the River Niger is erroneously mapped and there are many ficticious lakes in Central Africa, as it was really only during the mid and late nineteenth century that the interior could be mapped with any accuracy.

46

Map 37

SANSON, Nicolas (1600-1667)
Afrique. Engraved by A. Peyrounin. (Paris, N. Sanson, 1656.)
Map, 20 x 27,5 cm, uncoloured.
In: Marmol, Caravajal L. *De L'Afrique*, vol. 1. Paris: Louis Billaine, 1667, fold p. 1.

This map of Africa is typical of Sanson's technique and naming, although smaller in size than his 1650 map (see Map 35). It is a late map and although the same in shape it contains similar cartographic details and the same (though unnamed, in this map) peculiar mountain range in the south. The cartouche in this 1656 edition is also differently placed and different in shape and design.

The original Marmol volumes appeared in Spanish in 1573 and 1599 and are some of the most important and rare early works on Africa. In the French edition, published in 1557, all the maps are by the French cartographer, Sanson.

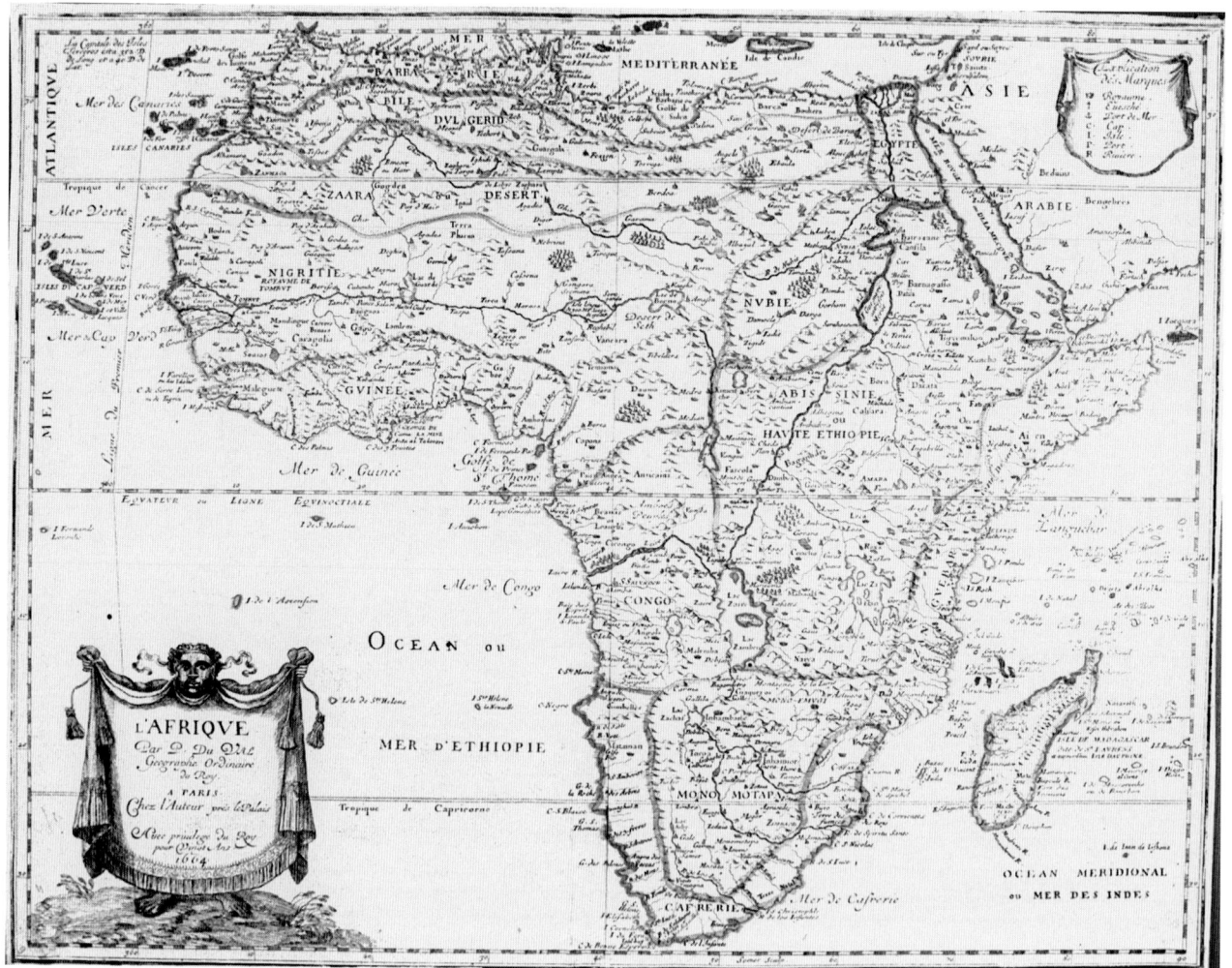

Map 38

DU VAL, Pierre (1618-1683)
L'Afrique. Engraved by Somer. (Paris, P. du Val, 1664.)
Map, 40 x 53 cm, coloured outline.
Prime meridian through Ferro Island.

This map of Africa was reissued in 1684 and, in the same year, as a large four-sheet map. The cartouche containing the title is made up of a fringed cloth held up by a native standing on a mound of earth and the legend explaining the abbreviations appears in the top right corner on a piece of drapery somewhat resembling a continental coat-of-arms. The general shape of Africa is squat because it has been drawn somewhat wider from east to west than usual. The usual seventeenth-century decorations are omitted but all the lakes rivers, bays and towns of maps of this period are here.

Pierre du Val was born in Abbeville in 1618 and died in Paris in 1683. He was a relative and pupil of Nicolas Sanson, and the author of several atlases of all parts of the world and of geographical treatises.

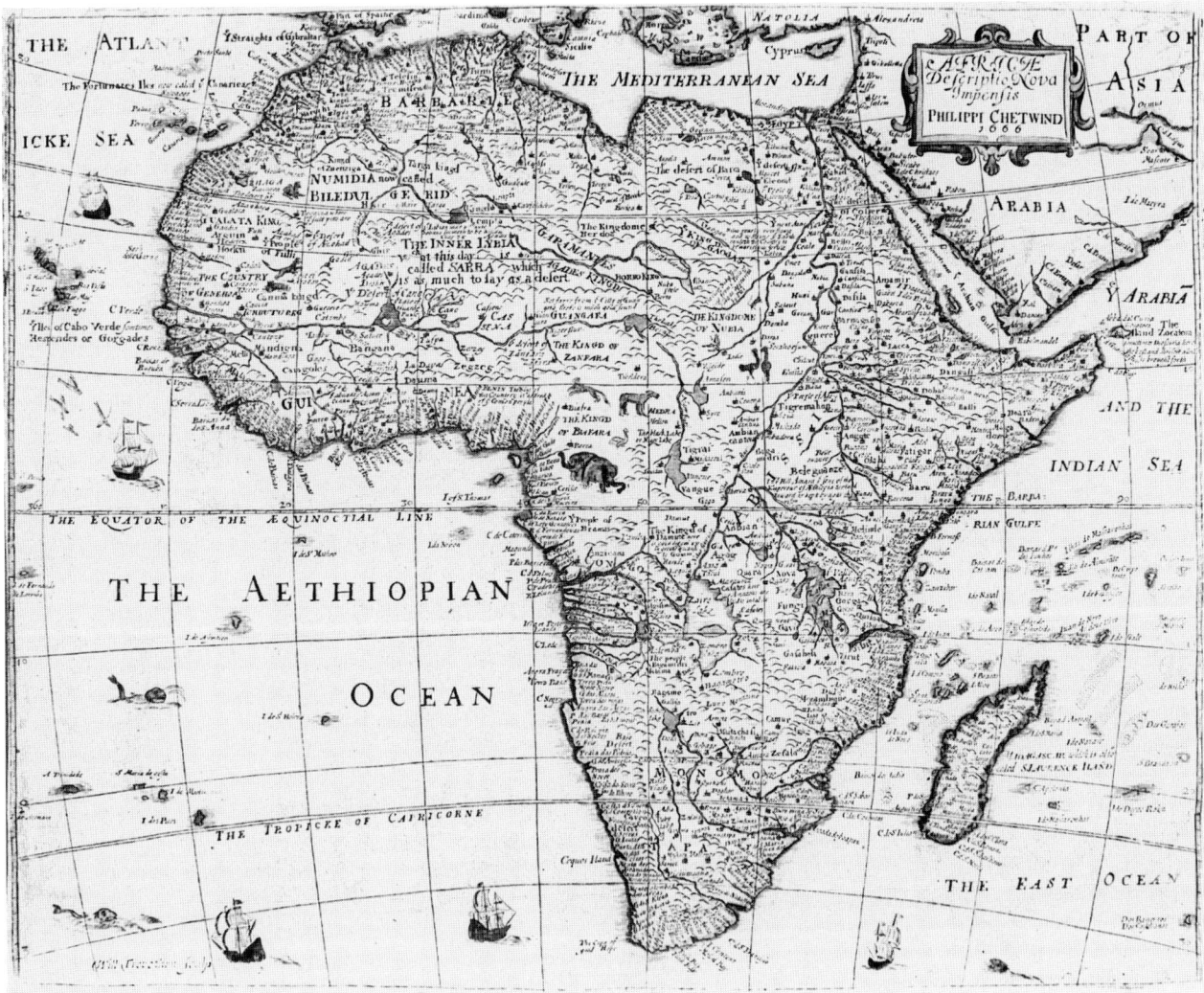

Map 39

CHETWIND, Philip (fl. 1670s)
Africae Deseriptio Nova Impensis. (Latest description of Africa.) Engraved by
William Trevethen. (London, P. Chetwind, 1666.)
Map, 33 x 43 cm, coloured.

Philip Chetwind was the publisher of the four maps of the continents in Heylin's *Cosmographie* of 1670, 1674 and 1677. William Trevethen engraved these maps, which included that of Africa reproduced above. There is a slightly decorated scrollwork title cartouche in the upper right corner. There are four ships in the Atlantic and one in the Indian Ocean as well as a number of flying fish and sea monsters. On the mainland the usual seventeenth-century decorations – elephants, lions, ostriches and a giraffe – appear, and large lakes and rivers are shown in South and Central Africa. No tribal names appear in the south, but many present-day names of bays can be recognised. Vigiti Magna and Monomotapa, for which explorers searched for many years, are both marked as well as Zimboa (Zimbabwe?). The Nile, as appears often in seventeenth-century maps of Africa, rises from the two lake tributaries – Lake Zaire near the Congo, and Lake Zaflan towards the east. The course of the Niger is inaccurately mapped.

This map is identical to that of Henry Seile (Map 36) of 1652. The cartouche in the upper right corner is also identical in shape and design, except that Seile's name is replaced by that of Philip Chetwind and the map is now dated 1666.

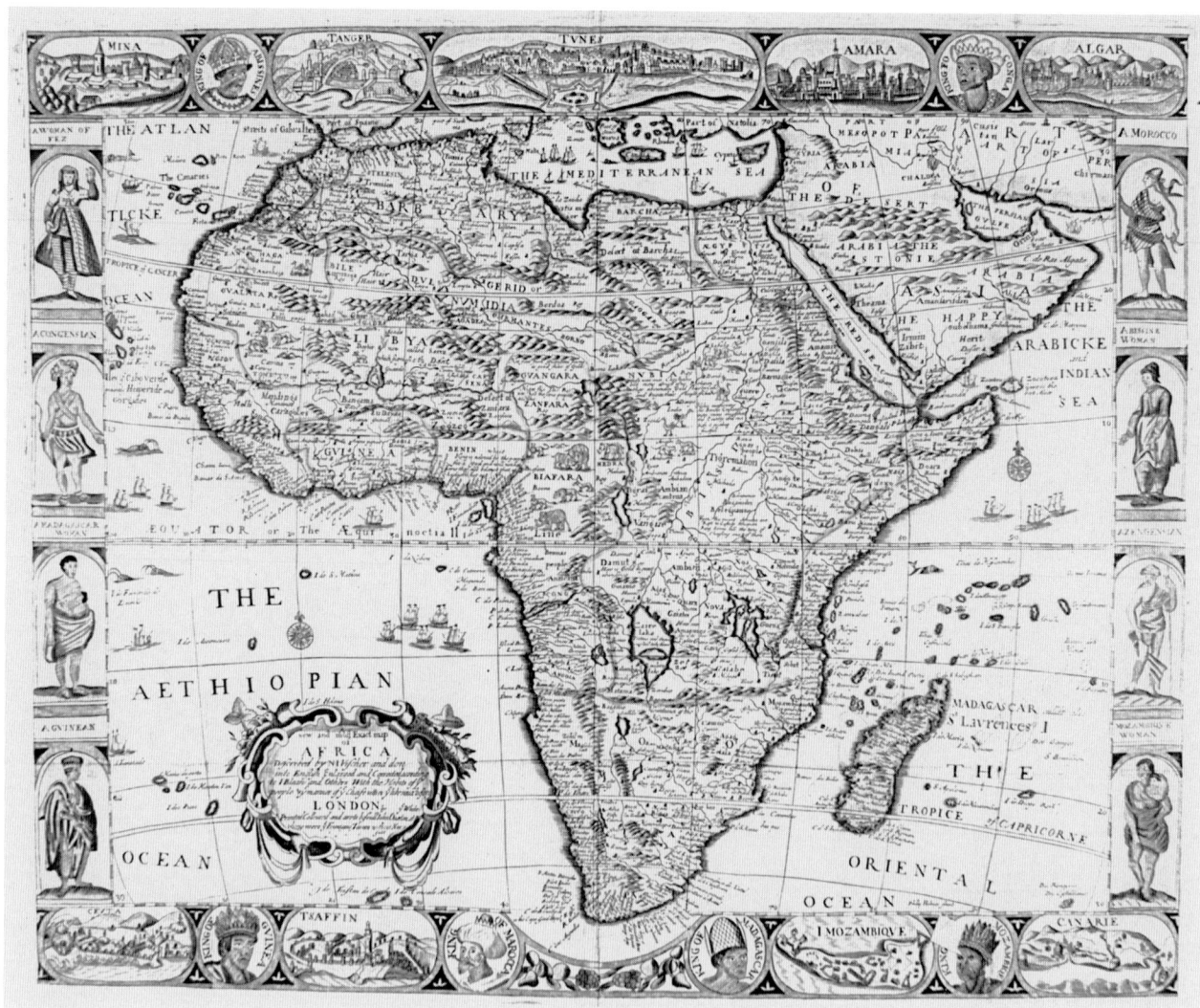

Map 40

OVERTON, John (1640-1713)
A new and most Exact map of Africa Described by N.I. Vischer and don into English; Enlarged and Corrected according to J. Blaeu and Others, With the Habits of ye people, and ye manner of ye Cheife sitties ye like never before.
Engraved by Philip Holmes. (London, J. Overton, ye White Horse neere ye Fountaine without Newgate, 1666.)
Map, 42 x 51 cm, coloured.
Decorative frieze of towns and peoples.

The cartouche of this colourful map of Africa consists of decorative scrollwork adorned with fruit on either side near the top. Five figures reminiscent of those in Speed's map (see Map 30), representing the people of the various African countries, are to be found on the border on either side of the map. Both at the top and lower borders there are views of African cities, separated by portraits of African kings. The oceans are embellished with more ships than normally appear on maps of this period and the sea monsters are fewer and smaller. Central and West Africa are well populated by the usual animals, some unrecognisable, including a rhinoceros almost in the centre. There are numerous lakes, rivers and mountains throughout the continent and the whole of southern Africa is labelled Monomotapa.

John Overton, a map-seller and publisher, lived and worked at various addresses in London. He employed the engraver Hollar for copying maps and, in 1700, he acquired John Speed's plates.

50

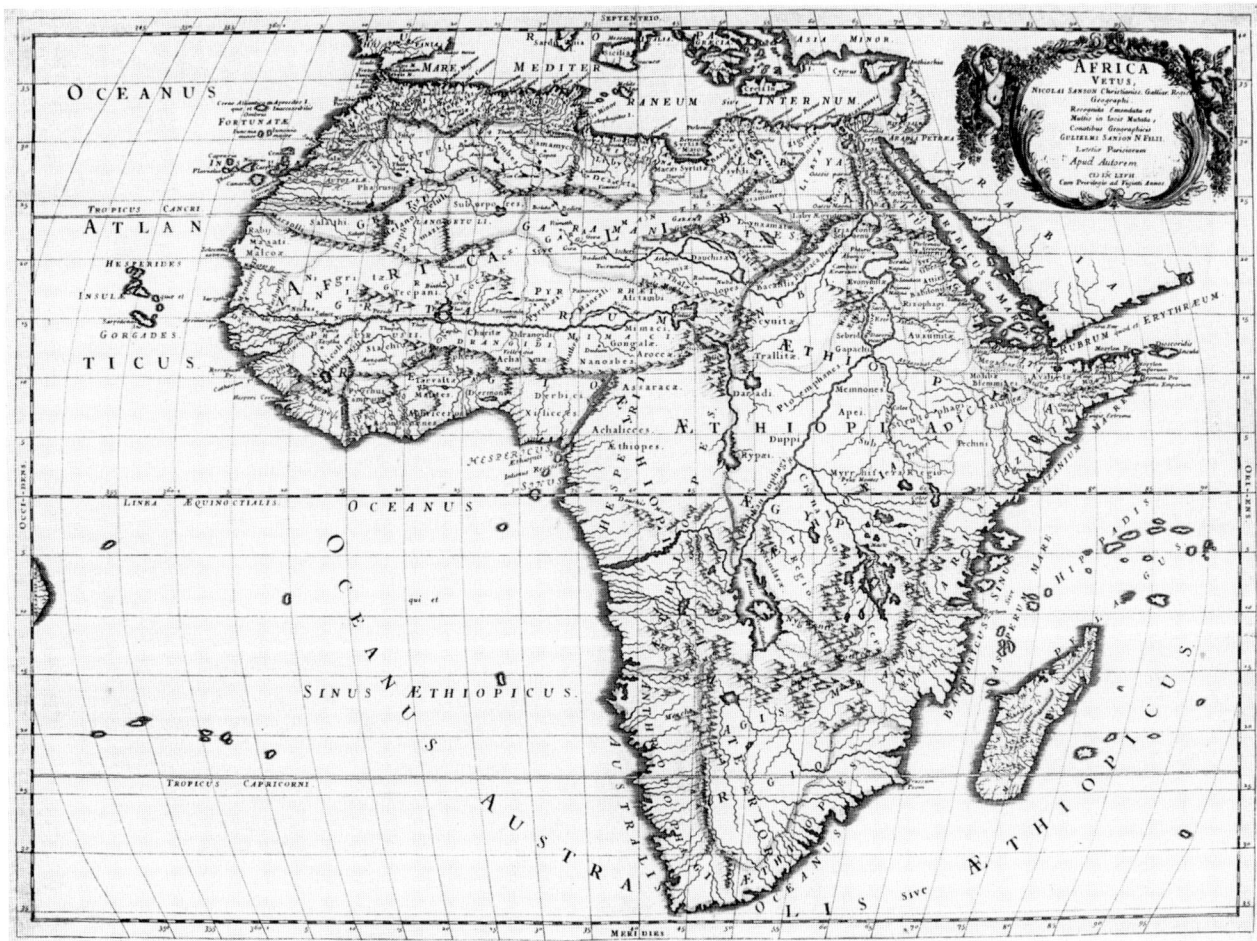

Map 41

SANSON, Nicolas (1600-1667)
Africa Vetus, Nicolai Sanson Christianis Galliar Regis Geographi Recognita Emendata et Multis in locis Mutata, Conatibus Geographicis Gulielmi Sanson N. Filii. (Africa in antiquity.) (Paris, Nicolas Sanson, 1667.)
Map, 40 x 56 cm, slightly coloured.

This map of 1667 has a simple wreath cartouche in the top right corner. The shape of the African continent is a little strange, with numerous coastal inlets. The rivers are distinct and noticeable and only the very large ones are named. No places, bays or rivers are named south of the Tropic of Capricorn, except for one cape on the east coast marked 'Prassum Prom.' Azania and Azanium Mare appear just below the Horn of Africa. All the inhabitants of southern Africa are labelled 'Anthropophagi Aethiopes' or cannibals. The origin of the Nile is once again based on the Ptolemaic concept of twin lakes.

This French family of cartographers, of whom the first was Nicolas Sanson, known as the founder of French cartographers, worked as map makers for three generations, over 100 years, when Hubert Jaillot succeeded to the business. Although their work was much less ornate than that of the Dutch and other schools their maps were very much copied, because geographically they were frequently superior and were also esteemed for their clarity and the delicacy of their engravings. Nicolas Sanson moved from his birthplace in Abbeville to Paris. In 1627 he met Richelieu, who was impressed by his map production and presented him to King Louis XIII, who was also favourably impressed and created him Geographe Ordinaire du Roi. Sanson produced over 300 maps and raised French cartography from insignificance to a prime position. Collections of his maps which appeared in his original *Cartes Generales...* generally do not correspond with the printed titles in the index of the atlas. His original atlas was re-issued by his sons in 1670 and again in 1676 with a new title, *Cartes de la Geographie Anciennes et Nouvelles.* Sanson employed Melchior Tavernier, Jean Boisseau, Peyrounin and others as his engravers. The folio edition maps of Africa first appeared in 1650 with further editions up to 1655. The small edition maps came out in 1656 with additional editions up to 1705.

51

Map 42

BLOME, Richard (fl. 1660-1710)
A New Map of Africa Designed by Mounsir Sanson, Geographer to the French
King. Rendered into English and Illustrated with Figurs by . . . by the Kings
Especiall Command. Engraved by F. Lamb (London, R. Blome, 1669.)
Map, 39 x 54 cm, coloured outline.
Dedicated to the Rt. Honourable Charles Howard.

This map of 1669 is almost identical to Sanson's map of Africa of 1650 (see Map 35), including the title cartouche, except for its mention of Blome, who greatly increased the number of place names and also included small ships and sea monsters, adding an heraldic dedication to the Right Honourable Charles Howard, Earle of Carlisle, in the lower left corner.

Richard Blome was a publisher, engraver and heraldic writer who flourished in London from 1660 to about 1710. He issued a description of the world in 1670 and supplied the maps for Heinrich Varenius's *Cosmography*. He issued a set of English county maps in his *Britannia* of 1673. Although his maps lack accuracy, they were popular at the time for their fine decorative charm.

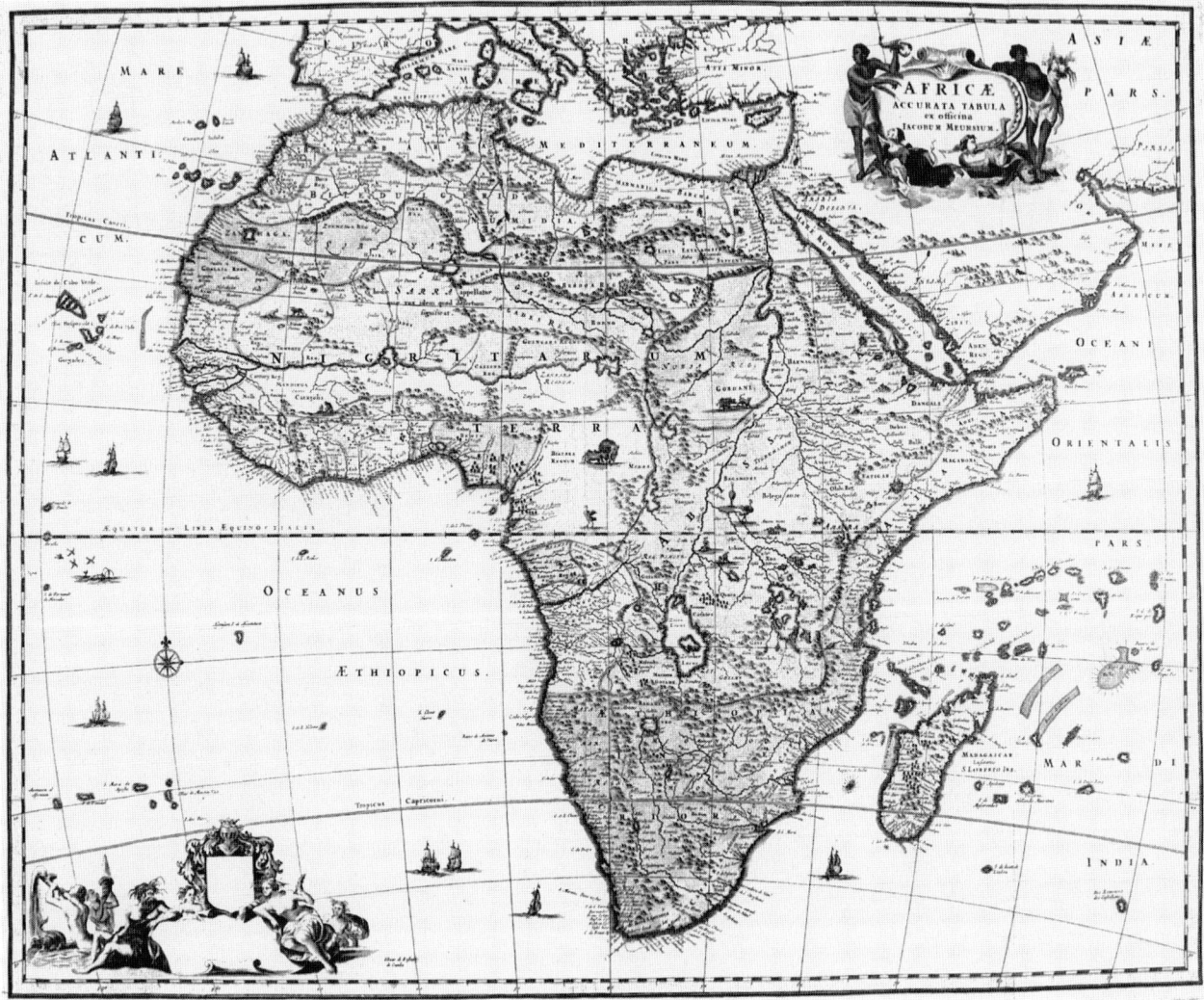

Map 43

MEURS, Jacob van (1620-1680)
Africae Accurata Tabula. (Accurate map of Africa.) (Amsterdam, J. van Meurs, 1670.)
Map, 43 x 54 cm, coloured.
Prime meridian through Tenerif.
Blank shield in cartouche at lower left of map.

This map of Africa appears in Ogilby's volume on Africa of 1670. The author of the map is recorded as Jacob van Meurs, an Amsterdam map publisher, engraver and bookseller, who published (for Dapper and Montanus) a map of America in 1641, and one entitled 'Wandering Israel' in 1677. The business was continued after his death by his widow.

This map is essentially the same as Visscher's 'Africae Accurata Tabula' of 1690 (Map 55), in which Nicolas Visscher's name has been substituted on the cartouche and the dedication renamed, although its design is practically the same. The fact that the embellished dedication at the lower left is blank, in this map, has prompted map dealers in single sheets to describe it, erroneously, as a proof copy. This map is also found in the various editions of Dapper's *Africa* in Dutch, French and German.

53

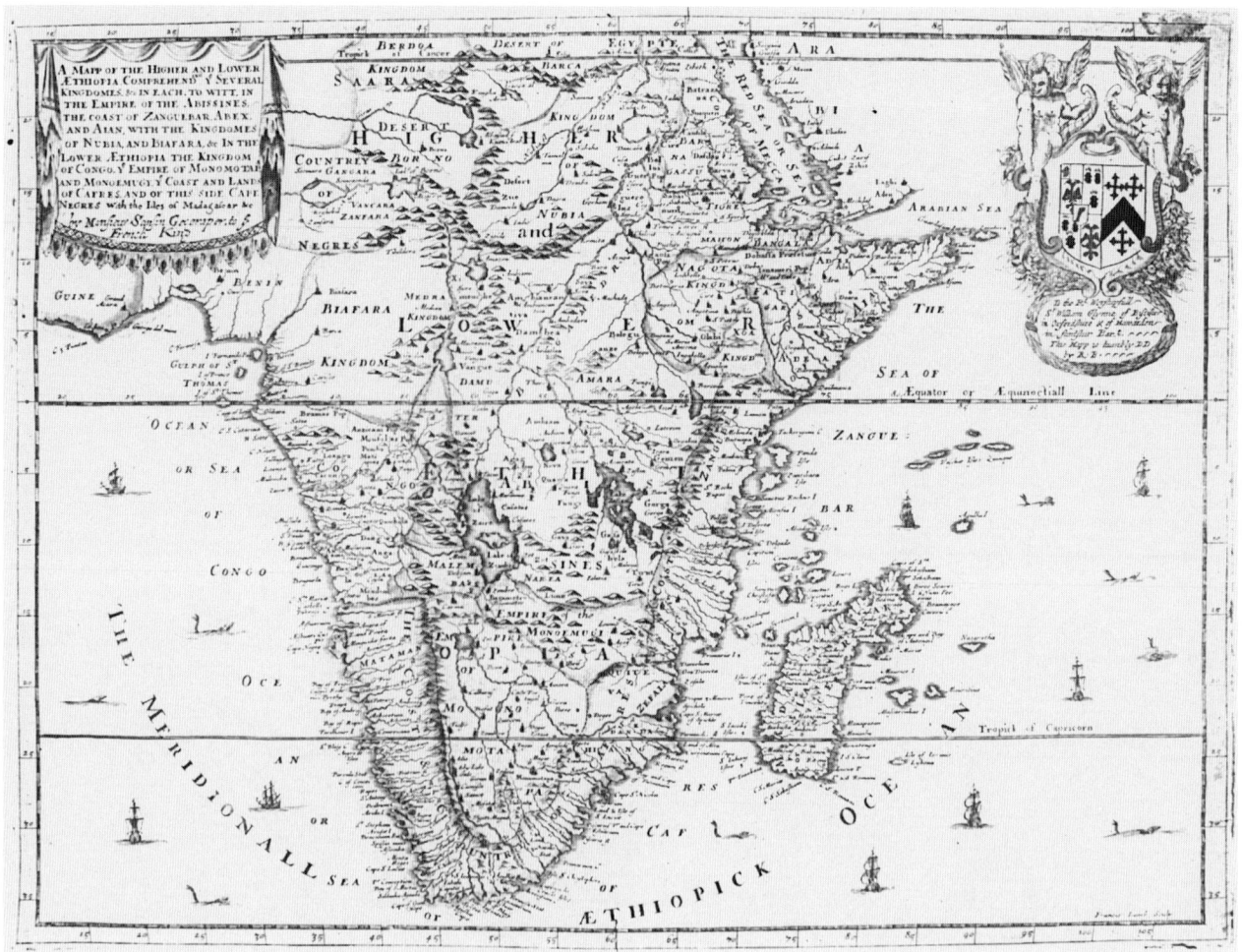

Map 44

SANSON, Nicolas (1600-1667)
A Mapp of the Higher and Lower Aethiopia. Comprehending Ye Several Kingdomes &c. in Each, to Witt, in the Empire of the Abissines, the Coast of Zanguebar, Abex, and Aian, with the Kingdomes of Nubia, and Biafra &c. In the Lower Aethiopia the Kingdome of Congo, Ye Empire of Monomotapa and Monoemugy, Ye Coast and Lands of Cafres and of this Side Cape Negres with the Isles of Madagascar &c.

Map, 30 x 40 cm, coloured.
Dedication: 'To the Rt. Worshipful Sir William Glynne of Bissister in Oxfordshire . . .'

The cartouche at the upper left corner of this map is obviously derived from the drapery of Sanson's 'Basse Aethiopie' (See Map 158). In the top right corner is the dedication to Sir William Glynne, surrounded by a coat of arms on which are perched two angels with cornucopias. The English publisher of this version has improved on the French edition by adding three ships and three whales to the Atlantic and five ships and five sea monsters to the Indian Ocean. The information on this map is much the same as is found on other Sanson maps of Africa; the larger lakes, the broken west coast with numerous rivers and the clearly defined boundaries between the states of the interior of southern Africa are characteristic.

The source of the map has not been traced. Although William Berry is known to have published English versions of Sanson's maps, it is most unlikely that this is one of his issue.

54

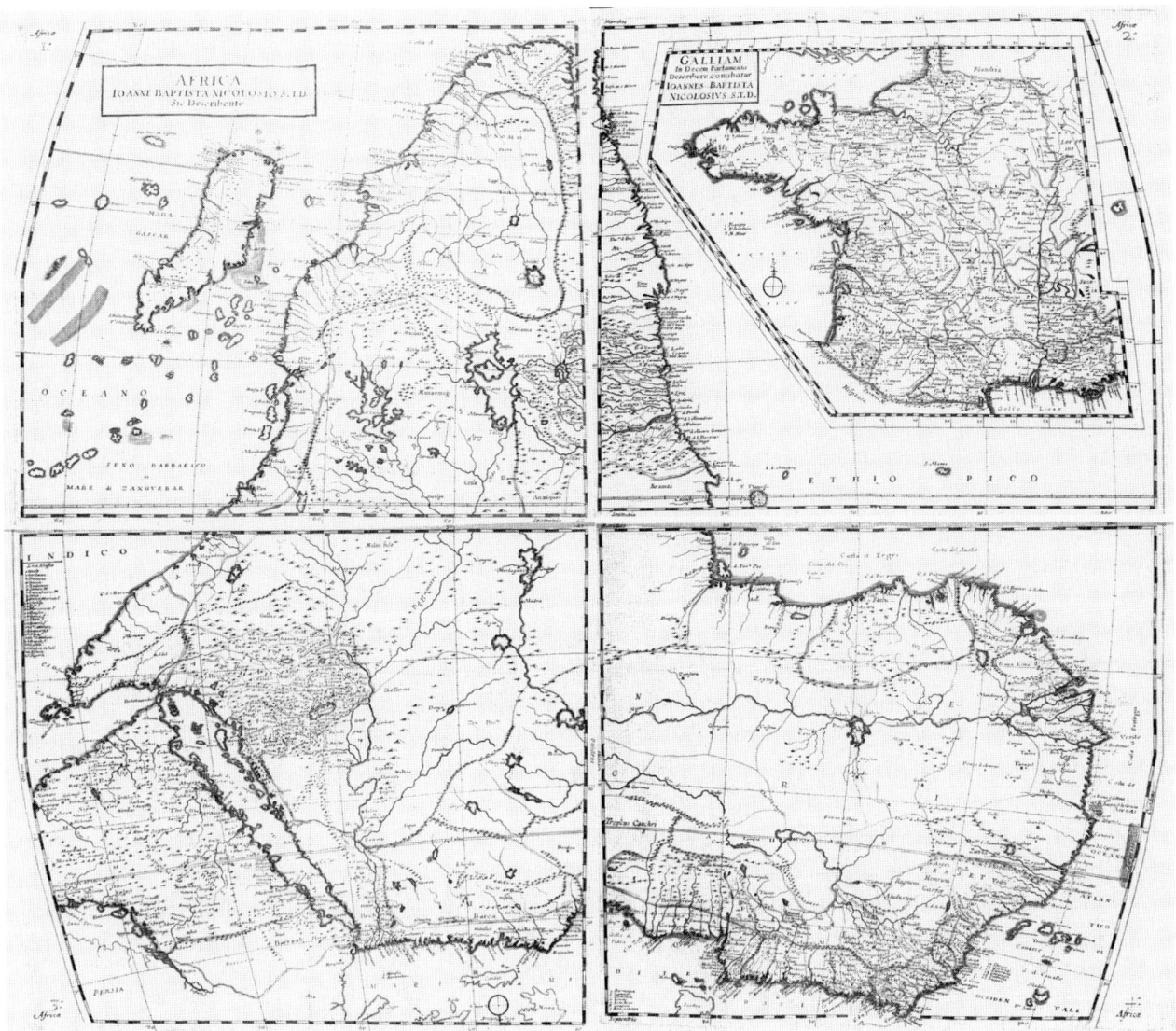

Map 45

NICOLOSIO, Joanne Baptista (1610-1670)
Africa... sic describente. (Africa described.) (Rome, J.B. Nicolosio, 1671.)
Map, four sheets 38 x 44 cm each, slightly coloured.
Sheets marked 1 and 3 on one mount, 2 and 4 on second mount.
Inset on map 2: Galliam (France).

This folio Italian map of Africa appears in four sheets marked 1-4, with an inset on sheet 2 of Galliam (France). It appears in an unusual atlas, *Hercules siculus sive studium Geographicum*, published by a Roman cartographer, Nicolosio, who was commissioned to make maps by Propaganda Fide in 1652. In addition to this atlas he was the author of *Dell' Hercole e studio geografico* with maps and directions for making globes (1670-71).

The atlas in which this four-sheet map appears is the second edition of 1671, the first edition having appeared in 1660. The front map of the atlas is a double-page map of the world in two hemispheres. Thereafter each continent is given four double-sheet pages (including separate maps of North and South America). Each sheet is folio size.

Topographical features included on the map reflect seventeenth-century knowledge of the continent: Abyssinia and Arabia are shown in some detail, with much less in southern Africa. The map appears upside down, as was the custom with other Italian cartographers. This atlas, with its remarkably finely engraved title page, frontispiece and armorial insignia, was seen by the author in the Vatican Library.

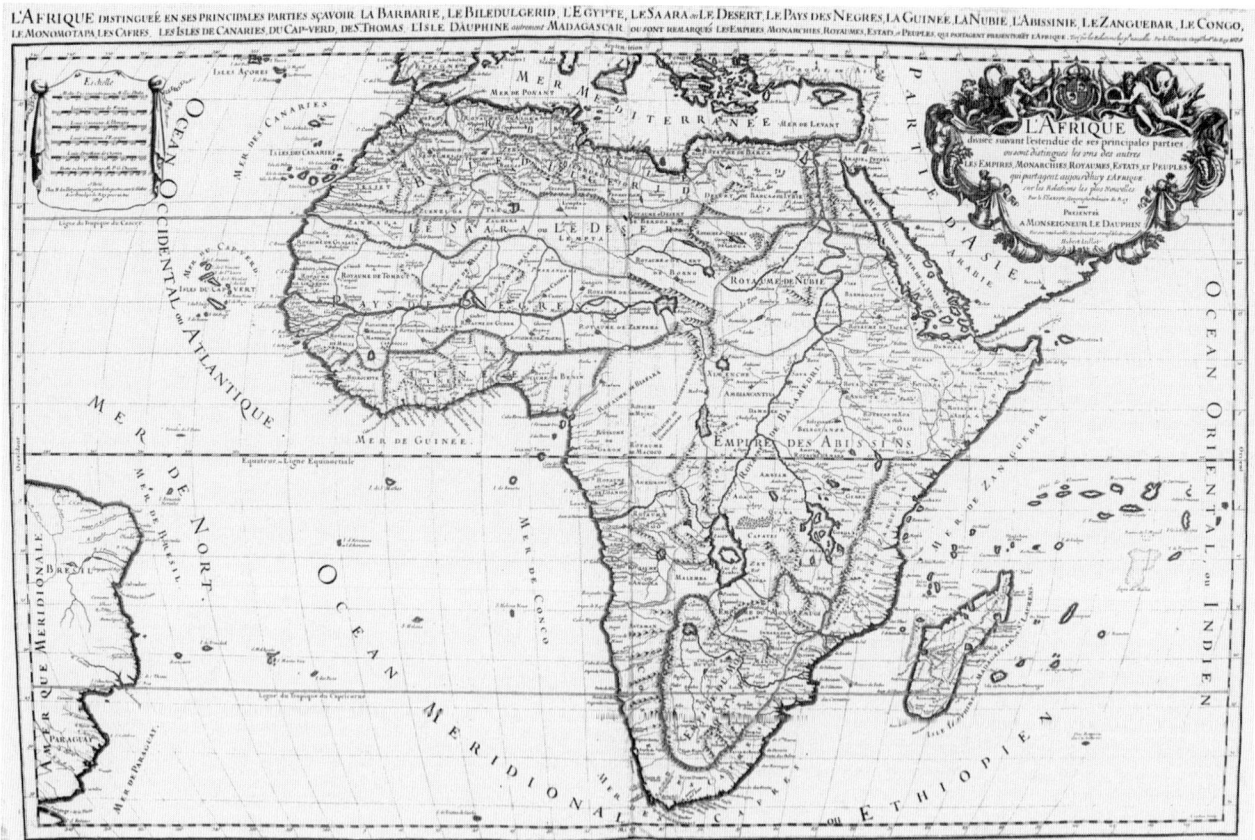

Map 46

JAILLOT, Alexis Hubert (1632-1712)
L'Afrique divisée suivant l'estendue de ses principales parties ou sont distingués les uns des autres, les Empires, Monarchies, Royaumes, Estats et Peuples qui partagent aujourd'huy l'Afrique sur les Relations les plus Nouvelles par le Sanson.
(Africa divided into its principal parts . . .)
Engraved by Robert Cordier. (Paris, A.H. Jaillot, 1674.)
Map, 54 x 88 cm, slightly coloured.
Scale in Italian miles, French, German and Spanish leagues.
Dedication: 'Presentée à monseigneur le Dauphin.'
Title across top: *L'Afrique distinguée en ses principales parties sçavoir la Barbarie, le Biledulgerid, L'Egypte, le Saara ou le Desert, le Pays des Negres, la Guinée, La Nubie, L'Abissinie, le Zanguebar, le Congo, le Monomotapa, Les Cafres, Les Isles de Canaries, du Cap-Verd, de S. Thomas, L'Isle Dauphine autrement Madagascar . . . Tiré sur les Relations les plus nouvelles Par le Sanson...*

This large map of Africa was published in 1674 by the Frenchman Alexis Hubert Jaillot, who succeeded to Sanson's publishing house in Paris. He was born in the small hamlet of Avignon in Franche Comté and in 1657 he and his brother Simon went to Paris, where they worked as sculptors. Hubert became friendly with some Flemish engravers, among them Nicolas Berey, publisher and map-maker to the Queen. In 1664 he married Jeanne Berey (by whom he had seven children) and joined his father-in-law in his trade. In 1676, after the death of his first wife, he married Charlotte Orbane, by whom he had eight children. He was appointed geographer to Louis XIV in 1678 and died in 1712, a wealthy and respected personality.

Ten years after Jaillot arrived in Paris, Sanson, the foremost geographer of that period, died, leaving behind a vast quantity of unpublished material. Sanson's sons, wishing to continue their business, asked Jaillot to become their associate. Thus Jaillot began his task of re-engraving Sanson's work on a larger scale, producing very fine issues from 1674 onwards. The great fire of 1672 which destroyed Blaeu's empire in Holland left a large gap in the publishing of maps, which Jaillot readily filled. He published increasing numbers of single maps and, later, atlases – among them the *Atlas Nouveau*. All these maps were printed in France by the French engravers and many editions were reprinted in Amsterdam, under contract with the Dutch publisher

Mortier and, later, Covens. In 1693 he was requested to undertake the task of publishing the *Neptune Français,* which was to be his most important work, and in 1694 he published the *Atlas Français,* which went into many editions, continued by his heirs for fifty years. His sea charts appear to have been more accurate than his terrestrial maps.

His large map of Africa (after Sanson) of 1674 was his first edition of this continent. The cartouche in the right-hand top corner is oval in shape and made up of elaborately detailed drapery, figures and cornucopias, with, in the top centre, the arms of the Dauphin. It is flanked on both sides by a male figure pouring water from a jar. An ostrich and a lion appear on the left and on the right an elephant and a crocodile. A larger section of South America than is usually found in maps of Africa can be seen on the left and two St. Helena islands appear – a common error in maps of this period. The two large lakes, Zaire/Zembre and Zaflan, in Central Africa, are conspicuous as the origin of the Nile. The River Zaire (Congo) also arises from Lake Zaire. The Kingdom of Monomotapa is depicted with its capital on the Rio de Spirito Sante, and Vigiti Magna also appears. A number of rivers have been marked in the south, including R. du Cap de Bonne Esperance, but most of this country is marked 'terre deserte.'

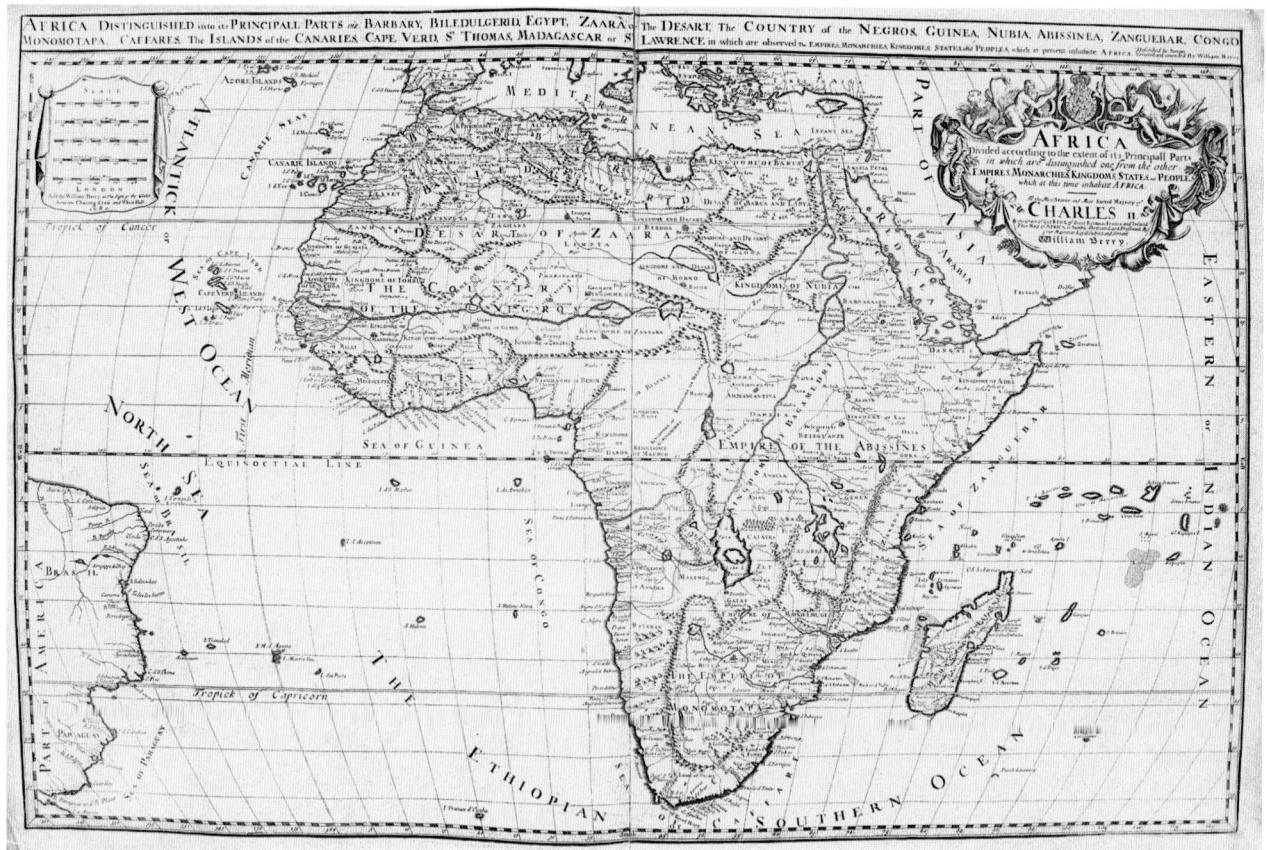

Map 47

BERRY, William (fl. 1669-1708)
Africa divided according to the extent of its Principall Parts in which are distinguished one from the other Empires, Monarchies, Kingdoms, States and Peoples which at this time inhabite Africa. (London, William Berry at the sign of the Globe between Charing-Cross and Whitehall, 1680.)
Map, 54 x 89 cm, slightly coloured.
Scale in English and Italian miles, French, German and Spanish leagues.
Prime meridian through Ferro Island.
Dedication: 'To the Most Serene and Most Sacred Majesty of Charles II.'
Title across the top: *Africa distinguished in its principal parts viz. Barbary. Biledulgerid, Egypt, Zaara or The Desert, the Country of the Negros, Guinea, Nubia, Abissinea, Zanguebar, Congo, Monomotapa, Caffrares, the Islands of the Canaries, Cape Verd, St. Thomas, Madagascar or St. Lawrence . . . Described by Sanson. Corrected and amended by William Berry.*

A large map of Africa with a majestic cartouche in the top right corner and a dedication to Charles II, originally described by Sanson (see Map 46), and amended by William Berry in 1680. Berry was an engraver, globe-maker, publisher and bookseller, who flourished from 1669 to 1708 in London. He was sometimes referred to as the English Sanson, having published an atlas after Sanson in 1680-89; and also collaborated with Robert Morden in the production of his *Globe* of 1683.

A large section of the western bulge of Brazil appears, as so often found in maps of Africa. There are still two St. Helena islands, as was the custom until it was finally established that one was fictitious. The two large lakes in Central Africa appear prominently almost side by side, the eastern one marked as L. d'Zaire in the north and Lake d'Zembre in the south with an island separating the names. The western lake is unnamed but conforms to Lake Zaflan in Sanson's map. The Empire of Monomotapa features prominently, with Monomotapa marked on the Rio d'Spiritu Sancta. In southern Africa many rivers are marked, including 'River of Cape of Good Hope.' Most of this southern country is marked 'Land of Desert.'

This map is based on Jaillot's map of Africa of 1674 (see Map 46) but not merely a copy of it as suggested by some writers.

58

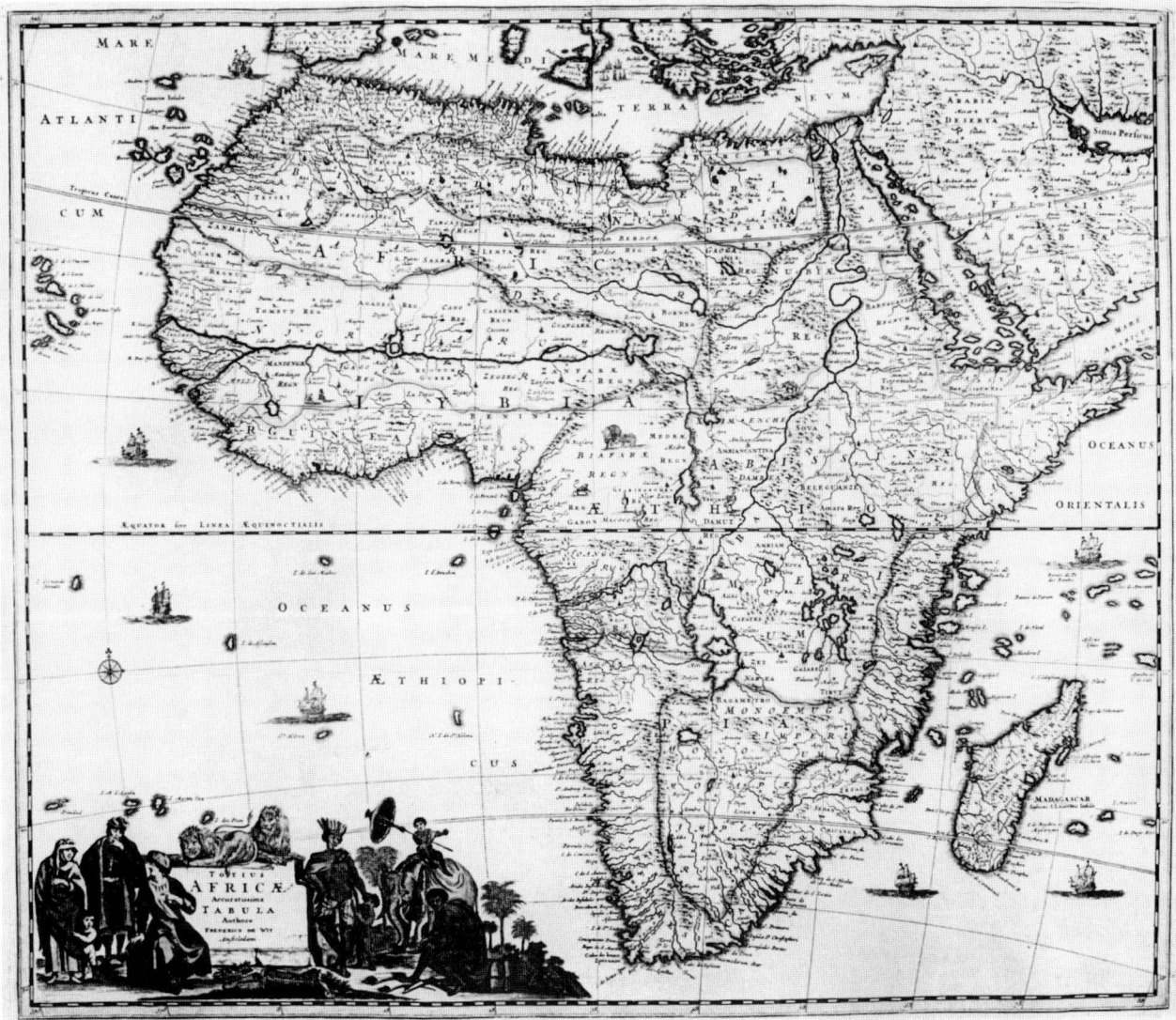

Map 48

DE WIT, Frederick (1616-1689)
Totius Africae Accuratissima Tabula. (Most exact map of the whole of Africa.)
(Amsterdam, F. de Wit 1680.)
Map, 48 x 58 cm, slightly coloured.
Prime meridian through Ferro Island.

The map of Africa described here has a large title piece on the lower left with three Moors and a child on the left, two negroes and two children on the right, with an elephant; two lions above the title and seven ships in the surrounding seas. Small vignettes of different animals appear within the interior. A noticeable feature of this map is the large number of deep inlets on the south-east coast. The large lakes and numerous rivers are also a feature of this map. The origin of the Nile is Ptolemaic in design; the kingdom of Monomotapa comes down as far south as the Rio de Infante, and there are few names in the interior of the Cape.

Though it shows no real geographical advance this map became extremely popular and had a wide appeal through many publishers. Danckerts published it under his name without any other change, and the same cartouche is to be found in the de Ram map (see Map 50).

59

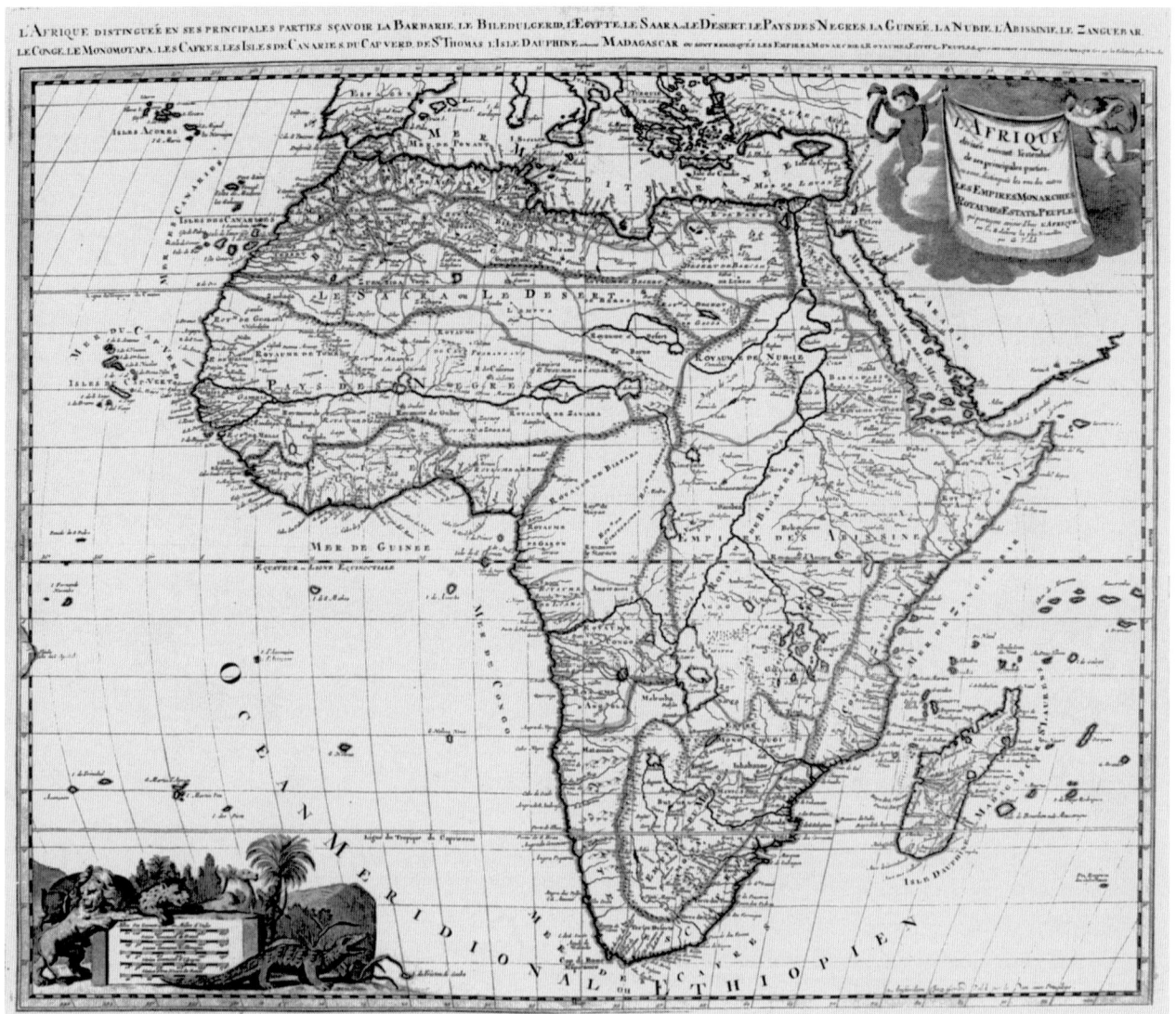

Map 49

VALK, Gerard (c. 1650-1726)
L'Afrique divisée suivant l'estendue de ses principales parties, ou sont distingué's les uns des autres, les Empires, Monarchies, Royaumes, Éstats et Peuples qui partagent aujourd'hui L'Afrique sur les relations les plus nouvelles... (Africa divided into parts which are further divided into empires etc.)
(Amsterdam, G. Valk. 1680.)
Map, 46 x 57 cm, slightly coloured.
Scale in Italian miles, French, German and Spanish leagues.
Title across the top: *L'Afrique distinguée en ses principales parties sçavoir la Barbarie, le Biledulgerid, l'Egypte, le Saara ou le Desert, le Pays de Negres, la Guinée, la Nubie, l'Abissinie, le Zanguebar, le Conge, le Monomotapa, les Cafres, les Isles du Canaries, du Cap Verd, de St. Thomas, l'Isle Dauphine autrement Madagascar . . .*

This map of Africa has a cartouche in the right upper corner consisting of two cherubs holding a fringed square of material on which the title appears. At the left lower corner are the scales in Italian, French, German and Spanish, surrounded by a lion, an elephant, a leopard, a camel and, in the foreground, a crocodile. In the background is a palm tree and other foliage, with an African about to shoot an arrow from a bow. The details of the map itself are typical of the late seventeenth century. The large rivers are marked and the Nile appears according to the ancient pattern, arising

60

from two lakes side by side. No tribal names appear in the south but the Monomotapa kingdom with the town of that name appears extending well down into southern Africa. There are two islands of St Helena, as so often in maps of this period.

Gerard Valk was a Dutch publisher and engraver working in Amsterdam. In partnership with Schenk he acquired some of Blaeu's and Jansson's copperplates. He issued globes in 1700-1715 and an atlas in 1702, the Caspian Sea (1721), and large maps of the world and four continents in 1680, among which this map probably appeared. He was joined by his son Leonard, who continued the business on his father's death in 1726.

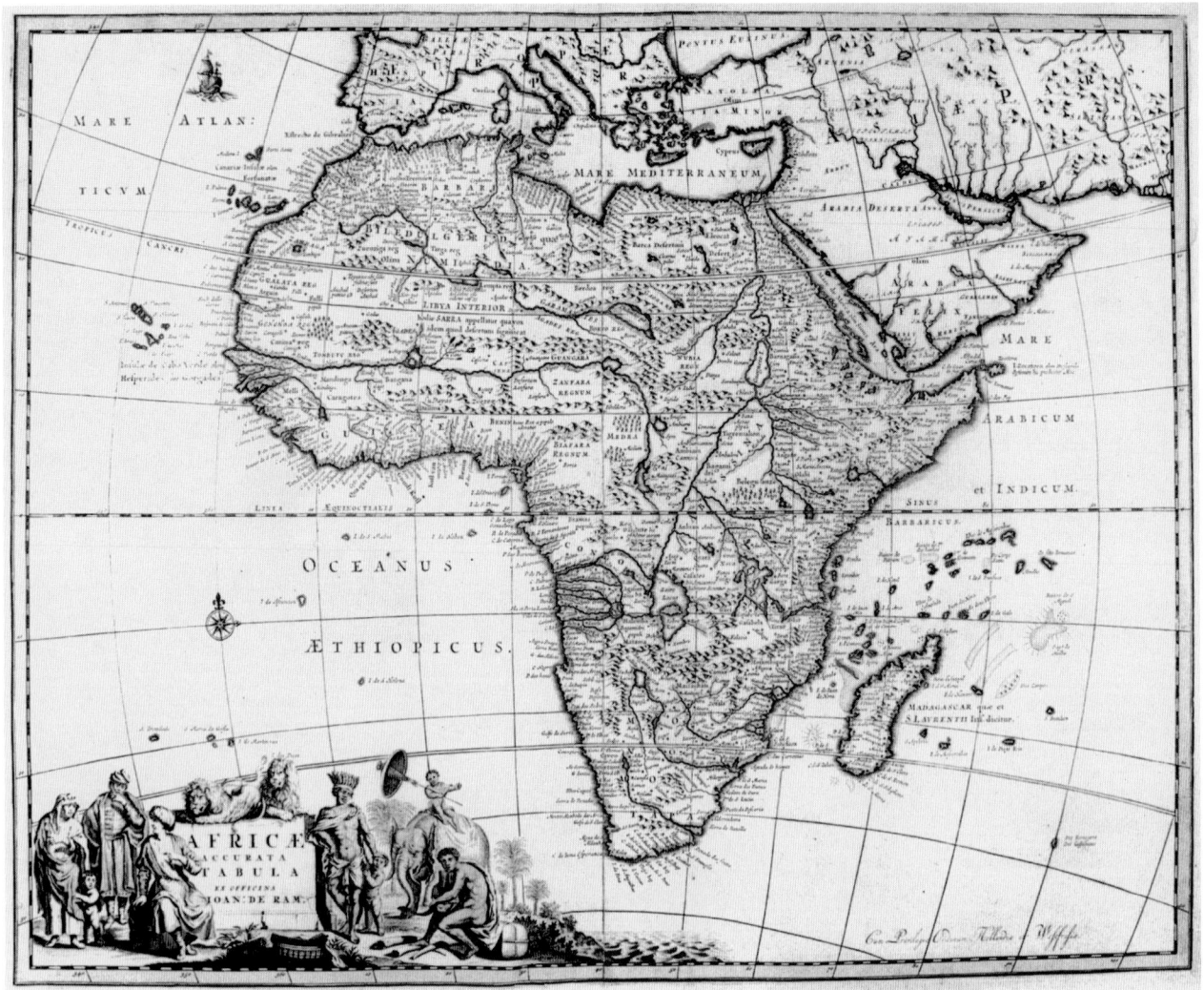

Map 50

DE RAM, Joan (Johannes) (1648-1693)
Africae Accurata Tabula. (Detailed map of Africa). (Amsterdam, J. de Ram, 1680-1690.)
Map, 44 x 56 cm, slightly coloured.
Prime meridian through Tenerif.

The cartouche of this map is identical to that of the de Wit (Map 48) and the Danckerts (Map 57) except for the variation in text. The rest of the map is similar in its somewhat fictitious detail, but it lacks the surrounding ships in the eastern and western oceans. The detailed animal representations have also been omitted in this de Ram edition.

Johannes de Ram was an engraver, publisher, globe-maker and art dealer based in Amsterdam. He appears to have issued atlases with maps by various geographers as well as those of his own design, including one of the world. His widow, who married J. de la Feuille, reissued his plates.

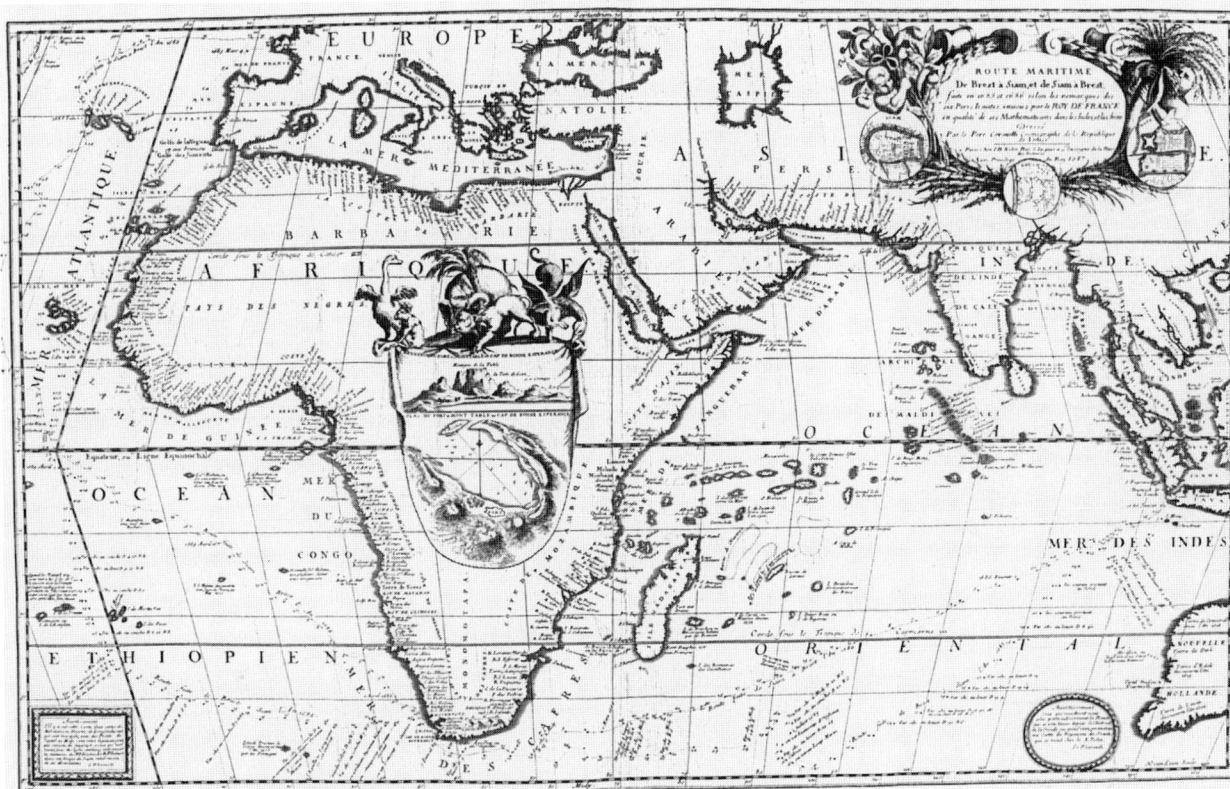

Map 51

CORONELLI, Vicenzo Maria (1650-1718)
Route maritime de Brest à Siam, et de Siam à Brest, faite en 1685 et 1686 selon les remarques des six Pères Jésuites, envoiez par le Roy de France en qualité de ses Mathematiciens dans les Indes, et la Chine. Dressé par le Père Coronelli . . .
(Sea route from Brest to Siam and back taken by the six Jesuit fathers in 1685 and 1686...) Engraved by H.van Loon. (Paris, J.B. Nolin, Rue S. Jacques, 1687.)
Map, 44 x 73 cm, uncoloured.
Prime meridian through Ferro Island.
Inset: Elevation of Table Mountain. Plan of Table Bay. Plans of Siam, Louvo and Batavia.

This large French map featuring Africa prominently, with the Middle and Far East, was published by Coronelli and Nolin to describe the sea route taken by the Jesuit Fathers sent out by Louis XIV to accompany the embassies to Siam. Tachard, one of these Jesuits, describes his visits and records his experiences in two volumes, during four visits to the Cape between 1685 and 1688. The first book was issued in 1686 and the second in 1689.

The map traces the routes taken by these Jesuit Fathers in 1685 and 1686, logging the places and dates of arrival. At the top right is a highly decorative oval cartouche, surrounded by scrollwork, trees and bushes, with a cherub on each side holding a tree and branches. At its lower border are three inset plans of Siam, Louvo and Batavia.

The continent of Africa is accurately shaped, with only coastline names and places marked. A large area of the interior is taken up by a decorative cartouche, at the top of which are various large animals of which a monkey and an

elephant are identifiable. One unusual animal has a dragon-like appearance and another resembles a large bird or ostrich. Three of the animals are holding the upper border of a hanging drapery in which are depicted two insets of the Cape of Good Hope – one showing a view of Table Mountain and the other a coastline plan of the bay and its surroundings. At the lower end of the map are two legends describing the Jesuit Fathers and Nolin, the publisher.

Coronelli, the son of a tailor, entered the Franciscan order as a young man and became a famous theologian, mathematician and cartographer. He is particularly famous for his construction of terrestial and celestial globes, particularly a large five-metre pair he made for Louis XIV. He founded the first Geographical Society in 1680 and was appointed Cosmographer to the Venetian Republic in 1685. One of his main productions was a large atlas, *Atlante Veneto*, (1691-1696) in two volumes, with finely engraved copperplate designs and maps.

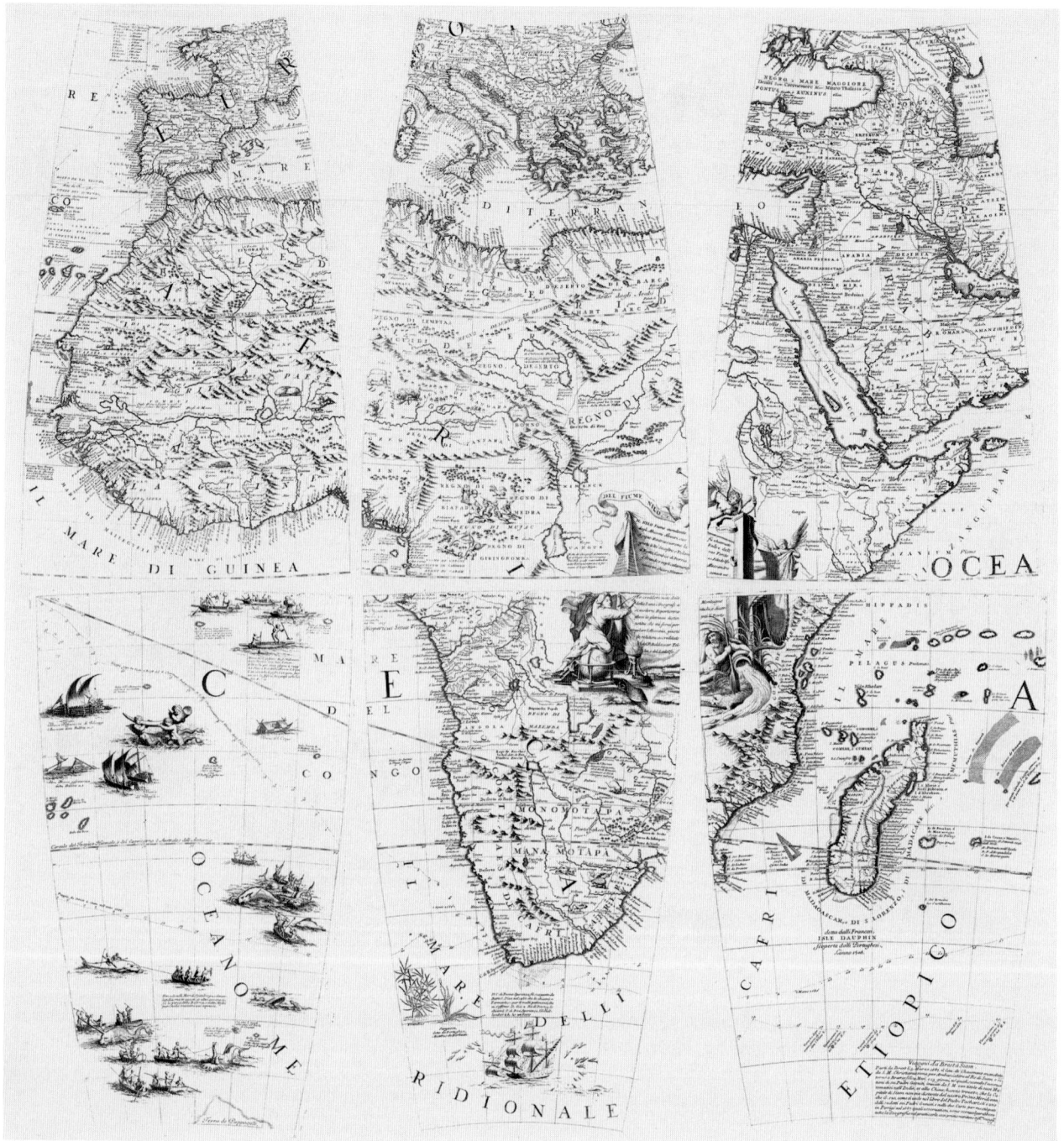

Map 52

CORONELLI, Vicenzo Maria (1650-1718)
(Africa.) (Venice, P. Coronelli, 1688.)
Six gores assembled to form one map, 91 x 75 cm, uncoloured.

These six globe gores assembled to form a general map of Africa were published by V.M. Coronelli, the most illustrious globe-maker of the seventeenth century, in *Libro dei Globi*, the tenth volume of his *Atlante Veneto*.

In each generation leading cartographers and map publishing houses had applied themselves to the production of globes, as well as atlases and maps. Gerard Mercator's globes of 1541 and 1551 had been famous in their day, to be superseded in the 1590s by those of Jodocus Hondius. By 1640 these Hondius globes were outrivalled by the larger and more splendid globe of Willem Blaeu. The golden age of Dutch cartography was over by the 1680s, and Father

64

Coronelli took up map-making in 1685. His atlas series, the *Atlante Veneto*, was designed as the continuation of Blaeu's *Atlas Major*. In the *Libro dei Globi* (published as volume ten of the thirteen-volume *Atlante Veneto*), Coronelli brought together a record of all the globes he had made, from the smallest, two inches in diameter, to the largest, the great globes of five metres in diameter which he had constructed for Louis XIV of France. Coronelli seems to have made every endeavour to produce maps which should omit nothing of real interest and value to geographers, navigators and explorers. The gores from which the globe was assembled were thereafter bound in copies of the atlas and thereby preserved.

These six gores assembled here represent the African portion of Coronelli's globe of 1688, which had a diameter of three and a half feet, and are a typical late-seventeenth-century depiction of Africa. This is one of the most detailed of early maps of Africa, full of topographical features and place names, and surprisingly accurate for the period. Stevenson, writing on the Coronelli globe, quotes Coronelli as stating that 'besides outlining the Monomotapa and Abyssinia Countries we have been the first to describe correctly the source of the Nile River, correcting by many degrees the errors of the Ancients'. He adds: 'Coronelli's depiction of the Zambesi River was apparently the best to date and it must be deduced that this Venetian geographer had been able to consult the early Portuguese documents which today are lost to us.' (E.L. Stevenson, *Terrestrial and Celestial Globes*, vol. 2, p. 102.)

In the centre of the map, accompanying a portion of four of the six gores, is a beautiful cartouche consisting of a personification of the sources of the Nile and other figures holding a legend which contains comments by Coronelli on his sources, Father Baldassar Tellez and Hiob Ludolf. The entire bottom left gore consists of an empty sea with a host of beautiful vignettes of whale-hunting, sailing ships, fishermen and informative notations. E.L. Stevenson also states that Coronelli added a rather unusual number of explanatory and informative legends to his maps, but never seemed to crowd the space that he had at his disposal. So exquisitely engraved were his maps that he was able to avoid the appearance of confusion noticeable on certain other globes of his century.

The map shows in detail the route of Tachard and the French Jesuit Fathers from Brest to Siam in 1685 with a detailed logging of places and distances.

Reference:

Stevenson, Edward Luther. *Terrestrial and Celestial Globes, their History and Construction*, 2 vols. Published for the Hispanic Society of America by the Yale University Press, London, 1921.

Map 53

DELACROIX, Louis Antoine Nicolle (fl. c. 1750)
Afrique. (Paris, L.A.N. Delacroix, 1688.)
Map, 13,5 x 15,5 cm, uncoloured.
Prime meridian through Ferro Island.
In: De la Croix, P. *Relation Universelle de L'Afrique Ancienne et Moderne*, vol. 1. Lyons: T. Amaulry, 1688, opp. p.35.

A small map of Africa that appears in the *Relation Universelle de l'Afrique Ancienne et Moderne* by Sr. (P.) de la Croix. It presents an unusual shortened southern half. The southern extremity appears divided by a vertically flowing river, into the Cape on the west and Agulhas to the east. There is not very much information in the interior of southern Africa except for the usual places and names used by many of the map makers of the time.

Abbé Delacroix was a French engraver whose information was used by de L'Isle and Brion de la Tour. He also engraved for Robert de Vaugondy in 1750 and he was associated with the *Nouvel Atlas Portatif* of 1778.

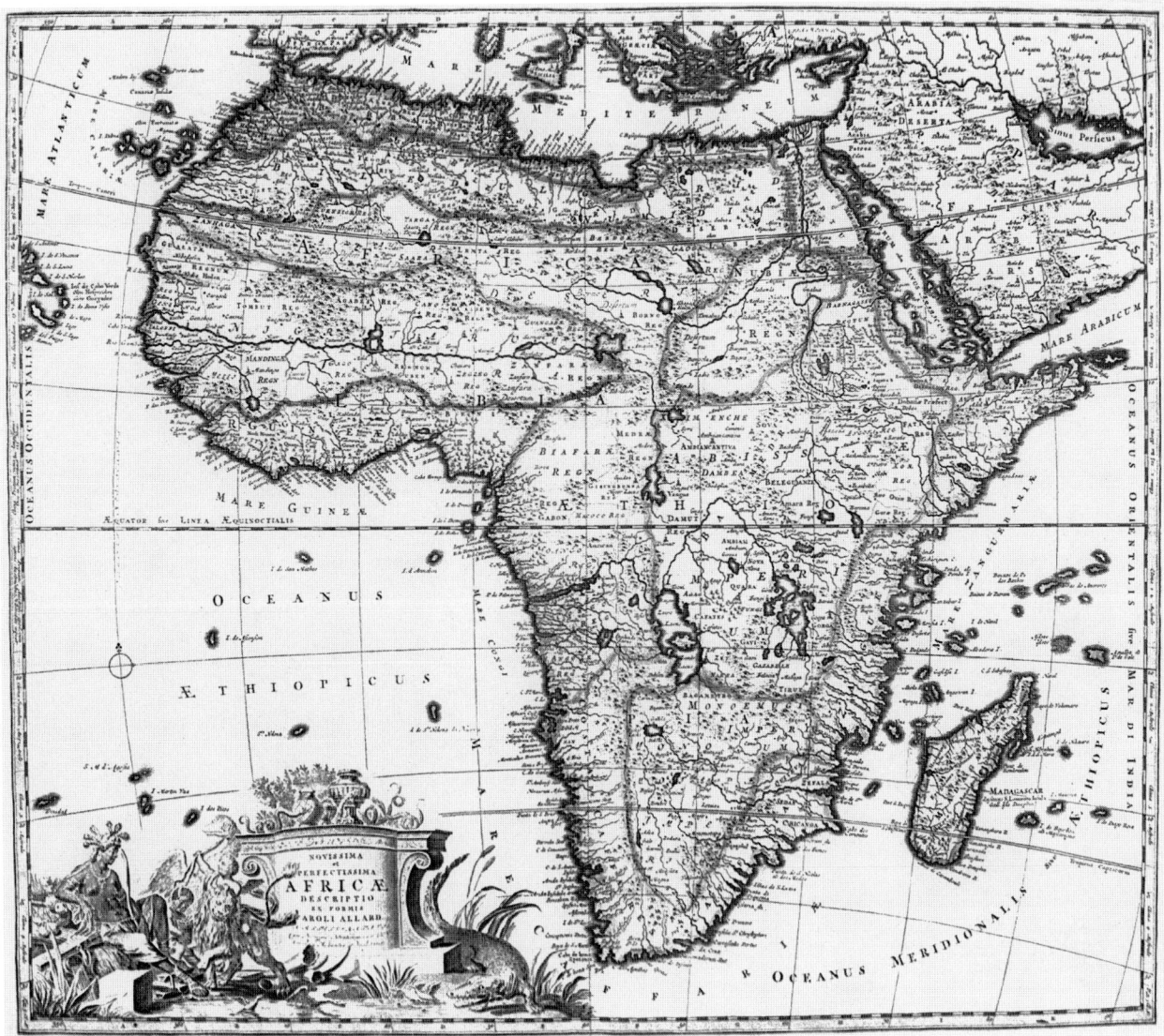

Map 54

ALLARD, Caroli (1648-1706)
Novissima et Perfectissima Africae Descriptio ex formis Caroli Allard.
(Newest and most detailed Africa.) Engraved by Ph. Tideman del G.V. Gouven.
(Amsterdam, C. Allard 1690.)
Map, 49 x 57 cm, coloured.
Prime meridian through Tenerif.

This attractive map of Africa was published by Karel Allard, son of Huych (Hugo) Allard, founder of the map-producing family. Karel worked as an engraver and publisher in Amsterdam, producing an *Atlas Minor* in 1694, an *Atlas Major* c.1765 and a hundred-plate *Orbis habitabilis oppida* in 1698. He handed over his stock in trade to his son Abraham in 1706.

The cartouche is attractively African in style and content, while the map itself contains a mass of fictitious cartographical details, especially within the southern interior.

Map 55

VISSCHER, Nicolas II (1649-1702)
Africae Accurata Tabula. (Detailed map of Africa.) (Amsterdam, N. Visscher, 1690.)
Map, 43 x 54 cm, coloured.
Prime meridian through Tenerif.
Dedicated to D. Gerardo Schaep.

The oval cartouche rests on two angels with an African as supporter on either side, one holding a scorpion-like animal and the other a horn of plenty. The dedication in the lower left corner is surmounted by what appears to be Gerardo Schaep's coat of arms. Neptune and a female figure, in company with mermaids and tritons blowing through shells, surround the dedication. Eleven small ships in full sail are scattered throughout the sea but only one whale and a few flying fish represent creatures of the sea. Elephants, ostriches, monkeys, lions and rhinoceros are used to decorate the empty spaces on the continent. The Nile has its origin in the two large lakes in Central Africa.

The Visscher family worked in Holland as map makers for most of the seventeenth and part of the eighteenth centuries. Their publications were extensively used and copied by other firms because of their high quality and accuracy, probably inferior only to those of Blaeu.

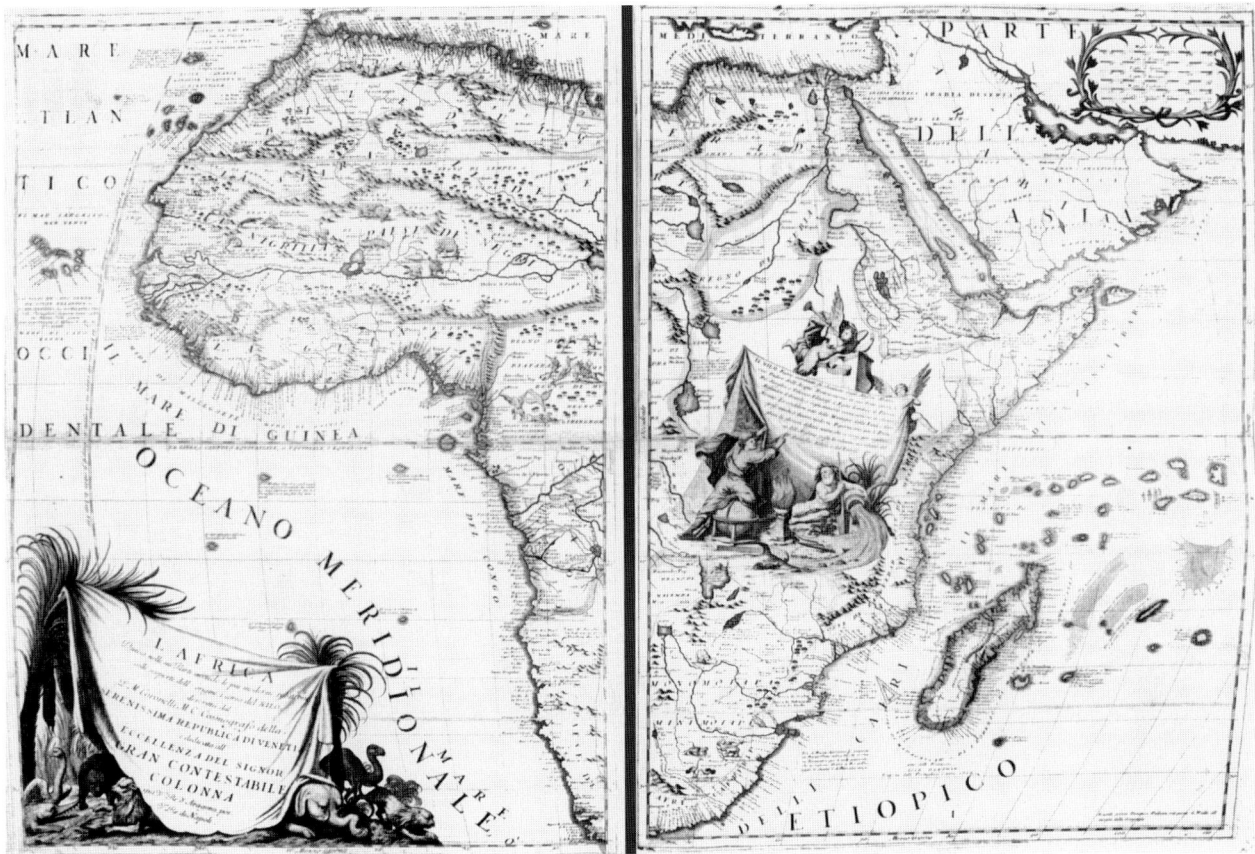

Map 56

CORONELLI, Vicenzo Maria (1650-1718)
L'Africa divisa nelle sue Parti secondo le più moderne, relationi colle scoperte dell'origine e corso del Nilo... (Africa divided into its several parts according to recent information with the discoveries of the source of the Nile.) (Venice, V.M. Coronelli, 1691.)
Map, 2 sheets 59 x 44 cm each, slightly coloured.
Scale in Italian miles, French, Spanish, German and English leagues, sea leagues.
Prime meridian through Ferro Island.
Dedicated to Eccelenza del signor Gran Contestabile Colonna.

Vicenzo Maria Coronelli entered the Franciscan Order as a young man and eventually became a famous theologian, mathematician and cartographer. After residing for a short time in Ravenna he settled in Venice, where he eventually became General of the Franciscan Order. In all he drew and engraved over five hundred maps and founded the first geographical society, the Academica Cosmographica degli Argonaute. He is particularly famous for his construction of terrestrial and celestial globes and his *Libro dei Globi* (Venice, 1697) was a major event in the history of globe making. He constructed, by order of Louis XIV, a very large terrestrial and celestial globe, which was seen in 1901 in the Bibliothèque Nationale in Paris. Thereafter it was revived and stored in the Orangerie at Fontainbleau until recently when it was erected on the ground floor of the Pompidou National Gallery in Paris – a truly majestic sight (seen by the author in 1980) towering on its base approximately ten metres. Coronelli also constructed many other globes, equally beautiful but smaller.

This double-sheet map of Africa is very colourful and first appeared in his atlas, the *Atlante Veneto*. The large cartouche in the lower left-hand corner of the west sheet consists of an animal skin hanging from palm trees. On the left are an ostrich, a crocodile, a leopard and a lion. A camel and a lion appear in full view on the right with an elephant's head peeping out from under the skin. The sea is not decorated but the mainland is embellished with all kinds of animals, together with scenes of horsemen in combat with lions and archers aiming at a large ostrich-like bird. A most noticeable feature of the decoration on the mainland is a scene consisting of a large piece of drapery containing information about the source of the Nile, at the very top of which there is an angel blowing a trumpet, while another holds one corner of the drapery on which the recording angel has just written the legend. To the right is a bearded figure – Father Nile – reclining against a bowl from which water is rushing over the back of a crocodile.

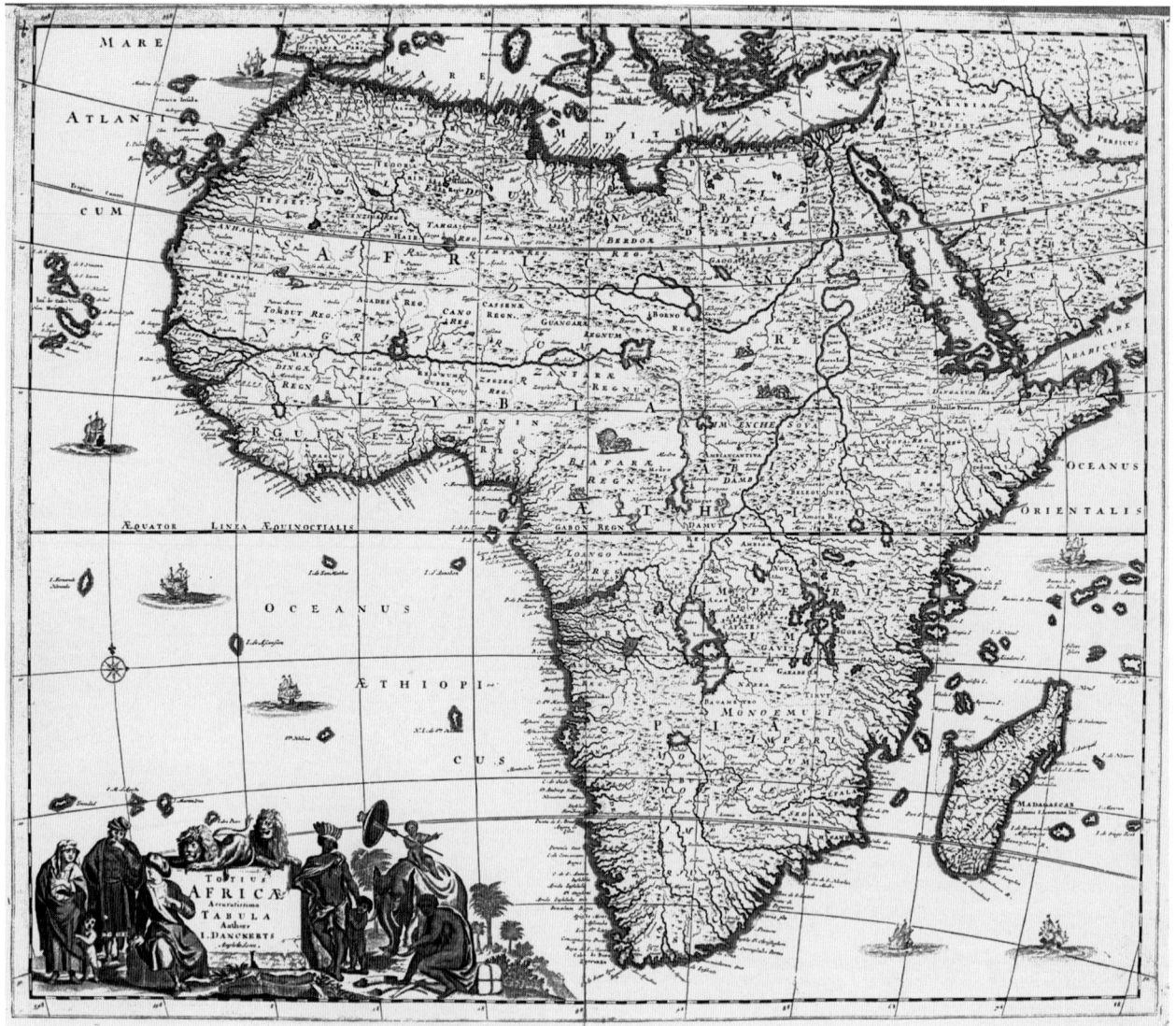

Map 57

DANCKERTS, Justus the Elder (1635-1701) or the Younger (d. 1692)
Totius Africae Accuratissima Tabula. (Detailed map of the whole of Africa.)
(Amsterdam, J. Danckerts.)
Map, 49 x 57 cm, slightly coloured.
Prime meridian through Ferro Island.

A map of Africa emanating from the family Danckerts of Amsterdam, publishers and engravers. Of this family there were two with the above name and it is difficult to establish which of them published this map.

The cartouche is a colourful one with people in varied costume: Moors, Black Africans, children and a mounted elephant. There are two lions seated on masonry; the title appears on this masonry. This cartouche, and the map itself, are identical to that published in 1680 by the prominent Dutch mapsellers the de Wits (see Map 48) with all its fictitious details, especially those of Central and South Africa. The same three sailing ships appear in the Oceanus Ethiopi (Atlantic) and in the Oceanus Orientalis (Indian) with an added one off the coast of Spain. The same animals decorate the various regions and the origin of the Nile conforms to the Ptolemaic two-lake system.

70

Map 58

ANON

Africae tabula. (Map of Africa.)

Map, 27 x 34 cm, coloured. Prime meridian through Paris. Top right: 'Numéro six.'

Neither author nor publisher of this map has been identified. Mrs Gillian Hill, professional research assistant at the Map Library of the British Library, was unable to identify the map from catalogues and reference works in the Library's possession. She placed it as: 'rather earlier than 1740, perhaps about 1700 or earlier. This is borne out by a study of the plates in Tooley's *Collector's Guide to Maps of Africa* which most nearly resemble yours in general appearance and in shape of the coastline (e.g. plate 57 : 93) are all from the seventeenth century.'

Mr R.V. Tooley has also seen a photograph copy of the map and he wrote: 'If it is a large wall map (roughly the size of a folding map) it could possibly be by Meurseum who did a series of wall maps of the four continents about 1670 which it resembles in style, but I could not verify this.'

The map is not wall size. It has a decorative cartouche in the lower left corner (in contemporary colours), the main features of which show two elephants each mounted by a

person in what appears to be oriental dress. On the left is a mountain in front of which a smaller elephant is depicted and in the left foreground an ostrich, a lion and a male figure carrying a box under his right arm and a bow in the left, with arrows carried on his back. In the right foreground are two semi-nude natives – a female carrying a child and the male a spear in his left hand. In the right background is the sea with sail and rowing boats in the distance. On shore are three men who appear to be holding a flat object and are possibly panning for gold. The coastal and inland geography is typical of the eighteenth-century period with lakes, rivers and place names. An unusual feature on the left of the map, below the equator, is a colour chart indicating whether areas on the map itself are Christian, Jewish, Mohammedan or heathen. At the top of the map is 'numéro six' in faded script, suggesting that it may have appeared in an atlas. Dr Anna Smith, of the Africana Museum, Johannesburg, especially knowledgeable on maps of Africa, is also unable to provide any further information concerning its provenance.

Map 59

L'ISLE (Insulanus), Guillaume de (1675-1726)
L'Afrique dressée sur les observations de Ms. de l'Academie Royale des Sciences, et quelques autres, et sur les memoires les plus récens. (Africa according to the observations made by the Royal Academy of Science...)
Engraved by N. Guérard. (Paris, chez l'auteur sur le Quai de l'Horloge à la Couronne de Diamans, G. de l'Isle, 1700.)
Map, 45 x 58 cm, slightly coloured.
Scale in French, Spanish and German leagues.
Prime meridian through Ferro Island.

This map of 1700 from the geographer Guillaume de l'Isle is distinguished – like all his maps – by his scientific approach to his subject. De l'Isle was at first taught by his father, then tutored by the astronomer and geodesist at Bologna, Jean Dominique Cassini. Working closely with the Académie des Sciences he became the foremost geographer of his age. He has been called the father of modern geography and ranks after Ptolemy and Mercator as one of the major forces in the progress of cartography.

De l'Isle's maps were copied and reprinted for well over a hundred years, both in his own country and abroad. Most of the maps of Africa issued in the eighteenth century were based on de l'Isle. He was appointed Premier Geographer to Louis XV in 1718, and his work is distinguished by its scientific value, with minute care and constant revision. After his death the business was continued by his widow, in partnership with Phillipe Buache, who married de l'Isle's daughter and whose name appeared on some of de l'Isle's posthumous maps.

The map featured here, first published in 1700, remained the standard map until 1722 when he issued a revised edition. Contrary to general practice de L'Isle's maps were dated and, although reissued many times, retained their original date of compilation. This map is one of the landmarks in the history of mapping of Africa. De l'Isle was the first to give the correct longitude – 42° – to the Mediterranean, thus correcting the northern littoral of the continent. He was the first to discard Lakes Zaire and Zaflan, inherited from Ptolemy, giving the correct source of the Blue Nile in Lake Tzana (Tana) in Abyssinia. On the other hand, according to some authorities, he shows the Senegal joined to the Nile by a hypothetical line. This is corrected later.

The Congo or River Zaire is a marked feature and further south the R. du Saint Esprit and the Zambeze appear, with the Portuguese settlements at Sena, Tete and Chicora. The Dutch Fort and Hellenbok (Stellenbosch) are shown at the Cape and native kingdoms such as Monomotapa, Sofala, Monoamugi, etc. are named. He also gives the site of various mines of gold, silver, emeralds and salt, wells in the desert and gum in Senegal.

In the top right corner is a typical de l'Isle cartouche consisting of a shield surrounded by trees and animals, including three elephants, an ostrich and two natives, one of whom is on horseback behind the ostrich and the other in the foreground riding a crocodile.

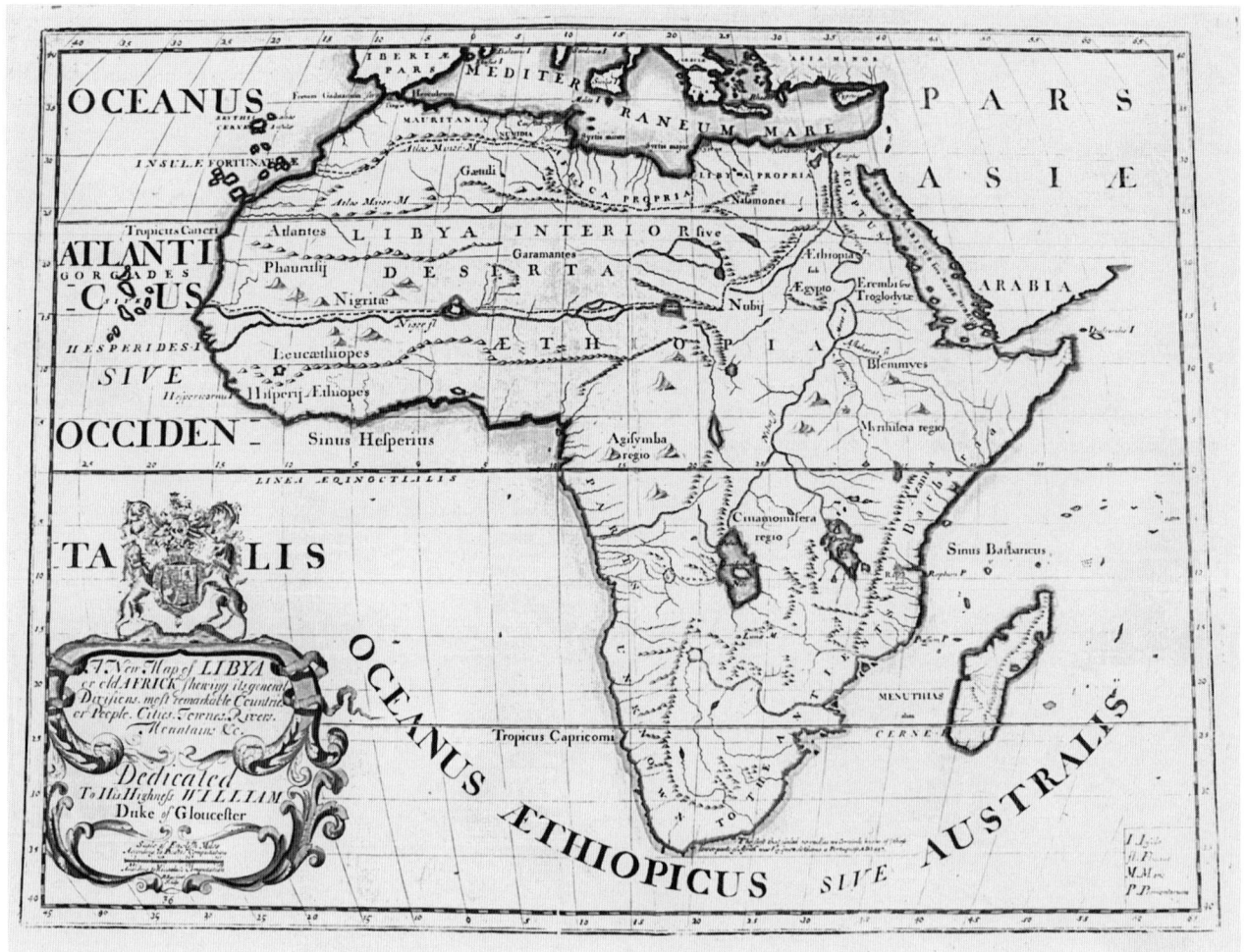

Map 60

WELLS, Edward (1667-1727)
A New Map of Libya or old Africk Showing its general Divisions, most remarkable Countries or People, Cities, Townes, Rivers, Mountains &c.
Engraved by R. Spofforth. (London, E. Wells, 1700.)
Map, 37 x 49 cm, slightly coloured. Scale in English miles.
Dedication: 'To His Highness William Duke of Gloucester.'

Edward Wells, mathematician and geographer, who published this map of Africa, was educated in London and Christchurch, Oxford. His atlas, *A New Sett of Maps*, was first published in 1700 and was reissued in England several times from 1701 to 1738. Wells is recorded as an accurate geographer of his time. A feature of his atlas is that each map appears twice, first in its then current form and again with ancient and classical names. It thus contains two maps of Africa.

The map featured here is the 'ancient' one, with a decorative scrollwork cartouche surmounted by the coat of

arms of the Duke of Gloucester. The continent of Africa has been left surprisingly blank, particularly the south, with the statement 'Unknown to the Antients.' The only other note in the southern part is 'The first that sailed around these lower parts of Africa was Vacquez de Gama, a Portuguese A.D. 1497.' The map is crudely drawn and does not appear to have any real geographic value, though it does have some decorative attraction.

74

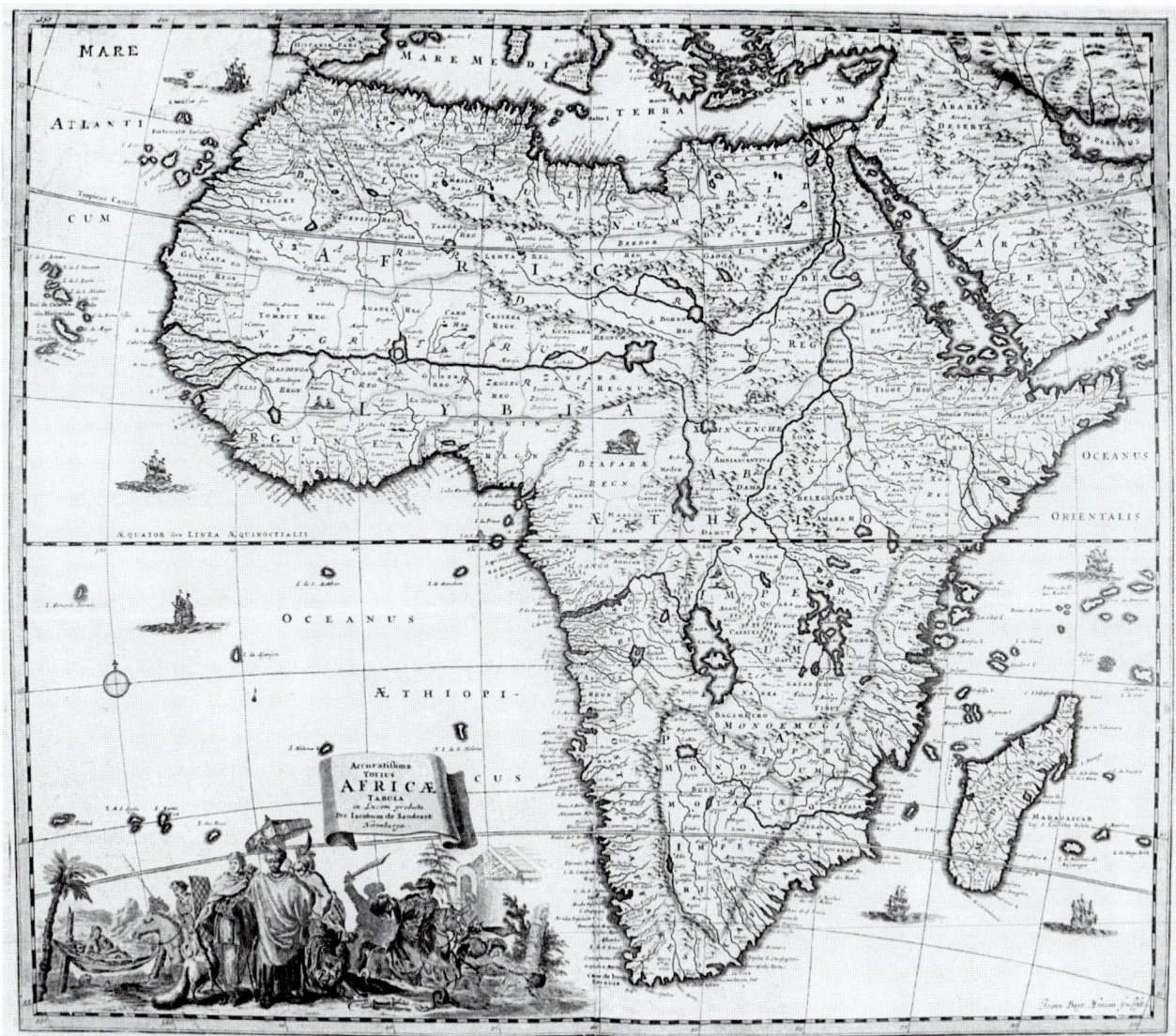

Map 61

SANDRART, Jacob von (1630-1708)
Accuratissima Totius Africae Tabula in Lucem producta ... (Exact map of the whole of Africa) Engraved by Johann Baptist Homann. (Nürnberg, J. von Sandrart, 1700.)
Map, 48 x 56 cm, coloured outline.
Prime meridian through Ferro Island.

A typical late seventeenth-century map of Africa with a highly decorative cartouche in the lower left corner. The description is placed in a simple scroll above the picture, which shows a woman seated in a hammock, nursing a baby, while a man seated on a donkey is looking on. In the left foreground is the rear view of a fat-tailed sheep (reference is made to this unusual feature in other maps of this series). A large native potentate dominates the scene with a lion at his left side, a figure holding an umbrella over his head and another with a bow in his hand appearing to guard him. In the foreground are snakes and a lizard, while four men, two on horseback, appear to be involved in a battle in the middle distance. There are three ships in the Atlantic and three in the Indian Ocean.

The map is a close copy of the Dutch map of de Wit (Map 48), with the same decoration and fauna on the interior of the map, and also relies on Ptolemy for its description of the Nile basin.

Sandrart, born in Frankfurt, was a painter and engraver. It is interesting to note that this map is an early example of Homann's work as an engraver, before he set up on his own.

75

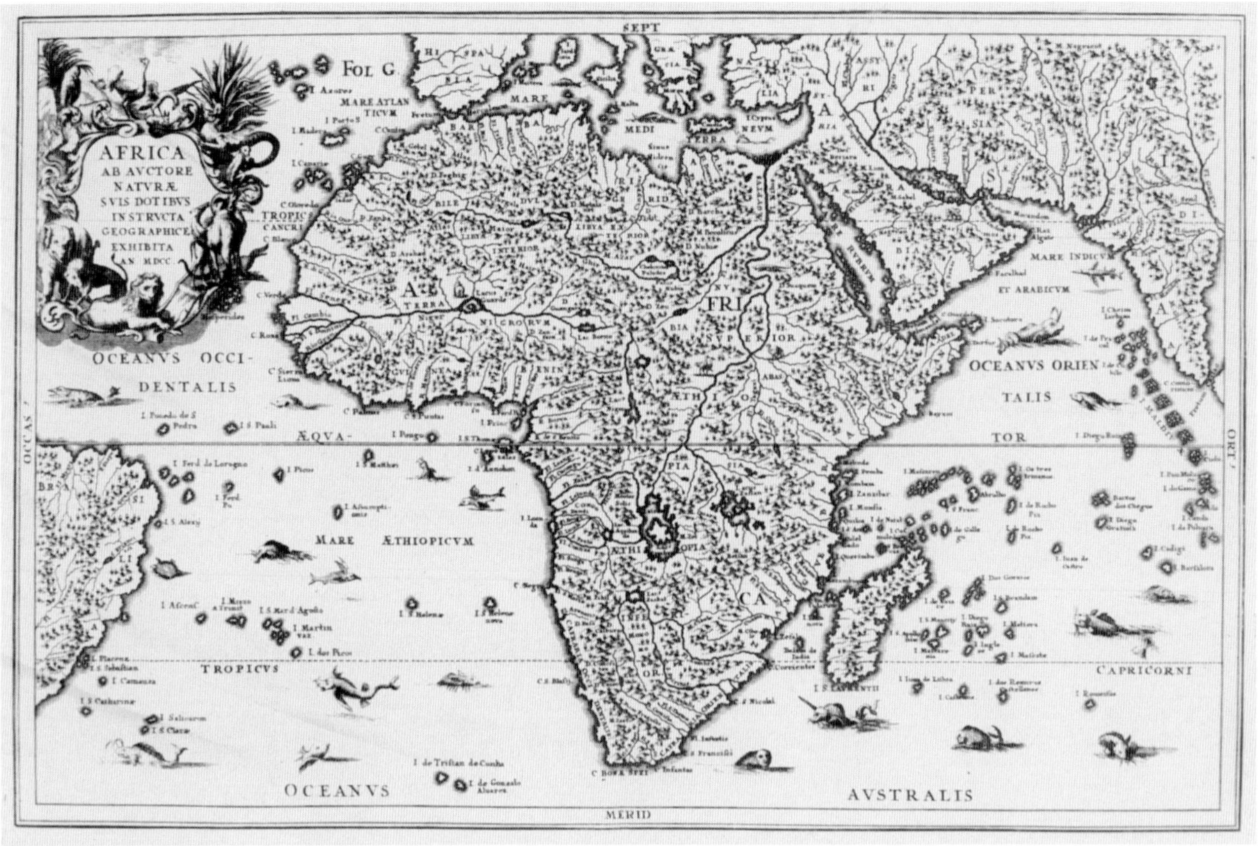

Map 62   SCHERER, Heinrich (1628-1704)
Africa Ab Auctore Naturae Suis Dotibus Instructa Geographice Exhibita An
MDCC. (Africa geographically represented in 1700.) (Munich, H. Scherer, 1700.)
Map, 22 x 35 cm, uncoloured.

Heinrich Scherer, a Jesuit, was a geographer and professor of mathematics in Munich. He published his *Atlas Novus* in 1710 with about 200 maps (including the above) engraved by Homann. It appeared in parts variously entitled *Geographica Naturalis, Atlas Marianne* (1702), *Geographica Politica, Geographica Artificalis, Tabellae Geographicae* (1703) and *Critica Quadruportita* (1710).

The map includes Arabia, India and the Middle East and the eastern bulge of Brazil. The seas abound with islands, sea monsters, one of which has a human face, and flying fish. The cartouche at the top left, with its title, is surrounded by detailed engravings of animals and birds of Africa, artistically executed with foliage and leaves. The interior of the continent is full of detailed geography of the time, with the large lakes and rivers depicted as inaccurately as in its cartographical predecessors.

76

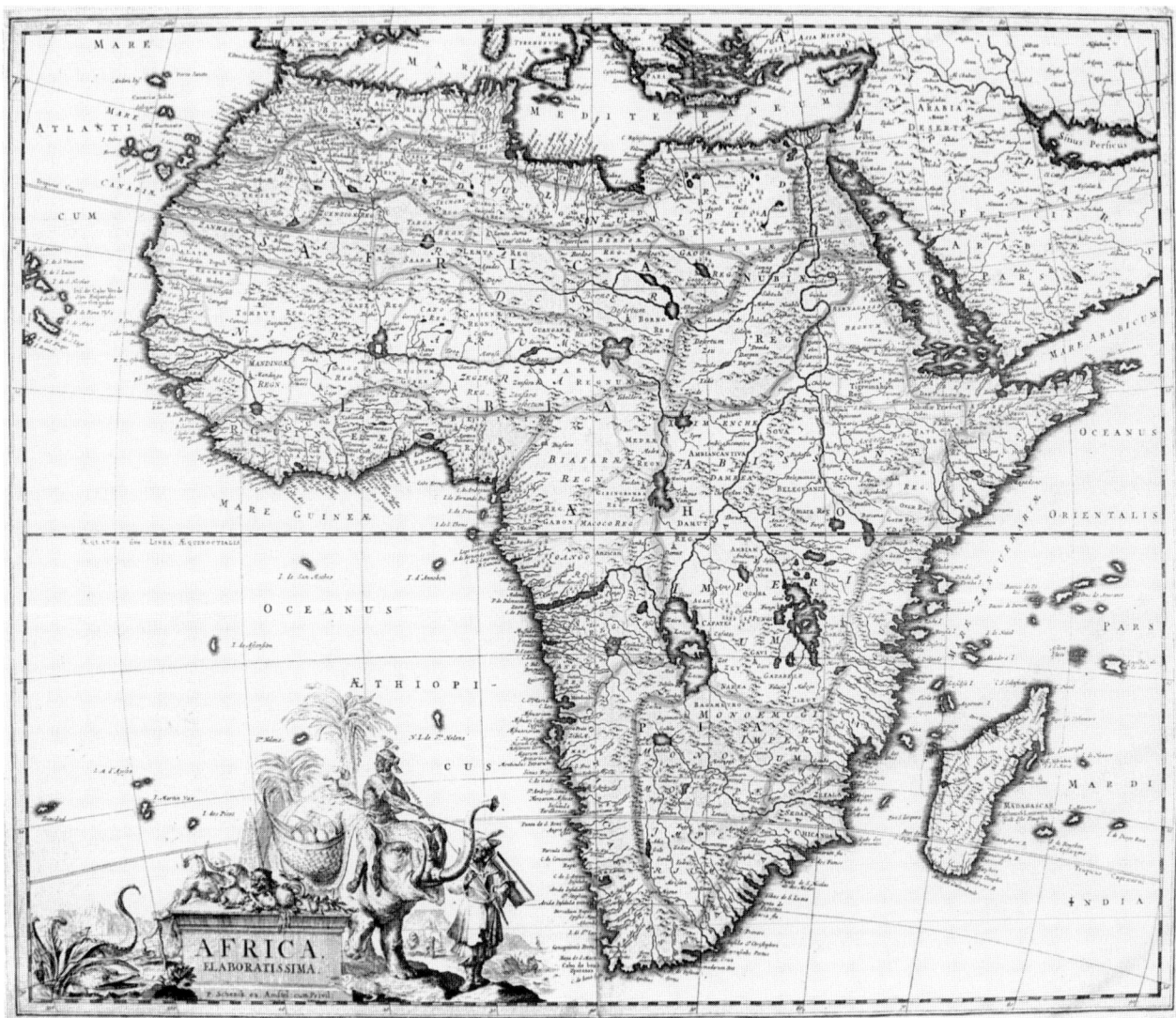

Map 63

SCHENK, Pieter (1645-1715)
Africa Elaboratissima. (Detailed Africa.) (Amsterdam, P. Schenk.)
Map, 48 x 58 cm, coloured.
Prime meridian through Tenerif.

This eighteenth-century Dutch map features a decorative and powerful cartouche in the lower left corner. It consists of masonry on top of which is fruit, a monkey, a mouse and a snake. On the left a crocodile appears against a background of mountains and on the right is a large elephant, mounted by a native, bearing a basket of fruit. Next to the elephant is a Moor armed with shield and spear. In the background is a flat-topped mountain (Table Mountain), in front of which two ships appear, and in the middle distance there is a small boat with three men. On the map itself large

lakes and rivers appear but the decorative animals have disappeared from the mainland. The coastline of the southern part of the continent is markedly broken, with many large rivers.

Pieter Schenk, a well-known publisher and engraver of Amsterdam, acquired plates from Blaeu, Jansson and Visscher. He published atlases of England, Hungary, the *Atlas Contractus* and *Theatre de Mars*. The atlas from which this map was taken may be Schenk's *Atlas Contractus* as it contains maps dated between 1703 and 1709.

Map 64

FER, Nicolas de (1646-1720)
L'Afrique Dressée selon les dernières Relations et suivant les Nouvelles decouvertes dont les Points principaux sont placez sur les Observations de Mrs. de L'Académie Royale des Sciences. (Africa according to the latest discoveries made by the Royal Academy of Science.) Engraved by H. van Loon. (Paris, chez l'auteur dans l'Isle du Palais sur le Quay de l'Orloge a Ia Sphere Royale, N.de Fer, 1705.)
Map, 46 x 59 cm, coloured. Scale in French leagues.
Dedication: 'A Nosseigneurs Les Enfans de France.'

The author of this map of Africa, French-born Nicolas de Fer, was a geographer, engraver and publisher of considerable ability who created several atlases. After his death in 1720 he was succeeded by his sons-in-law J.F. Benard and G. Danet. His atlases include *France Triomphante* (1693), *Forces de l'Europe* (1696) and *Atlas Royal* (1695). He produced approximately 600 maps, which were sought after more for their decorative quality than their geographical content.

In this map Africa has a curious bulge in the west coast between the Cape and the Congo. False Bay does not appear although Robben Island is noted. Unusual islands appear in the Atlantic and Indian Oceans including the fictitious second island of St. Helena. Monomotapa appears in the south and in Central Africa there is a paragraph of information on the Nile, which no longer has its origin in southern Africa. The dedication appears in the lower left corner and the cartouche in the lower right corner depicts what appears to be a lime kiln in operation. The names of some Hottentot tribes appear in the south. This map appeared, with slight variations in the cartouche, in 1700, 1705 and 1722 (according to the author's information).

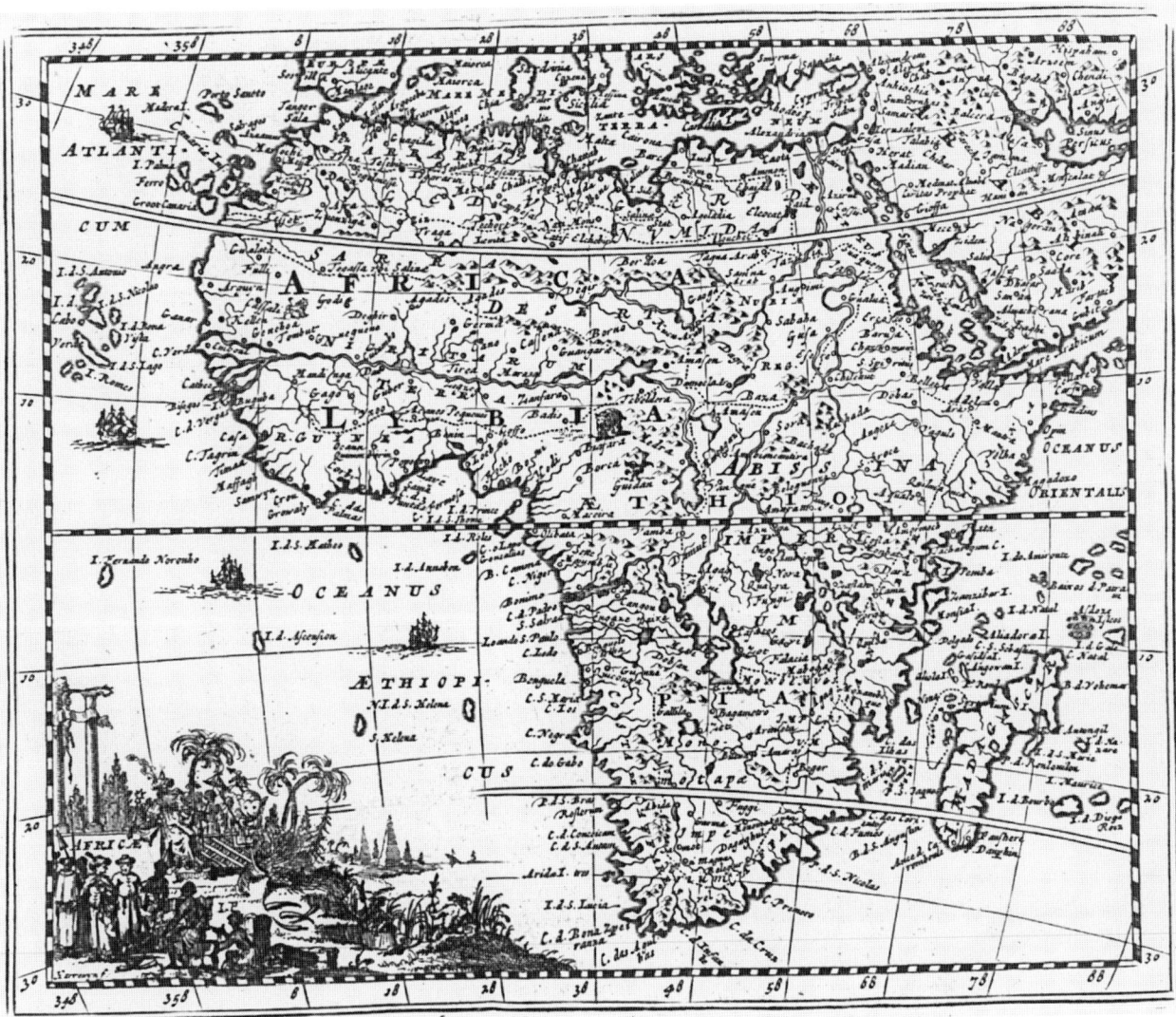

Map 65

FERNANDEZ DE MEDRANO, Sebastian

Africæ. Engraved by Jacques Harrewyn. (Antwerp, 1709.)
Map, 15 x 18 cm, uncoloured.

This map appears in a volume entitled *Geographia O Moderno Descripcion Del Mundo...* published by Fernandez de Medrano in Antwerp in 1709.

The cartouche bearing the title, 'Africæ', shows the ruins of a building at the lower left against a background of palms with pyramids in the background. A group of native Africans and Moors are depicted in the foreground. There are four ships in the Atlantic Ocean and a few unusual looking animals scattered on the mainland. The coastline is very broken up with bays. The Nile, as usual for the period, has its origin in the large lakes in Central Africa. Many fictitious cities are named and the mapping is generally poor.

Harrewyn was one of a family of engravers of Brussels, including Jacques, Jacques-Gerard (son of Jacques), François and Jean Baptiste, who engraved for Afferden in 1709. (Afferden was a Belgian cartographer who produced an atlas, *El Atlas Abbreviado*, in Amberes (Antwerp), in many editions including one in 1709.) Harrewyn is known to have engraved for him in 1709 and Afferden's name has been associated with this map.

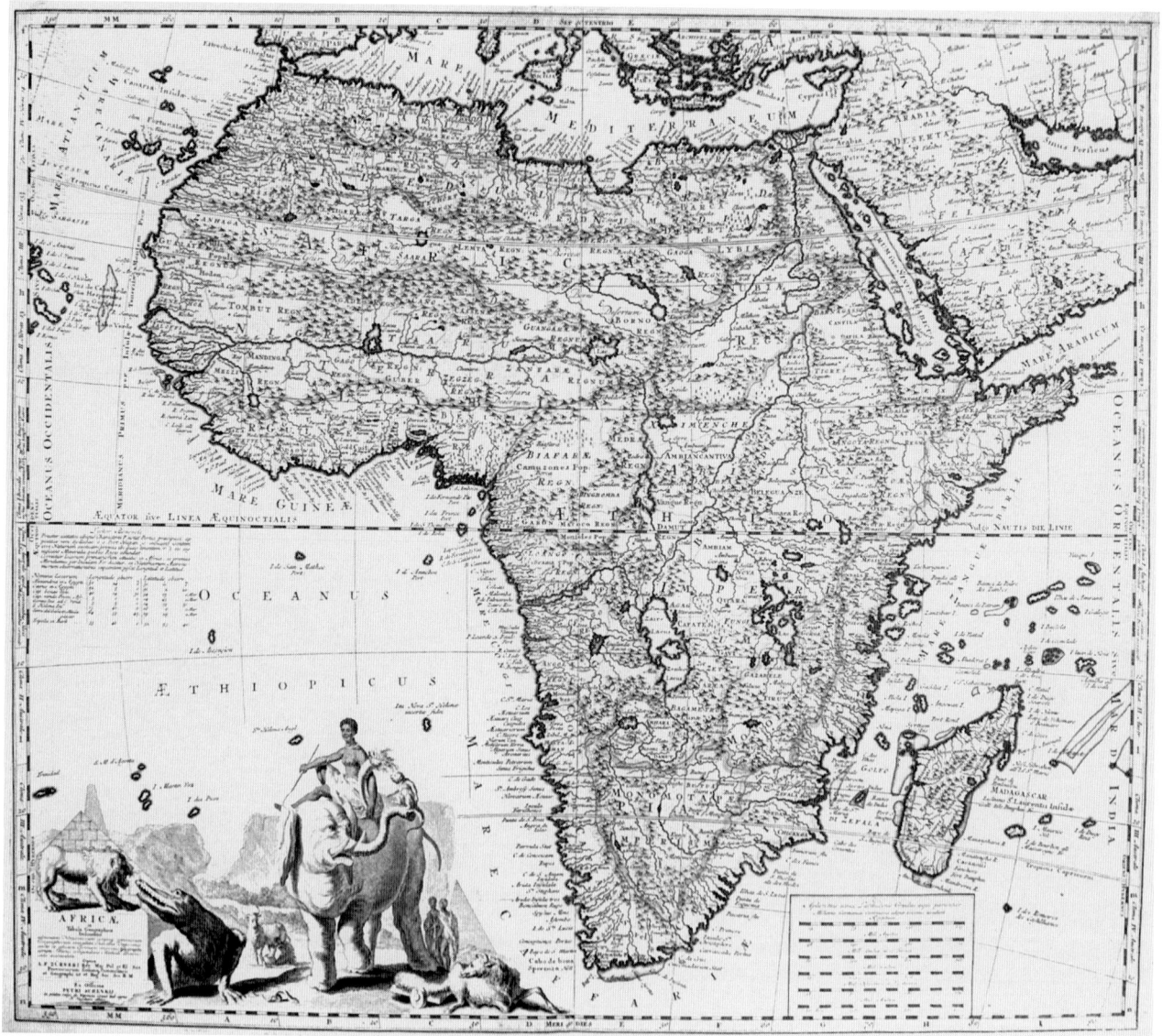

Map 66

ZÜRNER, Adam Friedrich (1680-1742)
Africæ in tabula geographica delineatio admentem novissimorum eorumq.
optimorum geographorum emendata, indicibus utilissimis aucta et adusum
tyronum imprim is geographicorum. (Geographical map of Africa drawn and
measured according to the best and most recent information...) (Pieter Schenk, 1709.)
Map, 50 x 58 cm, coloured.
Scale in English, Italian, French, German, Spanish and Arabic miles.
Prime meridian through Tenerif.

Zürner, a geographer and publisher, produced maps of Africa (in 1700), America (in 1709) and nearly 1000 other maps used by Schenk, Valk and Weigel, among others. The map probably first appeared in *Atlas Contractus*, published by P. Schenk in c.1709 or in Valk and Schenk's *Atlantis Syllage Compendiosa* published in Amsterdam in 1709.

The impressive cartouche at the lower left border shows, on the masonry bearing the inscription, a lion snarling at a very large crocodile which is standing near a snake. In the background are two pyramids and in the middle distance, a cow being milked. Next to the crocodile is a large elephant ridden by an African woman holding a horn of plenty. Near the elephant's feet is a crouching fat-tailed sheep with its tail carried on a little cart with wheels to prevent it being dragged on the ground (see other fat-tailed sheep in other cartouches). In the distance on the right is another pyramid and two native Africans.

80

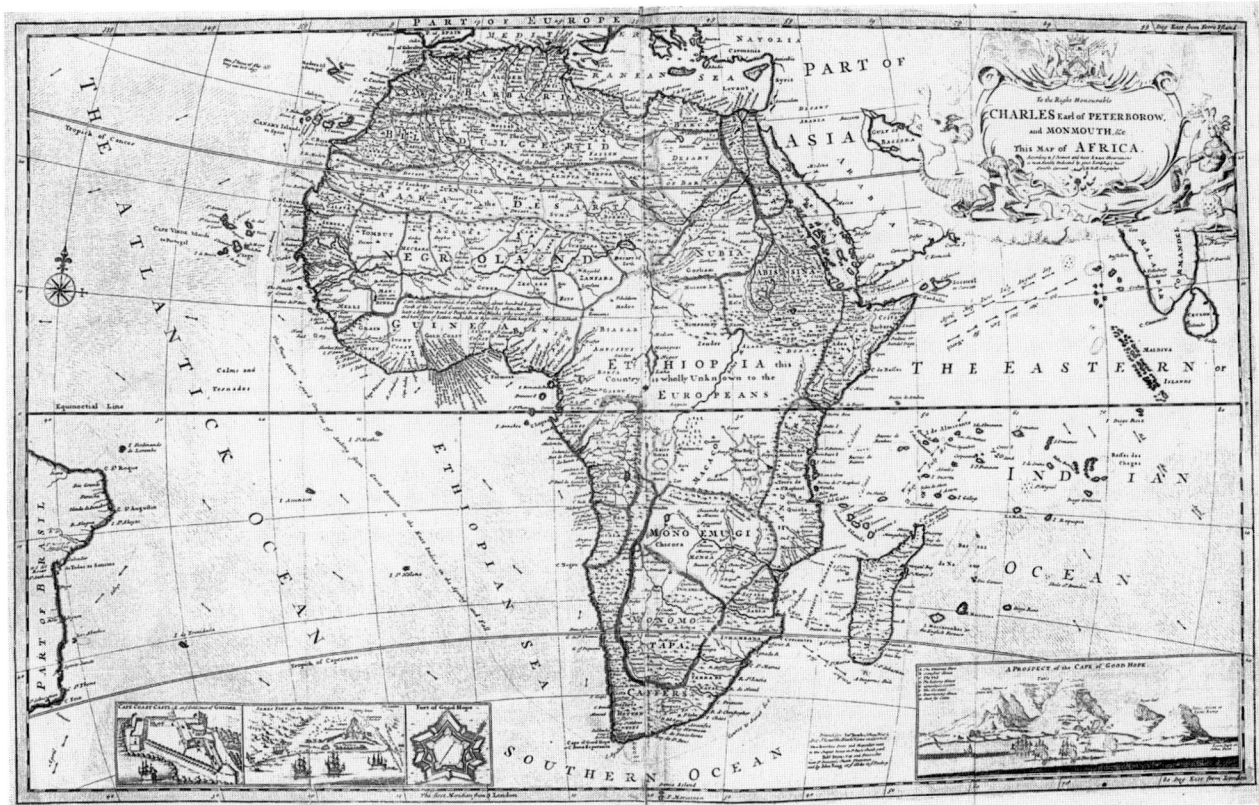

Map 67

MOLL, Herman (d. 1732)
To the Rt. Honourable Charles Earl of Peterborow and Monmouth. This Map of Africa. According to ye newest and most exact observations is most humbly dedicated... (London, printed for In° Bowles, print and map seller next to the Chapter House in St. Paul's churchyard by Philip Overton, map and print seller near St. Dunstan's Church, Fleet Street and by John King at ye Globe, 1710.)
Map, 58 x 97 cm, slightly coloured.
Prime meridian through London and Ferro Island.
Dedication: 'To the Rt. Honourable Charles Earl of Peterborow and Monmouth.'
Inset: A prospect of the Cape of Good Hope, Cape Coast Castle – Guinea. James Fort – St. Helena, Fort of Good Hope.

Herman Moll, engraver, geographer and bookseller of Dutch origin, went to England in 1678 and worked there as an engraver for Moses Pitt. He also engraved maps for Grenville Collins, John Adam Seller and Charles Price. He commenced selling maps in Vanleys Court, Blackfriars, but later settled at an address in the Strand. On his death in 1732, his heir was his only daughter Henderina Amelia.

Moll engraved two maps (of America and Europe) for *Moore's Geography* in 1681, but he made his mark in the eighteenth century when, in the first two decades, he became the foremost map publisher in England. He published more than twenty-five atlases and geographical works with reissues and loose maps. The same maps were used in different publications with minor alterations, something which is particularly known of this large folding map of Africa. He had a good reputation and was highly conscientious. A notable feature of his maps are the long legends he used in blank spaces.

The dedication title to the Earl of Peterborow at the top right features a coat of arms, presumably of this nobleman. There are figures of mounted natives chasing an ostrich, a Black African riding a crocodile, an elephant, a lion and a pyramid. It has been suggested that this cartouche was partly inspired by Guerard's design for de l'Isle's map of Africa.

Inset views along the base of the map show, on the left, Cape Coast Castle on the coast of Guinea: James Fort, St. Helena; Fort of Good Hope (Cape) and on the right a detailed 'Prospect of the Cape of Good Hope,' showing the different parts of Table Mountain. No settlements other than the 'Fort of Good Hope' are marked and the Hottentot tribes are not specified. Various names on the southern shore marked on this map are not known today. The boundaries and cities of Monomotapa are marked and both Vigiti Magna and Zimboae appear as do the Delagoa River and the Zambezi. Further north Moll attempts to compromise with ancient geography, retaining the two great lakes of Zaire and Zaflan but turning them into morasses and dissociating them from the Nile, which is shown rising in Lake Tzana in Abyssinia, a feature of de L'Isle's 1700 map.

The winds and the calms are indicated; the route from Britain to the East is marked in the Atlantic and, in the Indian Ocean, mention is made of the course to Surat. Within the map, in Guinea, a legend appears referring to white men inhabiting an area one hundred leagues from the coast of Guinea, who wear clothes which differ from those worn by the blacks and some of whom keep the Christian Sabbath.

This is a popular map of Africa and has been reissued many times with variations, particularly of the names of the printers or map-sellers. According to Tooley there are seven such variations, one of which does have a changed dedication title.

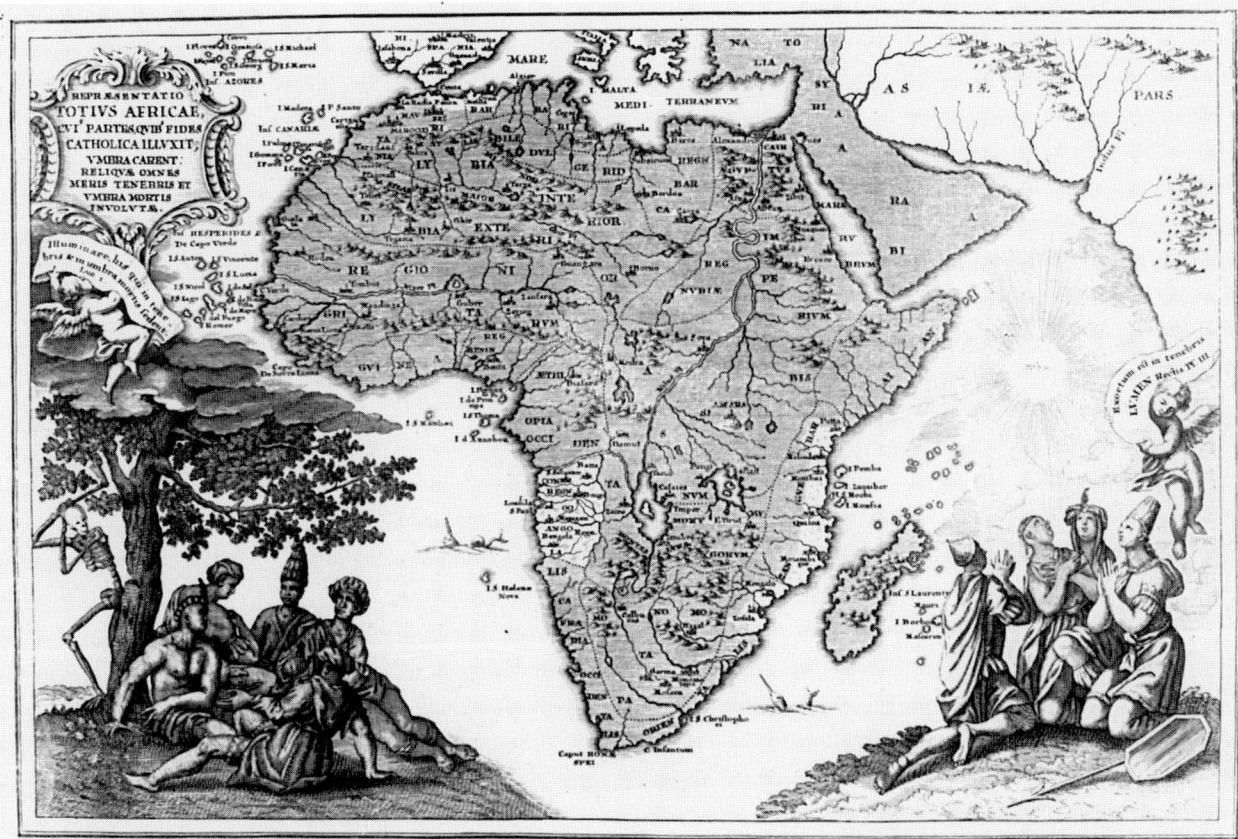

Map 68

SCHERER, Heinrich (1628-1704)
Representatio Totius Africae... (Representing the whole of Africa...) (Munich, H. Scherer, 1710.)
Map, 22 x 34 cm, uncoloured.

This is another version of Africa published by Scherer, and presumably engraved by Homann. The map itself is somewhat darkened by transverse, closely parallel lines and is less detailed than Scherer's other map of Africa (Map 62). The Nile features prominently as a very long river with its main tributary coming out of Lake Zaire, and minor branches joining it from around Lake Zaflan to the east.

At the top left there is a title cartouche with foliage surrounding it. Below this is an angel carrying a scroll on which is written in Latin an extract from the Bible (Luc. I).

Below the angel are clouds surrounding a leafy tree with, sitting in its shade, five figures in native African and oriental dress, appearing to be in conversation. Behind the group is a skeleton wielding an axe, felling the tree. The group under the tree are unaware of his presence. In the lower right corner are four figures kneeling and praying up to the heavens with their heads turned towards a brightly shining Hebrew word meaning 'God.' Above them and to the right is another angel carrying a scroll on which is written another biblical extract.

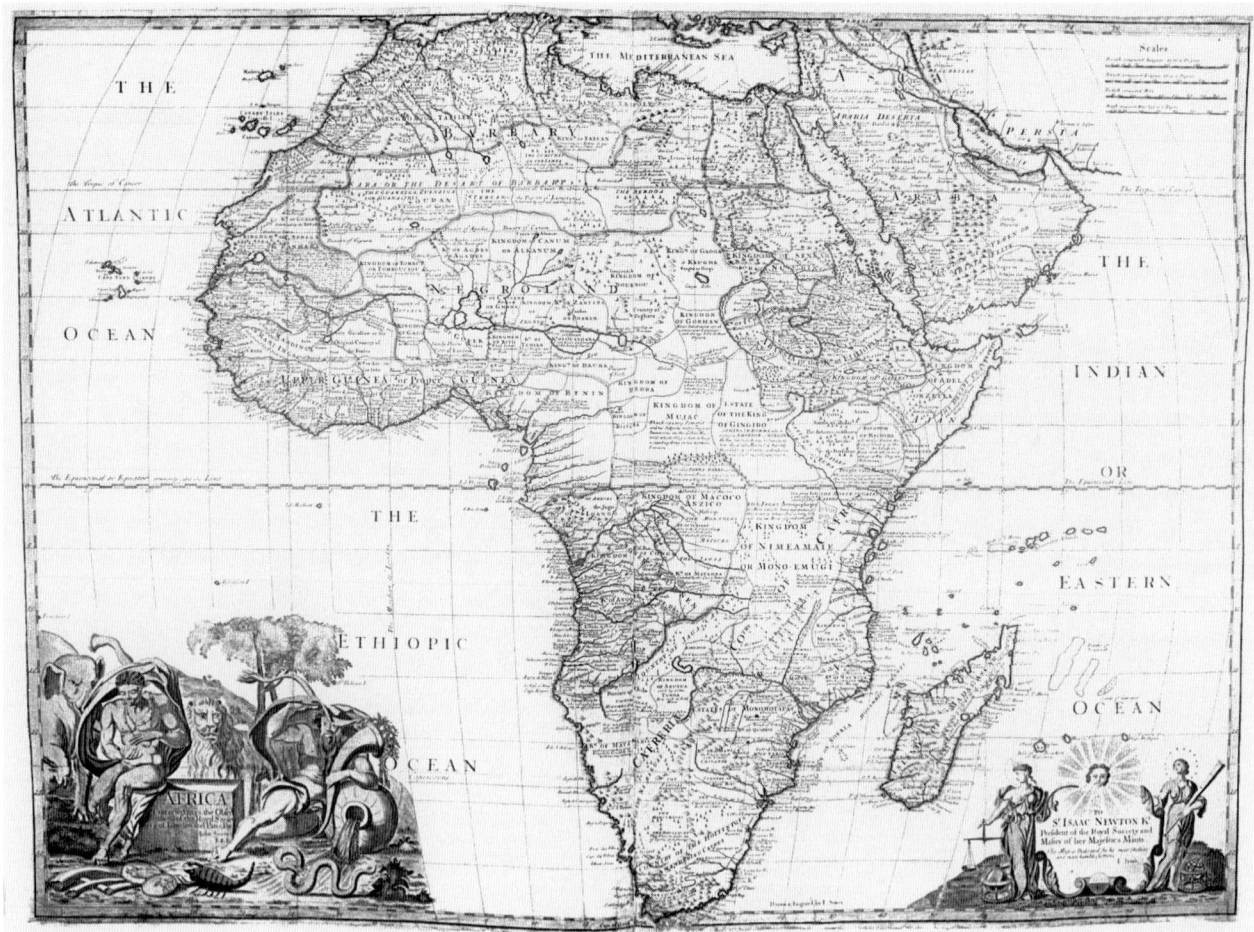

Map 69

SENEX, John (d. 1740)
Africa Corrected from the Observations of the Royal Society of London and Paris. (London, J. Senex, 1711-1714.)
Map, 69 x 97 cm, coloured.
Scale in British and French leagues, Turkish and British miles.
Prime meridian through London.
Dedicated to Sir Isaac Newton.

The cartouche at the lower left is very similar, but not identical, to the large C. Price map of 1711 (see Map 77). On the left is a large elephant moving towards a draped native sitting on the masonry bearing the inscription. He is wearing beads and holding some object in his hand. Behind the masonry is a lion, and in the background two trees and a hill. Adjacent to the lion is a seated male figure holding a horn of plenty and leaning his arm on an urn from which water is flowing towards a snake. In the right lower corner the dedication to Sir Isaac Newton is colourfully illustrated and it is mentioned that he is President of the Royal Society and Master of Her Majesty's Mints.

The general shape of the map is fairly correct but some fictitious names still persist in the interior, although the author has placed notes at different spots to the effect that the country has not yet been explored. In the south there are various Hottentot names as well as 'Hellenbok' and a few known rivers.

John Senex, publisher and engraver, produced two fine maps of Africa, one dedicated to the Marquis of Annandale and another larger one dedicated to Sir Isaac Newton. This map is presumably one of the sheet maps Senex issued separately.

84

Map 70

AA, Pierre van der (1659-1733)
L'Afrique selon les nouvelles observations de Messieurs de l'Académie des Sciences etc. (Africa according to the latest observations made by the Academy of Science.) (Leiden, P. van der Aa, 1713.)
Map, 47 x 65 cm, partly coloured.
Title across the top: *Africa in praecipuas ipsius partes distributa...*

A large map of Africa with a Latin description across the top border and a French title within the cartouche in the lower left corner. This cartouche is colourful, with a nude woman sitting on a rectangle of masonry situated in a rural scene with foliage and trees in the background, with elephants at the back around the trees. In the foreground is a partial view of an elephant and, adjacent to the masonry on the right, a lion looking to its right towards a coiled snake appearing to be moving towards it.

There appears to be a compromise in the mapping of the Nile basin, showing the river rising from Lake Tzana, in an area marked Demba, though the Ptolemaic lakes Zaire and Zaflan are retained, unconnected to the Nile. The map shows the Cape settlements in the south and the Portuguese ones on the lower eastern coastline and mainland. The centre of the map is blank with a Latin legend stating: 'I would rather show this part of Africa as unknown and uninhabited than rely on my imagination.'

The map is most decorative, illustrating Pierre van der Aa's predilection for embellishment. Although the map first appeared in 1713, it was reissued in *Galérie Agréable du Monde*, an atlas of sixty-six volumes published in 1729.

85

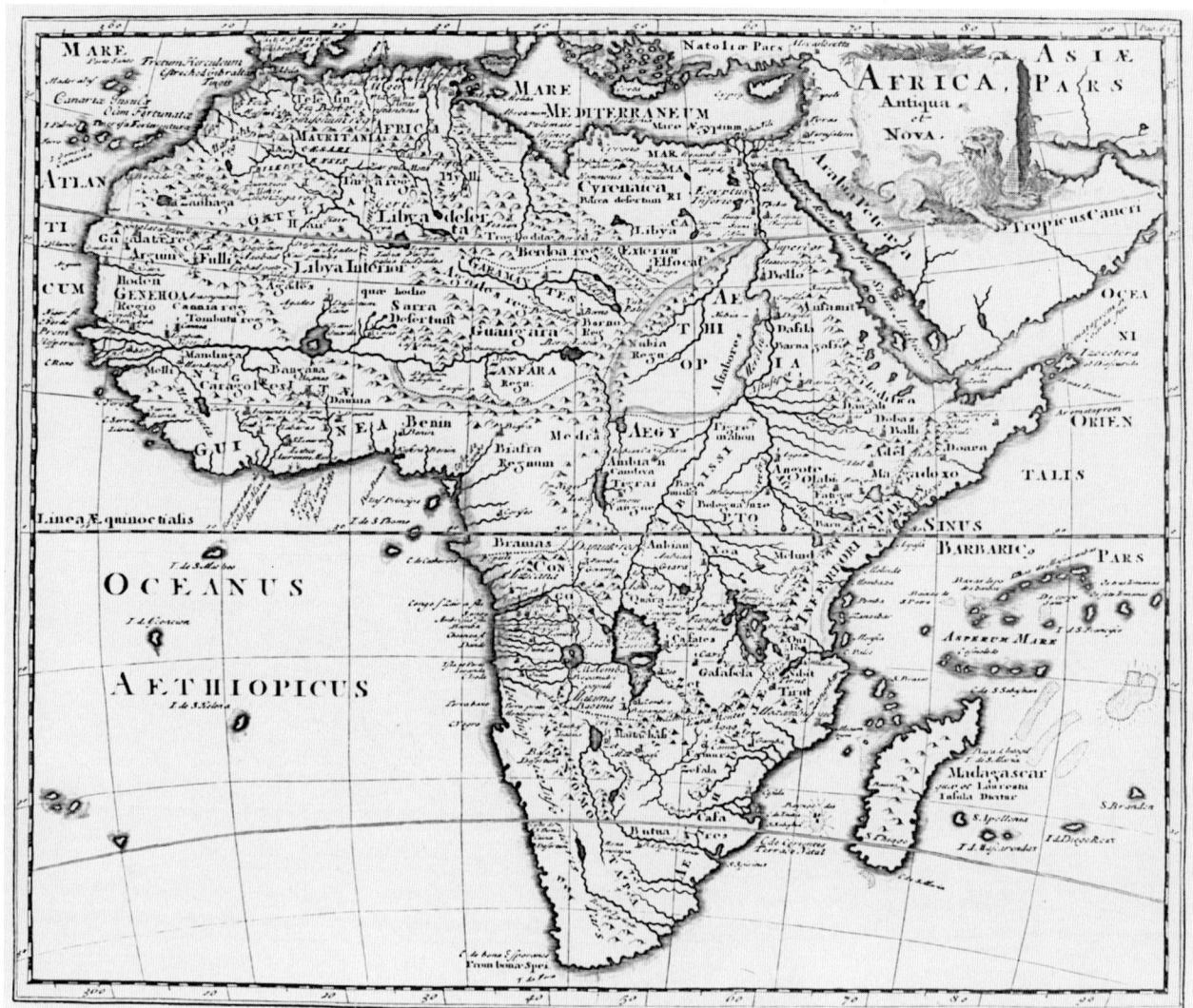

Map 71

MOLL, Herman (d. 1732)
Africa Antiqua et Nova. (Africa ancient and modern.) (London, H. Moll, 1714.)
Map, 21 x 26 cm, slightly coloured.
Top right: Pag. 623.

This small map of Africa came from the fourth volume of *Atlas Geographus: or A Complete System of Geography* (1714) by Herman Moll. Item 51 in Anna Smith's *Exhibition of Decorative Maps of Africa up to 1800* is identical to the copy shown here, in that the Rio de Spirito Sancto and Rio do Infanto are the only rivers named in the south. This copy is numbered 623 whereas the number of the copy in Anna Smith's series is 395. In addition, this copy is slightly larger and does not include a reference to John Senex as engraver on the left lower border. Otherwise the map has the usual features of eighteenth–century maps, particularly in its large lakes and rivers.

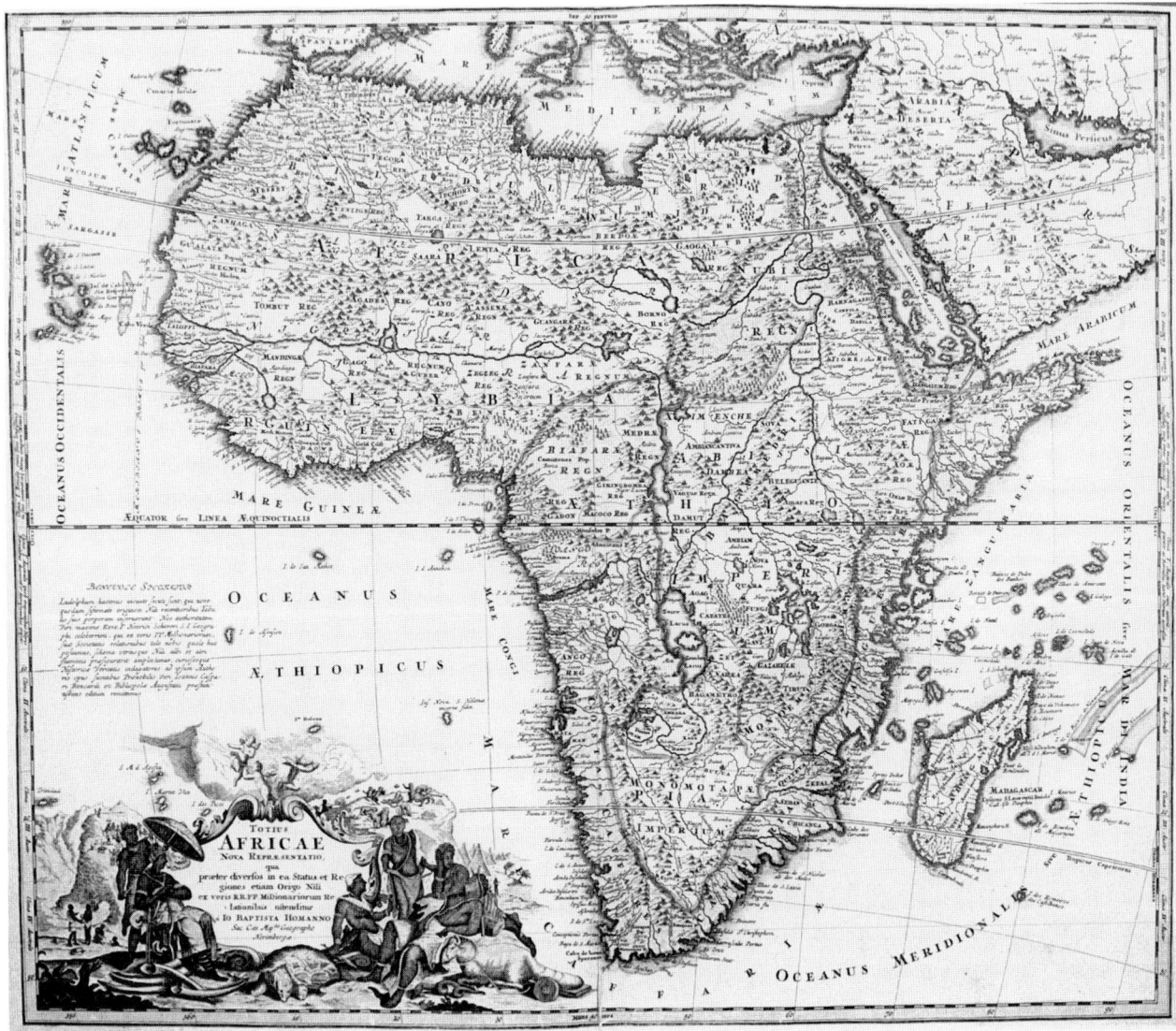

Map 72

HOMANN, Johann Baptist (1663-1724)
Totius Africa Nova Repraesentatio qua practer diversos in ea Status et Regiones, etiam Origo Nili et veris RRPP Missionariorum Relationibus ostenditur... (The whole of Africa divided into states and regions...) (Nürnberg, J.B. Homann, 1715.)
Map, 47 x 56 cm, coloured. Prime meridian through Tenerif.

This map has a lengthy note on the supposed accuracy of the source of the Nile, probably the first of its kind in the history of map making of Africa. This interest in the Nile is reflected in the cartouche, where at the far left is a gorge labelled 'Fontes Nile.' A crowd of natives can be seen on the Nile bank. In the middle distance another four natives hold tusks. Prominent in the foreground, a native potentate is seated under an umbrella. In the clouds above the cartouche are three cherubs, one holding a crucifix. In the right background is a cliff, with a troop of monkeys and two European hunters. Below the cliff is a desert-like landscape with six pyramids. In the right foreground are three natives sitting and one standing. By their side in the foreground is a fat-tailed sheep with its tail supported on a small wheeled cart, to support the weight and to prevent damage to it in dragging when walking.

This map is poor geographically in both shape and nomenclature. Both the real and fictitious St. Helena islands are shown. The Nile's origin, despite the lengthy note, conforms – as in so many other maps of this period – to the erroneous two-lake-side-by-side representation.

Johann Baptist Homann and his heirs dominated German map making for about a hundred years. Born in 1663 in Kammlach, Homann engraved for Funck, Jacob van Sandrart and Scherer. His first atlas was issued in 1704; the *Neuer Atlas* in 1707 followed by the *Grosser Atlas* (1731), Doppelmayer's Star Atlas of 1742.

Map 73

CHATELAIN, Henry Abraham (1684-1743)
Nouvelle Carte d'Afrique avec des Remarques et des Tables pour trouver sans peine les differents Peuples de cette Partie du Monde par les Renvois Alphabétiques suivant les plus nouvelles Observations de Messieurs de l'Académie des Siences. (New map of Africa with observations and index...)
(Amsterdam, 1718.)
Map, 46 x 58 cm, coloured.
Scale in French, Spanish and German leagues.
Prime meridian through Ferro Island.
Inset: Alphabetical index of kingdoms, provinces, towns, islands, capes, rivers and lakes.
Top right: No. D.

A French map of Africa – a co-operative work by Chatelain and Guedeville. Henry Abraham Chatelain, a geographic publisher, was responsible for an encyclopaedic seven-volume *Atlas Historique* which was first published in Amsterdam in 1705, and re-issued in 1718, 1721 and 1732-1739. Tooley dates this map 1705, basing his deduction on its similarity to the de L'Isle map of Africa of 1700, but the letter D which appears here in the top right corner is proof that the map was included in the 1718 edition. Guedeville (also Gueudeville), a geographer and printer, compiled the text assisted by Garillon. On each side border appears an index to the map's contents, and at the foot of the map on both sides appear 'historical remarks' describing various parts of the country. This map is based on the 1700 edition of de L'Isle's map; in both, the authors have deviated from the more usual two-lake origin of the Nile in favour of a single lake, Dambia, in northern Abyssinia.

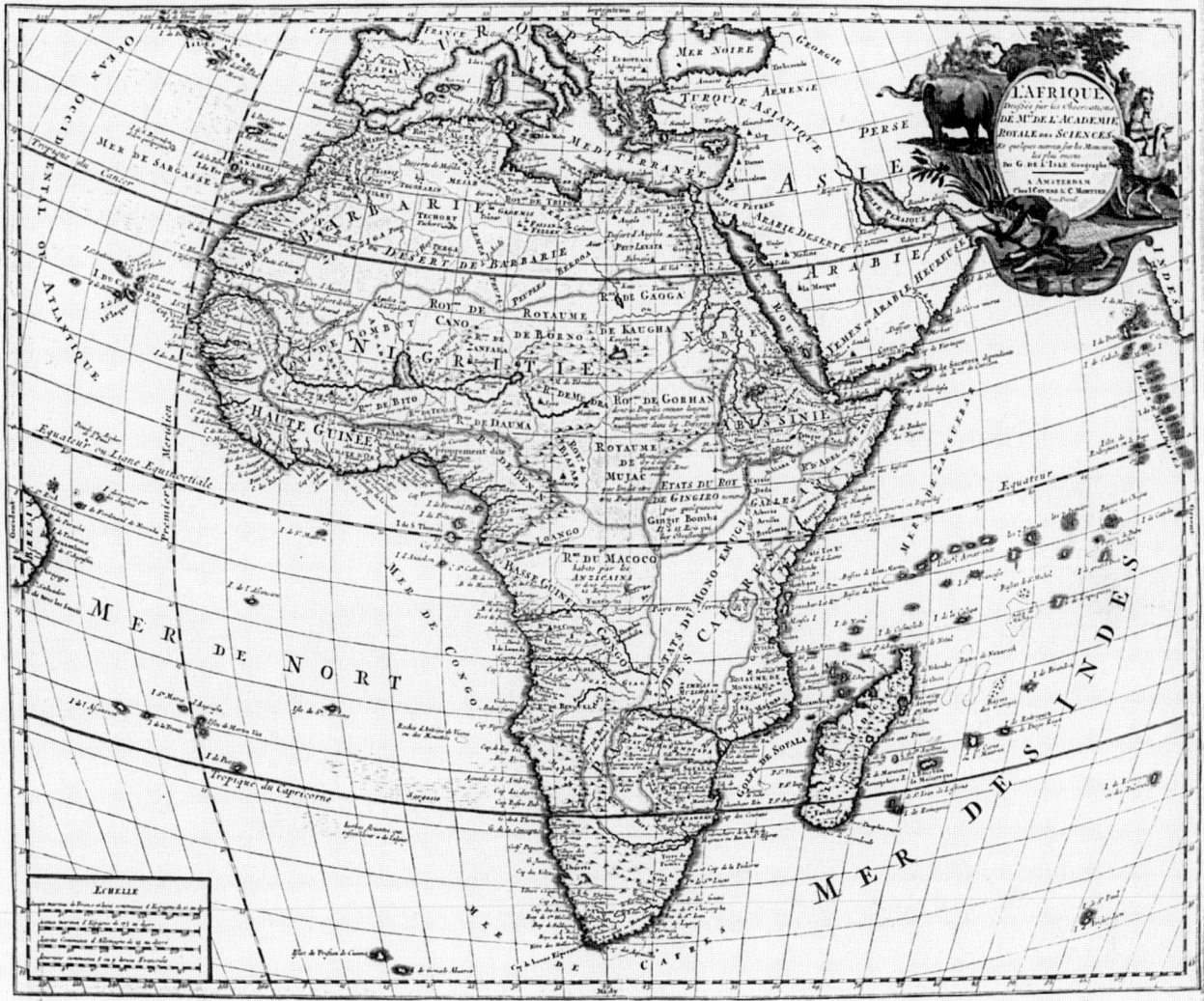

Map 74

L'ISLE, (Insulanus), Guillaume de (1675-1726)
L'Afrique Dressée sur les Observations de Ms. de l'Académie Royale des Sciences et quelques autres sur les Mémoires les plus recens. (Africa according to the observations made by the Royal Academy of Science.) (Amsterdam, J. Covens and C. Mortier, 1730.)
Map, 46 x 57 cm, slightly coloured.
Scale in French and Spanish sea leagues, Spanish and German leagues, travelling leagues.
Prime meridian through Ferro Island.
Title across top: *Africa Accuraté in Imperia, Regna, Status and Populos Divisa, ad Usum Serenissimi Bourgundiae Ducis.*
Dedicated to the Duke of Burgundy.

This is a 1730 edition of the de L'Isle map of Africa, published in Amsterdam by Covens and Mortier. Despite its later date, it does not incorporate any of the revisions of his post-1722 map of Africa, but retains the geographical features of the 1700 revision, which are described in the previous de L'Isle map. (See Map 59.)

The cartouche at the top right now displays the names of Covens and Mortier, a dedication to the Duke of Burgundy has been added across the top, and the 'Avertissement' above the scale at the bottom left has disappeared.

Map 75

VALK, Gerard (c. 1650-1726) and Leonard (1675-1755)
***Africa Mauro Percussa Oceano, Niloque admota tepenti . . . Cum Privilegio
Ordinum Hollandia et Westfrisia.*** (Amsterdam, G. and L. Valk, c.1720.)
Map, 49 x 59 cm, coloured.
Prime meridian through Tenerif.

This map of Africa has the typical features of the eighteenth-century Dutch maps of this continent. The cartouche at the left lower corner is identical, except for the insignia, to that of Allard's map (see Map 54), although the names of the artists have disappeared. The map, however, is different from the Allard, with captions now partly in Dutch and partly in Latin, and its shape and content show it to be a map of a later date. The typical bulge on the east coast near Madagascar has disappeared as well as some of the rivers, but Vigiti Magna and Monomotapa still appear.

The fort in Table Bay is marked as well as Hout Bay, but no Hottentot tribes are indicated. The author has two copies in different colours, one cartouche having been coloured and highlighted in gold.

Gerard Valk, a publisher and engraver based in Amsterdam, was in partnership with Schenk and acquired some of Blaeu's and Jansson's copperplates. He was joined by his son, Leonard, in compiling an atlas, and Leonard continued the business after his father's death.

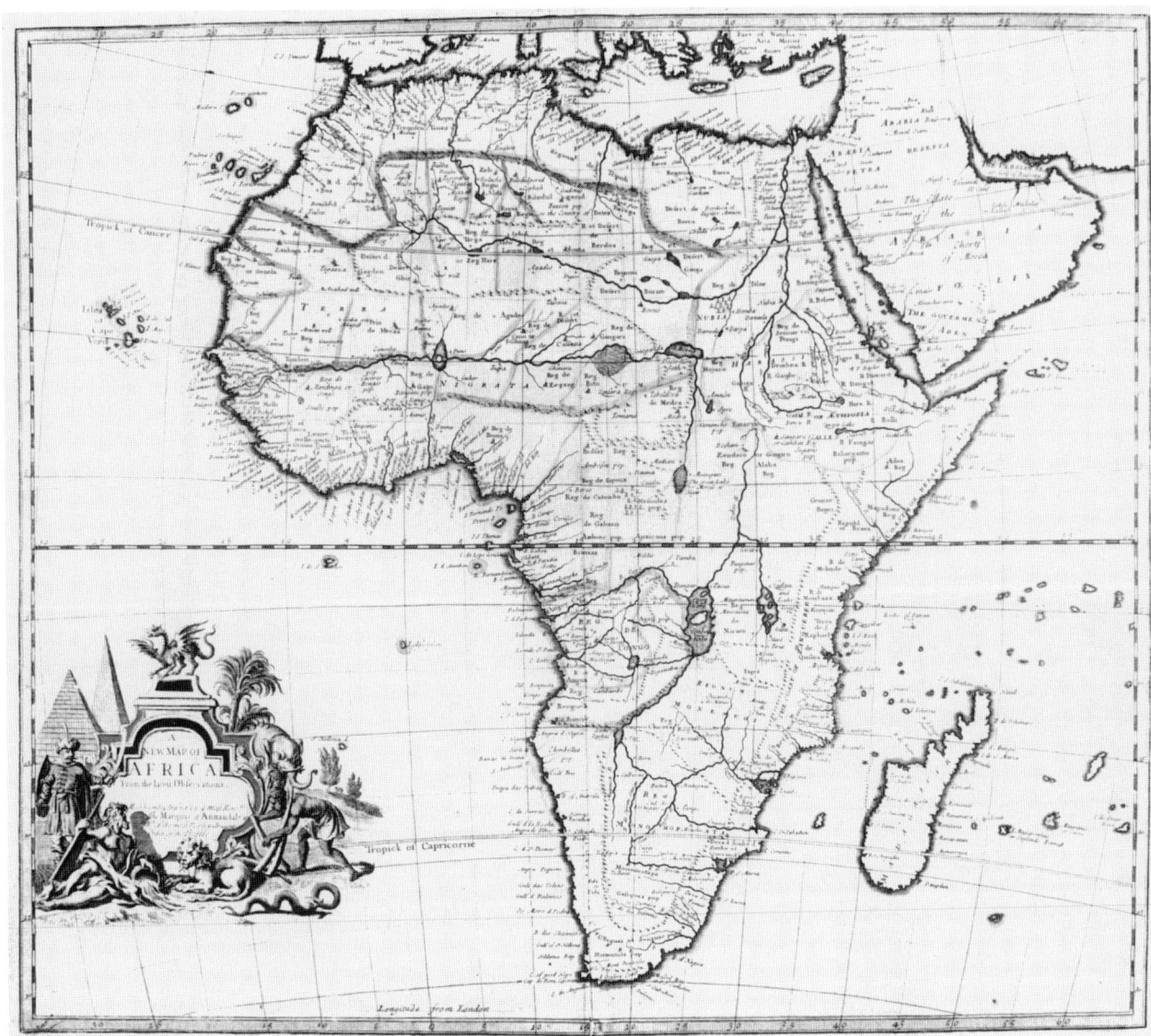

Map 76

SENEX, John (d. 1740)
A New Map of Africa from the latest Observations. (London, John Senex, 1721.)
Map, 48 x 55 cm, coloured outline.
Prime meridian through London.
Dedicated to the Marquis of Annandale.

The cartouche of this map, in the left lower corner, consists of a decorative shield bearing the inscription and surmounted by a dragon. To the left are two pyramids and a Moor brandishing a sabre and holding a shield, and below sits a male figure holding what appears to be an oar. In the centre a lion is lying down adjacent to a snake. On the right is a palm and two trees in the distance. An elephant is half concealed by the shield and in front is a native collecting tusks.

John Senex, engraver, publisher and Geographer to Queen Anne, was a contemporary of Herman Moll. He worked in London next to the Fleece Tavern in Cornhill. From about 1708 he began a partnership with C. Price of Ludgate Street, where they produced a series of large maps printed on two sheets and sold separately. By 1711 he had twenty maps of the world and the four continents, designed by C. Price and engraved by Senex. In 1714 Senex went into partnership with John Maxwell in London and published the *English Atlas.* In 1721 another improved work, the *New General Atlas of the World*, with 37 maps, was issued.

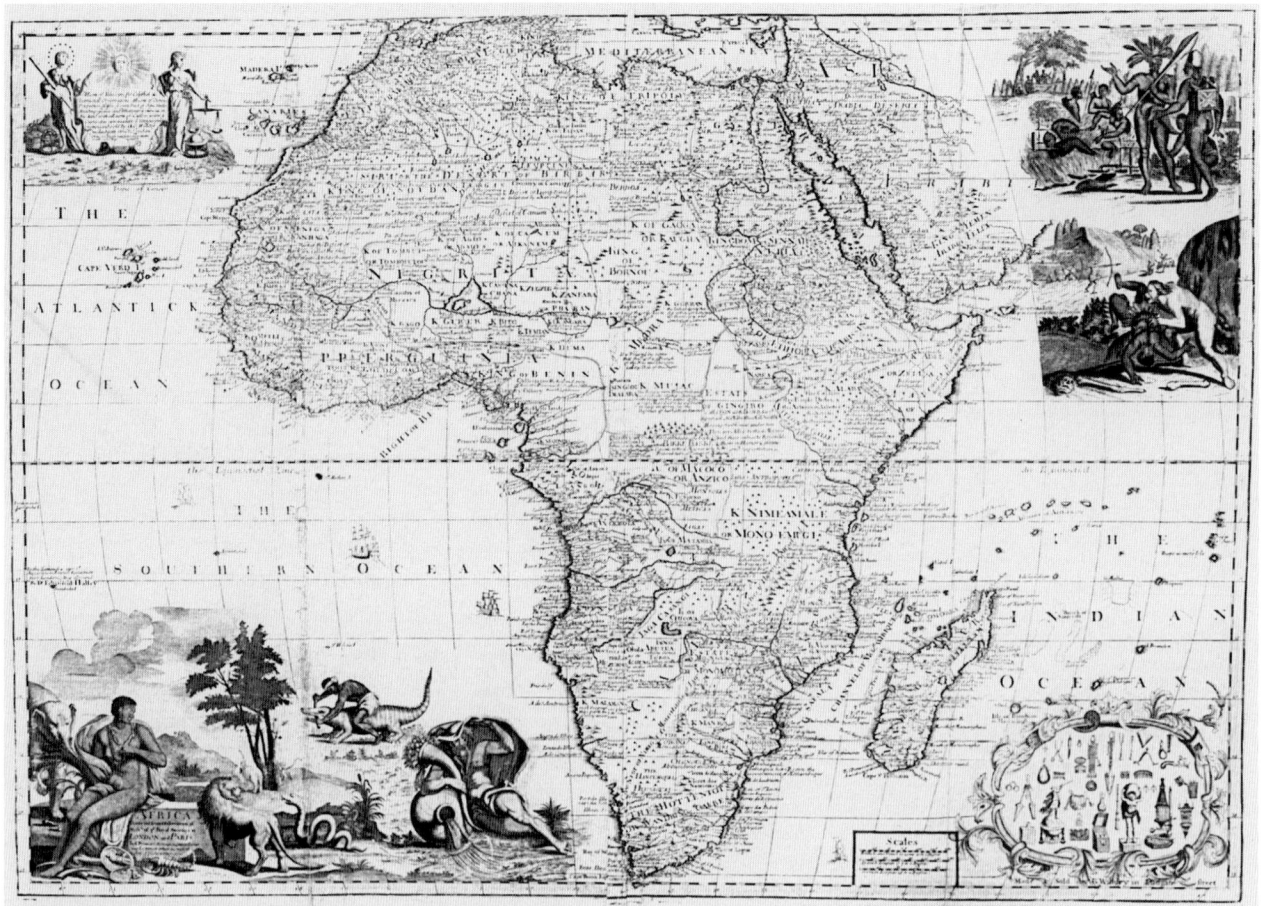

Map 77

PRICE, Charles (fl. 1680-1720)
Africa corrected from observations of Mess. of ye Royal Societies at London and Paris. London, C. Price, 1721.
Map, 56 x 95 cm, coloured. Scale in British, French, Spanish and German leagues.
Inset: Instruments used in map making. Scenes of cannibalism.

This large, colourful and fanciful map of Africa has an elaborate cartouche at the lower left. A draped native woman is seated on masonry bearing the inscription, with corn at her side and at her feet some elephant tusks, medallions and a scorpion. On the other side of the masonry is a lion looking at a snake. In the background is a mountain and a tree under which two elephants are sheltering. On the right is a native riding a crocodile. There is also a figure of a man sheltering under some drapery and resting his arm on an urn from which some liquid is flowing. In his hand he holds a horn of plenty. The left half of the cartouche is practically identical to that of the large van der Aa map of Africa (see Map 70) except for the title on the masonry, while the figure at the right is a mirror–image copy of the figure in the cartouche of Senex's map (see Map 69). Also copied (though reversed, left to right) is the scene at the top left, which in the Senex map illustrated the dedication to Sir Isaac Newton. The scales appear towards the right and in the corner, above the words 'made and sold by G. Willdey in Ludgate Street', is a miscellaneous collection of articles surrounded by decorative scroll work. Presumably these are the telescopes,

glasses, spectacles, etc., which are referred to in the advertisement for G. Willdey under the left–hand corner.

In the right hand top corner two scenes depict different stages of cannibalism – the upper scene shows people engaged in cooking parts of the human body on an open fire. The other scene is of a fight in the rocks between two groups, one of which is armed with bows and arrows while the other has spears. In the foreground is a man (who appears white–skinned) overwhelming another. On the ground various human bones are lying around.

Three ships are in full sail in the Atlantic; one south of Madagascar. Contrary to usual cartographical practice, facts or legends about certain places, rivers and mountains in the interior are noted. Stellenbok appears in the south, as noted on de L'Isle's map, more or less where Stellenbosch is situated. Fort Hollandia and the Hottentot tribal names are noted in addition to the usual coastal places and a few rivers.

Charles Price was a publisher, draughtsman, surveyor and globemaker working in London in partnership, at times, with Senex, Maxwell, Willdey, and also with Jeremiah Seller.

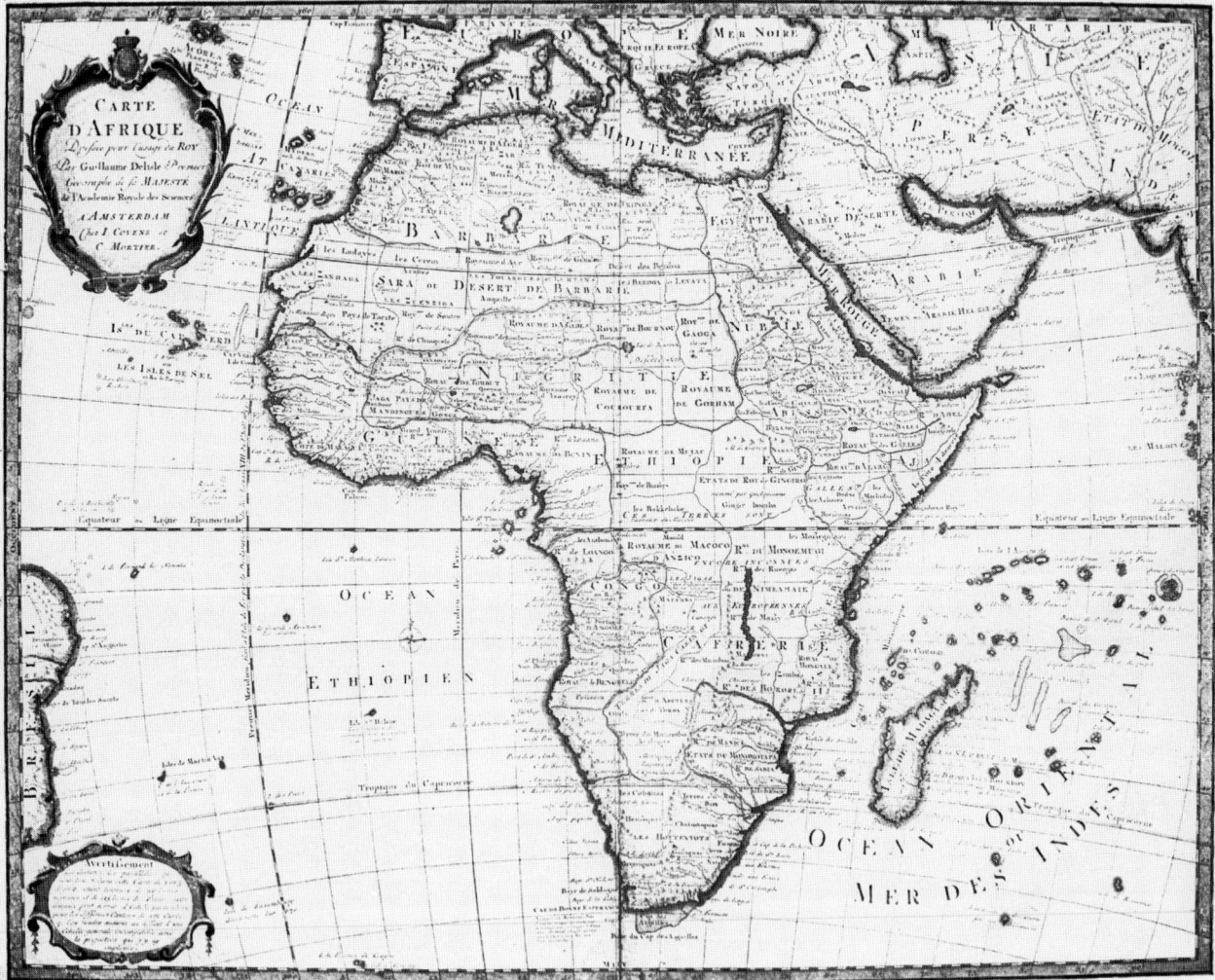

Map 78

L'ISLE (Insulanus), Guillaume de (1675–1726)
Carte D'Afrique Dressée pour l'usage du Roy... (Map of Africa for the King...)
Engraved by J. Kondet. (Amsterdam, J. Covens and C. Mortier, 1722.)
Map, 48 x 62 cm, coloured.
Title across top: *Africa Accurate in Imperia, Regna Status e Populus Divisa ad Usum Lodovici XV*
Galliarum Regis. (Africa divided into kingdoms, states and peoples...)
Prime meridian through Ferro Island and Paris.
Dedicated to Louis XV.

This map of Africa is one of the series of revised maps that de l'Isle first issued in 1722, published by the two foremost Dutch publishers, Covens and Mortier. In this redrawn version de L'Isle makes further improvements. For the first time he separates the Senegal and Niger rivers, making the latter rise in the Kingdom of Tombut and flow eastwards to Lake Bournou. Inland from Zanzibar he inserts

a large lake, Moravi (Nyasa), with the mountains of Lapata – the spine of the world. A typical de l'Isle cartouche is drawn at the top right corner and a dedication title appears boldly across the top border of the map.

This is a Dutch example of de l'Isle's world-wide expansion of map publications.

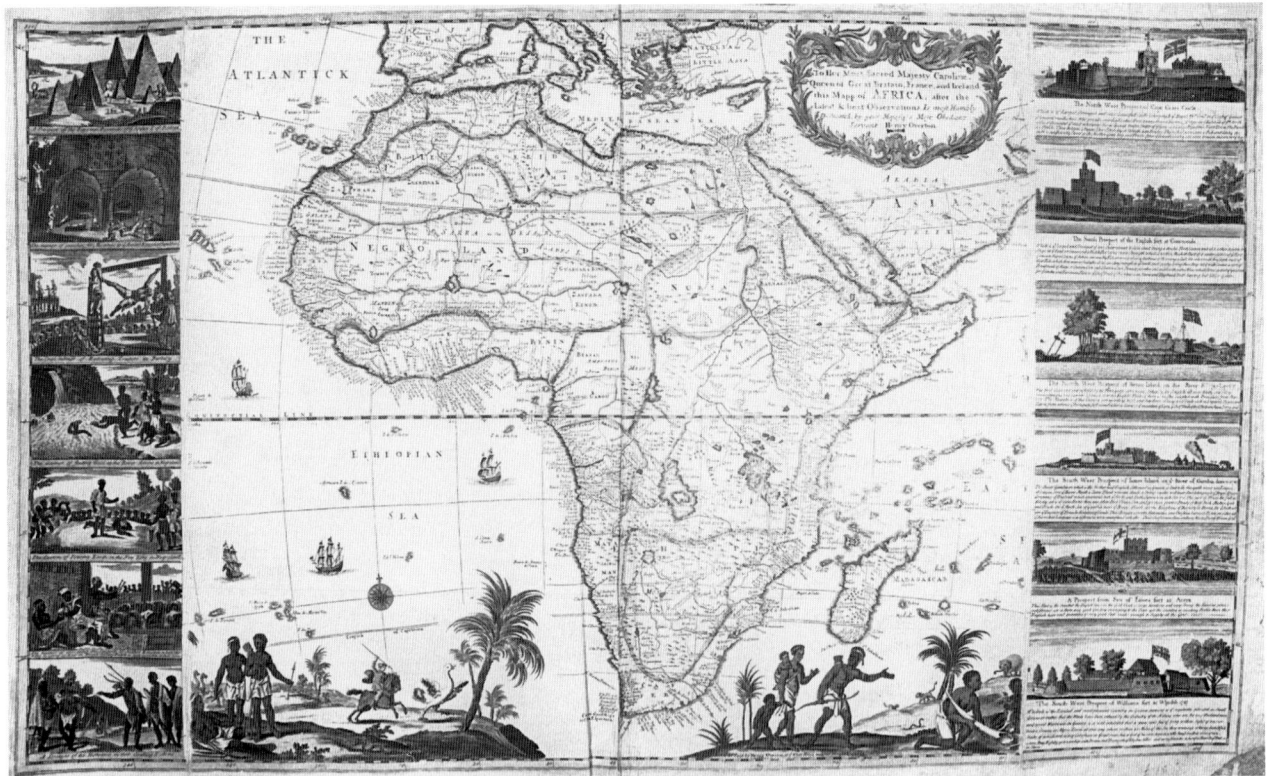

Map 79

OVERTON, Henry (fl. 1706-1764)
To Her Most Sacred Majesty Caroline, Queen of Great Britain, France and Ireland this Mapp of Africa, after the latest and best Observations is most humbly dedicated... (London, Henry Overton, ye White Horse without Newgate, c. 1727.)
Map, 57 x 97 cm, slightly coloured.
Decorative borders depict scenes from African life and African forts.

This large map of Africa notes at its lower border that it was 'Edited and sold at ye White Horse without Newgate, London.' The cartouche on the top right border contains the dedication title, surrounded by a floral wreath and the map itself is flanked on each side by decorative borders. On the right border there are six English African Forts: 'The North West Prospect of the Cape Coast;' 'The South Prospect of the English Fort at Commenda;' 'The North–West Prospect of Bense Island on the River Sierra Leone'; 'The South-West Prospect of James Island on the River of Gambia drawn 1727'; 'A Prospect from sea of James Fort at Accra'; 'The South West Prospect of Williams Fort at Whydah in 1727.' Each fort is flying a Union Jack, with the exception of James Fort at Accra, flying St. George's flag. Below each fort is a geographic and historic description.

On the left decorative border there are seven scenes of African places and native habits and customs: 'Prospect of the Egyptian Pyramids of ye Spyn and of Radope'; 'The manner of searching for Mummys in ye Ancient Egyptian sepulchers'; 'The cruel manner of executing traitors in Barbary'; 'The manner of finding gold at the River Alzine in Negroland'; 'The Custom of pouring earth on the New King in Negroland'; 'The manner of informing the subjects of Ethiopia when the King is going to eat or drink, it being death to look upon him at that time'; and 'A draught of the Hottentots in their clothing and arms.'

Henry Overton, who was responsible for the issue of this map, succeeded his father John Overton (1630–1713) in 1707, when he bought his father's stock, and both worked in London at the White Horse without Newgate near the Fountain Tavern. Henry reissued the anonymous *County Maps* of 1708 and reprinted Speed's *County Maps* with additions.

94

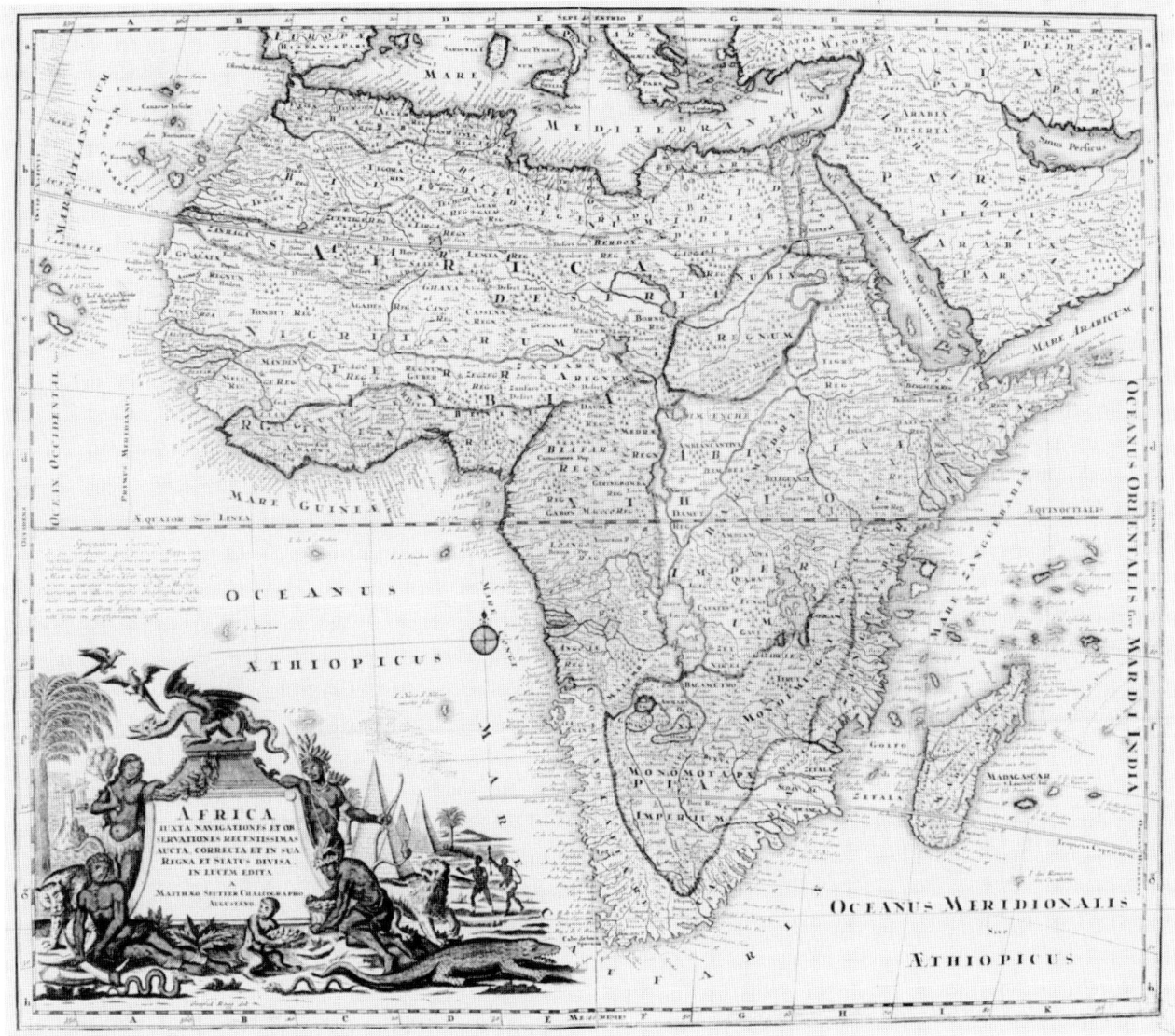

Map 80

SEUTTER, George Matthaus (1678-1757)
Africa Juxta Navigationes et Observationes Recentissimas Aucta, Correcta et in Sua Regna et Status Divisa in Lucem Edita... (Africa according to the latest voyages and observations.) Engraved by Gottfried Rogg. (Vienna, M. Seutter, 1728.)
Map, 49 x 57 cm.

This map of Africa was published by George Matthaus Seutter, a German cartographer and publisher of Augsburg.

In the lower left corner is a large decorative title cartouche engraved by Gottfried Rogg, with natives, pyramids, animals, lighthouses and ships. Although all the decorative animals have disappeared from the mainland the enormous lakes are shown in Central Africa and the information about the southern extremity of the continent is largely fictitious. The Nile is shown not only originating in the south at lakes Zaire and Zaflan, but also continuing further south, and the Abyssinian province of Amhara is shown in the kingdom of Monomotapa. This map is in fact crowded with erroneous detail.

The son of a goldsmith, Matthaus was apprenticed to Homann, the map publisher at Nürnberg, in 1697. He set up his own business in Augsburg where his map publishing flourished, issuing an *Atlas Geographicus* in 1725, an *Atlas Novus* in 1728 in Vienna (in which the above map appeared) and a *Grosser Atlas* about 1735. He was appointed geographer to the Imperial Court. With his son Albrecht Karl, son-in-law Tobias Conrad Lotter and the engraver Silbereisen he also issued a large number of town plans.

Map 81 GIBSON, John (fl. 1750-1792)
Old Map of the Continent according to the greatest diametrical Length from the Point of East Tartary to the Cape of Good Hope. (London, T. Jeffery, 1731.)
Map, 21,5 x 18 cm, uncoloured.
In: *The Gentleman's Magazine or Monthly Intelligencer*, vol. 1, 1731. London, F. Jeffery.

This miniature map of Africa, together with Europe, Asia and the Far East, is shaped to provide a global effect. The cartouche in the lower right corner is oval, with an ornate floral border on which is skilfully included a female face wearing a wide-brimmed hat. The title appears within the cartouche.

The map of Africa, geographically, provides somewhat sparse information. In the south, Monomotapa is noted with the unusual word 'Empire' in the centre of the map. Isolated mountain ranges appear irregularly and inaccurately. Mention is also made in the south of the 'Country of Cafres.'

This map was engraved by John Gibson, an engraver and draughtsman in Clerkenwell, London. He is known to have engraved for the long-established and well-known *Gentleman's Magazine* from 1758-1763.

Map 82

INSLIN, Charles (fl. 1699-1735)
Carte Generale de L'Afrique Contenant les Principaux Etats qui y sont Contenue Dressée sur les Nouvelles Observations. (General map of Africa showing the principal states according to the latest observations.) (Paris, Rue St. Jacques a St. Pierre, Crepy, 1735.)
Map, 51 x 64 cm, coloured outline.
Prime meridian through Ferro Island and Paris.

A highly decorative French map of Africa issued by Charles Inslin, a Parisian geographer and engraver for many other cartographers such as De Fer, Froger, Desnos and Jaillot. The map was published by Crepy, a publisher of Paris.

The map itself conforms to the usual observations of the eighteenth century, with rivers, lakes and towns as well as the conventional origin of the Nile from the two–lake system in lower Central Africa. The cartouche at the lower left depicts a semi-nude female standing over the title wearing a large flat hat. She is being observed from the right by an elephant with large flapping ears. Surrounding the cartouche and all the way up the left border of the map is a series of coats of arms of all the main cities of Europe – twenty-nine in all – making the map most colourful and decorative.

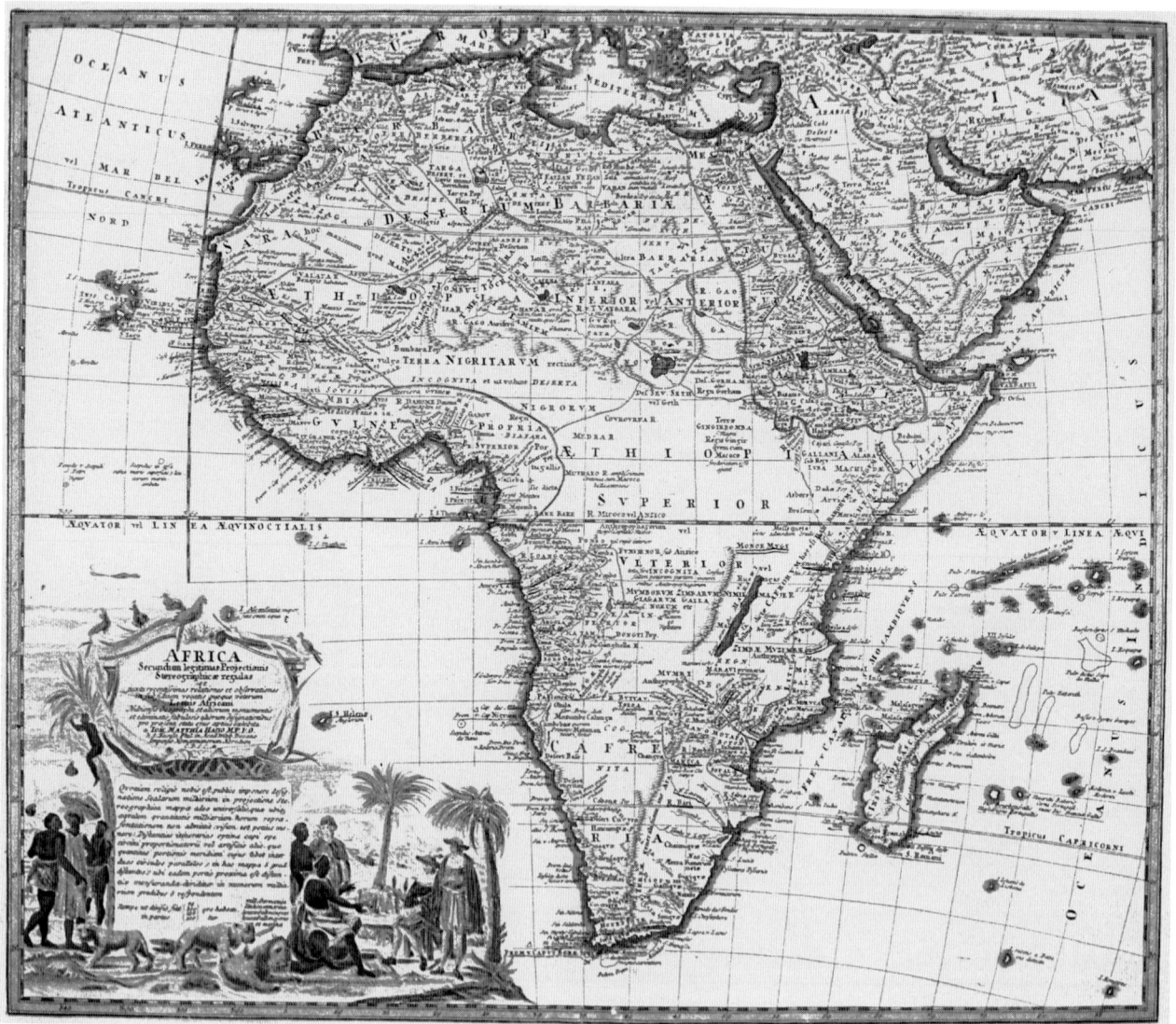

Map 83

HAAS, Johann Matthias (1684-1742)
Africa Secundum legitimas Projectionis Stereographicae regulas et juxta recentissimas relationes et observationes in subsidium vocatis quoque veterum Leonis Africani Nubiensis Geographi et aliorum monumentis et eleminatis fabulosis aliorum designationibus pro praesenti statu ejus aptius exhibita.
(Africa according to the most recent reports and observations.) (Nuremberg, Homann's Heirs, c.1737.)
Map, 45 x 54 cm, coloured.
Scale in Italian miles, German and nautical leagues.
Prime meridian through Ferro Island.

Johann Matthias Haas, also known as Hasse, Haase, and Hasius, became Professor of Mathematics at Wittenberg and worked for Homann and his heirs. 'Africa', produced in 1737; 'Guinea,' 1743; *Grund Staedton*, 1745 and *Atlas Historicus*, 1750, were some of his works. This cartouche contains no date but the copy in the Johannesburg Public Library's *Descriptive Catalogue* (1952) dates the map as 1737.

An ornate title consisting of a panel surrounded by elephant's tusks rests on a piece of masonry containing a long description of the scale used. On the left are four Africans, one of whom has climbed to the top of a palm tree. In front of the masonry are members of the lion family, a snake-like lizard and a tortoise. On the right an interview is being conducted between Europeans and Africans under a palm tree. The African chief is sitting on the back of another who is crouching on a mat, while others look on. Two Europeans are standing at this interview and a third is seated. In the background is a view of Table Bay and Table Mountain.

Most of the usual ornamentation has disappeared from the map as well as most of the fictitious lakes and rivers. 'Castell Batov' indicates the Dutch settlement at the Cape and Stellenbosch is marked. Some tribal names, a few unnamed rivers and place names on the coast bear a fairly accurate resemblance to their actual position in the south.

The Homann family published an identical map but for the addition 'Noribergae in officina Hommaniana' below the frame. This family issued numerous atlases containing maps by different cartographers and engravers, making it difficult to state definitely from which atlas this map was taken, but it probably comes from one of the editions of their *Atlas Geographicus Maior.*

Map 84

L'ISLE (Insulanus), Guillaume de (1675-1726)
Carta Generale Dell' Africa. (General map of Africa.) (Italian ed., 1740-1750.)
Map, 33 x 42 cm, uncoloured.
Scale in Italian miles.

In attempting to establish the provenance of this Italian map, reference is made by Tooley (*The Map Collector Series*, no. 6), to a similar map of the Cape of Good Hope which he states appeared in the Italian edition of de l'Isle's *Atlante Novissimo che contiene tutti le parti del mondo*, published in Venice between 1740 and 1750. This he confirms in his *Dictionary of Mapmakers*.

The cartouche in the top right corner is an animated scene of a youth spearing a crocodile with, behind him, another person riding a horse. On the right is an elephant and a large ostrich-like bird between two trees. Geographically the map outline is accurate, with many rivers, mountains and bays, illustrated in the manner typical of eighteenth-century maps of Africa.

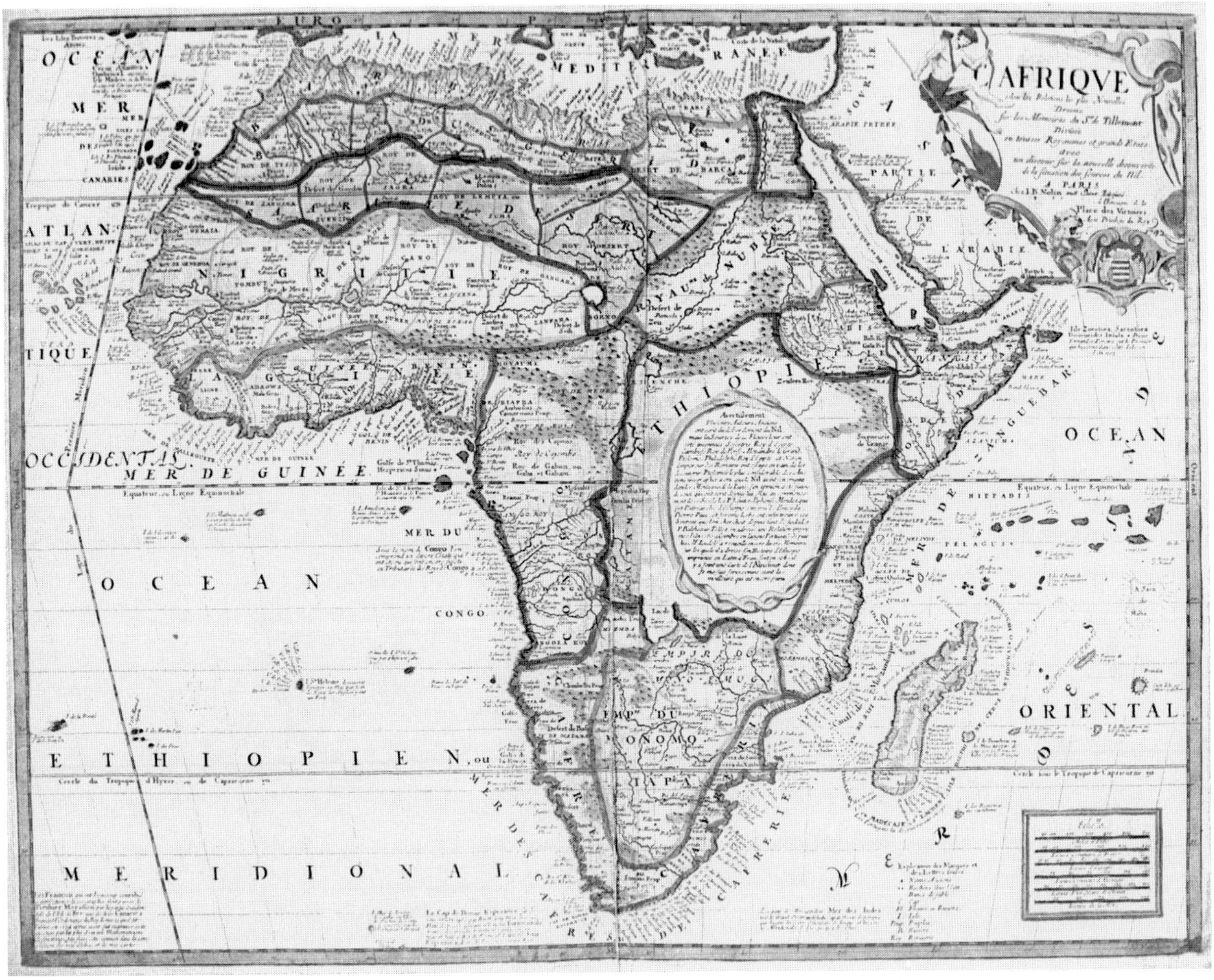

Map 85

TILLEMONT, Nicolas du Trallage (d. 1699) and NOLIN, Jean Baptiste (1686-1762)

Afrique selon les Relations les plus Nouvelles Dressée sur les Mémoires du... Divisée en tousses Royaumes et grands Etats avec un discours sur la nouvelle découverte de la situation des sources du Nil. (Africa according to recent findings... with a note about the latest discovery regarding the sources of the Nile.) (Paris, J.B. Nolin, 1742.)

Map, 45 x 58 cm, coloured.
Scale in Italian miles; French, Spanish and German leagues; nautical leagues.
Prime meridian through Ferro Island.

This attractive French map of Africa, dated 1742, with its cartouche in the top right corner, referring to Sr. de Tillemont and J.B. Nolin as the publishers, is really identical, except for the name in the cartouche, to one originally published by Vincenzo Coronelli in 1689.

It is known that Tillemont, as he is called in the cartouche, is really Jean Nicolas, Sieur de Tillemont du Trallage. He is well known for his review of Coronelli's maps for Nolin in 1688-1690, of which this map is one. Jean

Baptiste Nolin the elder (1657-1725), was a French geographer and engraver who worked extensively in Paris on maps and globes of many countries. His son, Jean Baptiste Nolin junior (1686-1762), continued the firm as a map maker and became Royal Geographer. As this map is dated 1742, whereas Coronelli's original was published in 1689, it must have been issued by Nolin junior. This is further substantiated by the fact that the address on the title is Rue St. Jacques, to which he eventually moved and worked.

101

Map 86

HEYDT, Johann Wolfgang
Africa. Drawn and engraved... J.W. Heydt. (1744.)
Map, 22,5 x 26 cm, uncoloured.
Top right: pag. 340.

A map of Africa from J.W. Heydt's *Alterneuste Geographisch und Topographische Schau–Platz van Africa und Oost–Indien*, published in 1744.

Whales, sea monsters and sailing ships appear in the seas with tritons and mermaids in the lower right corner blowing horns. Scattered throughout the mainland are trees, the usual animals and men. Large lakes appear in the centre with the Nile arising from the two-lake system. Saldanha Bay, Table Bay, 'Muschel Bay' and 'Bay de Algoa' are in the south as well as the 'C. das Angul' (Agulhas). The east coast has a queer bulge jutting out near 'Coffala,' opposite Madagascar, the lower corner of which is labelled 'C. des Coriendas.' The rest of the interior includes very little detail.

102

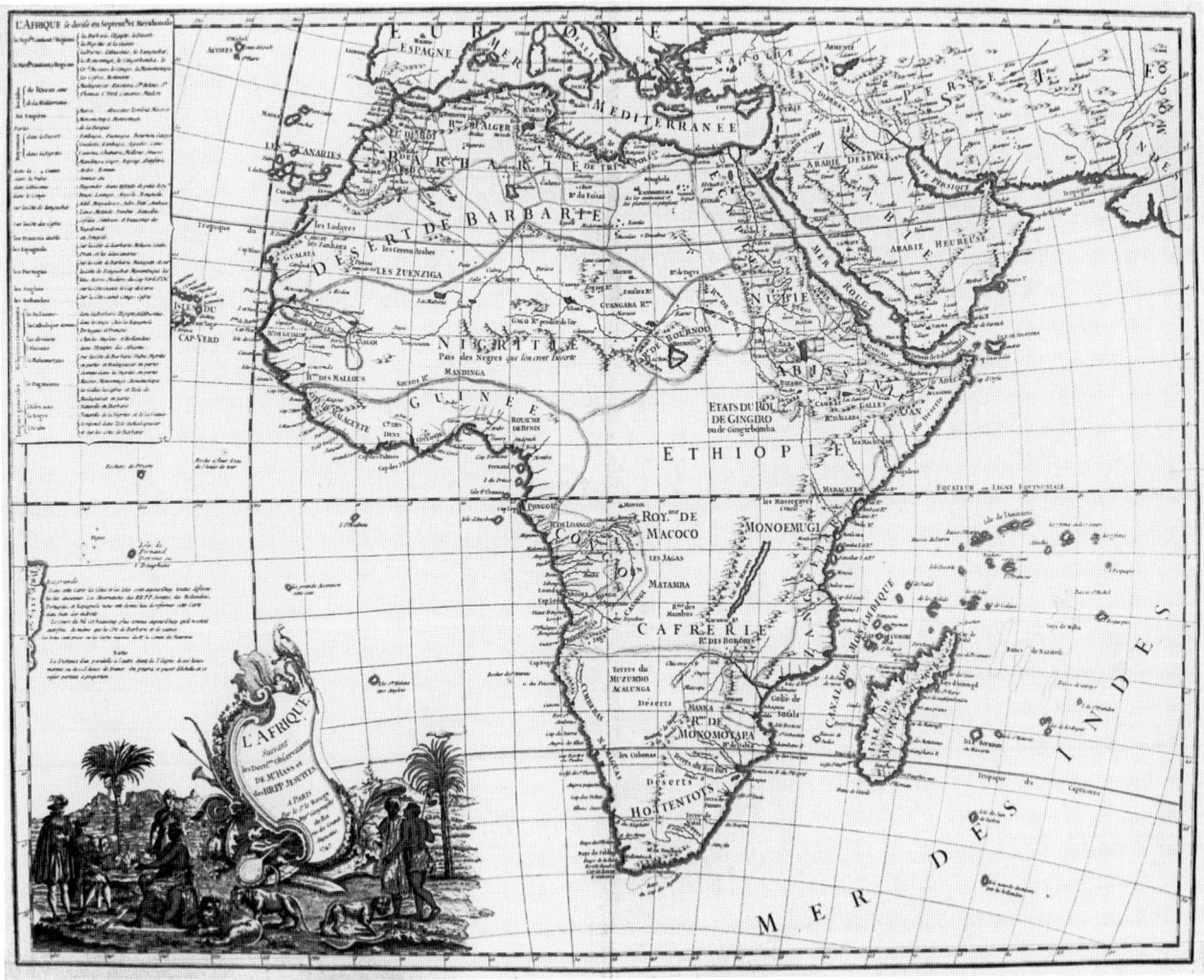

Map 87

LE ROUGE, George Louis (fl. 1740-1780)

L'Afrique Suivant les Derniers Observations de Mr. Hass et des RRPP Jesuites. (Africa according to recent observations made by Mr. Hass and the Jesuits.) (Paris, G.L. le Rouge, 1747.)

Map, 49 x 63 cm, coloured outline.

Prime meridian through Ferro Island.

This map of 1747 has a large decorative title cartouche at the lower left corner very similar to that of the map of Africa (see Map 83) published by Homann, assisted by Johannes Matthias Haas or Hasius, in 1737. The cartouche in this map shows white persons talking to a native who sits on the back of another native. There is a fairly faithful sketch of Table Mountain appearing in the background. In the foreground are three lions – two female and one male. The title is surrounded by an oval curved frame surmounted by

an eagle and flowers. On the right are two other natives, one carrying an umbrella.

George Louis le Rouge was a French geographer and engraver. He was appointed Geographer to the King and published many atlases, among which were the *Nouvel Atlas Portatif* of 1748, the *Theatre de la Guerre en Allemagne* in 1733-35, and the *Curiosités de Paris*, 1778.

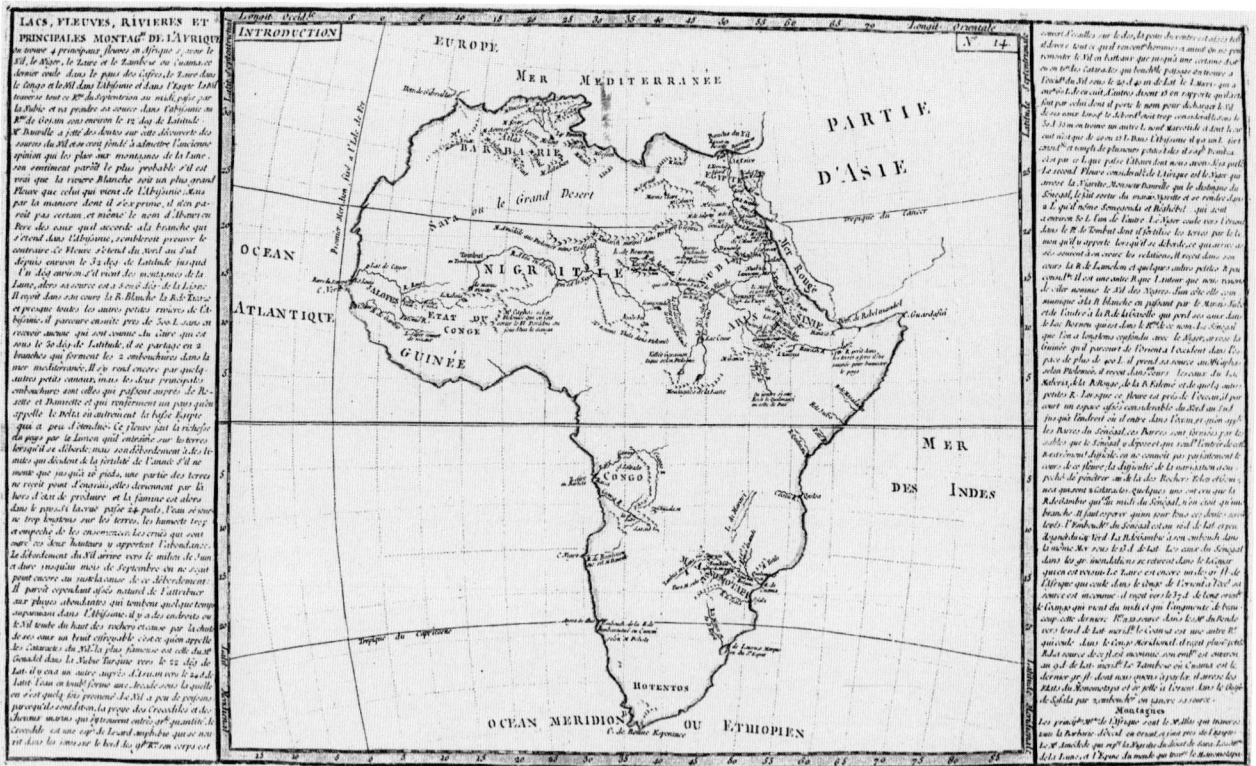

Map 88

CLOUET, Jean Baptiste Louis (fl. 1730-1793)
Lacs, Fleuves, Rivières et Principales Montagnes de L'Afrique. (Lakes, rivers and principal mountains of Africa.) (Paris, J.B.L. Clouet.)
Map, 41 x 39 cm, coloured outline.
Prime meridian through Ferro Island.
French text in margins.
Top right corner: No. 14.
Top left corner: Introduction.

The descriptive legends on each side of this French map of Africa are neatly engraved and form an integral part of the image.

The map has an accurate outline with the minimum of geographical detail except in the areas known to the geographer. The centre and south are almost devoid of any information except for the Cap de Bonne Esperance and the word 'Hotentos.' The part of the east coast explored by the Portuguese, 'Manomotapa,' is shown and the Mountains of the Moon appear as the origin of the Nile, with the two-lake system just above the mountain range. More is recorded of North Africa. The features of the map resemble those of d'Anville's map of 1727. This map has the merit of removing many of the fictitious and inaccurate features of so many of the eighteenth-century period, as well as attempting to describe the lakes, rivers and principal mountains of the continent.

The Abbé Jean Baptiste Louis Clouet was a member of the Académie des Sciences of Rouen as a geographer. He compiled wall maps of the four continents (1788-1793) and he also issued an atlas, *Geographie Moderne*, in 1747, with a corrected edition in 1793.

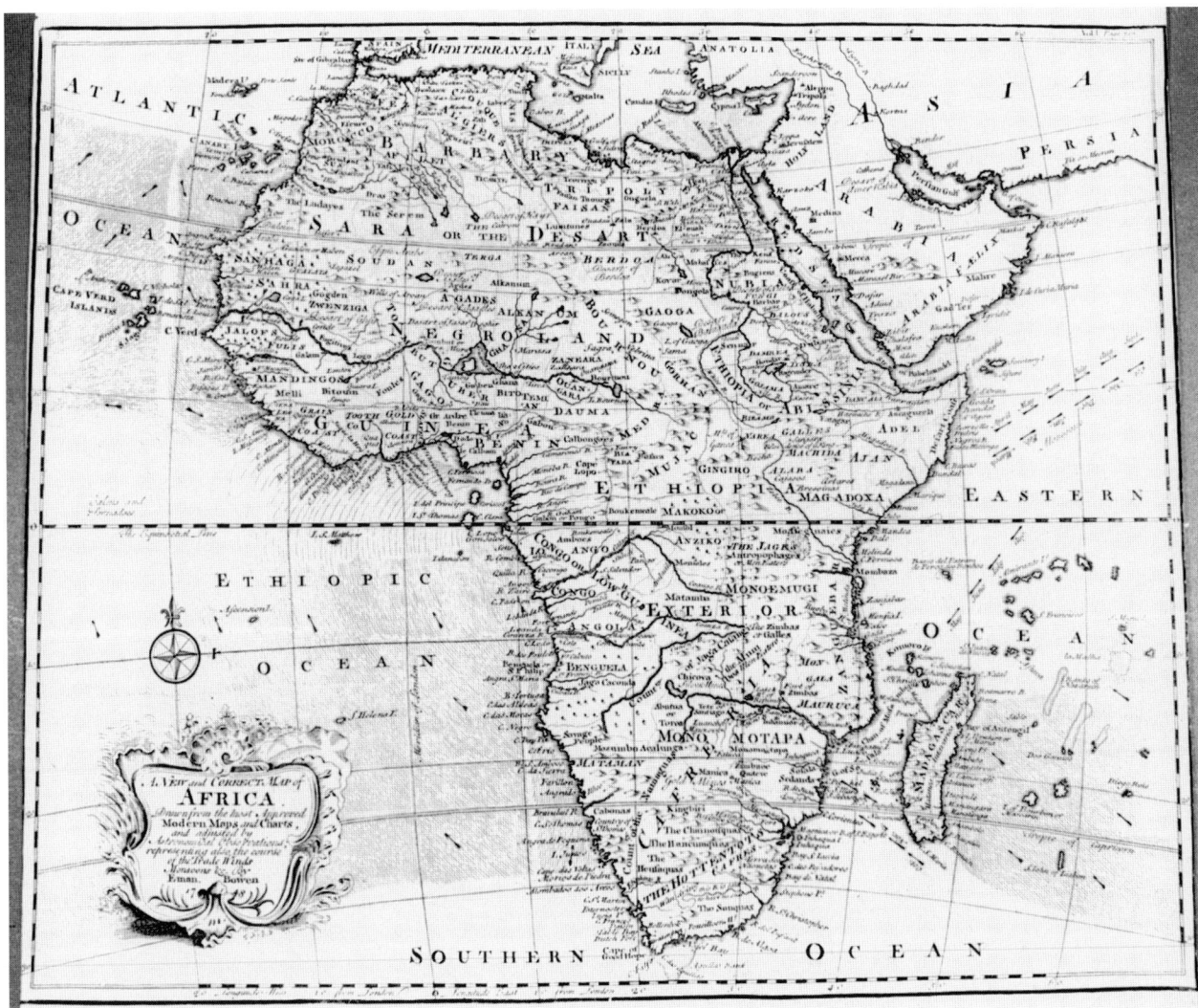

Map 89

BOWEN, Emanuel (fl. 1720-1767)
A New and Correct Map of Africa. Drawn from the most Approved Modern Maps and Charts, and adjusted by Astronomical Observations representing also the course of the Trade Winds, Monsoons &c. (London, E. Bowen, 1748.)
Map, 37 x 45 cm, uncoloured.
Prime meridian through London.
Top right: Vol. 1. Page 717.

This map of Africa was drawn for Harris's *Collection of Voyages* (1748), identified by the volume and page number. It is interesting because it indicates the various winds, and is almost entirely free of ornament except for a rococo cartouche at the lower left. In the southern part Hellenbok, Drakenstein and the Dutch fort appear as well as a river which is said to have no end. The Hottentot tribes have moved further north and Monomotapa still appears. Zimbaoe and the Fort of Zimbas are both marked and gold and silver mines appear in Zimbabwe. The larger lakes have gone and large rivers are not as prolific as in other maps of the period.

The Nile appears to arise from Lake Dambea, taking a markedly curved course before joining other tributaries.

Emanuel Bowen, map and print seller and engraver to George II of England and Louis XV of France, published English county maps from various addresses in London. He also made maps for Harris's *Collection of Voyages* and many other world atlases, which all contained maps of Africa and southern Africa. A feature of his maps was his engraved paragraphs of text which were inserted in various parts of the maps. After his death in poor circumstances he was succeeded by his son Thomas.

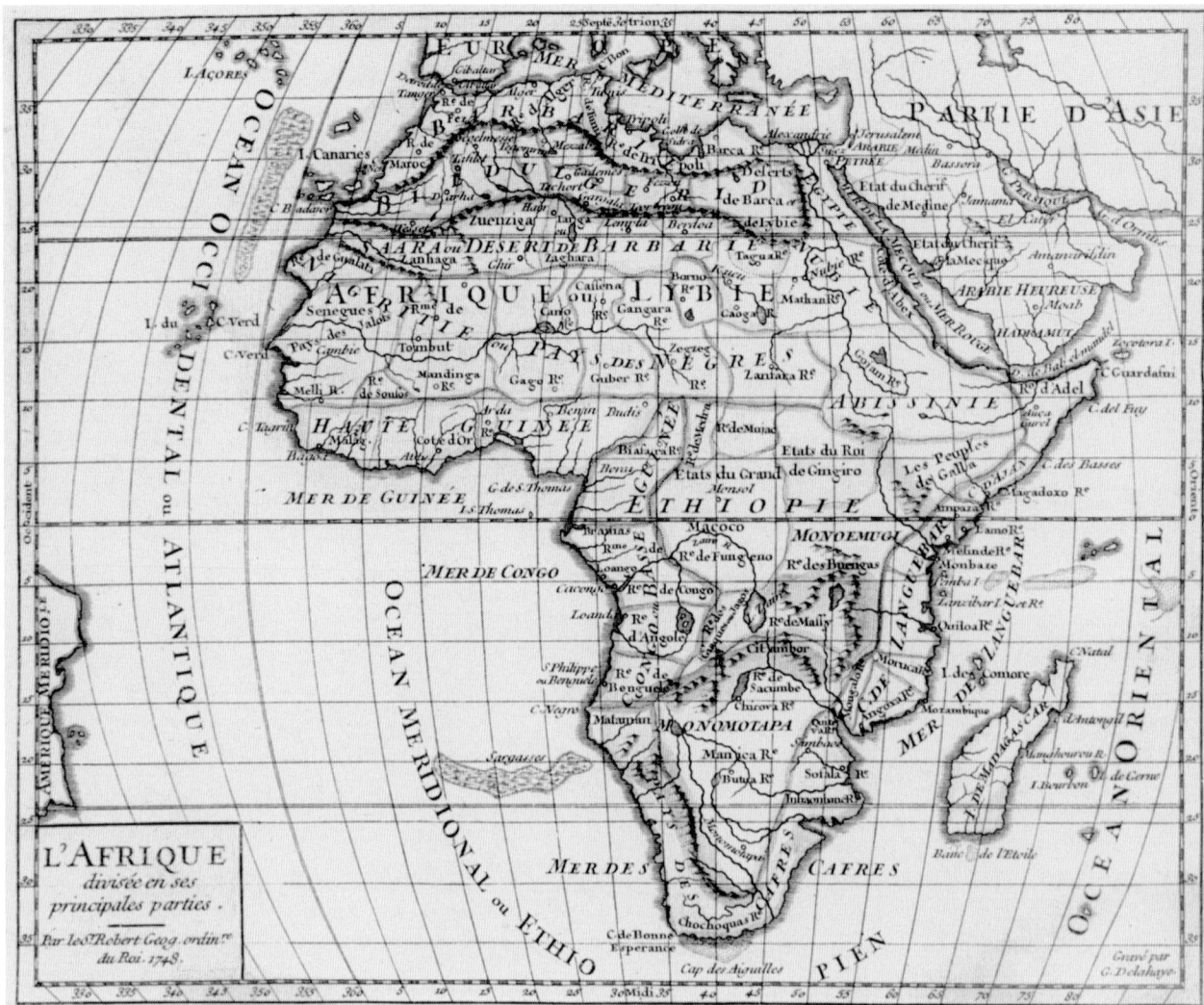

Map 90

ROBERT DE VAUGONDY, the family of
L'Afrique divisée en ses principales parties. (Africa divided into its principal
parts.) Engraved by G. Delahaye. (Paris, Robert de Vaugondy, 1748.)
Map, 15,5 x 19,5 cm, coloured.
Top right: 171.

This small map of Africa comes from a French family of cartographers. The father, Gilles Robert de Vaugondy, was born in Paris in 1688 and died there in 1766. He succeeded his uncle Pierre Moulard Sanson and was thus able to utilize the accumulated records of the Sanson family. His son, Didier Robert de Vaugondy (1723-1786), assisted his father and in 1757 they were joint publishers of an *Atlas Universel*, of which 601 copies were printed on large paper and 517 on smaller paper, the atlas taking fifteen years to produce. They were eminent geographers, and their maps were finely engraved

by such well-known engravers as Haussard and Delahaye. Both father and son were appointed geographers to the King.

This map, although medium in size, provides a moderate number of place names throughout the interior, engraved by Delahaye. The cartouche at the left lower corner is a simple box, dated 1748, indicating that it appeared at the time of their *Atlas Portatif*.

106

Map 91   ANVILLE, Jean Baptiste Bourguignon d' (1697-1782)
Afrique publicée sous les auspices de Monseigneur le Duc d'Orléans Primier Prince du Sang. (Africa published under the patronage of the Duke of Orleans.)
Engraved by Guillaume Nicolas Delahaye. (Paris, J.B. d'Anville, aux Galeries du Louvre, 1749.)
Map, 97 x 97 cm, slightly coloured.
Scale in French leagues, sea leagues, arabic miles, Giaman arabian sea measure, travelling hours for caravan.
Prime meridian through Ferro Island.
Inset: Azores.

107

This large wall map is of two long double-folding sheets, joined and linen-backed. The top right corner has a large oval emblematic title piece surmounted by a seated Black woman, adorned with an elephant head, with a fish in her right hand and flower on her left shoulder. At her sides are a lion, a camel and an ostrich. The French title appears below her and below the title a river-god, a crocodile and pyramids. Below the cartouche the map is described as having been published under the patronage of the Duke of Orleans.

Apart from some detail in the Congo and a small area on the east coast, practically the whole of southern Africa is left blank. At the Cape, the Dutch Fort, Stellenbo, Drakensteen, Waveren and Hout Bay are depicted as well as various native tribes. The captions are in Dutch, French and Portuguese. The Nile arises from a range of mountains above which are two lakes smaller than normally seen in the Ptolemaic concept, from which two tributaries join to form the Nile.

Jean Baptiste d'Anville was born in Paris and spent his adult life studying geography, becoming the foremost exponent in Europe. His ambition was to record original work based on astronomical observation and to avoid unverified material. This led him to leave large blank spaces on his maps that had been filled, by others, with fictitious and hearsay material. His maps were delicately engraved and his attention to detail was renowned.

He became Royal Geographer and Cartographer at the age of twenty-two and a member of the Académie des Sciences in 1773. He produced a large number of maps under the title of *Atlas Général*, mainly folding ones, and as new discoveries appeared, the maps were accordingly revised. His work was largely copied. He amassed a large personal collection of maps of great value which were acquired by Louis XVI and housed in the Bibliothèque Nationale.

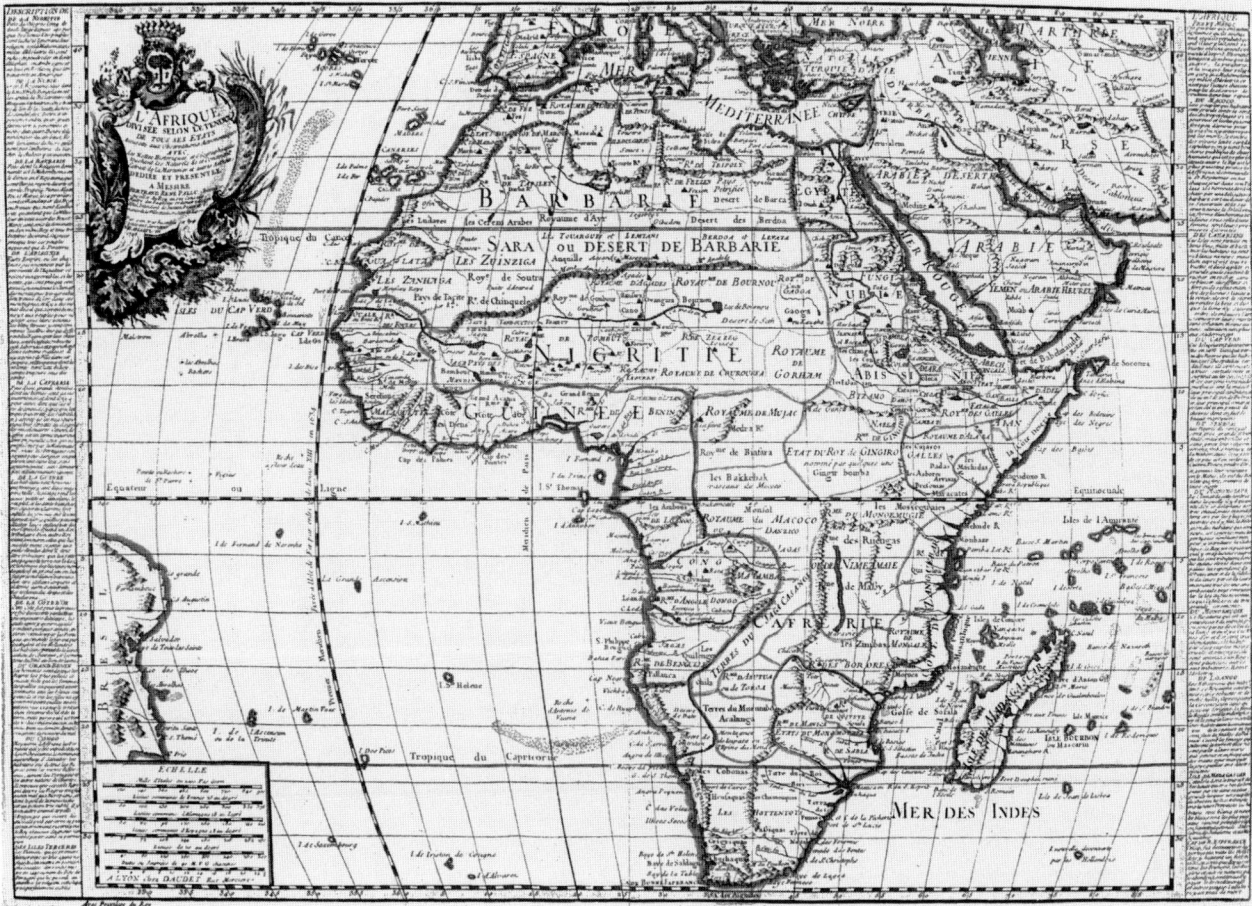

Map 92

BAILLEUL, Gaspard (1703-1781)
L'Afrique Divisée Selon le Tendue de Tous Ses Etats Assujetti aux Observations Astronomiques avec des Nottes Historiques et Geographiques touchant les Naturels de ce Continent. (Africa divided into regions according to astronomical observations...) (Lyon, Daudet, 1752.)
Map, 47 x 61 cm, slightly coloured.
Scale in Italian miles, French, Spanish and German leagues.
Prime meridian through Ferro Island.
Dedication: *Dediée et presentée a Messire Bertrand René Pallu*
Borders consist of columns of historical and geographical descriptions.

This decorative French map of Africa includes the Mediterranean, Turkey, Arabia and the Middle East, with a finely engraved floral and armorial dedicatory cartouche at the top left. The prominent bulge of Brazil is included, as with so many maps of Africa of the eighteenth century. The borders contain descriptions of many parts of Africa from the Cape of Good Hope to the northern state of Morocco, together with many islands small and large encompassing the continent of Africa. The commonly depicted 'two lakes side by side' origin of the Nile does not appear here.

Gaspard Bailleul was a Parisian engineer and geographer who published a number of atlases. Daudet was a map seller of Lyons, especially known for his connection with Bailleul.

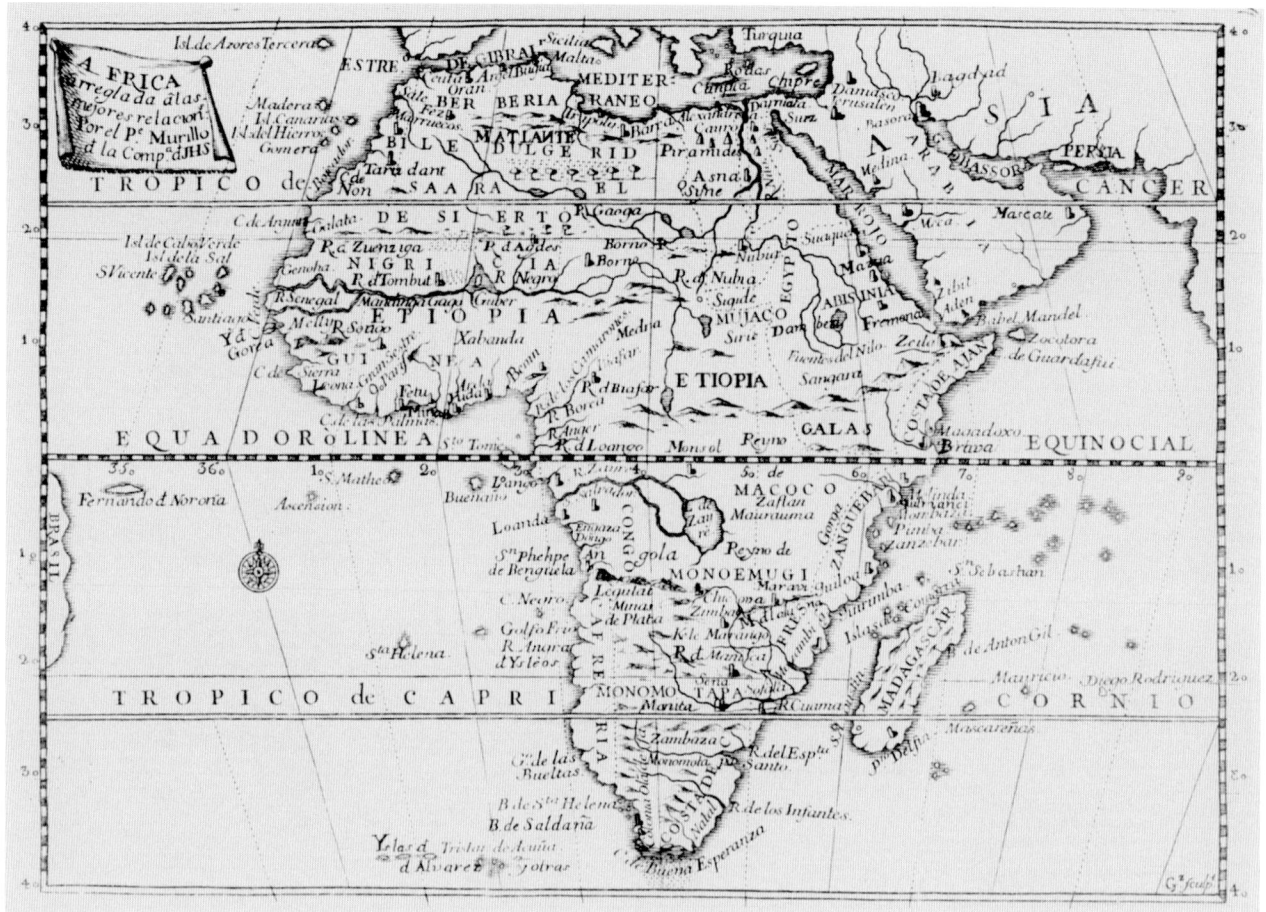

Map 93

MURILLO, Velarde Pedro de (1696-1753)
Africa arreglada a las mejores relacion[e]s... (Africa agreeable to the best descriptions). Engraved by GZ. (Spain, P. Murillo, 1752.)
Map, 16 x 23 cm, uncoloured.

A miniature map of Africa by the Spanish Jesuit cartographer Murillo, from his *Geographia Historica Africa Arreglada a las Mejores Relaciones* (1752). The title appears in the top left corner in a draped cartouche. The shape of the south is unusual, with a bulge on the western coast. There is limited geographical detail on the map, especially in the south, where a few bays, capes and rivers are mentioned. Monomotapa is featured and Caffraria is misplaced on the western coast. A small portion of the bulge of Brazil is noted.

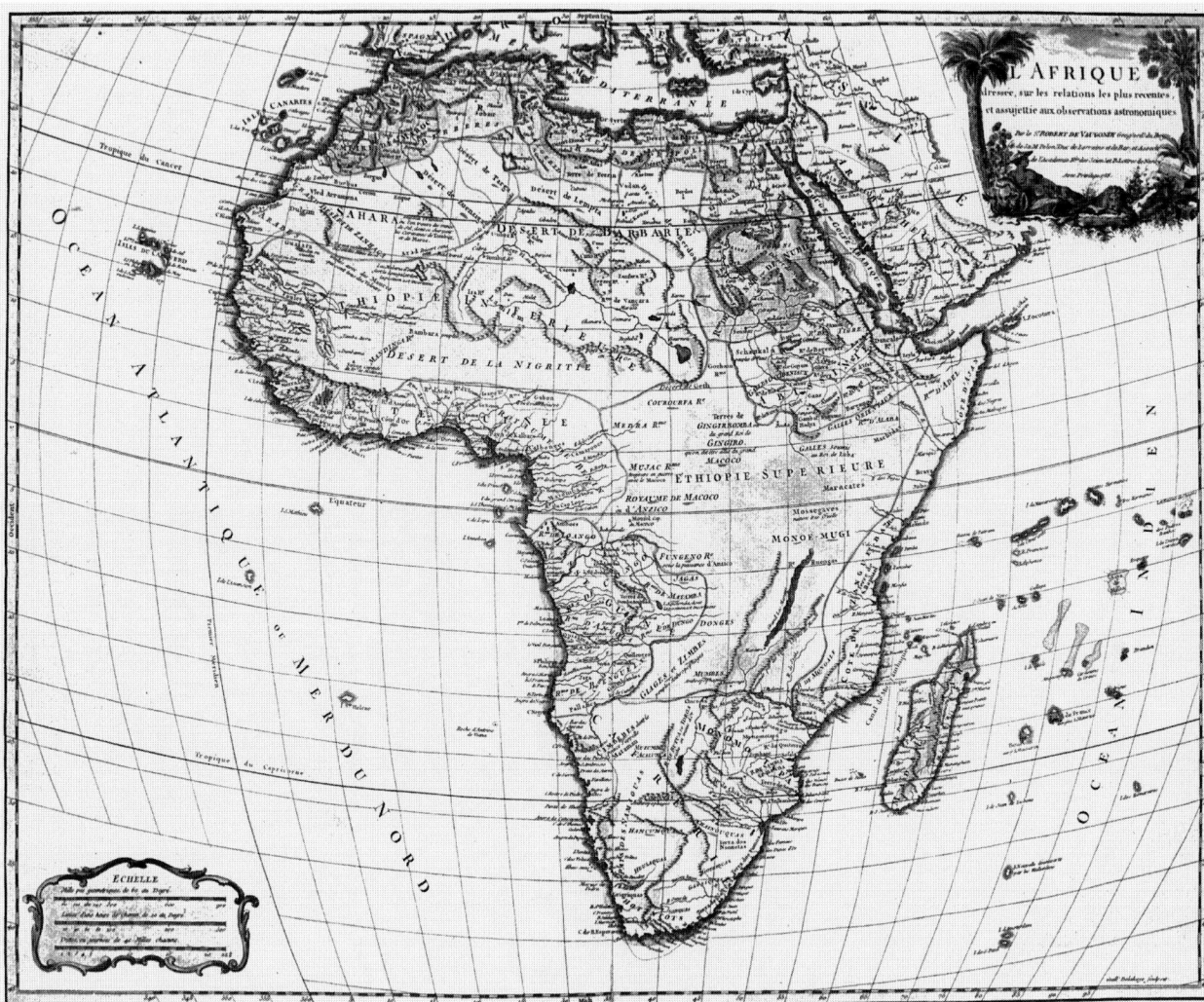

Map 94

ROBERT de VAUGONDY, Gilles (1688-1766) and Didier (1723-1786)
L'Afrique dressée sur les relations le plus recentes, et assujettie aux observations astronomiques par... Géog. du Roy, de Sa M. Polon, Due de Lorraine et de Bar et Associé de l'Academie Royale des Sciences et B. Lettres de Nancy, avec privilege. (Africa drawn according to recent reports.) Engraved by Guill. Delahaye. (Paris, Robert de Vaugondy, 1756.)
Map, 46 x 59 cm, coloured.
Scale in travelling leagues.
Prime meridian through Ferro Island.

The cartouche of this map of Africa of 1756 consists of a somewhat romantic scene depicting a native woman reclining between two palms. In the lower right-hand corner is marked 'Guill Delahaye Sculpsit.' The interior of southern Africa has very little other than tribal names, but Monomotapa is still included and the lakes and rivers appear quite accurately.

Gilles Robert de Vaugondy of Paris succeeded his uncle, Pierre Moulard Sanson, and was thus able to exploit the large volume of accumulated records of the Sanson family. He published a small quarto *Atlas Portatif* in 1748 and his son

Didier assisted him in the joint publication in 1757 of an *Atlas Universel* (in which this map probably appeared), of which 601 copies were printed on large paper and 517 on small paper, the atlas taking 15 years to produce. The maps of Gilles Robert de Vaugondy and his son were finely engraved, mainly by Haussard and the Delahays, and they form fine examples of eighteenth–century French engraving. The first series of maps of Africa, a folio edition, was issued in 1749 and subsequently reprinted and reissued from 1749 to 1778, now including southern Africa.

111

Map 95

LENGLET-DUFRESNOY, Nicolas (1674-1755)
Africa Dividida Em Suas Rigioes Principais Estados. (Africa divided into its principal states.) (Coimbra, Joseph da Costa, 1757.)
Map, 14,5 x 17,5 cm, uncoloured.
Scale in leagues.

Abbé Nicolas Lenglet-Dufresnoy was a French diplomat and geographer who published three books on geography in 1716, 1764 and 1768. This map appeared in *Descripcao da Terra* (1757), a small geographical volume, published in Coimbra, Portugal.

This miniature map of the African continent depicts the regions and states as they were in 1762, when this map was published. The book included seven folded maps: of the world, of Europe, Asia, North America, South America, Portugal and Africa. This map is simple in content but quite accurate in shape, with a single title in a box in the upper right corner. The geographical details are sparse, especially in the south, where mention is made only of Caffraria and Monomotapa. The surrounding islands are placed in their correct positions in relation to the mainland, with only one island of St. Helena.

112

Map 96

TIRION, Isaak (1739–1769)

Kaart van Afrika door den Heer d'Anville... (Map of Africa after d'Anville.)

(Amsterdam, I. Tirion, 1763.)

Map, 35 x 35 cm, coloured.

Scale in French, Italian, Arabian and sea miles, days journey per caravan.

Prime meridian through Ferro Island.

Inset: Vlaamsche Eilanden.

Isaak Tirion was an Amsterdam bookseller. This particular map of Africa appears in his *Hedendaagsche Historie of Tegenwoordige Staat van Afrika...* published in Amsterdam in 1763.

The name of the map appears in Dutch, in a ruled box in the right-hand top corner and the scale is also surrounded by miles, in the lower left-hand corner. An inset appears in the top left-hand corner titled: 'Vlaamsche Eilanden.' As stated in the title, the map is based on d'Anville.

Stellenbosch, Drakenstein and Waveren are indicated as well as the rivers Berg, Olyphantin, Breede and Riv. Zonde End in the southern part of Africa. Zimbabwe is marked

and also 'Groote Waterval.' As the latter is somewhere in the vicinity of Victoria Falls it is possible that information about these falls was made known at this early date. Gold and silver mines are also indicated but further north than in actual fact. Large tracts of country are left vacant as they had not yet been explored – rather than have them filled with inaccurate, fictitious geography. The Nile is shown according to Paez, as well as the older version of the Ptolemaic tradition, with the two lakes and the Mountains of the Moon. This map also appears in Tirion's *Nieuwe en Beknopte Land Atlas*, first published in Amsterdam between 1730 and 1740.

113

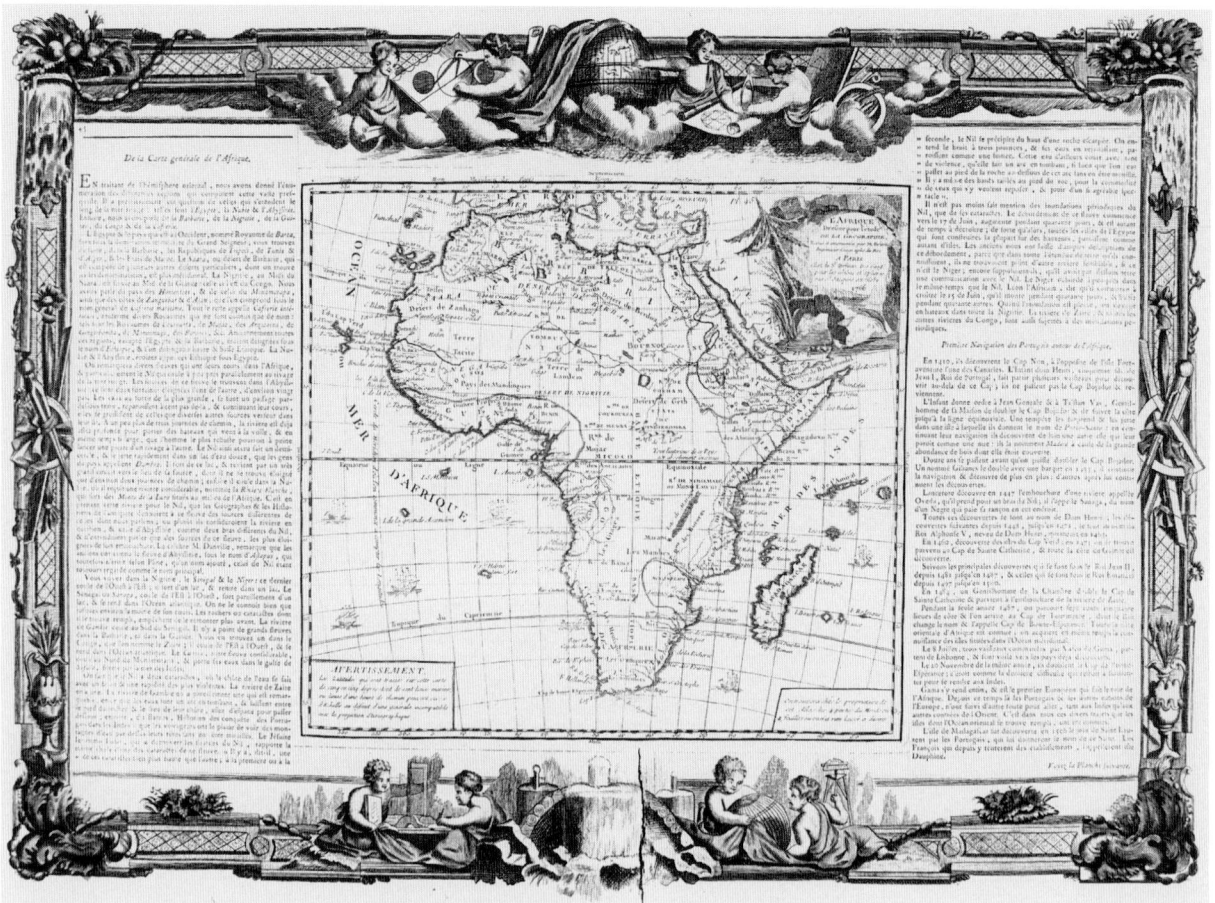

Map 97

BRION DE LA TOUR, Louis (fl. 1757-1800)
L'Afrique Dressée pour l'étude de la Géographie. (Paris, L.C. Desnos, rue St. Jacques, au Globe, 1766.)
Map, 23 x 26 cm, slightly coloured.
Prime meridian through Ferro Island.
French text appears within decorative borders.

A map of Africa within a highly decorative, originally coloured printed frame, by Brion and Desnos, both French publishers of maps – the former Geographer to the King of France and the latter to the King of Denmark. Brion de la Tour published many atlases and this map may have originally appeared in his *Atlas Général* of 1766.

Within the map area itself at the top right corner is a decorative cartouche of trees and the title surrounded by what appear to be mountains and water leading to a cataract with a cavern exit. The interior of this map contains little geographical material. Large rivers are evident but the larger lakes are not marked. The Nile appears once again to arise from two small lakes lying closely adjacent to each other. In the south there is a minimum of place names and rivers, with sparse evidence of the native tribes.

On each side border is an extensive text with a detailed description of Africa, its states, rivers and cataracts, and added to this there is a detailed description, with dates, of the early Portuguese discoveries, particularly on the west coast. The highly decorative top and lower borders consist of ornamentation, fruit and foliage, dominated by four cherubs using trigonometrical instruments, designing figures and admiring globes – on one of which (at mid-top) appear a number of animals being pointed to by one of the cherubs. The side borders are also attractively decorated with vases and instruments.

114

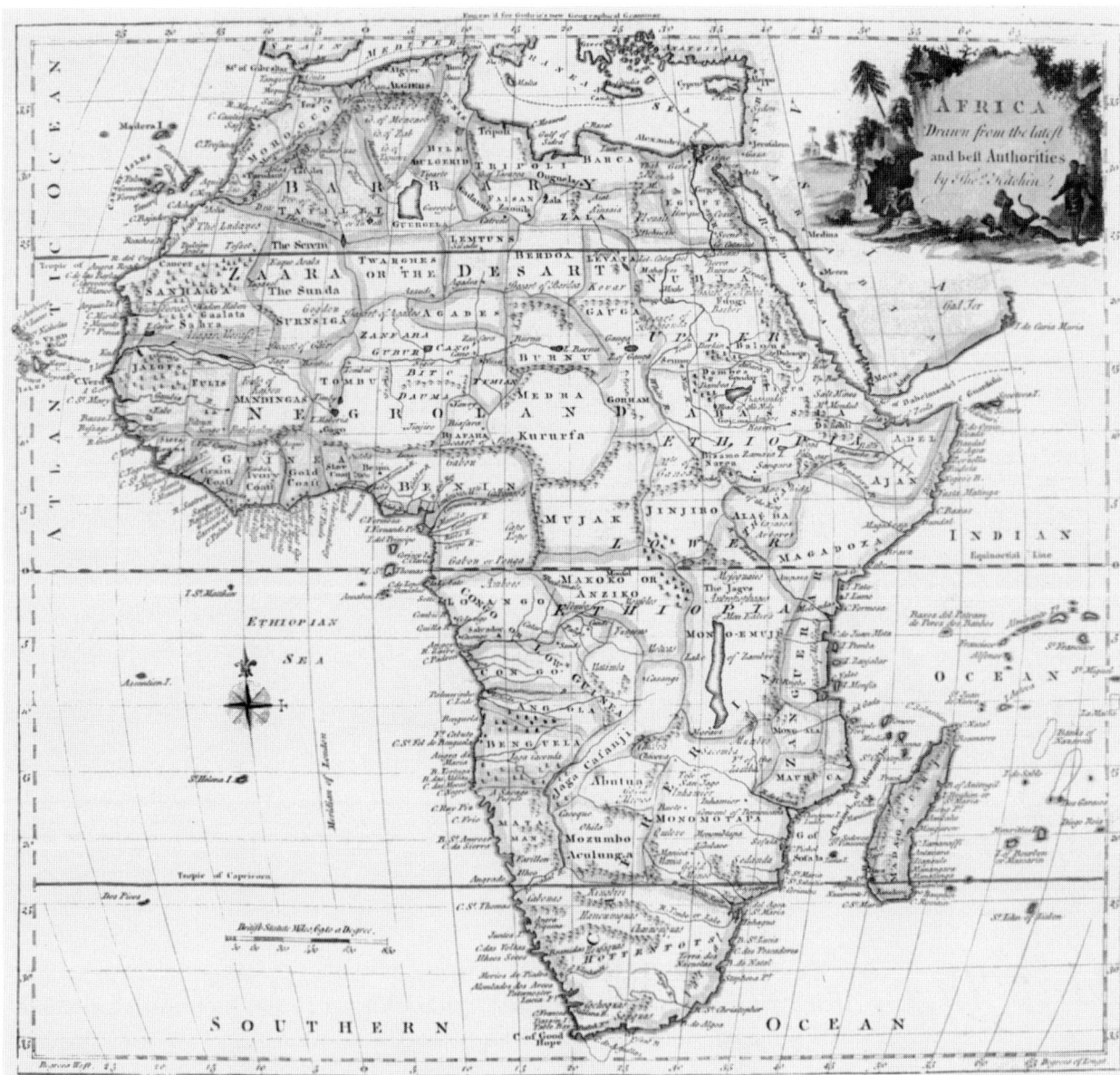

Map 98

KITCHIN, Thomas (1718-1784)
Africa Drawn from the latest and best Authorities. (London, T. Kitchin, 1770.)
Map, 34 x 37 cm, coloured.
Scale in British statute miles.
Prime meridian through London.
Title across the top: *Engraved for Guthrie's new Geographical Grammar.*

This map of Africa was drawn and engraved by Thomas Kitchin (Kitchen) for the Scottish geographer William Guthrie's *New Geographical Grammar*, 1770 with subsequent editions, and also appears in an atlas to Guthrie's *Systems* (1785 and 1820).

A colourful cartouche features in the upper right corner showing the title on a squarish rock-face surrounded by shrubs and small trees. At its lower border appear two individuals, one standing with a child. The other sits, carrying a quiver of arrows on her back and a bow in her hand. A frontal view of what seems to be a lion appears on the left.

The map shows various well-defined boundaries of the southern tribal countries. The Dutch Ft. (fort) is the only indication of a settlement at the Cape. Some capes and bays are marked. Monomotapa appears as a city and as a kingdom, with the legendary gold mines appearing at its southern boundary well south of the Fort of Zimbas and the present-day Zimbabwe. In this map the origin of the Nile begins to get away from the Ptolemaic concept of the two-lake system. The 'head of the Nile' is depicted as starting from an unnamed lake and then curving to enter Lake Dambea, whence it assumes a circuitous route northwards.

115

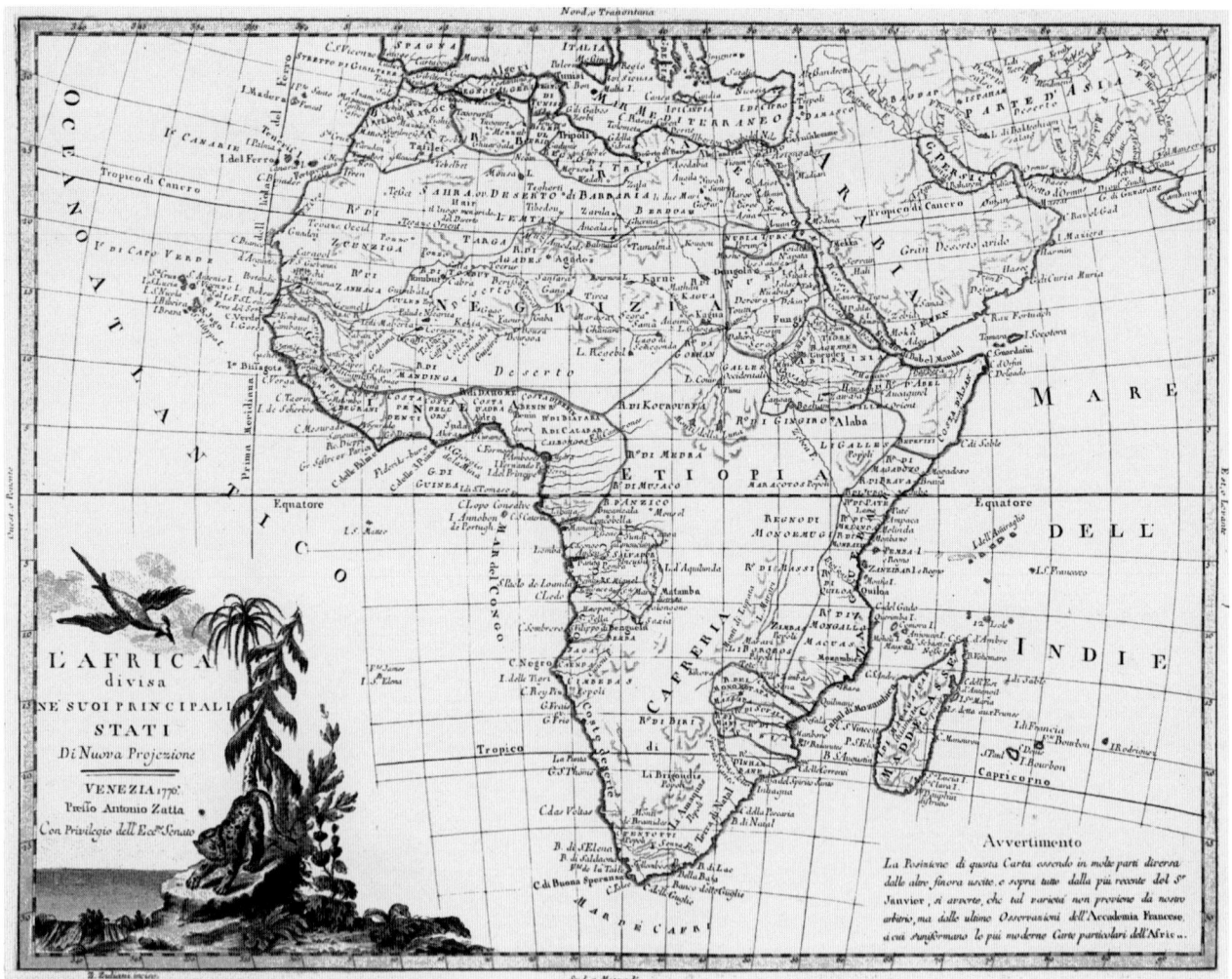

Map 99

ZATTA, Antonio (fl. 1757-1797)
L'Africa divisa ne'suoi Principali Stati Di Nuova Projezione. (Africa divided into its principal states.) Engraved by Z. Zuliani and Ab. V. Formaleoni. (Venice, A. Zatta, 1776.)
Map, 30 x 39 cm, coloured outline.
Prime meridian through Ferro Island.

The cartouche consists of a cataract over which a pheasant-like bird is flying. On the side of the river is a palm, another tree and some foliage amongst which a leopard can be seen. According to the 'Avvertimento' in the lower right corner this map is based on that of Janvier (see Map 104), and the same information about southern Africa can be found on the two maps. This map is fairly accurate geographically and unknown parts of the country are marked as such instead of being dotted with fictitious places. Stellenbosch is the only inland place shown in South Africa.

Antonio Zatta was a Venetian publisher who issued his *Atlante Novissiomo* of four volumes (in which this map appeared) in Venice from 1775-1785.

116

Map 100

BONNE, Rigobert (1727-1795)
L'Afrique divisée en ses principaux états (Africa divided into its principal states.) Engraved by Jean Lattré. (Paris, J. Lattré, 1778.)
Map, 70 x 98 cm, coloured.
Scale in Turkish and Arabic miles, hours march by caravan, French leagues and sea leagues.
Prime meridian through Ferro Island.
Inset: Isle Bourbon and Isle de France.

A large map of Africa issued by the French cartographer, Bonne, an engineer and cartographer, Hydrographer to the King, who published many atlases – among which were the *Atlas Maritime* (1762), *Atlas Portatif* (1781) and *Atlas Encyclopédique* in two volumes (1787-1788).

The map is accurately drawn with information reflecting the known geography of the east and west coasts at that time. Southern Africa is almost devoid of detail which was, however, known to other publishers. Some tribal names are mentioned. There are two insets – one of Isle Bourbon and the other of Isle de France, on the lower right border. The

cartouche, at the lower left, is particularly attractive. A large rectangular block of masonry gives the French title, with trees behind and, to the left and covering the rest of the border, attractively drawn shrubs. In the foreground, reclining on a supine lion, are a person in oriental turban and dress, holding an umbrella, and a child. Beyond the left border of the title piece and partly hidden by it is a large elephant. At the bottom of the map are five scales – an unusual number – giving Turkish and Arabic miles, hours march by caravan, French and sea leagues.

117

Map 101

PROBST, Johann Michael (d. 1809)
Africa. (Augsberg, J.M. Probst, 1778.)
Map, 16 x 24 cm, coloured.
Scale in German miles.
Prime meridian through Tenerif.
German description of Africa on right.

This small map of Africa has a decorative cartouche in the lower left corner showing a seated native wearing a headdress and holding a bow. To his left is a crocodile and on his right are two objects which may be quivers, and a long-stemmed pipe. The shape of the mainland is strange, with a large inlet in the middle of the Congo and a protruding part of the east coast in the vicinity of Mozambique. The south is rather broad with a flat extensive tip. There is a legend panel on the right of the map consisting of a description of Africa in German.

Johann Michael Probst was a publisher and engraver of Augsberg, who worked with Seutter and Lotter.

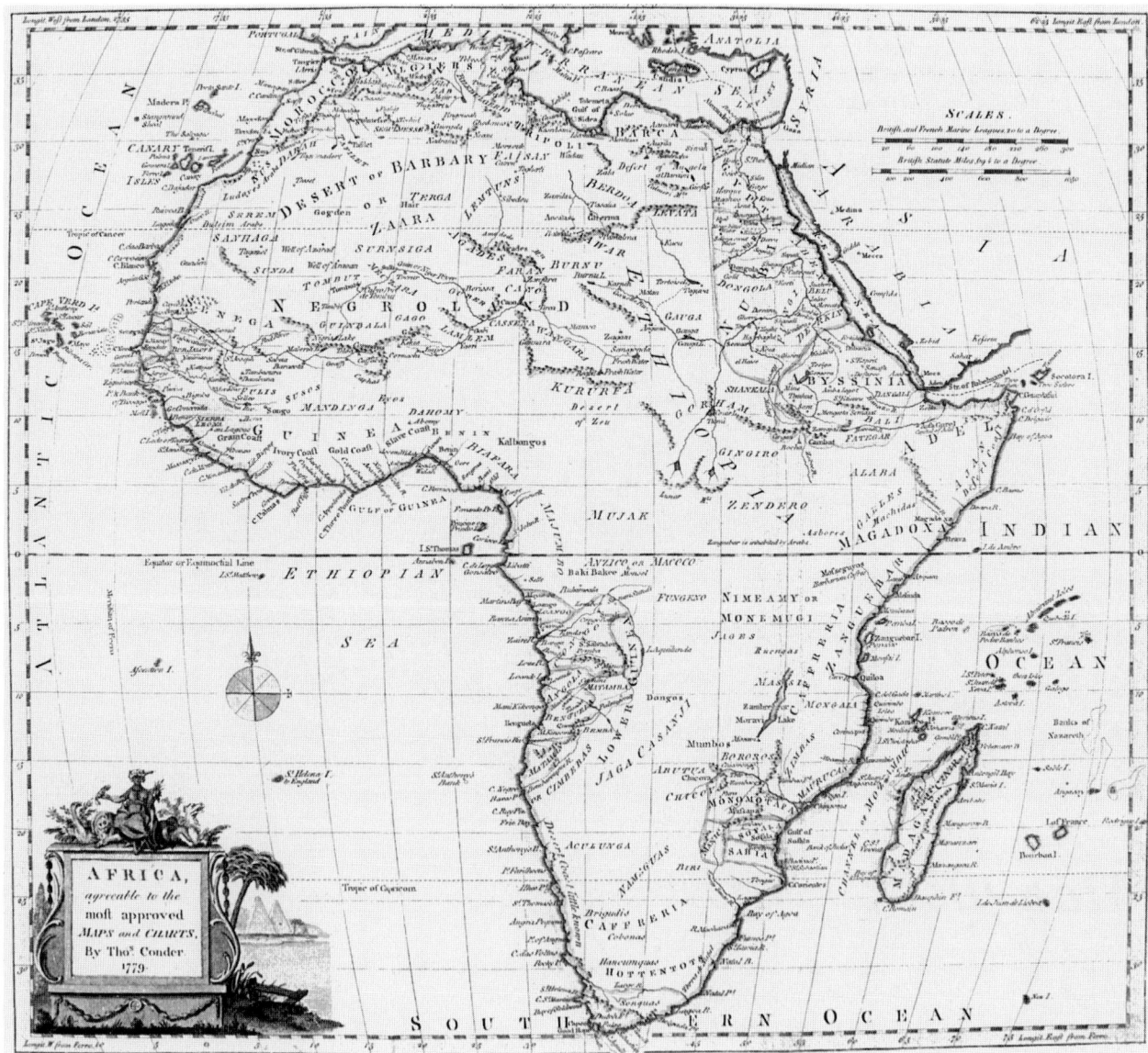

Map 102

CONDER, Thomas (fl. 1775-1801)
Africa, agreeable to the most approved maps and charts. (London, T. Conder, 1779.)
Map, 33 x 37 cm, coloured outline.
Scale in British and French marine leagues, British statute miles.
Prime meridian through London and Ferro Island.
Title across the top: *Engraved for Moore's New and Compleat Collection of Voyages and Travels.*

There is a cartouche at the lower left corner, square in outline, surrounded by floral scrollwork, trees and foliage and surmounted by a seated native in costume flanked on his right by a lion and an unidentifiable bird, and on his left by a camel. In the background on the right is a seashore on which two pyramids stand.

The interior of the map depicts the geography known at that time, with very little in the central and southern parts. The west coast bears evidence of the Portuguese settlements, especially in the Congo. The Nile arises from two adjacent small lakes with tributaries joining in from all sides.

This map of Africa with its surrounding islands was published by Thomas Conder in 1779. He was an engraver and cartographer who issued a number of atlases and travel books, among which were Moore's *New and Compleat Collection of Voyages and Travels* of 1778. He also engraved a large folding map of southern Africa which is illustrated in Lt. Paterson's *Narrative of Four Journeys . . .* (1777 and 1778), depicting Paterson's route in that country.

119

Map 103

BONNE, Rigobert (1727-1795)
Afrique. Engraved by André R. Bonne (1782).
Map, 21 x 32 cm, coloured.
Scale in Portuguese and French leagues and sea leagues.
Prime meridian through Ferro Island and Paris.
Top left: Liv. I,III, et XI.
Top right: No. 31.

This small map of Africa provides the minimum of cartographical detail, especially of southern Africa. Some bays and the ancient kingdom of Monomotapa are named with a few rivers. The only native tribe to be mentioned is the Hottentots. Arrows in the oceans probably indicate the monsoons because those depicted at the top right are named by months.

Bonne, a French engraver and cartographer, was associated particularly with the publication of the Atlas Maritime, 1762, *Atlas Portatif* (1781) and the *Atlas Encyclopédique* of two volumes, 1787-1788.

120

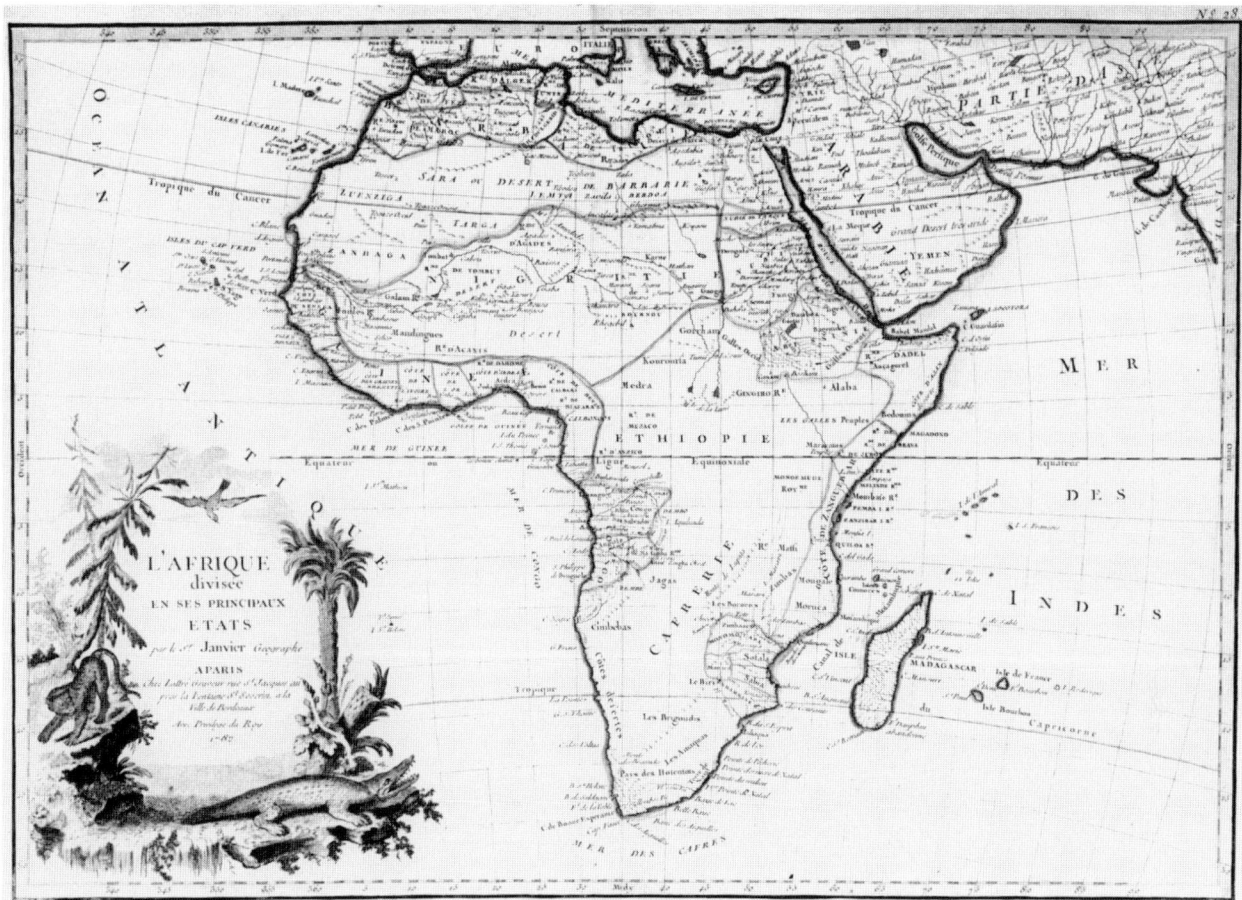

Map 104

JANVIER, Jean (fl. latter 18th C)
L'Afrique divisée en ses principaux Etats. (Africa divided into its principal parts.) Engraved by Jean Lattré. (Paris, J. Janvier, 1782.)
Map, 30 x 43,5 cm, coloured outline.
Top right: No. 28.

Janvier, a French geographer who worked in Rue St. Jacques, Paris, in the second half of the eighteenth century, produced a series of maps in 1760 and collaborated in an *Atlas Moderne* in 1762 and 1771. This map of Africa of 1782 was issued in a similar edition in 1762, and in subsequent editions, in addition to two Italian issues in 1776 and 1784, the latter by Remondin.

The cartouche in the lower left corner is made up of pines and other trees under which a leopard-like animal is stalking.

Above the title a bird is shown in flight. In the right foreground is a crocodile. 'C. de Bonne Esperance' and 'Stellembos Pic' are the only settlements marked in southern Africa. A few Hottentot tribes are indicated. 'Cafrerie' is noted as a territory but further north than the area later called Kaffraria. This map is numbered 28 and probably comes from the *Atlas Moderne*.

121

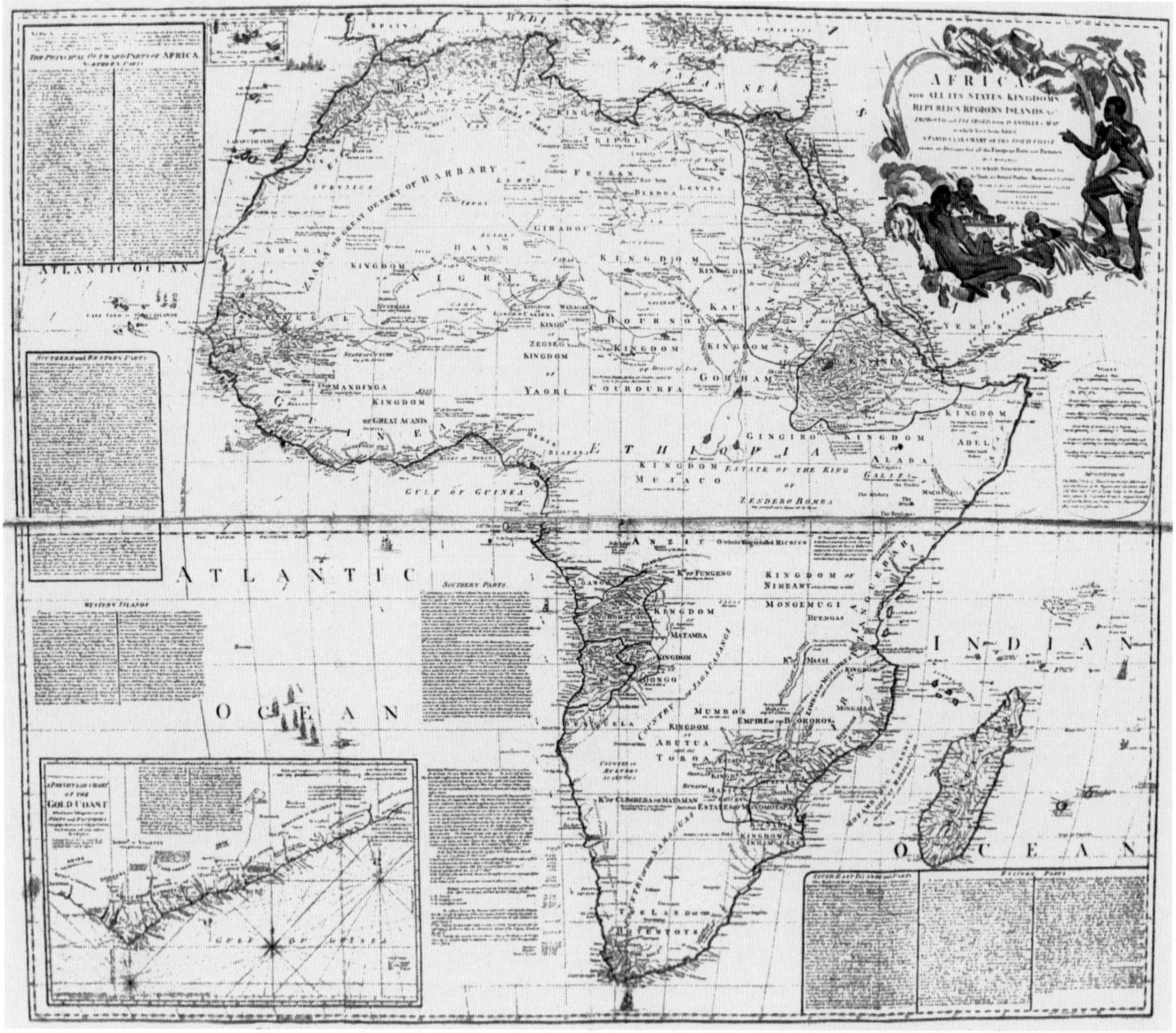

Map 105

BOULTON, S. (fl. late 18th C)
Africa With All Its States, Kingdoms, Republics, Regions, Islands, etc, Improved and Inlarged from d'Anville's Map; to which have been Added A Particular Chart of the Gold Coast, wherein are Distinguished all the European Forts and Factories. (London, Robert Sayer, Fleet Street, 1787.)
Map, 2 sheets each 51 x 123 cm, coloured outline.
Scale in English miles; French land leagues, Arabian miles; Giaman Arabian sea measure; travelling hours for caravan; British and French sea leagues.
Prime meridian through Ferro Island.
Inset: Azores; *A Particular Chart of the Gold Coast; A summary description relative to the trade and natural produce, manners and customs of the African continent and islands.*
Sheets numbered 26 and 27.

122

This large two-sheet map of Africa was originally published by J.B.B. d'Anville (1697-1782), the French geographer and cartographer. He became very important as a cartographer in Europe when he reformed unconfirmed printed cartography and rejected plagiarism, which led him to leave blank spaces on maps where before they had been filled with figments of imagination and hearsay evidence. In his maps of Africa and southern Africa, the earliest of which appeared in 1727, it is interesting to note that apart from some detail in the Congo, the upper eastern coast and the Cape Colony, southern Africa is left blank. This two-sheet map of the whole of Africa was first issued by Boulton under d'Anville's name in 1772 and reissued in 1787. A new title was designed for this English edition. A large inset of the Gold Coast is sited in the lower left corner. A smaller one appears on top of the Azores. The sea is engraved with legends, including an account of the Hottentot woman and a long account of the experiences and ways of the Dutch East India Company at the Cape.

Geographically the map is practically the same as the d'Anville's map of 1749 (see Map 91). The cartouche shows a native man and woman reclining on seats covered with skins, with a child looking on holding a bird. On the other side a native stands and points to the wording of the cartouche.

In the southern extremity the Dutch Fort, Stellenbosch and Drakenstein are indicated. It is interesting to note the legend 'Hot Waters' in the vicinity of the present–day Caledon with its hot springs. Gold and silver mines are shown in what is now Zimbabwe.

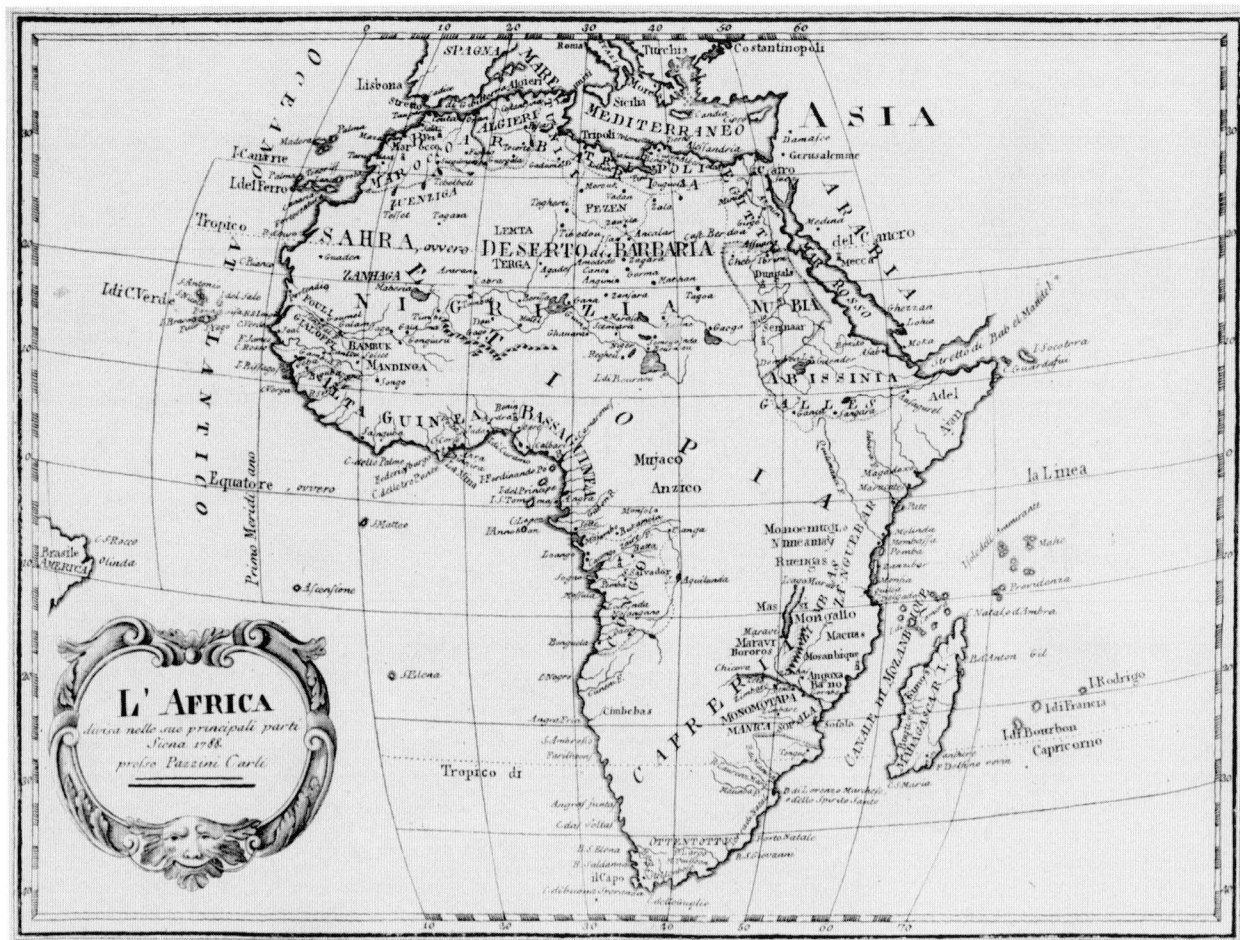

Map 106 CARLI, Pazzini (fl. late 18th C)
L'Africa divisa nelle sue principali parti. (Africa divided into its principal parts.)
(Siena, P. Carli, 1788.)
Map, 22,5 x 31 cm, coloured outline.
Prime meridian through Ferro Island.

This small Italian map of Africa is by Pazzini Carli, an eighteenth-century publisher in Siena.

The map presents the barest details geographically, and it may have been used for educational purposes at school. The cartouche at the lower left corner is a colourful wreath of flowers with a face at the lower centre. The Cape of Good Hope is noted, and also the early settlement of Stellenbosch.

124

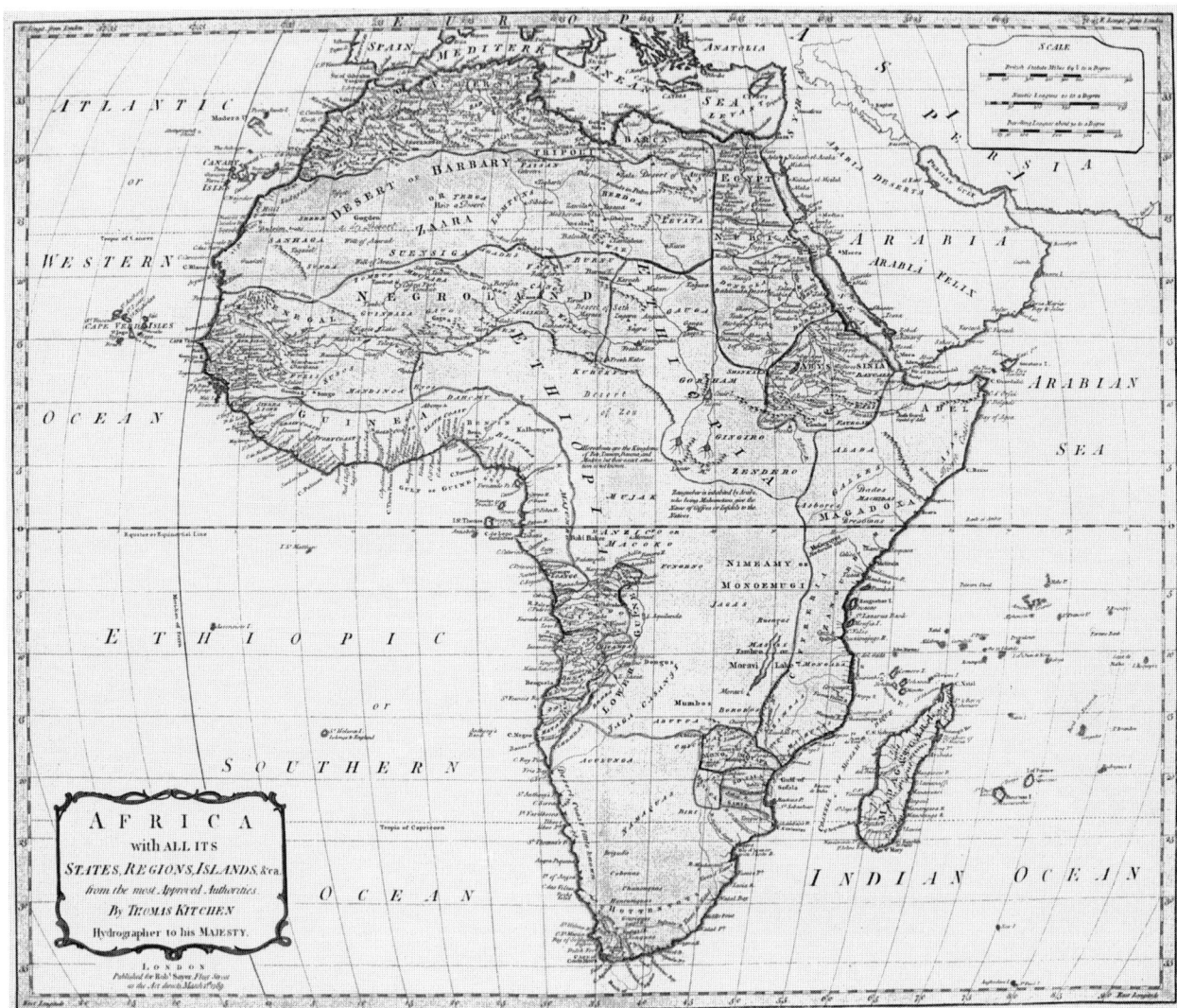

Map 107   KITCHEN, Thomas (1718-1784)
Africa with All Its States, Regions, Islands &ca. from the most Approved
Authorities. (London, Robert Sayer, Fleet Street, 1789.)
Map, 44 x 53 cm, coloured.
Scale in British statute miles, nautical leagues and travelling leagues.
Prime meridian through Ferro Island.

Many issues of this map were published, varying in size and with some slight geographic differences. One edition appeared in William Guthrie's *Grammar of Several Kingdoms of the World*, published in 1785. Another issue was published in 1787 in which 'gold mines' appear at the site of the present day Witwatersrand.

The map gives only limited detail in Central Africa. Various boundaries of the southern regions are shown and the Dutch Fort is noted as an indication of a settlement at the Cape. Certain native names are mentioned, and the southern coastline has some names of bays and capes, such as Mossel B., and rivers referred to as the Elephant and the Infante (Great Fish River of today). Monomotapa is shown

as a state and a city. The Nile and its origin is inaccurately depicted, which is understandable as the famous explorers Speke and Burton et al. had not yet arrived on the scene. Zimboae is marked in the east, and west of it is a range of mountains with a place called Fura, probably an indication of gold having been found there.

Thomas Kitchen (Kitchen) was an engraver and publisher as well as being Hydrographer to the King. He worked mainly in London, in Charing Cross and Holburn Hill. His prolific output was published mostly by Bowen and Dalrymple. This particular edition was published by Robert Sayer, the well-known publisher of Fleet Street.

125

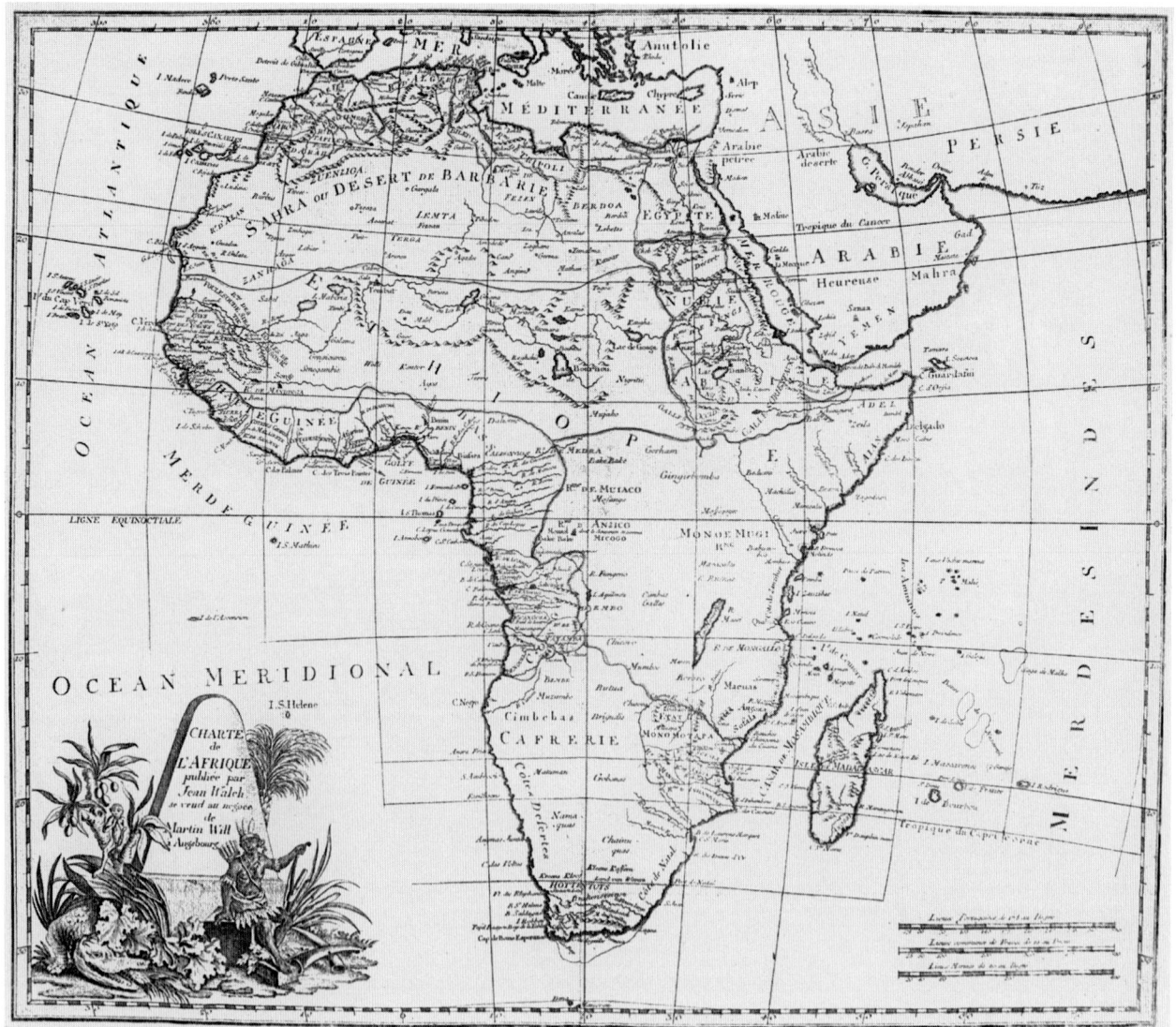

Map 108

WALCH, Jean (fl. late 18th C)
Charte de L'Afrique. (Map of Africa.) Engraved by Martin Will. (Augsberg,
J. Walch, 1790.)
Map, 46 x 55 cm, slightly coloured.
Scale in Portuguese and French leagues, sea leagues.

A French map of Africa published by Jean Walch of Augsberg in 1790. The outline is fairly accurate with geographical details limited to regions known at that time, especially in the northern and southern extremities of the continent. Very few rivers are noted and the Nile appears to arise in a single river from Lac Dembe in Abyssinia. The ancient states of Monomotapa and Sofala are still illustrated. The cartouche appears at the lower left corner and includes an African with a headdress of feathers. A monkey and a crocodile are on the left among a tree and other foliage, and in the right foreground are two tusks.

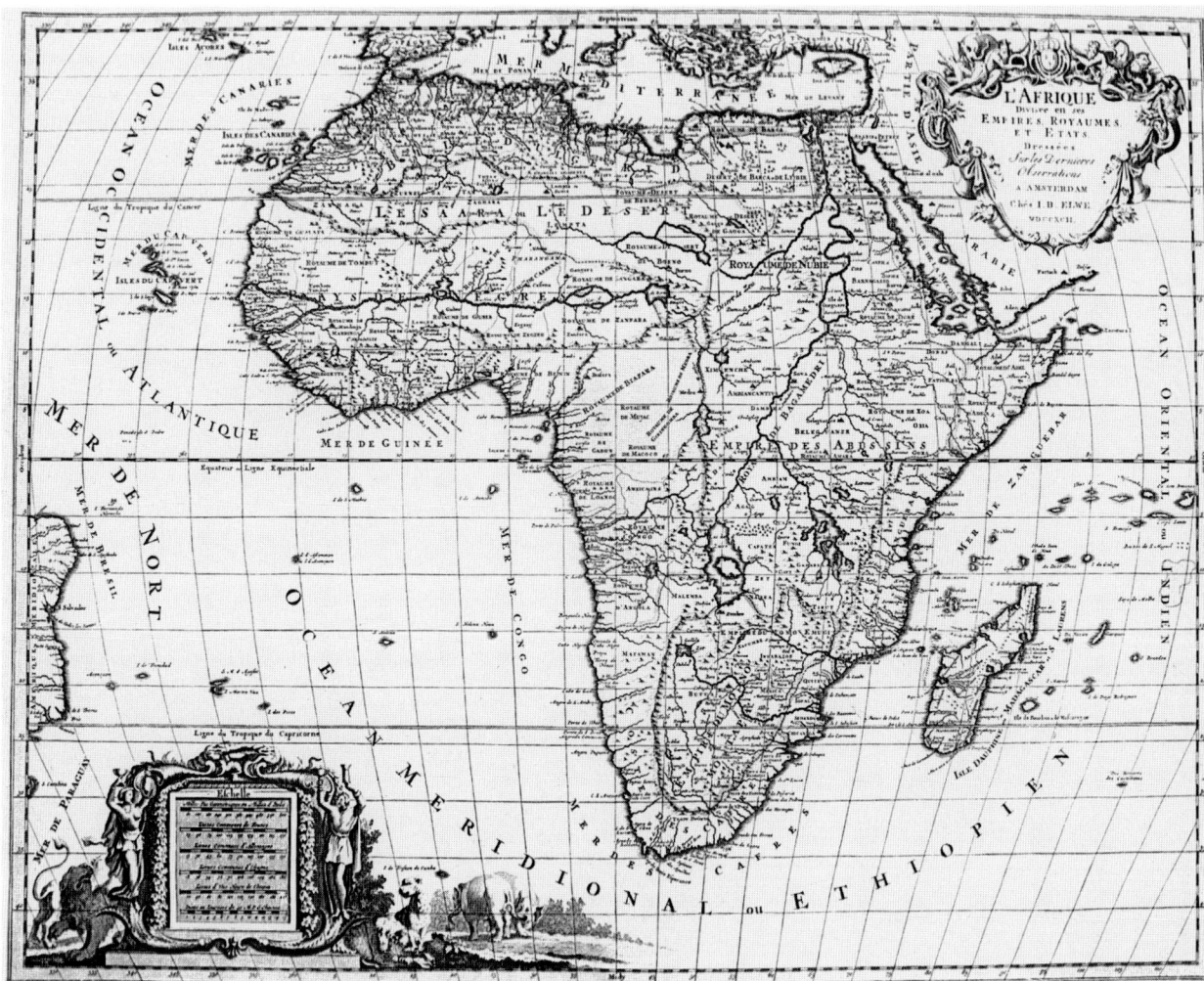

Map 109

ELWE, Jan Barend (fl. late 18th C.)
L'Afrique Divisée en ses Empires, Royaumes et Etats Dressees Sur Les Dernières Observations. (Africa divided into empires, kingdoms and states...) (Amsterdam, J.B. Elwe, 1792.)

Map, 45 x 59 cm, coloured.
Scale in Italian miles, French, German and Spanish leagues.
Prime meridian through Ferro Island.

This map of Africa is identical to that of A.H. Jaillot of 1674 (see Map 46), though the decorative scales have been embellished. The title within the top right cartouche has been altered so that the map is attributed to J.B. Elwe, a publisher of Amsterdam. Jaillot's map of Africa appears to have been popular, having been published (in addition to Elwe) by Covens and Mortier, William Berry and R. and J. Ottens at varying times.

Map 110   PAYNE, John Willet (1752-1803)
Africa, 1792. Engraved by Ferguson. (1792.)
Map, 33,5 x 36 cm, slightly coloured.
Scale in miles.
Prime meridian through London.
Along the bottom: *Engraved for Jackson's Edition of Payne's new System of Universal Geography.*

This map appeared in 1792 and was engraved for Jackson's edition of *Payne's New System of Universal Geography.* The shape of the continent appears reasonably accurate but reflects very little European penetration into the interior. In the south reference is made to Hottentot tribal names, the Dutch Fort and 'A Savage People'. Lake Moravi (Malawi) assumes a more elongated appearance and two rivers in the south are marked R. of Elephants and Large R. The origin of the Nile is referred to as the 'Head of the Nile,' arising in Lake Dambea.

128

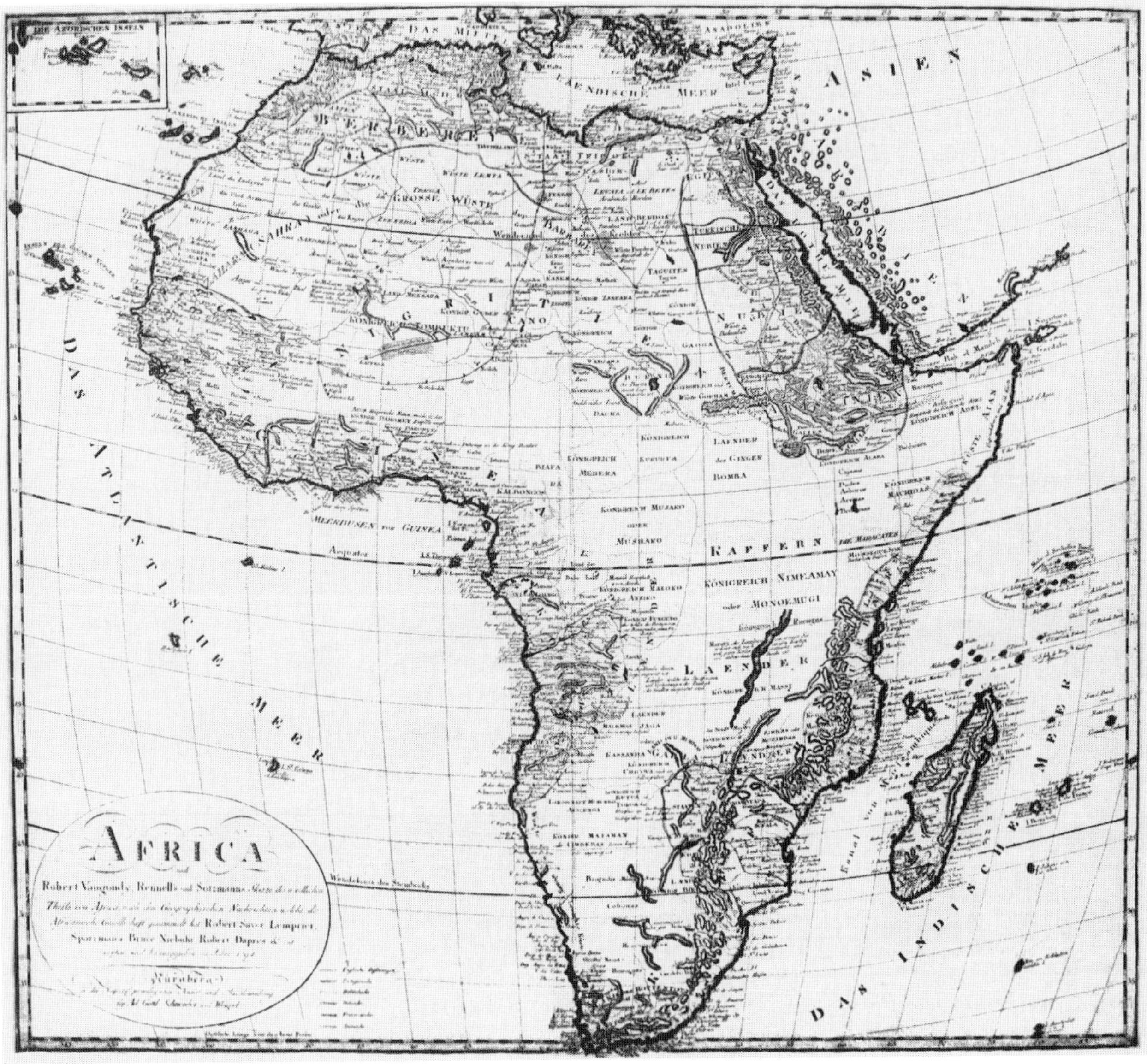

Map 111
SCHNEIDER, Adam Gottlieb (1745-1815) and WEIGEL, Johann Christoph (d. 1746)

Africa nach Robert Vaugondy, Rennell's und Solzmann's Skizze des nordlichen Theils von Africa... (Africa according to...) (Nürnberg, Schneider and Weigel, 1794.)
Map 52 x 58,5 cm, slightly coloured.
Prime meridian through Ferro Island.
Inset: *Die Azorischen Inseln.*

This map shows a mass of accurate geographic detail in its southern parts, inland as well as on the coast. Places such as Stellenbos[ch] and Drakenstein are marked. The 'Kau Boeke Veld,' the 'Varme Bocke Veld,' 'Langen Cloof' and 'Cango' are some of the names that have not appeared before. The colour index next to the cartouche indicates the power responsible for particular parts of Africa, marked English, Portuguese, etc., but the parts of the map so coloured do not seem to be historically accurate – the Cape, for instance, has never been Danish. The map comes from

the *Atlas der Geographie von der bekannten ganzen welt . .* (Nürnberg, A.G. Schneider and Weigel, 1794-1805).

This map of Africa comes from the joint publishing firm of Schneider and Weigel. Schneider was a cartographer-publisher of Nürnberg and, working with Weigel the younger, founded a Nürnberg publishing house in 1746, working on maps of the western hemisphere and the Netherlands. Weigel the younger had published three atlases before joining Schneider.

129

Map 112   GÜSSEFELD, Franz Ludwig (1744-1807)
Charte von Africa nach astronomischen Beobachtungen, auch alten und neuen
Nachrichten, ingleichen den Charten von Sayer, Rennel, Arrowsmit... (Map of
Africa according to astronomical observations and information found on the maps
of Sayer, Rennel and Arrowsmith.) (Nürnberg, Homannischen Erben, Homann Heirs),
1797.
Map, 46 x 15,5 cm, coloured.
Scale in German, French, English and Portuguese miles, nautical miles.
Prime meridian through Ferro Island.

A late eighteenth-century German map of Africa. The
inscription is in the lower left corner in a box and below it is
a note in German stating that accurate geographical
knowledge of the interior will only be available after further
exploration. No inland settlements are named. In addition
to the usual rivers in southern Africa the Orange (named
Grosse) and the Keureboom are shown. Many of the larger
rivers of Central and North Africa are omitted and what does
appear is quite inaccurate. The author refers to information
obtained from other cartographers but the name Arrowsmith
is out of place, as his map of Africa was first published in
1802. Güssefeld may, however, be referring to Arrowsmith's
world map of 1790. Güssefeld worked for the Homann Heirs
in Brandenbourg in 1773.

130

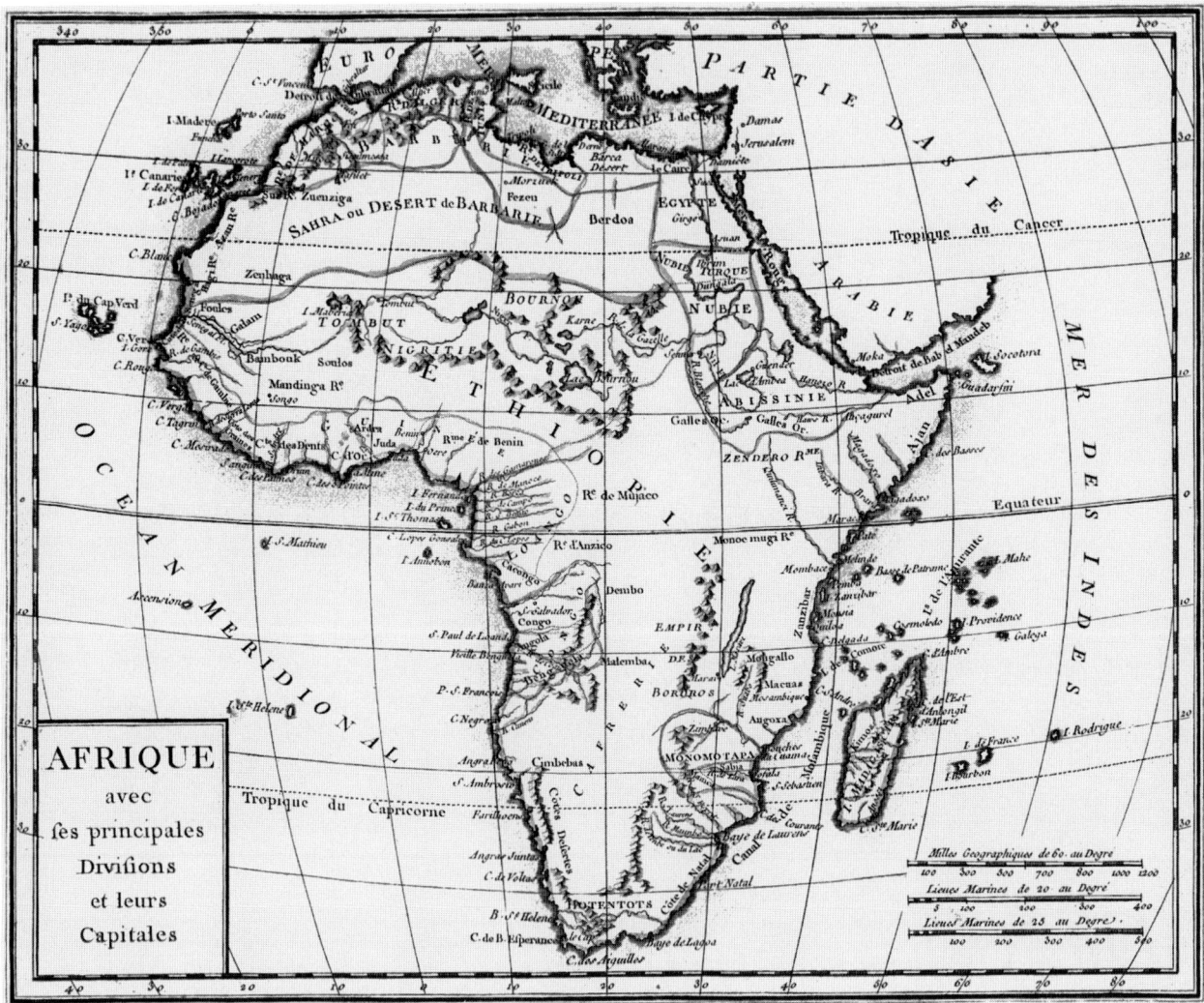

Map 113 ANON
Afrique avec ses principales Divisions et leurs Capitales. (Africa with its principal regions and their capitals.)
Map, 22,5 x 28 cm, slightly coloured.
Scale in geographic miles, sea leagues.
Prime meridian through Ferro Island and Paris.
Top right: 4.

A French map, maker unidentified, of the continent of Africa. As indicated in the simple box title in the lower left corner it shows the principal divisions of Africa and their capitals. The details on this map are sparse, with very little information in the southern part except for mention of the Cape of Good Hope, Cape Agulhas, the coast of Natal and Delagoa Bay, which is placed too far south. Mention is also made of the Kingdom of Monomotapa. This map, numbered 4 (in the top right corner), probably appeared in an atlas in the early or mid-nineteenth century.

131

Map 114
BONNE, Rigobert (1727-1795)
Mappe Monde, Sur Un Plan Horisontal. Situé à 45d de latitude Nord.
Hémisphère Oriental. (World map, 45° plane, eastern hemisphere.)
Map, 23 x 23 cm, uncoloured.
Prime meridian through Ferro Island.
Inset: *Effet de la pesanteur* (effect of gravity).
       Sphère oblique (oblique globe).

This French world map of the eastern hemisphere, with the Arctic Circle, includes almost the whole of Africa except for the southern tip. There is practically no geographical detail of southern Africa except for a few names. The central and northern parts are also very sparsely labelled with geographic information.

The origin of this map has been difficult to trace but reference to the Library of the British Museum has indicated that it is most probably from an atlas by Bonne.

132

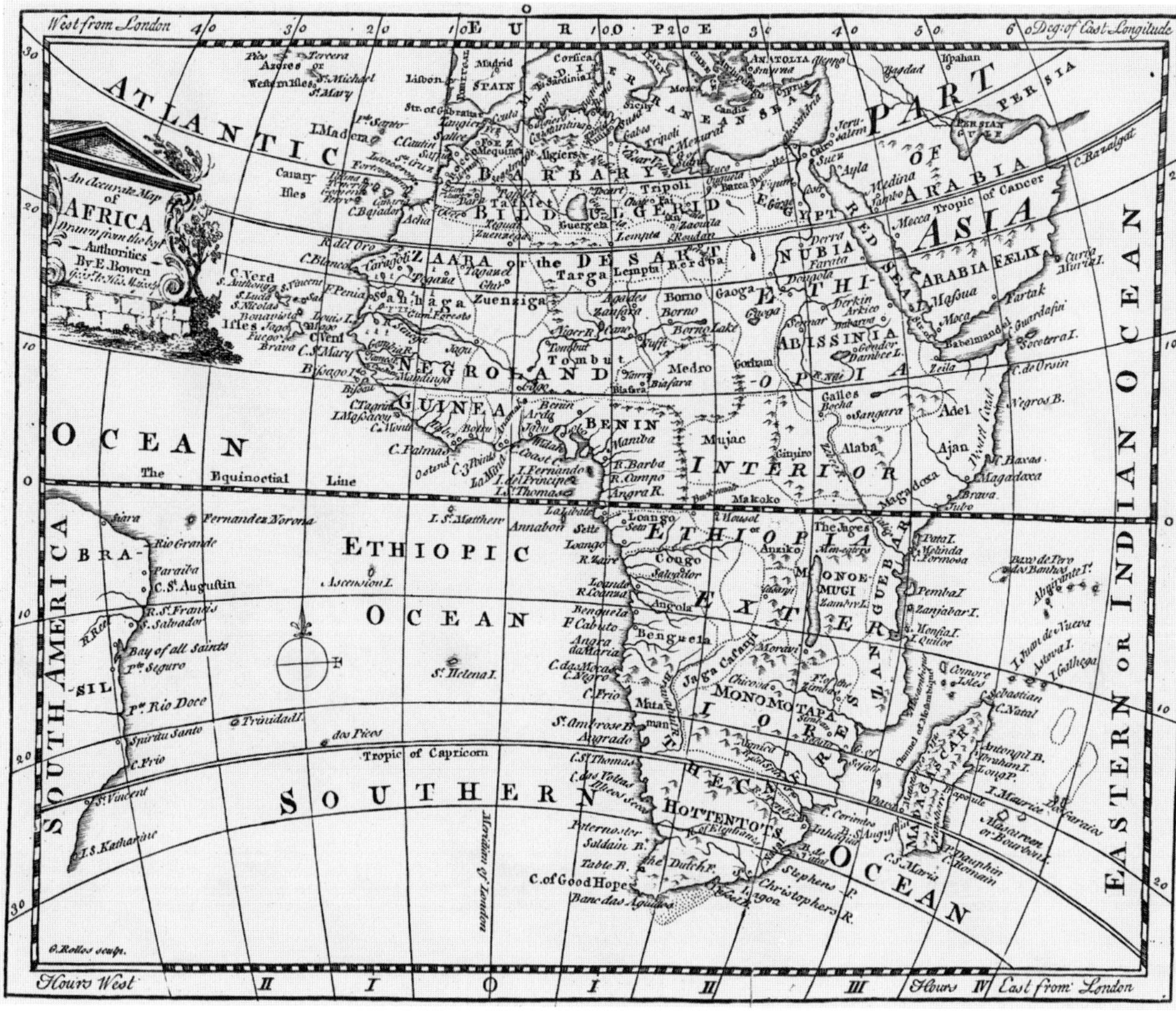

Map 115 BOWEN, Emanuel (fl.1720-1767)
An Accurate Map of Africa Drawn from the best Authorities. Engraved by G. Rollos. (London, E. Bowen.)
Map, 18 x 22 cm, uncoloured.
Prime meridian through London.

This map appears under the name of Emanuel Bowen, father of Thomas, both of whom printed and published maps and atlases in London. The engraver was G. Rollos, an engraver and map seller who flourished from 1754 to 1789, though his work is not featured on the other maps of Africa published either by Emanuel or Thomas.

This map illustrates a portion of the bulge of Brazil, and gives the Cape of Good Hope an unusually elongated and westerly tip. The title cartouche appears at the top left border in the form of an architectural pediment surrounded by flowers and resting on a base of bricks. The geographical information is scanty, especially of southern Africa, with mention, as in other Bowen maps, of 'Stephens P. and Christophers R.' below each other. 'R. of Elephants' is noted, and 'Hottentots.'

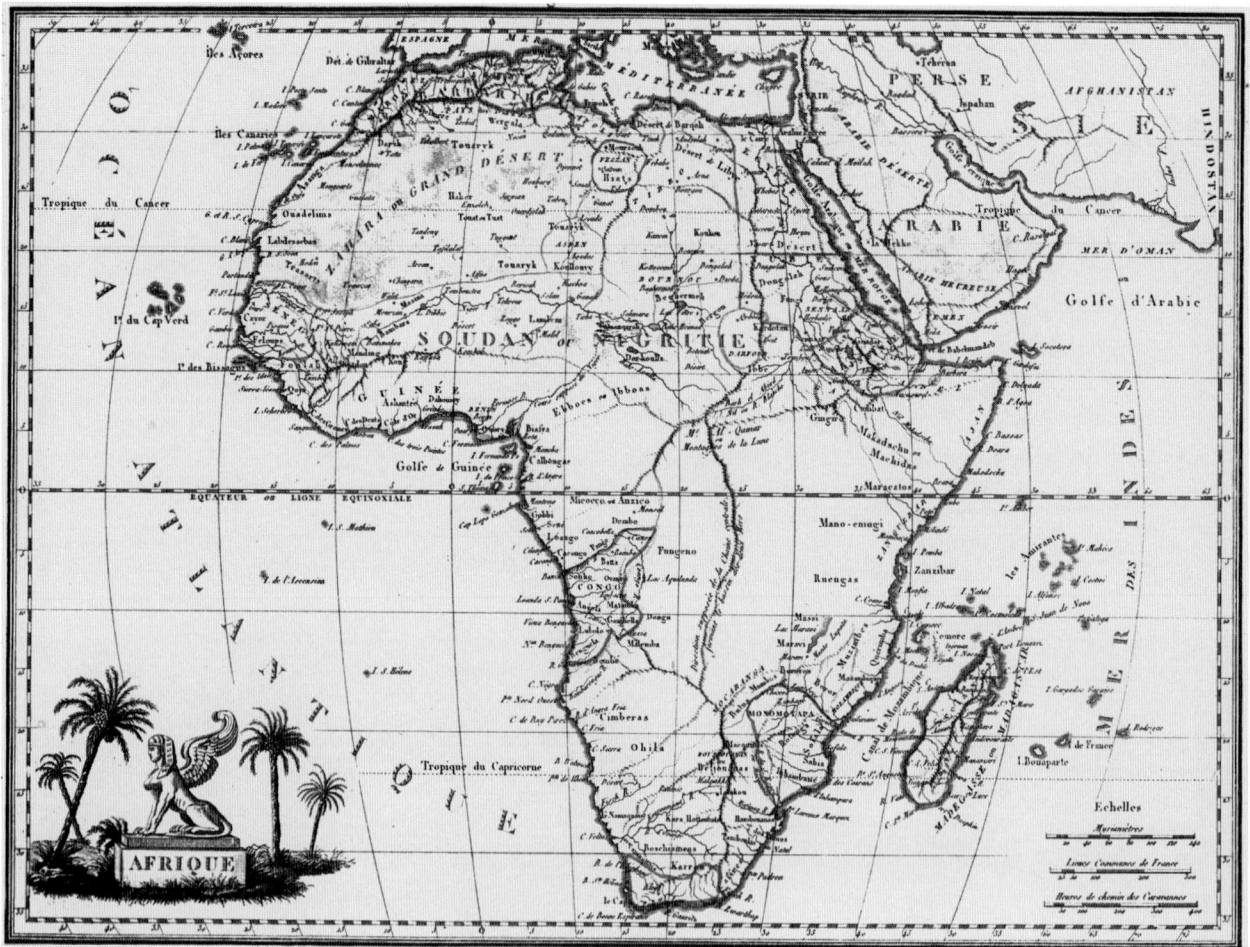

Map 116   CHAMOUIN, Jean-Baptiste-Marie (b. 1768)
Afrique. Engraved by Chamouin and Giraldon. (Paris.)
Map, 22 x 30 cm, coloured outline.
Scale in 10,000 metres, French leagues, caravan days.
Prime meridian through Paris.

A small French map of Africa engraved by Giraldon, and Jean-Baptiste-Marie Chamouin of Paris. Chamouin himself engraved for Lapie, another French geographer.

The cartouche at the lower left corner shows a sphinx-like animal sitting on a rectangular masonry block with the title 'Afrique' on it. The scene is surrounded by a number of palm trees. This is a moderately accurate map although somewhat sparse, with a number of tribal names. The surrounding islands appear to be fairly accurately placed. Rivers are plentiful but inaccurate, as in so many maps of this period.

134

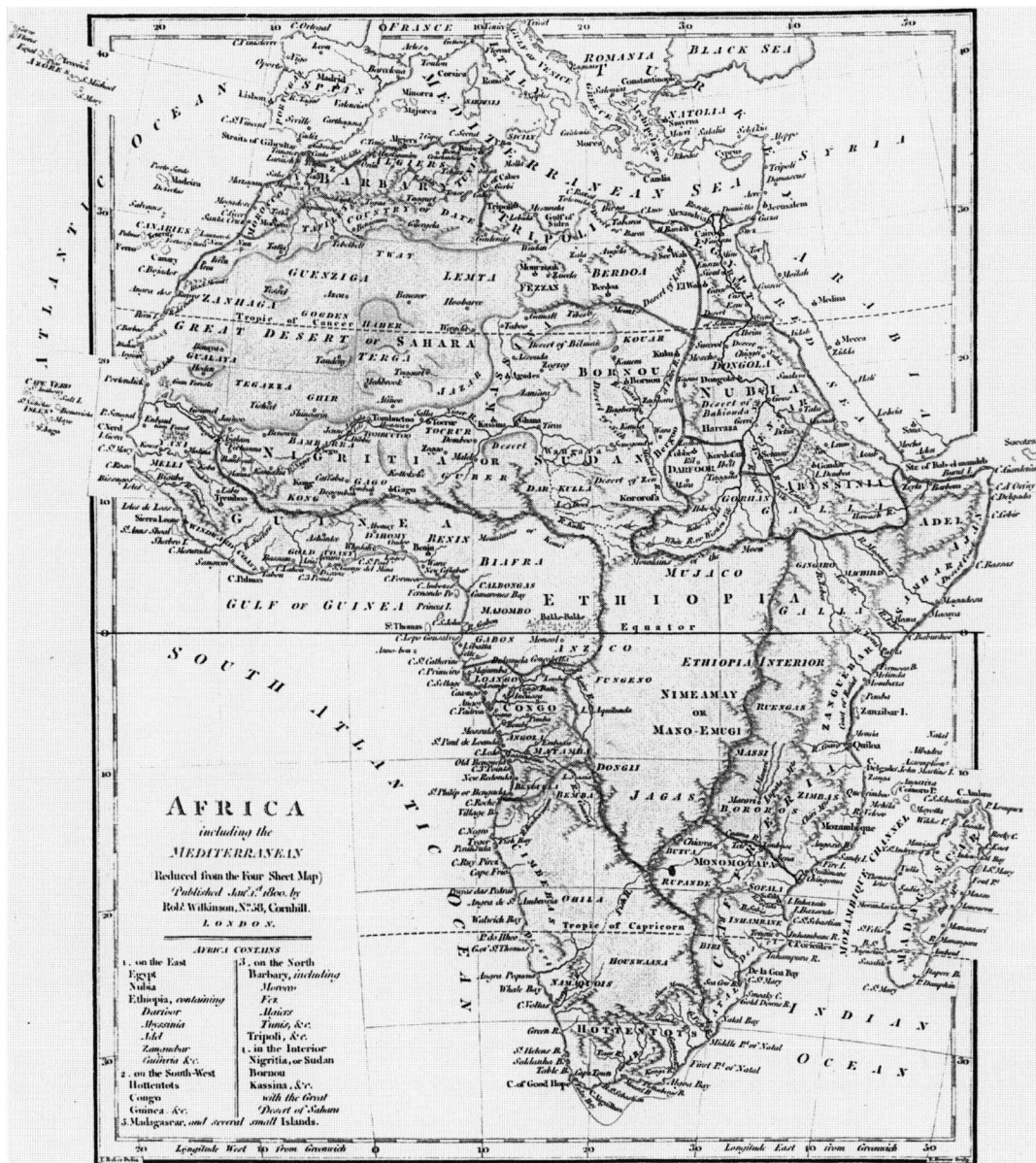

Map 117   WILKINSON, Robert (fl. 1785-1825)
Africa including the Mediterranean (Reduced from the Four-Sheet Map, Engraved by E.
Bourne; drawn by E. Baker. (London, Robert Wilkinson No. 58, Cornhill, 1800.)
Map, 28 x 22 cm, coloured.
Prime meridian through Greenwich.

A medium-sized map of Africa (reduced from a four-sheet edition) issued in 1800 by Robert Wilkinson of 58 Cornhill, London (later of 125 Fenchurch Street). He was a publisher who collaborated with Bowles and Carver, also publishers in London. Wilkinson published a number of atlases, of which the most important is his *General Atlas of the World*, in which this map appeared.

The map is coloured. The coastline appears to be quite accurate except for the southern tip which is rather narrow. For a map of 1800 it contains a moderate accuracy of detail,

and also a number of tribal names. Caffreria is depicted on the eastern aspect of the south, extending to and beyond Monomotapa. The White Nile arises from a centrally placed lake adjacent to a range of mountains named Mountains of the Moon. The Blue or Eastern Nile arises from a more easterly lake to join the White further north. There is a table of legends in the lower left corner headed by 'Africa contains' and then a double list of place names, including Madagascar and several small islands.

135

Map 118   LAURIE, Robert (1755-1836) and WHITTLE, James (d. 1818)
Africa, and Its Several Regions and Islands according to the most recent
Descriptions. (London, Laurie and Whittle, No. 53 Fleet Street, 1800.)
Map, 45,5 x 57 cm, coloured.
Scale in British and geographical miles, sea leagues and travelling leagues.
Prime meridian through Ferro Island.

This map was published in 1800 by the well-known publishers and engravers Laurie and Whittle, originally of 53 Fleet Street, London. In 1794 they acquired Sayer's geographical business and in 1803 they amalgamated with Imray, Noris and Wilson. Robert Laurie retired in 1812, replaced by his son Richard. The firm published for many other geographers and worked in close association with other publishers, especially d'Anville and Kitchin. The business passed to Findlay and Kettle, in 1873 and 1875 respectively, who produced an *American Atlas*, an *Oriental Pilot* and a *General Atlas*.

This map has the shape and outline of many of the early nineteenth-century maps of Africa, showing a fairly extensive geographic knowledge of the interior and the surrounding islands. Nontheless, the great rivers are inaccurately placed and once again the Nile erroneously appears to originate in two adjacent lakes in Central Africa. In the south a large number of fictitious rivers appear to run from north to south with imaginative names. More accurate coastal names appear especially around the Cape of Good Hope and on the eastern Natal coastline. Further north are the kingdoms of Monomotapa, Manica, Sabia and Sofala, the latter described as 'supposed to be the capital of the ancient land of Ophir.' Subsequent research and writings proved this positioning of Ophir to be historically and geographically incorrect (see Map 310). Silver is noted as abounding in Chekoya, sited in 'Manomotapa,' and gold in Butwa, adjacent to Manica, both areas in present-day Zimbabwe and Zambia.

A MAP of AFRICA for C.F. DAMBERGER'S Travels

Map 119 GOLDBACH, C.F.
A Map of Africa for C.F. Damberger's Travels. Engraved by S.J. Neele.
(London, Longman and Rees, 1800.)
Map, 35 x 34 cm, uncoloured.
Scale in days journey with camels and in days journey of Damberger.
Prime meridian through London and Ferro Island.
In: Damberger, Christian Frederick, *Travels in The Interior of Africa from the Cape of Good Hope to Morocco from 1781 to 1797*, vol. 1. London: Longman and Rees, 1801, fold p. 1.

137

A map of Africa of 1800 which appears in the well-known bogus travel book compiled by C.F. Damberger and printed in German, with a translation into English. Mendelssohn, in his *South African Bibliography*, remarks: 'one of the cleverest volumes of fabricated travels ever produced. The details are so circumstantial, and the mixture of fact and fiction is accomplished with so much skill that it is not uncommon to find people who do not know that it is nothing more than a well-contrived literary deception.' There is no doubt that the author, Damberger, a Leipzig cabinet-maker, never left his native land, yet he was able to have this map printed for the book. The engraver of the map was S.J. Neele, well-known as such in London. The contents and geographical features were laid down by Major Rennel (1742-1830), identified as Surveyor General to the Dutch East India Company by his map of North Africa, by Forster's map of South Africa and by the cartographers D'Anville and de Vaugondy, etc. The final compilation of the map was done by C.F. Goldbach, author of the *Neuester Himmels Atlas* (1803).

The map has the usual features of that period, based mainly on Portuguese and European sources. An unusual and interesting feature on the southeast coast is Padrao Point – one of the stone crosses laid down on the South African coastline by the early Portuguese. It is too far west to be identified as the Diaz Cross. Across Central Africa is marked 'Doubtful part of C.F. Damberger's routes.' The hoax thus extended even to providing the cartographer with a fictitious route.

Map 120

REICHARD, Christian Gottlieb Theophil (1758-1837)
Atlas Des Ganzen Erdkreises in der Central Projection. (Map of part of the
world...) (C.G. Reichard, 1803.)
Map, 43 x 42 cm, coloured outline.

This map includes the whole of the Mediterranean, the
Middle East, Arabia and part of Asia and India. Southern
Africa is geographically sparse but it does include the V. der
Guten Hofnung, Saldanha Bay and other surrounding places.
The interior is poorly represented and the Nile appears to
arise from a multitude of rivers. The title and date appear in
a simple oval at the lower right.

Reichard, a German cartographer, was the author of the
Atlas Ganzen Erdkreises (1803) in which this map of Africa
appeared. He published a *Hand Atlas* in 1832 and one of
North America.

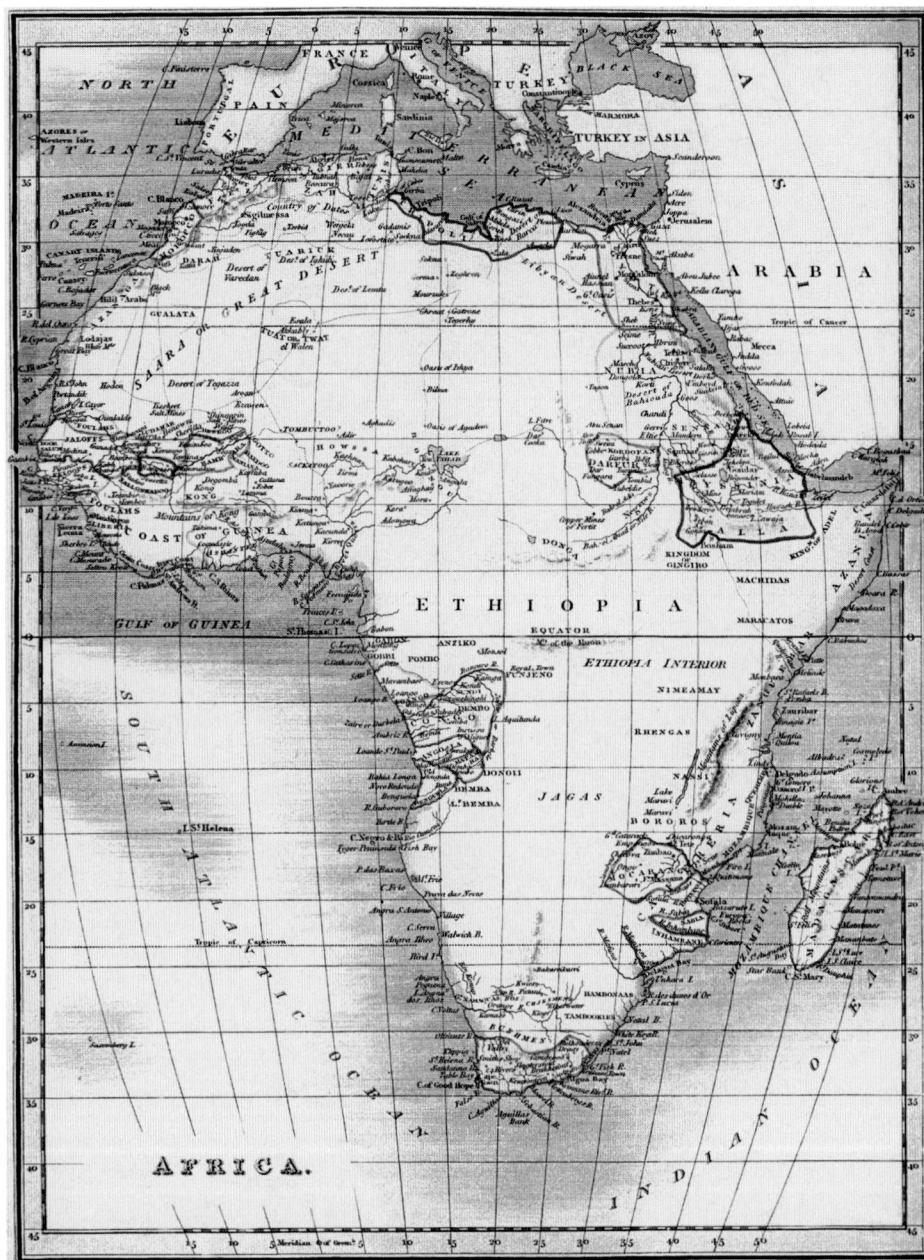

Map 121

SMITH, Charles
Africa. (London, C.S. Smith, No. 172 Strand, 1808.)
Map, 35 x 26 cm, coloured.
Prime meridian through Greenwich.
Top right: 42.

This map, including the Mediterranean and part of Arabia, is in colour and gives very little information on the northern central and southern parts of Africa. In the south the Cape Colony is demarcated but named with only a few inland legends. The coastal capes are featured and many tribal names are indicated, some in the empty spaces. The west coast of Africa illustrates the Guinea, Ivory and Grain Coasts. Although the map is undated it probably comes from the *General Atlas* of 1808.

Charles Smith was a publisher, map and globe seller and engraver extraordinary to H.R.H. the Prince of Wales. The business was later known as C. Smith and Son (1845).He and his son published many atlases of England and London, maps of English canals, rivers and roads, and a *General Atlas* in 1808. In 1836, for Samuel Arrowsmith, this establishment published a large-scale map of Africa, for school use, on four sheets of 61 x 79 cm each.

140

Map 122

PINKERTON, John (1758-1826)
Africa. Drawn by J. Herbert, engraved by S.J. Neele. (Philadelphia, Dobson.)
Map, 69 x 50 cm, slightly coloured.
Scale in British statute miles.
Across the top: *Pinkerton's Modern Atlas* (issued 1815).

This map is modern in character, without the usual antiquarian cartouche, but with a box title simply marked 'Africa.' The geography, however, is based on the ancient maps – particularly the east coast with the kingdoms of Sofala, Sabia, Manica, etc. The southern area notes the Hottentot and Bushmen tribes. The rivers Zaire and Congo are placed with inaccurate detail and the origin of the Nile is recorded as coming from a range of mountains referred to as Mountains of the Moon. Lake Morave, however, does appear. Large tracts of the interior of Africa are marked 'unknown.'

It was drawn under the direction of John Pinkerton by J. Herbert and engraved by S.J. Neele (1758-1824), foremost engraver of London. Pinkerton, a geographer and publisher, worked in Edinburgh. His atlases included *Scotland, Voyages and Travels*, in sixteen volumes (1807-1814) and a *Modern Atlas* (published in two parts 1809-1814 and as a whole in 1815) in which this map appeared, as indicated at its top. However, at the bottom this particular map is marked 'Published by Dobson Philad' making it an American issue. Tooley lists Thomas Dobson as a map maker in Philadelphia around 1799.

141

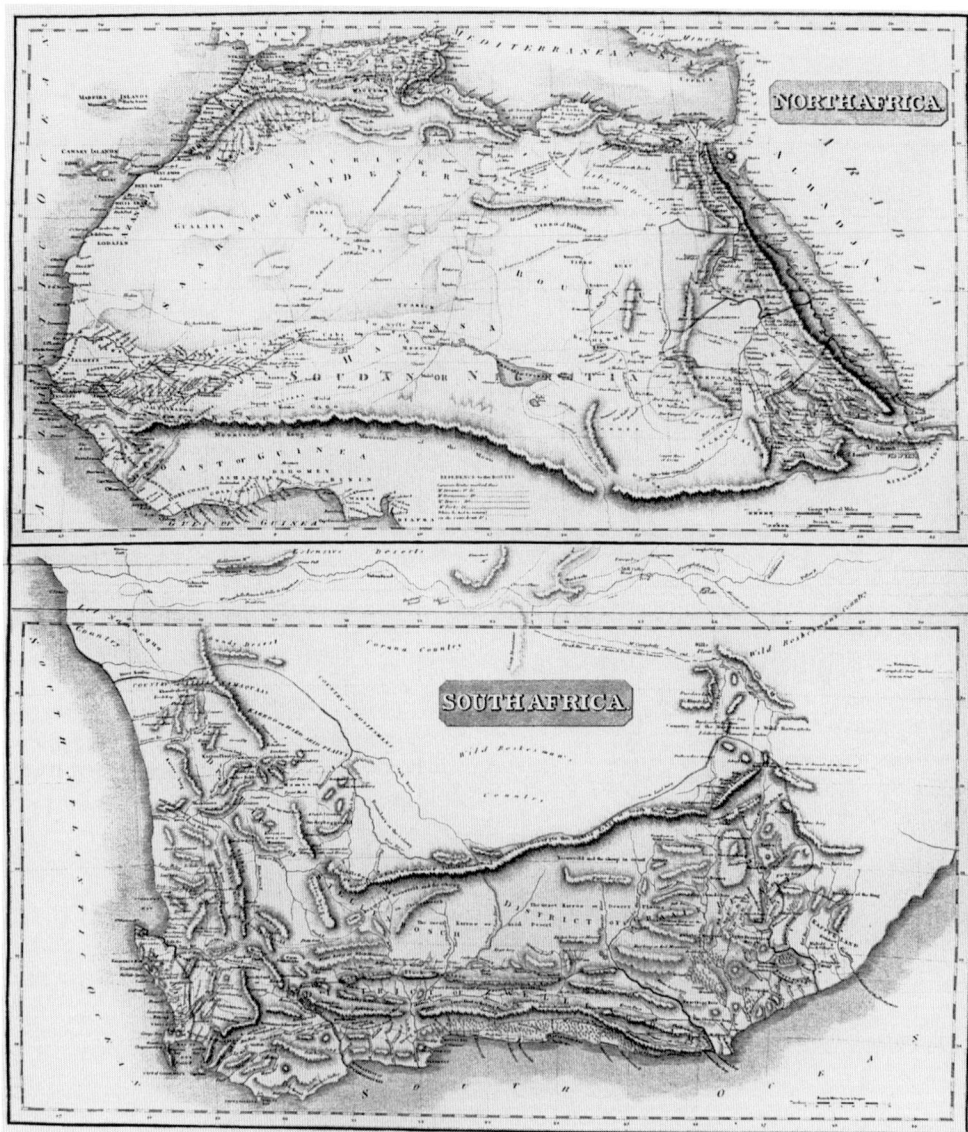

Map 123 THOMSON, J. and Co. (fl. 1810-1860)
North Africa and *South Africa*. (Edinburgh, J. Thomson and Co., 1814.)
Map, 27 x 51 cm and 26 x 51 cm, coloured.
Scale in geographical and British miles.
Bottom right: No. 49.
Along the bottom: *Drawn and engraved for Thomson's New General Atlas.*

John Thomson and Co. were publishers in Edinburgh who issued a *New General Atlas* in 1814, 1819 and 1828, a *Classical and Historical Atlas* in 1824 and an *Atlas of Scotland* in 1831.

These maps, labelled No. 49 (in the lower right corner), come from the *New General Atlas*. They are not a composite map of the whole of Africa but of North and South Africa separately, the one below the other. In the north section a legend appears indicating in different colours the routes of various explorers, namely Browne, Hornem, Bruce and Park. Caravan routes appear and also the journeys of salt caravans, noting the various salt mines. Just above the coast of Guinea

in the upper map appears an unusually placed range of mountains referred to as the Mountains of Kong or Mountains of the Moon.

Mountain ranges appear to dominate the map in the lower south, with many placenames and rivers. An interesting feature is the route taken by Campbell in what appears to be an unmarked area of the northern section of this map. This is presumably the John Campbell who, in 1815 and 1822, undertook two journeys in South Africa at the request of the London Missionary Society. The map shows his route through Pella on his way back to Cape Town.

142

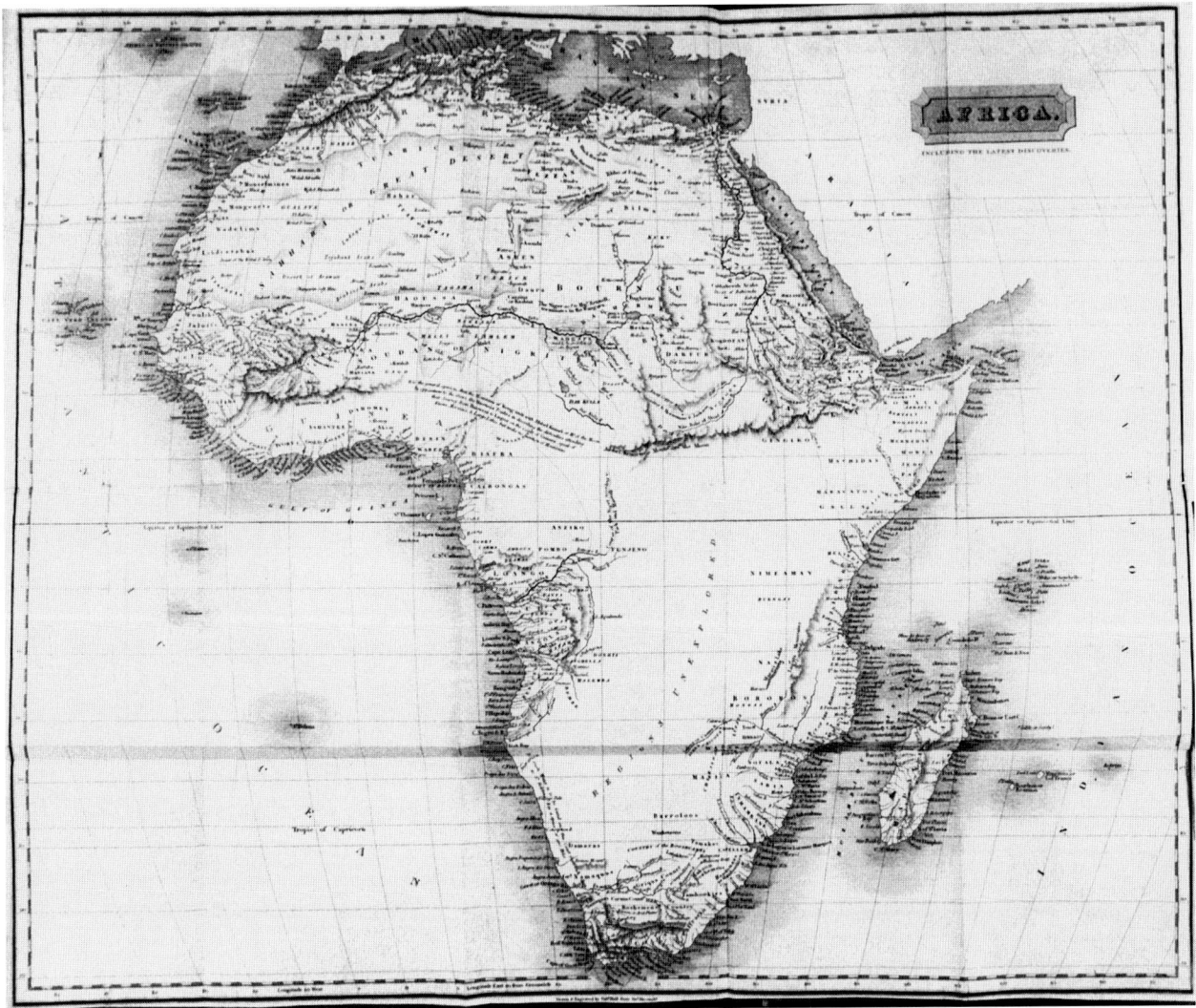

Map 124   LEYDEN, John
Africa Including the Latest Discoveries. Engraved by Sydney Hall. (Edinburgh, Archibald Constable and Co., 1817.)
Map, 40 x 50 cm, uncoloured.
In: Leyden, John. *Historical Account of Discoveries and Travels in Africa... vol. 1, enlarged by Hugh Murray.* Edinburgh, Archibald Constable, 1817, fold in front.

A map of the continent of Africa which appears as an illustration in the two-volume travel book by Murray and Leyden. The map was drawn and engraved by Sydney Hall, the London engraver and publisher.

The geographical features are concentrated mainly in the north, which at that time was better known. A large tract of Central Africa is left bare as 'regions unexplored' – a more faithful situation than filling it with numerous fictitious places, mountains and rivers. The south has the usual Cape Colony and coastal names, together with tribal boundaries and names.

Map 125 HÉRISSON, Eustache (b. 1759)
Carte de L'Afrique Divisée en ses principaux Etats pour servir à l'instruction de la Jeunesse. (Map of Africa divided into states...) (Paris, Basset, Rue St. Jacques No. 64, 1821.)
Map, 50 x 54 cm, coloured.
Scale in French and Spanish leagues, sea leagues and 10,000 metres.
Prime meridian through Paris and Ferro Island.
Borders consist of columns of descriptions of African states.

This map, issued in 1821, is from one of Eustache Hérisson's later atlases. On each side border are descriptive panels on the geography of the interior of Africa and some of the islands. The interior is devoid of the many fictitious rivers and lakes seen in other maps of this and earlier periods. The state of Monomotapa is noted, as well as many tribal names and colonies of southern Africa. At the Cape, Stellenbosch, Graaf Reynet and Zwellendam are noted. It has an interesting feature in the depiction of the route of Cook's voyages of 1771, 1775 and 1780 around the Cape of Good Hope to Australia.

Hérisson was a French hydrographical engineer, geographer and pupil of Bonne who issued maps for Grenet in 1785 and engraved for Bonne and Basset of Paris. He issued a *World Atlas* in 1818, an *Atlas ou Dictionnaire de Géographie Universelle* in 1806 and an *Atlas Portatif* in 1806, 1807 and 1811.

144

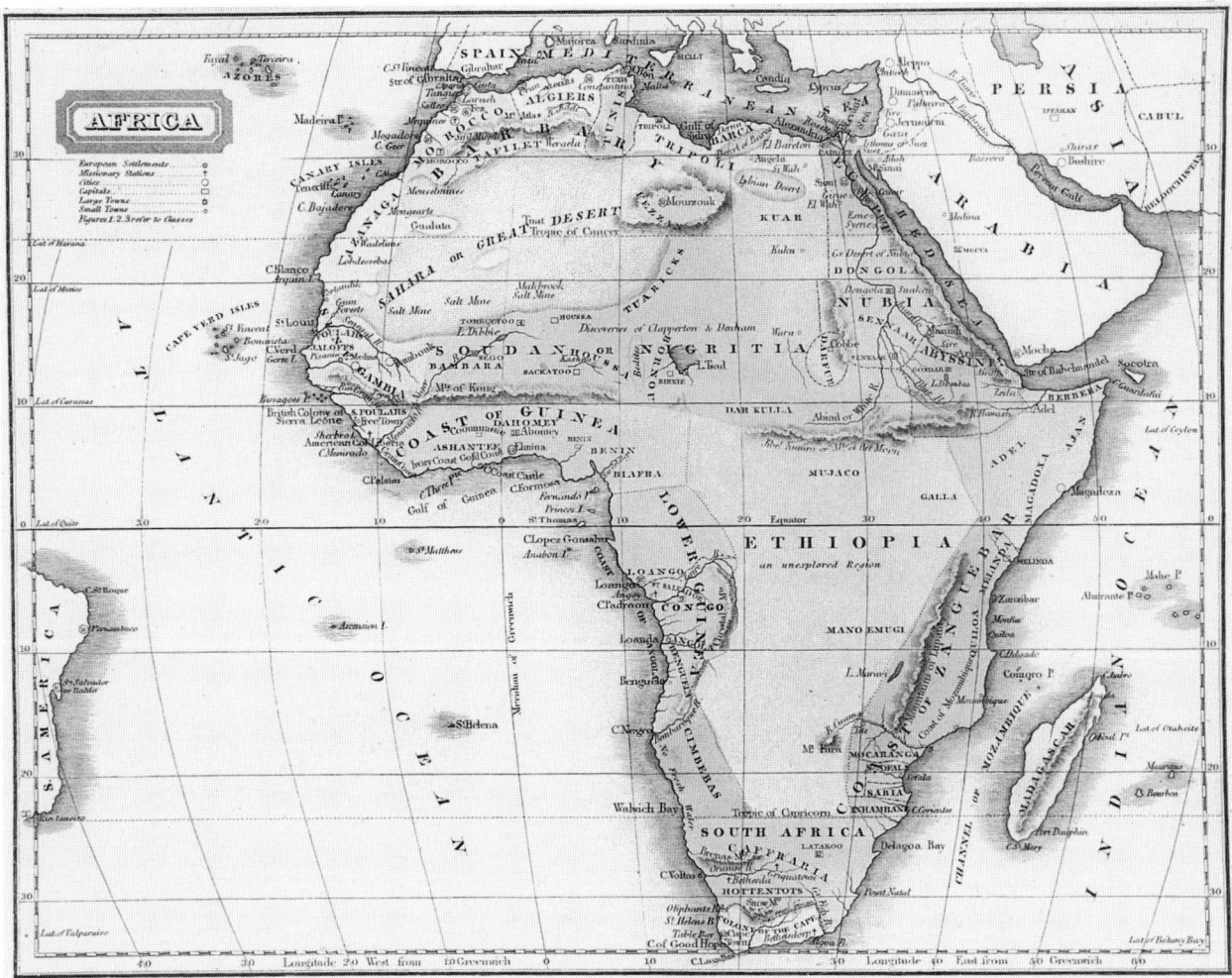

Map 126 WOODBRIDGE, William Channing (1794-1854)
Africa. (Hartford, Connecticut, W.C. Woodbridge, 1821.)
Map, 20 x 26 cm, coloured.
Prime meridian through Greenwich.

A small coloured map of Africa with a simple title at the upper left corner. There is a minimum of geographical detail in the map itself apart from its accurate shape. The whole of the central part is referred to as 'Ethiopia – an unexplored region.' The south mentions only a few of the known places at the Cape, with an unusual reference to a mountain range called Snow Mts. The coastal names on the west and east coasts are sparse. North Africa also contains very little information and the Blue and White Niles are illustrated as two separate east and west tributaries forming a main river.

William C. Woodbridge, an American publisher and geographer of Hartford, Connecticut, issued this map in 1821, probably as part of a school atlas to accompany his *Rudiments of Geography*. He issued a larger atlas in 1831 and a political map of the United States in 1845.

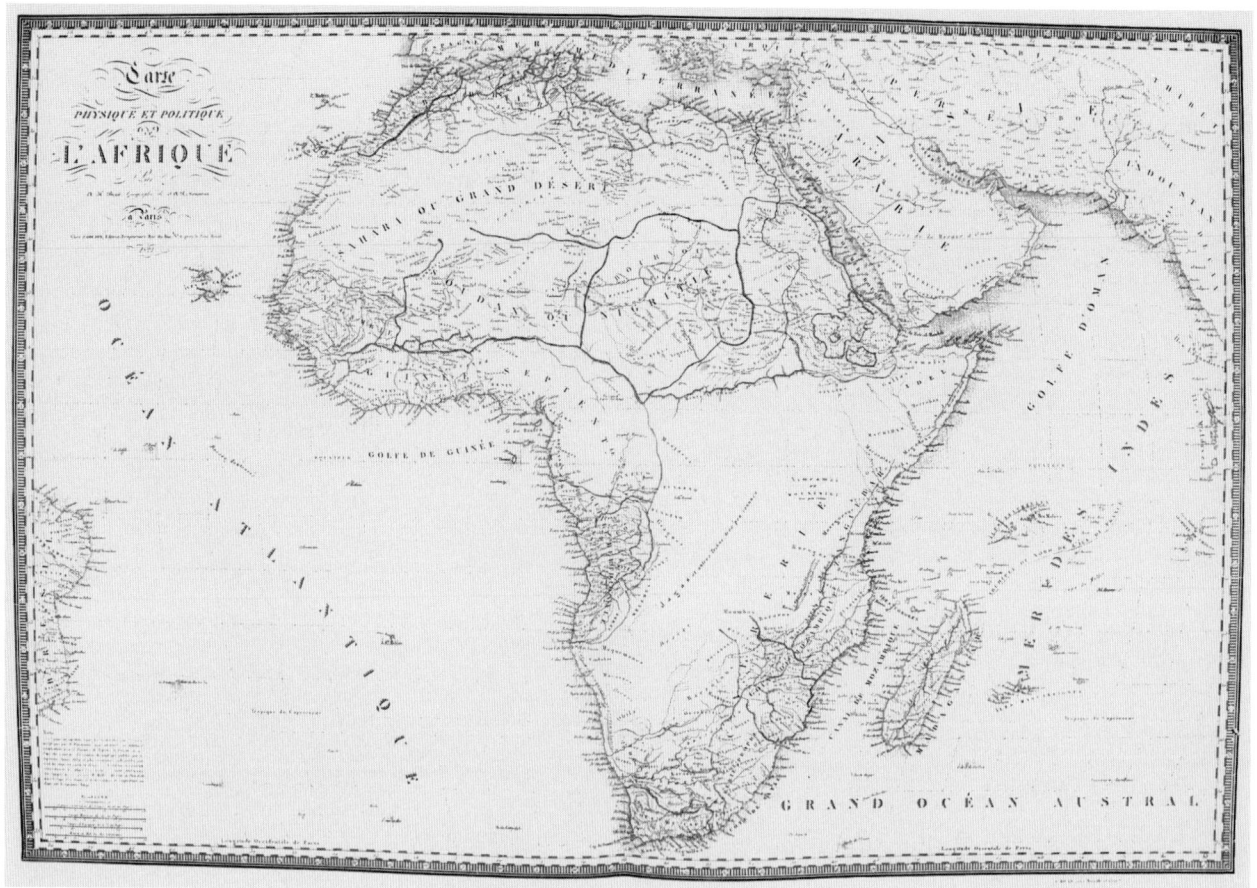

Map 127

BRUÉ, Adrien Hubert (1786-1832)
Carte Physique et Politique de L'Afrique. (Physical and political map of Africa.)
(Paris, J. Goujon, 1822.)
Map, 57 x 85 cm, coloured outline.
Scale in French, Spanish and sea leagues, days journey per caravan.
Prime meridian through Paris.

The publisher of this map, Adrien Hubert Brué, worked in Paris and was Geographer to the King. He produced the *Atlas de France* (1820-1828), *Atlas Classique* (1830), *Atlas Universel* (1816) and others.

This large French map of Africa includes Arabia, a portion of India and of the bulge of Brazil, with an attractively lettered title at the top left. The shape and outline appear to be accurate, with a fair amount of geographical detail in the north and southeast of the continent, but with a vast blank space in the centre. The large rivers are evident and the Nile appears to have its origin in a number of tributaries arising from an extensive east-to-west range of mountains labelled Mountains of the Moon. The south is filled with many geographical details, with names such as Elephant river. The areas belonging to different native tribes are shown more accurately than in many other maps at this time. The Mozambique area and coastline are fairly accurate. Madagascar and the surrounding islands appear with their original French names, namely I. de France, Isle Bourbon, etc.

146

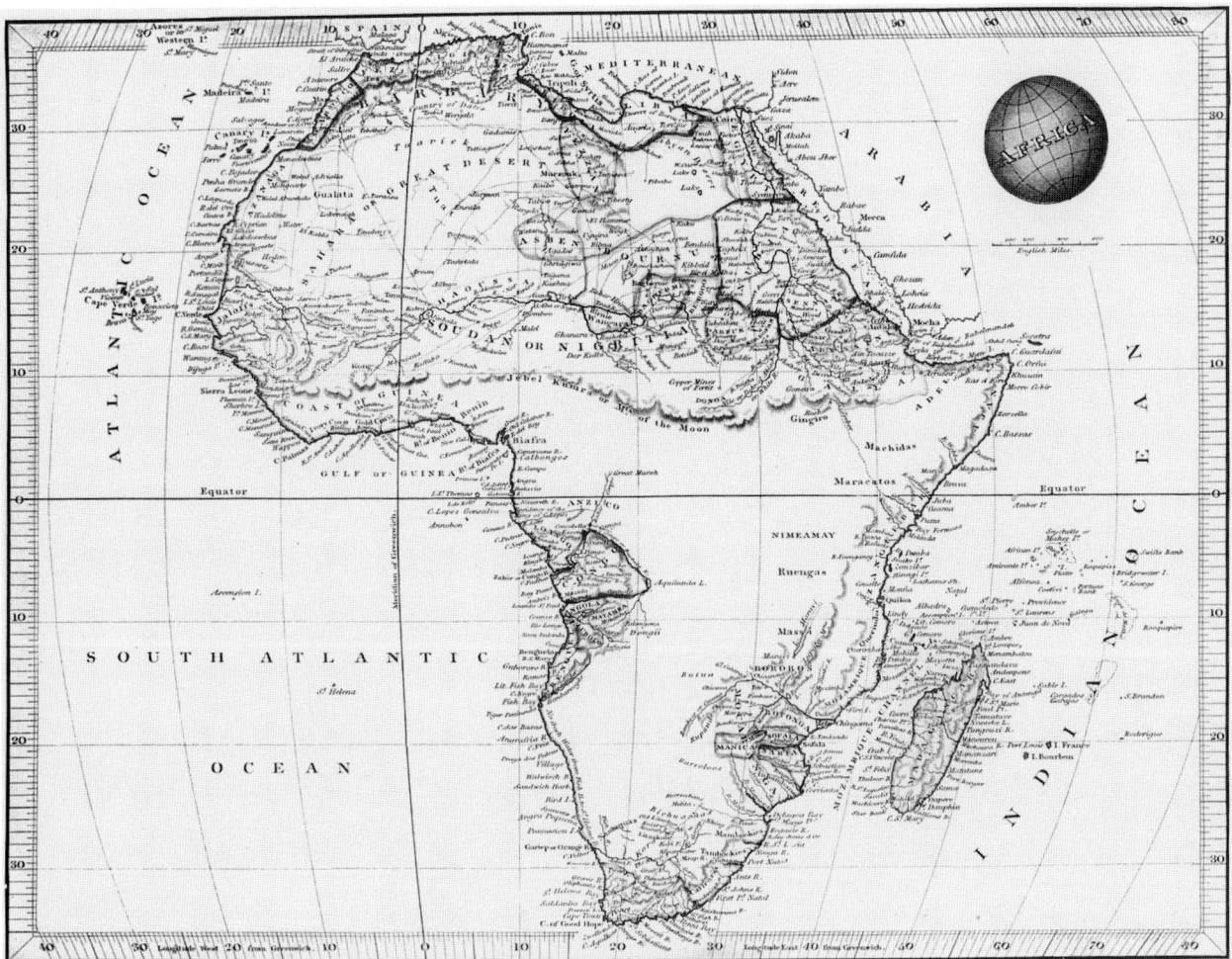

Map 128

ARROWSMITH, Aaron (1750-l833) and Samuel (d. 1839)
Africa. (London, A. and S. Arrowsmith, 1825.)
Map, 22,5 x 30 cm, outline coloured.
Scale in English miles.
Prime meridian through Greenwich.
Bottom right: 33.

This map comes as a combined imprint dated 1825, and appeared in *Outlines of the World* in the same year. Central Africa is devoid of any geographical features except, in the north, 'Jebel Kumra or Mountains of the Moon,' and only the known lands are depicted, around the Congo complex and on the west coast. Southern Africa features the eastern Portuguese possessions and the Cape, with Paarl, Clanwilliam, Tulbagh, Zwellendam, George and Graaf–Reinet.

The Arrowsmiths were the foremost British map publishers of the early nineteenth century, with an output between them of 750 maps. The founder of the family was Aaron I, who came to London in about 1770, established himself as a land surveyor and worked for other publishers, certainly Carey and possibly Faden. He set up his own

publishing establishment and his first production was a *Chart of the World upon Mercator's Projection showing all the new Discoveries*, in 1790. Although he published the majority of his own maps he occasionally worked for others or jointly with them. As well as being a geographer and publisher Arrowsmith was an engraver and, apart from his own family who assisted him, he employed other engravers, namely Foot, Pickett, George Allen, Palmer, Cooper, Lowy and Sidney Hall. After his death in 1833 he was succeeded by his two sons Aaron and Samuel, who continued the business from the family house in 10 Soho Square, London. They followed their father's tradition of producing new maps and revising old ones and between them published over 100 maps. They sometimes used the same imprint but also used their names separately.

147

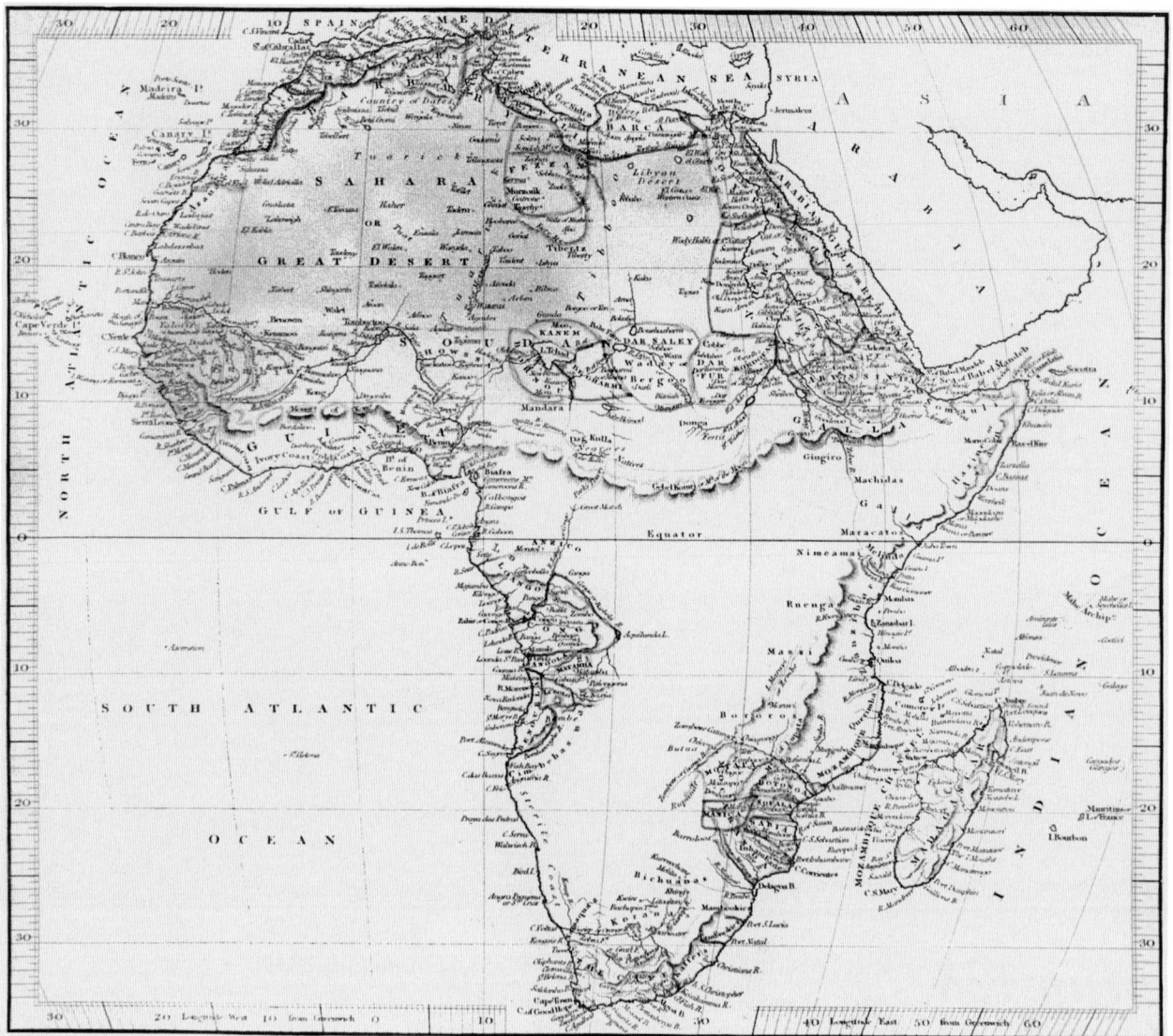

Map 129

ARROWSMITH, Samuel (d. 1839)
Africa. (London, S. Arrowsmith, No. 10 Soho Square, 1828.)
Map, 24 x 27 cm, coloured outline.
Scale in geographical and English miles.
Prime meridian through Greenwich.
Top right: 4.

This map of Africa of 1828 was published by Samuel Arrowsmith, son of the founder of the leading British map publishers of the early nineteenth century. Samuel was then the Hydrographer to His Majesty and, with his brother, Aaron II, issued maps from their establishment, No. 10 Soho Square.

'Africa' appears above the top border and the map is numbered 4. The southeast coast is marked 'Sterile Coast.'

Genadendal, Tulbagh, George and Bachapin Town are shown. The south is accurately depicted although somewhat sparsely filled in. The Portuguese settlements on the eastern coast are illustrated, and also the Congo. The rest of Central Africa is left barren and unoccupied.

148

Map 130

CAREY, Matthew (1760-1839)
Africa According to the best Authorities. (Philadelphia, M. Carey.)
Map, 33,5 x 35,5 cm, coloured outline.
Scale in statute miles.
Prime meridian through Philadelphia.
Title across the top: *Engraved for Carey's American edition of Guthrie's Geography Improved.*

This map of Africa, described as 'according to the best authorities,' was published in Philadelphia as a line engraving. This is the first folio-sized map of Africa to be engraved and printed in America. It appeared in Carey's *General Atlas* and in the general atlas for Carey's edition of *Guthrie's Geography Improved* (Wheat and Brun).

Geographically the map is not very accurate, with a host of fictitious tribal names in southern Africa. The Cape of Good Hope includes a Dutch Fort with inaccurate naming of bays and rivers, one named 'the endless river.' The Kingdom of Monomotapa and its capital town of the same

name are recorded. The Lake of Zambre appears in what is the proper siting of today's Lake Malawi, and is probably an ancient name from the time of Ptolemy's writing, often described in connection with the origin of the Nile. In West Africa many fictitious names, in what is termed Negroland, are used to describe tribes and kingdoms.

Matthew Carey was an American publisher of Irish descent working in Philadelphia. He published the earliest American *Atlas of the United States* in 1795, a *General Atlas* in 1794, an *American Pocket Atlas* and a *Scripture Atlas* in 1817.

149

Map 131 MENTELLE, Edmé (1730-1815)
Carte Genérale de L'Afrique. (General map of Africa.) (Paris, J.B. Delaval, Rue Geoffrey-Langevin No. 7,1829.)
Map, 31 x 42 cm, coloured outline.
Scale in Dutch miles, Portuguese and French leagues, sea leagues.
Prime meridian through Ferro Island and Paris.
Top left: No. 150.

In this map there is not a great deal of cartographical detail. The northern, central and southern areas are comparatively empty, which is more commendable than placing a mass of fictitious kingdoms, rivers and mountain ranges in the spaces. In the south the 'Guvt de Cap' (Cape Colony) and 'Cote de Natal' (Natal coast) show some modern place names and inlets, together with a number of tribal names and kingdoms. The map retains some features of older maps, such as Monomotapa, Sofala and some names on the Mozambique coastline. Further north 'Lac Murari' (Lake Malawi) appears with a cut-off northern boundary. The origin of the Nile appears as a mass of fictitious north-flowing rivers

with one comparatively small lake adjacent to a place named Enfras. The Congo and Zaire rivers are not marked as such but the Niger River shows its source in West Africa, not too far distant from the coastline. The cartouche is a simple rectangular box at the lower left corner, with the scales in the upper right corner.

This early-nineteenth-century French map of Africa comes from the publishing house of Edmé Mentelle, a geographer and historian who issued a number of atlases. He was a member of the Royal Institute of Science and Professor of the Central Schools Department of Seine.

150

Map 132　ARROWSMITH, Samuel (d. 1839)
Eastern Hemisphere. (London, S. Arrowsmith, No. 10 Soho Square, 1828.)
Map, 22,5 x 22,5 cm, coloured outline.
Scale in geographical miles, English miles.
Top left: 1.

This small eastern hemisphere of a world map, on a typical Mercator projection, was issued by Samuel Arrowsmith, one of the two sons of the founder of this Arrowsmith family, Aaron. It was issued in 1828 from the family residence in Soho Square and was probably map number one in one of their atlases, possibly *Outlines of The World* (1828). The map itself provides very little geographical detail in Africa, except for a number of names of bays and

towns in the southern part. The northern sector is equally devoid of geographical information. Around the periphery of the map the varying temperature zones are marked. The North and South Poles with their Arctic and Antarctic Circles are included.

[The improved depiction of the course of the Niger suggests that this map is later than 1828, since the Lander brothers only traced the Niger to its mouth in 1830 (JCS).]

151

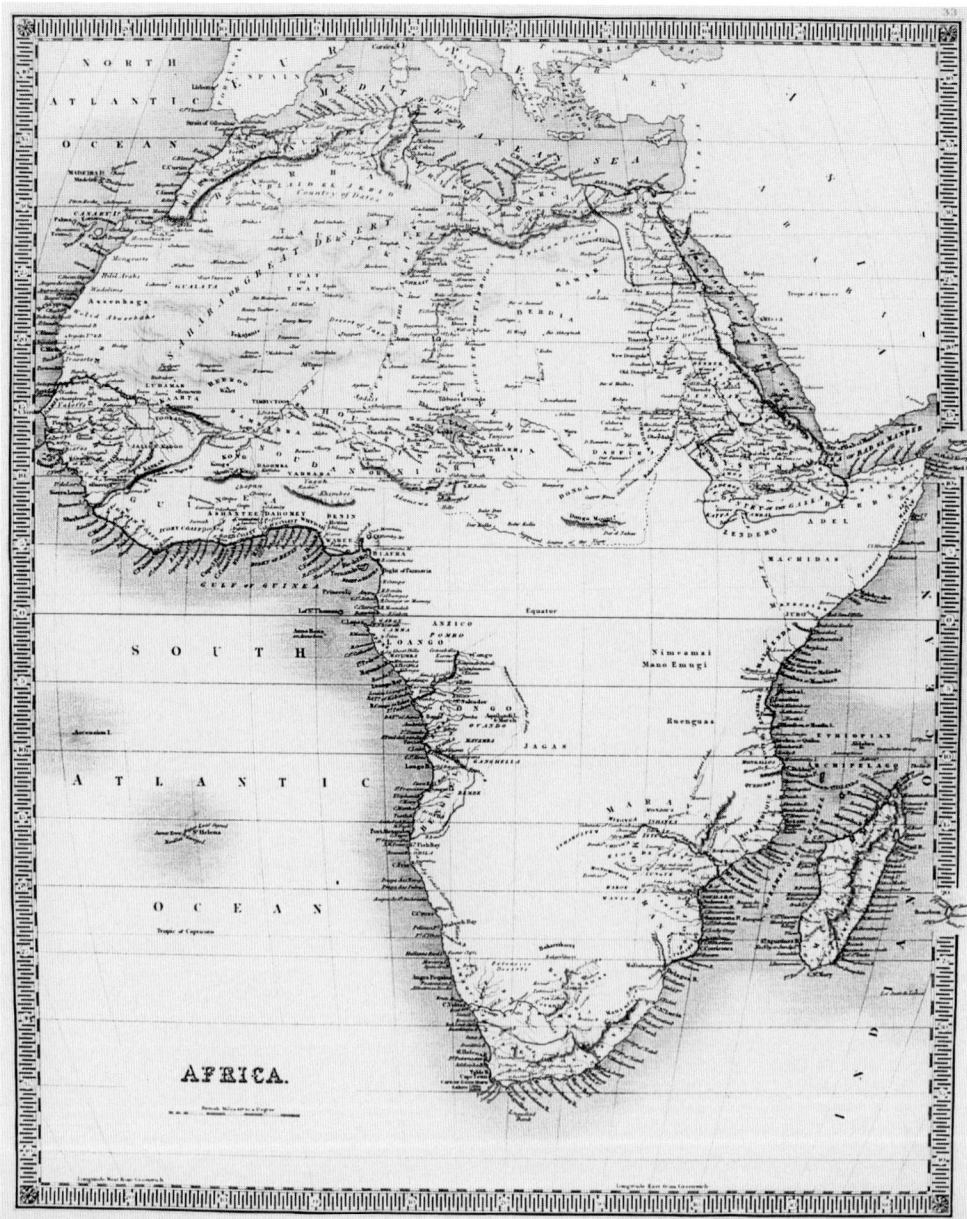

Map 133

TEESDALE, Henry

Africa. Engraved by J. Dower. (London, H. Teesdale, 302 High Holborn, 1831.)
Map, 41 x 33 cm, coloured.
Scale in British miles.
Prime meridian through Greenwich.
Top right: 33.

This is a scientifically accurate map of the continent of Africa, published in colour. Whole tracts of land are left unoccupied, most probably because of lack of geographical knowledge. The known geography of southern Africa at the Cape and on the east and west coasts is accurately drawn. In the Cape an unusual 'Zak River' appears, but Stellenbosch, Zwellendam, Graaf Reinet and Grahamstown are accurately placed. In the north a peculiar range of mountains, referred to as 'Donga,' in an area by the same name, appears to be the source of the Nile. Below this range an interrupted line runs across upper Central Africa which eventually joins the River Niger and which is referred to as the 'supposed source of the Niger.'

Henry Teesdale published atlases and maps of English counties and a *New British Atlas* in 1829, a *New Travelling Atlas* in 1830 and a *General Atlas* in 1831, in which this map of Africa appeared.

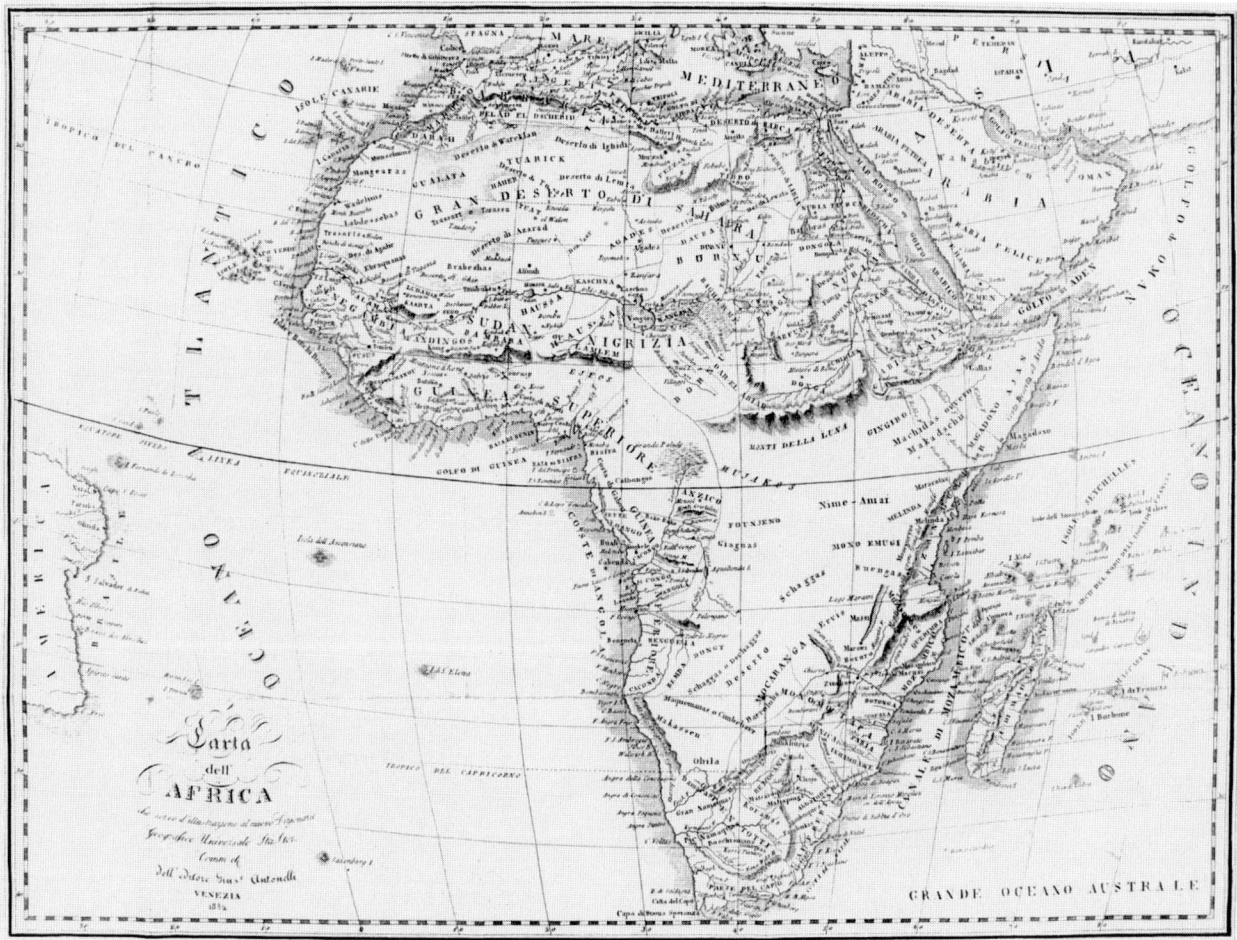

Map 134 ANTONELLI, Giuseppi
Carta dell' Africa che serve d'illustrazione al nuovo Dizionario Geographico Universale... (Map of Africa forming part of a new gazetteer...) (Venice, G. Antonelli, 1832.)
Map, 34 x 46 cm, uncoloured.
Prime meridian through Ferro Island.

This Italian map of Africa includes the Mediterranean, Arabia and the bulge of Brazil. As with other maps of that date the Cape Colony and the bays are depicted fairly accurately. None of the present-day colonies appear on the map but African tribes are noted in their respective regions.

The author of this map, Giuseppi Antonelli, was a Venetian publisher who put out the *Dizionario Geographico* in 1833, *America* in 1834 and *Oceania* in 1839.

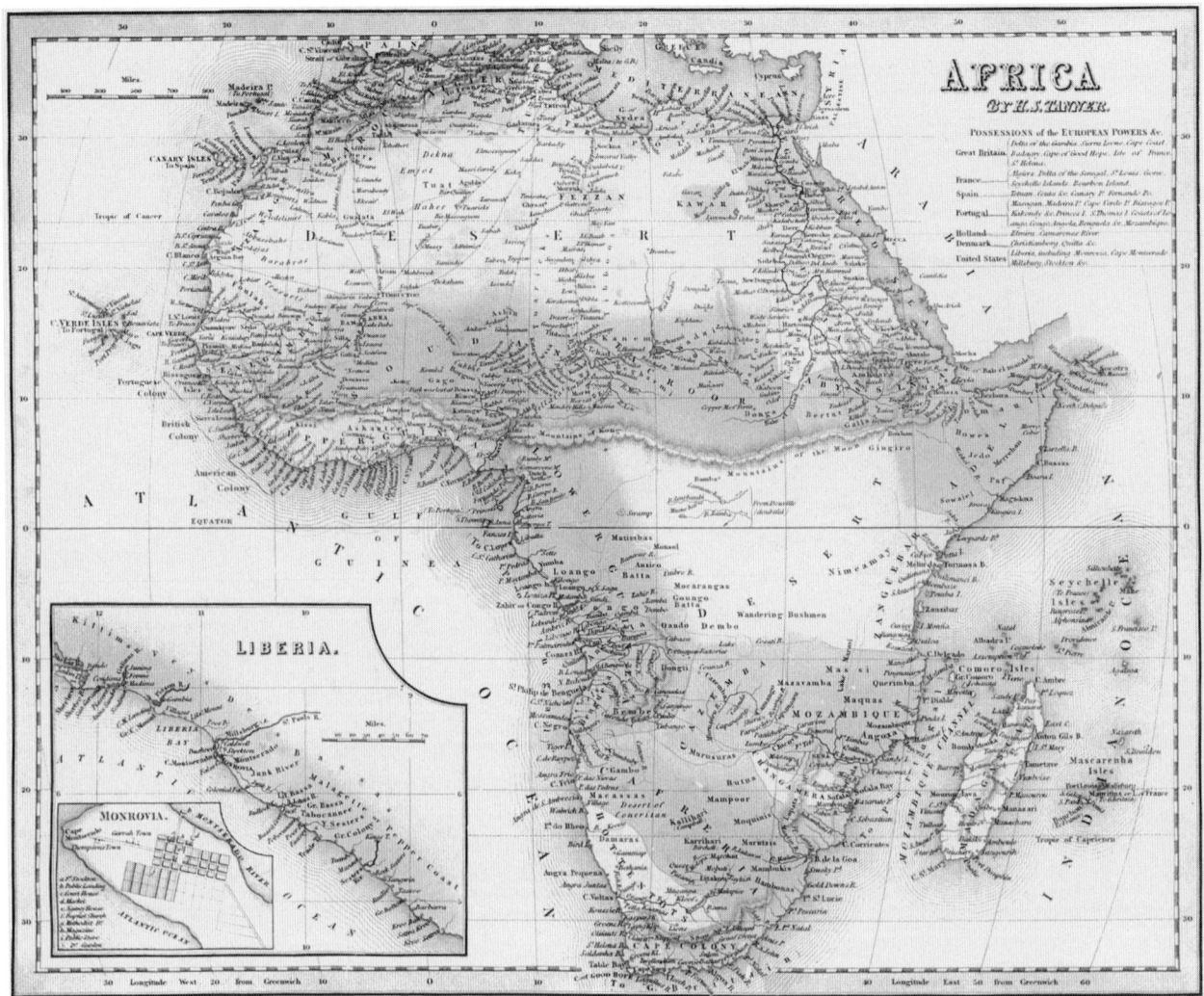

Map 135

TANNER, Henry Schenk (1786-1858)
Africa. (Philadelphia, H.S. Tanner, 1834.)
Map, 29 x 36 cm, coloured.
Scale in miles.
Prime meridian through Greenwich.
Inset: Liberia and Monrovia.
Title across the top: *Tanner's Universal Atlas.*
Bottom right: 61.

This map is coloured according to a legend in the upper right corner (below the title), indicating the possessions of the European powers – Great Britain, France, Spain, Portugal, Holland, Denmark and the United States. Such a map of Africa for the American public had to provide extra information on Liberia, which is listed as a possession of that power. The inset clearly designates the major towns and rivers of this 'American Colony' and also includes an orderly plan of Monrovia. In the south the Cape Colony is named, but with very sparse geographical information. 'Snow Mountain' refers most probably to the Drakensberg range. There are a number of tribal names. The Nile arises from a long truncated range named Mountains of the Moon in the east and Mountains of Kong in the west. The Blue Nile (marked Blue R.) is shown arising from Lake Demba in Abyssinia.

This map was published in the United States of America by Henry S. Tanner, an engraver, draughtsman and publisher of Philadelphia, known for his atlases of America.

154

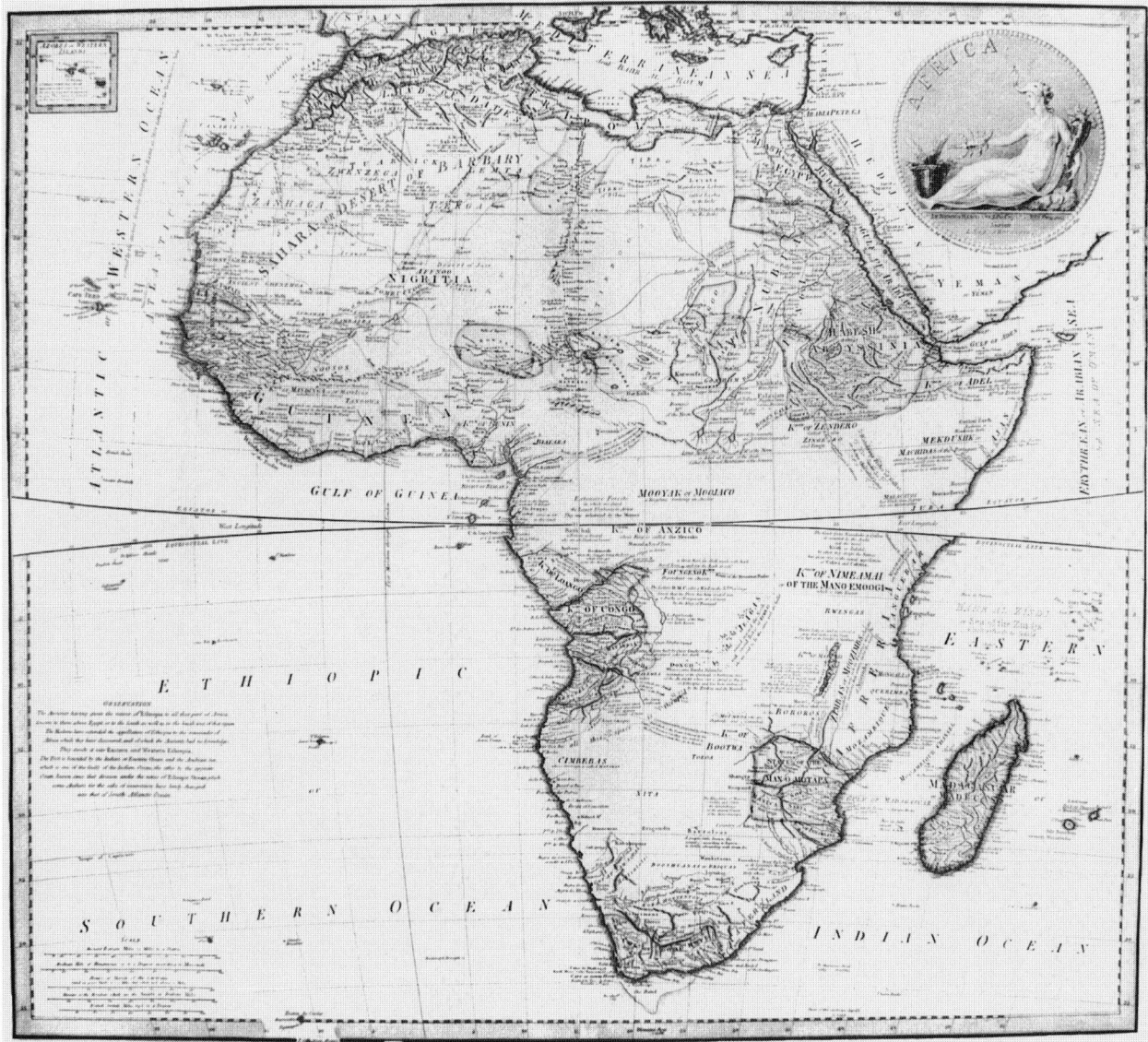

Map 136 DE LA ROCHETTE, Louis Stanislas d'Arcy (1731-1802)
Africa. Engraved by W. Palmer. (London, J. Wyld, 1835.)
Map, 51 x 57 cm, coloured outline.
Scale in ancient Roman miles, Arabian miles, hours march of caravan, Italian miles, British statute miles.
Prime meridian through London.
Inset: Azores or Western Islands.

The title appears on a medallion at the top right, which features a seated female figure with an unusual hair style (a bird?), holding a sea animal in one hand and a horn of plenty in the other. The map bears many legends relating to the continent's history. In this edition reference is made to some findings of Barrow's travels. At the lower left of the map there is a short description of the State of Ethiopia and the Ethiopic sea. The Nile is once again featured as arising from two side-by-side lakes (smaller than normally seen), joining to form what is described as the White Nile.

De la Rochette, cartographer and engraver, was associated with Faden and Bowles, both well-known publishers, printers and map sellers. His various editions of the Cape of Good Hope are particularly well-known to collectors and historians of southern Africa. This map of the continent of Africa was first published in 1803 and republished in 1833, 1835 (this copy), 1838 and 1840.

155

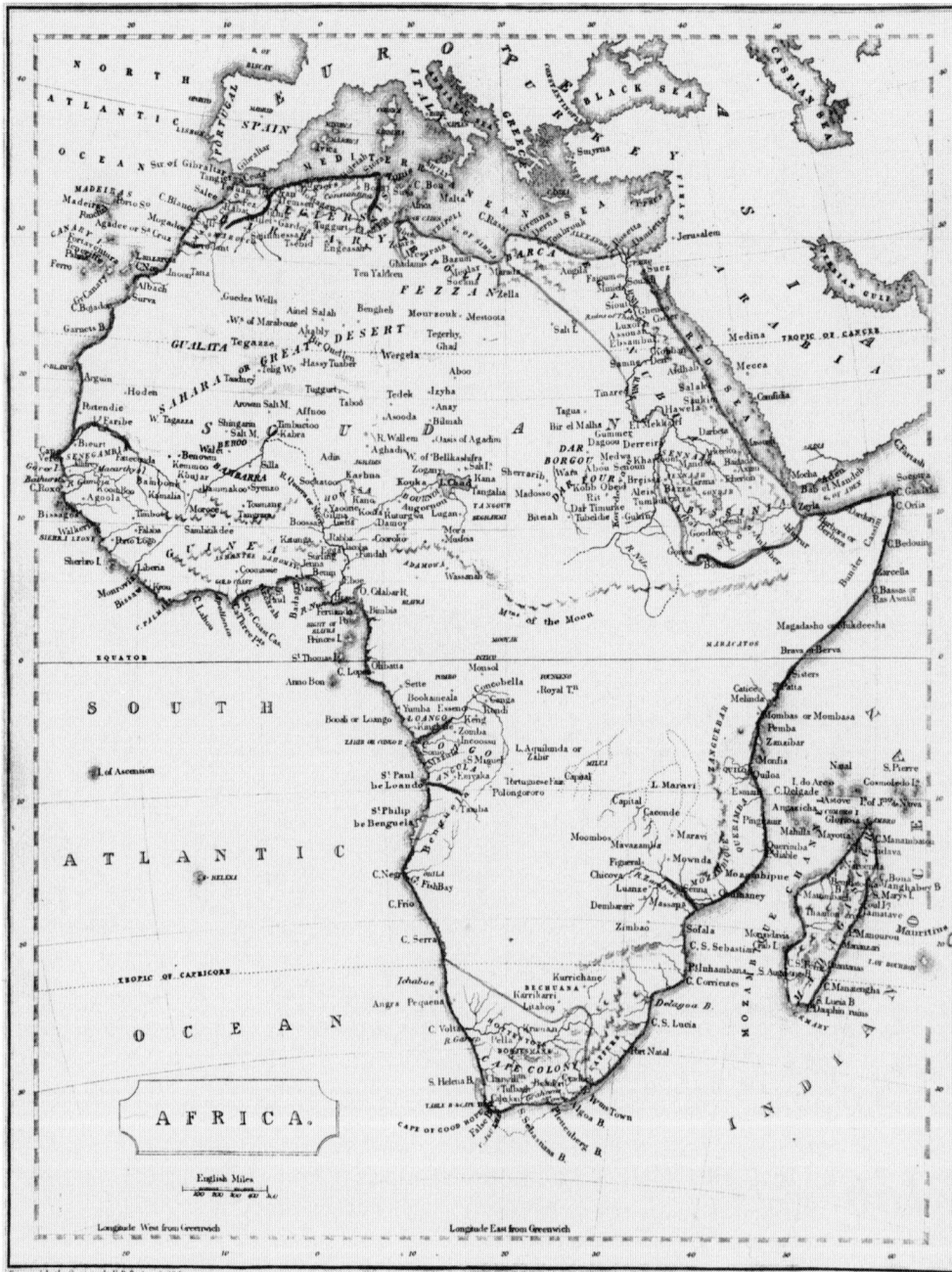

Map 137

VIRTUE, George

Africa. Engraved by F.P. Becker. (London, C. Virtue, Ivy Lane, 1836-1839.)
Map, 24 x 19 cm, coloured outline.
Scale in English miles.
Prime meridian through Greenwich.

A small map of Africa engraved by F.P. Becker and Co., and issued by Virtue, a London publisher. Among Virtue's publications was an atlas (1836-1839) in which this map probably appeared.

The outline of the continent is accurate. In southern Africa, the early Cape Colony is demarcated with an accurate northern border. The geographical information is very limited, particularly in the central part of the map. A few tribal names do appear, and the coastline lacks some legends that did appear in other maps of the period. The Nile appears without any link to the great lakes and there is a faint mountain range across Central Africa labelled :Mtns of the Moon," which the Nile appears to cross.

156

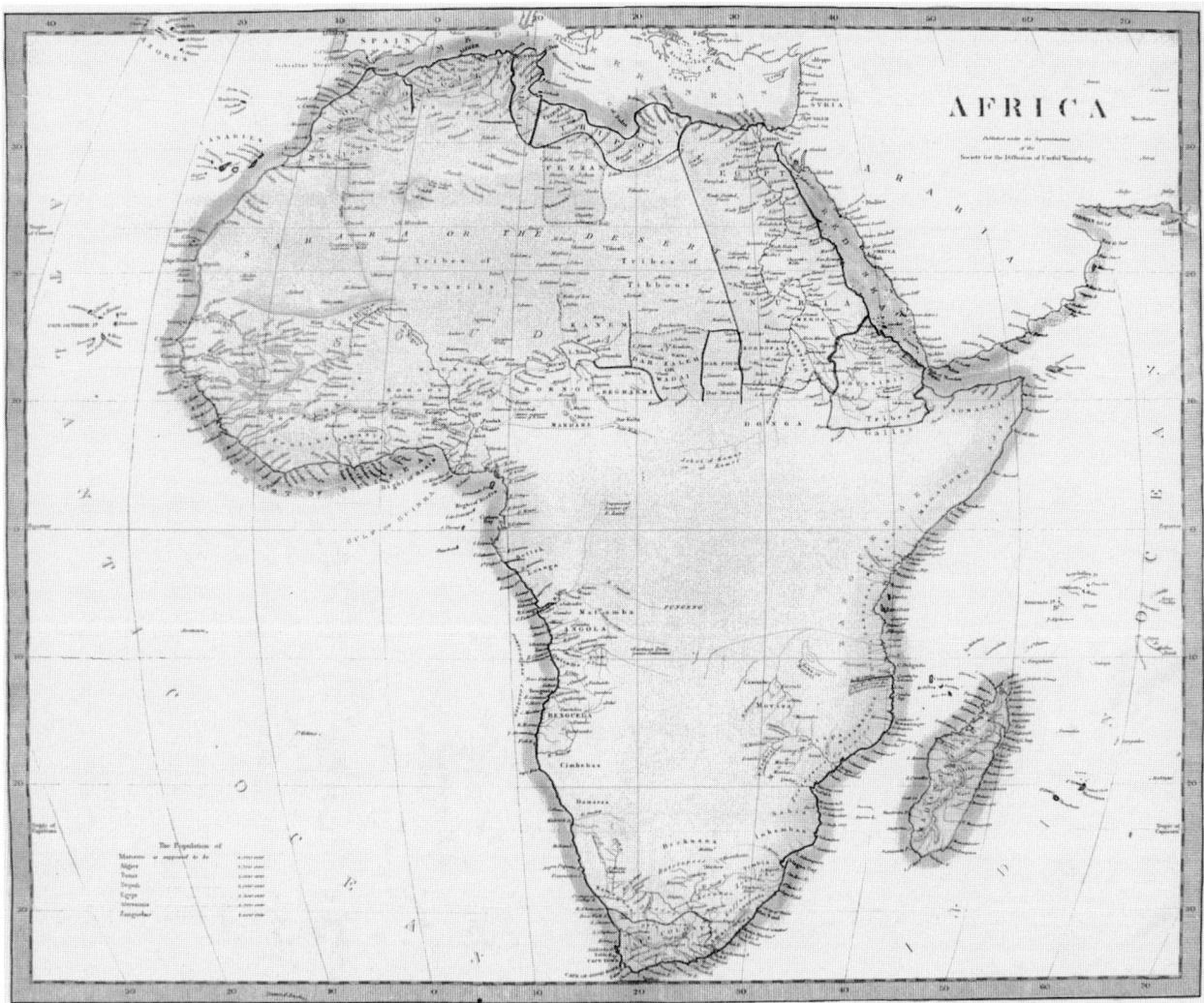

Map 138 SOCIETY FOR THE DIFFUSION OF USEFUL KNOWLEDGE
Africa. Engraved by J. and C. Walker. (London, Society for the Diffusion of Useful Knowledge, 1839.)
Map, 31 x 39 cm, slightly coloured.
Prime meridian through Greenwich.

This map of Africa was issued by the Society for the Diffusion of Useful Knowledge in 1839. It is a fine line-engraved coloured map, the interior and coastlines of which give information then known and not a mass of fictitious geography, as was the case with many of the sixteenth-and seventeenth-century maps preceding this period. The Cape Colony is accurately mapped with a number of tribes correctly depicted. The known features of Portuguese Mozambique, the western coast and the interior of Angola are also correctly described. A table of population numbers is noted in the lower left corner. The S.D.U.K. atlas was in print for many years and the maps underwent several revisions. The high standards to which its maps were held made it very influential.

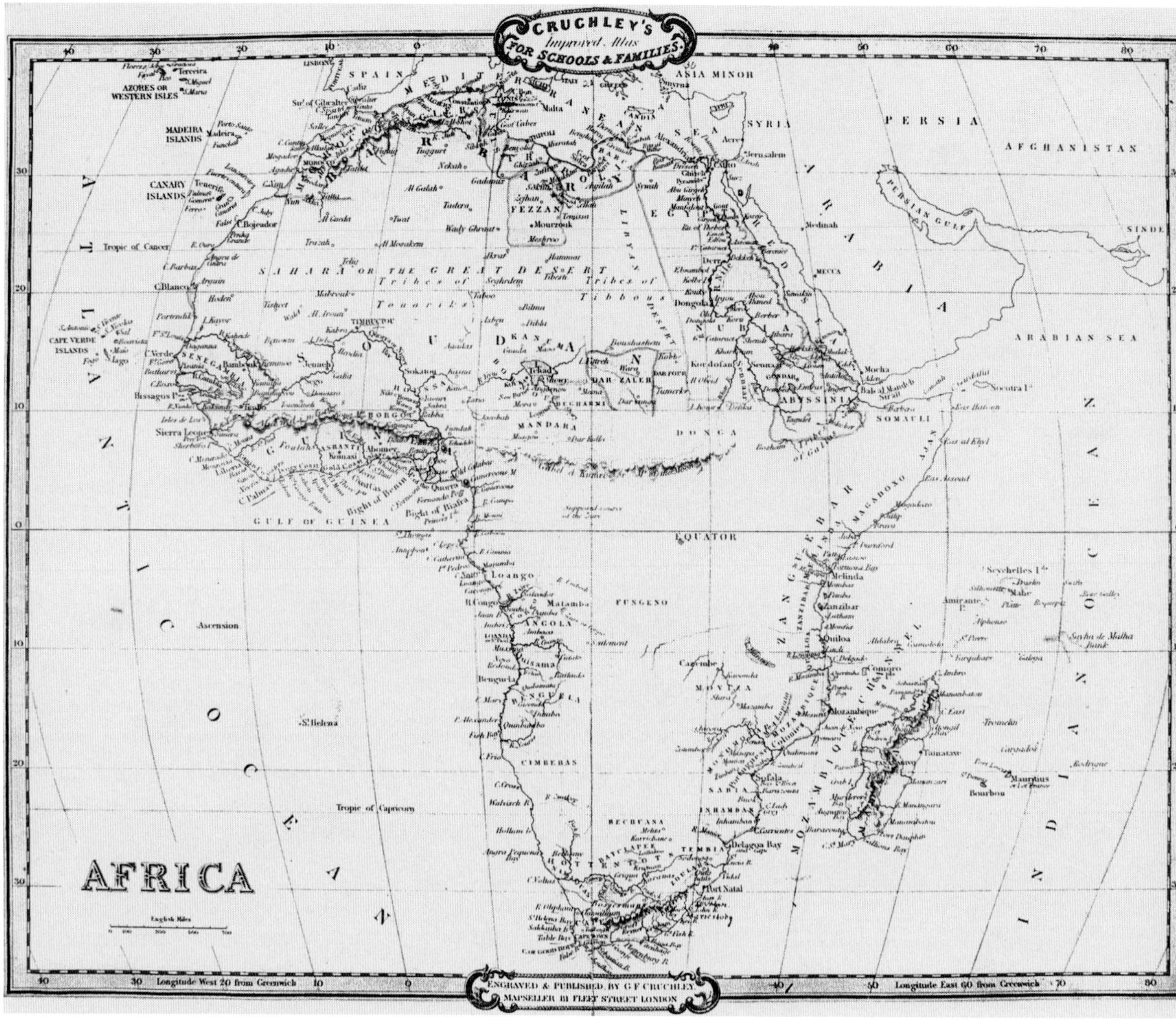

Map 139

CRUCHLEY, George Frederick (fl. 1823-1876)
Africa. (London, G.E. Cruchley, 81 Fleet Street 1841.)
Map, 34,5 x 44,5 cm, coloured.
Scale in English miles.
Prime meridian through Greenwich.
Title across the top: *Cruchley's Improved Atlas for Schools and Families.*
Top right: 27.

This particular map is dated 1841 and appears to have come from an atlas because of the number 27 displayed at the top right corner – probably from the *General Atlas* of 1843. The reason for assuming this date is the lack of geographic detail in southern Africa, especially the non-existence of the early Transvaal Republic and its associated towns and settlements. There is a good deal of the Cape Colony, with tribal names, and one of the first references to the 'Zoulahs' in what was early Natal. The northern and central areas are left almost free of any features rather than having to include a mass of unknown or fictitious details.

This map of Africa was engraved and published by G.F. Cruchley, an English engraver, map seller and globemaker who worked at various addresses in London. He was trained by Aaron Arrowsmith, founder of the Arrowsmith family of publishers. Cruchley reissued some of the Arrowsmith maps, notably the 1835 Cape of Good Hope, in 1845, showing Burchell's route of his travels. In 1844 he bought the engraved plates of John Carey, one of the most prolific English cartographers. Cruchley issued many atlases, including the *Environs of London*, in 1824, and a *General Atlas* in 1843.

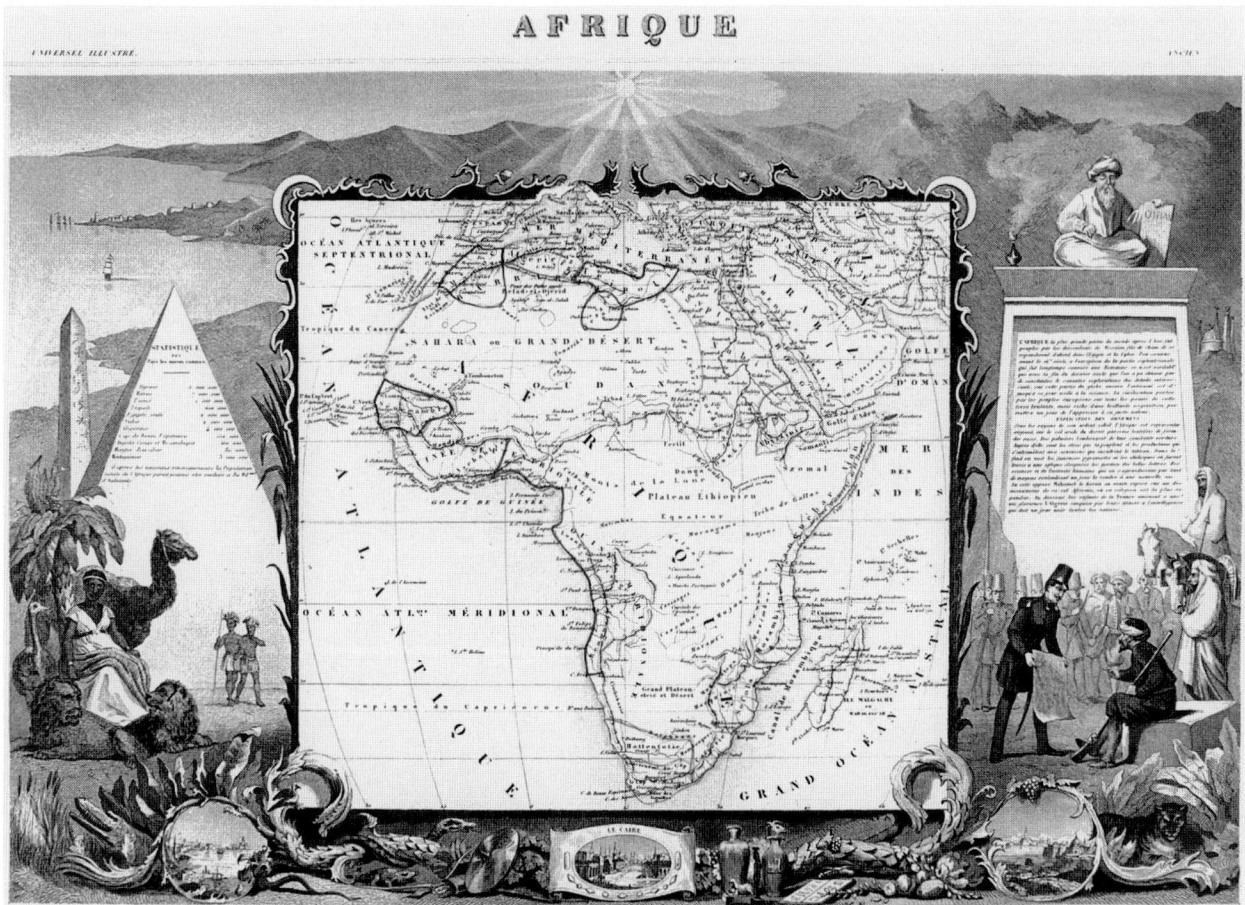

Map 140

LEVASSEUR, Victor
Afrique. Engraved by Laguillermie, illus. by Raimond Bonheur. (Paris, A. Combette, ca. 1840's.)
Map, 29 x 43 cm, coloured outline.
Inset: Views of Alexandria, Cairo and Algiers.
Top left: 'Atlas Universel Illustré.'
Top right: 'Ancien continent.'

This is an attractive map of Africa set within a colourful and descriptive framework of landscape, fruit, foliage and animals, with descriptive text and statistical information on towns and islands. The left border contains a camel, two recumbent lions and an ostrich, surrounding a seated African woman. In the background are two black warriors. The right border shows a uniformed officer (French) showing a document to a seated, turbaned Arab holding a gun. In the background is a party of other soldiers and Arabs, one riding a horse. On the lower border there are small insets containing fine steel engravings of the towns of Alexandria, Cairo and Algiers, interspersed with small animals and floral decoration.

One interesting feature is the appearance of a turbaned and bearded seated figure on a shelf of masonry above the descriptive legend on the right border. He is holding a tablet with the lettering CORAN inscribed on it. On the left of the shelf is a small oval vase which appears to be burning incense.

The map appears to be an adjunct to the more impressive pictorial representations surrounding it, and although its shape is accurate, the geographical detail (in French) is poor and sparse, considering its date (ca. 1840's).

159

Map 141   LAPIE, Alexandre Emil and LAPIE, M.
Carte d'Afrique. (Map of Africa.) Engraved by Lallemand. (Paris, Rue Mazarine No. 50,
Eymery Fruger Cie, 1851.)
Map, 39 x 54 cm, slightly coloured.
Scale in French leagues; nautical leagues; days travel per caravan.
Prime meridian through Paris.

This mid-nineteenth-century French map of Africa offers a wealth of geographical detail, particularly in North Africa, in attractive and varied type. Southern Africa is extensively marked with accurate detail of that period, especially that of the Colonie du Cap. There are a few inaccuracies – Plettenberg is placed in the interior towards the east of the map. Bushmen and Hottentot names are marked as well as capes, harbours and rivers. Portuguese place names appear on the Natal coastline, and Monomotapa appears as a kingdom.

Madagascar and the other surrounding islands are accurately sited. Réunion, with its original European name, Bourbon, and Ile de France, appearing in the left lower corner, are very similar to the Brué map (see Map 127).

The great rivers of Central and North Africa are marked but not with great detail or accuracy. The Nile and its origin appear, as in so many other maps of Africa of this period, quite inaccurately and without the placing of the important lakes of the northern part of the continent. Across the middle of the continent appear the Mountains of the Moon, dividing known from unknown territory. A fine ornamented oval blind stamp appears on the map just to the right of the title with the initials in the centre and surrounding it the lettering *Atlas Universel de Geographie* (Lapie).

The authors of this map are father and son Lapie of Paris. The father, Lieutenant-Colonel Alexandre Emile, who was Geographer to the King, issued his *Atlas Universel* in 1829 and two subsequent editions in 1837 and 1851. The son, Captain M. Lapie, was in turn Geographer to the Dauphin.

160

Map 142

ANON
Africa. (Barcelona, 1850-1860.)
Map, 34,5 x 42 cm, slightly coloured.
Scale in Spanish leagues, French leagues, Arabic miles, English leagues, Portuguese miles, geographic miles.
Prime meridian through Madrid.

An undated map of Africa published in the Spanish language, from a Spanish atlas. It appears, from the geographical features presented, that this map would probably have been published about 1850. The outline of the continent appears to be accurate, with vast areas, particularly in Central Africa, devoid of geographical detail.

In the south the usual mid-nineteenth century features appear, particularly at the Cape, with the then known rivers and bays. A number of tribes are noted, together with the main rivers flowing eastwards and westwards. The lake basins and the origin of the Nile were unknown at the date of publication of this map.

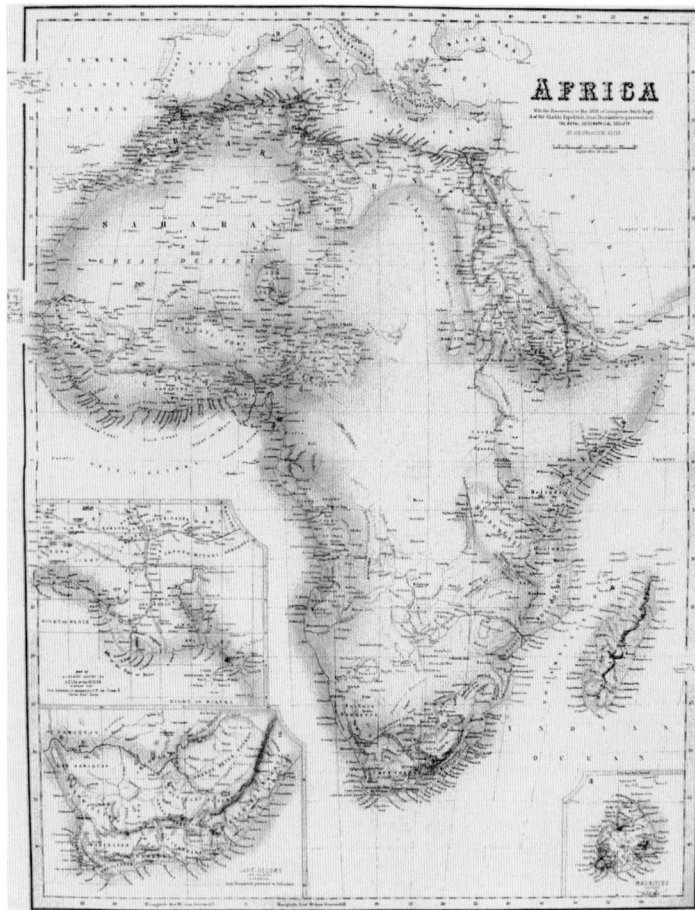

Map 143

SWANSTON, George
Africa with the discoveries to May 1858 of Livingstone, Barth, Vogel and the Chadda expedition from documents in possession of the Royal Geographical Society. (Edinburgh, A. Fullarton and Co. 1860.)

Map, 49 x 38 cm, slightly coloured.
Scale in English miles.
Prime meridian through Greenwich.
Insets: Cape Verde Islands.
Delta of the Niger.
Cape Colony.
Mauritius.
Right hand side (top and bottom): 64.

This nineteenth-century map of Africa is one of the more modern maps based on documentary evidence from the travels and expeditions of Livingstone and Barth up to 1858, which had been submitted to the Royal Geographical Society.

The map is printed in a modern colour, with a simple title, in the top right corner and four insets: of the Cape Verde Islands, the delta of the Niger, the Cape Colony and Mauritius. There are still large empty areas of land obviously unknown to the publisher, but additional information now appears, especially of the great lakes, and the presence of Nyanza-Victoria as a source of the Nile. In the south,

additional colonies appear: the Orange River sovereignty and, on the inset map, the southern aspect of the Vaal Republic, the forerunner of the South African Republic. More accurate tribal names and areas are noted but the ancient Monomotapa still exists.

Swanston was an engraver and draughtsman of Edinburgh responsible for a number of atlases, among which was *Africa*, 1858 of which this map is number 64 (as appears on the map). Fullarton and Co., publishers and engravers of Edinburgh, London and Dublin, were very active in the field of atlases, mainly of Great Britain.

162

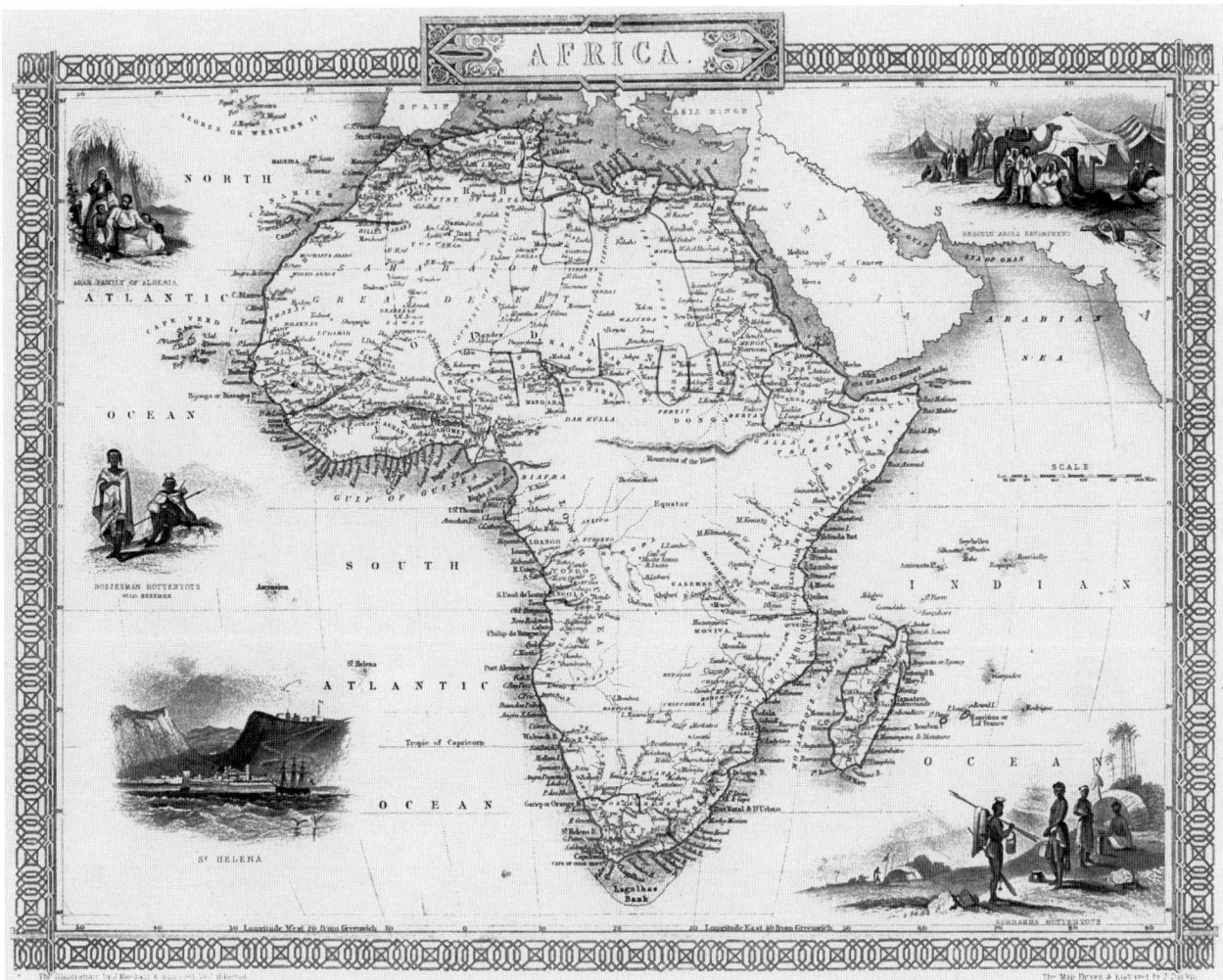

Map 144

TALLIS, John

Africa. Drawn and engraved by J. Rapkin; illus. drawn by J. Marchant and engraved by J.H. Kernot. (London, John Tallis and Co. 1880.)
Map, 24 x 32 cm, coloured outline.
Prime meridian through Greenwich.

This map of Africa is a line engraving of 1880 with a great deal of information in the southern part demarcating the Cape Colony, but with less information towards the north. The coastal geography is well endowed with accurate names.

The engraved vignettes, five in all, show an Arab family of Algeria, Bosjeman Hottentots, a view of St. Helena, a Bedouin Arabs' encampment and Korranna Hottentots. These form a colourful picture surrounding the continent. The illustrations were drawn by J. Marchant and engraved by J.H. Kernot. The map was drawn and engraved by J. Rapkin.

The firm of Tallis and Co. flourished from 1835 to 1900 with varying imprints. During this period they operated from various addresses in London, Edinburgh, Dublin and New York. Their illustrated atlas of 1850-1851 was one of the last decorative atlases. All maps were engraved on steel and adorned with small vignette views. Africa and parts thereof appeared in different sections as the atlas was published in parts.

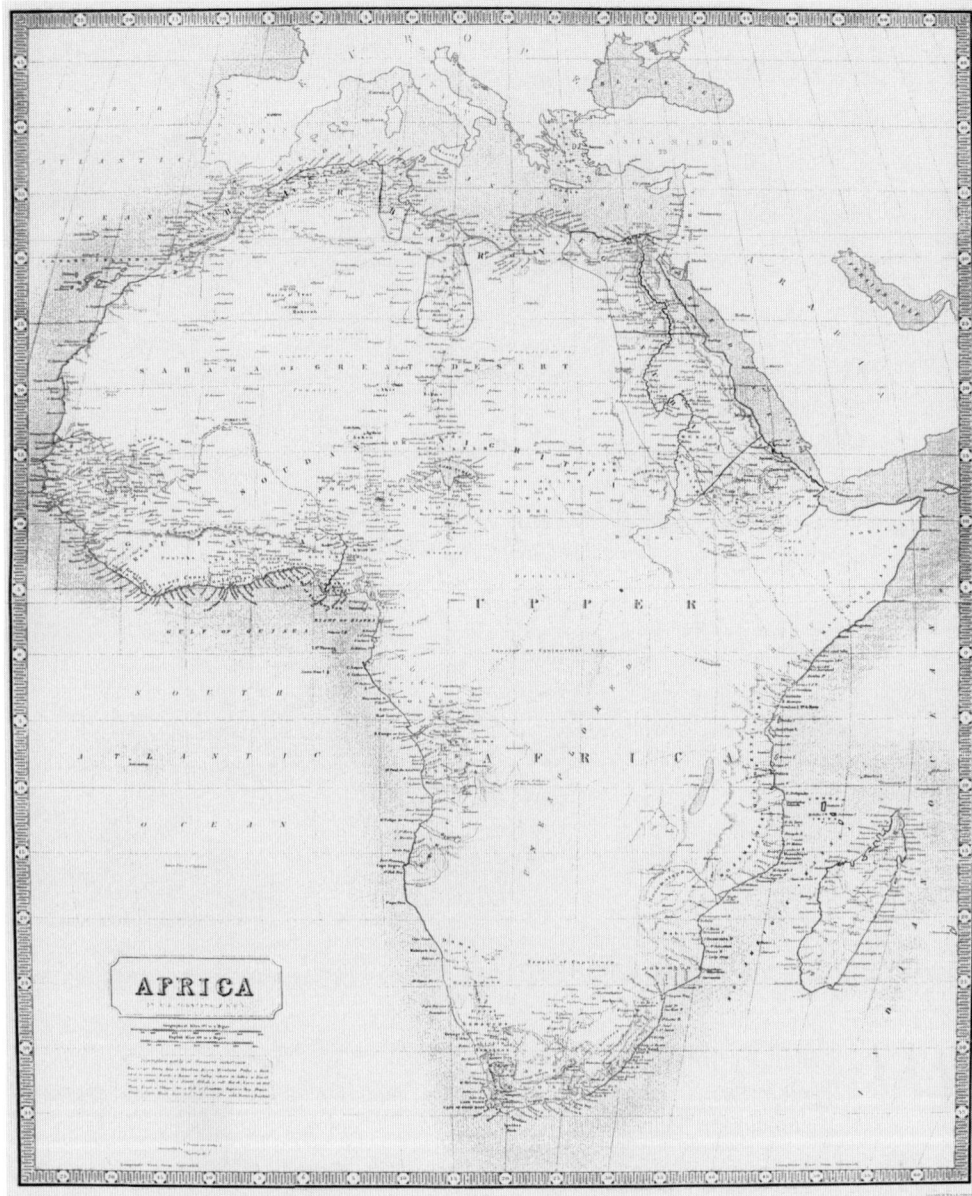

Map 145 JOHNSTON, Alexander Keith (1804-1871)
Africa. Engraved by W. and A.K. Johnston. (Glasgow, Robert Weir and James Lumsden and Son., 1843.)
Map, 58,5 x 49 cm, coloured outline.
Scale in geographical and English miles.
Prime meridian through Greenwich.
Bottom right: National Atlas 34.

This is a large early-nineteenth-century map of Africa. There is a conspicuous absence of geographic detail in the northern central and southern interior. South Africa and especially the then known Cape Colony and its coastline are noted with greater and fairly accurate detail. North of the Cape Colony there are a number of tribal names. The rivers and mountain ranges are depicted fairly accurately. This map appeared prior to the discovery of the source of the Nile, as Lake Victoria Nyanza does not appear. Monomotapa appears, and the Mozambique coastline is

accurately labelled with the names given to the rivers and inlets by the early Portuguese explorers.

The map publishing and printing firm which produced this map was that of W. and A.K. Johnston, founded by Sir William Johnston, of Hill Square, Edinburgh. Robert Weir and James Lumsden of Glasgow both had their own publishing firms and they collaborated with W. and A.K. Johnston.

164

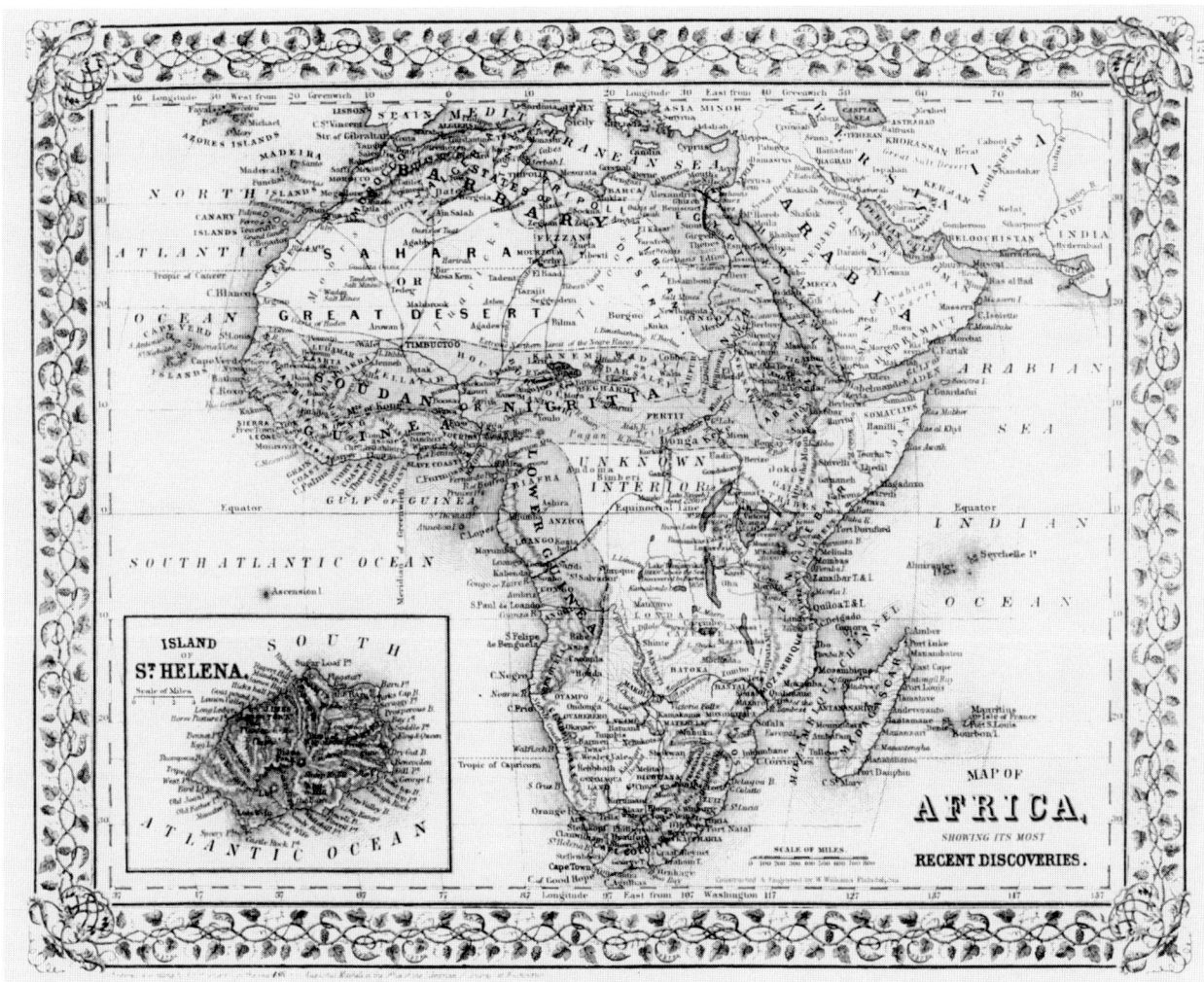

Map 146   MITCHELL, Samuel Augustus
Map of Africa Showing Its Most Recent Discoveries. Engraved by W. Williams.
(Philadelphia, S.A. Mitchell, 1881.)
Map, 27 x 34 cm, coloured.
Scale in miles.
Prime meridian through Washington.
Inset: Island of St Helena.
Top right: 120.

This map of Africa was printed and published in Philadelphia and first appeared in Mitchell's *New General Atlas* in 1867. Its engraver, W. Williams, was a geographer and draughtsman in Philadelphia, working for Mitchell from 1839-1853.

The map and its contents are accurate for that period. Southern Africa is now divided into the Cape Colony, the Transvaal Republic and the Orange River (Colony), showing Bloemfontein. Natal is not depicted as such but Durban and Port Natal and 'Zulu' obviously refer to this colony. Victoria Falls is marked in its correct position. Further north the lake systems now appear more correctly mapped showing the elongated Lake Nyassa (now Malawi), Lake Tanganyika and Victoria Nyanza. Lake Tanganyika is noted as having been discovered by Burton in 1859. The map also shows Speke's route via Victoria Nyanza to the Ripon Falls. From here northwards the Nile is depicted arising from Lake Victoria. The west coast includes up-to-date information, especially around Liberia, where American influence prevailed. The great era of immigration of African-Americans to Liberia was in the first half of the nineteenth century and this post-Civil War map reflects other American interests in Africa. The great explorers of Africa and their shaping of the history of this continent were intriguing to the reading public. The routes of Stanley and Livingstone are included, and there is a detailed inset of St. Helena, where Napoleon died and was buried.

165

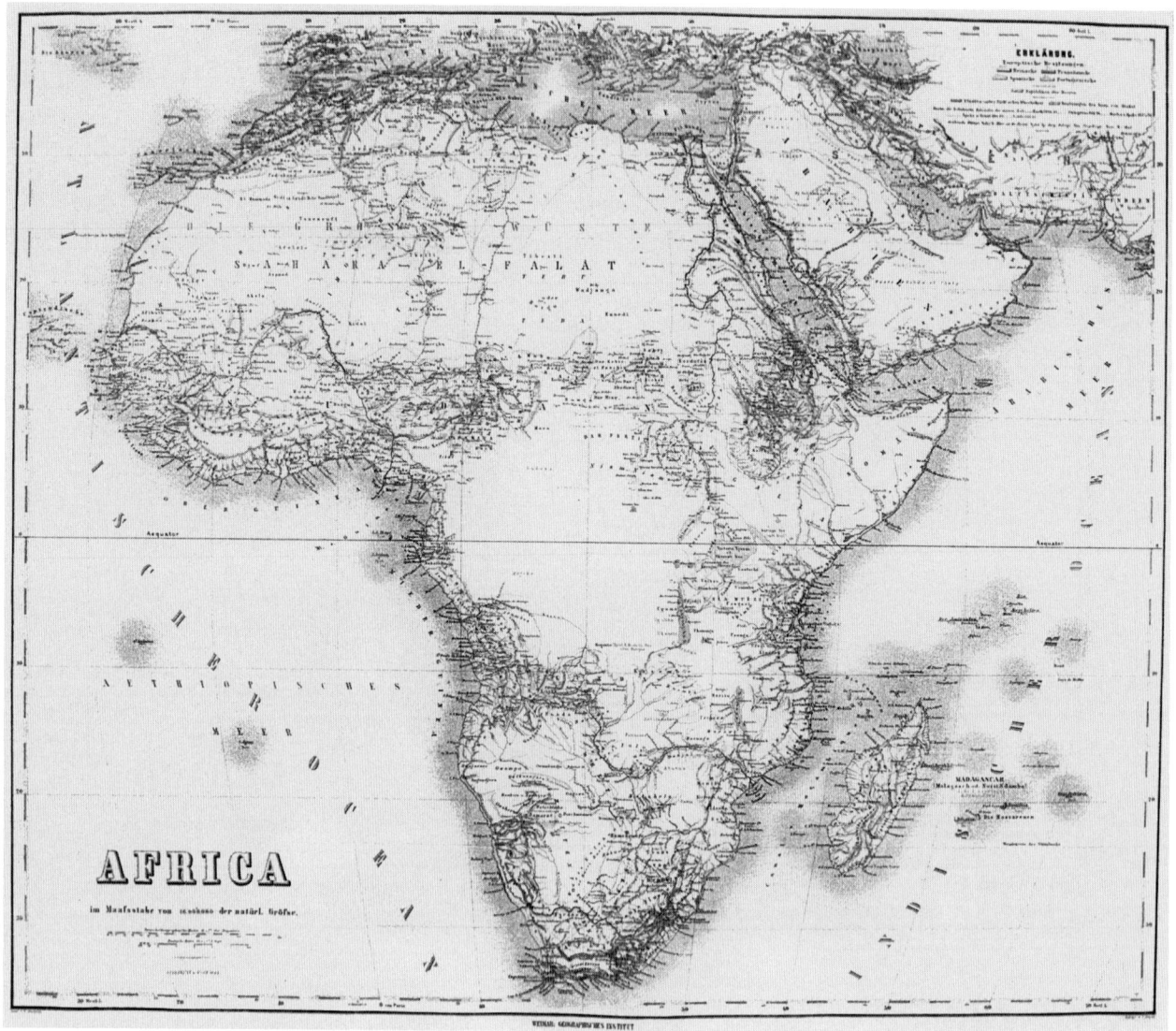

Map 147

GRÄF, Adolph

Africa. Drawn by A. Gräf and engraved by G. Haubold. (Wiemar, Geographisches Institut, V. Geyer, 1867-1881.)
Map, 53 x 64 cm, slightly coloured. Scale in German and English miles.
Prime meridian through Ferro Island and Paris.
Top right: 55.

This map of Africa is extremely detailed, with accurate, up-to-date information. It is not dated, but was probably made between 1867 and 1881, as details on the map suggest that diamonds had been discovered in the northern Cape (1867), though the Transvaal Republic has not yet been annexed by Britain (1881).

At the top right there is a colour-coded list of the geographic areas in the possession of Britain, France, Spain, Portugal, Turkey and the Iman of Maskat, which also includes the 'Republica der Boeren.' Below this is a list of all the early explorers, especially in Central and North Africa, with the dates of their travels: Barth, 1850-55; Livingstone, 1841–56; Burton and Speke, 1857-58; Speke and Grant 1861-72 and Rohlfs, 1861-67. The origin and course of the Nile is based on the discoveries of these pioneer travellers.

In southern Africa the Cape Colony, the Orange Free State and the Transvaal are accurately delineated, with a red border around the Cape to indicate British possession. A blue border around the Transvaal Republic indicates Boer possession, and marked within it are Prätoria and other early Transvaal place names, though the map is too early to include Johannesburg (1886). On the northern Cape border Hopetown – the site of the discovery of the Kimberley diamond fields – appears, though Kimberley itself is not marked.

166

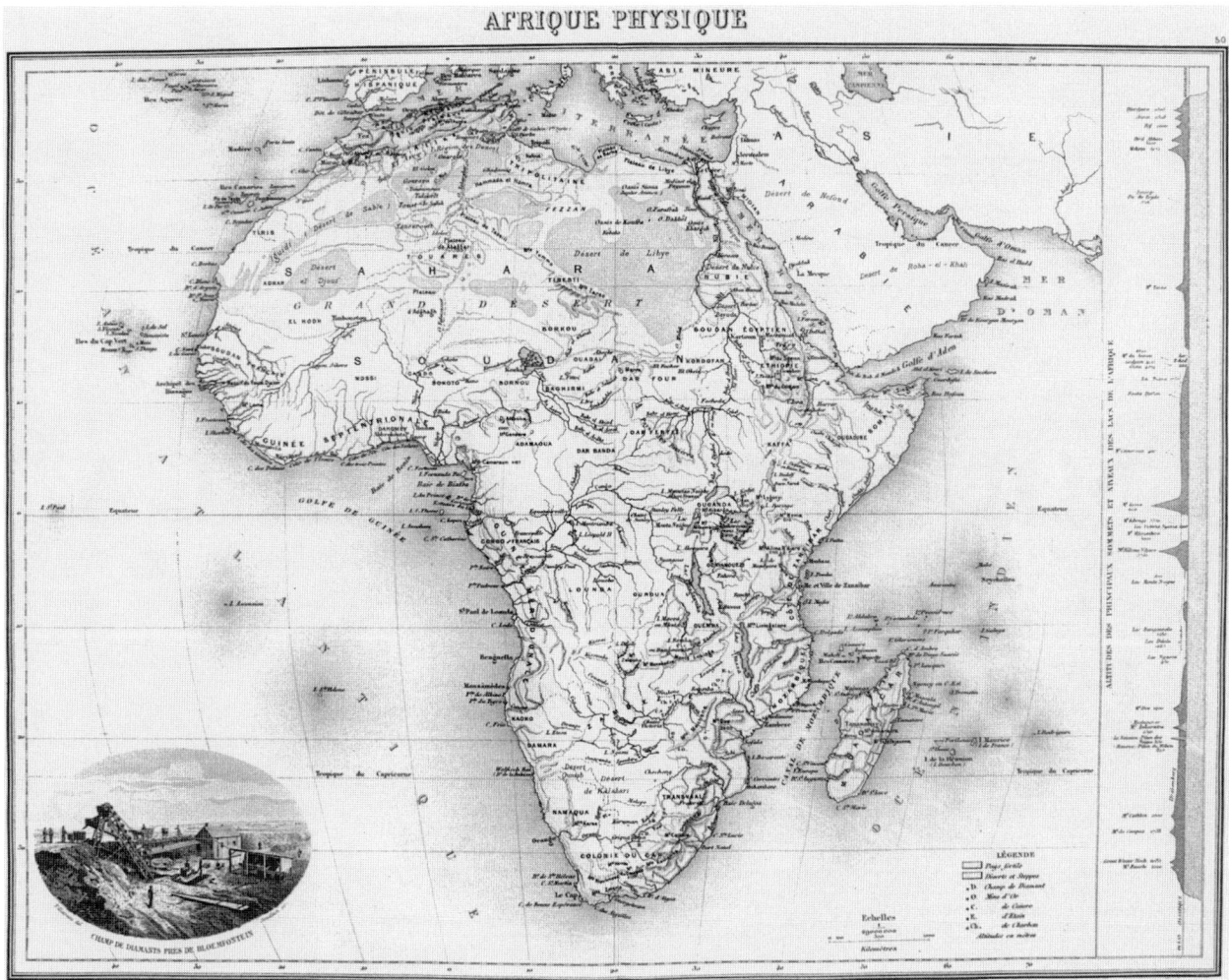

CHAMP DE DIAMANTS PRES DE BLOEMFONTEIN.

Map 148

MIGEON, J.
Afrique Physique. (Physical map of Africa.) Engraved by G. Lorsignol. (Paris, J. Migeon, du Moulin Vert, 1891.)
Map, 31 x 42 cm, coloured.
Scale in kilometres.
Inset: Heights of principal peaks and depths of principal lakes in Africa.
Engraving of diamond fields near Bloemfontein.

A coloured, late-nineteenth century French map of Africa, with an unusual oval engraved inset, in the left lower corner, of a mining operation in the veld referred to as 'diamond-fields near Bloemfontein.' The exact site should, of course, be described as Kimberley, which was well known as a diamond-digging town in Griqualand at the date of this map. Griqua Town is labelled but not Kimberley. This engraving was executed by two artists other than those responsible for the map, Fillatreau and Soudain.

The map is fairly accurate in shape and geographical content. In the south the Transvaal is marked, including Pretoria, though not a great deal of detail is given. An interesting feature in central Africa is the appearance of a better outline and depiction of the great lakes. Lake Tanganyika, Lake Nyassa, Victoria Nyanza, Albert Nyanza and Stanley Falls appear, and the Nile is more realistically shown emerging from Lake Victoria Nyanza.

167

SOUTHERN AFRICA

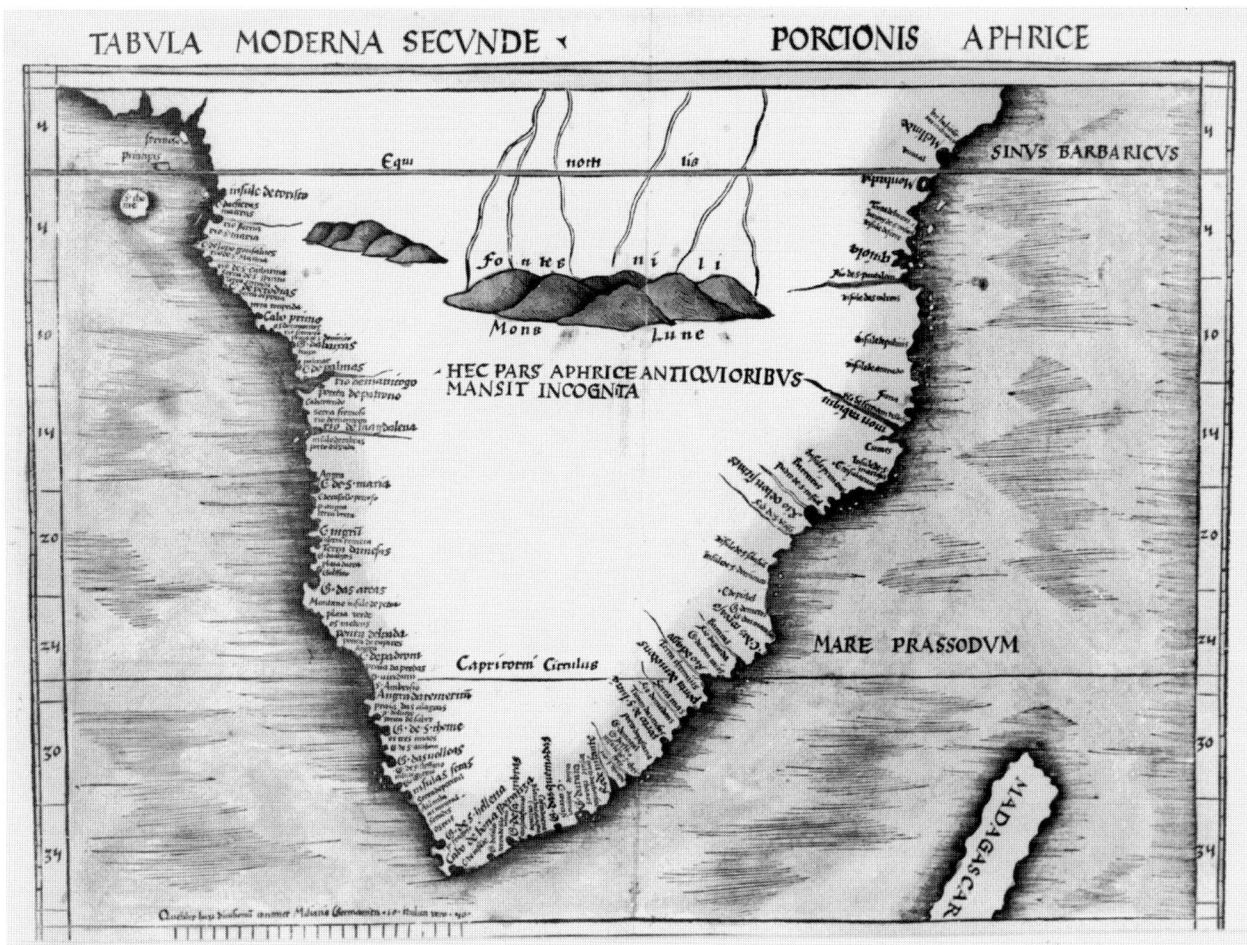

Map 149

WALDSEEMÜLLER, Martin (1470-1518)
Tabula Moderna Secunde Porcionis Aphriee. (Modern map of second half of
Africa – South Africa.) (Johannes Schott, Strasbourg, 1513.)
Map, 35 x 50 cm, coloured.
Scale in German and Italian miles.

This fine woodcut map of South Africa depicts the area
from the equator to the Cape of Good Hope, together with
most of Madagascar. The coastline is full of names of rivers
and bays, printed inland from the coast, using many of
Portuguese origin. The centre is left blank, with the
statement that this part of Africa remains unknown. The
prominent central mountain range is called Mons Lune and
described as the origin of the Nile. The Indian Ocean is
labelled Mare Prassodum. Three islands appear off the upper
west coast marked Formosa, Principis and S. Thome.

This map is from the 1513 Strasbourg edition of the
Waldseemüller atlas. It was reissued in slightly smaller form
in Strasbourg in 1520, 1522 and 1525; in Lyons in 1535, and
in Vienna in 1541.

Martin Waldseemüller, who was the first to suggest the
name America for the new world, was the principal

geographer of the Alsatian School. He finally settled in St.
Die, Lorraine. It is thought that because the Portuguese
authorities would not allow printed maps of their discoveries
in the Cape to be published, some of their particulars and
manuscript maps were secreted out of Portugal to St. Die to
be used by Waldseemüller. Thus, in 1513, in association with
a coterie of geographer friends, he produced an edition of
Ptolemy with twenty-seven ancient maps and twenty maps
based on contemporary knowledge – the first modern atlas.
These maps were compiled by Waldseemüller and included
this, the first separately printed map of South Africa. This
work was so authoritative that no other version of Ptolemy's
work appeared for the next twenty-seven years. As an
influential geographer of the early sixteenth century, his work
was copied by other important map publishers, as was that
of Mercator and many others. No significant advance in the
mapping of Africa was made on his work for almost fifty years.

169

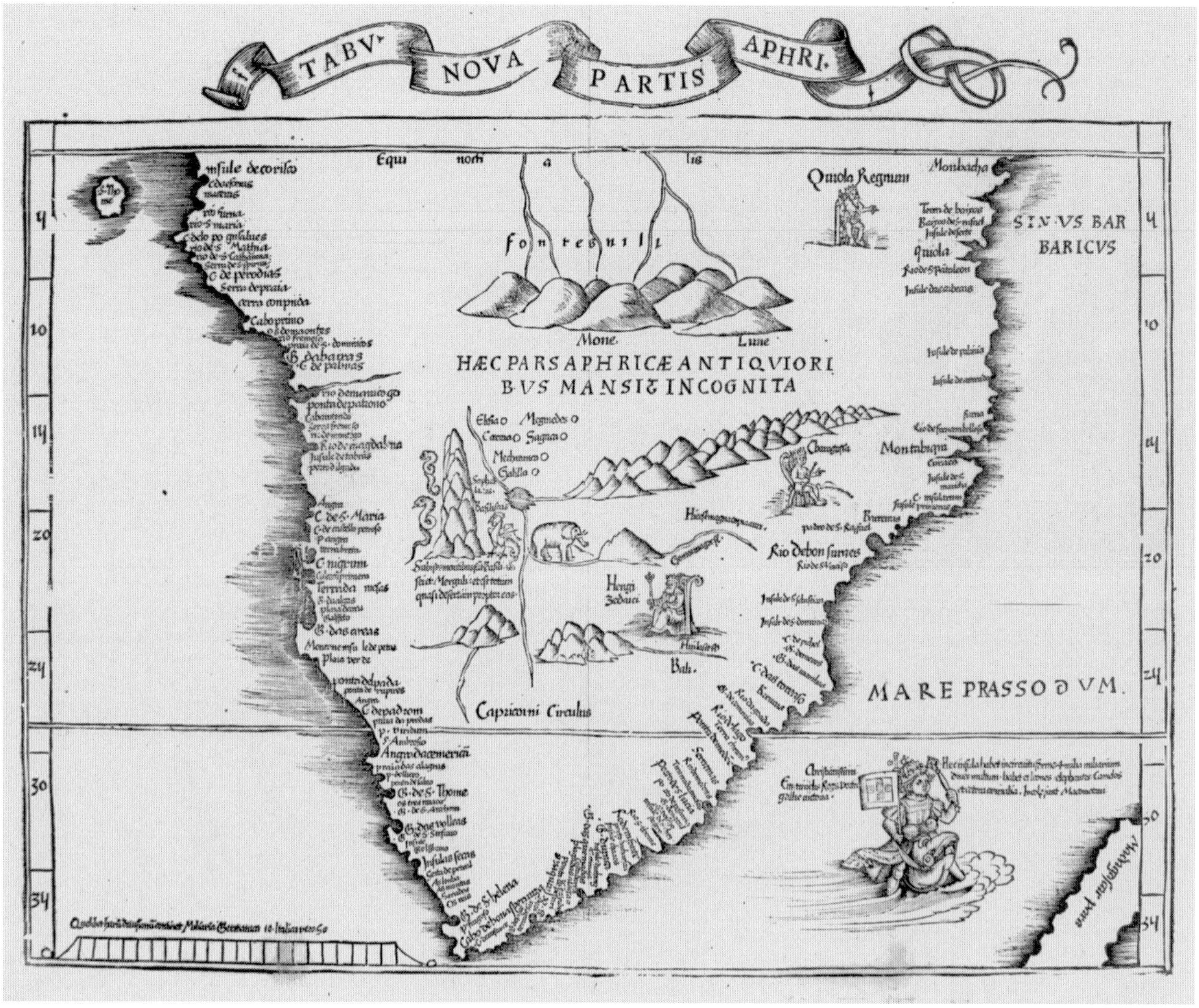

Map 150

WALDSEEMÜLLER, Martin (1470-1518)
Tabu. Nova Partis Aphri. (New map of part of Africa.) (Lyons, Melchior and Gaspar Treschel, 1535.)
Map, 30 x 42 cm, uncoloured.
Scale in German and Italian miles

This map of southern Africa is a woodcut published in Lyons by Melchior and Gaspar Treschel in 1535. The woodcut border and ornament are by Hans Holbein and Graf. The map itself is based on the Waldseemüller map of 1513 (see Map 149) which was reissued in 1520, 1522, l525, 1535 and 1541. This map is characterised by the title and scrollwork above the map which identify Lyons as the place of publication.

This reissued map now has three kings on their thrones, an elephant, a cockatrice and two serpents next to a sugarloaf mountain, while the King of Portugal rides a bridled sea monster on the Mare Prassodum, holding the banner of Portugal in his right hand and the sceptre in his left. Mountains are added and inland rivers appear to the south of the Mountains of the Moon. The reverse of this reissue is blank.

170

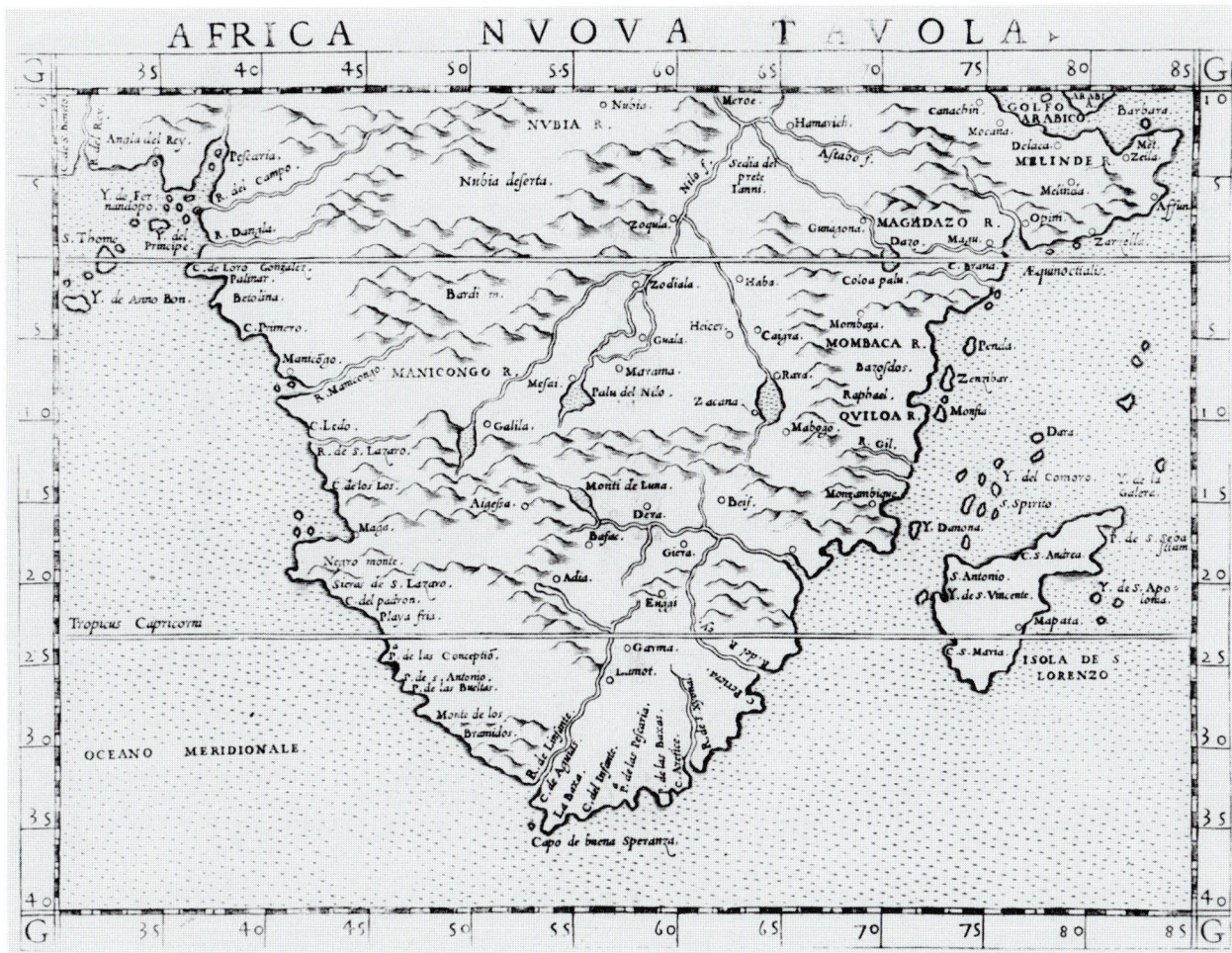

Africa Nuova Tavola. (New map of Africa.) (map image with title "AFRICA NUOVA TAVOLA")

Map 151

RUSCELLI, Girolamo (c.1504-1566)
Africa Nuova Tavola. (New map of Africa.) (Venice, V. Valgrisi, 1561-1562.)
Map, 18 x 24 cm, uncoloured.

This map is one of a series produced by two Italian printers, Vincenzo Valgrisi, who printed Ruscelli's Italian translation of Ptolemy in 1561 and 1562, and Giordano Ziletti, who also printed editions of Ruscelli's Ptolemy in 1564 and 1574. This map is probably from one of the Valgrisi editions. It is a typical Italian copperplate engraved map with decisive outline, stippled seas and anthill mountain ranges. The origin of the Nile conforms to the Ptolemaic concept and certain important rivers are depicted. The final editions of this work were printed in 1598 and 1599 in Venice by the Heirs of Melchior Sessa.

171

Map 152

SANUTO, Livio (1520-1576)
Africae Tabula X. (Venice, L. Sanuto, 1588)
Map, 38,5 x 51,5 cm (top) and 41,5 cm (below), irregular shape, uncoloured.
Scale in millaria.
In: Sanuto, Livio. *Geographia*. Venice, L. Sanuto, 1588.

'Africae Tabula X' is a finely engraved map of southern Africa, showing the course of the Limpopo and the Zimbabwe Ruins, which Sanuto calls 'the work not of humans but the devil,' because the granite walls were of much greater perfection than the Portuguese fortresses by the sea. There are also accounts of Benomatapa (Monomotapa), Simbaoe (Zimbabwe), Manica and Golfo do Natal, etc. Table Bay appears as Saldanhe Agoada and Crucis Insule (the Isle of St. Croix) is shown at the entrance to what is now known as Algoa Bay. For further details on Sanuto's *Geographia* and the maps it contained, see Map 15 and Map 304.

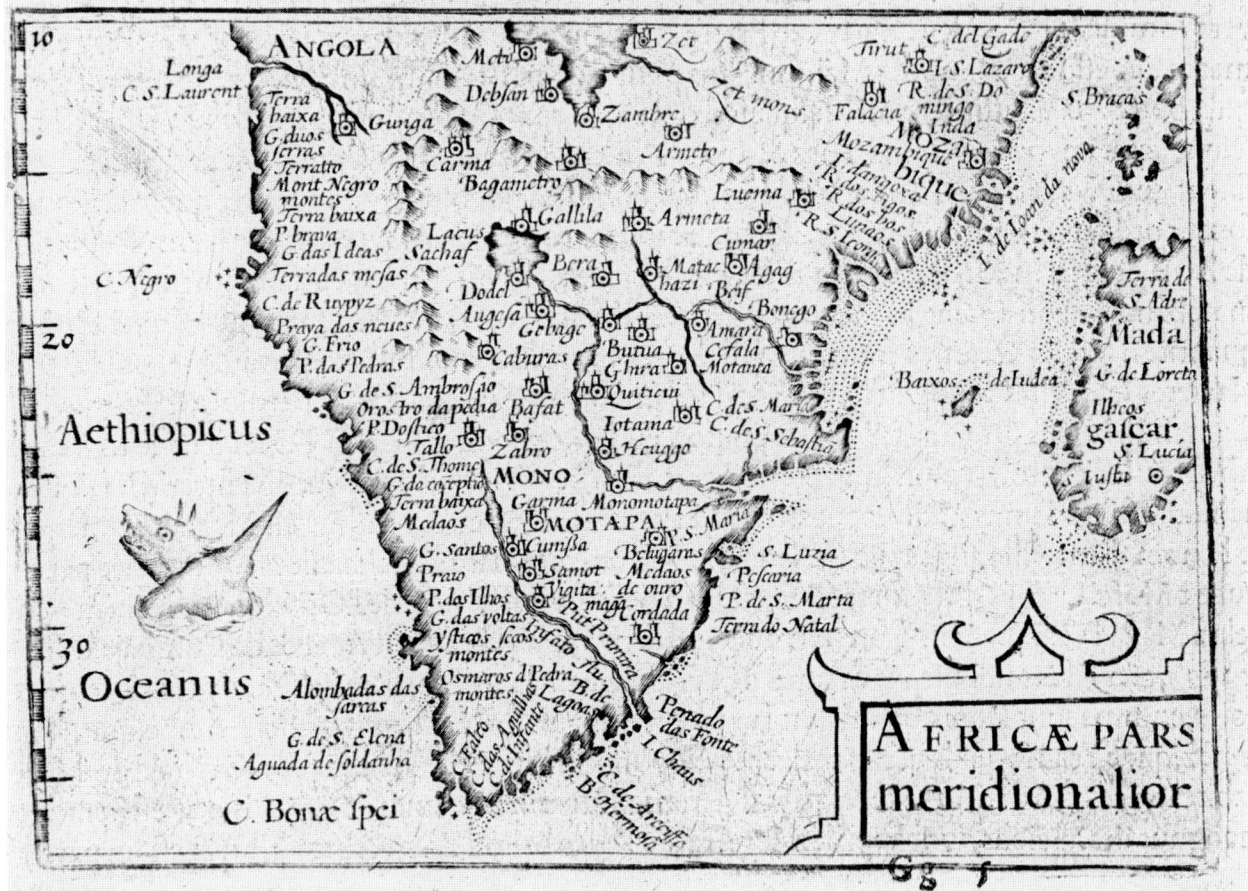

Map 153 BERTIUS, Petrus (1565-1629)
Africæ Pars meridionalior. (Southern Africa.) (Leiden, 1606.)
Map, 8,5 x 12 cm, uncoloured.
Title across top: *La Pointe d'Afrique.*
Top right: 105.
Bottom right: Cgs.

This map of the southern part of Africa was issued originally in 1598, and then at six different periods until 1618. The one featured here appeared in 1606, in Barent Langenes's *Caert Thressor,* and is unchanged from the original, except for the lettering above the map in French and the French text at the back, with the page number 105 and with the signature Cgs (the initials of Christoffle Guyot, the printer) below the title cartouche.

In spite of its small size this map provides a great deal of finely engraved information on the coastline and in the interior, repeating much of the geographical knowledge of this early seventeenth-century period. There is a very unusual sea monster in the Aethiopicus Oceanus. The title cartouche appears in the lower right corner in a box-like geometric design.

Petrus Bertius (also known as Pierre or Pieter Bert or Berts) was born in Beveran in Flanders. He was the brother-

in law of Jodocus Hondius and Pieter van der Keere (Petrus Kearius). He studied at Leiden University, and then visited Germany and Russia. On his return to Leiden he was appointed Director of the University Library and Professor of Mathematics. In 1620 he became Cosmographer to Louis XIII, and he died in 1629 in Paris.

Bertius is perhaps best known for his miniature *Atlas Tabulorum Geographicum Contractorum,* first issued in 1600, of which most of the maps were engraved by van den Keere. Bertius's early maps first appeared in Langenes's *Caert Thressor* of 1598. Langenes was a Dutch cartographer and publisher who set the fashion for miniature atlases. This particular series by Bertius are often referred to as the Langenes-Bertius maps and were issued from 1598 to 1618, illustrating the various regions of Africa. They were reissued in various editions, with modifications and improvements.

173

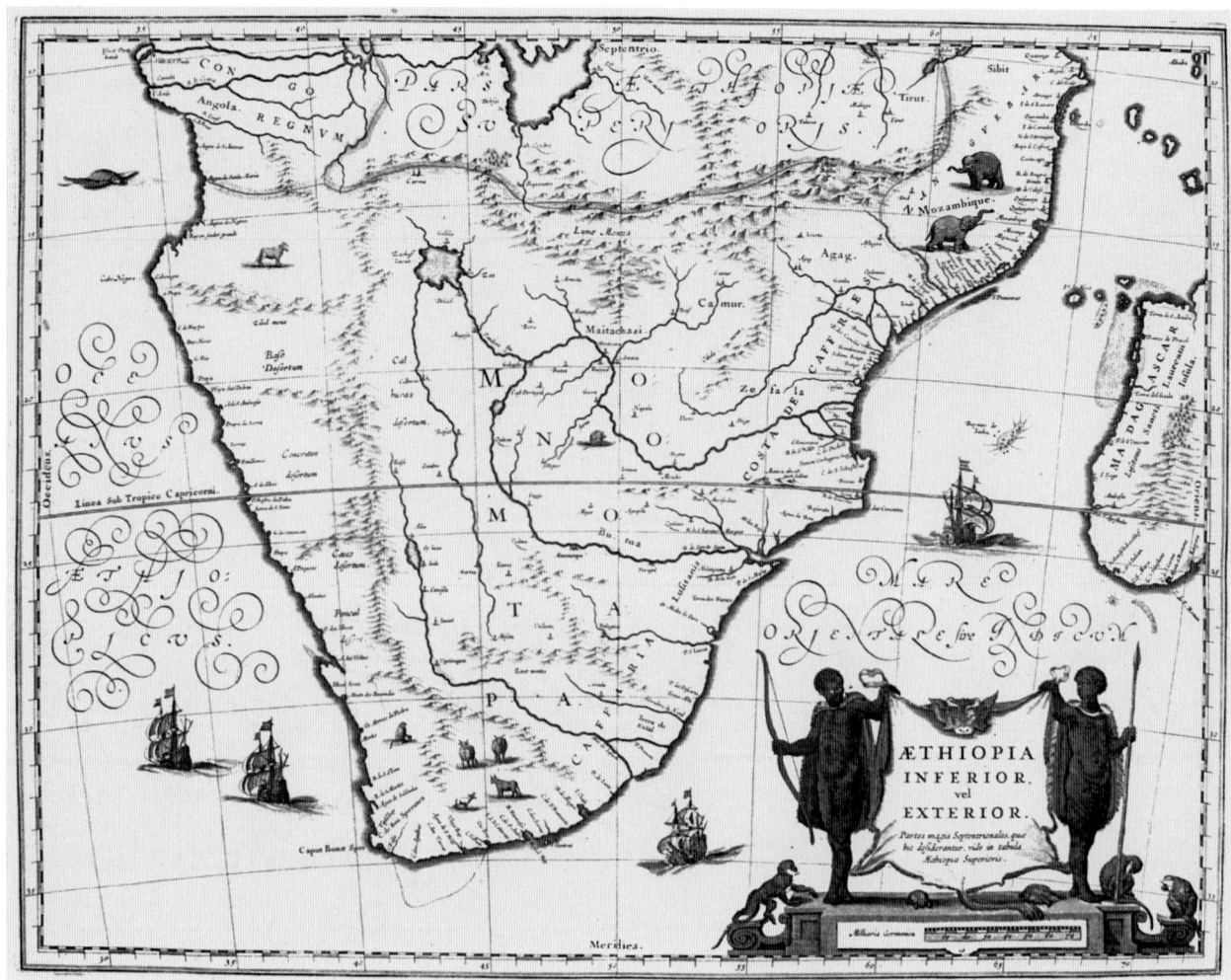

Map 154
BLAEU, Willem Janszoon (1571-1638)
Aethiopia Inferior vel Exterior. Partes magis Septentrionales, quae hic desiderantur, vide in tabula Aethiopiae Superioris. (Southern Africa...)
(Amsterdam, W.J. Blaeu, 1635.)
Map, 38 x 50 cm, coloured outline.
Scale in German miles.
On the verso a description of Kaffraria, Monomotapa, Congo, Zanzibar, Quiloa, Mombaza and Ajana.

This handsome map of southern Africa features an appropriate decorative cartouche in the right lower corner with the title displayed on an ox skin being held up by an African on either side. There are monkeys and tortoises around the base. Various types of animals are depicted on the mainland, the most recognisable of which are two elephants near the Mozambique coast. Two sailing ships appear in the Atlantic Ocean and one in the Indian Ocean between the mainland and Madagascar.

The map covers an area from Congo-Zanzibar to the Cape. It was published before Van Riebeeck settled at the Cape, and such knowledge of South Africa as cartographers had was based on prior Portuguese exploration, travellers' tales and rumours about the interior, which were used to fill in the blank spaces on the maps. This was a standard map throughout the seventeenth century and it went into many

174

editions, and was copied by other publishers. Joannes Jansson published an almost identical copy except that he removed the ship above the title cartouche. It was also issued by Merian, on a reduced scale, with one less monkey and an extra ship. Frederick de Wit reproduced it exactly in 1694 without the text on the back.

The firm of Blaeu was founded by Willem Janszoon and his sons Joan (John) and Cornelis in 1599. Blaeu was a scientist, trained by the celebrated Danish astronomer Tycho Brahe. After returning from his studies with Brahe he was appointed examiner of navigation and hydrographer to the powerful Dutch East India Company. On his death in 1638 the business passed into the hands of his two sons.

During the lifetime of Willem Janszoon the Blaeu family published a prodigious number of globes, wall maps, large folio atlases of town plans, and sea and land atlases, the last of which appeared as the famous twelve-volume large folio atlas issued not only in Dutch but also in Latin, French, German, Italian and Spanish. Their maps were issued either plain or in colour, special examples being heightened with gold. The more colourful ones were eagerly sought after by connoisseurs and collectors, while their sea charts were prized by navigators and seamen.

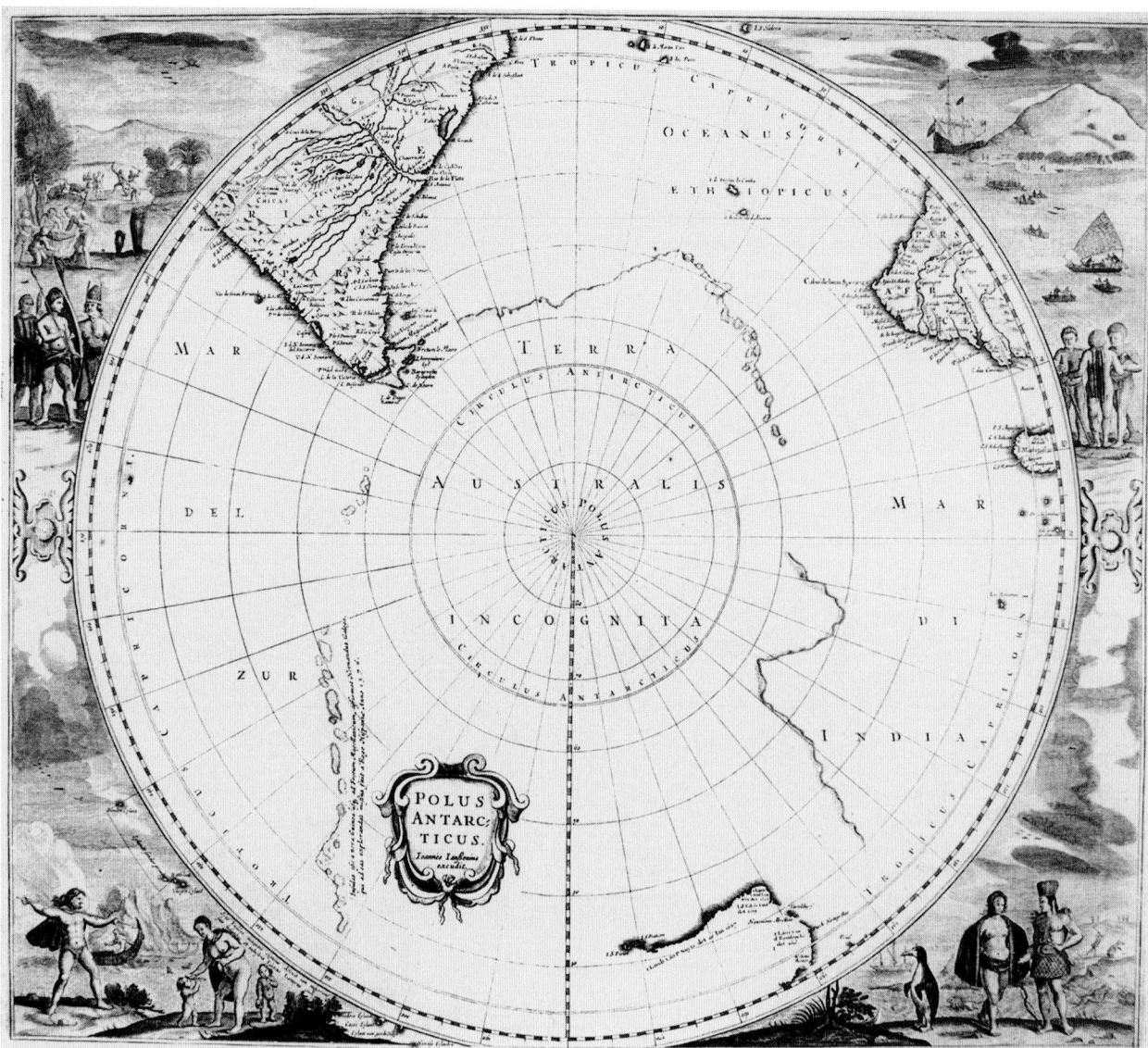

Map 155

JANSSON, Joannes (1588-1664)
Polus Antarcticus. (The Antarctic Circle.) (Amsterdam, J. Jansson, 1637.)
Map, 43 x 49 cm, slightly coloured.

This map was executed by Joannes Jansson in 1637, and has been chosen to be included in this series of maps of Africa because it depicts a portion of the southern part of the continent. Joannes Jansson was born in Arnhem in 1588, and married Elizabeth Hondius, daughter of Jodocus Hondius, himself a recognized geographer, engraver and publisher who settled in London. Jansson and his wife settled in Amsterdam. Upon the death of Jodocus Hondius II, Joannes Jansson and Henry Jansson together published a series of atlases, one of which is the rare *Appendix Nova Atlanti* (Amsterdam, 1647) in which this map appears.

This South Polar hemisphere includes Africa, South America and part of the Australian coastline, as well as an archipelago which the Latin text calls New Guinea and attributes to the reports of Magellan.

This is a strikingly decorative map. In the lower left corner a native feeds a fish to one of her children. In the background a native paddles a canoe and in the upper left hunters and warriors congregate, while a feast is being prepared. At the upper right three Africans stand on the shore of a busy harbour. In the lower right a woman and a man stand with a penguin to their right. In the far distance there is another figure with a penguin. Further coloured illustrations line the borders.

In plate vii of Tooley's *Map Collectors Series* no. 2, a variation of this map appears. It is identical except for a blank oval cartouche above and to the left of the usual square-like cartouche, in which is printed the title 'Polar Antarcticus.' In the variation the name Hondius appears in the cartouche, whereas in the above map the name Jansson appears.

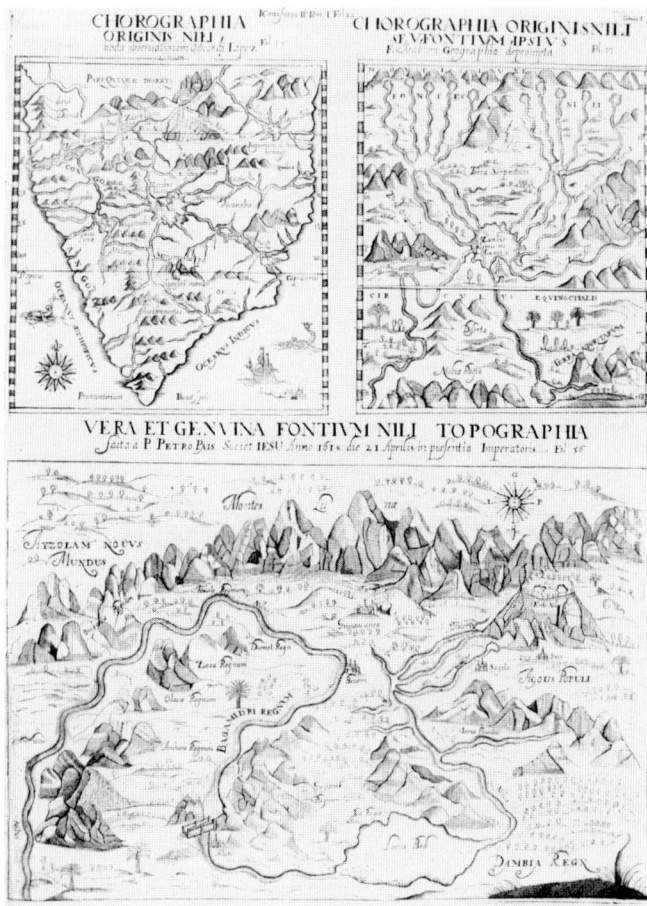

Map 156

KIRCHER, Athanasius (1602-1680)
Chorographia originis Nili juxta observationem Odoardi Lopez. Fol. 55. Chorographia Originis Nili... ex Arabum Geographia deprompta. Fol. 53. Vera et Genuina Fontium Nili Topographia facta a P. Petro Pais... 1618. Fol. 56. (Three maps of the origins of the Nile according to Duarte Lopez, to Arab geography and Peter Pais.) (Amsterdam, A. Kircher, 1652.)
Three maps on one sheet, 46 x 36 cm, uncoloured.
Title across the top: *I Conismas II Tom. I Fol. 53.*
Upper right: Tomus I.

This three-map sheet first appeared in *Oedipus Aegyptiacus* in Rome, 1652. The map at the top left represents the origin of the Nile as explored by Lopez and described by Pigafetta in his well-known map, which was drawn especially to depict this origin of the Nile from two lakes, one below the other, thus discarding the Ptolomaic concept of the two lakes side by side.

The top right map represents the Montes Lunae origin of the Nile, north of Lake Zambia, while the wider map at the bottom shows in enlarged detail the mountains and rivers as understood by the geographers of that period.

All three maps are embellished in their interior by mountains, citadels, trees and unusual animals. An attempt is made to demarcate the various North African kingdoms.

Kircher is reputed to be the first to have described the finding of the Nile as recorded by the Jesuit traveller Paez. An account appears in vol. 3 of Bruce's *Travels to discover the source of the Nile in the years 1768 to 1773.*

Athanasius Kircher, a celebrated Jesuit scholar, was born in Geyseen and educated in a Jesuit college in Fulda, where he studied medicine, mathematics and natural history. He travelled extensively and became interested in translating oriental scripts, both Coptic and hieroglyphic. He eventually settled in Rome, where he died. He published a number of works, characterised by rather curious maps, among which was *Mundus Subterraneus* in 1665 (where this map appeared), with editions in 1668 and 1678. Kircher's maps were engraved in Amsterdam by Jansson and Weyerstrat, whom the Jesuit never met, but to whom he supplied rough sketches.

177

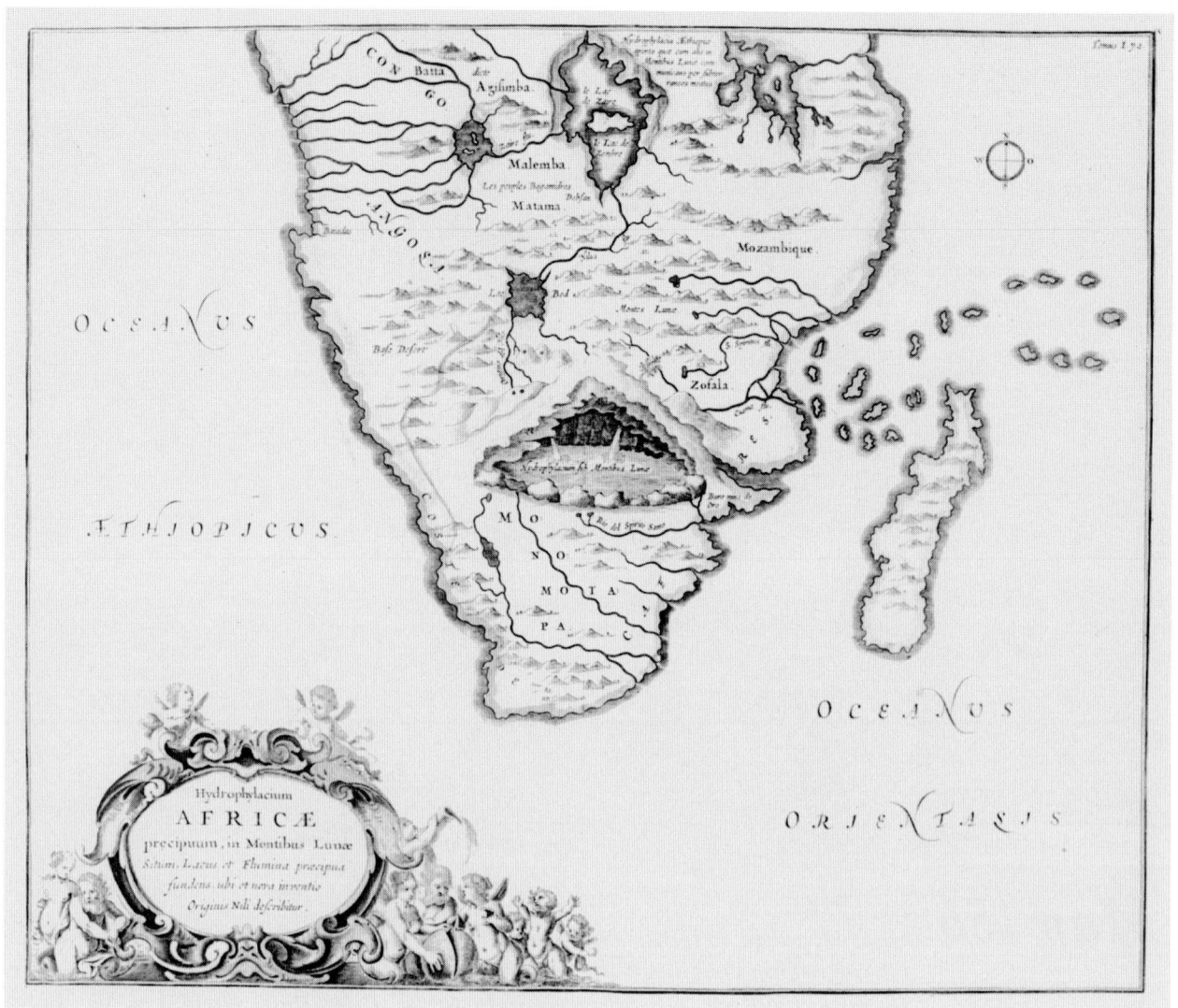

Map 157

KIRCHER, Athanasius (1602-1680)
Hydrophylacium Africae precipuum, in Montibus Lunae Situm, Lacus et Flumina praecipua fundens ubi et nova inventio Originis Nili describitur.
(Map of southern Africa showing sources of the Nile in a cavern beneath the Mountains of the Moon.) (Amsterdam, A. Kircher, 1665.)
Map, 34 x 41 cm, coloured.
Upper right: Tomus 1.72.

Athanasius Kircher, a Jesuit missionary, was well known for his extensive learning and voluminous writings on the natural history of the Kingdom of Abyssinia. He published an account of the 'fountains of the Nile' from a journal left by Peter Paez, a contemporary Jesuit colleague of Kircher's.

Bruce's *Travels to discover the source of the Nile in the years 1768-1773* (vol. 3, 1790) records Kircher's account, in which Paez reports: 'On the 21st April in the year 1618, being there with the King and his army, I ascended the place and observed everything with great attention: I discovered first two round fountains each about four palms in diameter and saw with the greatest delight what neither Cyrus, the King of the Persians, nor Cambysis, nor Alexander the Great, nor the famous Julius Caesar could ever discover. The two openings of these fountains have no issue in the plain on the top of the mountain but flow from the root of it. The second fountain lies about a stone-cast west from the first: the inhabitants say that this whole mountain is full of water and add that the whole plain about the fountain is floating and unsteady, a certain mark that there is water concealed under it.' He also refers to this place as Geesh.

Kircher's description of the cavern is clearly shown above. He refers to it as 'Hydrophylacium Africae precipuum in Montibus Lunae'; and shows the Rio del Spirito Santo coursing parallel to the lower border of the cavern, entering the sea at a spot referred to as a place of gold.

Geographically there is very little accurate detail, with a complete absence of coastal place names, though large lakes and rivers appear, including Lac Zaire and an unnamed lake to the east of it, together with numerous mountain ranges. The Cape of Good Hope is not mentioned.

There is an attractive cartouche in the lower left corner in which the title appears in an oval surrounded by a scroll design surmounted by two cherubs. To the right of the cartouche is a group of angels, three of whom are looking at a partially opened sphere. Above this group is another cherub drinking from a horn of plenty.

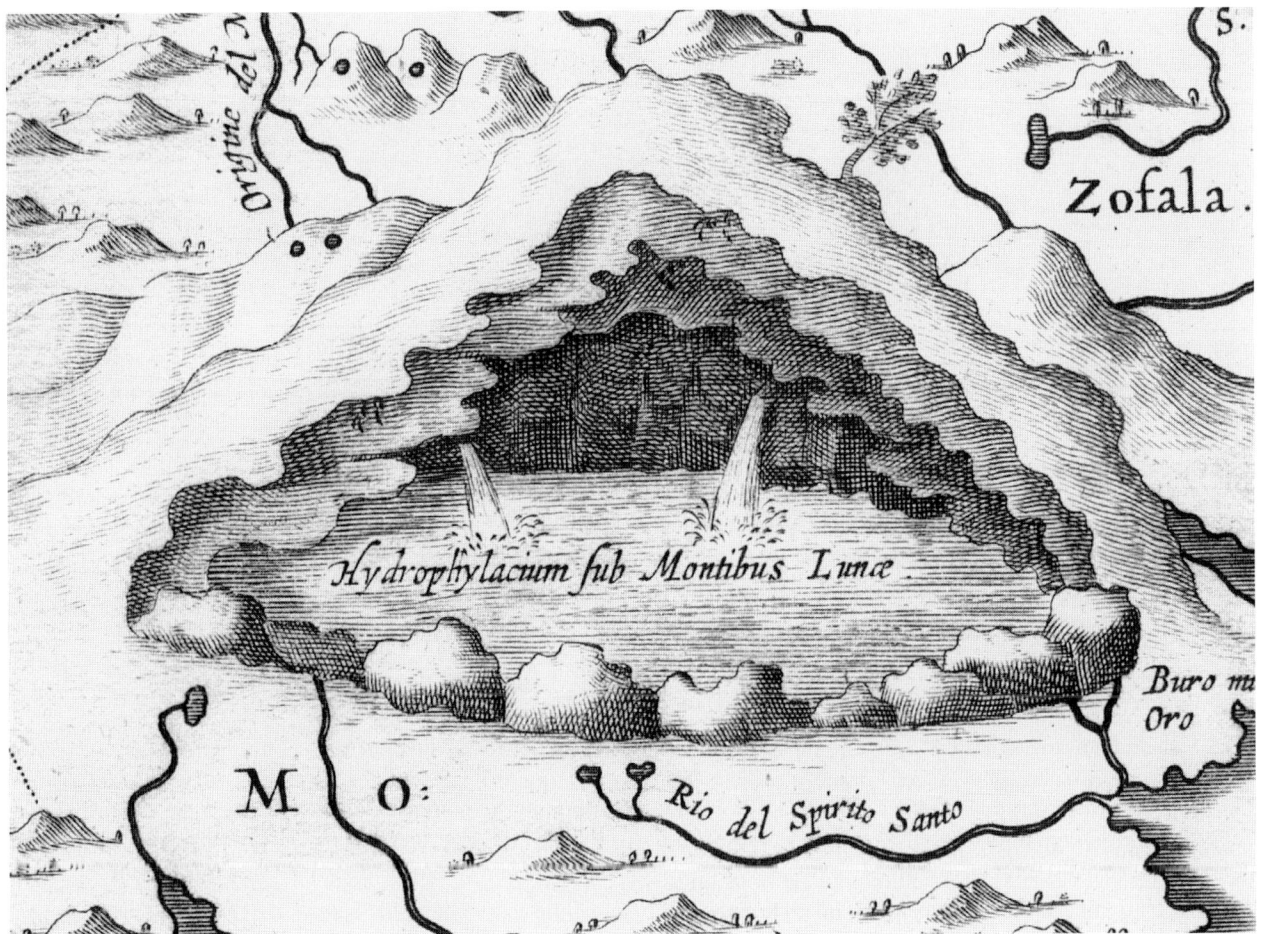

179

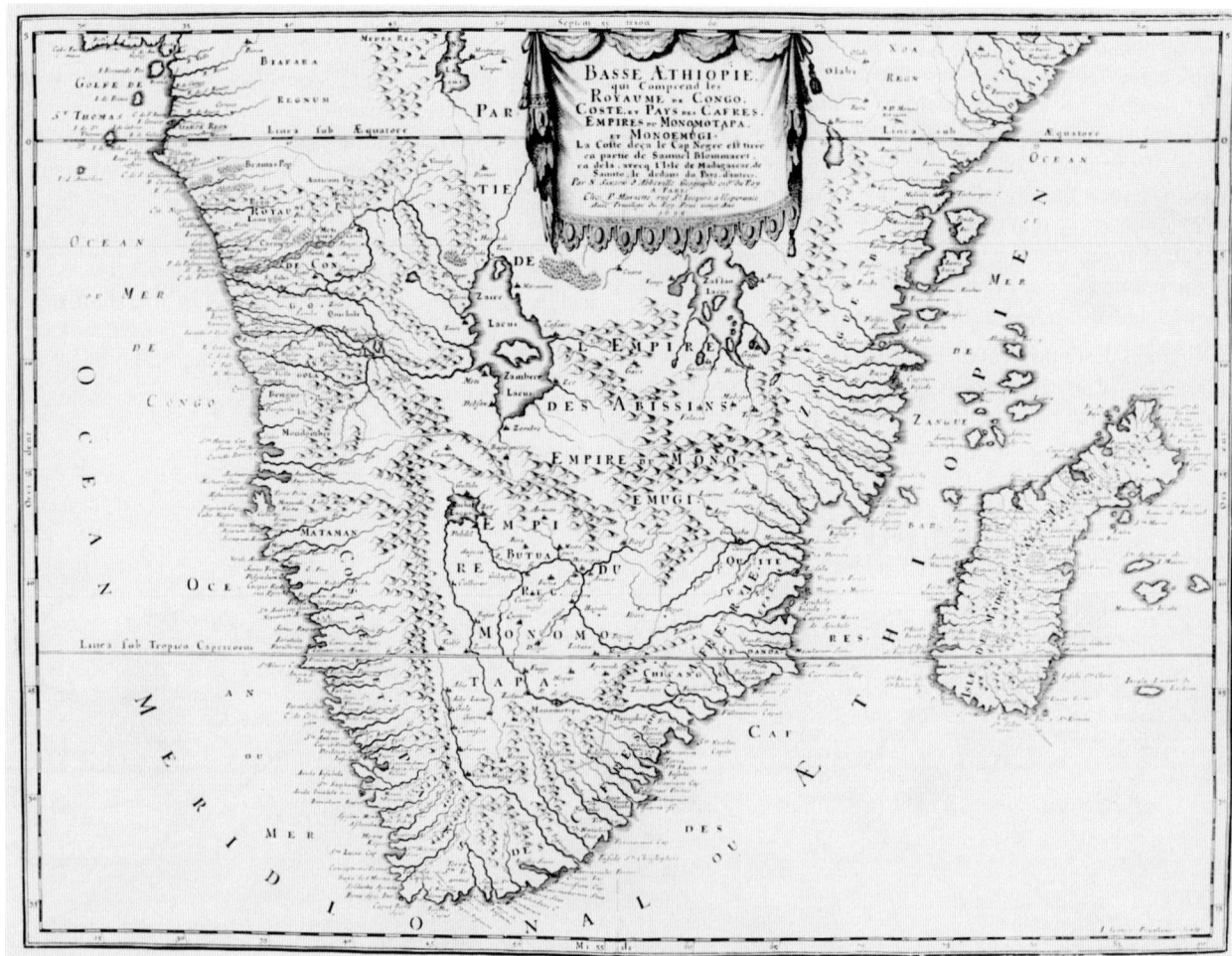

Map 158

SANSON, Nicolas (1600-1667)
Basse Aethiopie qui Comprend les Royaume de Congo, Coste, et Pays des Cafres, Empires du Monomotapa, et Monoemugi. (Southern Africa comprising the kingdom of Congo, coast of Kaffraria, empires of Monomotapa and Monoemugi.) Engraved by Jean Somer Pruthenus. (Paris, P. Mariette, 1655.)
Map, 42 x 56 cm, cartouche slightly coloured.

A Sanson map of southern Africa. The only decoration is the cartouche, made up of fringed drapery with tassles. The title within the cartouche states that the coast below Cape Negro is derived from Samuel Blommaert, Madagascar from Sanuto and the interior from other sources.

This map, although a fine example of cartography, is based on very little information; for example, the desert–like coast of South West Africa is presented as a land rich in rivers. The fictitious states are given fictitious boundaries and the interior is marked with large cities such as Vigiti Magna and Monomotapa. This information led to subsequent exploration, in search of these fabulous cities. Although the map is dated three years after the foundation of a settlement at the Cape by Jan van Riebeeck, no mention is made of any European settlement anywhere in the south.

This map is probably taken from the work *Cartes Générales de Toutes les Parties du Monde* by N. Sanson, published in Paris by P. Mariette in 1658 with maps dating from as early as 1632.

180

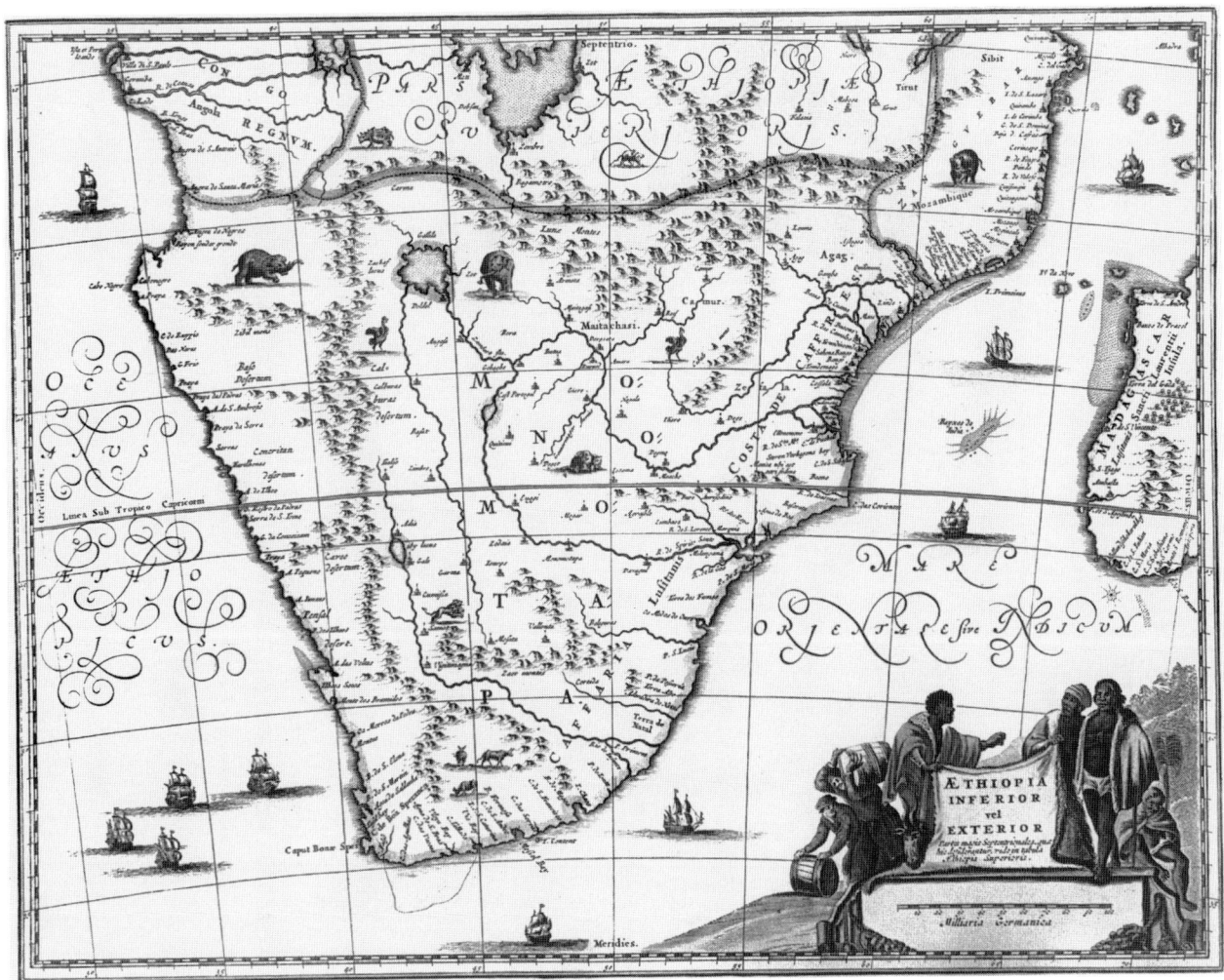

Map 159

OGILBY, John (1600-1676)
Aethiopia Inferior vel Exterior. (Southern Africa.) Engraved by Jacob van Meurs.
(London, J. Ogilby, 1670.)
Map, 28 x 36 cm, coloured outline.
Scale in German miles.
In: Ogilby, John. *Africa, being an Accurate Description of the Regions of Egypt, Barbary, Lybia and Billedulgerid...* London: J. Ogilby, 1670.

This decorative map of southern Africa bears a close resemblance to Blaeu's South Africa, having the same geographic content, but with more ships and a greater variety of animals on the mainland. The same florid script is evident.

The cartouche at the lower right corner consists of a group of natives surrounding an animal skin with the inscription on it. Two men on the left are handling casks. On the right there are two further natives, one a woman, seated on the masonry bearing the scale in 'Milliaria Germanica.' This map appeared in the Dutch, French and German editions of Dapper's *Africa*, but nowhere is a cartographer or engraver named. It is assumed that Jacob van Meurs (Mersius) was responsible for the engraving because of its similarity to a map on which his work is identified. These two maps both appear in Ogilby's *Africa*.

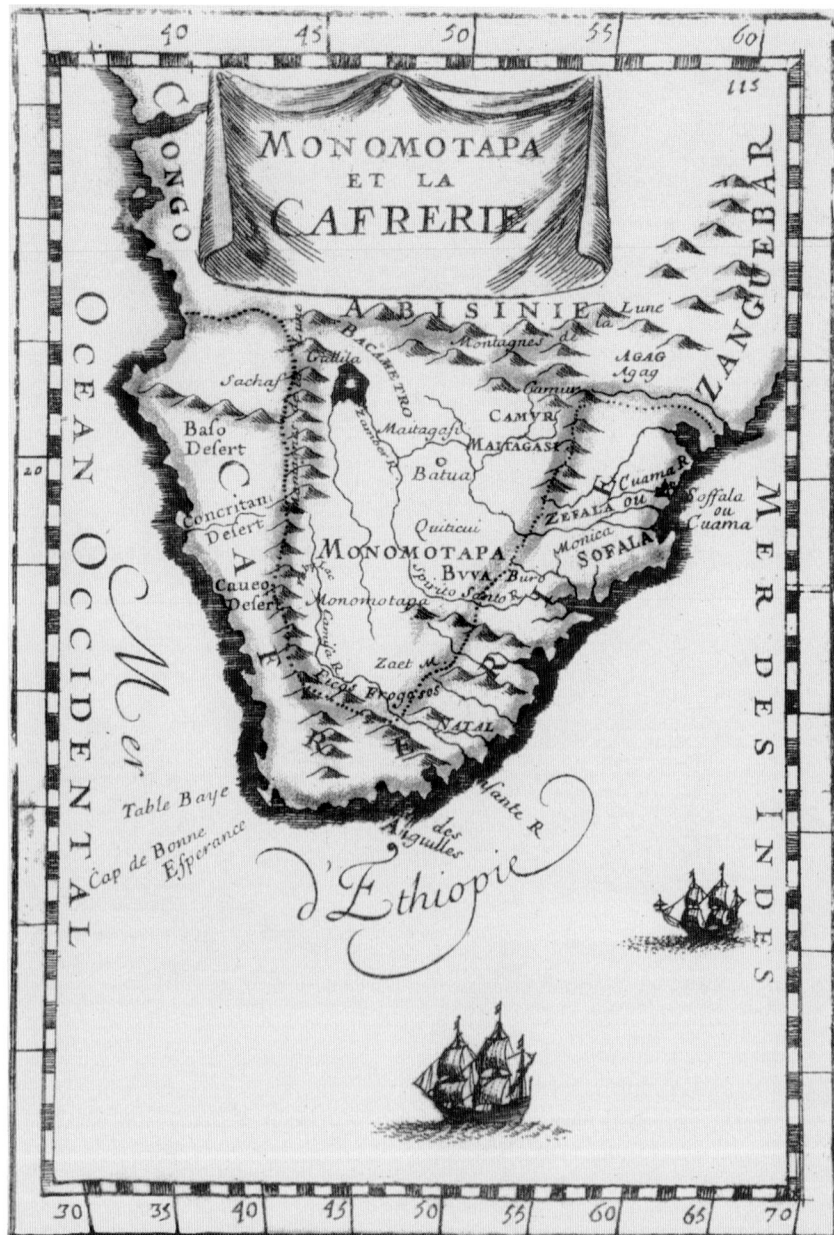

Map 160   MALLET, Alain Manesson (1630 1706)
Monomotapa et la Cafrerie. (Monomotapa and Kaffraria.) (Paris, D. Thiery, 1683.)
Map, 14 x 10 cm, slightly coloured.
Across the top: *De L'Afrique. Figure XLIV. 115.*

A miniature map of southern Africa described as 'Monomotapa and Cafrerie' on a drapery cartouche at the top of the map. It was issued by Mallet, a Parisian engineer. He took service in the Portuguese army for a while, before returning to settle in Paris, where he wrote on mathematics, fortification and practical geometry. His *Description de L' Univers*, in five volumes, with numerous copperplate maps, plans and views was published by Denys Thiery in 1683. As Mallet had travelled widely many of the plans were surveyed by himself. It was a popular publication because of the extensive illustrations, and it was reprinted in 1686 with German text.

The interior is sparse, with only a few place names indicated, for example, Monomotapa. There are two sailing vessels in the southern seas.

In addition to this map Mallet published two other general maps relating to Africa - 'Afrique Ancienne - Afrique Moderne' and 'Ancienne Ethiopie - Lybia à la Cap.'

182

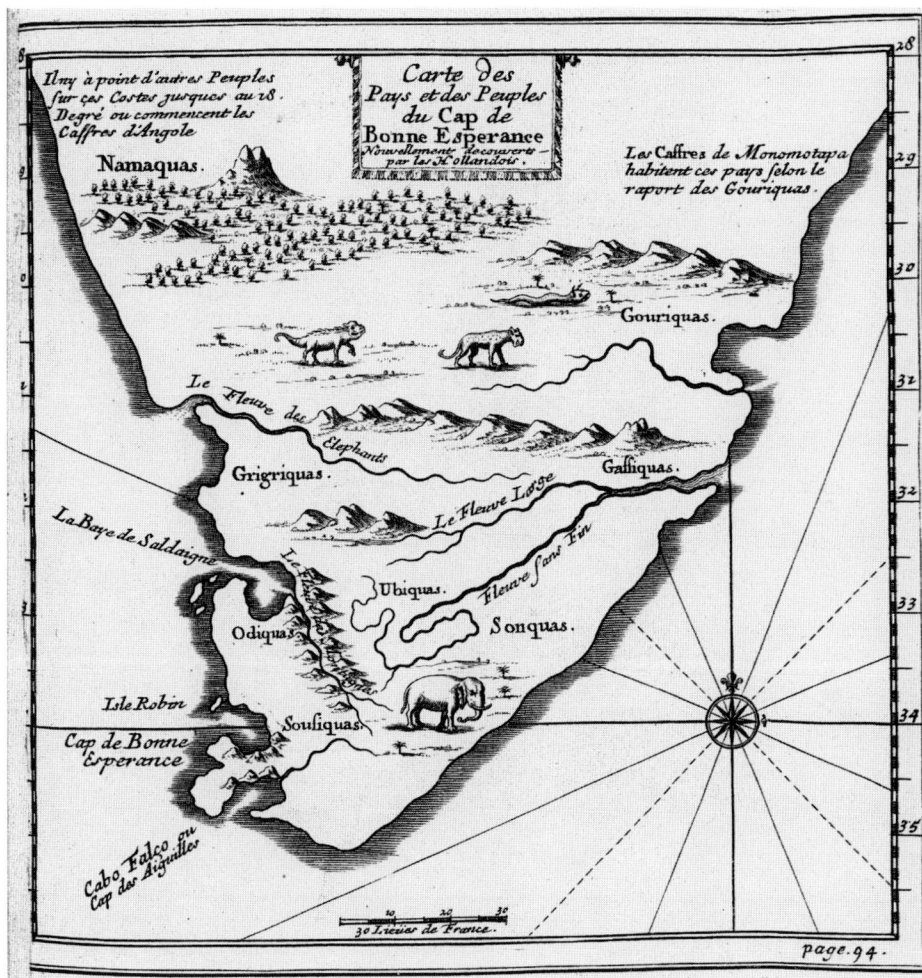

The map image contains the following labels:

Ilny à point d'autres Peuples sur ces Costes jusques au 28. Degré ou commencent les Caffres d'Angole

Carte des Pays et des Peuples du Cap de Bonne Esperance Nouvellement decouverts par les Hollandois.

Namaquas.

Les Caffres de Monomotapa habitent ces pays selon le raport des Gouriquas.

Gouriquas.

Le Fleuve des Elephants

Grigriquas.

Gassiquas.

Le Fleuve Losse

La Baye de Saldaigne

Fleuve Sans Fin

Ubiquas.

Sonquas.

Odiquas.

Isle Robin

Souliquas.

Cap de Bonne Esperance

Cabo Falco ou Cap des Aiguilles

30 Lieues de France.

page.94.

Map 161

TACHARD, Guy (d. 1714)
Carte des Pays et des Peuples du Cap de Bonne Esperance Nouvellements Decouverts par les Hollandois. (Map of the Cape of Good Hope, and the people thereof.)
(Paris, Seneuze, 1686.)
Map, 16,5 x 16,5 cm, uncoloured.
Scale in French leagues.
Bottom right: page. 94.
In: Tachard, G. *Voyage de Siam des Pères Jésuits...* Paris: Seneuze, 1686, fold p. 94.

This small, unusual map of southern Africa appears in the French work *Voyage de Siam des Pères Jesuits*, by Guy Tachard, published in Paris in 1686. The same map appears in his Dutch version of 1687, but it is smaller and differs in detail. This is the 'little map made with his own hand' which Tachard says the German apothecary Hendrik Claudius gave to him (p. 63, English edition). Claudius, who was in the employ of the Dutch East India Company, could draw so well that he was sent to the Cape at the time of Simon van der Stel, then Governor, to draw plants and to study their medicinal properties. He drew not only plants, but animals, landscapes and the indigenous people. He also drew this map. He took part in the historic expedition to Namaqualand with van der Stel. Many of his drawings and watercolours were acquired by the Johannesburg City Council from an owner in Holland in the early 1950s. They formed the basis of the first Frank Connock publication of the Johannesburg Africana Museum, under the able editorship of Anna H.

Smith, in 1952 – entitled *Claudius – watercolours of the Africana Museum.*

As this map was made a little over twenty years after the founding of the settlement at the Cape, very little was known of the country and consequently the information conveyed is very incomplete. Because of some contact wfth the Hottentot tribes their names have been inserted. As they found strange animals on the Namaqualand expedition, Claudius inserted an elephant, a snake, a hyena and a chameleon. They had come across rivers, the Berg, the Elephant and a few others, and Claudius added them to the map. The map has the merit of not illustrating anything that was purely fictitious. The coastline was, of course, inaccurate compared say with the 'Paskaart' published by Hondius – an indication of how much superior coastal charts were in comparison with land maps, even from the earliest days. Claudius was deported by Commander Simon van der Stel for having given this map to the Jesuits.

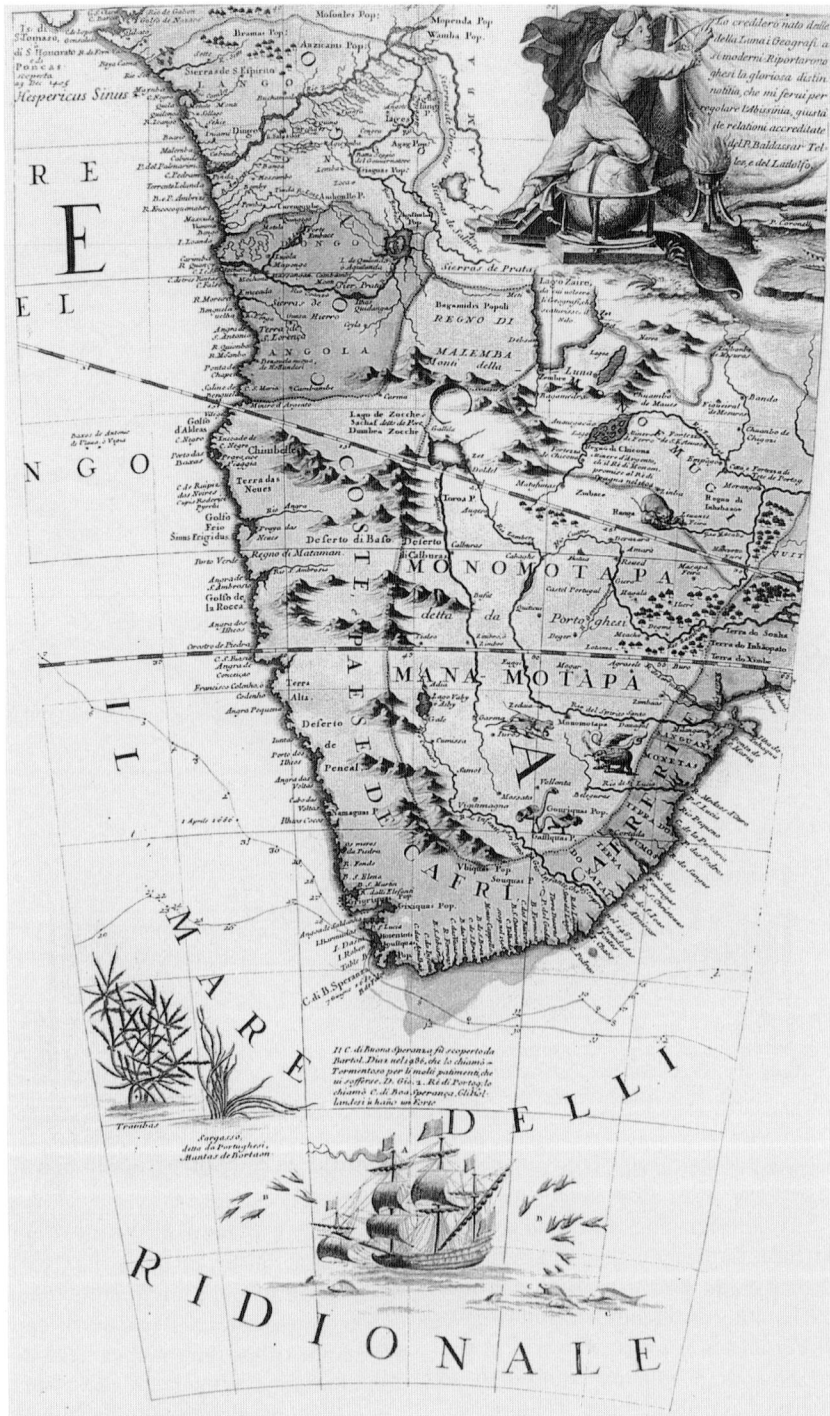

Map 162 CORONELLI, Vicenzo Maria (1650-1718)
Lo creddero nato dalle della Luna i Geografi a si moderni Riportarono ghesi la gloriosa distinnotitia, che mi ferui per regolare l'Abissinia, guista le relationi accreditate del P. Baldassar Tellez, e del Ladolfo. (Born of the most recent geographical findings.) (Venice, V. Coronelli, 1688.)
Map, 45 x 25 cm gore, coloured.

    This Coronelli map of southern Africa is recognisable as the gore for that portion of Africa on the Coronelli globe.

(For further biography of Coronelli and detailed description of the map see Map 52.)

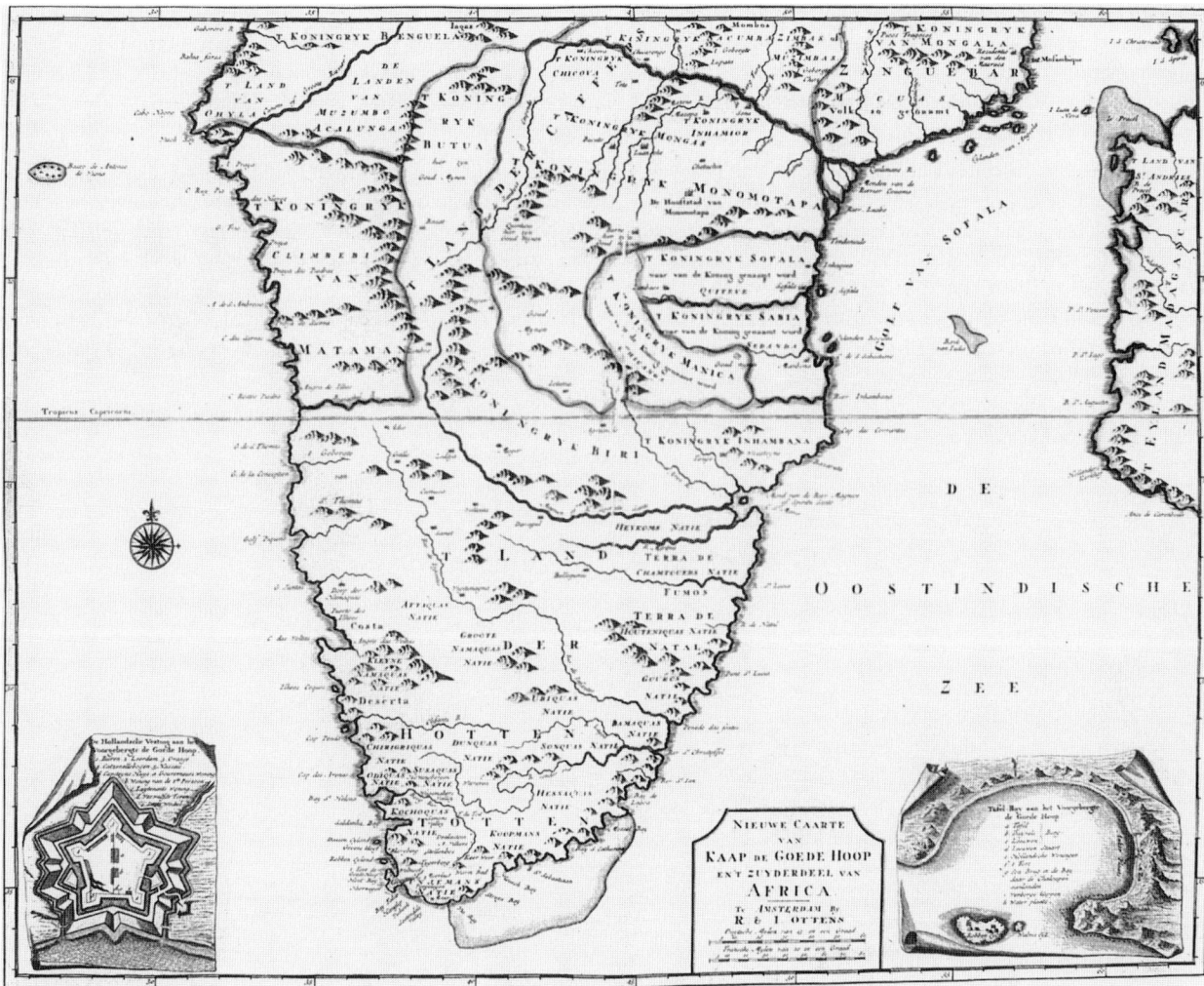

Map 163

OTTENS, Reinier and Josua (fl. 1725-1750)
Nieuwe Caarte van Kaap de Goede Hoop en't Zuyderdeel van Africa. (New map of Cape of Good Hope and southern part of Africa.) Amsterdam, R. and J. Ottens, 1700.)
Map, 44 x 56 cm, slightly coloured.
Scale in Dutch and French miles.
Inset: *De Hollandsche vesting aan het voorgebergte de Goede Hoop* (the Castle).
Tafel-Bay aan het voorgebergte de Goede Hoop.
Top right: Pag. 48.

The Ottens family were publishers of maps in Amsterdam for many generations. This map of the Cape of Good Hope and southern Africa was issued by Reinier and Josua Ottens, sons of the founder Joachim.

A plan and description of the Fort, started in 1666 and completed in 1679, is one of the insets, with a table of identification of the various parts of this stronghold. Another inset, with a map of Table Bay and its surrounding mountains, is featured in the lower right corner, enclosing an explanatory legend. The only decoration on the main map is a conventional compass rose with an additional cross pointing to the east. (The use of such a cross is due to its association with Jerusalem in the east.)

The map covers the area between Benguela on the west and Mozambique on the east. The legends within the map are of great historical interest, referring to the larger settlements such as Stellenbosch, Drakenstein and Waveren, and also to the historic farms of Meerlust, Vergelegen and Constantia. Some of the Hottentot tribes appear to be misplaced even for the eighteenth century, for example 'Houteniquas Natie' appears between the two lines of the heading 'Terra de Natal.' The courses of the rivers are somewhat inaccurate.

Further inland are the usual imaginary cities including 'Vigitimagna' and the capital of Monomotapa, labelled as such. The Limpopo is here named 'Magnice of R. Spiritu Santo' but the Zambezi is now marked 'Rio de Zambeze of Empondo.'

It is uncertain for which of the many Ottens atlases this map was issued but 'pag. 48' appears on the top right corner. This map also appears in P. Kolbe's *Naaukerige en Uitvoerige Beschryving van de Kaap de Goede Hoop* (vol. I.) under the authorship of Balthazar Lakeman (see Map 168 and Map 217), and is also stated by the Catalogue of the Library of Congress to appear in N. Visscher's work of c. 1700, *Variae Tabulae Geographicae.*

CAP DE BONNE ESPERANCE

A. LA RADE.
B. LE IARDIN.
C. Les Maisons de Village.
D. Cabanes des Mourres.
E. Reservoir où les Navires font de l'Eau.
V. Sommet de la Montagne du Lion.
G. Gorge de la dite Montagne.
M. Moulin.
I. L'endroit où etoit N°... quand il a fait se Vessin.

VUE du CAP de BONNE ESPERANCE.

Montagne de la Table
M. de Vent
Montagne du Lion
LA RADE

OCEAN MERIDIONAL

MER DES CAFRES

MER DES CAFRES

PAYS DES CAFRES

CAP DE BONE ESPERANCE

Aguada de Saldanha

C. das Agulhas

Mortier. Map 266   A

Hottentots Habitans du Cap de Bonne Esperance

Description du Cap de bonne Esperance & des environs.

Carte des Pays et des Peuples du...

Des Hottentots.

Du Rhinoceros.

Zembras ou Anes Sauvages du Cap.

Des Anes Sauvages.

Vache marine.

Le Ceraste ou Serpent Cornu.

Sentimens des Hottentots sur la Religion.

Cameleon.

...ce Nouvellement decouverts par les Hollandi

Description des Environs du Cap de bonne Esperance.

Chacune des Nations qui habitent aux environs du Cap a son chef ou Capitaine auquel elle obeit. Cette charge est hereditaire & passe des pères aux enfans. C'est aux ainés qu'apartient le droit de Succession, & pour leur conserver l'autorité & le respect, ils sont les seuls heritiers de leurs pères. Les Cadets n'ayant point d'autre heritage que l'obligation de servir leurs ainés. Leurs habits ne sont que de simples peaux de moutons avec la laine, preparées avec de l'excrément de vaches & une certaine graisse qui les rend insuportables à l'odorat & à l'écorat. Ils en frottent aussi leurs cheveux, qui se reduisent par ce moïen en...

Des Namaquas.

La 2.de de ces Nations est celle des Namaquas, découverte en 1683...

Du Lezard.

Ce que le grand Lezard du Cap a de plus remarquable, c'est que quand on le frappe, il se plaint comme un enfant qui pleure, & que se mettant en colère, il dresse les écailles dont il est tout herissé. Sa langue est bleuatre & fort longue, & lorsqu'on s'en approche, en l'entend souffler avec beaucoup de violence.

On trouve aussi au Cap un autre Lezard, marqué de trois croix blanches, dont la morsure n'est pas si dangereuse que celle du premier...

Grand Lezard du Cap.

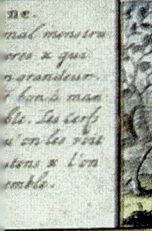

Serf.

Coutumes des Peuples Barbares du Cap.

Quand une femme a perdu son mari, elle doit dans la suite se couper autant de jointures de doigts, qu'elle se remarie de fois...

Bonnes qualitez des Hottentots.

La Barbarie n'a pas tellement effacé dans ces peuples toutes les traits de l'humanité qu'il n'y reste quelque reste de vertu. Ils sont fideles, & les Hollandois les laissent entrer librement dans leurs maisons sans crainte d'en être volez...

Petit Lezard du Cap de Bonne Esperance.

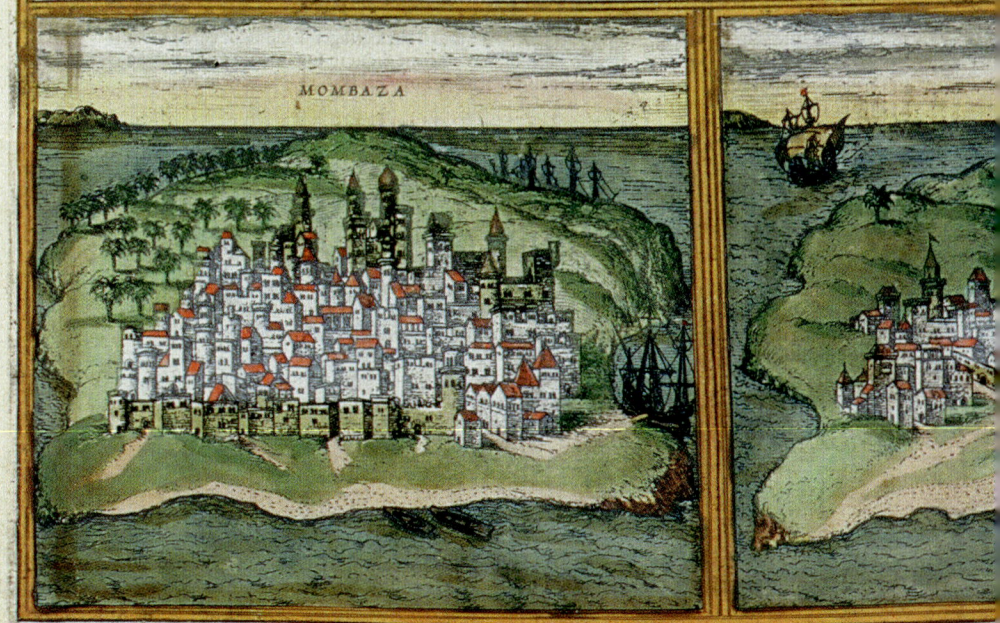

MOMBAZA

C   Braun and Hogenberg. Map 332

Cum Priuilegio

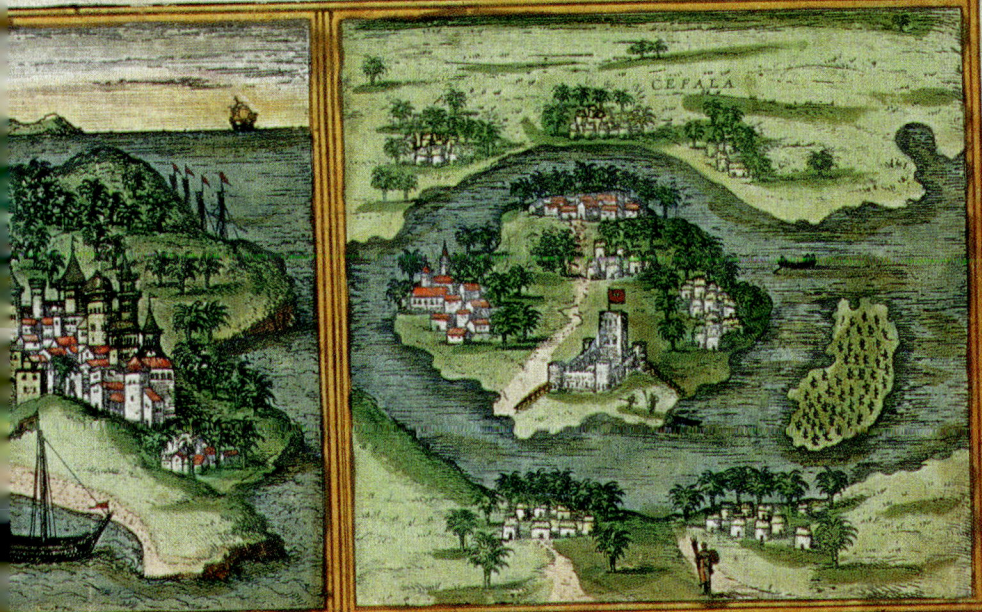

CEFALA

D   van Meurs. Map 43

ASIÆ

PARS.

ASIA MINOR

PERSIA

AFRICÆ
ACCURATA TABULA
ex officina
IACOBUM MEURSIUM

MARE

ARABICUM.

OCEANI

ORIENTALIS

PARS.

MADAGASCAR
Lusitanis
S. LAVRENTII INS.

MAR DI

INDIA.

Map showing the southwestern coast of Africa with the following labels:

Longitude markings: 35, 40, 45 (top and bottom)
Latitude markings: 10, 15, 20, 25, 30, 35

CON GO REGNVM

Isla et Porto leaõe
Villa de S. Paolo
Corimba
R. de Coanza
Loboe
R. Lozzi
C. Brios

Angola.

Angra de S. Antonio
Angra de Santa Maria

P A R S S V

Mti.
Deblui
R. Zembre
Baganeuro
Carma

Angra de Negros
Bayon fondos grandis

Galfila
Zachaf Lacus
Zot
Doldel
Angofa
Zambere flu.
Gobobbe
Arnika
Bora
Butus
Cast Portugal

Cabo Negro
Cabonegro
Praya

Zibil mons

Basô Defertum

Cal
buras
defertum.

M
N

C. de Roppu
Das Nevos
G. Fria
Praya

Praya das Pedras
A. de S. Ambrosii
Praya da Serra

Seros
Farilhones

Concritan
defertum

Bafat
Talfo
Zimbre

Aduu
dy lacus
S. Golo
S. Cruzilla

Caft Portugal
Quilcei
Deger

Enggi

M

Zodaia
Leuos
Mafuuhega

T

Linea Sub Tropico Capricorni.

O   C   E   A   N   V   S

A. de Ilheo
O. Refibo da Pedra
Serra de S. Tome

G. de Conjicam

Praya
P. Pequeno

**Caveo
defertum.**

Garma
Samet

Mofata
Vallents

Zaet montes

Æ   T   H   I   O   P   I   C   V   S

Mantas

Poncal
defert

P. dos Ilheos
das Voltas

Ilheos Secos
Monte dos Bramilos

Occidēns.

Os Morros de Pedra
Flankos

P A

Os. Vochingena

B. de S. Elena
Ra de S. Martin
Agoa de Saldanha
Tafelbay
C. de Boä Speranza
Golfo

Caput Bonæ Spei

ÆTHIOPIÆ ORIS.

Tirut.

Sibit

Agag.

Ca mur.

Ze fa la

COSTA DE CAFFRES

Mozambique

MADAGASCAR Sancti Laurentii Insula.

Oriens

Baxos de India

MARE
ORIENTALE sive INDICVM

ÆTHIOPIA
INFERIOR,
vel
EXTERIOR.

Partes magis Septentrionales, quæ
hic desiderantur, vide in tabula
Æthiopiæ Superioris.

Milliaria Germanica

Terra Sancta

PERSIAE PARS

Ierusalem

Suez

Sinus

Persicus

GYP

Oriens

IMPERII

Aswan

Medina Talnabi, vbi Mahumetis
sepulcrum magna frequentia visitur.

Mare

Bello.

Mecha, patria
Mahumetis.

Fuing

Snachen

Suachem

Aiman, que olim
ARABIA FELIX.

Gan
fila.

Dafila.

Ex Arabia Felice Thus ad
nos deferter; quod hic, &
non alibi nascitur. Incole sua
lingua Louan vocant.

Giebel

Dafila

Rubrum

Barnagasso.

guere

Abarah

Lacce.

Zibit

Aden

Dangali.

Adel

Zocotora insula, que olim Dioscoruda

dicta; optima aloe, que inde Zocotori=
na appellatur, ad nos vehitur. Incole
eam Catcomar vocant Turci, Perse et
Arabes Cehar; Hispani Acchar; et La=
sitani Azure nominant; vti author:
est Garcias ab Horto in sua aro=
matum historia.

Dobas.

Tigre

Doar

Balli.

Angote

Fatigar.

Magadazo.

Olabi.

BARBARICVS
SINVS

Baru

90

Presbiteri

Amara

70

Agola

Azel

ORIENS

Melinde

Canze.

Gemen

Quiloa

Gorga

Sibit.

Tirut
Mozambique

PRESBITERI
IOHANNIS, SI
VE, ABISSINO
RVM IMPERII
DESCRIPTIO.

S.

Map panel labels (top city vignettes):

TANGER · CEVTA · ALGER · TVNIS

Goleta nunc destructa

Side panel figure captions (top to bottom):

Marocchi

Senagenses

Meriostores in Guinea

Cab: lopo Gonsalvi Accola

Milos Congensis

Map text:

MARE ATLAN

TROPICUS CANCRI
TICUM

S. Antonio
S. Vincente   S. Lucia
S. Nicolao   I. de Sal
S. Iago   Boa Vista
I. Bravo   I. de May
                     I. del Fuego   C. Verde

Insulæ de Cabo verde olim
Hesperides sive Gorgades

Canariæ Insulæ olim Fortunatæ
I. Palma
Ferro
I. Gomera
Canaria

HISPANIÆ PARS

Estrecho de Gibraltar

BARBARIA

BILE   DULGERI   qu
HAGA   Zuenziga reg.   NUMIDIA

GUALATA REGN
Arguin
Hoden

LIBYA INTERIOR
quæ hodie
SARRA appella
idem quod desertum

GENEHOA
REGIO

TOMBVTV
REGN

Mandinga   Bangana
Caragoles   Zegzeg
             Duma

GVINEA   BENIN

C. Roxo
Baixas de
Bocabu
C. Verga
C. Serra Liona
Baixas de S.
Anna
C. de Palma

Melli
Mandinga
Maluel

I. Fernando de P
I. del Principe
I. de S. Thome

ÆQUATOR   sive   LINEA ÆQUINOCTIALIS

Vega
Abrolho
S. Paulo
I. de Fernando
de Loronho
Rocas

OCEANUS

I. de S. Matheo
I. de Nobon

I. d. Ascension

AETHIOPICUS

J. de S. Helena

A. Trenidad
S. Maria de
Ascension
I. Ian Pisos
I. de Ascension

TROPICUS CAPRICORNI

Cum privilegio
ad decennium

I. de Tristan de cunha
I. de Gonçalo
Alvares

G   Blaeu. Map 32

CARTA
Geografica de Africa,
del Capo de bon Esperanze Sino al
Regno et Imperio Monamotapo
e Rio de la Goa, Scoperto dell'
Citadino e Commandante
Alvise Pisani

BRIQUAS of BINAS of DAMARAQUAS Na

LA NATION de DAMORAQUES
Scoperto nell 1785
de L.A. Pisani

Twee Broeders
Koper Berg
Monte del Cuivero

Terra del REGNO de Grand.ᵈˢ
NAMAQUAS

Precle Namaquers

TERRA de BOSMANILNATION
ROGGEVELD of HANTOM

Olifantehime

NIEUWEVELD of TERRA
Scoperta Nell A 1771

BORTEVELA

Helena Bay

Pikat Berg

Saldana Bay

Cabinet Topographique
Cae Riti

Hennrigh Berg

ZWART
LAND

Rabben Eiland
Kaap Stadt

CAPO de BONE
ESPERANZA

HOUT NEGRAAS LA

LANDTOGT gedaan door de ZUID-KUST, v
Louis Almaro Pisani Burger Comma

H   Pisani. Map 183

REGNO del IMPERIO de Monamotapa
Que si troeva gli Monti d'Oro Scoperti dell' Pisani
ne ll'Anno 1788=1789.

NATION of TSONGQUAS of COBONAS

SAMBIS VLAKTE

A l'het woody Country

HAMBONAS Nation

TAMBOUQUAS Natie

MABUQUAS Natie

KAFFERSBERGE

TERRA DEI CAFFRI
met Land der Caffers

Gebalinik

de Groote Bergen

Renoster Berg

Rio das St. John

Rio das St. Christophoro

Swart Kops Reviers Bay
Porto de la Reyene Negr

Skrumme Reviers Bay

Bay St. Lorenzo Marques
o. Rio di de Goa
I Unbuka

Porto de St. Lucie

PORT NATAL
Scoperto del Cittadino
Alvise Pisani A.o 1782

26 Nov. 1730

Schale of English Miles 69½ tot. Degr.

Former Travellers arrest
their capatain service of
the Groezveners people
2 Aug. 1781. Kop. Cokson

Questo porto e
Stato Scoperto nell'A. 1782
del Commandante Alvise Pisani

PORT NATAL

tutti gli mineralli
sono Segnato suo
grado Goride Bergen

FRICA in den Jaare 1781 tot 1793 door
en Ingezeetenen der Colonie Zwellendam

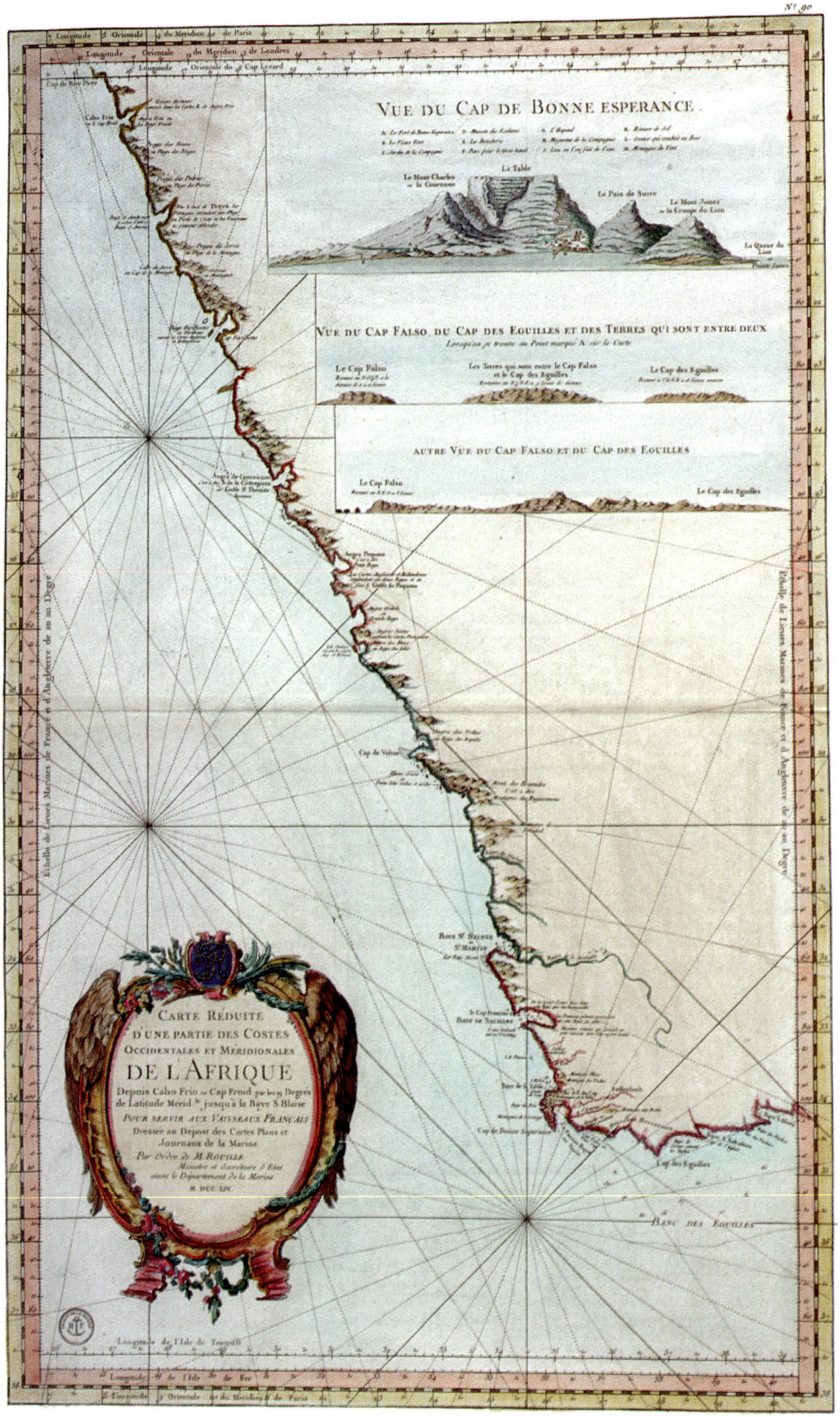

VUE DU CAP DE BONNE ESPERANCE.

La Table

Le Mont Charles
ou la Couronne

Le Pain de Sucre

Le Mont James
ou la Croupe du Lion

La Queue du Lion

VUE DU CAP FALSO, DU CAP DES EGUILLES ET DES TERRES QUI SONT ENTRE DEUX
Lorsqu'on se trouve au Point marqué K sur la Carte

Le Cap Falso

Les Terres qui sont entre le Cap Falso
et le Cap des Eguilles

Le Cap des Eguilles

AUTRE VUE DU CAP FALSO ET DU CAP DES EGUILLES

Le Cap Falso

Le Cap des Eguilles

Cap de Voltas

Baye Sᵗᵉ Helene
Sᵗ Martin

CARTE REDUITE
D'UNE PARTIE DES COSTES
OCCIDENTALES ET MÉRIDIONALES
DE L'AFRIQUE
Depuis Cabo Frio ou Cap Freud par les 15 Degrés
de Latitude Mérid.ᵉ jusqu'à la Baye S. Blaise
POUR SERVIR AUX VAISSEAUX FRANÇAIS
Dressée au Depost des Cartes Plans et
Journaux de la Marine
Par Order de M. ROUILLÉ
Ministre et Secrétaire d'Etat
ayant le Département de la Marine
M.DCC.LIV.

Cap de Bonne Esperance

Cap des Eguilles

BANC DES EGUILLES

I   Bellin. Map 279

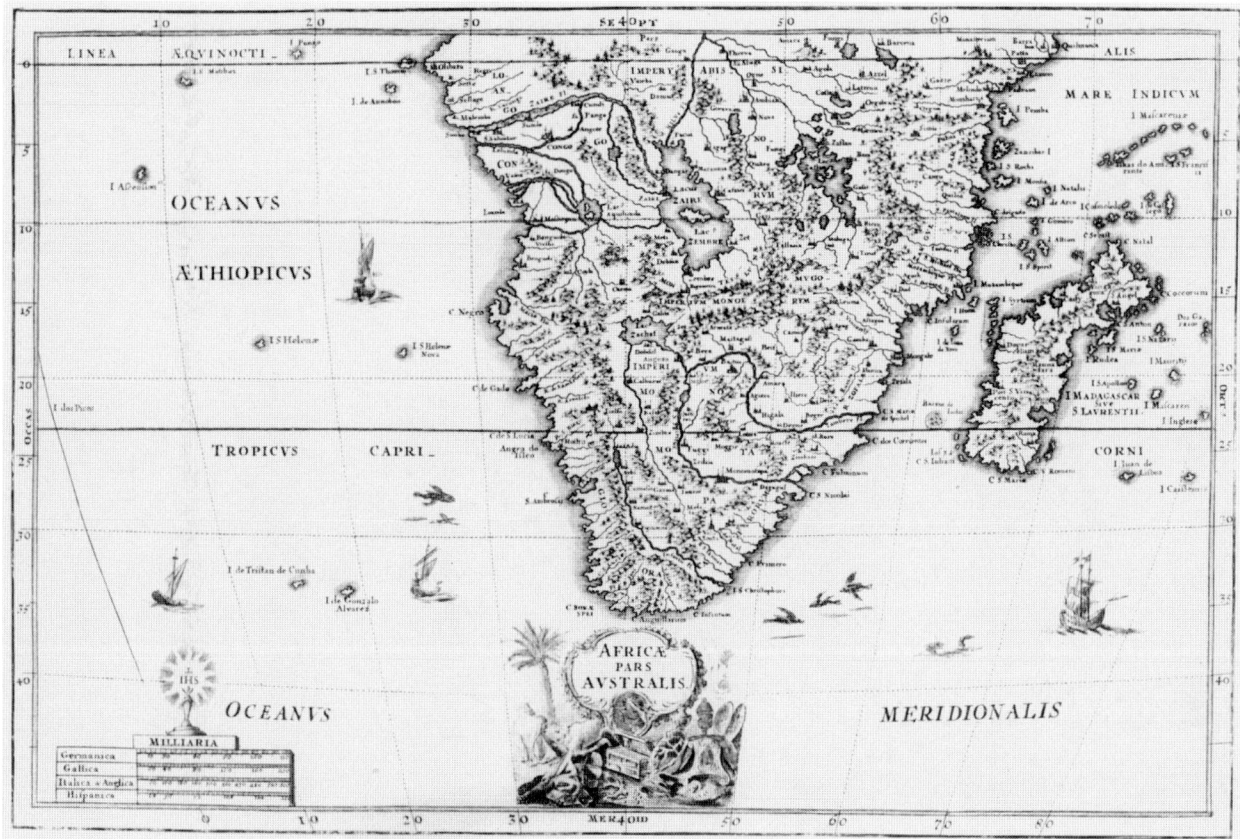

Map 164   SCHERER, Heinrich (1628-1704)
Africae Pars Australis. (Southern Africa.) (Munich, H. Scherer, 1702.)
Map, 23 x 35 cm, coloured.
Scale in German, French, English, Italian and Spanish miles.

This coloured map of southern Africa was issued by Scherer with two other maps in the same style, one of Africa (see Map 62) and one of the world. Scherer was a Jesuit father, a geographer and Professor of Mathematics at Munich.

This is an attractive map with a centrally placed cartouche on the lower border, with an oval scroll surrounding the title. Below the title are two animals (possibly lions) and a palm tree. To the right is an open box adjacent to two ormolu plates resting on a table covered by a cloth. In the east and west oceans flying fish, a sailing vessel and a sea serpent are illustrated. Rivers, lakes and mountains appear, and are typical of the maps of the eighteenth century.

187

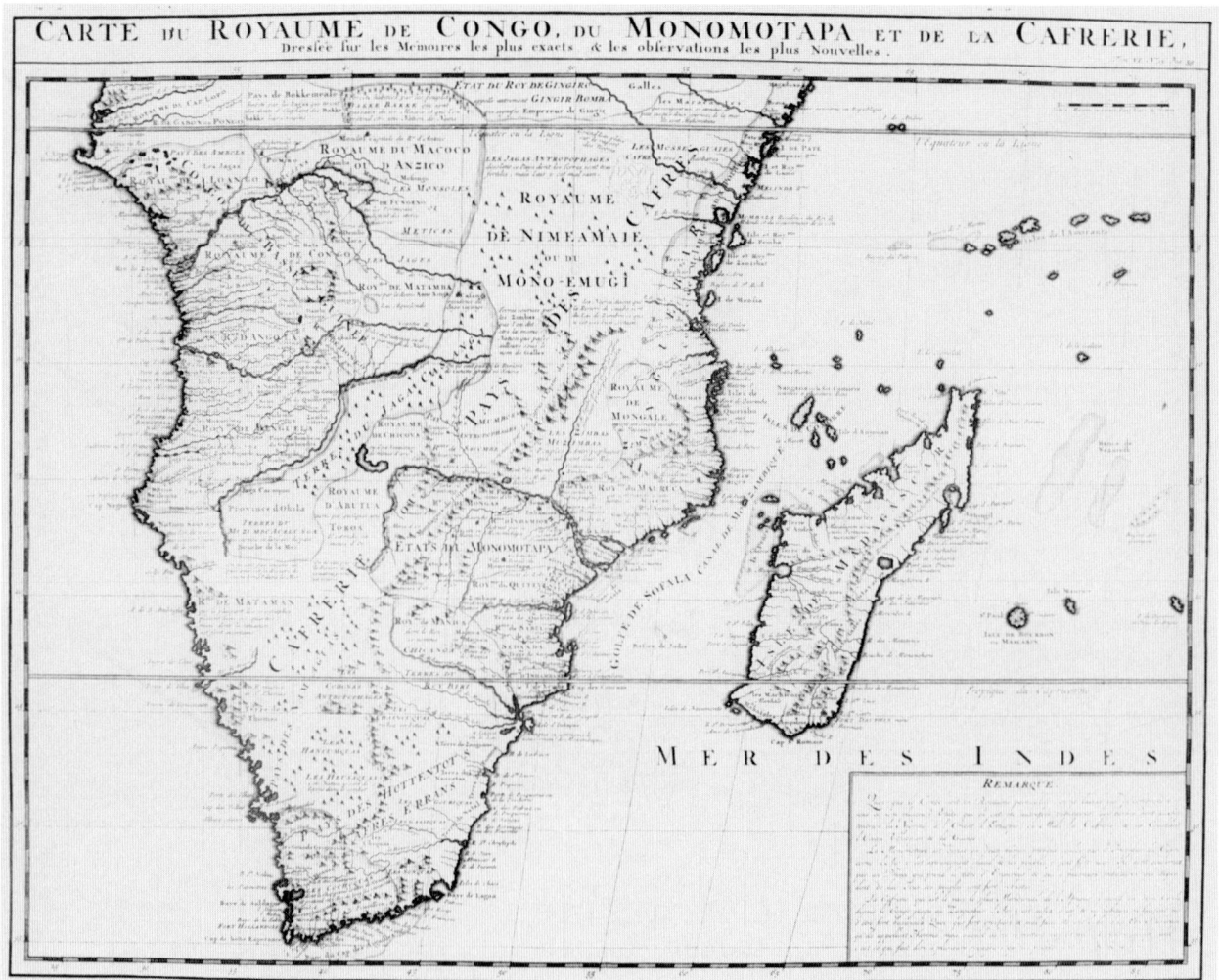

Map 165

CHATELAIN, Henry Abraham (1684-1743)
Carte du Royaume de Congo, du Monomotapa et de la Cafrerie, Dressée sur les Mémoires les plus exacts et les observations les plus Nouvelles. (Map of the kingdom of the Congo, Monomotapa and Kaffraria based on recent observations.) (Amsterdam, Gueudeville, 1719.)
Map, 39 x 52 cm, coloured.
Scale in sea leagues.
Top right: Tom. VI, No. 15, Pag: 59.

A map of southern Africa issued in vol. 6 of the *Atlas Historique et Méthodique* which was published in seven volumes, compiled by Gueudeville and Garillon, with a supplement by H.P. de Limiers and maps by Henri Abraham Chatelain, engraver and geographer of Amsterdam. First published in 1705-1720, with a second edition in 1732, it contains 300 fine engraved plates, many of them large folding. The maps are largely based on de L'Isle. There are many maps and views of Africa and especially of the Cape.

In the lower right-hand corner of this map there is a long note describing the Congo, Monomotapa and Kaffraria. The interior of the map is full of detail, especially the tribal kingdoms such as Monomotapa. It is interesting to note many references to gold, in the form of 'Mines d'Or.' Associated with one such area is M. Fura, which the author of *The Eldorado of the Ancients*, Carl Peters, refers to as the 'fons et origo' of the region of the Zambezi in the biblical voyages of Solomon to Ophir. This has since been completely disproved.

Anna Smith, in her extensive *Exhibition of Decorative Maps of Africa up to 1800* (1952), makes reference to an identical watermark in the paper of map 73 in Heawood's volume.

188

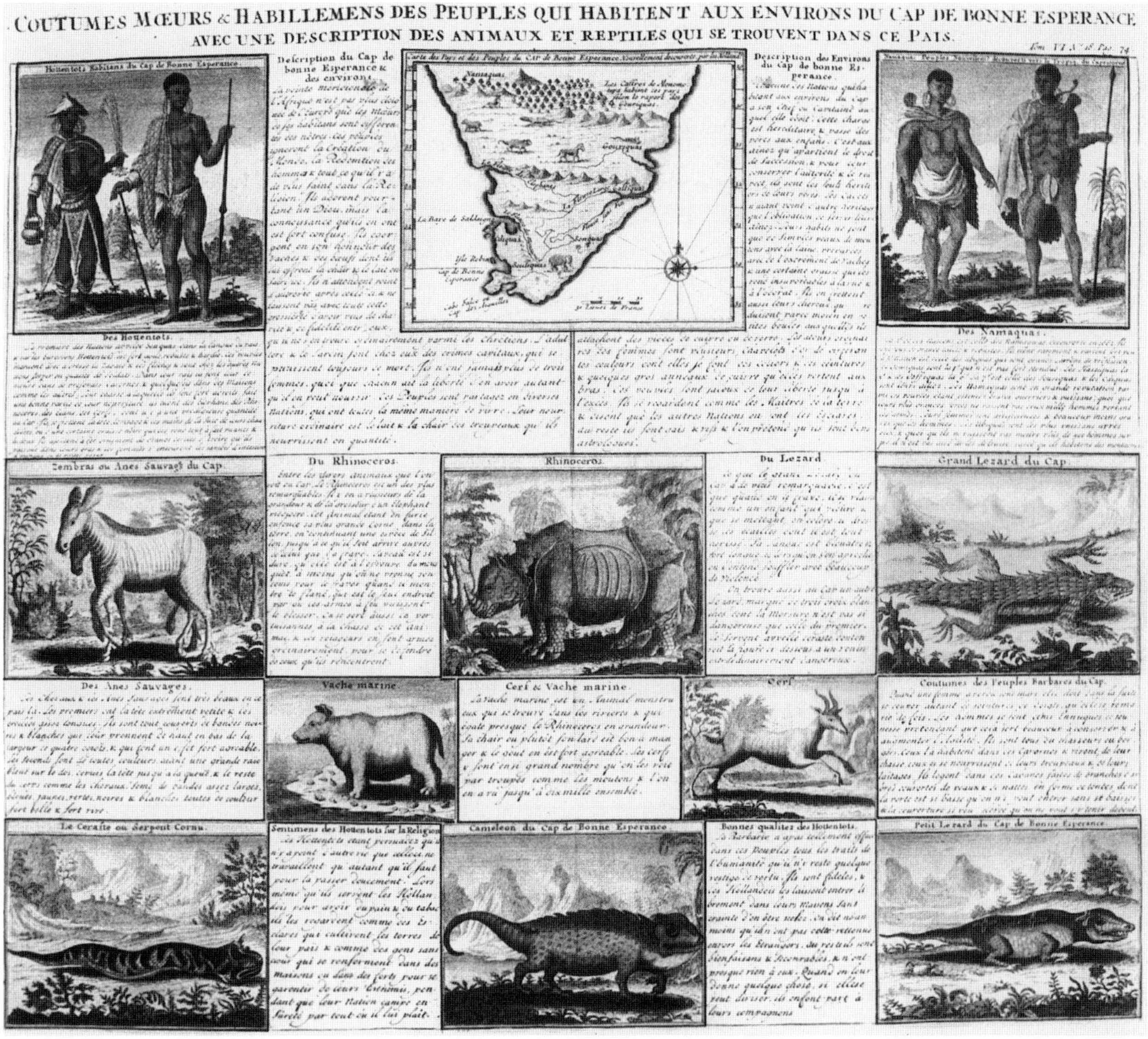

Map 166

CHATELAIN, Henry Abraham (1684-1743)
Coutumes, Moeurs et Habillemens des Peuples qui Habitent aux Environs du Cap de Bonne Esperance avec une Description des Animaux et Reptiles qui se Trouvent dans ce Pais. (Dress and customs of the inhabitants of the Cape of Good Hope.) (Amsterdam, Gueudeville, 1719.)
Map, 10,5 x 13 cm, coloured.
Ten engravings, 36 x 44 cm, coloured.
Top right: Tom. VI. No. 18. Pag. 74.
Title of map: *Carte des Pays et des Peuples de Bonne Esperance. Nouvellement decouverts par les Hollands.*

This map, which contains ten engravings of natives and animals of southern Africa, comes from the *Atlas Historique* by Henry Abraham Chatelain, a seven–volume folio atlas first published in 1705-1720 and compiled by Gueudeville. It also depicts a small, unusual, centrally placed map of Africa almost identical to that which appears in the Tachard travel book of Africa published in 1686. Many of the animals illustrated here also appear in this volume. The provenance of these illustrations and the map are described in the Tachard item relating to Hendrik Claudius, the Dutch apothecary, after whose drawings these engravings were executed (see Map 161).

189

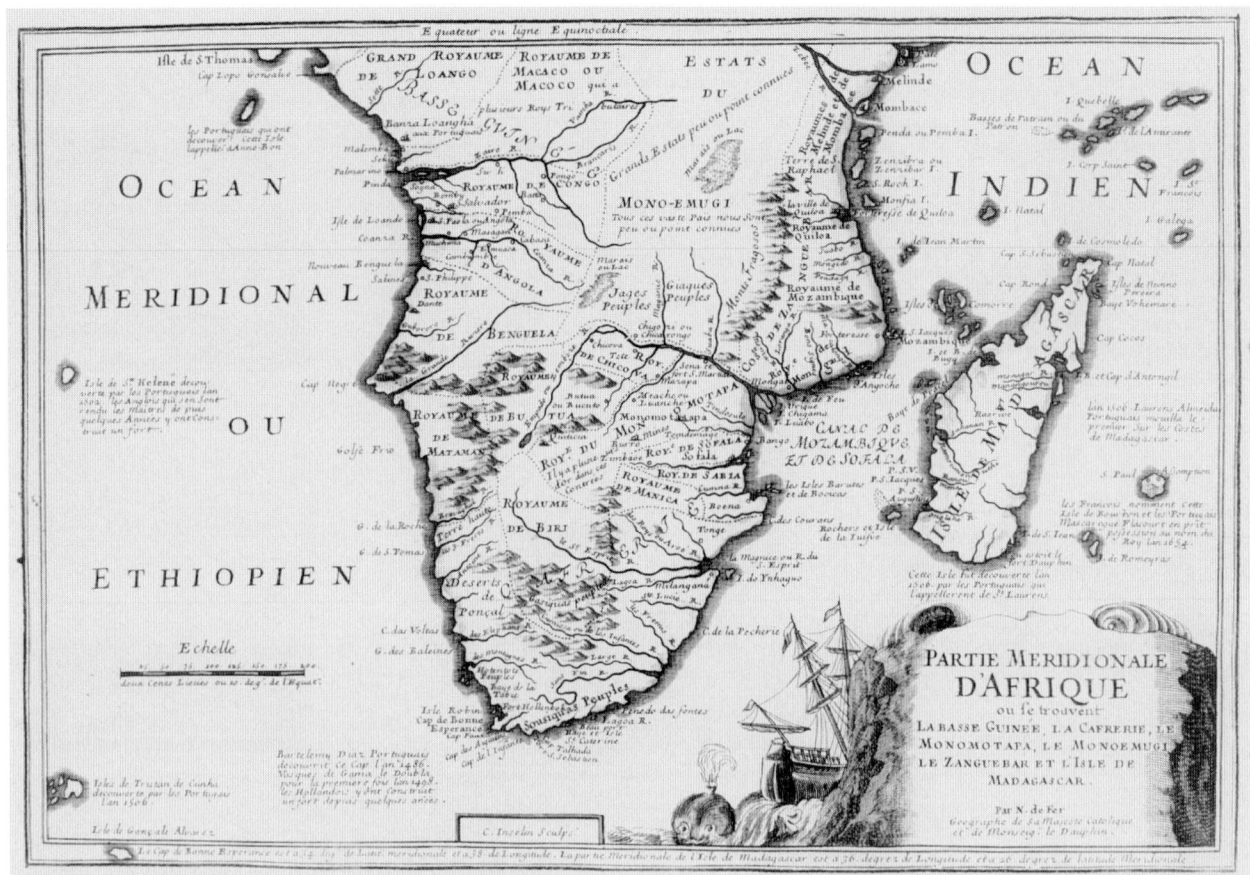

Map 167

FER, Nicolas de (1646-1720)
Partie Meridionale D'Afrique ou se trouvent La Basse Guinée, La Cafrerie, Le Monomotapa, Le Monoemugi, Le Zanguebar et L'Isle de Madagascar.
(Southern part of Africa . . .) Engraved by C. Inslin. (Paris, N. de Fer, 1715.)
Map, 21 x 31 cm, slightly coloured.
Scale in leagues.

A map of southern Africa published by Nicolas de Fer, a French geographer, engraver and publisher of considerable creative ability. He produced several atlases as well as large, very decorative wall maps. He travelled widely in Europe and was extremely prolific in map-making, having some six hundred to his credit which appear to be sought after more for their decorative quality than their geographical content. During his creative period he was geographer to the French King.

The geographical content of the sub-continent is poor. The bulge of the south-west coastline is not as marked as in de Fer's other maps of Africa. There are more rivers and place names around the extreme southern coastline than elsewhere. The Fort and 'Hollenbok' appear, as on so many other maps of the period. There is a cartouche in the lower right corner displaying the title, to the left of which is a galleon, partly hidden by the edge of the cartouche. In the foreground a sea animal is depicted, probably a whale because of water spurting from the head. There is a small legend at the lower border describing the discovery of the Cape by Bartholomew Diaz in 1486 and the construction of a Dutch fort.

190

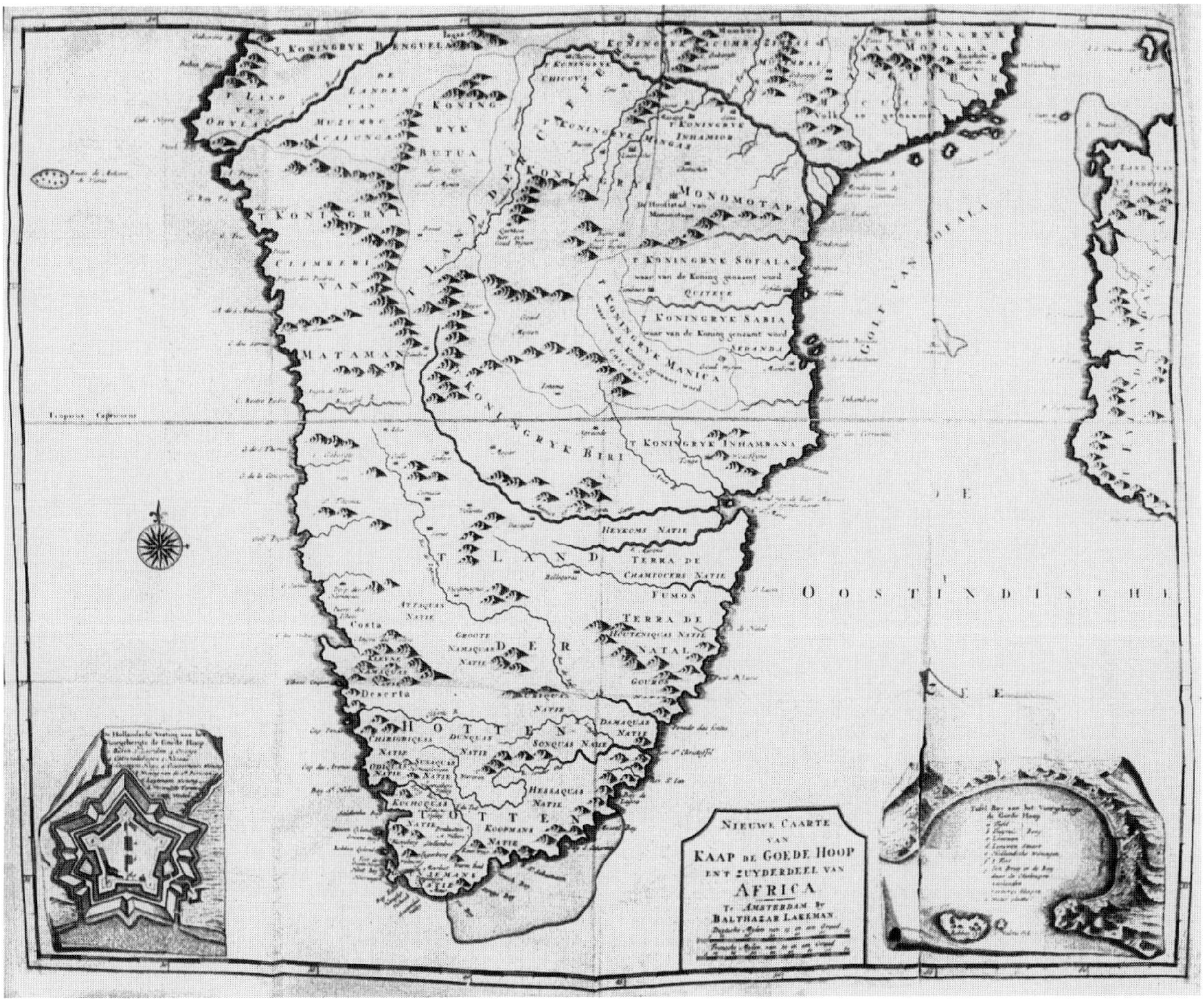

Map 168 LAKEMAN, Balthazar
Nieuwe Caarte van Kaap de Goede Hoop en't Zuyderdeel van Africa.
(New map of Cape of Good Hope and southern part of Africa.) (Amsterdam,
B. Lakeman, 1727.)
Map, 44 x 56 cm, uncoloured.
Scale in Dutch and French miles.
Inset: Table Bay. The Fort.
In: Kolben, Peter. *Naaukerige en Uitvoerige Beschryving van de Kaap de Goede Hoop.* Amsterdam: B.
Lakeman, 1727. Dutch edition, vol. 1 p. 48.

This map of southern Africa appears in Peter Kolbe's
Naaukerige en Uitvoege Beschryving van de Kaap de Goede Hoop...
(1727), first published in German in 1719. It is identical to
that issued by Reinier and Josua Ottens (Map 163) except
for the imprint of Balthazar Lakeman. (For a description of
this map see Map 163.)

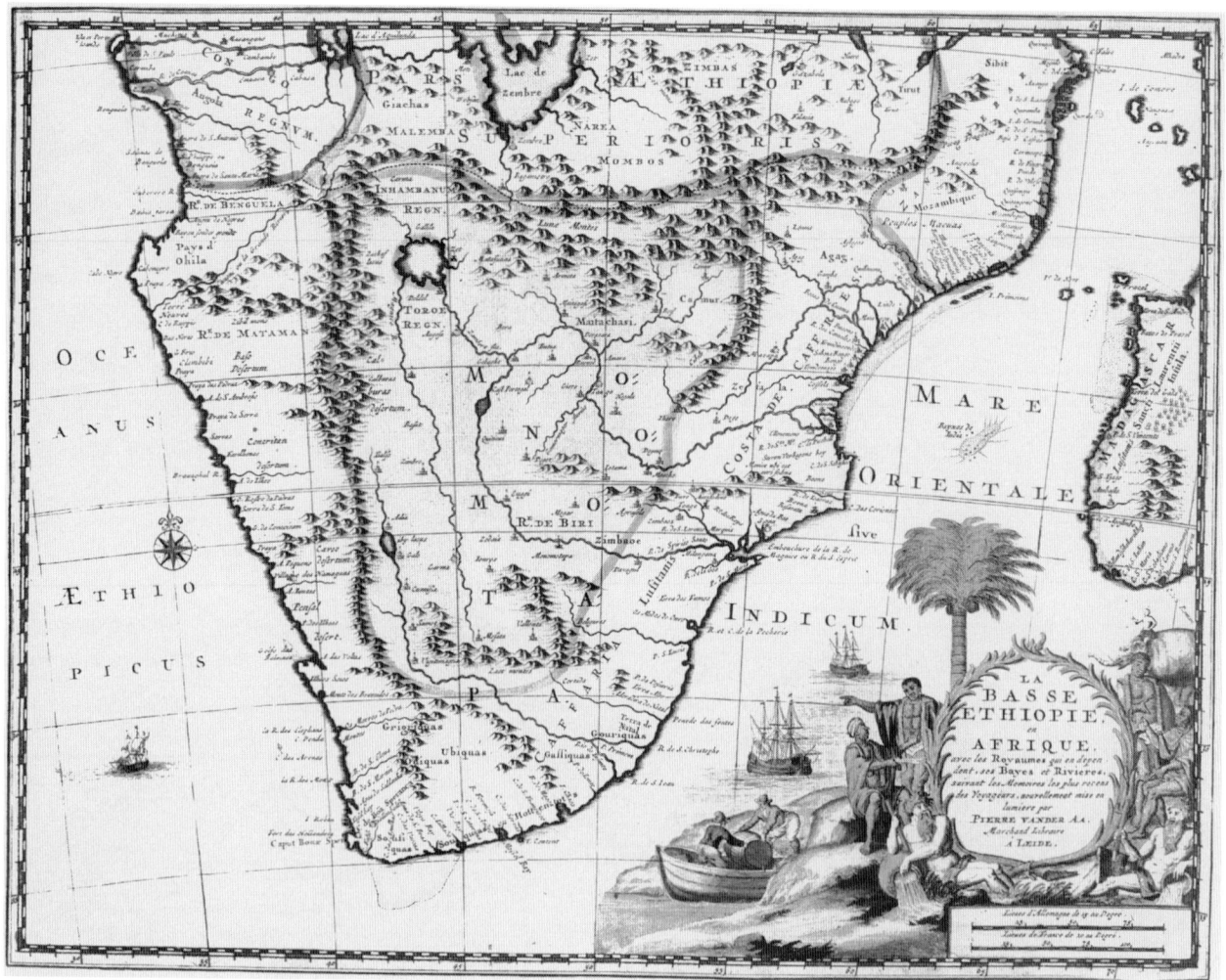

Map 169

AA, Pierre van der (1659-1733)
La Basse Ethiopie en Afrique avec les Royaumes qui en dependent, ses Bayes et Rivieres, suivant les Memoires les plus recens des Voyageurs, nouvellement mise en lumiere par... (Southern Ethiopia and dependent kingdoms...) (Leiden, P. van der Aa, 1729.)
Map, 28 x 36 cm, coloured.
Scale in French and German leagues.

This map has a colourful cartouche in the lower right corner with natives, two river gods and European sailors handling a barrel in a small rowing boat. A mounted elephant appears at the top of the cartouche and sailing vessels are depicted in the surrounding seas. Hellenbok and Fort des Hollandois are depicted and large rivers enter the Indian and Atlantic Oceans, with the known places and bay names around the southern Cape coast. Monomotapa is spread over a wide area of the southern continent.

This coloured map is one of several maps of Africa and South Africa published by Pierre van der Aa, who became a book and map seller at twenty-three years of age, in Leiden. His output of maps and atlases was varied and prodigious, and included the *Atlas Nouveau* and a twenty-eight volume *Atlas Zee en Land Reysen*. The *Galérie Agréable du Monde*, a sixty-six volume atlas compiled in 1729, contained over three thousand maps. Volumes 61-63 deal solely with Africa. Van der Aa was a successful map dealer but a comparatively poor geographer and his maps are typical of the early eighteenth-century style. Luiken, Goerée and Stoopendal were employed as his engravers.

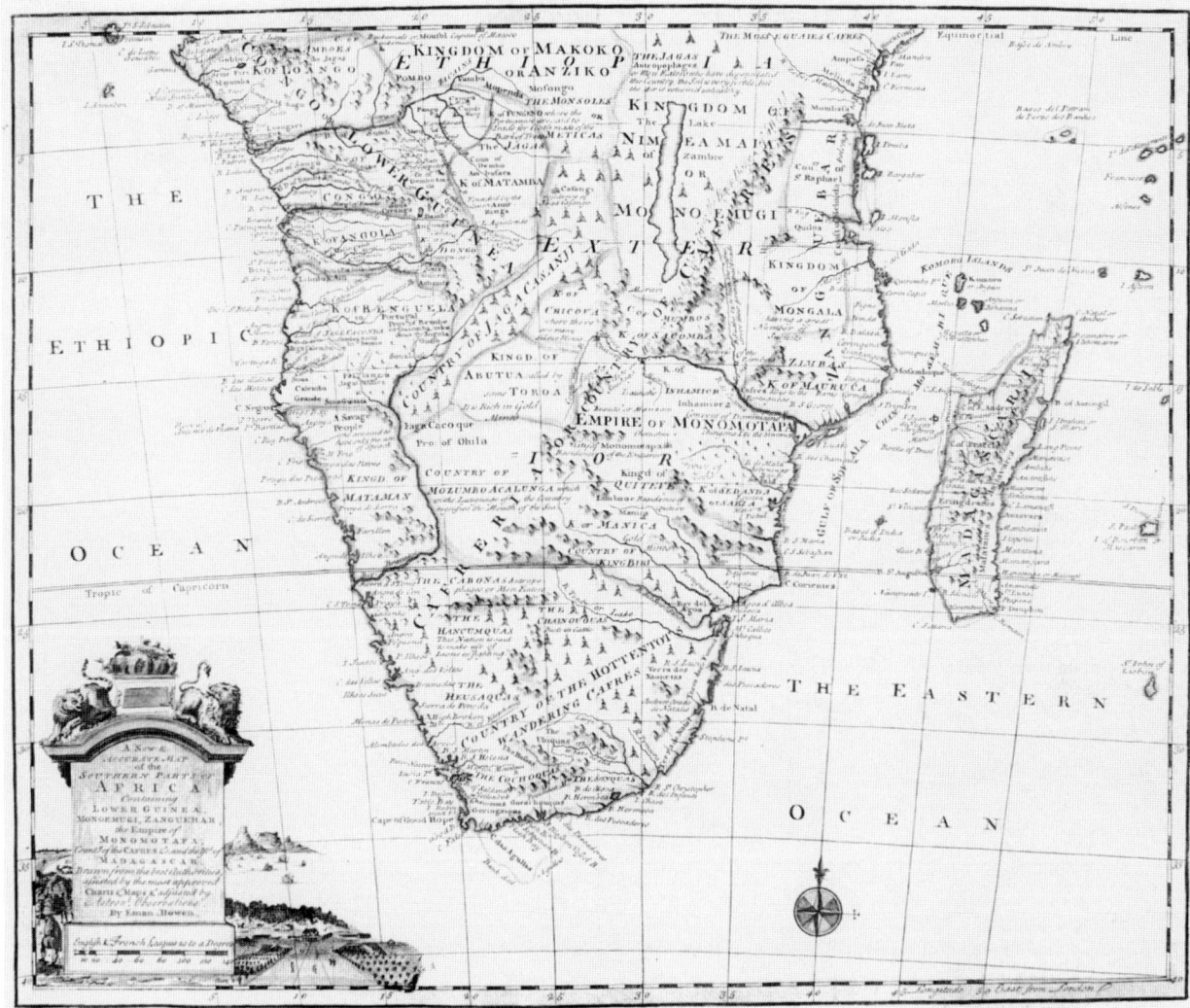

Map 170

BOWEN, Emanuel (fl. 1720-1767)
A New and Accurate Map of the southern parts of Africa Containing Lower Guinea, Monoemugi, Zanguebar, the Empire of Monomotapa, Country of the Cafres &c. and the Island of Madagascar. Drawn from the best Authorities, assisted by the most approved Charts and Maps and adjusted by Astron. Observations. (London, E. Bowen, 1747.)
Map, 35 x 43 cm, slightly coloured.
Scale in English and French leagues.
Prime meridian through London.
Bottom left: No. 56.

This map was issued by Emanuel Bowen in 1747. The cartouche is in the form of a headstone with the scale in English and French leagues on the base. On the top of the stone on the left is a tiger and on the right a lion. The background to the left shows a group of huts with two Hottentots and on the right a distant view of Table Bay with ships. In the foreground, very similar to Vergelegen, is the farm of Willem Adriaan van der Stel (one of the early Governors of the Cape). There is an additional decoration of a compass rose in the Indian Ocean. The names of the various Hottentot tribes appear all over the southern

extremity of the mainland and the Dutch fort and Hellenbok, presumably Stellenbosch, appear in the Cape. South of Natal is a river marked 'this river is said to have no end.' The geographical features of the map closely resemble those of de L'Isle's map of South Africa.

Tooley, in his *Collectors' Guide to Maps of Africa*, refers to three states of the map, and says that this edition was reissued in 1752 and 1766. This map probably comes from Bowen's *Complete Atlas* of 1752 and also appears in his *Systeme Geographicæ* of c.1750. A Dutch edition was published in 1764 and reissued in 1782 (see Map 176).

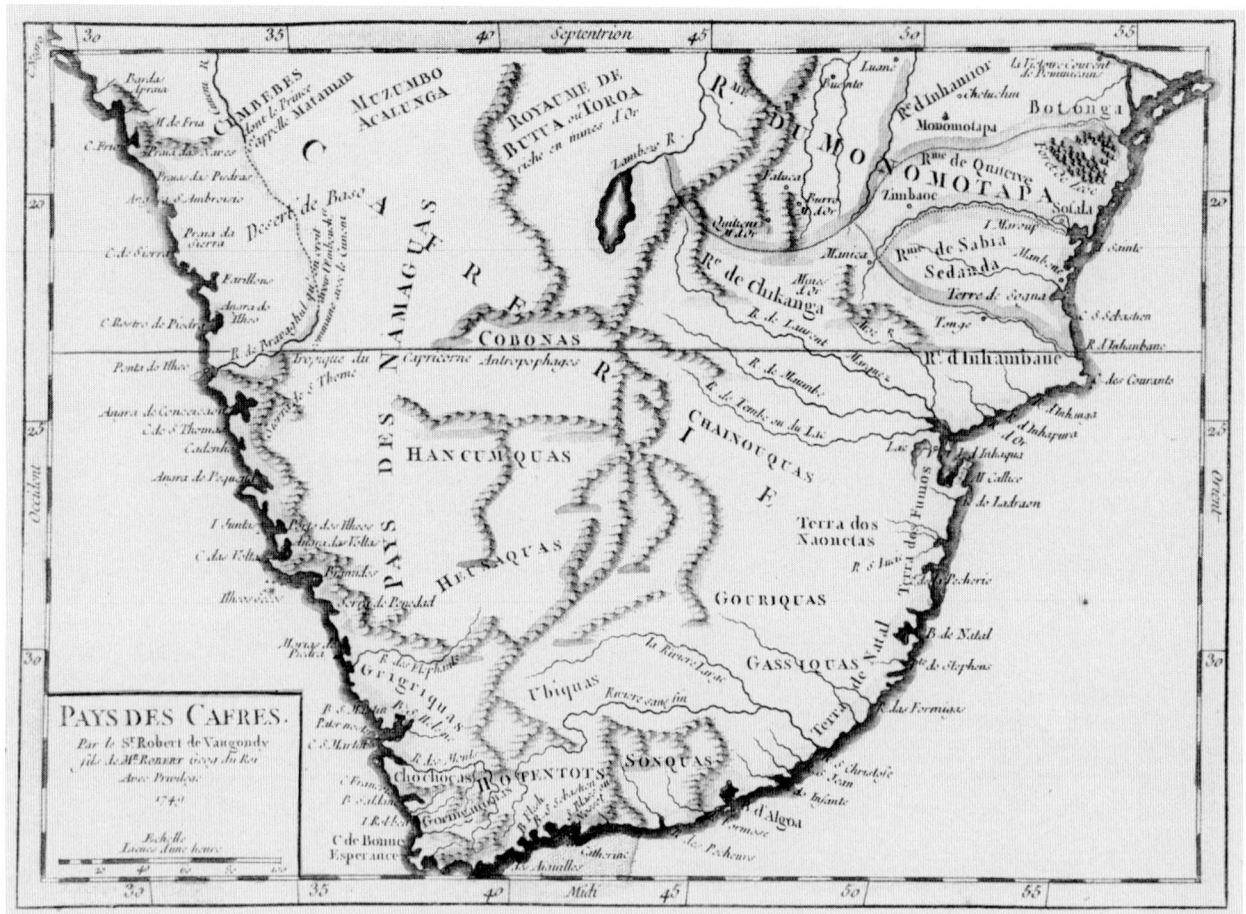

Map 171　ROBERT DE VAUGONDY, Didier (1723-1786)
Pays des Cafres. (Caffraria.) (Paris, D. Robert de Vaugondy, 1749.)
Map, 16 x 23 cm, coloured.
Scale in leagues.
Top right: 178.

A small map of southern Africa issued by Didier, son of Gilles Robert de Vaugondy, in 1749. The map concentrates on mountain ranges and tribal kingdoms. There is very little detail inland except for rivers, but the coastline features a number of the bays and inlets known at that time. The title is within a simple box situated at the lower left corner.

194

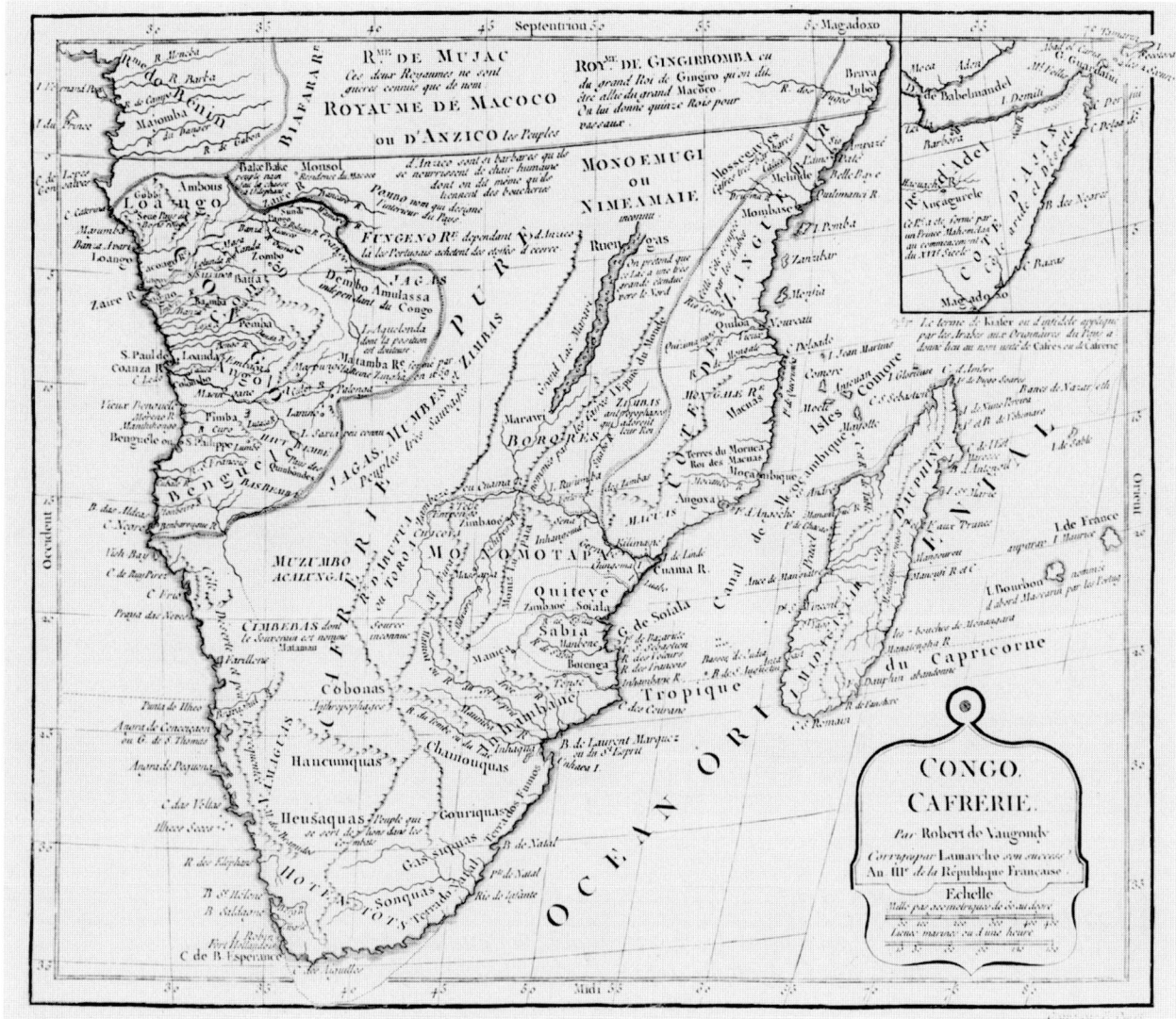

Map 172   ROBERT DE VAUGONDY, Gilles (1686-1766) and Didier (1723-1786)
Congo Cafrerie par... Corrigé par Lamarche son successr. (Congo, Caffraria.)
Engraved by E. Dussy. (Paris, Robert de Vaugondy, 1762-1795.)
Map, 24 x 28 cm, coloured outline.
Scale in miles and sea leagues.
Inset: Cote d'Ajan.

This small French map of southern Africa entitled 'Congo Cafrerie' was engraved by E. Dussy. The map, first issued in 1762, was corrected by Delamarche, who was the successor to Robert de Vaugondy, in 1794/5, 'An 111ᵉ de la République Française.' The map itself provides a moderate amount of information on southern Africa and the many tribes extending up to Monomotapa. The Congo contains much detail. The elongated lake in Central Africa is illustrated as the 'Grand Lac Marari' which became Lake Nyassa. In more recent times it reverted back to its present name, Lake Malawi.

There are numerous editions of the Robert de Vaugondy family maps of Africa and South Africa in atlases of varying sizes published from 1748 to 1779.

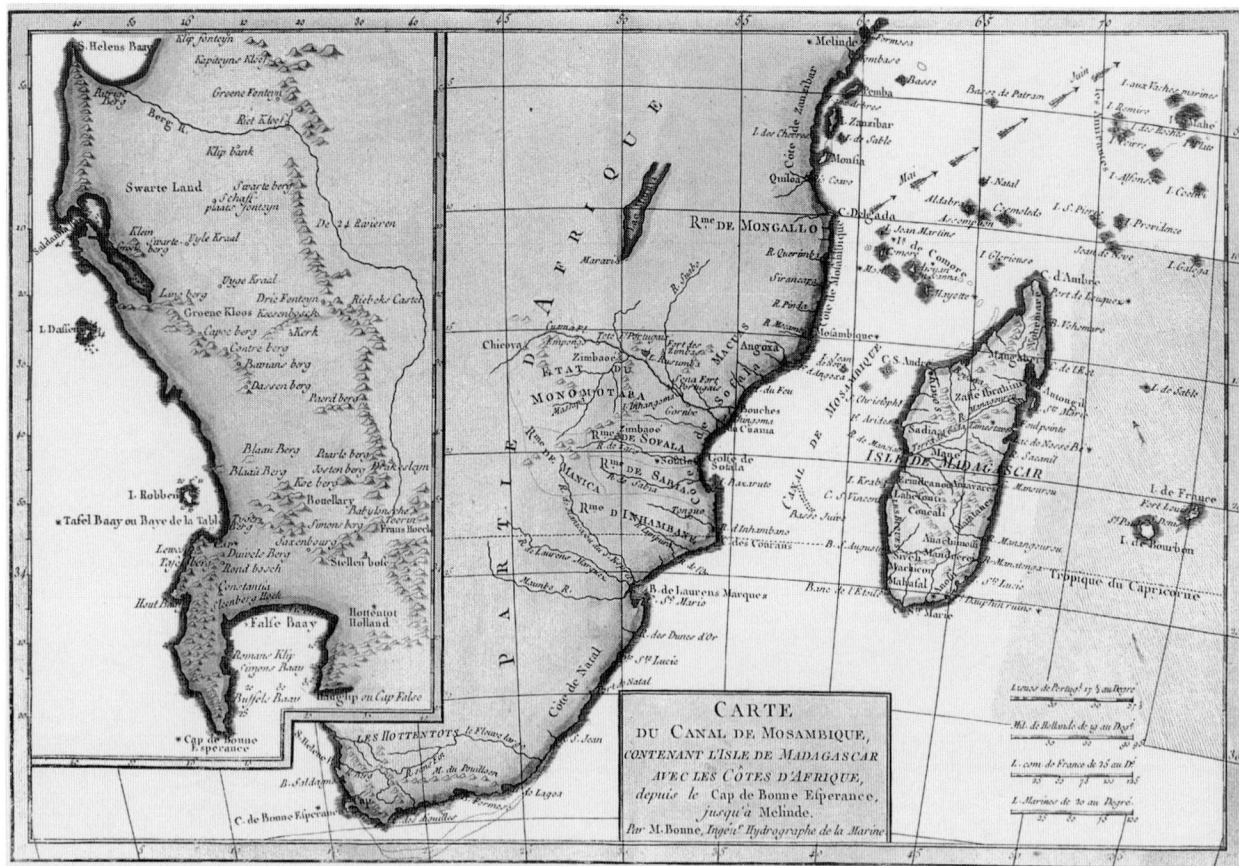

Map 173 BONNE, Rigobert (1729-1794)
Carte du Canal de Mosambique, contenant L'Isle de Madagascar
les Cotes D'Afrique, depuis le Cap de Bonne Esperance jusqu'à Melinde. (Map
of the Mozambique Channel, Madagascar and the coast of Africa from the Cape
of Good Hope to Melindi.) Engraved by André. (Paris, R. Bonne, 1762.)
Map, 21 x 32 cm, coloured.
Scale in Portuguese and French leagues, Dutch miles, sea leagues.
Prime meridian through Ferro Island and Paris.
Inset: Cape coast from St. Helena Bay to False Bay.
Top left: Livre I, II, III, IV.
Top right: No.11.

Rigobert Bonne, a French engraver and hydrographer, probably published this map of southern Africa in his *Atlas Maritime* of 1762. On the top left reference is made to book numbers. There is a detailed inset of the Cape of Good Hope giving a fairly accurate indication of the bays and the place-names in the interior. Adjacent to Stellenbosch appears the name Saxenbourg, which is known to be the name of a farm. In a sea chart of this area by John Bew (see Map 285) mention is also made of the name Saxenbourg, which was thought in this map to be Stellenbosch.

Map 174
TIRION, Isaak (d. 1769)
Kaart van het Zuidelykste Gedeelte van Afrika of het Land der Hottentotten.
(Map of southern part of Africa...) (Amsterdam, I. Tirion, 1763.)
Map, 33 x 37 cm, uncoloured.
Scale in Dutch and French miles.
In: Tirion, Isaak. *Hededaagsche Historie of Tegenvoordige Staat van Afrika... volgens de waarneemingen van de Heeren Shaw, Adanson, de la Caille...* Amsterdam: I. Tirion, 1763, fold p. 625.

This map of the southern portion of Africa and the land of the Hottentots comes from Tirion's *Hededaagsche Historie* of 1763, and depicts the area to the south of the Tropic of Capricorn. It is fairly accurate in the Cape Town vicinity but from Cape Agulhas to the east the coast is distorted. Further inland is largely conjectural and the large fictitious city of Vigitimagna is indicated. The 'Kobonas' are recorded as cannibals; the 'Heusaquas' as being 'used in the struggle against the lions.' The cartouche is on the right encased in a box and below it is the scale in Dutch and French miles. Immediately above these two boxes there is a compass rose.

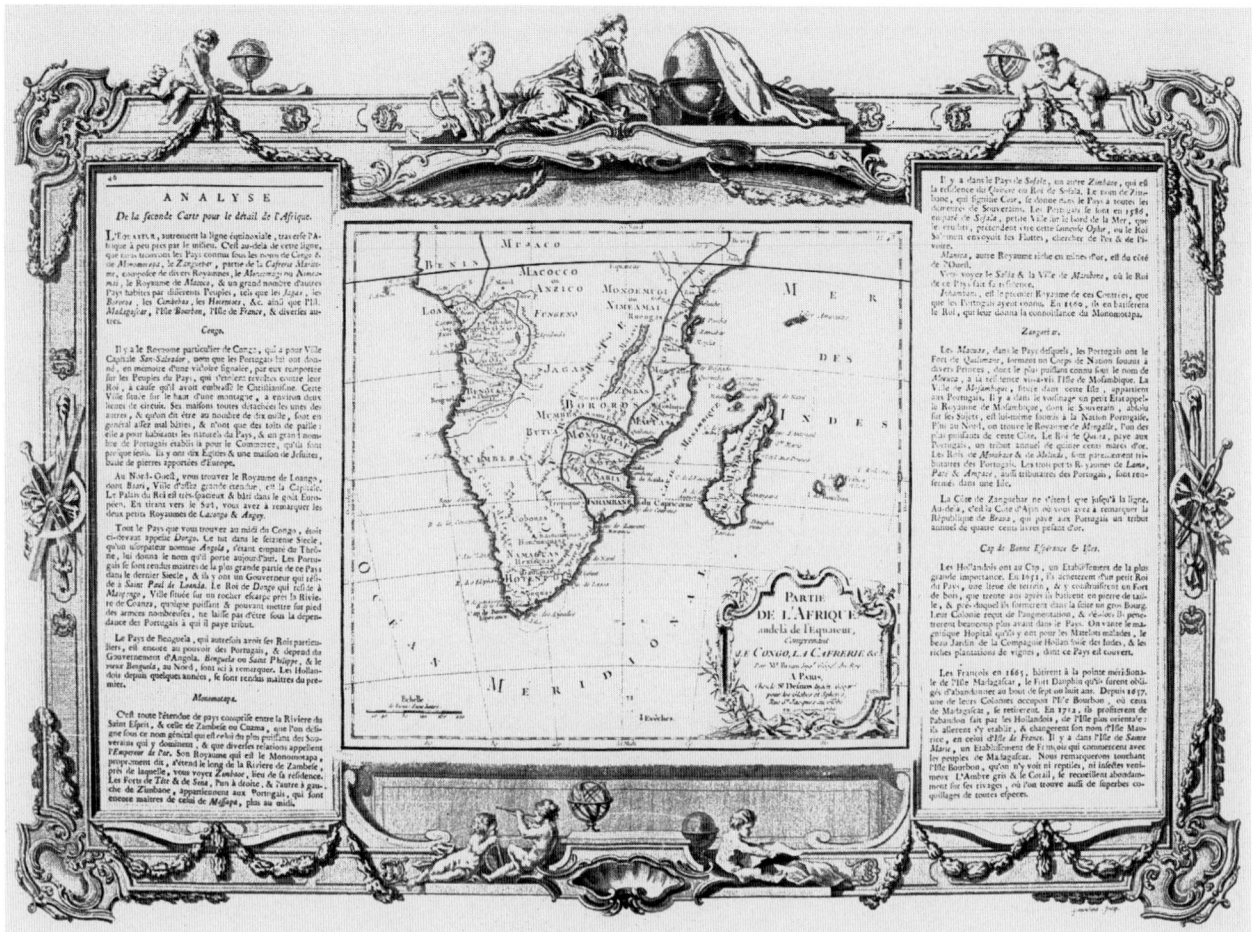

Map 175 BRION DE LA TOUR, Louis (fl. 1757-1800)
Partie de L'Afrique audelà de L'Equateur, Comprenant Le Congo, La Cafrerie &c. (Africa south of the Equator comprising the Congo, Cafraria, etc.)
(Paris, Louis Charles Desnos, ingèn., Géogr. pour les Globes et Spheres, Rue St. Jacques au Globe, 1766.)
Map, 22 x 25 cm, slightly coloured.
Scale in leagues.
Map surrounded by a decorative frame with a description of the Congo, Monomotapa, Zanzibar, the Cape of Good Hope and the islands.

A decorative map of southern Africa with the cartouche on the lower right. The whole map is surrounded by a decorative frame with scrollwork and figures. There are two side panels with descriptive information on the Congo, Monomotapa, Zanzibar, Cape of Good Hope and the islands. The decorative appearance is more striking than the geographic content of a somewhat insignificant map. The cartographical information is moderately accurate. Hottentot tribes are mentioned and the Fort at the Cape noted. The west coast is referred to as 'Côte Deserte.' The Orange River has not yet become known.

This map may have formed part of the Brion de la Tour *Atlas Général*, published by Desnos of Rue St. Jacques. Brion de la Tour, engraver and Geographer to the King, was the author of many atlases from 1757-1800.

198

Map 176 BOWEN, Emanuel (fl. 1720-1767)
Nieuwe en Naauwkerige Kaart van het Zuidlyk Gedeelte van Africa. . . .
verbeeterd door W.A. Bachiene. (A new and accurate map of the southern parts of
Africa, improved by W.A. Bachiene.) Engraved by J. van Jagen. (E. Bowen, 1782.)
Map, 34 x 42 cm, uncoloured.
Scale in English and French miles.
Prime meridian through London.

This is a Dutch version of the 1747 English edition of the southern parts of Africa by Emanuel Bowen (see Map 170). Geographically it is identical except for a less decorative cartouche in the left lower corner – a square ornamented scroll containing the title – with the addition of 'improved by W.A. Bachiene,' a Dutch preacher and geographer of Maastricht.

Map 177

LOTTER, Tobias Conrad (1717-1777)
Africae Pars Meridionalis cum Promontorio Bonae Spei Accuratissime Delineato Opera... (Southern part of Africa with Cape of Good Hope.) (Vienna, T. Lotter, 1778.)
Map, 46 x 55 cm, coloured.
Dedicated to Domino Wilhelmo.
Inset: *Ager Promonotorii Bonae Spei. Sinus Saldanhae of Saldanha Bay. Castellum Bavatorum in Promontorio Bonae Spei of Het Castel de Goede Hoop.*

The dedication to Domino Wilhelmo, on the lower left border, appears to be on material blowing in the wind. There are three detailed insets: the Cape Fort, Saldanha Bay and the Cape of Good Hope, all with a detailed identifying legend. The compass rose in the Atlantic is the only decoration in the seas. The coastline is not very accurately depicted. Stellenbosch, Riebek's Casteel, Swellendam, Roode Sand, Wagenmakers Valley, etc. appear but their geographical positions are not very accurate. We note the legends 'Ruinen van een Portugusisch Fort' and 'Ruinen van het Hollands Fort' on the Spiritu Santo and Marquis rivers. Hottentot tribes are noted all over the country.

According to Dr Anna Smith *(Exhibition of Decorative Maps of Africa up to 1800)* this map could be from the *Atlas Géographique de cent et huit cartes générales et spéciales par les géographes Tobies Conrad Lotter,* published in 1778 (according to the Library of Congress).

Tobias Conrad Lotter had been apprenticed to Funk, Professor of Natural History at the University of Leipzig. He married the daughter of M. (George Matthäus) Seutter in 1740 and succeeded him in 1756.

200

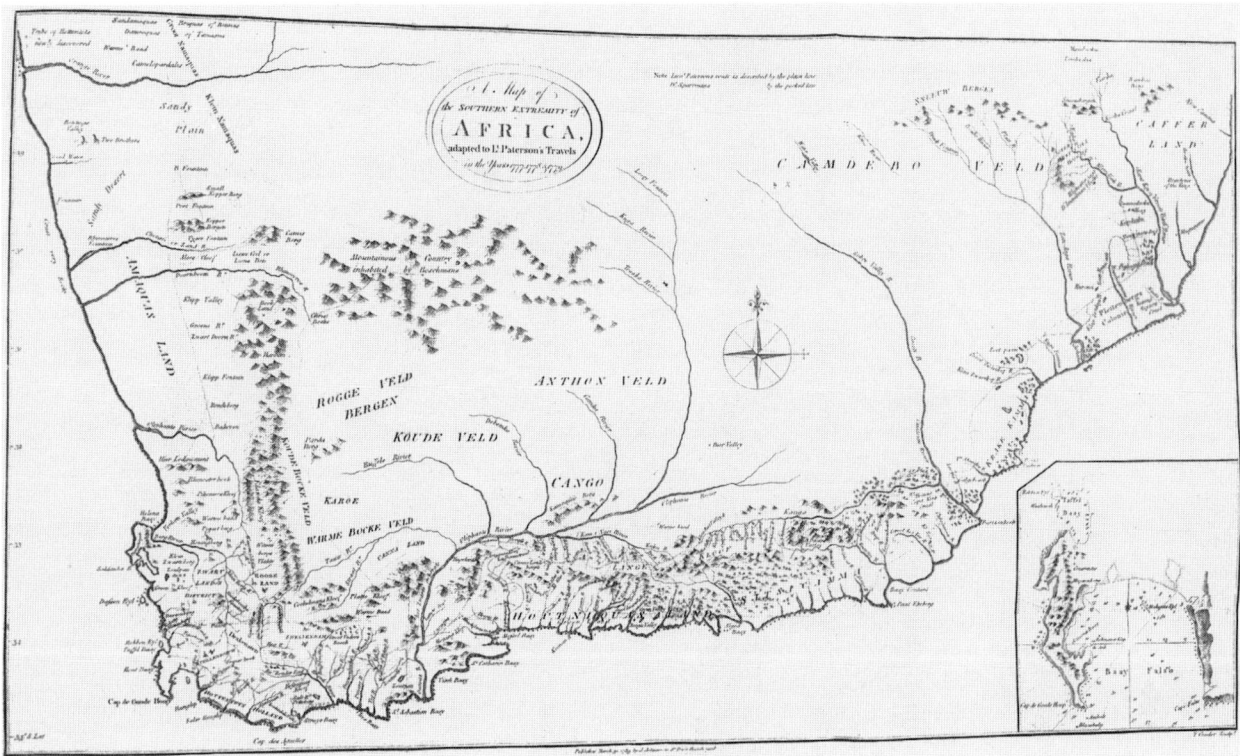

Map 178

PATERSON, William

A Map of the Southern Extremity of Africa adapted to Lt. Paterson's Travels in the Years 1777, 1778 and 1779. Engraved by T. Conder. (London) J. Johnson, St. Paul's Church yard, 1789.
Map, 32 x 56,5 cm, uncoloured.
Inset: Cape Peninsula.
In: Paterson, William A. *Narrative of Four Journeys into the Country of the Hottentots and Caffraria...*
London: J. Johnson, 1789, fold at back.

A map of southern Africa adapted to Lt. Paterson's travels, narrated in his book *Four Journeys into the Country of the Hottentots and Caffraria.* The title of the map is within a simple oval and the map is undecorated. In the lower right corner is an inset without a title of Table and False Bays showing soundings. The Cape Colony is shown with as much detail as was known at that date. Paterson accompanied Colonel Gordon (Commander of the troops of the Dutch East India Company) and Jacob van Reenen on several trips to the interior. In the course of his travels he penetrated and recorded as far as Namaqualand in the west and the Great Fish River in the southeast.

This map is an exact copy of Sparrman's (see Map 223), with the same errors.

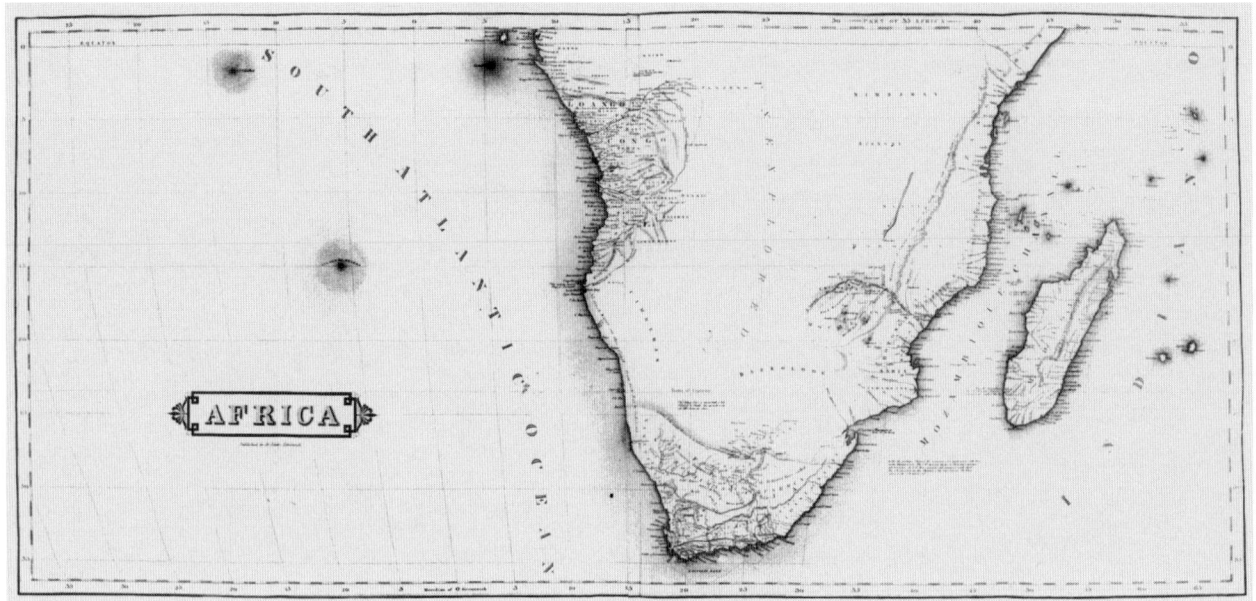

Map 179

LIZARS, Daniel (fl. 1776-1812)
Africa. (Edinburgh, D. Lizars.)
Map, 39 x 86 cm, coloured.
Prime meridian through Greenwich.
Across the top on the left: 'Part of Africa.'
Top right: LVI.

A map of southern Africa extending from just above the equator, and labelled 'Africa' in a rectangular decorative box in the lower mid-Atlantic Ocean. The part unoccupied by Africa itself is unusually wide. This is because it consists of the two lower plates only of a four-plate map of the whole of the African continent, engraved and published by Daniel Lizars Sr. as part of his *A New and Elegant General Atlas of the World...*, the maps with imprints and watermarks dating from 1808 to 1812. The plates were amended and reissued monthly from c.1826-31 by Daniel Lizars, Jr. to form *The Edinburgh geographical and historical atlas...* with 68 plates. Africa's plate numbers were changed to 53-56, in Roman numerals, as per this copy. It is printed in colour and concentrates on the known areas, especially the Cape Colony, the Congo and Mozambique. The Island of Madagascar is featured prominently, providing a good deal of detail on both the coastline and the mainland.

The known Portuguese places such as Sofala, Inhambane and the associated rivers, with the Zambezi to the north, are illustrated. An unusual range of mountains extends on the eastern aspect, marked as the 'Mountains of Lupala or the spine of the world covered with perpetual snow.' The Mozambique Channel is noted with its various monsoons and their respective periods of the year.

Southern Africa has the usual placenames known in the late eighteenth and early nineteenth centuries, together with a number of indigenous tribes. Caffraria is erroneously placed almost completely across the subcontinent north of the Cape Colony.

The Congo shows a number of Portuguese placenames and on its coastline appears 'Pillar with the Arms of the King of Portugal' – presumably a reference to a padrão erected by the early Portuguese explorers around the southern African coastline.

The map was published, as is stated below the title, by engraver and publisher D. Lizars of Edinburgh. He was apprenticed to Andrew Bell, also an engraver of Edinburgh, and was succeeded by his sons William Home and John Lizars. W.H. Lizars was well known as the engraver of the first ten plates of Audubon's *Birds of America*.

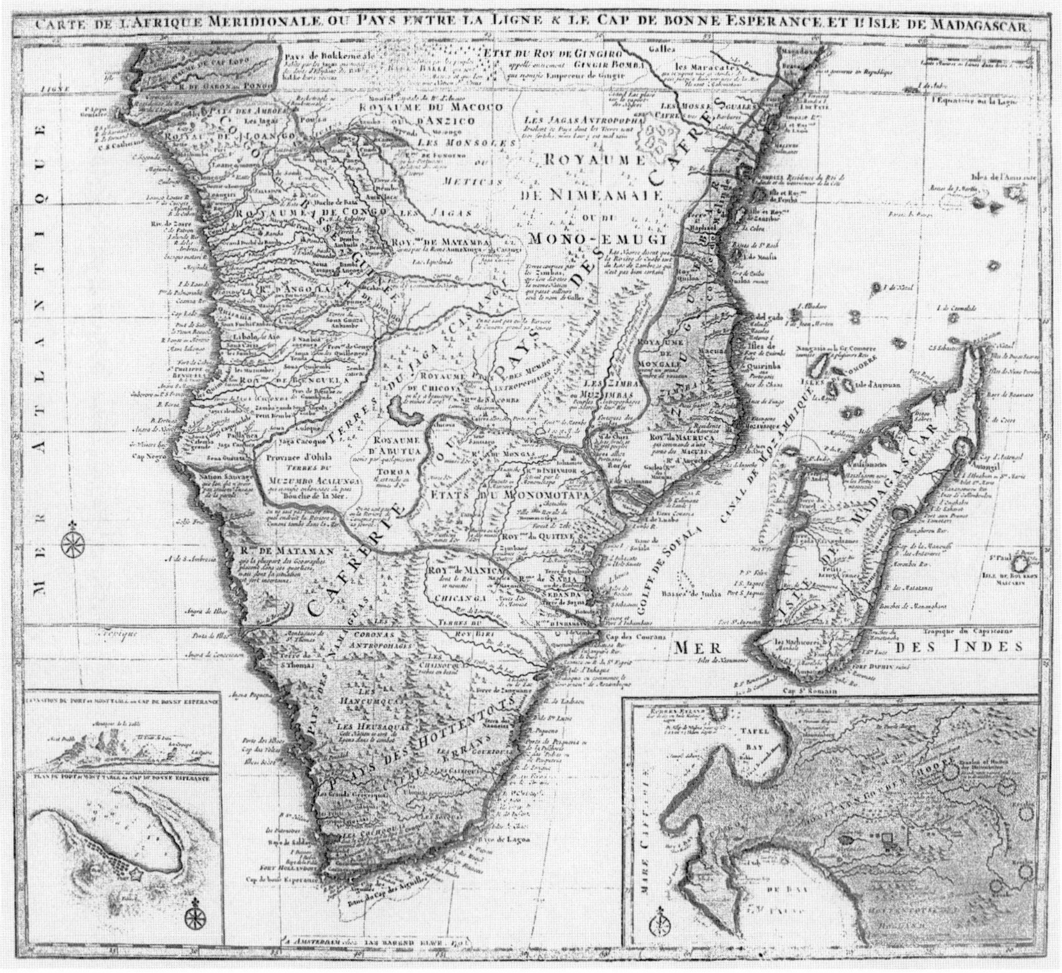

CARTE DE L'AFRIQUE MERIDIONALE OU PAYS ENTRE LA LIGNE & LE CAP DE BONNE ESPERANCE ET I! ISLE DE MADAGASCAR

Map 180

ELWE, Jan Barend
Carte de L'Afrique Méridionale ou Pays Entre La Ligne & Le Cap de Bonne Esperance et L'Isle de Madagascar. (Map of southern Africa or countries between the equator and the Cape of Good Hope and Madagascar.)
(Amsterdam, J.B. Elwe, 1792.)
Map, 49 x 57,5 cm, coloured.
Scale in nautical and travelling leagues.
Inset: The Cape Peninsula. Plan and elevation of Table Bay.

A map of southern Africa with two interesting insets, one of Table Mountain with a plan of Table Bay, and the other, somewhat larger, and of considerable historic and geographical interest, of the Cape of Good Hope. Extensive detailed and fairly accurate topographical information is noted together with the names of owners of farms and plots of that period. Amongst these are de Beer, Swart Pieter, Martin de Smit, Roelof Pasman, Gerrit Kloeton, and the French Huguenot settlement. These place names add considerably to the knowledge of the history of the Cape Colony and especially the Cape of Good Hope. Present research on the hereditary genetic disease of Huntington's Chorea in the Afrikaner population of South Africa notes that the names of Roelof Pasman and Gerrit Kloeton,

mentioned on the map, appear among the early sufferers from the disease. (*South African Medical Journal*, Aug. 20th, 1980.) The rest of southern Africa includes the various native kingdoms of the interior and what was known then of the coastline and rivers and ports, many of which are fictitious and highly imaginative.

This map was originally issued by the Visscher family of Dutch mapmakers in 1710, eighty years earlier, and is one of many examples of the issue of identical maps by different publishers under their own names. Hendrik de Leth, engraver and publisher of Amsterdam, married into the Visscher family and eventually took over the business. He too had his name imprinted on this identical map.

203

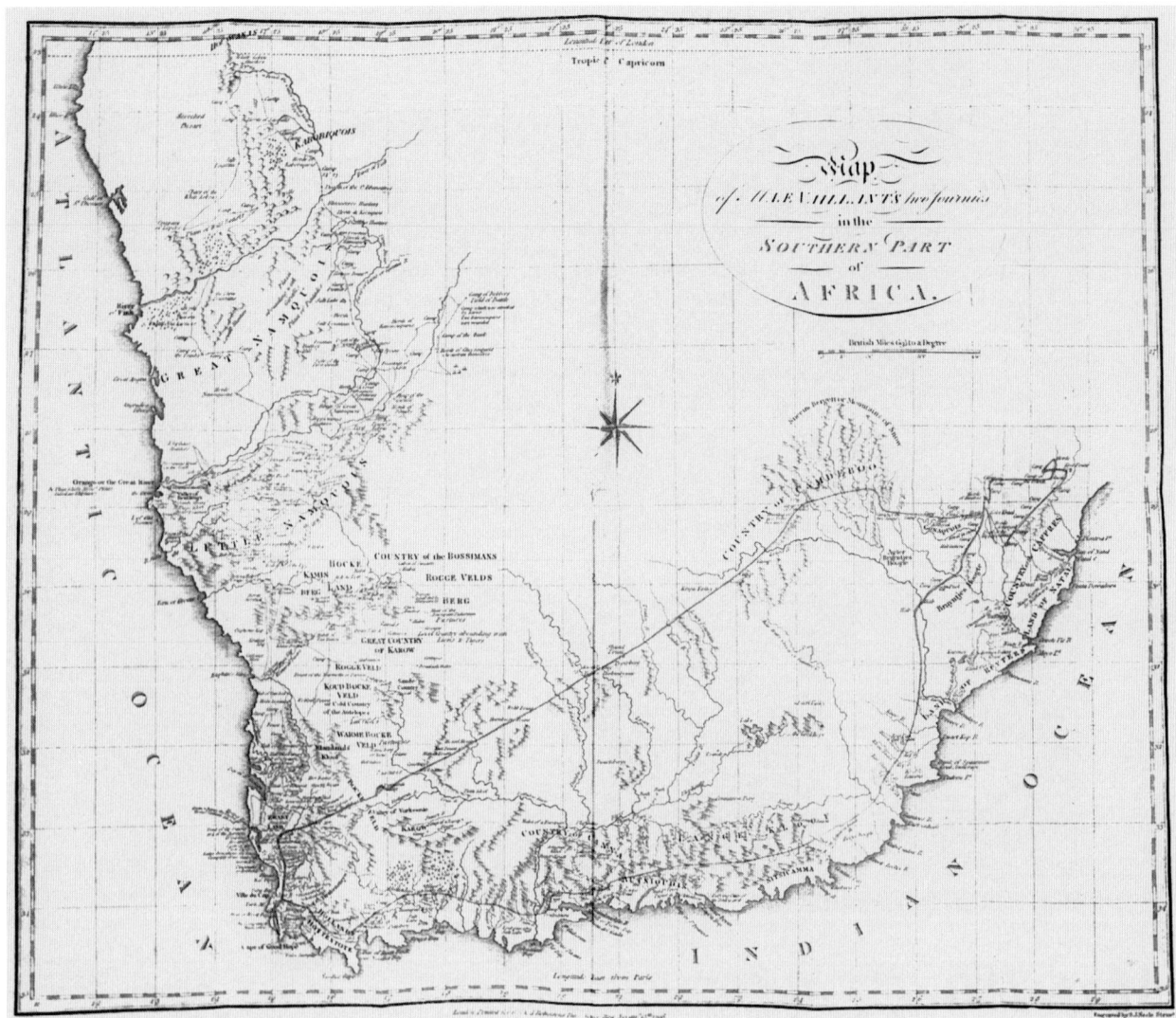

Map 181

LE VAILLANT, François (b. 1753)
Map of M. Le Vaillant's two Journies in the Southern Part of Africa. Engraved by
S.J. Neele. (London, G.G. and J. Robinson, 1796.)
Map, 33,5 x 41 cm, uncoloured.
Scale in British miles.
Prime meridian through London.
In: Le Vaillant, F. _New Travels into the Interior Parts of Africa by way of the Cape of Good Hope in the
years 1783, 1784 and 1785_, vol. 1. London: G.G. and J. Robinson, 1796, fold p. 1.

François Le Vaillant was born in Parimariba in Dutch Guiana in 1753 and after an education in Holland, France and Germany he proceeded to Paris where he studied natural history collections and, full of enthusiasm and ambition, decided to travel into the interior of Africa in order to further his opportunities of gaining information by observing the specimens in their native countries. This led to two journeys which he undertook in the years 1783, 1784 and 1785.

This map depicts his two journeys. His first journey started on December 18th, 1781 and took the route via Hottentots Holland to Swellendam and Muscle (Mossel) Bay. He continued eastwards to Agoa (Algoa) Bay. At the end of 1782 he reached Klein Vis (Fish) River and penetrated into the interior, remaining for some time with a tribe known as the 'Gonaqua' Hottentots. He returned to the Colony via the 'Sneeuw Bergen,' passing through the Eastern Province and then traversing the Karroo, crossing the Buffalo and Touws Rivers, to finally arrive at the plantation of his friend Slater in June 1783. He went north and for over a year traversed Namaqualand, Damaraland, parts of Bechuanaland and the Kalahari Desert.

This map illustrating Le Vaillant's two journeys has an oval title in which the only ornamentations are the flourishes round the word 'map.' A centrally placed compass rose is the only other decoration. The map extends from the Tropic of Capricorn southwards, although the Bay of Natal is the most northerly feature on the east coast. The coast is largely fictitious except around Table Bay and repeats the distorted eastern coastline as seen in Sparrman's map (see Map 223.) The route through Namaqualand and along the southern coast marks the halting places, and the adventures encountered make interesting reading.

Map 182 BARROW, John
A Chart of the Southern Extremity of Africa. Engraved by S.J. Neele. (London, Cadell and Davies, 1806.)
Map, 24,5 x 36 cm, partly coloured.
In: Barrow, John. *A Voyage to Cochinchina in the Years 1792 and 1793.*
London: Cadell and Davies, 1806, fold in appendix.

A chart of southern Africa essentially the same as Sir John Barrow's earlier map (see Map 228) with boundaries of districts, rivers and remarks relating to relevant sites. It has the addition of the route of a journey to Leetako (Kuruman) and the residence of the Bechuana chief (Kuruman area). The coastline is still rather exaggerated. Names of tribes adorn the interior and stories about them are repeated. The Hamboona in the Swaziland area are 'descended from three European Women.' To the west, it is mentioned that 'the mountains of this part of the Country said to abound with Copper Ores.'

The map is based on the travel and experience of Truter and Somerville, who made one of the earliest journeys north of the Orange River, an account of which appears as an appendix to Barrow's *A Voyage to Cochinchina.*

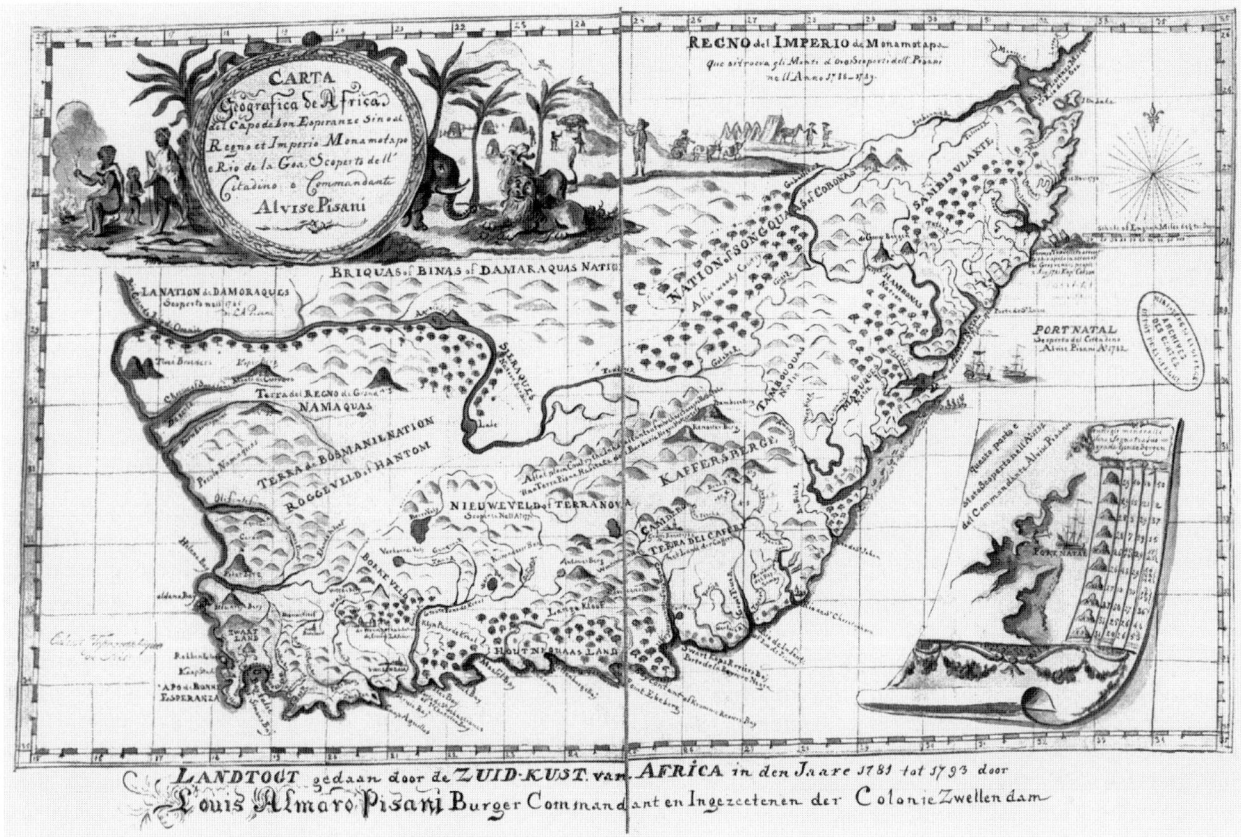

Map 183

PISANI, Alvise Louis Almaro
Carta geografica de Africa del Capo de bon Esperanze sino al Regno et Imperio Monamotapo e Ria de la Goa. Scoperto dell' Citadino e Commandante... (Map of Africa from Cape of Good Hope to the kingdom of Monomotapa...) (L.A. Pisani, 1793.)
Manuscript map, 36 x 61 cm, coloured.
Scale in English miles.
Inset: Port Natal.
Across the bottom: *Landtogt gedaan door de Zuid–Kust van Africa in den Jaare 1781 tot 1793 door Louis Almaro Pisani Burger Commandant en Ingezeetenen der Colonie Zwellendam.*

A coloured manuscript map of southern Africa of the late eighteenth century, extending to latitude 25° north, with a title in Dutch across the whole of the lower border, translated as 'Land journey performed of the Southern Coast of Africa in the years 1781 to 1793 by Louis Almaro Pisani, Burger Commandant and inhabitant (citizen) of the Colony of Swellendam.'

A colourful flowery oval cartouche in the upper left border contains another title, in Italian, translated as 'Geography of Africa from the Cape of Good Hope up to the Kingdom and Empire of Monomotapa and the river of Goa discovered by the citizen and Commander Alvise Pisani.' Surrounding the cartouche are colourful scenes of natives sitting and standing behind a fire on the left with lions and an elephant to the right and beehive huts in the background. Further to the right background are men in dress of the time with a small cart driven by another figure and pulled by what appear to be oxen. Above this scene is another caption in Italian translated as 'Kingdom of the Empire of Monomotapa in which there are mountains of gold discovered by Pisani in the years 1788-1789.'

207

A scale in English miles is sited at the top of the east coast surmounted by a compass rose with radiating rhumb lines and the usual north-pointing fleur-de-lys. A sinking ship is illustrated just to the lower left of the scale in an inlet of the coast, with the caption 'Former travellers' arrivet to this apels in servis of the Grosvenor's people 2nd August 1781, Kap. Cokson,' written in poor English and obviously referring to the wreck of the *Grosvenor*. Further south on this coastline appear two sailing vessels in the Bay of Natal, with a caption in Italian alongside Port Natal translated as 'Port Natal discovered by the citizen Alvise Pisani in 1782.'

An inset in the lower right corner of the map, in the form of a scroll with a floral border, shows a sailing vessel in an inlet marked 'Port Natal' with an oblique title in Italian translated as 'This port was discovered in the year 1782 by the Commander Alvise Pisani.' Alongside this scene is a vertical table at the top of which is a caption, translated as 'All the minerals are marked in grades (quality) – golden mountains,' showing a mountain in each square with, alongside, four columns of figures headed by the words 'long. and lat.'

The details of the map itself are as to be expected at that period both on the mainland and the coastline, some fairly accurate and others quite incorrect. An attempt is made to map the major river systems of southern Africa. Native tribes and their kingdoms are named. Across the centre of the map, running obliquely, is a mountain range with, at its base, captions in Italian and English, noting this to be 'A flat country the Inhabitants of which are Savages, Hottentots.' At the top left of the map is an area described in Italian as 'The nation of the Damoraques discovered in 1785 by L.A. Pisani.'

Who was this Pisani, the author of this somewhat unusual manuscript map not strictly the work of a cartographer/publisher, yet having an accurate outline with many geographical and historical features drawn in it? Pisani has different forenames (Alvise, and Louis Almaro), in different places, suggesting that two different persons may have worked on creating this map. However, on perusing the literature available concerning Pisani and Swellendam, it appears here too that different names for this personality are used.

Alvise Pisani was born in Venice and migrated to South Africa, where it is recorded that he married a Miss Dina Joh. Crafford in Paarl and settled in Swellendam. Due to certain grievances, the early Dutch burghers in the Swellendam

Republic formed themselves into 'Cape Patriots.' One of their grievances was the drastic way a commission terminated the farmers' wheat agreement with the Dutch East India Company, and another was the issuing of paper money. The leader of the Patriots of Swellendam was Petrus Jacobus Delport, who was actively assisted by the Italian Louis Almaro Pisani. In June 1795 Delport, with some sixty armed men, occupied the Swellendam Drostdy and replaced the members of the Council with their own men. When the British attacked the Cape in August 1795 a detachment of burghers from Swellendam under its National Commandant, Delport, proceeded to the Cape to assist in the defences. After the surrender of the Cape, Delport refused to take the oath of allegiance to the British Crown and he was banished in 1798. Pisani was banished to the Netherlands, then fled to France, where he set himself up as an authority on Cape and South African affairs.

It was probably his boastful attitude to his contact with the Cape that prompted Pisani to have this map drawn up. It embodies many historical events and geographic features of the period but it also attributes to Pisani a host of discoveries, not one of which is historically correct. His use of the Dutch, English and Italian languages in describing the map and the events related to it was quite possibly due to his desire to attract the widest possible audience. Its real author is not revealed on the map itself but it is highly likely that it was created in France, most probably Paris, because it comes from the French Archives of Military Maps, which apparently is situated in Paris. If this suggestion is correct, then Pisani may have provided all the geographical and historcal details depicted on the map. Although some may be inaccurate, it is, nonetheless, full of known historical events.

References
Lichtenstein, Henry. *Travels in Southern Africa...* German edition, 1811-1812, pp. 613-619.
Lichtenstein, Henry. *Travels in Southern Africa...* Plumptree English edition, 1812, p. 371.
Beyers, Coenraad. *Die Kaapse Patriote*. Reprint, 1967, p. 268.
Wieringa, P.A.C. *Die Oudste Boeren Republiek*, p. 109.
Sluysken, A.J. *Verhaal Gehouden in den Commissaris van de Kaap de Goede Hoop*. The Hague: Isaac van Clef, 1797, pp. 250 and 257.
Standard Encyclopaedia of Southern Africa, vol. 6, p. 145; vol. 10, p. 388. Cape Town: NASOU.

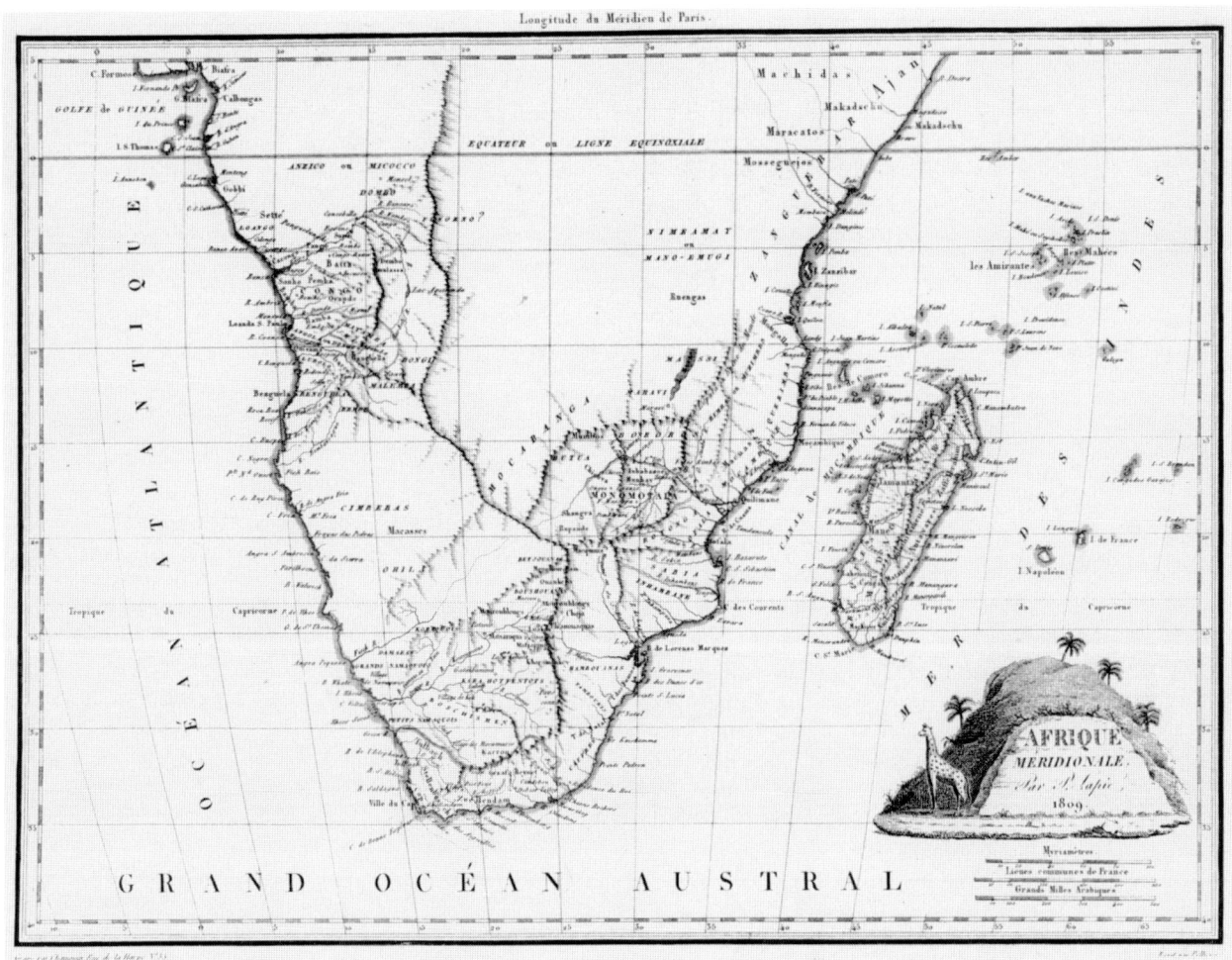

Map 184

LAPIE, Pierre (1779-1851)
Afrique Meridionale. (Southern Africa.) Engraved by Pellicier and Chamouin. (Paris, Lapie, 1809.)
Map, 21,5 x 29 cm, coloured outline.
Scale in kilometres, French leagues, Arabic miles.
Prime meridian through Paris.

A small map of southern Africa, issued by Pierre Lapie, a publisher of Paris and Geographer to the King, who produced a number of atlases.

The cartouche appears in the lower right corner with the title against part of a grassy hill topped by palm trees. There is an elegant-looking giraffe on the left. The Cape appears to be filled with accurate place names and there are a varied number of native tribal names extending particularly up the east coast as far as Monomotapa. Central Africa is devoid of place names, rivers and mountains.

Chamouin is noted as the general engraver and Pellicier the script engraver (who also worked for Delamarche, the French publisher).

209

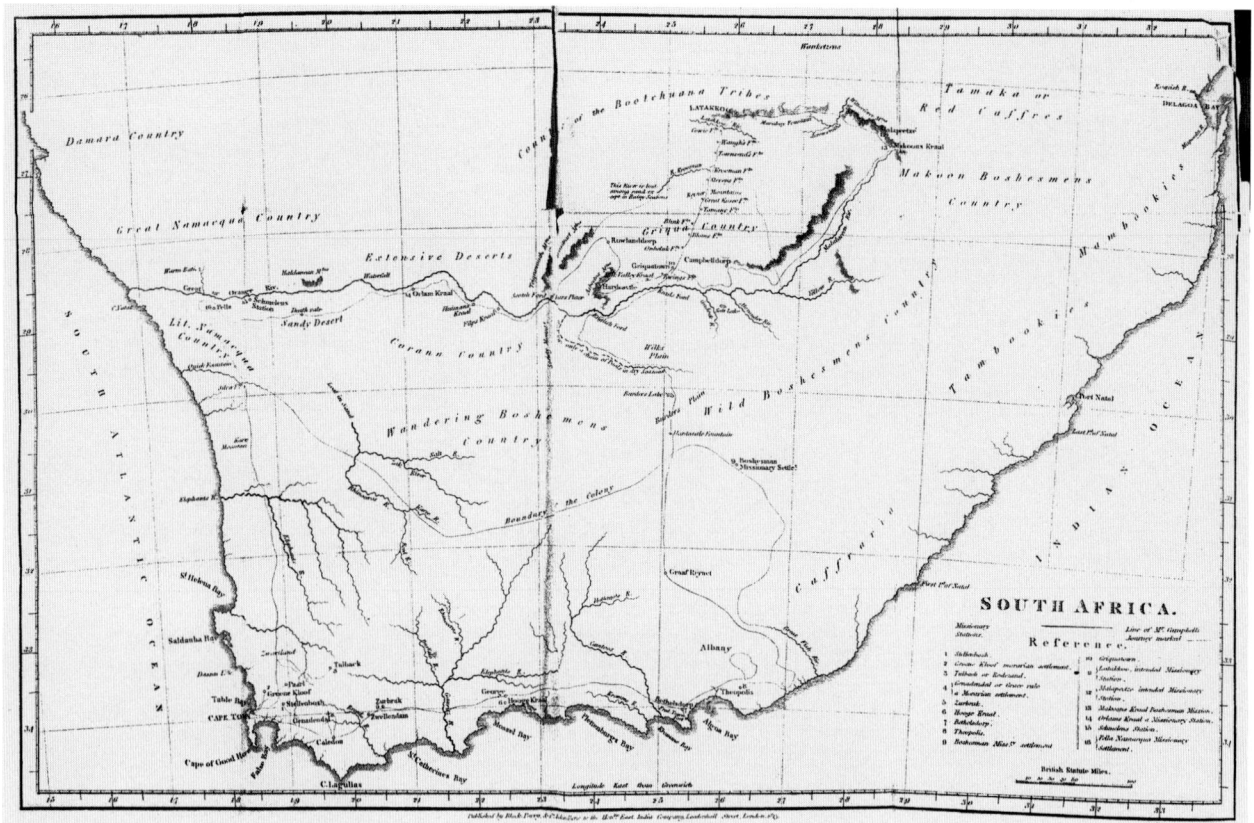

Map 185 CAMPBELL, John
South Africa. (London, Black, Parry & Co., 1815.)
Map, 25,5 x 40,5 cm, uncoloured.
Scale in British statute miles.
In: Campbell, Rev. John. *Travels in South Africa undertaken at the request of the Missionary Society,* 2nd.
ed. London: Black, Parry & Co., 1815, p. 1.

This map of South Africa was drawn to illustrate the first travels of the Rev. John Campbell in 1815. The directors of the London Missionary Society decided to send the author to inspect their settlements. He travelled in the Cape, to the Stellenbosch, Caledon and Genadendal mission settlements. He then extended his visits further north to the 'Bootchuana' country for missionary work there, and right through the Colony to Lattakoo.

The coastline is still slightly exaggerated, especially the south-east coast. The Orange, or Great River, as it is named on the map, is not very well known and its course below Prieska is straight to the west without any curves. In the lower right corner is a list of sixteen mission stations named on the map and a scale of English miles.

210

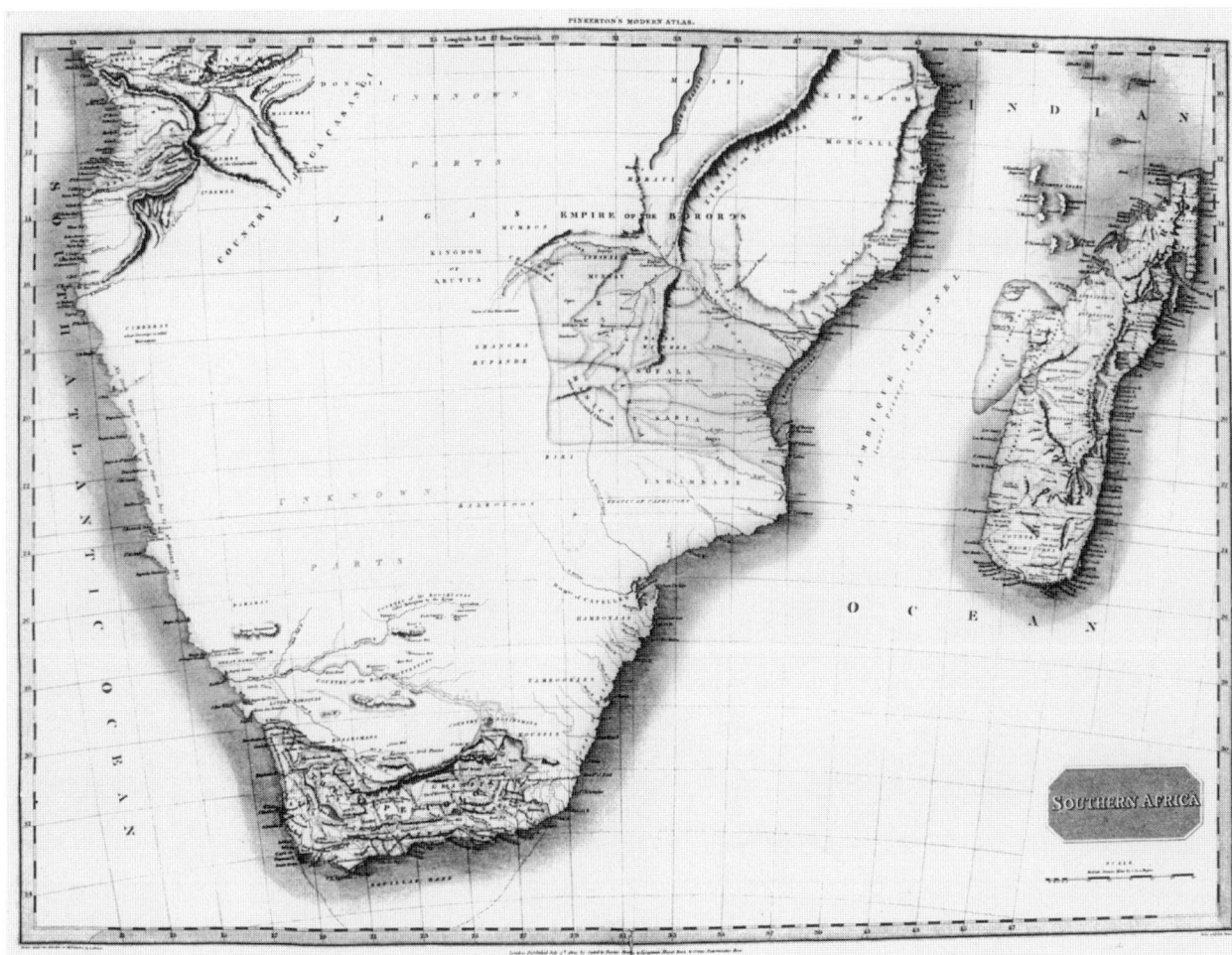

SOUTHERN AFRICA

Map 186   PINKERTON, John (1758-1826)
Southern Africa. Drawn by L. Herbert, engraved by S.J. Neele. (New York, Inskeep and Bradford, 1815; London, Cadell and Davies, Strand, and Longman, Hurst, Rees & Orme, Paternoster Row, 1809.)
Map, 50 x 69 cm, coloured.
Scale in British miles.
Title across the top: *Pinkerton's Modern Atlas.*

A map of southern Africa from *Pinkerton's Modern Atlas* of 1815, drawn by L. Herbert and engraved by S.J. Neele. This is an American edition, published by Bradford and Inskeep of New York and Thomas Moses of Philadelphia.

The map itself has no special cartouche, but a neat boxed title, 'Southern Africa' in the lower right corner. The interior is conspicious for the complete absence of geographical detail in wide areas of central and southern Africa, noted as 'unknown parts.' Madagascar features prominently and in greater detail. The ancient kingdoms of the east coast reappear as in the map of the continent of Africa as appears in the map of Africa in Pinkerton's *Modern Atlas.* The Cape Colony is depicted with fairly accurate details of places, rivers, mountain ranges and coastal inlets. The Great Fish River, however, is marked erroneously as a tributary of the Orange River.

A British edition identical to the American issue was published in 1809 by Cadell and Davies, Strand, and Longman, Hurst, Rees and Orme of Paternoster Row, as is noted along the bottom of the map.

211

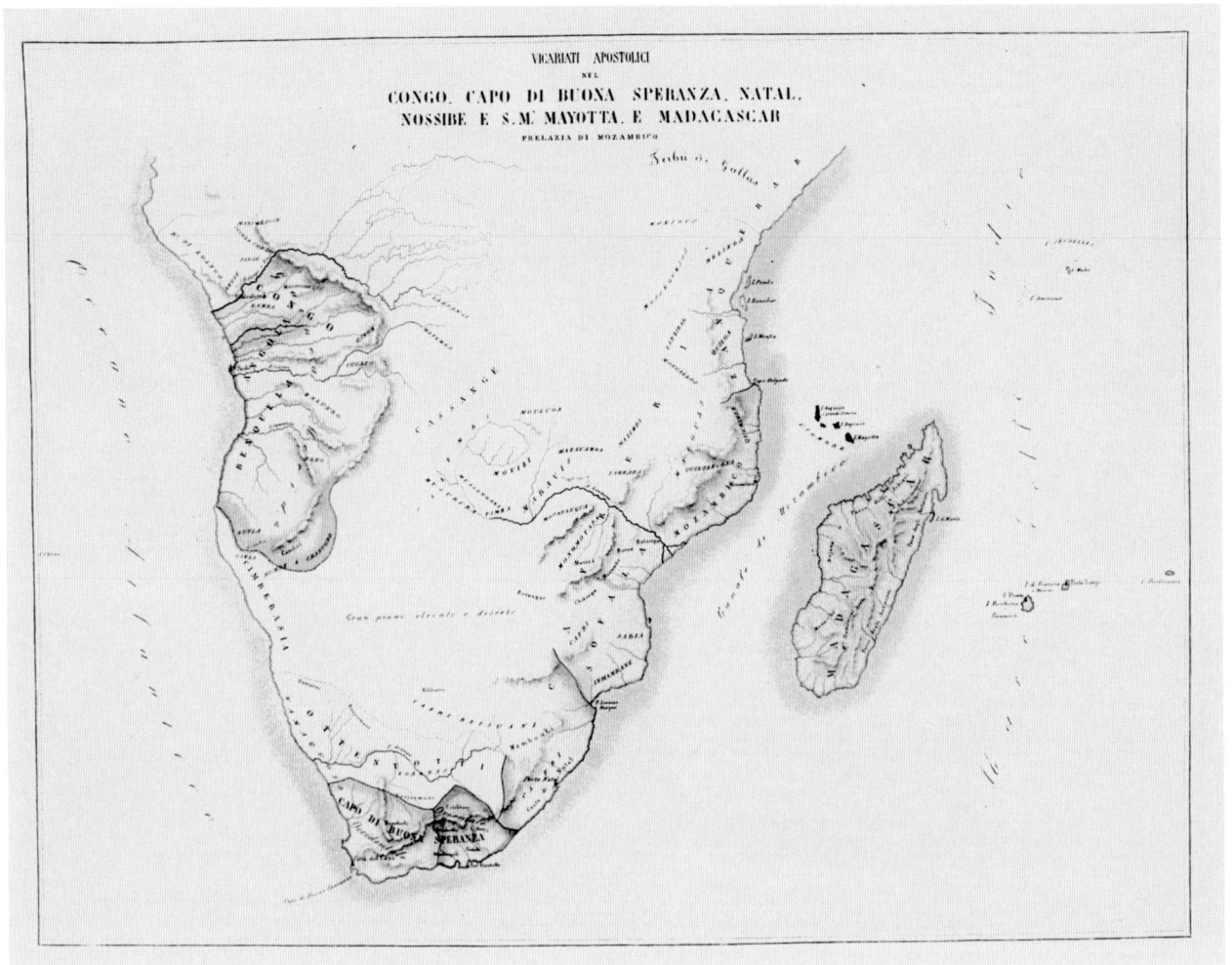

VICARIATI APOSTOLICI
NEL
CONGO, CAPO DI BUONA SPERANZA, NATAL,
NOSSIBE E S. M. MAYOTTA, E MADAGASCAR
PRELAZIA DI MOZAMBICO

Map 187 ANON
Vicariati Apostolici nel Congo, Capo di Buona Speranza, Natal, e S. Ma.
Mayotta, Nossibe Madagascar Prelazia di Mozambico.
Map, 47 x 62 cm, coloured.
Top left: TAV. CXII.

This coloured Italian map probably originated from the Vatican in Rome. It shows southern Africa divided into its various states and colonies. The Cape Colony is referred to as Capo Buona Speranza and very few towns are included. North of this area the Orange River appears to be fairly accurately illustrated. Sofala and Mozambique, with a few rivers, are noted on the east coast. Benguela, Angola and the Congo appear on the west coast. Madagascar and the surrounding islands are labelled by their French names. The title of the map appears at the top centre, naming some of the more important places. The Island of Mayotta, one of the Comoro group north of Madagascar, is also prominent.

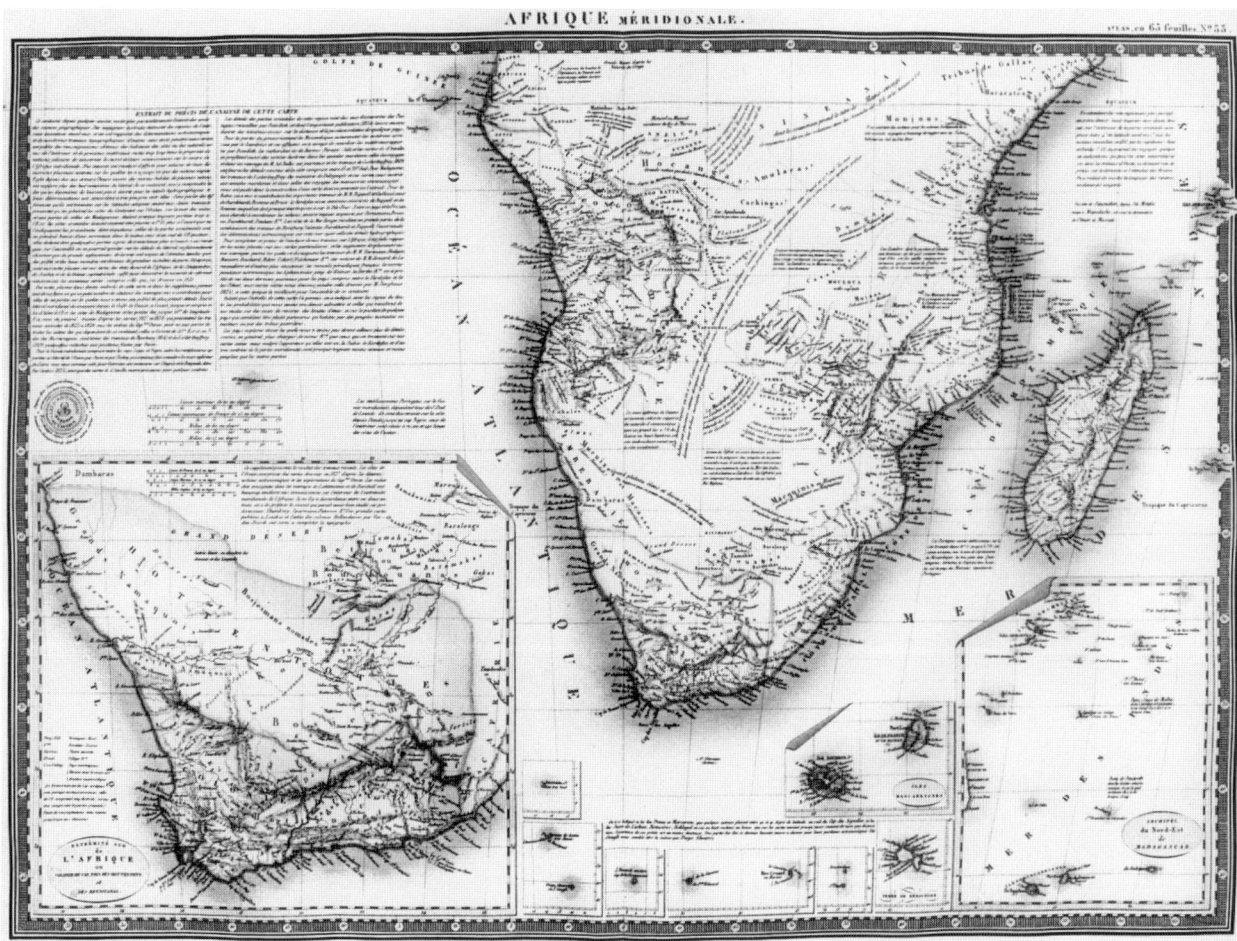

Map 188

BRUÉ, Adrien Hubert (1786-1832)
Afrique Méridionale (Southern Africa.) (Paris, A.H. Brué, 1828.)
Map, 36 x 51 cm, slightly coloured.
Scale in nautical leagues, French leagues and miles.
Prime meridian through London.
Inset: Enlargement of southern tip of Africa. Islands northeast of Madagascar.
Mascareignes islands. Kerguelen Island.
Top right: 'Atlas, en 65 feuilles, No. 53.'

The publisher of the atlas in which this map appeared was the Frenchman Adrien Hubert Brué, who published many atlases from the Rue des Macons, Sorbonne. His *Atlas Universel* appeared in 1816.

A feature of this map is its detailed insets. There is a large one of the south-west coast, the Cape of Good Hope and the Colony du Cap. The islands to the north-east of Madagascar also appear separately as an inset. The Mascareignes Islands feature on their own as the Ile Bourbon (Réunion) and Ile de France (Mauritius) with their coastal ports, towns and mountain ranges. The fourth inset is labelled Terre De Kerguelen, showing in detail the numerous coastal placenames. Six separate insets appear at the bottom of the map, each with its longitude and latitude and depicting numerous smaller islands of the southern Atlantic and Indian Ocean, among which are Marian, Ascension, Gough, Tristan d'Acune, Amsterdam and St. Paul.

213

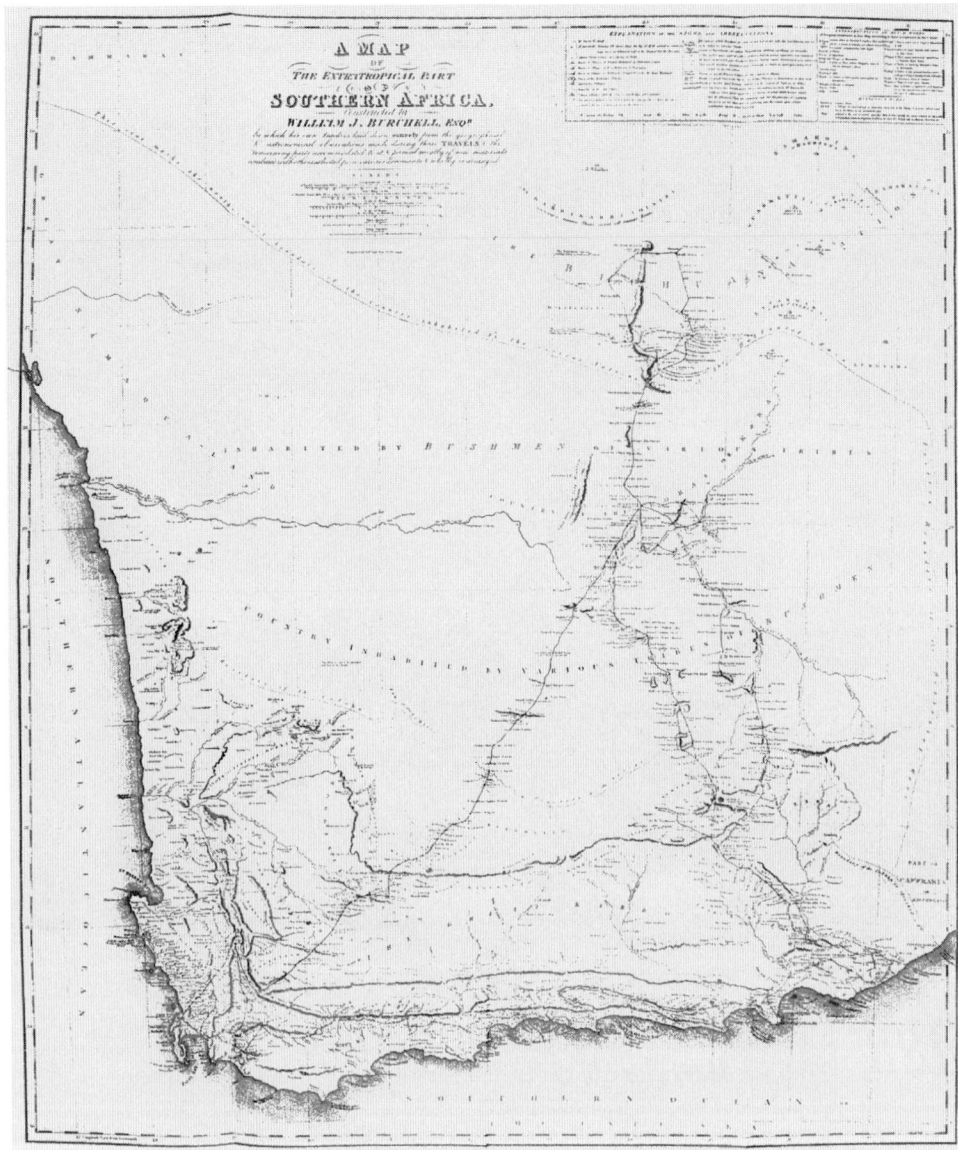

Map 189

BURCHELL, William J.
A Map of the Extratropical Part of Southern Africa. Engraved by Sidney Hall.
(London, Longman, Hurst, Rees, Orme and Brown, 1822.)
Map, 82,5 x 70 cm, uncoloured.
Scale in English statute miles, hours travelling, days journey.
Prime meridian through Greenwich.
Inset: *Explanation of signs and abbreviations; interpretation of Dutch and Hottentot words.*
In: Burchell, William J. *Travels In the Interior of Southern Africa*, vol. 1.
London: Longman, Hurst, Rees, Orme, Brown and Green, 1824, fold at back.

A map showing that part of southern Africa which the famous traveller and explorer William J. Burchell covered and described in his two-volume book of 1824, written after a four-year exploration.

This large and detailed map extends from 24° S as far as the Keiskama River on the east coast. The date of the founding of a village is written with its name. The author has also drawn a line marking the northernmost boundary of the Hottentots. The course of the Orange River as far as the Upington bend is more realistic, and an upper and lower waterfall is marked on it (the Augrabies Falls of today). Burchell notes that the Orange River was given its name in 1779 by General Gordon. For the first time the Cango Cavern appears on a map. This map is on the whole very informative as regards settlements, farms, etc. Burchell penetrated as far as Lattakoo, later known as Kuruman.

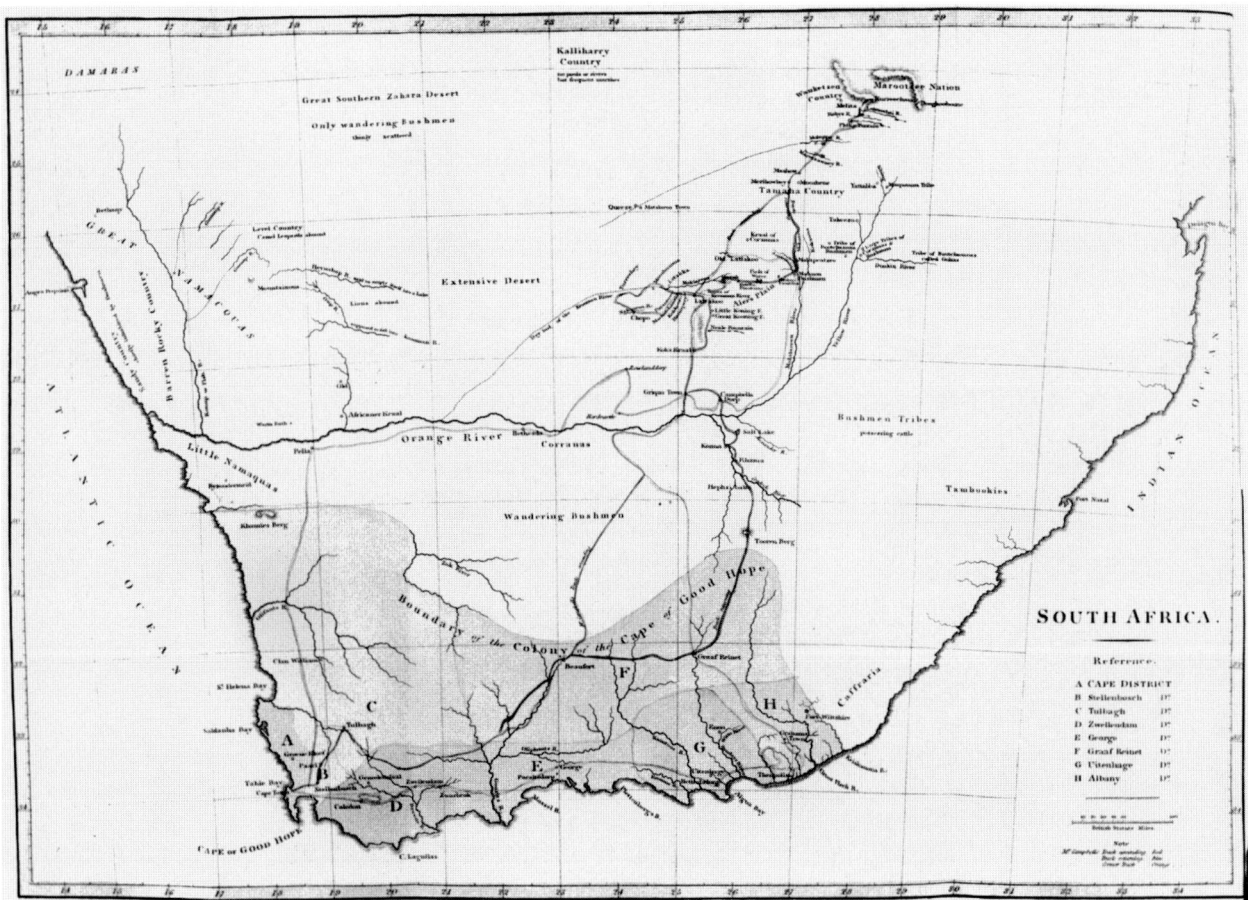

Map 190

CAMPBELL, John
South Africa. (London, F. Westley, 1822.)
Map, 31 x 44,5 cm, coloured.
Scale in British statute miles.
In: Campbell, Rev. John. *Travels in South Africa, undertaken at the request of the London Missionary Society, being a narrative of a second journey...* vol. 1. London: F. Westley, 1822, p. 5.

This coloured map depicts the second travels of the Rev. John Campbell through South Africa. He left London with Dr Philip, another well-known missionary, in 1818 and after a two months' stay in Cape Town they were joined by Messrs Evans and Moffatt and together proceeded to visit the missions in the Cape Colony and Kaffraria.

This map has a greater degree of accuracy on the west coast (compare Map 185), but the mouth of the Orange River is still not quite accurate. The various districts are illustrated in colour, with coloured tracks of the routes taken to and from the various missions. A further coloured route is illustrated showing Campbell's former route. There is not much detail in the Colony beyond the rivers, a few towns and tribal names. The greatest detail is in present-day Botswana, where fountains and villages are marked as well as Campbell's former tracks. An index reference to the various districts is situated in the lower right corner above a scale in English miles.

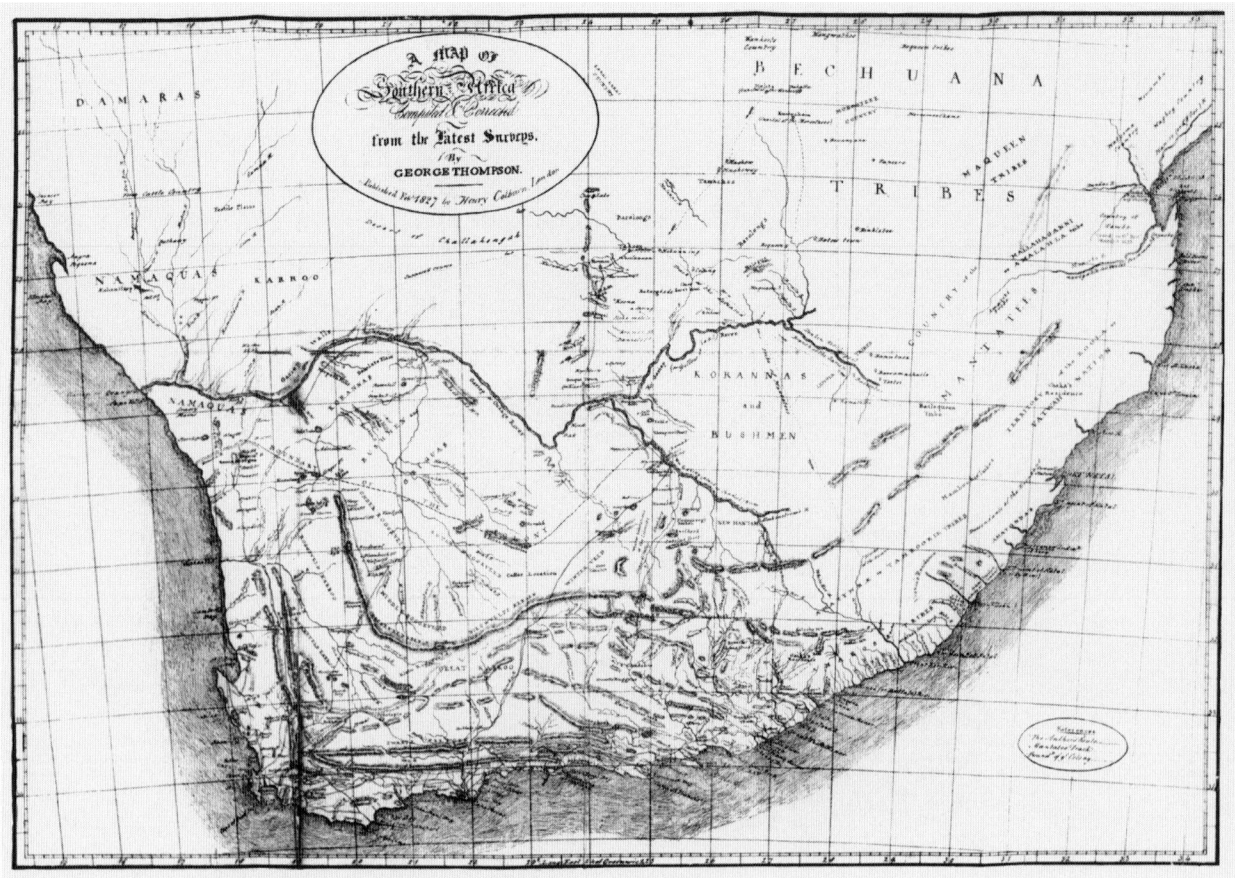

Map 191

THOMPSON, George
A Map of Southern Africa compiled and Corrected from the Latest Surveys...
(London, Henry Colburn, 1827.)
Map, 35 x 51 cm, uncoloured.
In: Thompson, George. *Travels and Adventures in Southern Africa*, 2nd ed. vol. II. London: H. Colburn, 1827, p.431.

This map of southern Africa has its title on the top left in a simple oval cartouche, the only ornamentation of which is in the flourishes around the words 'Southern Africa.' It is a typolithograph published by Henry Colburn in London in 1827. As such it is not very clear and the 1829 Dutch edition of the book has a better engraved map.

Thompson, a Cape Town merchant, resided in South Africa for many years and travelled throughout the greater part of the Colony. He marks the George IV cataract on the Orange River (Augrabies Falls). He depicts rivers but only the mouths are shown on the east coast. There is a note that the Maputa River is navigable for 40 miles from its mouth in Delagoa Bay. Tribal names are marked in the interior and the path of the defeated Matabele tribe is traced. Descendants of Europeans are noted to live inland from Port Natal. The Kalahari is given the unfamiliar name of 'Challahengah.'

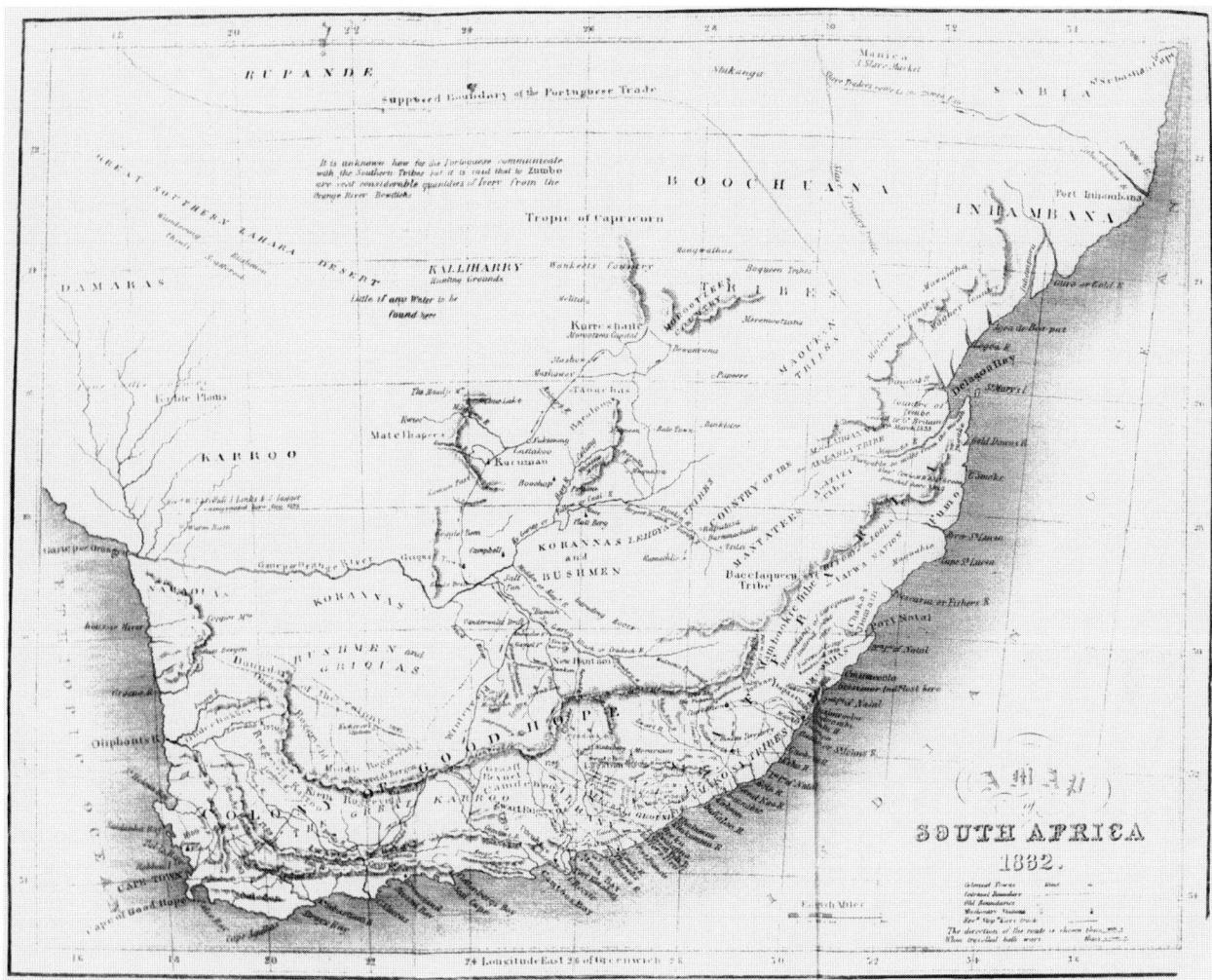

Map 192

KAY, Stephen
A Map of South Africa 1832. Engraved by Alex Findlay. (New York, Harper & Brothers, 1834.)
Map, 20 x 26 cm, uncoloured.
Scale in English miles.
Prime meridian through Greenwich.
In: Kay, Stephen. *Travels and Research in Caffraria...* New York, Harper & Bros., 1834, fold p. 1.

A map of South Africa which appears in Rev. Stephen Kay's book on his travels in that country, first published in London in 1833. The map illustrated here appeared in the New York edition a year later.

In his writings Kay, a corresponding member of the South African Institute, provides valuable information on the Bantu tribes and the Wesleyan missions. The coast immediately north of the Orange River is distorted by a sharp turn to the north-west and the course of the Orange River is inaccurate. The map extends on the east coast to St. Sebastian. Present-

day Botswana has more details than the rest of the interior. The routes of other travellers are depicted and also the place of assassination of Threlfall, Links and Jagger in South West Africa in 1825.

Within the map the colonial towns and boundaries, old boundaries, missionary stations and Rev. Kay's routes are depicted by specific symbols and lines. The routes of slave traders in the Delagoa Bay area are marked with reference to the Zumbo Fair on the Zambezi River.

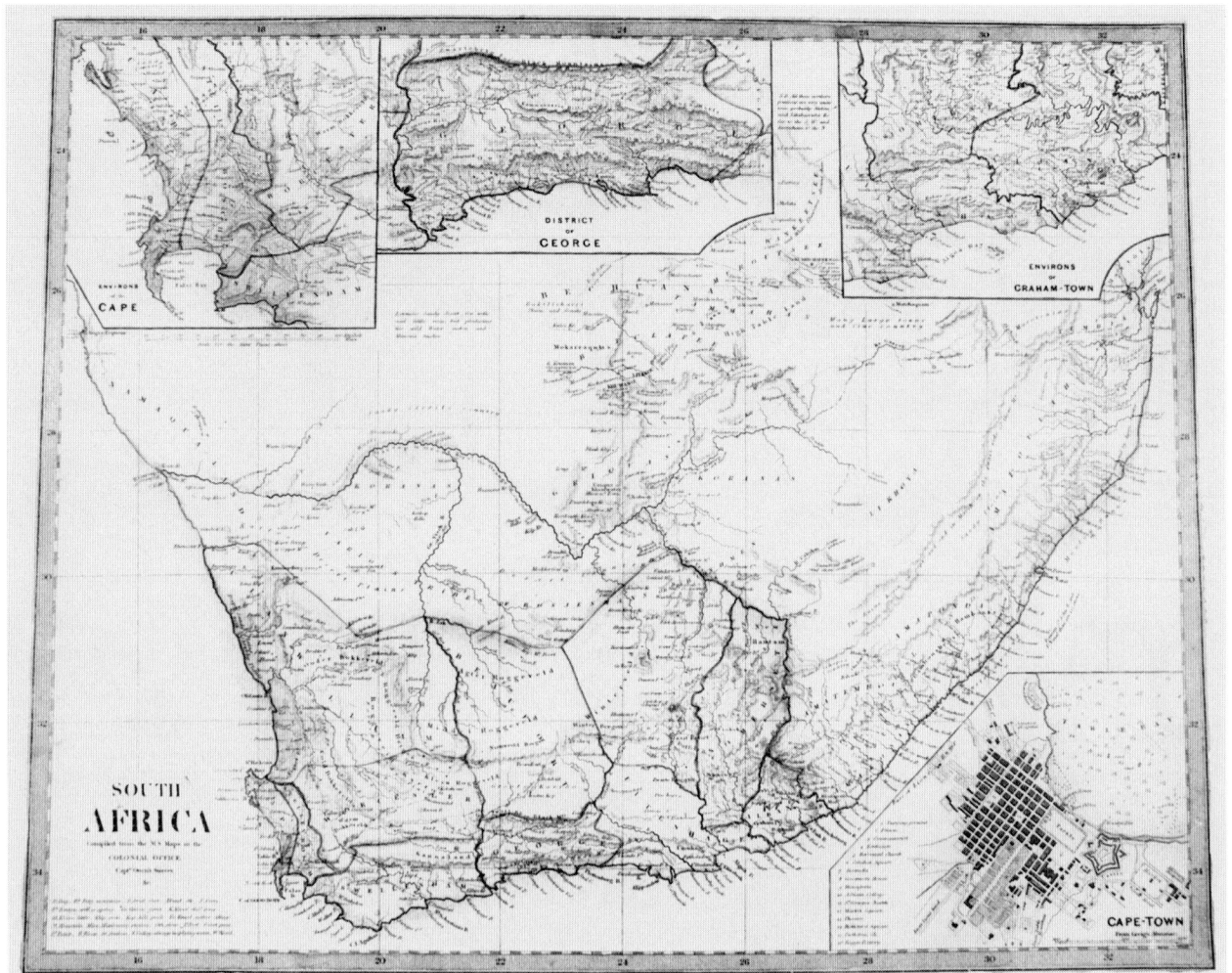

Map 193

OWEN, William Fitzwilliam (1774-1857)
South Africa Compiled from the M.S. Maps in the Colonial Office. Engraved by J. and C. Walker. London, Baldwin and Cradock, 1834.
Map, 31 x 39 cm, coloured.
Scale in English miles.
Inset: Environs of the Cape. District of George. Environs of Grahamstown. Cape Town.

A map of South Africa published under the supervision of the Society for the Diffusion of Useful Knowledge by Baldwin and Cradock in 1834. This society was responsible for a series of maps of the world, issued in parts by Baldwin and Cradock (1829-1832), Chapman and Hall (1844), C. Knight (1844-1852), G. Cox (1852-1853), and Stanford (1857-1870). Later editions were revised.

As well as making land maps, Capt. Owen RN supervised the charting and mapping of Africa and Arabia by command of the Lords Commissioner of the Admiralty. These were issued by the Hydrographical Office. A number of the names on the east coast of Africa appeared in the two-volume book *Narrative of Voyages to Africa, Arabia and Madagascar* (1833).

This map is highly accurate for 1834, showing in fairly small detailed insets the environs of the Cape district, George and Graham-Town (Grahamstown). In the lower right corner there is a detailed plan of Cape Town (from *Greig's Almanac*) with a numbered index of all important buildings. Table Bay is filled with soundings. An historic wooden pier is marked, as well as the Castle (Fort) and the original Somerset Hospital. The map itself shows an abundance of accurate detail of all geographical features and placenames. The tribal areas situated north of the Cape Colony are prominently placed.

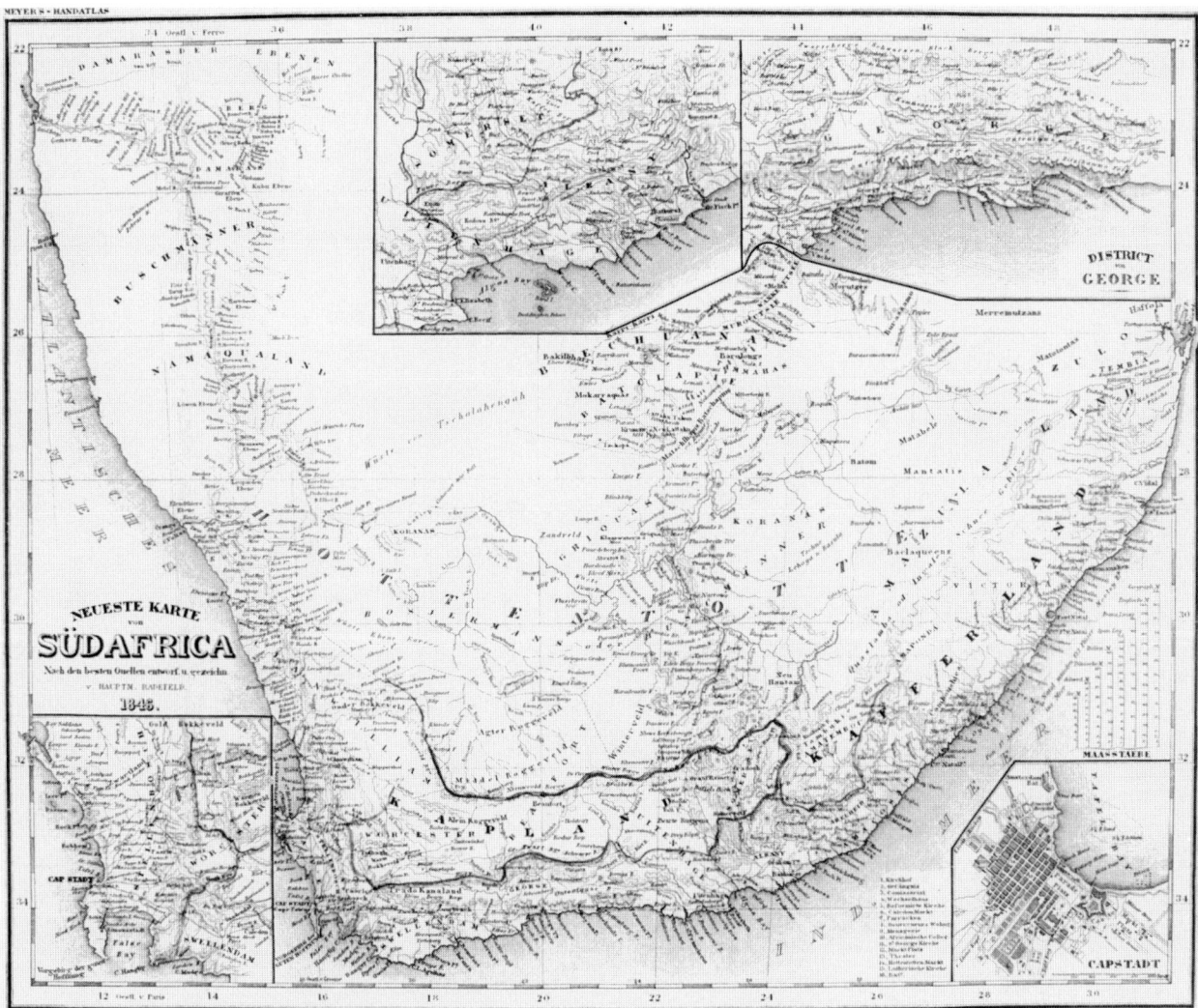

Map 194 RADEFELD, Carl Christian Franz (1788-1874)
Neueste Karte von Süd Africa Nach den besten Quellen entworf und gezeichn von... (Latest map of Africa drawn from reliable information by... (Amsterdam, Joseph Meyer) 1846.
Map, 29 x 36 cm, coloured outline.
Scale in geographical, English, Italian, Danish, Swedish and sea miles, French and Spanish leagues.
Prime meridian through Paris and Ferro Island.
Inset: Eastern Cape Coast. District of George. Cape Peninsula. Cape Town.
Top left: *Meyers – Handatlas.*
Top right: *No. 93.*

A fine steel-engraved German map of South Africa by Carl Christian Franz Radefeld, issued for *Meyer's Neueste Universal Handatlas.* In addition to an accurate and detailed map, especially of the Cape Colony and Kafferland, there are equally detailed inset maps of the Cape of Good Hope, a plan of early Cape Town, the district of George, Algoa Bay and the Eastern Province. This map provides detailed and accurate geographical data for its time.

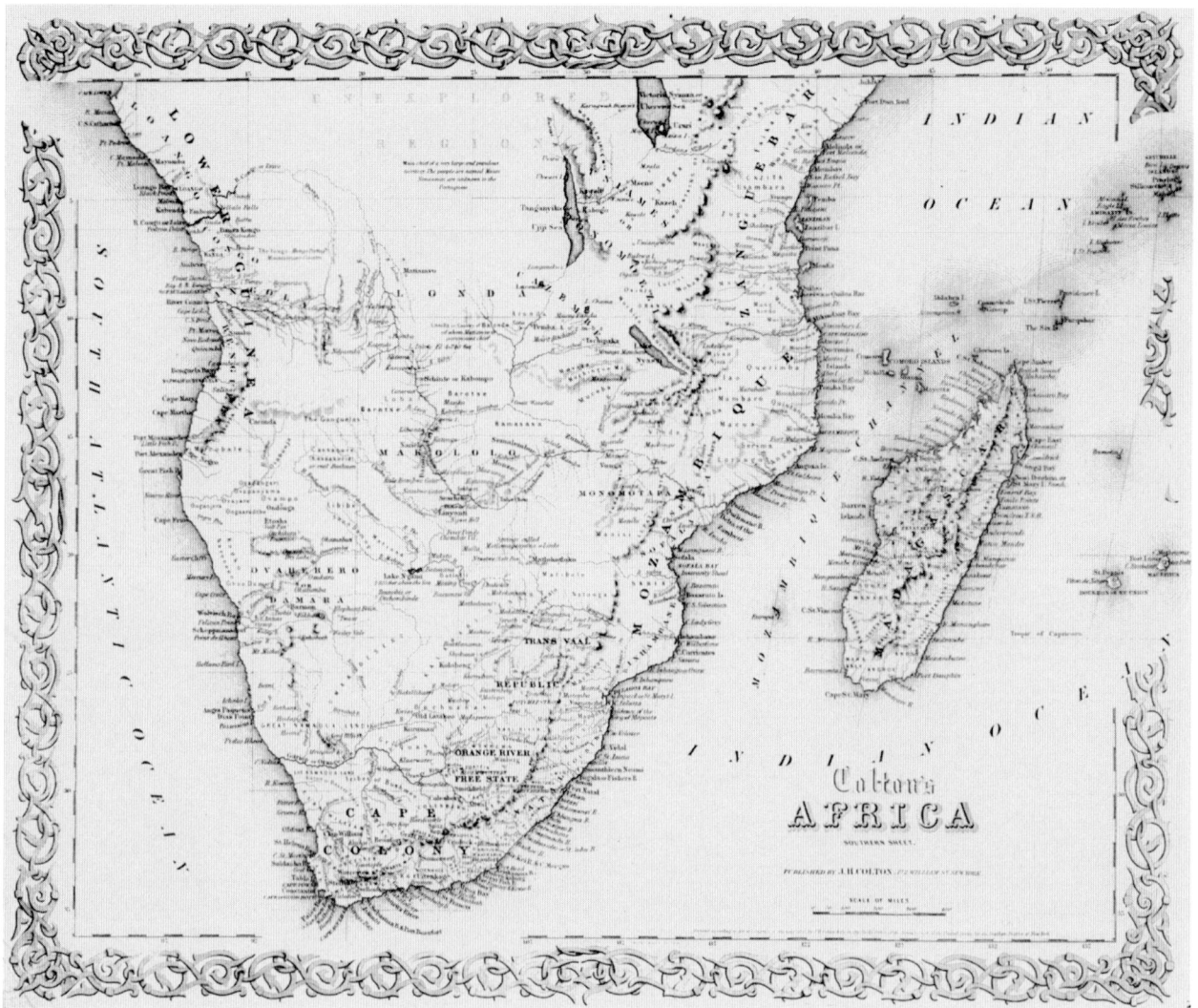

Map 195

COLTON, Joseph Hutchings (1800-1893)
Colton's Africa (southern sheet). (New York, J.H. Colton, 172 William Street, 1855.)
Map, 28,5 x 35,5 cm, coloured.
Scale in miles.
Prime meridian through Washington.
Bottom right: No. 36.

This map, referred to as 'Colton's Africa (southern sheet),' is dated 1855 and was published by Joseph Hutchings Colton, a publisher of 86 Cedar Street and 172 William Street, New York. His atlases include *South America* (1845), *World Atlas* (1855), general atlases, *Family Atlas* and *American School Geography* (1870).

This map of Africa appears to have been included in Colton's *World Atlas*, It is a fine colour printed map with an attractive border. For its date it provides a good deal of accurate information on southern Africa, especially the Cape Colony with its coastal geography. At this period the other provinces appear, such as the Orange River Free State – now known as the Orange Free State – with its capital, Bloemfontein. Natal is marked with its harbour, D'Urban, and other coastal names are accurately sited. The Transvaal Republic is noted without Pretoria, which did not appear on a map until the 1870s. The other smaller republics of Zoutpansburg, Potchefstroom and Leidenburg (Lydenburg) appear, and Mozambique and Madagascar with their associated islands are accurately sited. The great lakes now appear accurately sited for the date of the map – Lake Tanganyika and the southern portion of Victoria Nyanza. The Vaal and Orange Rivers are labelled, with the inclusion of the old name of Gariep.

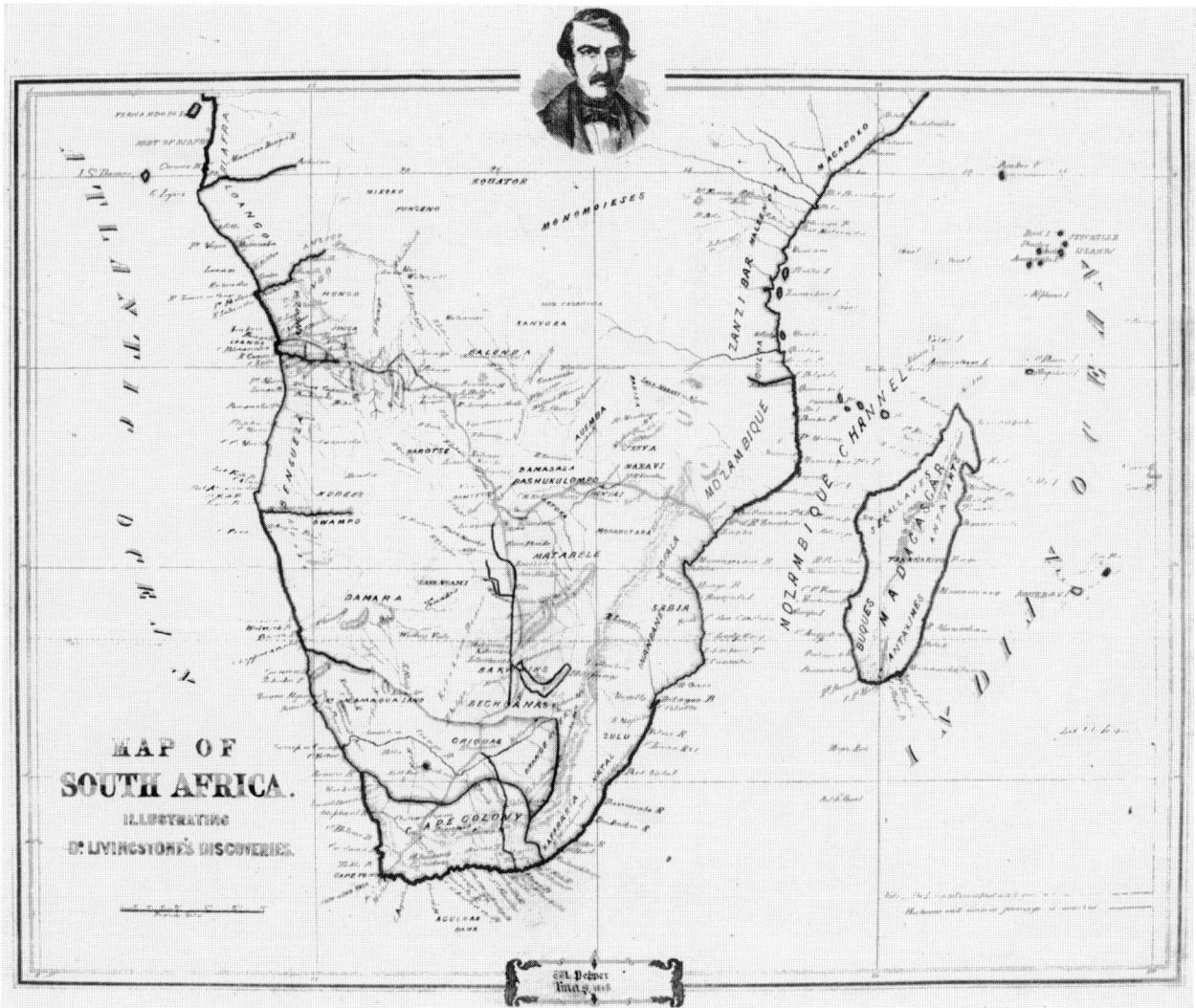

Map 196

PEPPER, W.
Map of South Africa Illustrating Dr. Livingstone's Discoveries. (London, W. Pepper, 1858.)
Map, 35 x 47 cm, coloured outline.
Scale in English miles.

This is an unusual manuscript map on which the title, lettering, portrait of Livingstone and the borders were apparently over-printed. David Livingstone's adventurous journeys in Central Africa put more on the map of the dark continent than any other explorer. He was the first European to cross the African continent, covering at least 30,000 miles and charting about one million square miles of unknown territory. He traced a part of the course of the Zambezi, including the Victoria Falls. He mapped some of the Central African river system and discovered Lake Malawi and the Shirwa Highlands.

A portrait of Livingstone dominates the map at the top. At the lower border a rectangular, slightly decorated box contains the name W. Pepper, Xmas 1858. The general geography of the map is sparse and it appears to be intended to show only the routes and discoveries of Livingstone.

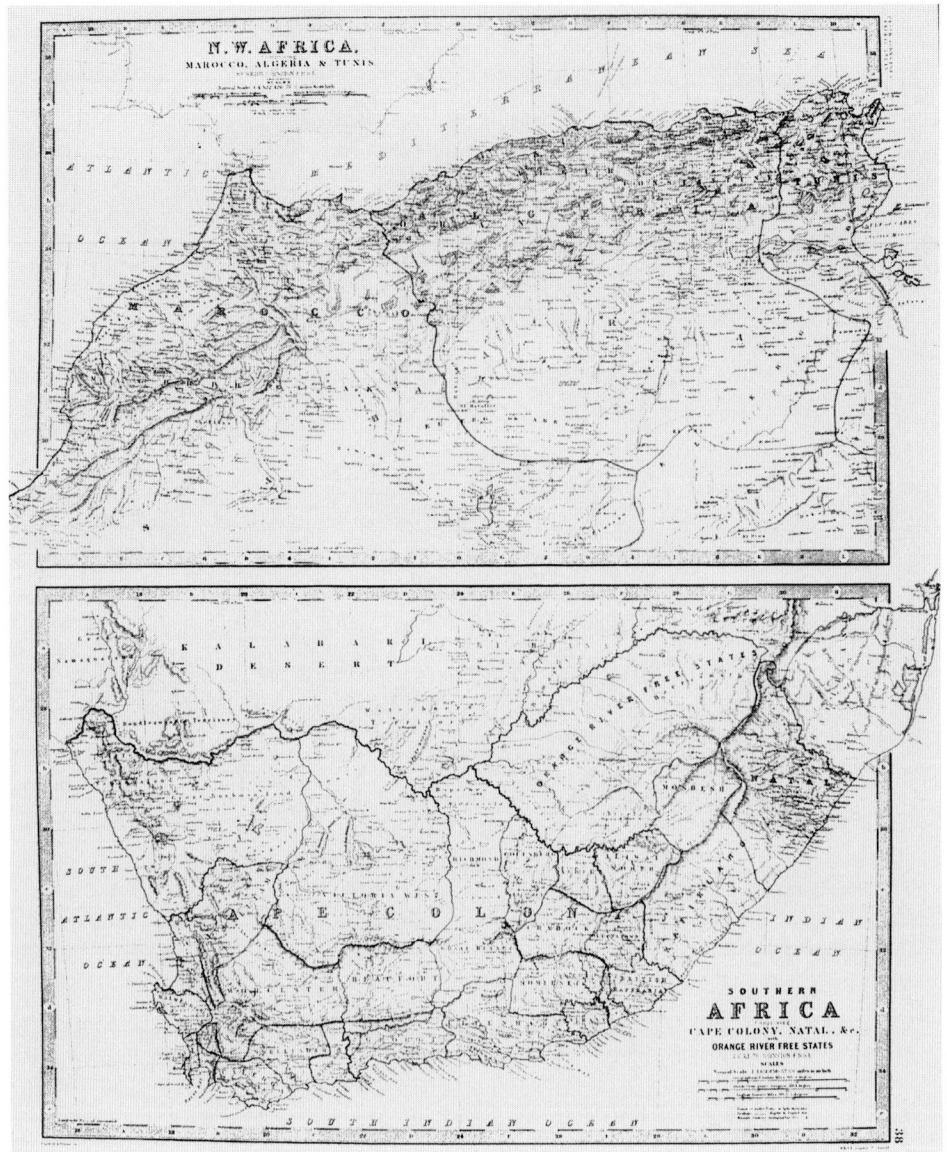

Map 197

JOHNSTON, Alexander
Keith (1804-1871)
*Southern Africa
Comprising Cape
Colony, Natal, &c. with
Orange River Free
States.* Engraved and
printed by W. and A.K.
Johnston. (Edinburgh, William
Blackwood and Sons, 1862.)
Map, 28 x 44 cm, slightly coloured.
Scale: 57,636 miles : 1 inch, geographical, Italian, Dutch and English miles.
Top right: *Keith Johnston's General Atlas.*
Bottom right: 38.

This map, in two parts, is made up of N.W. Africa –
showing in great and accurate detail its mountain ranges,
rivers and coastal inlets – and the southern portion of Africa.
The map of southern Africa shows greater detail than the
other Johnston single map of Africa (see Map 145).

Keith Johnston was the youngest brother of Sir William
Johnston, founder of the Johnston publishing and printing
establishment of Edinburgh. Keith styled himself
'Geographer at Edinburgh in Ordinary to the Queen.' He
issued a physical atlas in 1848, the first British atlas to provide
a synoptic view of physical geography. Subsequently he put
out other atlases and a *People's Pictorial Atlas* in 1873. This

map is marked as coming from Keith Johnston's *General Atlas.*
It was engraved by W. and A.K. Johnston and Blackwood
and Sons assisted in its publication.

The map of southern Africa provides a good deal of
geographical detail and, as the title indicates, it comprises
the Cape Colony, Natal and the Orange Free State. A portion
of the southern Transvaal is shown and referred to as the
South African Republic. The colonies are accurately marked
with their river boundaries. The Cape Colony was fully
established by then, with railway lines which are depicted
here. A 'Moshesh' area is marked, named after the Basuto
chief; later to become Basutoland and finally Lesotho.

222

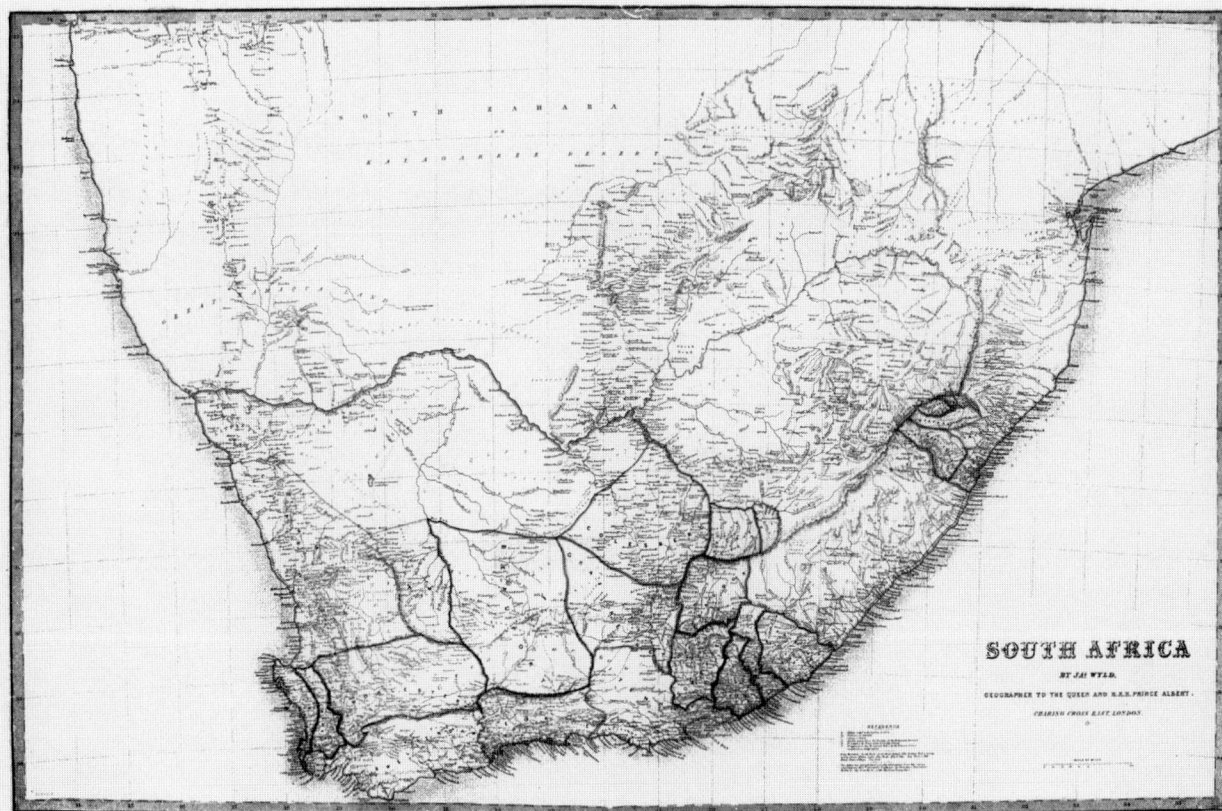

Map 198

WYLD, James (1812-1887)
South Africa by Jas Wyld, Geographer to the Queen and H.R.H. Prince Albert.
(London, Charing Cross East, 1844.)
Map, 52 x 82 cm, coloured outline.
Scale in miles.
On the verso: 32.

This and the three subsequent maps (199-201) of South Africa represent the various states of its mapping and settlements from c. 1846-1886. They all depict South Africa from 23°N to 34,5°S, showing the developing colonies and districts in coloured outline.

In all these maps the larger demarcating rivers as well as the Orange River (also known as the Gariep or Great River) and the Vaal (also known as the Ky Gariep River the Likwa or Likoua and the Yellow River), appear to subdivide the more northern area into various provinces, which eventually became the Transvaal, the Orange Free State and Natal. The Cape Colony, having been occupied since the mid-seventeenth century, is already full of more detailed geographical features.

The imprint on the map reads 'Geographer to the Queen and H. R. H. Prince Albert,' and as they were married in 1840, this map must have been published after this date. It can be dated as pre-1854 for various reasons, namely the non-presence of the Transvaal and Bloemfontein, the latter having just come into existence in 1854 (see detail on p. 272).

The region that was to become Natal was named by Vasco da Gama in 1487. In 1823 the port was known as Port Natal. The early travellers, John Cane, Alexander Biggar, Richard (Dick) King and Nathaniel Isaacs made contact with the great Zulu Chief Shaka, wishing to establish a settlement and a trading post. They received for this purpose a grant of 3500 square metres around the bay, and Port Natal was established. At the time of this map the area was divided into six separate districts, with 'Durban' one such district. On 23 June 1835, Allen Gardiner, the missionary, with fifteen residents of Port Natal, decided to name the town after Sir Benjamin D'Urban, Governor of the Cape. After the British annexation of Natal on 15 May 1854, the township of Durban (spelt now as such) was proclaimed a borough (see detail on p. 272).

On the south-east side of the map appears the important landmark of the Port of E. London, which was founded in 1848 by Sir Harry Smith. On the northern border of the Natal districts a range of mountains, the Zoutspanberg, appears, continuing into what was to be named the Drakensberg range.

223

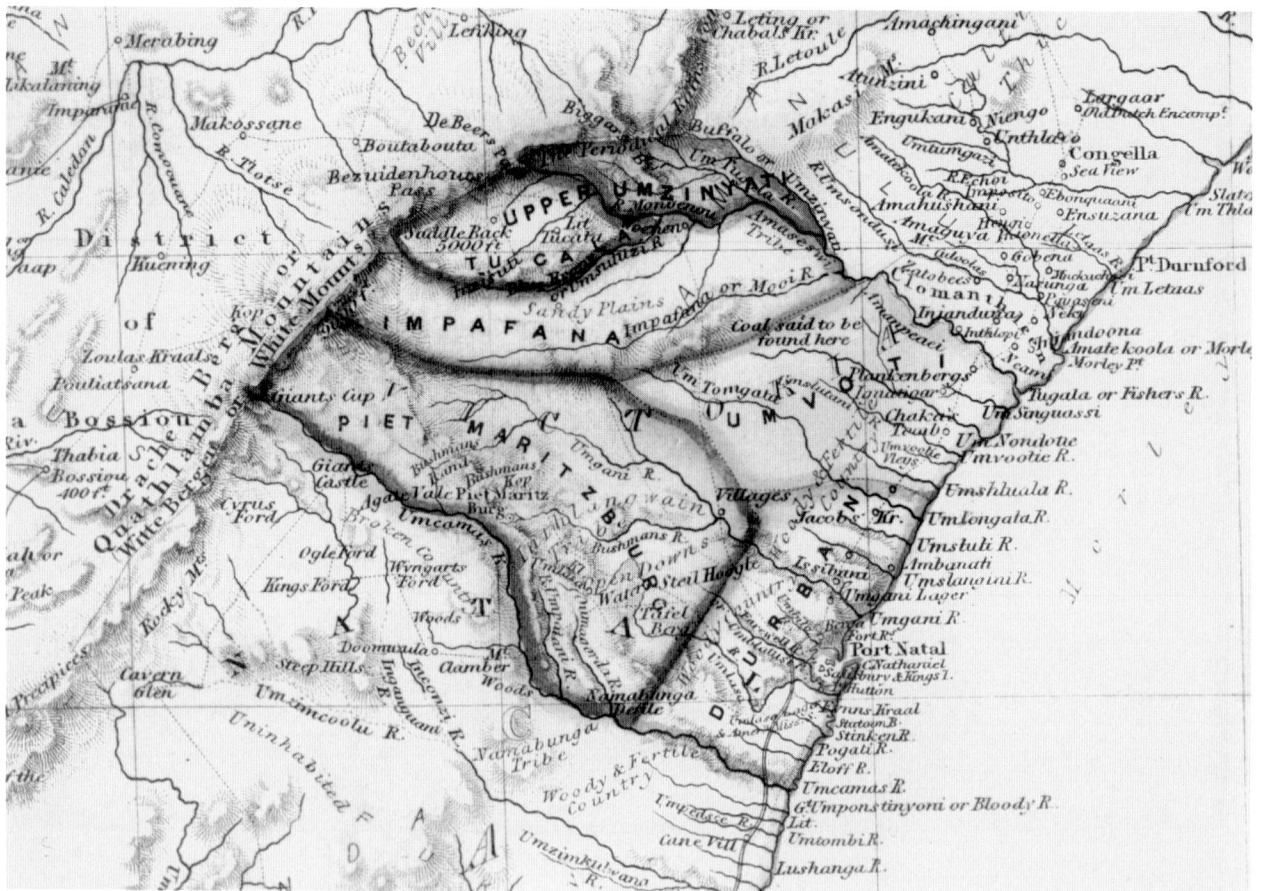

Detail Map 198 (above)

Detail Map 199 (below)

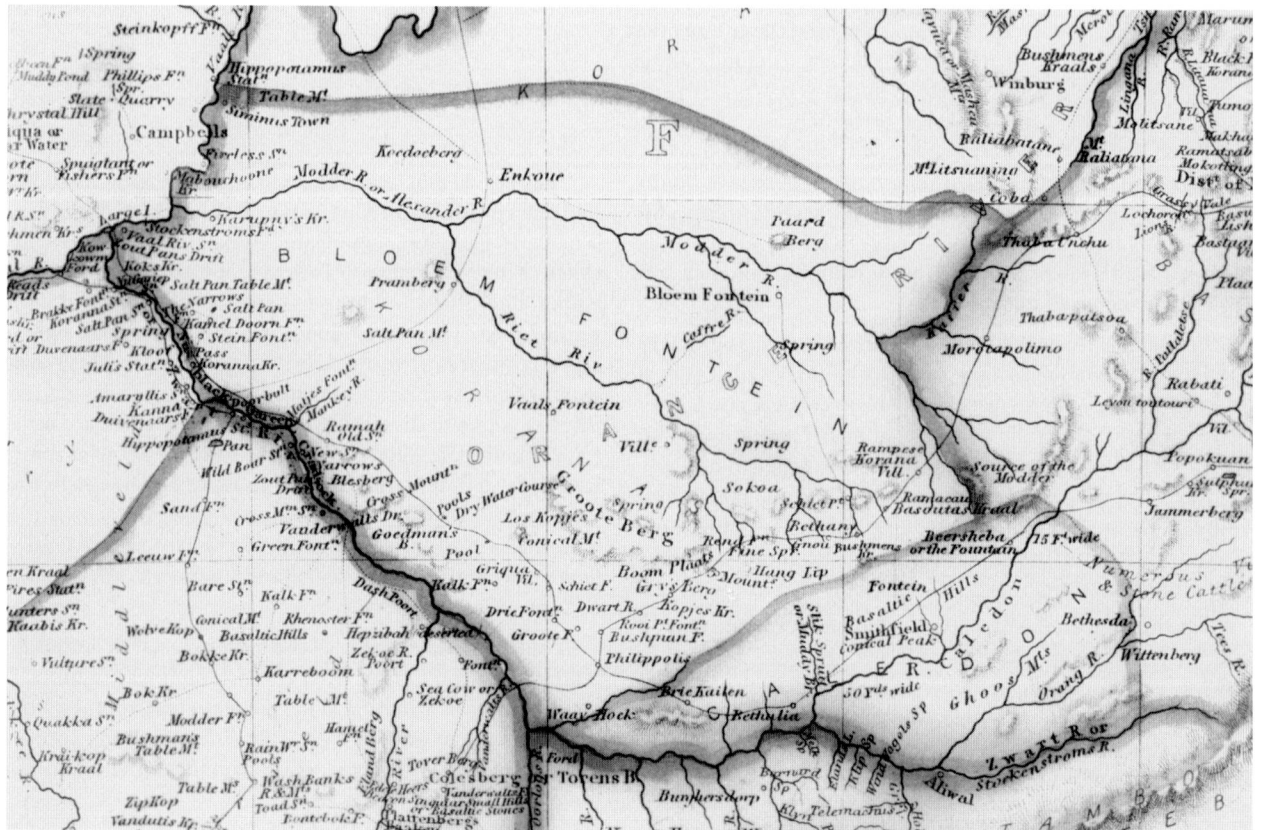

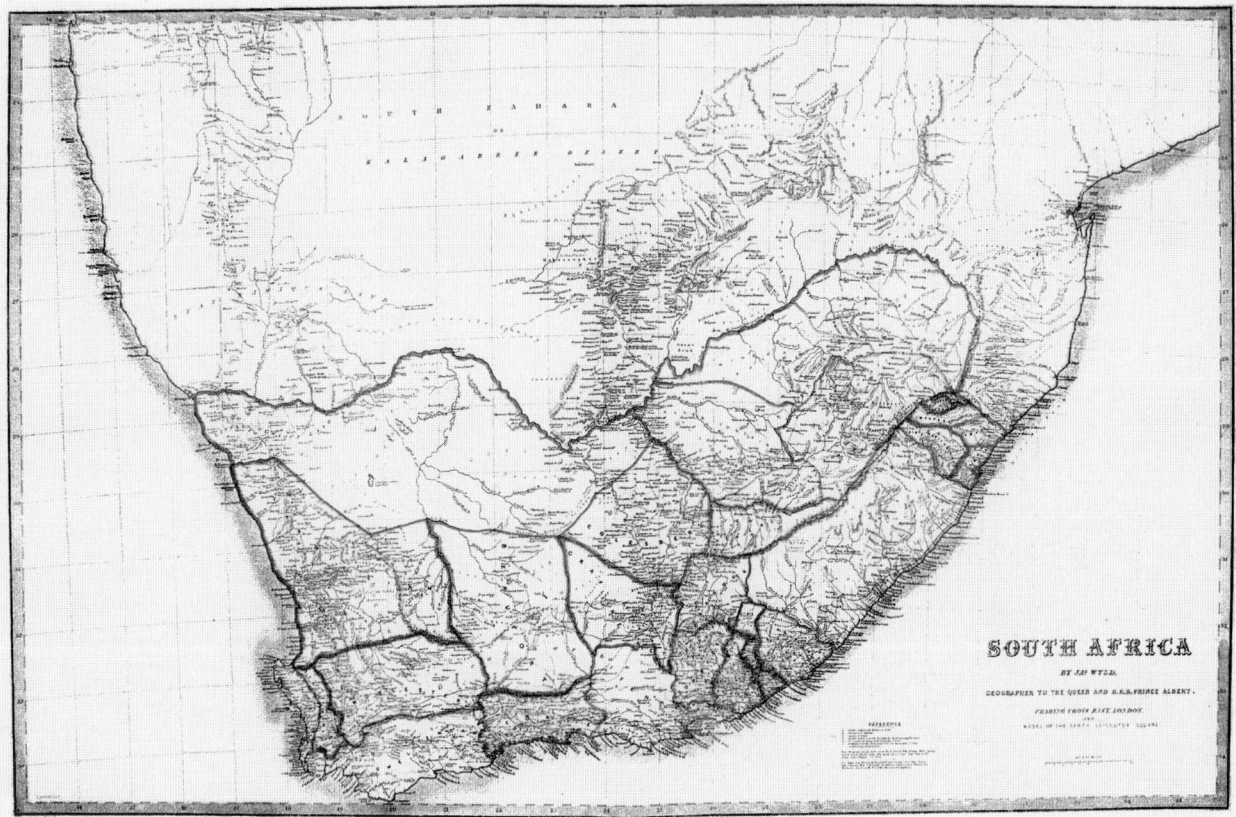

Map 199

WYLD, James (1812-1887)
South Africa by Jas Wyld, Geographer to the Queen and H.R.H. Prince Albert.
(London, Charing Cross East, and model of the Earth, Leicester Square, c.1854-1861.)
Map, 52 x 82 cm, coloured outline.
Scale in miles.
On the verso: 56.

This map of South Africa is a further state, with the additional reference, on the imprint, to the 'Model of the Earth, Leicester Square.' Wyld constructed his 'Monster Globe' between 1851 and 1861, dating the map between these periods. It can also be dated by the appearance of 'Bloem Fontein' (Dutch for 'Flower by the Fountain'). This town was founded when the original Free State Colony was established, in 1854, as its capital. Winburg, which was established as the town of the district of the same name, is situated 116 miles northeast of Bloemfontein, and, after Philipolis, is the oldest settlement in the Orange Free State.

Winburg was established in 1842 and the Voortrekker Hendrik Potgieter settled there. Further north, the positions of the Zoula, Matabeles, Moselekatse and many other tribal outposts are shown.

The six divisions of Natal are retained, with Port Natal indicating the bay. The Transvaal Colony does not yet appear except for a number of additional settlements and native names. The Vaal River still retains the additional names of Ky Gariep, Likwe and Yellow River.

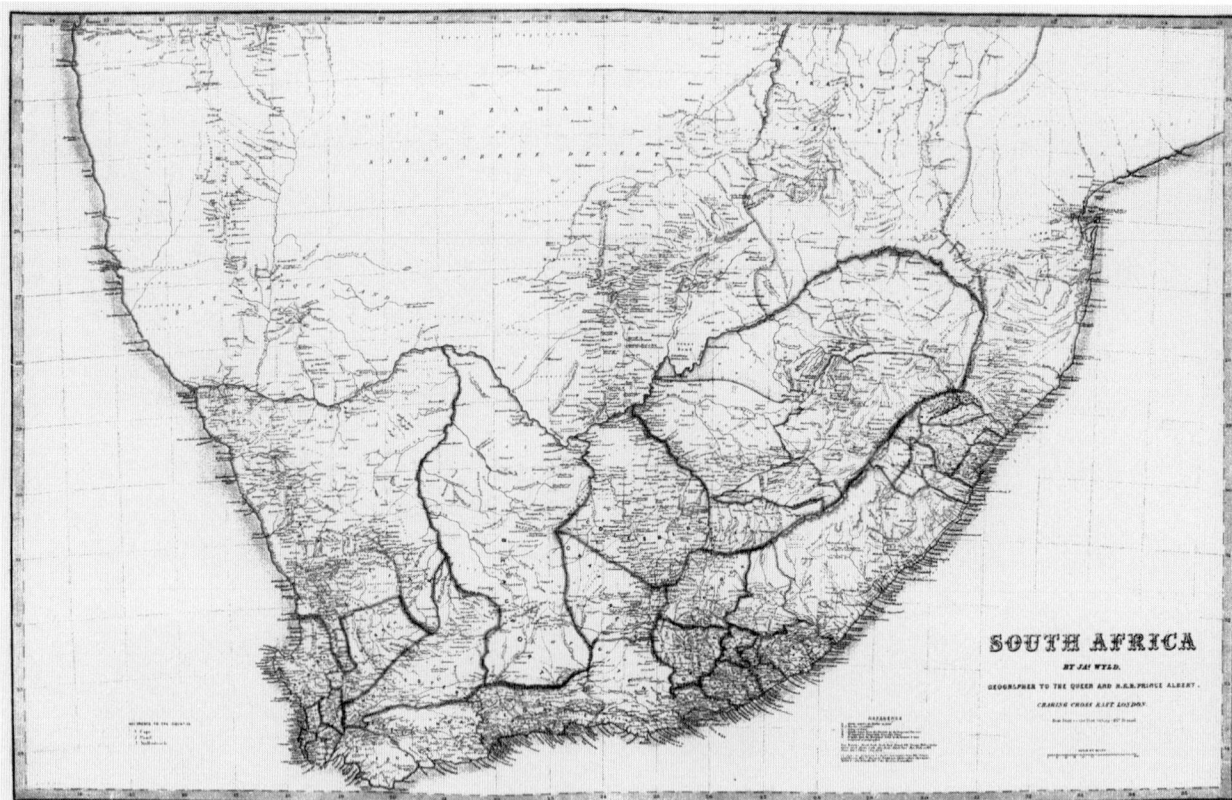

Map 200
WYLD, James (1812-1887)
South Africa by Jas Wyld, Geographer to the Queen and H.R.H. Prince Albert.
(London, Charing Cross East, next door to the Post Office – 457 Strand, after 1855.)
Map, 53 x 83,5 cm, coloured outline.
Scale in miles.
On the verso: 32

This is the third of the Wyld series of South Africa, with the title imprint changed to 'Charing Cross East, London, next door to the Post Office, 457 Strand.'

Additional settlements and more towns are added. The Transvaal Republic and Pretoria are named, the latter having been founded in 1855 with the arrival of the early Voortrekkers. The additional towns labelled in this map include Potchefstroom, Klerksdorp, Rustenburg and Leydenburg. Moselekatsie's residence is also noted. However, Kimberley, in the Vaal River district, does not yet appear.

The boundaries of the six Natal districts are now extended, with Durban still noted as one of them. From the details now shown on the map it can be dated as 1855-1856.

It is interesting to note that Potchefstroom is labelled as such. It is in the area which Andries Hendrik Potgieter, the Voortrekker leader, founded in 1838. The town of Potchefstroom has been named in various forms on maps dating from around this time, namely Mooirivierdorp and Potchefspruit. It was the early capital of the Transvaal Republic, with Pretoria the seat of the Government.

226

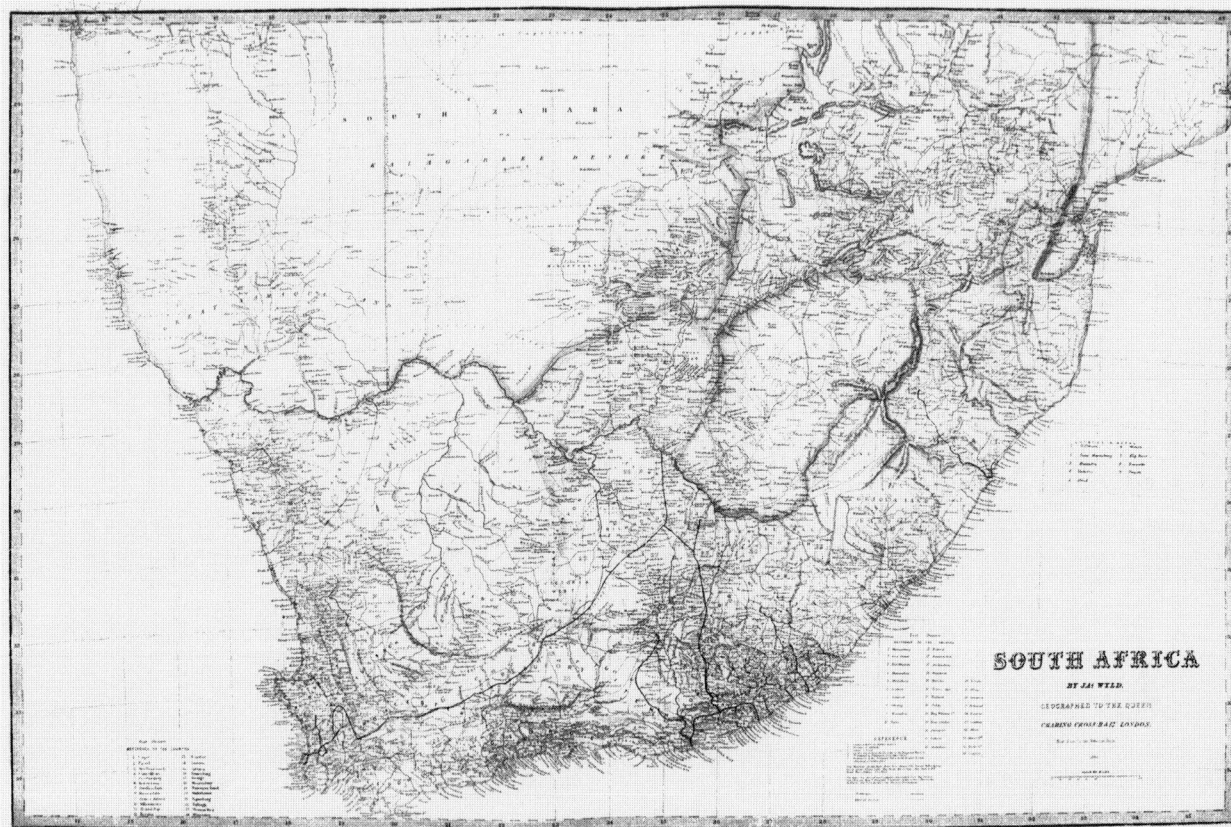

Map 201

WYLD, James (1812-1887)
South Africa by Jas Wyld, Geographer to the Queen. (London, Charing Cross (11 & 12) next door to the National Bank, 1886.)
Map, 52 x 82 cm, coloured outline.
Scale in miles.
On the verso: 33

This fourth state of the Wyld map of South Africa has an addition to the imprint, 'next door to the National Bank' and presents many geographic changes, especially in the northern area.

The area north of the Vaal River is now known as the Transvaal Territory, with an obvious increase of placenames, rivers and mountain ranges. A large area of the southern Transvaal is filled with Dutch names, including those of the owners of many farms and settlements. An interesting range of hills in the southern Transvaal is 'Gats Rand,' as it is still known today; this is one of its first appearances on a map. This is a range of hills extending from close to Potchefstroom for 120 miles in a northeasterly direction to a point southwest of what was to become Johannesburg, where it merges with what today is known as the Witwatersrand (Ridge of White Waters). The name 'Gats Rand' is a Dutch word meaning 'a ridge of holes.' This phenomenon was known to the early Dutch settlers as sinkholes and caves. Today it is part of the West Witwatersrand Line – the richest gold producing region in the world. It consists of three rock formations, interspersed with gold-bearing conglomerate. It was in this Witwatersrand area that the epochal discovery in 1886 of the largest goldfields known to mankind took place. This map is just too early to note the presence of Johannesburg.

The diamondiferous area of Hopetown, established in 1866, and the town of Kimberley (1877) are labelled.

The town of D'Urban is indicated, although its spelling had been changed to 'Durban' in 1854.

227

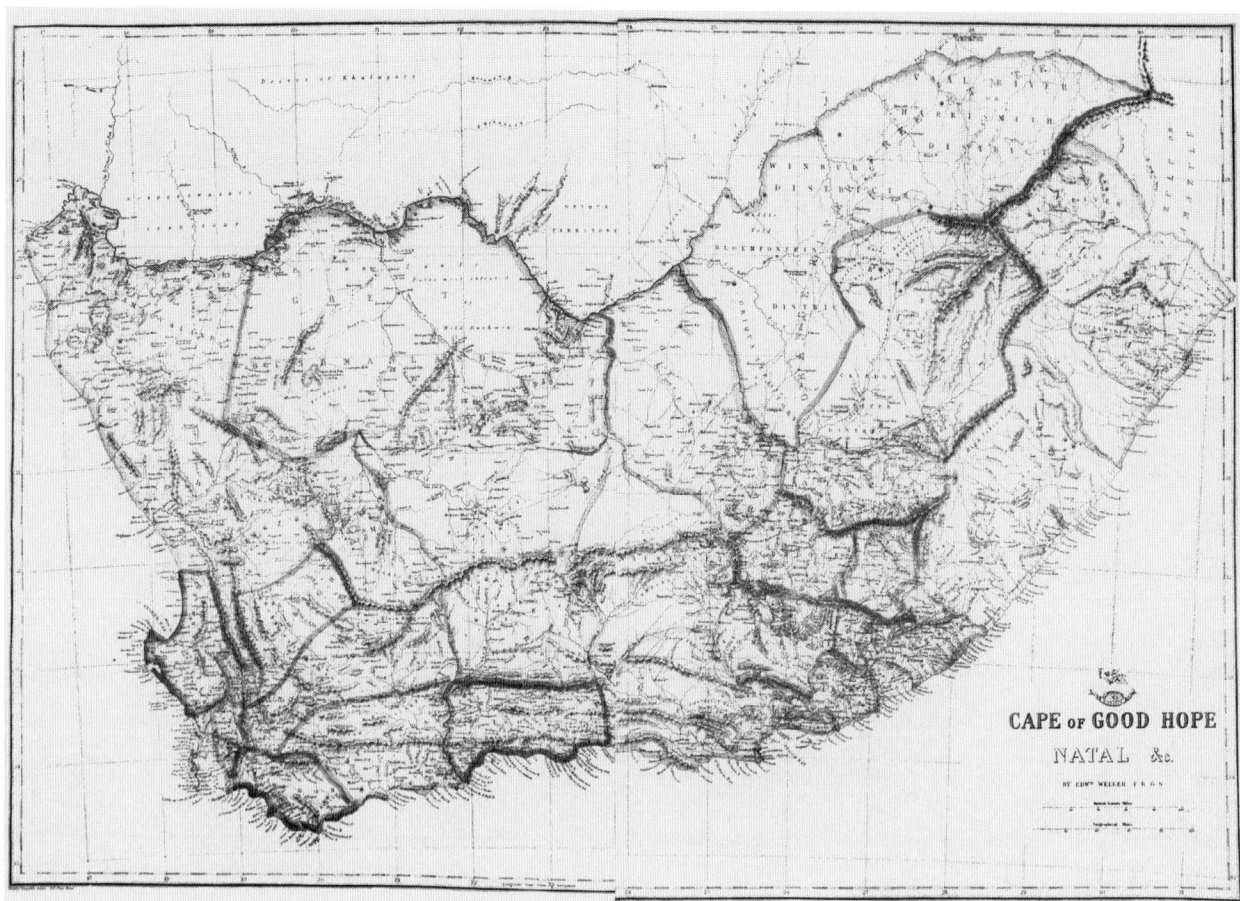

CAPE of GOOD HOPE

NATAL &c.

BY EDW^D WELLER F.R.G.S.

Map 202

WELLER, Edward (d. 1884)
Cape of Good Hope, Natal &c. Engraved by Edward Weller. (London, E. Weller, 1858.)
Map, 43,5 x 63,4 cm, coloured outline.
Scale in British statute miles and geographical miles.
Prime meridian through Greenwich.
Bottom left: *Weekly Despatch Atlas. 139 Fleet Street.*

A mid-nineteenth-century printed map of the Cape of Good Hope, Natal and surrounding areas, up to the Vaal River. The cartographer, engraver and publisher was Edward Weller of Duke Street, Bloomsbury, London, who was the author of the *Despatch Atlas* of 1858.

The title is placed in the lower right corner, surmounted by what appears to be the monogram of the *Despatch Atlas*: this name is on a narrow scroll across a portion of a globe with, above, a small flying figure with wings on his back, showing both heels.

The geography of this map is concerned mainly with the Cape of Good Hope, the Cape Colony and Natal. These areas are full of placenames, rivers and mountains, some accurate, others fictitious. The remarkable aspect of this map is the large number of subdivisions, the borders of which are shown in printed colours. These subdivisions conform

to the boundaries of certain districts of the Cape Colony and Natal. Some of the borders course along major rivers and mountain ranges but on the whole they do not reveal any true geographical demarcation. At the top of the map the Vaal and Orange Rivers form a prominent coloured border and, except for one historic site, Mooi River Dorp (present-day Potchefstroom), nothing of the Transvaal is shown. South of the Orange River and to the east of Namaqualand a large area is referred to as Bushmanland, with 'Parts most infested by Wild Bushmen,' The Orange Free State is included in the northern aspect of the map, with Bloemfontein depicted as a region and as a town – the latter also referred to as Queens Fort (Ft). Kimberley does not appear in the Orange River region, nor do any of the other early diamondiferous sites which were not discovered nor developed until 1867 in the Hopetown area of the Orange River.

228

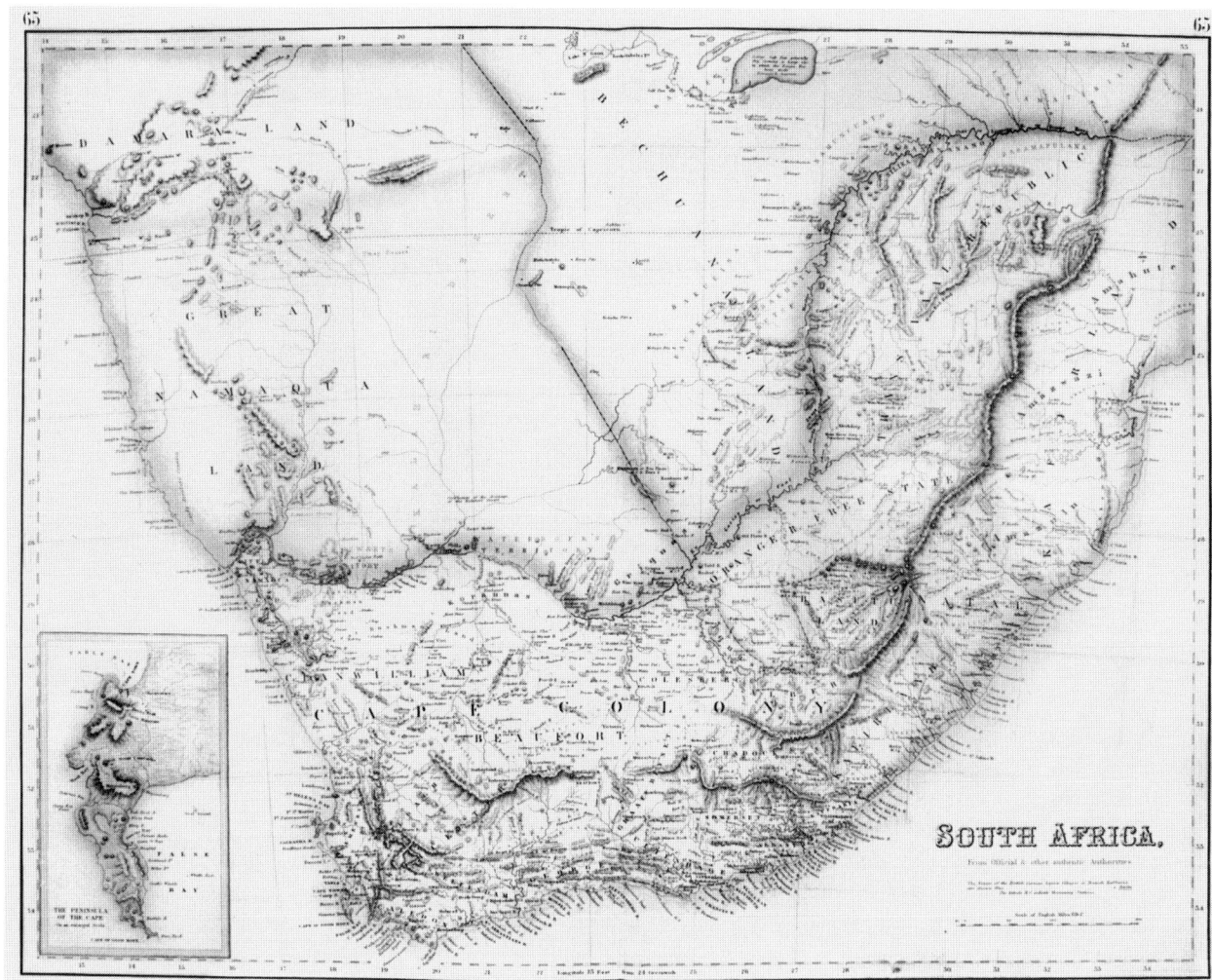

Map 203

FULLARTON, Archibald
South Africa, From Official and other authentic Authorities. Engraved by
Stanford's Geographical Establishment. (Edinburgh, A. Fullarton & Co., 1870-72.)
Map, 40 x 52 cm, slightly coloured.
Scale in English miles.
Prime meridian through Greenwich.
Inset: Peninsula of the Cape.
Top left and right: No. 65.

This map of South Africa was published by A. Fullarton & Co., a firm of publishers and engravers headed by Archibald Fullarton, operating in London, Edinburgh and Dublin in the nineteenth century. They issued many atlases of England, Wales and Scotland in various editions; a *Royal Illustrated Atlas*, published in 27 parts from 1854-1862, and a *Hand Atlas of the World*, 1870-1872. This map is probably from the later *Hand Atlas of the World* because of the great detail it contains from official and other sources.

The Cape Colony is depicted in great detail geographically, with its boundaries demarcated, and with a fine detailed inset of the Peninsula. The Orange Free State is similarly presented together with its capital, Bloemfontein. Basutoland is marked as well as Natal.

The Transvaal Republic appears with many of its early villages, settlements and towns, including Mooi River Dorp (Potchefstroom), Rustenburg and Klerksdorp. In the Eastern Transvaal Origstadt and Leydenburg are marked but Pilgrim's Rest and Barberton, where the earliest gold mines were established, are not featured. In these two latter places gold was found and mined in the mid-seventies and Pilgrim's Rest was established in 1873. This information may not have reached the publishers in time to be included in this map. The map has added information on the names of missionary stations.

229

Map 204

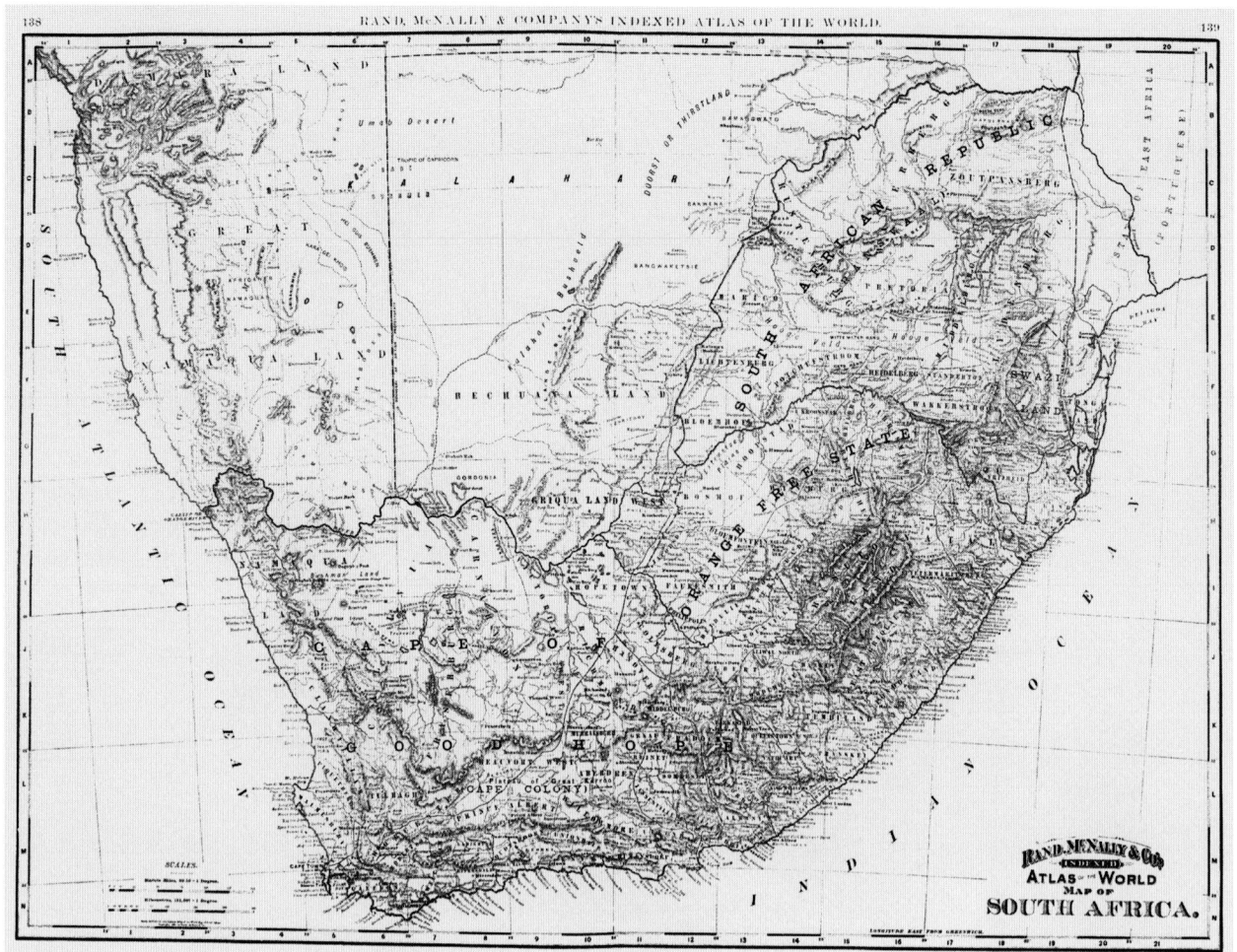

RAND, McNALLY & CO.
Map of South Africa. (Rand, McNally & Co., 1892.)
Map, 48,5 x 66,5 cm, coloured.
Scale in statute miles and kilometres.
Prime meridian through Greenwich.
Title across the top: *Rand, McNally & Company's Indexed Atlas of the World.*
Top left: 138.
Top right: 139.

A detailed map of South Africa extending as far north as the Limpopo border (22°S), to include what was then the South African Republic (now the Transvaal). The title in the lower right corner refers to it as *Rand, McNally & Company's Indexed Atlas of the World – Map of South Africa.* This map publishing firm in the United States of America has been firmly established since 1862 and since then has produced many accurate maps and atlases and geographic globes.

This map of 1892 is highly accurate for that period, featuring the four main provinces as they were then known – the Cape of Good Hope, Natal, Orange Free State and the South African Republic. These were all fully autonomous states with recognised borders until they were finally amalgamated in 1910 as the Union of South Africa. It is interesting to note that in the South African Republic the early goldfields of the Pilgrim's Rest region are marked. Many other historic sites are indicated, eg. Pretoria, one of the earliest large centres and the capital of the Republic, and Johannesburg, only six years after the world's largest goldfield was discovered and proclaimed in 1886. Kimberley and the surrounding diamond fields are depicted in Griqualand West, where large diamond discoveries were developed in 1867, well before gold was unearthed from the Witwatersrand.

230

CAPE OF GOOD HOPE

Map 205

BRY, Théodore de (1528-1598)
Delineatio Promontorii, Quod Cabo de bona Esperanca vulgo vocatur. (Map of the promontory known as Cape of Good Hope.) (Frankfurt, T. de Bry, 1598.)
Map, 14 x 17 cm, uncoloured.
Part of a folio page with Latin text below map.
Top right: III.
Bottom right: GK.
In: De Bry, Théodore. *Petits Voyages.* Frankfurt: T. de Bry, 1598.

This is a quaint early map of the Cape of Good Hope from Théodore de Bry's *Petits Voyages*, published in Frankfurt in 1598. The map is part of a folio page with text below relating to the area illustrated.

The seas have large, varying types of monsters, sailing ships and a paddling canoe. On land there are natives with bows and arrows, and reindeer. There is a cluster of trees surrounding rounded huts. At the top and bottom of the map there are two compass roses with the North correctly sited.

The flat summit of Table Mountain will be recognised at the bottom, near the centre. 'Aguada de Sadeyne' (Saldanha), 'Tafelberg' and 'C. de Bonesperance' are the only names to appear. Numbers up to six are placed at different spots on the map and are probably subsequently referred to in the text, which displays a finely engraved initial 'C'. (For further information on de Bry and his *Grand Voyages* and *Petits Voyages* see Map 334.)

Raven Hart, in his book *Before Van Riebeeck*, features this map with the addition, however, of a combined coat of arms and title – the map itself is identical. This source states that the map appeared originally in an account of Spilbergen's voyage which appeared in Isaac Commelin's *Begin ende Voortgangh...* a volume of condensed journals of the early Dutch travellers.

A third version of the map is featured in a book *Pictorial History of South Africa*, published by Odhams Press in 1940 with an unspecified author or editor. This third edition of the map now has two quite different cartouche titles at the top right and left, consisting of highly decorative armorial designs.

The map also appears in *Indiae Orientalis Pars, IX* (Verhoefs voyage to the Moluccas, 1607-1609), by de Bry, published in Frankfurt in 1612, with a Latin edition 1607-1612.

232

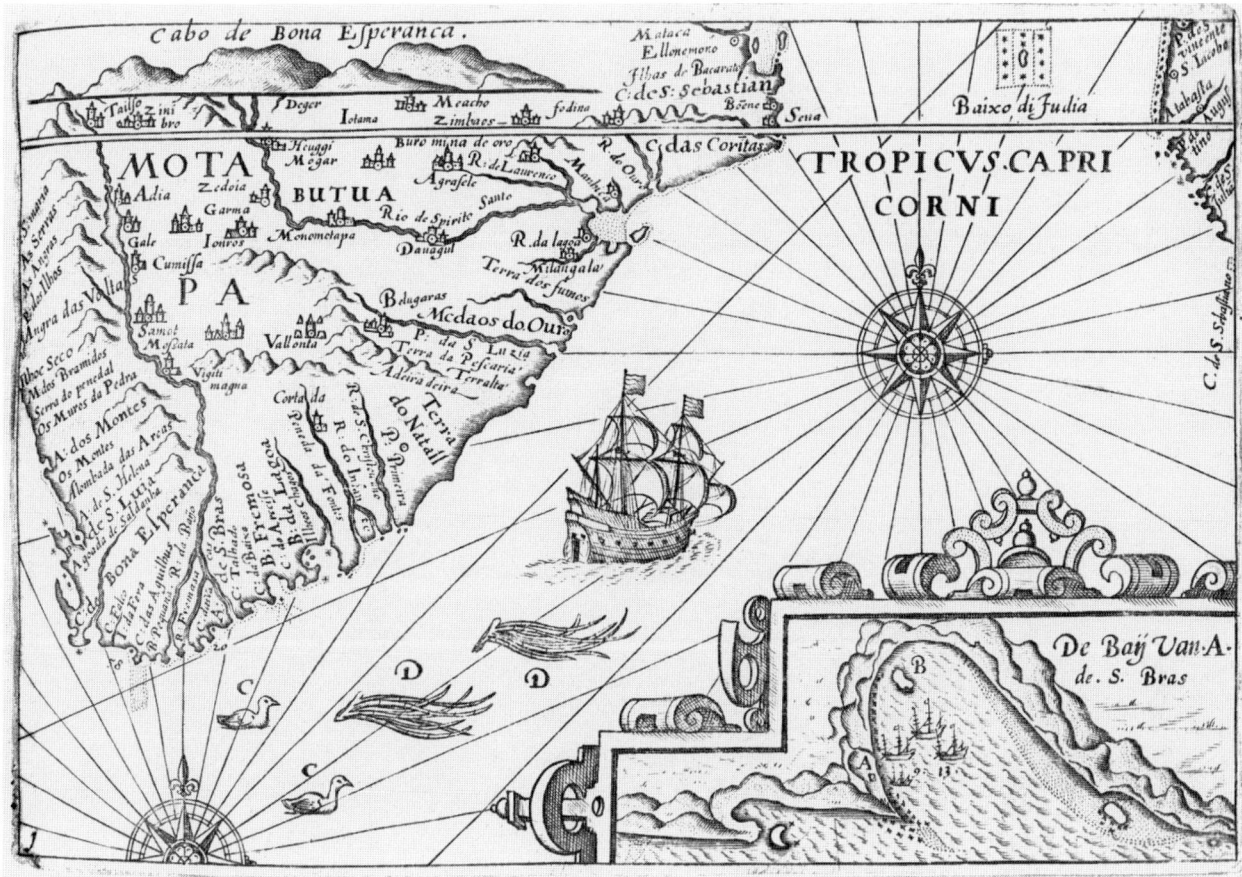

Map 206

ANON
Cabo de Bona Esperanca. (Cape of Good Hope.) (c.1595.)
Map, 14 x 22 cm, uncoloured.
Inset: *De Baij van A. de S. Bras.*

This very early map of the southern extremity of the African continent appeared in Isaac Commelin's famous collection of travels published in 1646 under the title *Begin ende Voortgangh van de Vereenighde Nederlandtsche Geoctroyeerde Oost-Indische Compagnie*, as an illustration of the earliest voyage of the Dutch to the East in 1595. The cartographer and publisher are not mentioned. The 'Cabo de Bona Esperanca' across the top left may be the heading of the map or of the mountains appearing as an inset.

In the lower right corner is an inset of 'De Baij Van A. de S. Bras' (identified with present day Mossel Bay), which received its Dutch name from P. van Caerden in 1601. A decorative scroll surrounds the bay, in which four ships lie at anchor.

The map itself has a compass rose with radiating lines in the Indian Ocean, as well as a large attractive sailing ship. To the south is another compass rose, two gulls and two seaweed-like objects which Godee-Molsbergen describes as

'trombas' in his *Pictorial Atlas of the History of South Africa* (these also appear in Coronelli's gore of southern Africa – see Map 162.) Reference letters C and D appear next to the seagulls and seaweed, and the letters A and B can be found in the inset, but without an identification table.

The peculiar shape of the southern tip of the continent is to be expected for that date and most of the coastal names are those introduced by the earliest Portuguese explorers. There are many rivers, which continued to appear on maps for over a century, including the large east-flowing Rio de Spirito Santo. Fabulous cities are present all over the interior and Vigiti Magna and Monomotapa can easily be identified.

This map provides us with an idea of how the Dutch sailors of the seventeenth century visualised South Africa. Maj R Raven-Hart, in his book *Before Van Riebeek*, refers to the map as appearing in Cornelis Houtman's *Voyages to the East* (Houtman left Texel in 1595.)

233

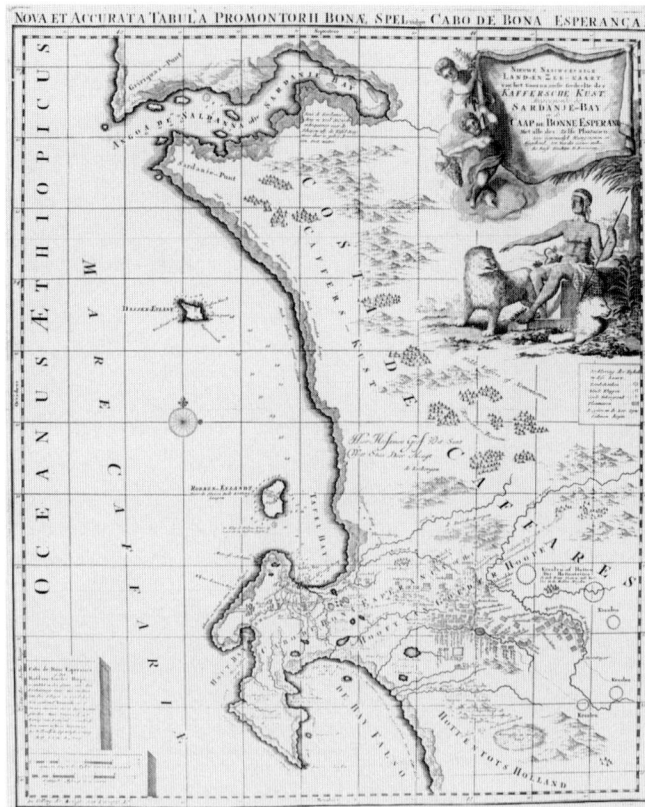

Map 207

OTTENS, Joachim (1663-1719)
Nieuwe Naaukeurige Land-en Zee-Kaart, van het voornaamste Gedeelte der Kaffersche Kust, Begrypende de Sardanje-Bay en de Caap de Bonne Esperanca met alle des Zelfs Plantazien. (New and accurate map and sea chart of the Cape coast from Saldanha Bay to the Cape of Good Hope.) (Amsterdam, J. Ottens, c. 1688-1702.)
Map, 57 x 48 cm, uncoloured.
Scale in Dutch and French miles.
Prime meridian through Tenerife.
Title across the top: *Nova et Accurata Tabula Promontorii Bonae Spei vulgo Cabo de Bona Esperanca.*

A fine engraved Dutch map of the Cape of Good Hope and Saldanha Bay. The cartouche at the top right corner illustrates a detailed title contained on a decorative drapery supported on its left by two angels, one blowing a trumpet and holding a bunch of grapes. Below this is a seated semi-nude figure holding a spear in his left hand and reaching to touch a lion. To his left is a hyena-like spotted animal. At the lower left corner is a legend in Dutch describing the discovery of the Cape, the origin of the name 'Cape of Good Hope' and its original name 'Tormentosa.'

The map itself is remarkably detailed around False and Table Bays and the Peninsula. The name 'Esselsteyn Bay' is marked adjacent to the Kalck Hoven (Lime Kiln). (Researches by the Simon's Town Historical Society brought to light the fact that Simon's Bay was named in the time of Governor Simon van der Stel. Subsequently a Dutch East India search party got as far as what is now Fish Hoek and named the bay Esselsteyn Bay.) The interior is particularly well detailed with names of many settlements and farms, with the owners' names included.

Although this map has been attributed to Ottens this has not been fully proven. Its date can be estimated between 1688 and 1702 because the French settlement at Fransch Hoek (established in 1688) is labelled, and although many farms and home owners are marked, the well-known Dutch home Libertas, owned and lived in by Adam Tas in about 1702, is not mentioned. It can thus be assumed that the map was drawn up between these two dates. Ottens, the Dutch cartographer-publisher, is associated with this map because it is listed in an atlas compiled by him, though it was not drawn up or published by him personally.

234

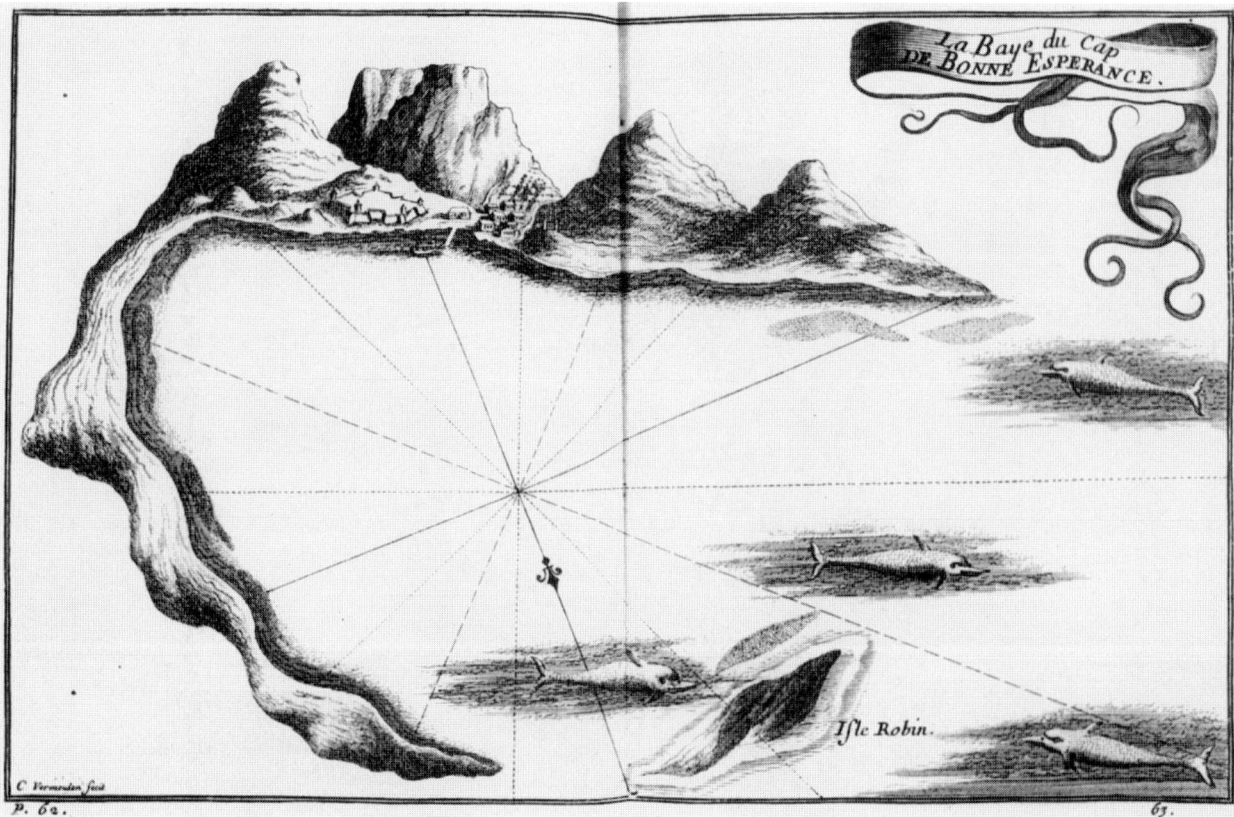

La Baye du Cap de Bonne Esperance.

Isle Robin.

C. Vermeulen fecit
P. 62.

63.

Map 208

TACHARD, Guy (d.1714)
La Baye du Cap de Bonne Esperance. (The bay of the Cape of Good Hope.)
Engraved by C. Vermeulen. (Paris, Seneuze, 1686.)
Sea chart, 18,5 x 29,5 cm, uncoloured.
Bottom left: P. 62.
Bottom right: 63.
In: Tachard, G. *Voyage de Siam des Pères Jésuits...* Paris: Seneuze, 1686, p. 62.

A map of the bay of the Cape of Good Hope which appears in the 1686 French edition of Tachard's *Voyage de Siam des Pères Jésuits envoyez par le Roy aux Indies et a la Chine.* Louis XIV, King of France, decided to send an embassy to the King of Siam with the Chevalier de Chaumont as ambassador. He thought it desirable that the College of Jesuits should accompany the expedition to make scientific investigations and observations and to assist any missionaries they might encounter in the course of their travels. They remained at the Cape for about a week and the author gives an interesting account of the settlement and the natives.

This chart of Table Bay has a title in the upper right corner, within a curved streamer-like scroll. The only ornamentation besides a compass rose and rhumb lines are four dolphins in the ocean. Robben Island is also illustrated. The Dutch Fort is depicted at the base of Table Mountain, and situated in a ravine between this and the adjoining peak is a series of buildings on an incline at the base of which are trees and foliage which could represent the Company's garden.

235

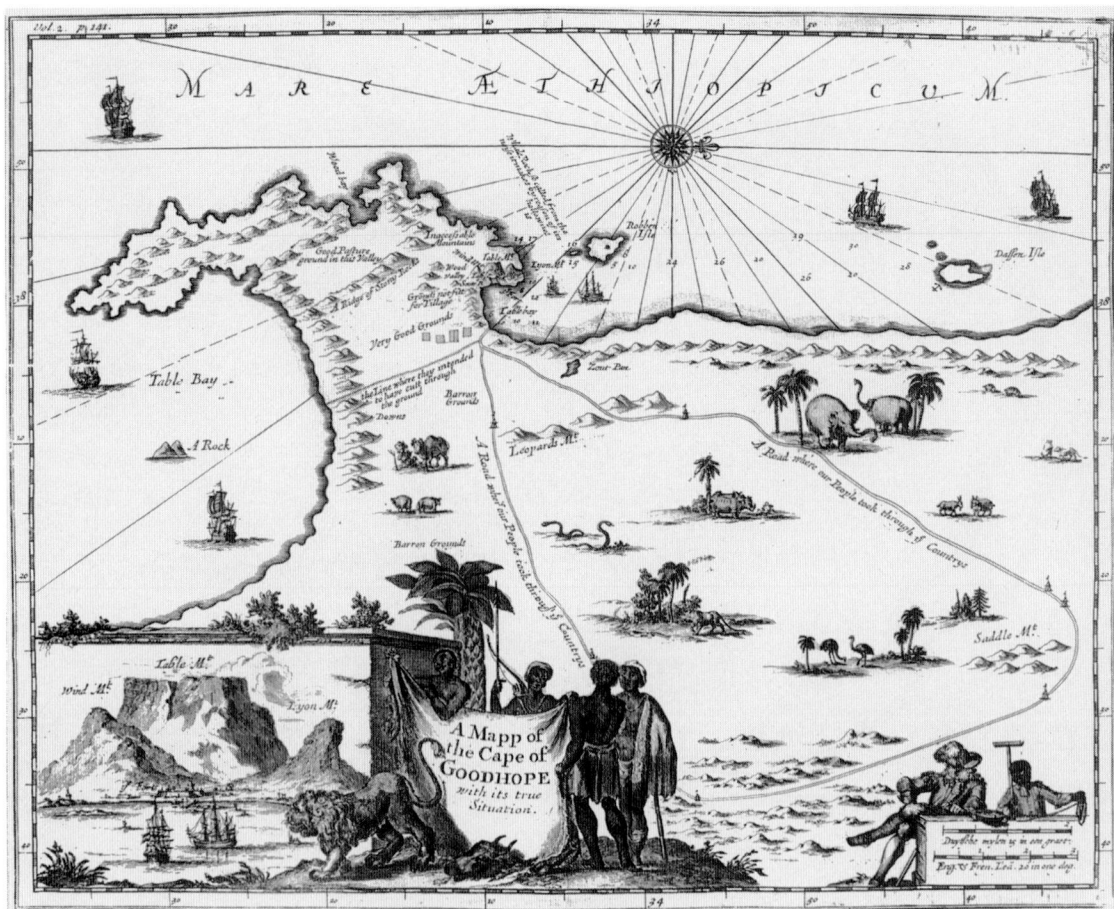

Map 209

NIEUHOFF, Johan (1618-1672)
A Map of the Cape of Good Hope with its true Situation. (London, 1703.)
Map, 27 x 35 cm, slightly coloured.
Scale in Dutch miles, English and French leagues.
Inset: View of Table Mountain.
Top left: Vol. 2. p. 141.

This early map of the Cape of Good Hope is the 1703 English version of the original Dutch map which first appeared in Johan Nieuhoff's travel book of 1682. In the Dutch version the initials J.N. (Johan Nieuhoff) appear after the title of the map but they are not present in the English edition.

The cartouche shows an animal skin supported by some natives. The one on the left stands in front of some palms with, to his right, a block of masonry which frames the Table Bay inset. Next to him, behind the title, another native stands holding a bow and a spear. On the right are two more natives. On the left in front of the masonry is a lion looking at a view of Table Mountain and Table Bay. There are two ships in the foreground, and the Fort and the various parts of the mountain are labelled.

In the lower right corner the scale is presented on what appears to be a low wall behind which a European with a plumed hat is seated, talking to a native carrying some kind of stick with a cross piece at the end.

The mainland is dotted with elephants, a lion, a rhinoceros, ostriches, snakes, tortoises, etc., but domestic cows and sheep also make their appearance. There are two ships in False Bay (labelled 'Table Bay') and in the Atlantic. The compass rose indicates that west and not north is at the top of the map.

In this version all the captions are in English except for the scale. There are fewer captions than in the Dutch edition and the most interesting one has disappeared – one referring to the land granted to the first free burgher at the Cape – which is marked 'Uytgedeelt land' (distributed land).

This map is of particular interest as it is the best-known early map of the Cape and because it shows not only the canal between Table Bay and False Bay proposed by Commissioner-General Rijckloff van Goens, but also the route of one of the early exploring expeditions. Although the above-mentioned canal was not constructed, an engraving of the Cape does exist showing ships coming out of the canal into the bay.

236

Map 210

CHATELAIN, Henry Abraham
(1684-1743)
Vue et Description du Cap de Bonne Espérance. (View and description of the Cape of Good Hope.)
 (Paris, Chatelain and Guedeville, 1705-1720.)
Engraving, 35,5 x 21 cm, uncoloured.
Inset: *Le fort des Hollandois au Cap du Bonne Espérance.*
Top right: Tom. VI, No. 17, Pag. 74.

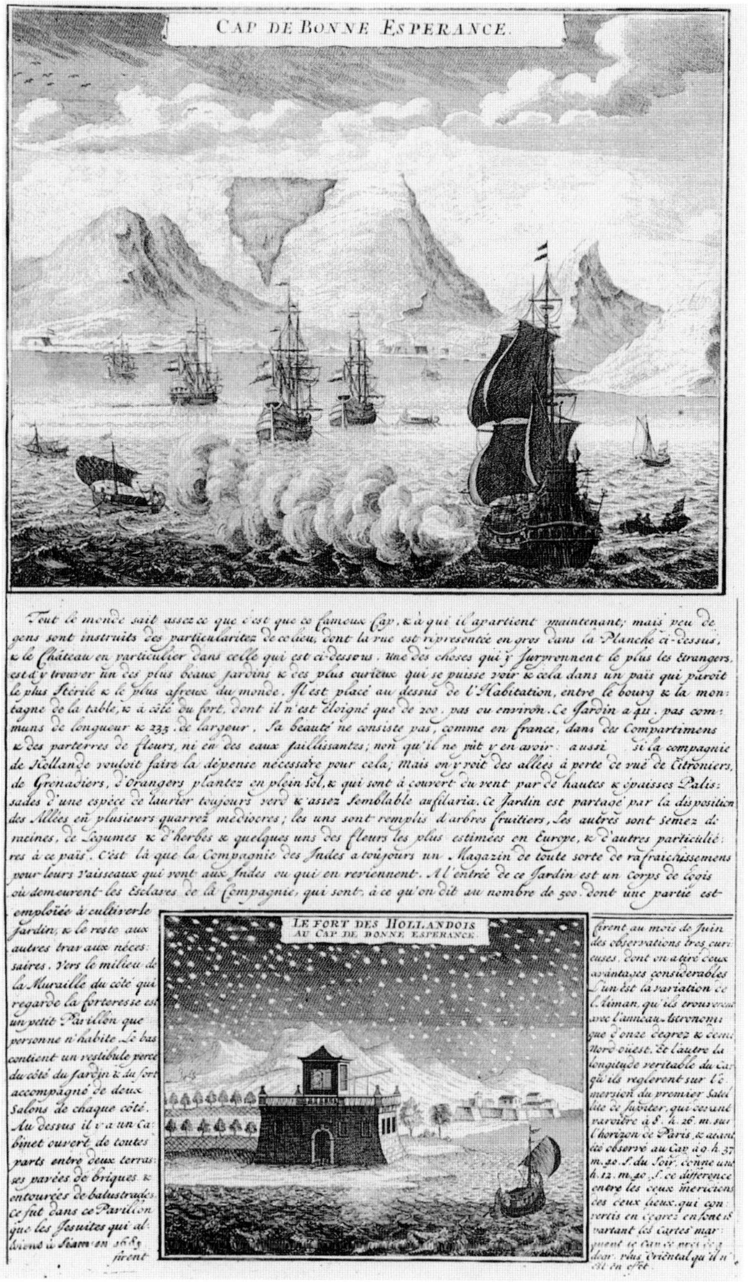

This copperplate engraving is of two views of the Cape. Above is Table Bay, with shipping in the foreground flying the Dutch flag, and separating the two views is an engraved descriptive text which refers mainly to the Company's gardens, 'the most beautiful and curious to be seen in a country that is sterile and frightful.'

The lower engraving shows the Dutch fort at the Cape and an observatory established by the Frenchman Tachard and the Jesuit Fathers to observe the southern skies.

The engraving comes from the seven-volume *Atlas Historique et Méthodique,* published by Chatelain and Guedeville in 1705-20 (with a second edition in 1732). These views are copies of earlier illustrations in Tachard's *Voyage de Siam des Pères Jésuits* in 1685.

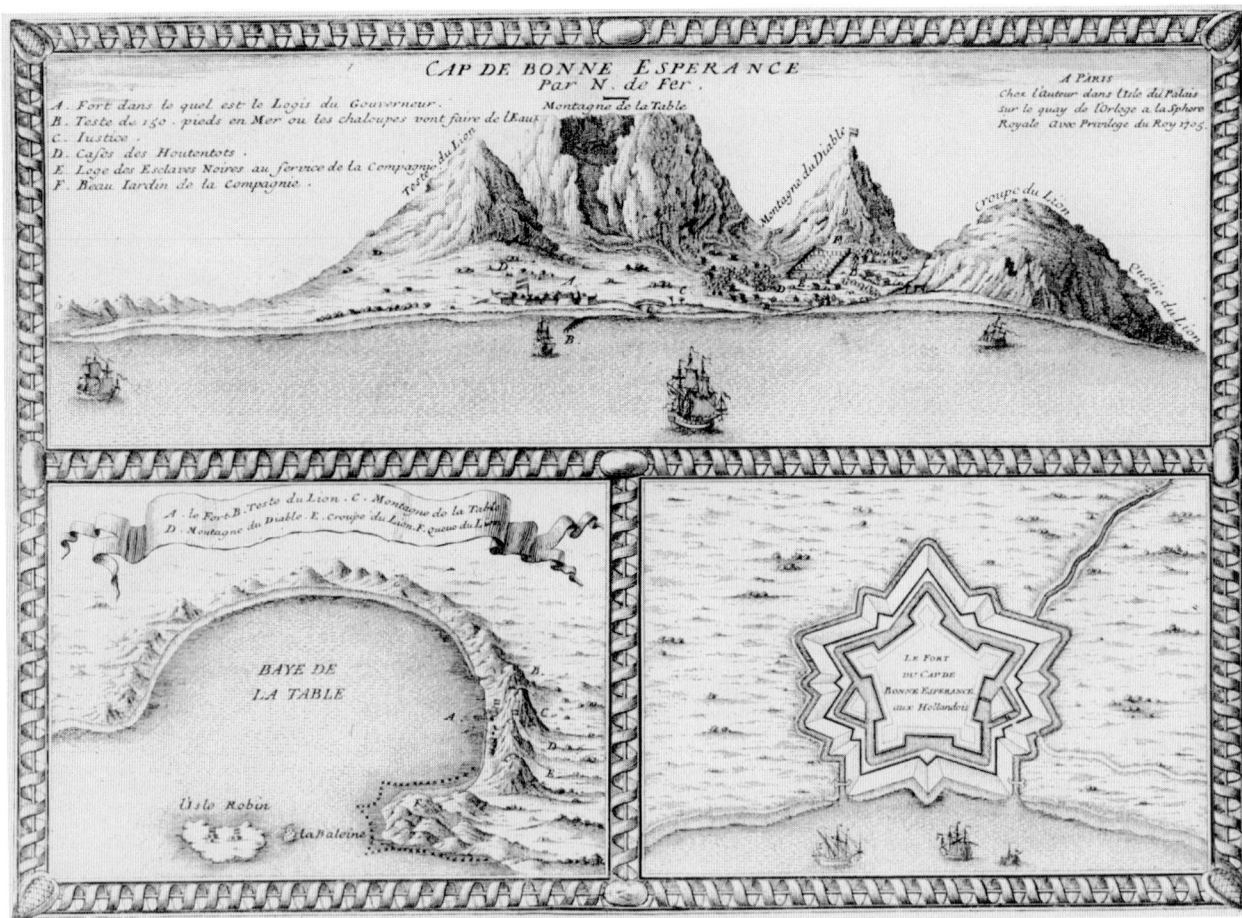

Map 211

FER, Nicolas de (1646-1720)
Cap de Bonne Esperance. (Cape of Good Hope.) (Paris, N. de Fer, sur le quay de l'Orloge
a la Sphere Royale, 1705.)
Engraving, 24,5 x 35 cm, uncoloured.
Top right: 327.

This view of the Cape from the sea is one of many such views that were published from 1666-1797; in atlases, as insets to maps in travel books of Africa or as separate maps. In the seventeenth century the Dutch ascendency in commercial power and maritime ability finally led to the establishment of permanent settlements at the Cape of Good Hope. They built a fort, installed a garrison and planted fruit and vegetables as a source of organised supply for the victualling of ships en route to the Far East.

This view is a copperplate etching, naming the different peaks of Table Mountain, and including a legend of important sites at the Cape. Below the view of Table Mountain are two insets – one showing Table Bay with Robben Island and a whaling station, and the other a ground-plan of the Fort, known today as the Castle.

This plate comes from the *Atlas Curieux* published in Paris by the French geographic publisher Nicholas de Fer in 1700-1705.

238

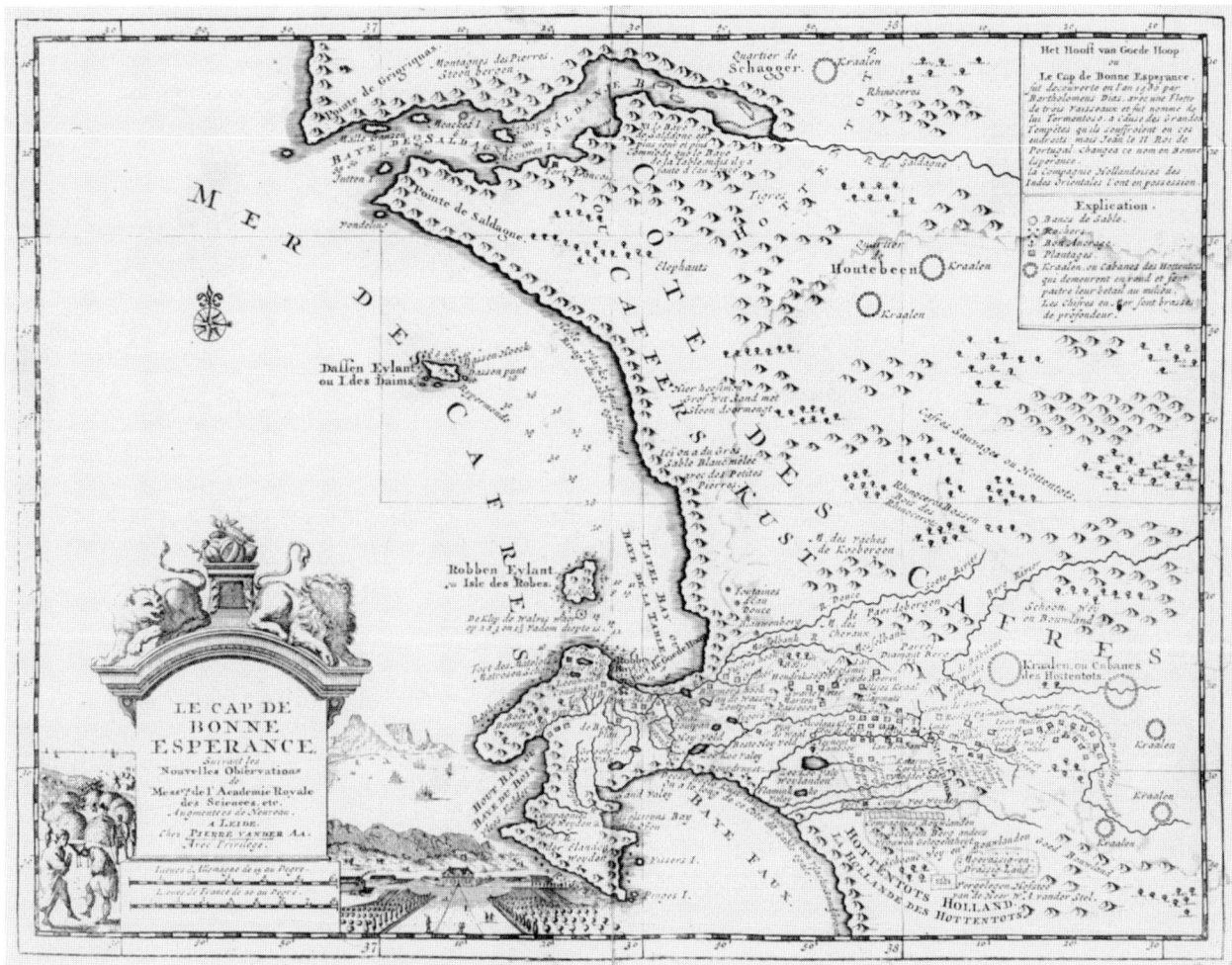

Map 212

AA, Pierre van der (1659-1733)
Le Cap de Bonne Esperance Suivant les Nouvelles Observations de Messrs. de l'Academie Royale des Sciences... (Cape of Good Hope according to the observations made by the Royal Academy of Science.) (Leiden, Pierre van der Aa, 1713.)
Map, 22,5 x 29,5 cm, uncoloured.
Scale in German and French leagues.

This is a small detailed map of the Cape from Saldanha to False Bay. The attractive title cartouche on the lower left is on a stone pediment surmounted by a lion and lioness, with a view of Table Bay in the background and the Governor's garden and homestead on the right. On the left is depicted a small collection of kraals and native inhabitants.

At the top right of the map is a short description of the Cape of Good Hope and below this a legend regarding the contents of the interior. Farms and homesteads and their owners are noted. A number of geographical inaccuracies occur, especially in place names and rivers.

Pierre (Pieter) van der Aa was a publisher, editor and bookseller of Leiden. He started as a bookseller at the age of twenty-three, becoming a prominent member of the Booksellers' Guild. As a publisher his output was varied and prodigious though his main interest was the production of maps and atlases, which he printed in all sizes. His publications include an *Atlas Nouveau* completed in about 1710, and the *Galerie Agréable du Monde,* of 66 parts in 27 volumes, completed in 1729 and containing over 3000 maps. Parts 61-63 deal solely with Africa. Though he was a successful merchant, as a geographer van der Aa was poor and his maps are collected mainly for their decorative qualities. Many of these maps were engraved by well-known Dutch engravers such as Luiken, Goeree and Stoopendael, and are typical examples of early eighteenth-century map productions.

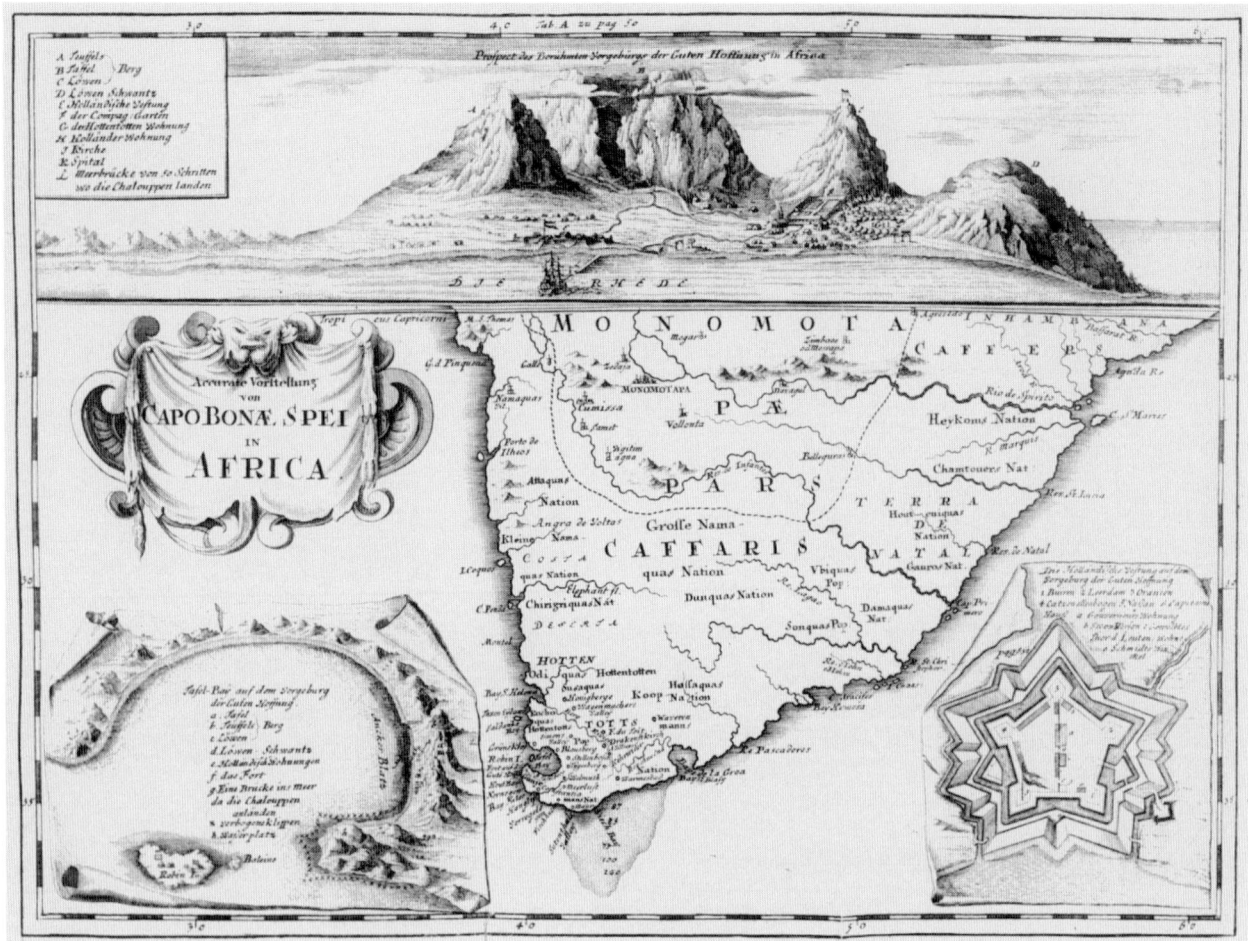

Map 213

KOLBEN, Peter (1675-1726)
Accurate Voorstellung von Capo Bonae Spei in Africa. (Accurate map of the Cape of Good Hope in Africa.) (Nürnberg, Peter Conrad Monath, 1719.)
Map, 28 x 38 cm, uncoloured.
Inset: Table Bay. The Fort. Table Mountain and surroundings.
Title across the top: Tab. A zu pag. 50.
In: Kolben, Peter. *Caput Bonae Spei Hodiernum* Nürnberg: P.C. Monath, 1719, fold p. 50.

This map with its attractive insets appears in the first German edition of Kolben's *Caput Bonae Spei hodiernum ...* published in Nürnberg in 1719 by Peter Conrad Monath.

The main map depicts the country south of the Tropic of Capricorn. Although this is a German map and allowing therefore for variations, most of the place names near the Cape are recognisable, although the coastline is inaccurate. Imaginary place names from earlier sources, such as Vigiti Magna and Monomotapa, are included. The publisher of the later Dutch edition decided to omit this map.

The cartouche with its title shows an animal skin hanging over some decorative masonry. Right across the top is an inset view of the Cape entitled 'Prospect des Berühmten Vorgeburgs der Guten Hoffnung in Africa.' It is accompanied by a table of identification corresponding to specific parts of the mountain and the town. In the lower left corner is an inset view of Table Bay described as 'Tafel Bay auf dem Vorgeburg der Guten Hoffnung' with a key drawing attention to the Castle and the landing stage. This inset is placed on a piece of parchment with the corners curled over. There is a similar curled parchment containing the third inset in the lower right corner, which curiously refers to 'page 632.' This inset shows a detailed ground plan of the Castle with an identification table similar to that found on the map issued by the Ottens family.

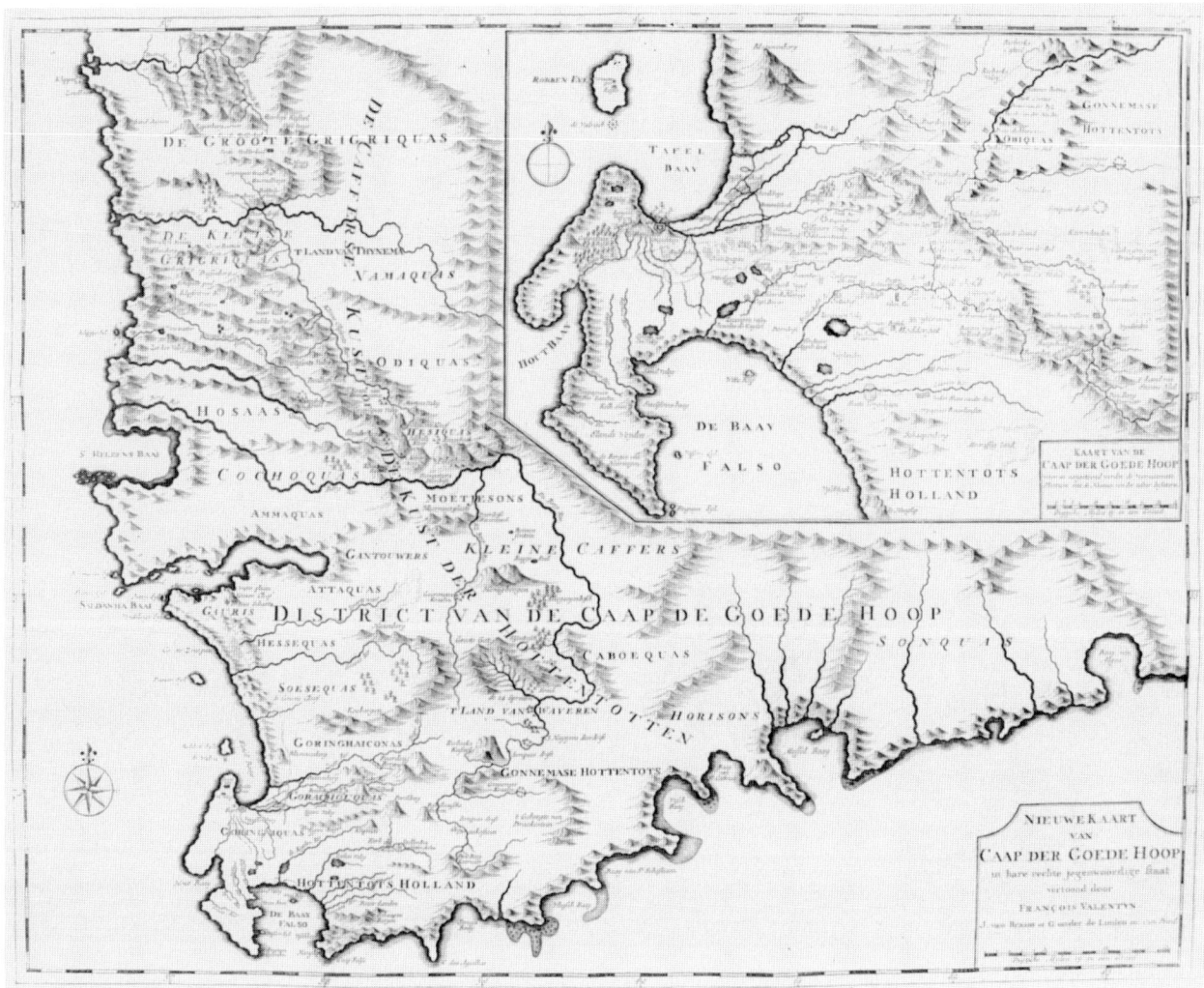

Map 214

VALENTYN, François (1666-1727)
Niuewe Kaart van Caap der Goode Hoop in hare rechte jegenwoordige staat vertoond door... (New map of the Cape of Good Hope.) (Amsterdam, J. van Braam and G. onder de Linden, 1724-1726.)
Map, 44 x 56 cm, uncoloured.
Scale in Dutch miles.
Inset: Detail of the Cape Peninsula with names of farms and their owners.
Top right: No. 43.
In: Valentyn, François. *Oud en Nieuw Oost Indien*, vol. 5. Amsterdam: J. van Braam and G. onder de Linden, 1724-1726.

A crisp map of the Cape of Good Hope which was published in Valentyn's *Oud en Nieuw Oost Indien* – an eight-volume travel book published in 1724 and 1726. François Valentyn, born in Dordrecht, one of the most celebrated Dutch travellers, went to Batavia in 1685 as an ecclesiastic in the employ of the Dutch East India Company. These volumes are particularly valuable for the researcher and are often referred to as the 'Encyclopaedia of the Dutch East Indies.' A large part of volume five, in which this map appears, relates to the Cape of Good Hope.

The cartouche is in the lower right corner with the scale in Dutch miles. In the upper right corner is a large inset map with its own title and scale as well as a circle surmounted by a fleur-de-lys to indicate north. This important inset features the names of the farmers, indicating clearly where their farms were situated, a valuable historical feature in tracing the development of the early Cape up to the time of the retirement of the renowned governor Simon van der Stel.

As this map is confined to parts of the Cape which were reasonably well known, it is fairly accurate and served as a source for other map makers over a number of years.

Map 215

VALENTYN, François (1666-1727)
Gezicht van Kaap der Goede Hope, als men op de Reede legt. (View of the Cape
of Good Hope...) (Amsterdam, J. van Braam and G. onder de Linden, 1726.)
Engraving, 27 x 36 cm, uncoloured.
Top right: No. 38. C.
Lower left border: 1. *'t Vlek.* 2. *de Tafel-berg.* 3. *de Wind-berg.* 4. *de Leeuwen-berg.*

This view of the Cape of Good Hope is similar to that
of Pierre van der Aa (Map 216). It appeared in *Oud en
Nieuw Oost Indien* by Valentyn, a well-known Dutch
explorer, who visited and described the Cape of Good
Hope while in the employ of the Dutch East India
Company 1685-1694 and 1706-1713. A cluster of buildings
appears in the background and there is another structure
to the left which resembles the Dutch Fort.

A scroll and title in Dutch appear at the top of the
engraving. Each of the prominent mountains is numbered
with an accompanying legend in the lower left border.

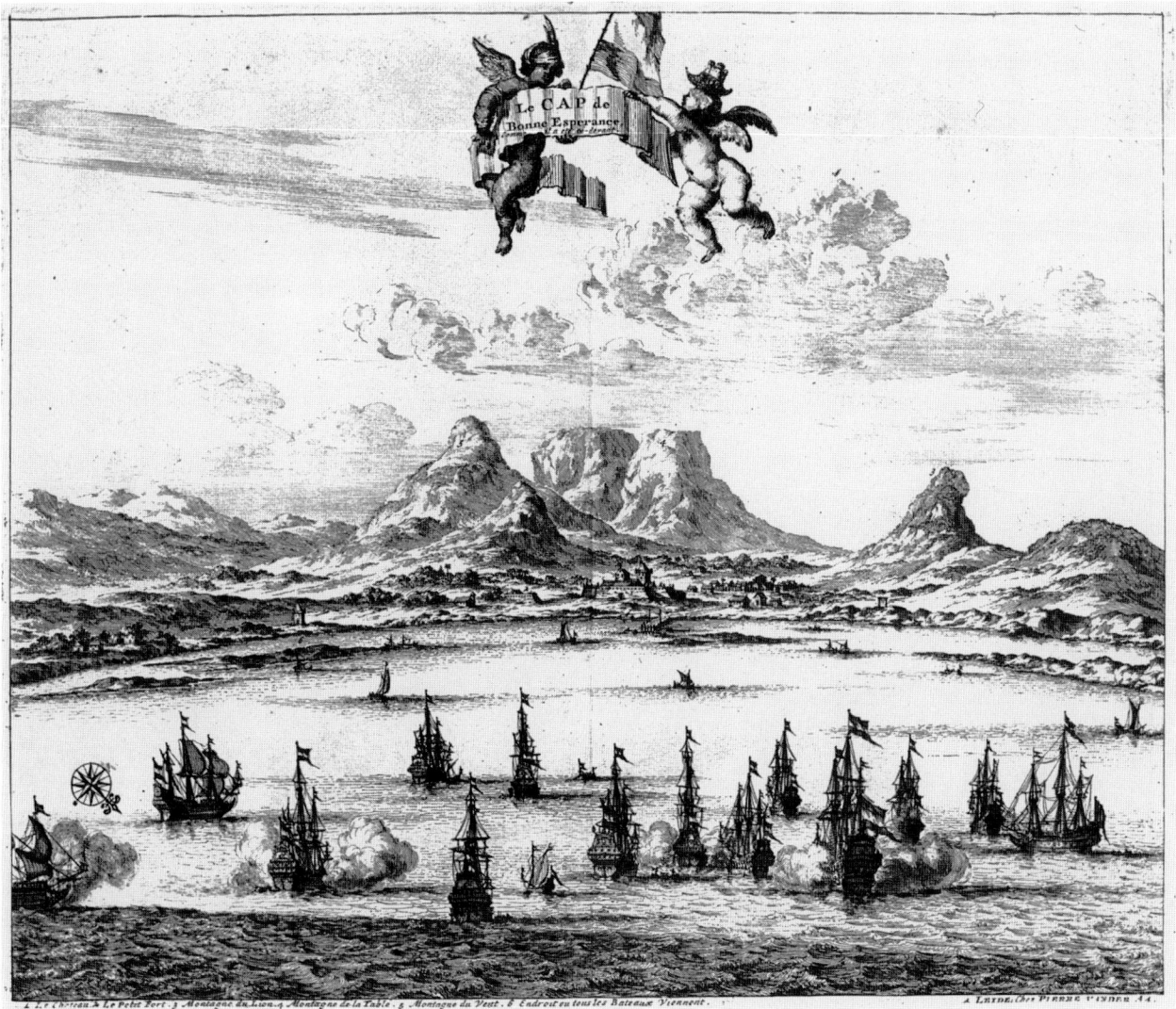

Map 216

AA, Pieter van der (1659-1733)
Le Cape de Bonne Esperance comme il a été ci-devant. (The Cape of Good Hope...)
(Leiden, P. van der Aa, 1727.)
Engraving, 28,5 x 34 cm, uncoloured.
Lower left border: 1. *Le chateau.* 2. *Le Petit-Fort.* 3. *Montagne du lion.* 4. *Montagne de la Table.*
5. *Montagne du Vent.* 6. *Endroit ou tous les Bataeux Viennent.*

An attractive copperplate view of the Cape of Good Hope with its fine mountain range in the background. The title is in French on a scroll held aloft by winged cherubs. A Dutch fleet is shown in the foreground, with the Fort, a wooden jetty and sparsely spread buildings. On the left foreshore is a building which appears to be the observatory set up by Tachard and the other Jesuit fathers who accompanied him to the East Indies. The individual mountains in the background are numbered and named in the legend below. This view is based on a similar plate in Ogilby's *Africa* which was in turn based on the work of Dapper.

243

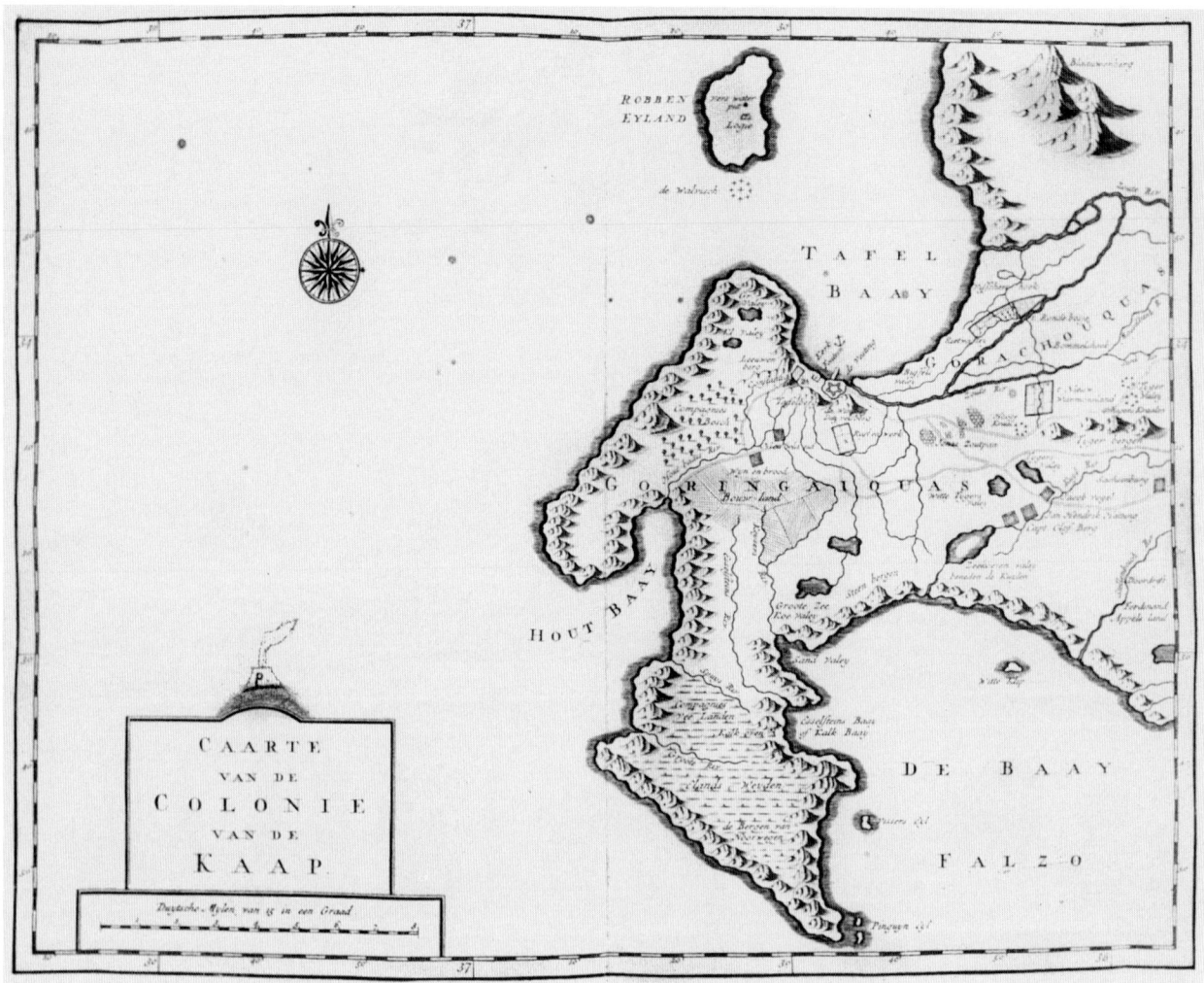

Map 217 LAKEMAN, Balthazar (fl. ca. 1700.)
Caarte van de Colonie van de Kaap. (Map of the Cape Colony.) (Amsterdam B. Lakeman, 1727.)
Map, 30 x 38,5 cm, uncoloured.
Scale in Dutch miles.
In: Kolben, Peter. *Naauwkeurige en Uitvoerige Beschryving van de Kaap de Goede Hoop...* Dutch ed., vol. 1. Amsterdam: Balthazar Lakeman, 1727, p.68.

This map of the Cape Colony appears in the Dutch edition of Peter Kolben's *Naaukeurige en uitvoerige Beschryving van de Kaap de Goede Hoop.* vol. 1, (of 2) published in Amsterdam in 1727. It extends from 'Blaauwenberg' and Robben Island to just past Gordon's Bay, as far as the Stellenbosch River. The Fort has plantations to the northeast of it. Owners of farms at that period are named – Captain Olaf Berg, Jan Hendrik Hattung, Jacob Vogel, etc. The land near the Hottentots Holland Mountains is called 'Ferdinand Appels Land.' In the peninsula the 'Kalk oven' is in 'Esselsteyn's Baay of Kalk Baay' which is very large; to the north are 'Sand Valey' and 'Groote Zee Koe Valey.' Near the 'Bergen van Noorwegen' in the south peninsula is the land where the elands grazed and just north of this are the 'vee landen' and forests belonging to the Company.

244

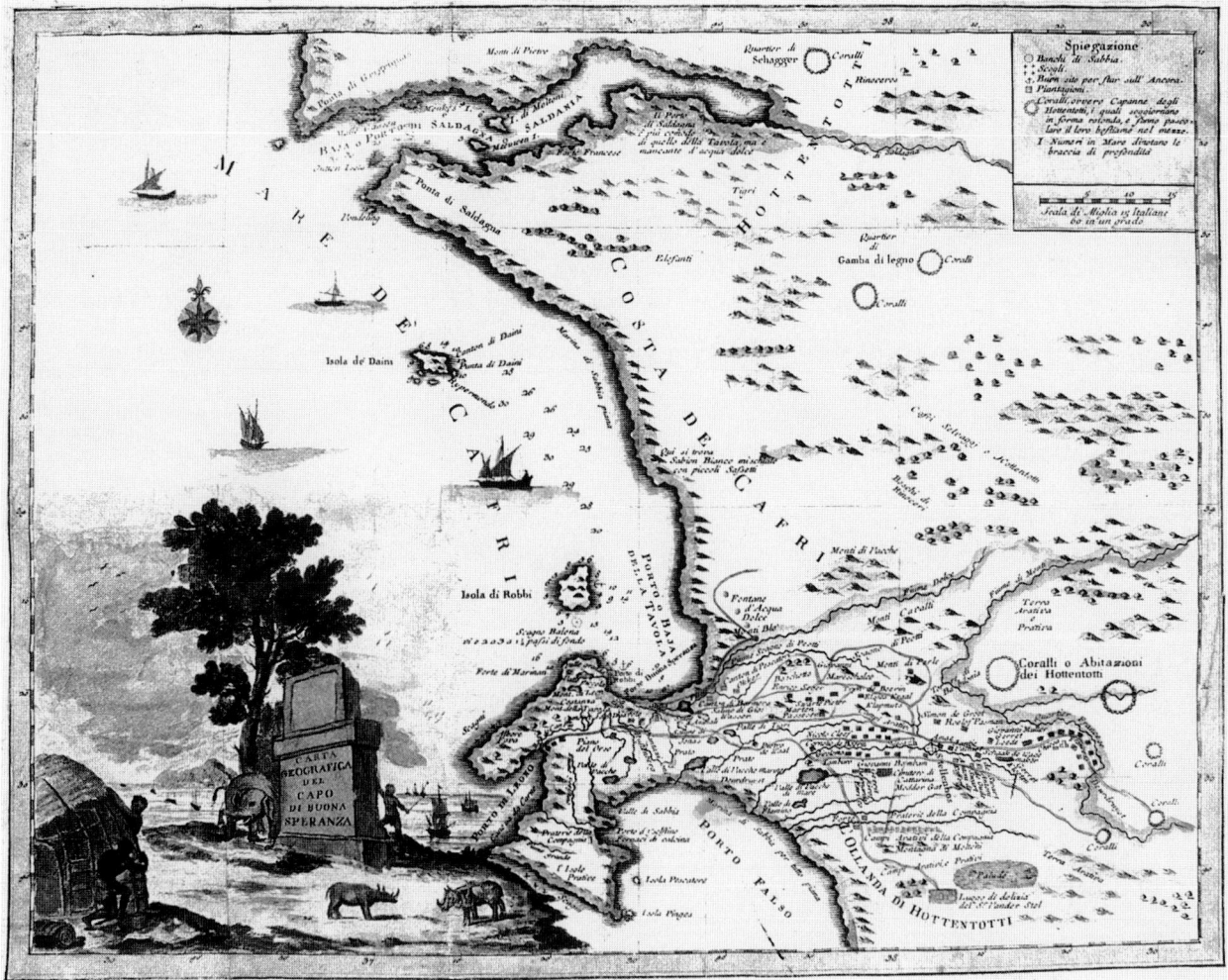

Map 218

L'ISLE, Guillaume de (1675-1726)
Carta Geographica del Capo di Buona Speranza. (Map of the Cape of Good Hope.) (Venice, G. de l'Isle, 1740-56.)
Map, 32 x 42 cm, coloured.
Scale in Italian miles.
Inset: *Spiegazione* (key).

As is well known, many countries reissued and copied de l'Isle's work, and this map (from the Italian version of his atlas, *Atlante Novissimo Che Contiene tutti le parti del Mondo*, published in Venice between 1740 and 1756) is an example of his international fame. It has a charming title piece of a native hut, natives, rhinoceros and an elephant, and embraces an area from Saldanha Bay to False Bay. It also shows the routes from the Cape inland to the kraals of the various tribes, land cultivated by the Company, sweet water wells, salt pans, woods, rivers and many of the early farming settlements with Italian versions of the names of the owners. Some twenty farmers are mentioned, the most important in the Cape's history being van der Stel, Swarte Pieter, Roelof Pasman, Gerrit Klotein and Simon de Groot.

245

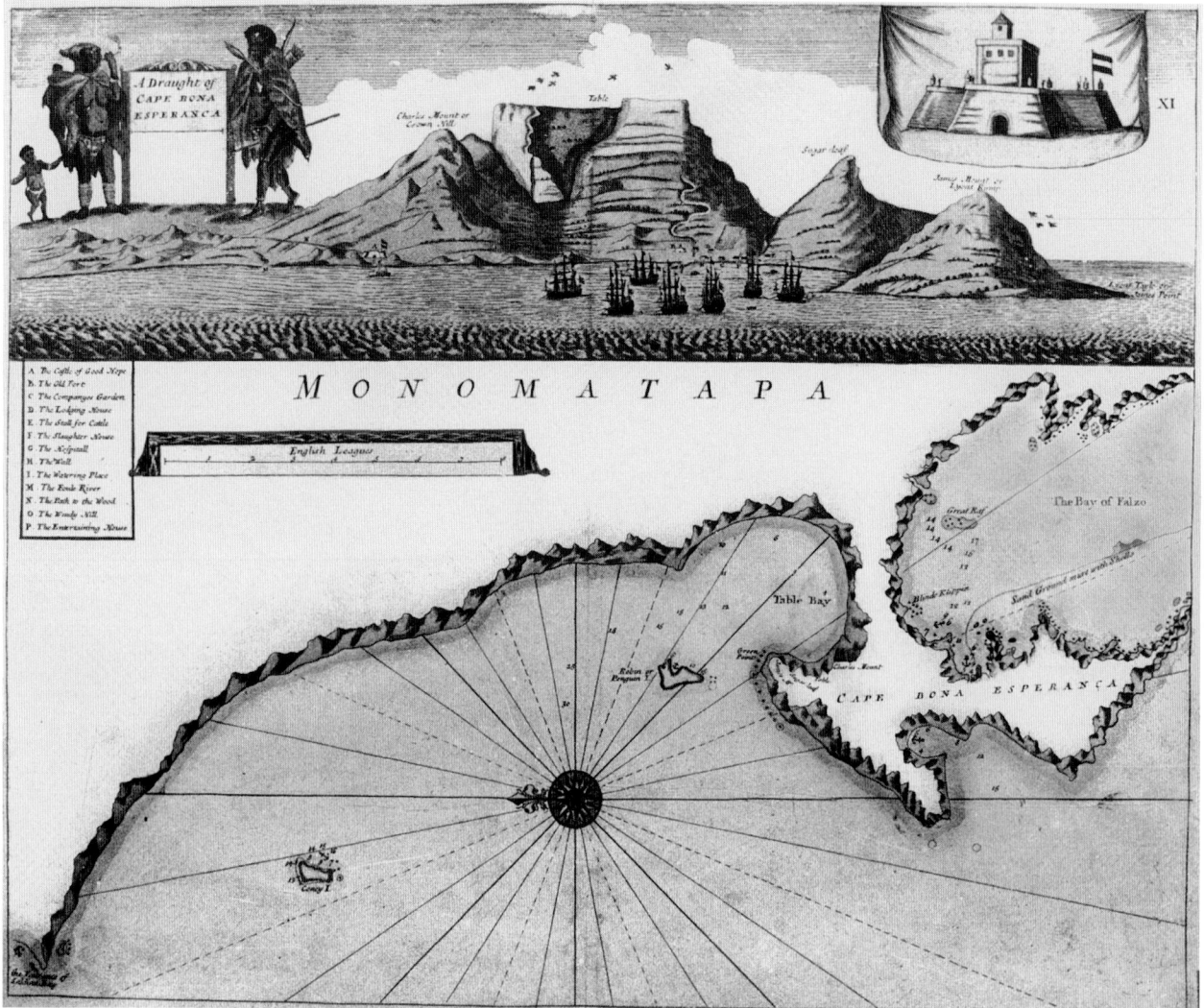

Within the illustration:

A Draught of CAPE BONA ESPERANCA

XI

Charles Mount or Crown Hill
Table
Sugar-loaf
James Mount or Lyon's Rump
Lyon's Tayle or Stone Point

A. The Castle of Good Hope
B. The Old Fort
C. The Companys Garden
D. The Lodging House
E. The Stall for Cattle
F. The Slaughter House
G. The Hospitall
H. The Wall
I. The Watering Place
M. The Ende River
N. The Path to the Wood
O. The Windy Hill
P. The Entertaining House

MONOMATAPA

English Leagues

Table Bay
The Bay of Falzo
Great Reef
Blinde Klippen
Sand Ground more with Shelfs
Robin or Penguin I.
Charles Mount
CAPE BONA ESPERANÇA

Map 219

SELLER, John (fl. 1669-1699.)
A Draught of Cape Bona Esperanca (1750)
Map, 43 x 53 cm, coloured.
Scale in English Leagues.
Inset: View of Table Bay. The Fort.

The title of this map is within a rectangular signboard, with a native clad in skins on each side. Alongside the cartouche is an elongated view of Table Mountain with each component named and a number of sailing vessels in the Bay. At the top right corner there is an unusual view of the Fort, illustrated on a piece of hanging drapery. Another title, Monomotapa, is printed above the scale. This map is thought to be by Seller (see Map 258 for biographical notes).

246

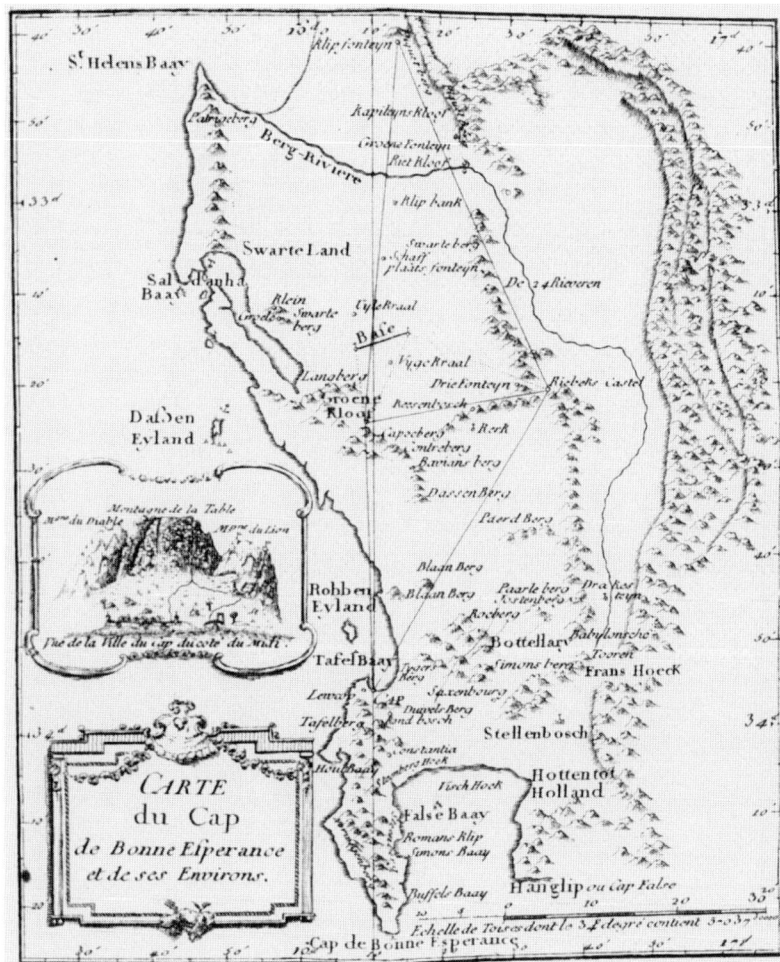

Map 220　LA CAILLE, Nicolas Louis de (1713-1762)
Carte du Cap de Bonne Esperance et de ses Environs. (Map of the Cape of Good Hope.) (Paris, N.L. de la Caille, 1752.)
Map, 15,5 x 12 cm, uncoloured.
Scale in toises.
Inset: *Vue de la Ville du Cap.* (View of the settlement at the Cape.)

The title of the map appears in the lower left corner and above it there is an inset marked 'Vue de la Ville du Cap du coté du Midi' in which Table Mountain (with certain points labelled) appears as a vignette with the Castle, the gallows, and in the distance the church and the Company gardens. In the main map topography is shown by anthill-like features. Gordon's Bay appears as 'Visch Hoek,' while Constantia, Rondebosch and Saxenbourg are inserted.

This map is of great interest because it shows de la Caille's triangulation for measuring an arc of the meridian. There is a very simple scale in toises at the foot of the map. Although it is a much smaller map than Tirion's Cape of Good Hope (see Map 221) there is no doubt that the latter is a close copy.

The Abbé de la Caille, a noted French astronomer and mathematician, published this map in his *Journal Historique*

du Voyage fait au Cap de Bonne Esperance in 1752. As a celebrated astronomer he proposed to the Academy of Sciences in Paris to follow up Halley's pioneer exploration of the southern skies from St. Helena to the Cape. This led him to be sent to South Africa in 1751 in the ship Le Glorieu and resulted in one of the most successful and useful scientific expeditions ever undertaken: still a landmark in the history of science. In his observations and investigations de la Caille established the parallax of the sun and moon with Mars adjacent, and embarked upon the measurement of an arc of meridian, the southern point of which was the Observatory in Strand Street, Cape Town, and the northern point Klipfontein. His original notes are in the Archival Department of the University of the Witwatersrand, Johannesburg.

247

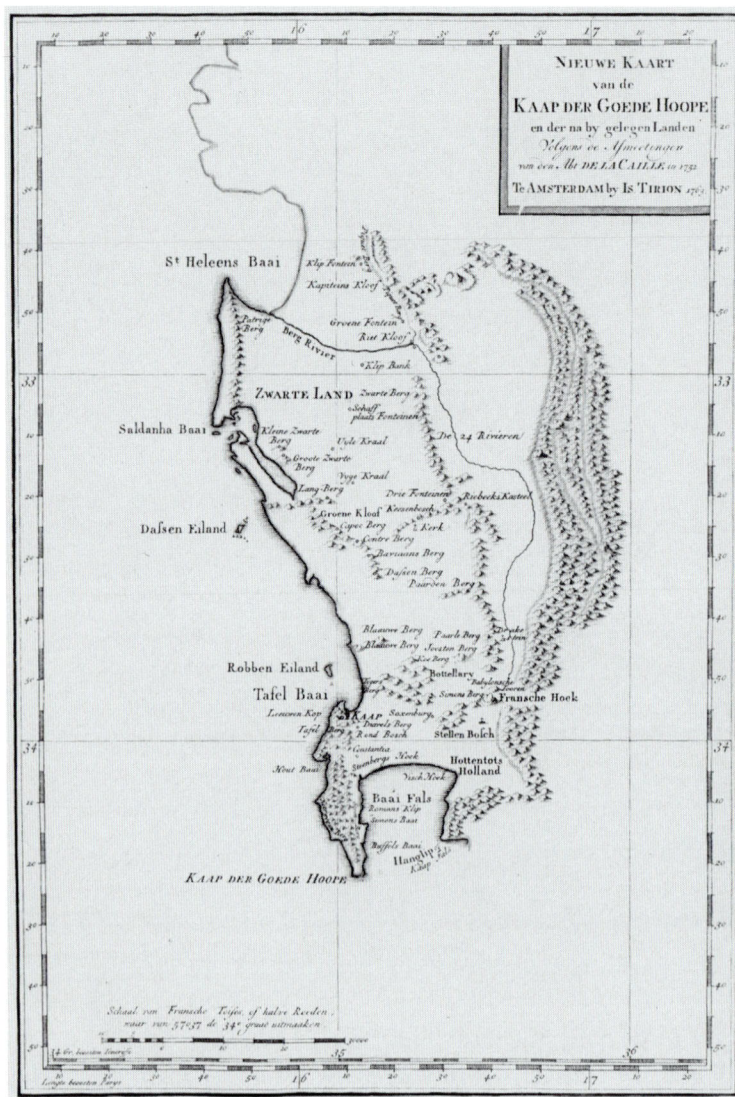

Map 221 TIRION, Isaak (d.1769)
Nieuwe Kaart van de Kaap der Goede Hoope en der na by gelegen Landen volgens de Afmeetingen van den Abt. de la Caille in 1752. (New map of the Cape of Good Hope and surrounding country...) (Amsterdam, I. Tirion, 1763.)
Map, 31,5 x 21,5 cm, coloured.
Scale in lineal measure of six French feet (toise).
Prime meridian through Paris and Tenerife.

This map of the Cape of Good Hope has its title in a box in the top right corner. It depicts the area between Cape Point and St. Helena Bay (St. Heleens Baai as described here) and Fransch Hoek and de Kaap make their appearance. The mountain ranges are fairly accurately named, more or less in accordance with modern nomenclature. On the Atlantic coast of the southern peninsula the mountain range is called Norweegen Berg. The Berg River course is accurate but the interior is not. The scale is shown in 'Fransche Toises of Halve Roeden,' and below this there is a note about the degrees east of Teneriffe.

In 1742, Isaak Tirion, the Amsterdam bookseller and publisher, issued an atlas of the Netherlands, and later, in 1744-69 a *Nieuwe en Beknopte Handatlas*, in which this map appears. It also appears in Tirion's *Hedendaagsche Historie of Tegenvoordige Staat van Afrika* of 1763. As the title indicates the map was issued in 1763 by Tirion and is based on the calculations of the Abbé de la Caille, a French mathematician and astronomer and a member of the Academy of Science, who visited the Cape and described his travels in a diary in 1752.

248

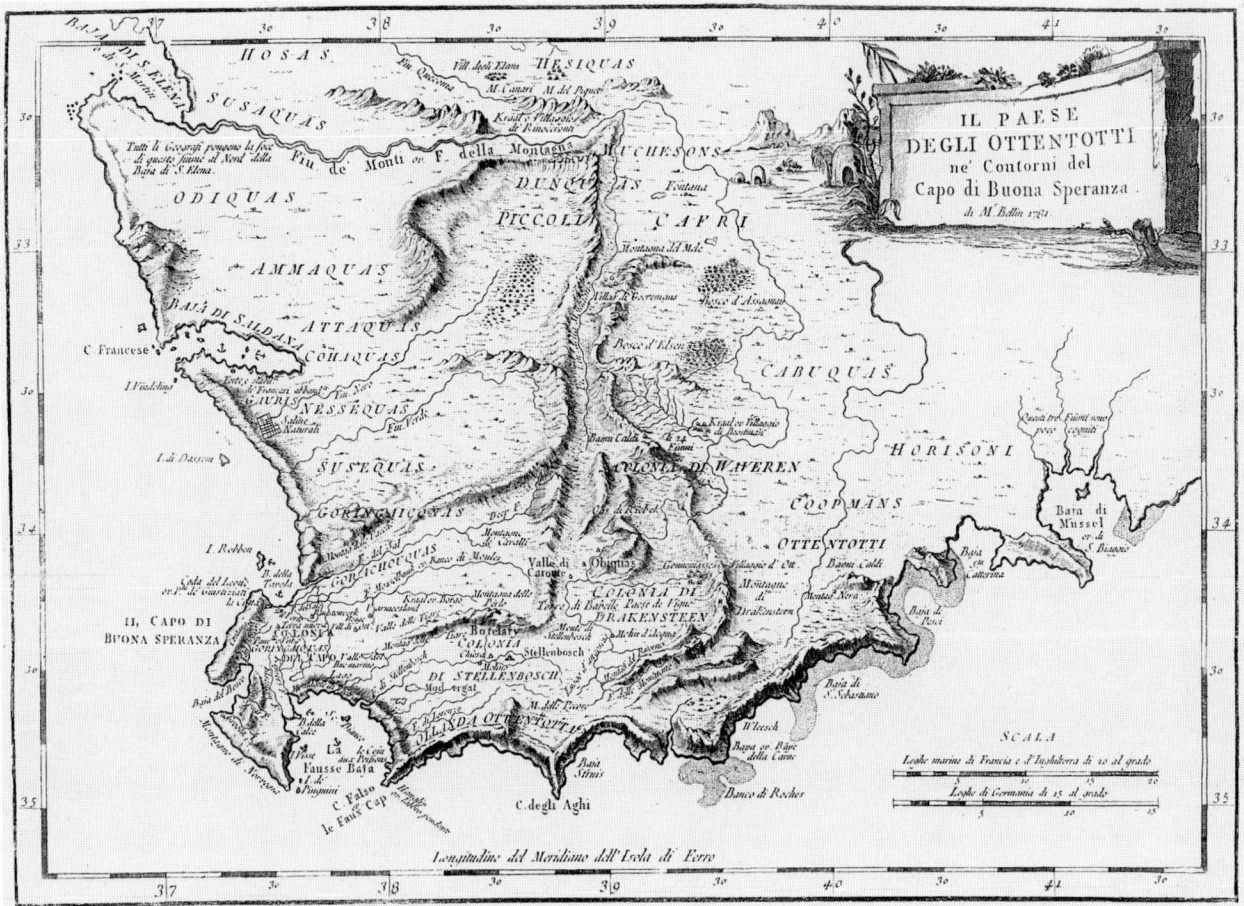

Map 222

BELLIN, Jacques Nicolas (1703-1772)
Il Paese Degli Ottentotti ne' Contorni del Capo di Buona Speranza. (The land
of the Hottentots... the Cape of Good Hope.) (Paris, J.N. Bellin, 1781.)
Map, 24,5 x 34,5 cm, uncoloured.
Scale in French, English and German leagues.
Prime meridian through Ferro Island.

Jacques Nicolas Bellin the elder was born in Paris in 1703 and died in Versailles in 1772. He was a French Royal Hydrographer, and the author of many atlases. This map appears in the Italian atlas *Teatro della Guerra Maritime* (Venice, 1781), one of Bellin's last works.

Bellin compiled many maps of Africa, including the Cape of Good Hope, Saldanha Bay, and large colourful sea charts. In this map of the Cape of Good Hope a curved rectangular masonry mass in the upper right corner forms the cartouche, within which an Italian title is described. Just to the left of the cartouche appears a mountain range, presumably Table Mountain, with a beehive hut in the foreground. The map is dated 1781 and the geographic features are typical of that period, with somewhat exaggerated bays, inlets and prominent mountain ranges. The west coast of the Cape Colony is depicted from St. Helena Bay, with a little of the east coast as far as Baya di Mussel (Mossel Bay).

249

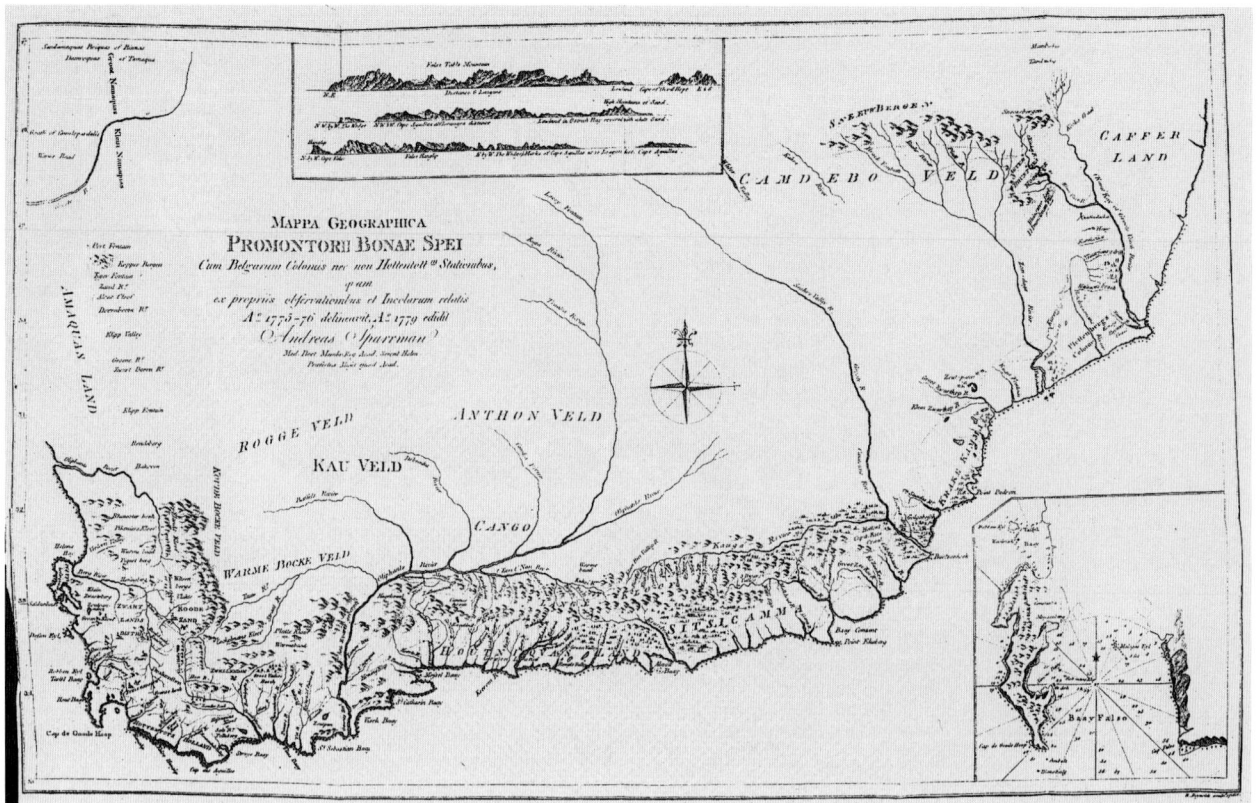

Map 223

SPARRMAN, Anders
Mappa Geographica Promontonii Bonae Spei... (Map of the Cape of Good
Hope...) Engraved by R. Reynolds. (London, G.G. and J. Robinson, 1785.)
Map, 33,5 x 54,5 cm, uncoloured.
Inset: False Bay. Elevation of coastline.
In: Sparrman, Andreas. *A Voyage to the Cape of Good Hope... from the year 1772 to 1776*, vol. 1. London:
G.G. and J. Robinson, 1785, fold at back.

This map appears in the 1785 English edition of Sparrman's travels of 1772-1776. The details of the settled parts of the Cape appear to be pretty accurate, but further east are not so reliable, probably because of his lack of accurate information. The Orange River is depicted but named Groote River. There is an inset at the top border of mountain ranges in three silhouettes of coastlines, and at bottom right of False Bay.

Sparrman's map is really a milestone in the history of South African cartography. In spite of its defects it was a great advance on its predecessors, many of which presented second-hand hearsay information and depicted little more

than a few rivers and some tribal names and kingdoms. Sparrman inserted a great number of authentic named features. His map was a valuable achievement which gave the world some authentic topography of the southeastern districts of the Cape. This map was copied almost unaltered (except for the title) for W. Paterson's book and was also used in le Vaillant's *Travels*. It was only really superseded in 1801 by the superior map of John Barrow, who made unkind remarks about it in volume two of his first edition of 1804. Considering the conditions under which he explored and the paucity of equipment, Sparrman did not do so badly.

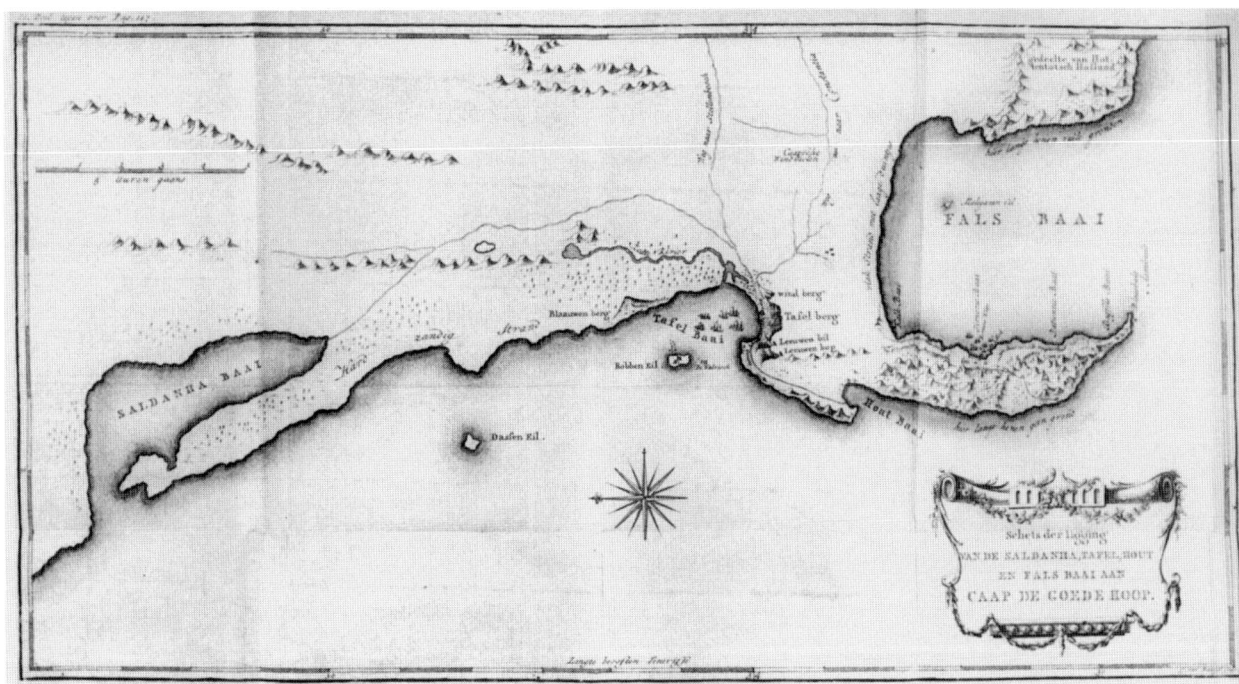

Map 224

STAVORINUS, J.S.
Schets der Ligging van de Saldanha, Tafel, Hout en Fals Baai aan Caap de Goede Hoop. (Sketch of Saldanha, Table, Hout and False Bays.) Engraved by C. van Baarsel. (Leyden, Honkoop, 1793.)
Map, 21,5 x 41 cm, uncoloured.
Prime meridian through Tenerif.
In: Stavorinus, J.S. *Reize van Zeeland over de Kaap de Goede Hoop, naar Batavia... 1768 tot 1771.*
Leyden: Honkoop, 1793, p. 117.

This map of Saldanha, Table, Hout and False Bays appears in Stavorinus's description of his travels in the years 1768 to 1771. A captain in the employ of the Dutch East India Company, he gave an accurate and valuable account of the Cape in the last quarter of the eighteenth century.

A rectangular floral scroll cartouche in the lower right corner contains the title. The top of the map is east and it extends from north of Saldanha Bay to include False Bay and the Hottentots Holland range. The topography is shown by anthill-like features, except for Table Mountain and 'Blaauen berg' which are shown as profiles. In both Table and Simon's Bay there are ships at anchor. The coast is not quite accurate as regards Hottentots Holland.

A road goes from Saldanha Bay to Cape Town, skirting the Salt River lagoon, then past the gallows at Green Point to Hout Bay. Another road runs from Simonstown over Red Hill to the coast at Olifantsbosch. There are two roads across the flats to Stellenbosch and Constantia and near the connecting road a windmill is noted (in the Eerste Rivier area).

251

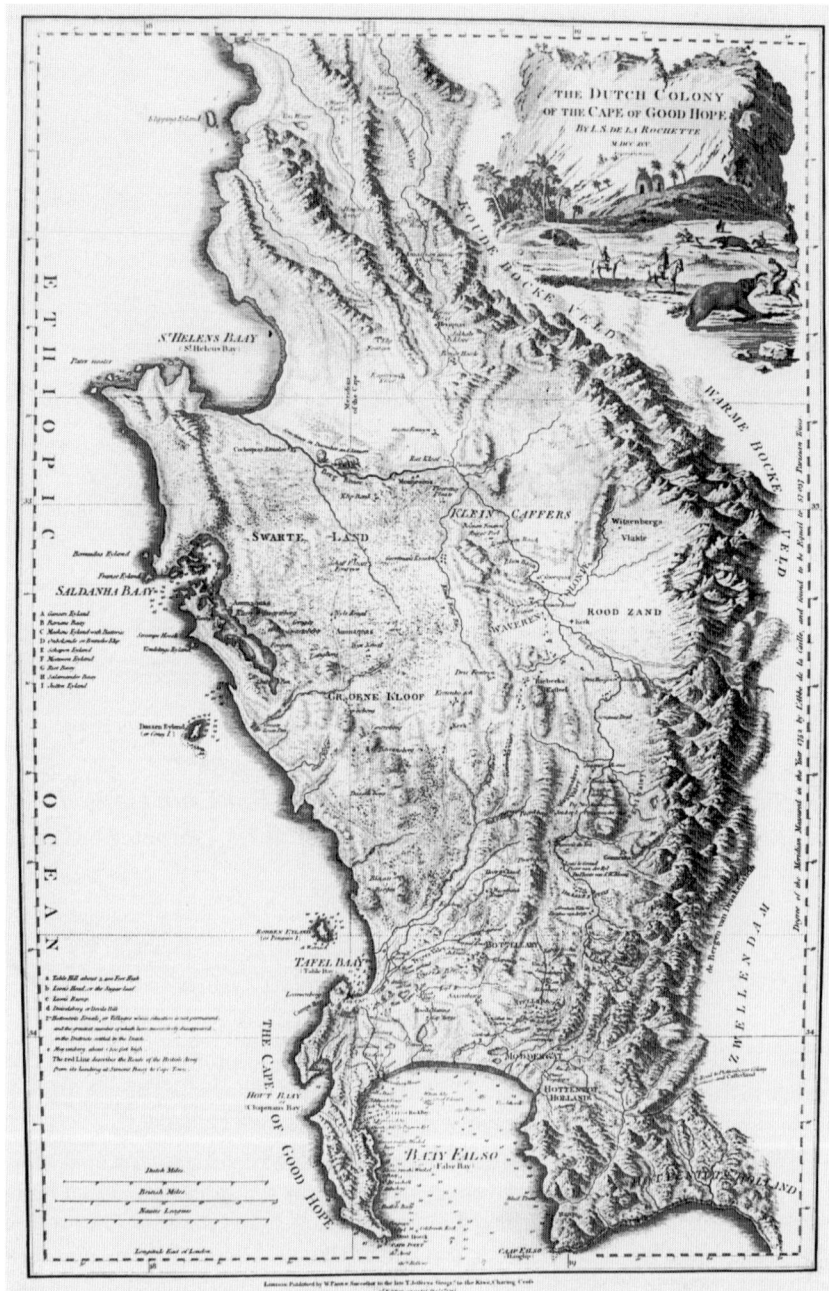

Map 225

DE LA ROCHETTE, Louis
Stanislas d'Arcy (1731-1802)
The Dutch Colony of the
Cape of Good Hope. Engraved
by W. Faden. (London,W. Faden,
1795.)
Map, 50 x 33 cm, uncoloured.
Scale in Dutch and British miles,
nautical leagues.
Prime meridian through London.

De la Rochette, an engraver associated with William Faden and James Wyld, was appointed publisher and cartographer to the King and Prince of Wales, working in St. Martin's Lane, London. With Faden he issued maps of North America, the Carribean Islands, South America and the Cape of Good Hope (the first edition of the above map) in 1782. This map is from William Faden's atlas of 1795 and gives an idea of contemporary knowledge of the area from 'Klipping Eyland' to 'Caap Falso,' with the Drakenstein Mountains as the eastern boundary of the Cape. Constantia, Stellenbosch, native kraals, fountains and river crossings are included in addition to the sites of the farms and settlements of approximately twenty-one settlers. The cartouche is made up of a cliff bearing the title, with palms and native huts and an elephant in the foreground. There is a key with legends identifying certain points on the mainland, and another identifying islands and bays. On the right is a note stating 'Degree of the meridian measured in the year 1752 by l'Abbé de la Caille and found to be equal to 57.037 Parisian Toises.'

Map 226

DEGRANDPRÉ, L.
Plan du Cap de Bonne Espérance et de ses Environs.
(Plan of the Cape of Good Hope and surroundings.)
(Paris, L. Degrandpre, 1793.)
Map, 48 x 28 cm, uncoloured.
Prime meridian through Paris.
In: Degrandpré, L. *Voyage a la Côte Occidentale de L'Afrique* vol. 2. Paris: Denut, 1801, p. 320.

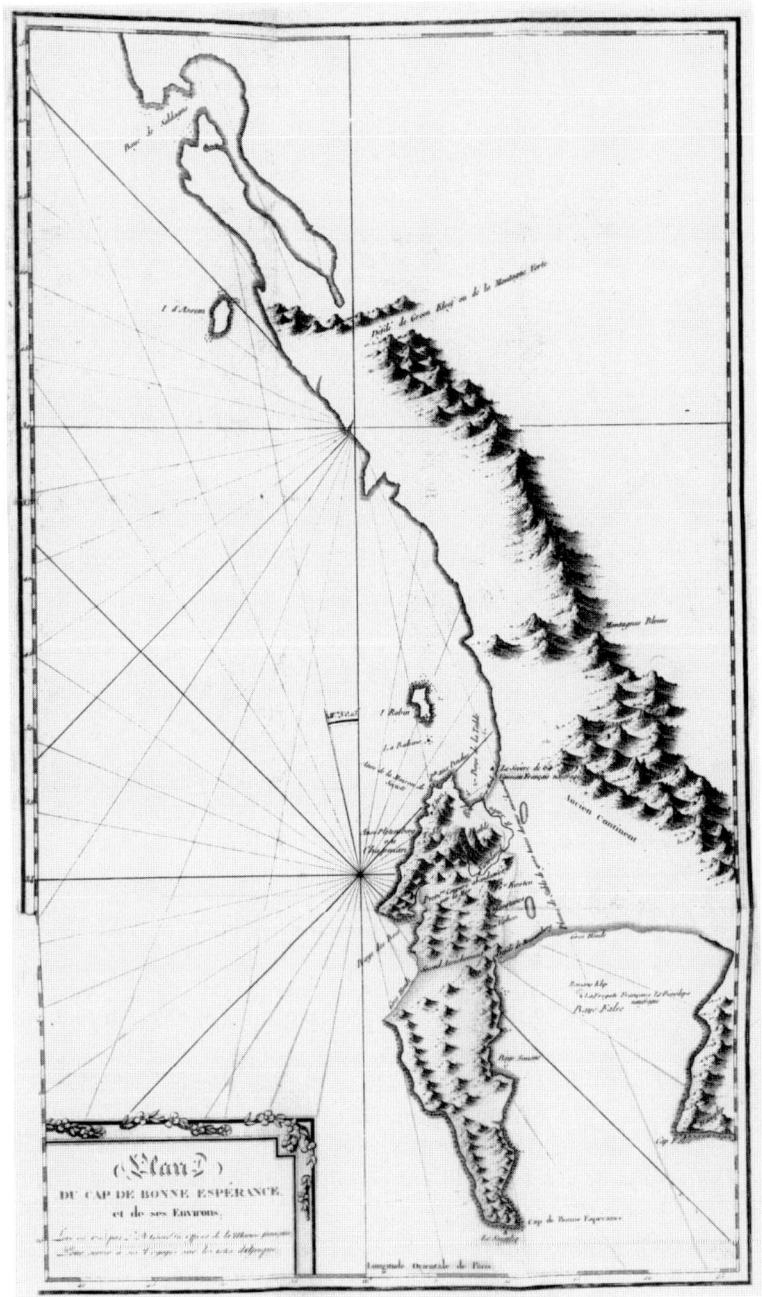

This map of the Cape of Good Hope appears in volume two of Degrandpré's account of his voyage to the west coast of Africa. He was an officer in the French Navy and drew many of his illustrations on the spot. The map is based on the meridian of Paris with radiating lines from a point off Llandudno. Saldanha Bay is distorted and brought too far south. The west coast mountains are depicted by a series of anthill-like features, broken only by the Groenekloof. Table Mountain is shown in profile. The farms of Alphen, Constance (Constantia), Newlands and Kirsten are indicated, as is also the Society House at Sea Point (possibly the site of the former Queen's Hotel there). There is a large gulf in the mountain at Constantia Neck, a second gulf from Muizenberg to Kommetjie and the third is the Cape Flats.

253

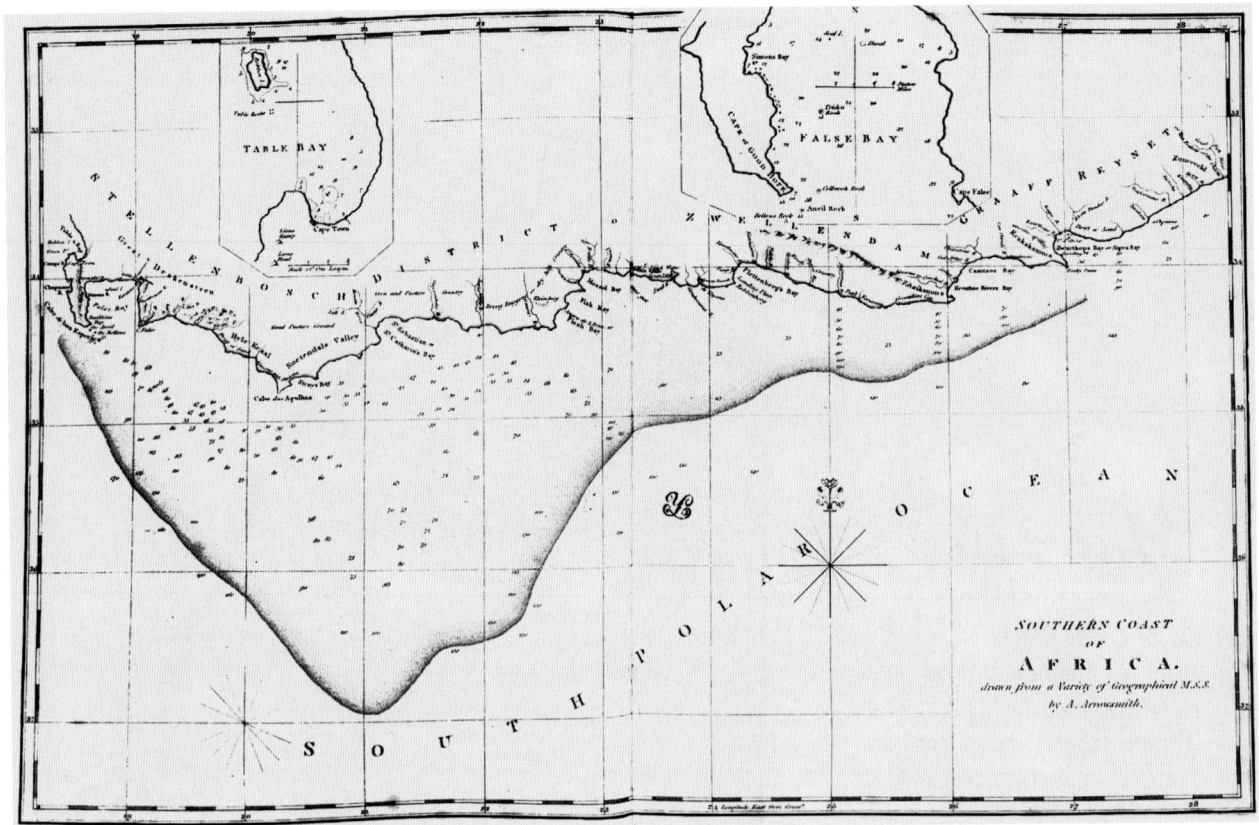

Map 227   ARROWSMITH, Aaron (1750-1833)
Southern Coast of Africa. Engraved by S.J. Neele. (London, Cadell and Davies, Strand, 1803.)
Map, 25,5 x 41 cm, uncoloured.
In: Untitled atlas containing maps of Africa. London: Cadell and Davies, 1803.

This atlas of 1803 appears to have been put together to contain a series of four double-page linen-backed coastal maps, mainly of western and southern Africa, with a single map depicting the Calabar and Bonny rivers of the Nigeria coast (see Maps 330-333).

This map is the fourth in the series, and extends from Table Bay to the Great Fish River. There is an inset of Table Bay and another of False Bay, and soundings appear all along the coast.

These maps were drawn from a variety of geographical information by A. Arrowsmith, the founder of the well-known family of British map publishers who produced, between them, approximately 750 maps. Arrowsmith held the position of Hydrographer to the Prince of Wales and later to His Majesty in 1820. He was assisted by his sons Aaron II and Samuel.

254

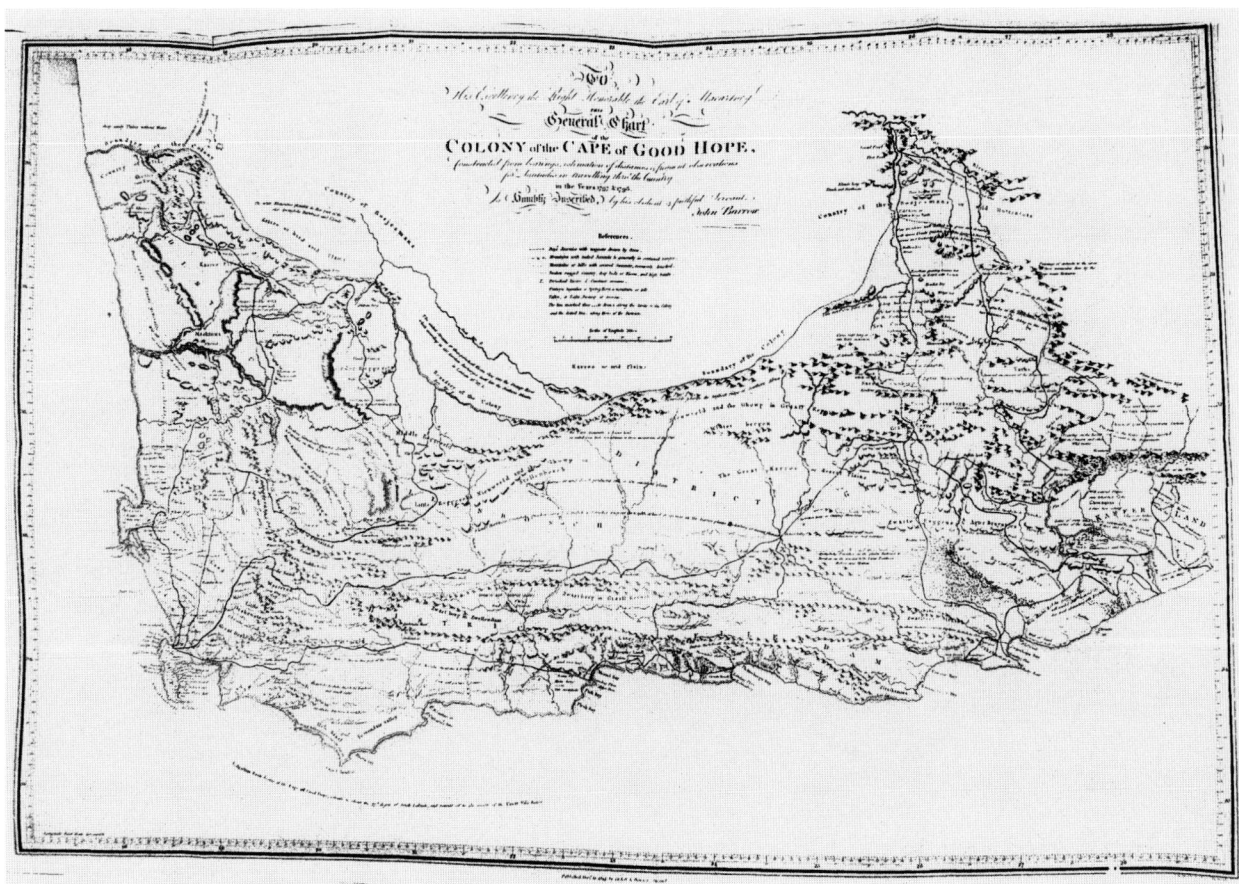

Map 228

BARROW, John (1764-1848)
General Chart of the Colony of the Cape of Good Hope... Engraved by S.J. Neele.
(London, Cadell and Davies, 1805.)
Map, 46 x 69 cm, coloured outline.
Scale in English miles.
Prime meridian through Greenwich.
Dedicated to His Excellency the Right Honourable the Earl of Macartney.
In: Barrow, John. *Travels into the Interior of Southern Africa...* vol. 2. London: Cadell and Davies,
1806, fold in front.

This is the first of the more accurate maps typical of the nineteenth century. The coastline is more realistic but with some inaccuracies especially on the False Bay coast. Some of the topography is represented by the newer technique using hachures – short lines being drawn in the direction of the slope, but the majority of ranges are shown by the old anthill method. The colonial boundary is from the Koussie River in the west to the Great Fish River in the east. The district boundaries are all drawn in. The routes taken by

Barrow in his travels, 1797-8, are well illustrated in this map. He travelled beyond the boundary of the Orange River and his map is covered with interesting remarks on the country, fauna and customs of the people. Some names are unfamiliar to us, eg. Little Europe for the Strand area in False Bay. This map is the best map of the Colony up to that date and the information it contains was used by later publishers for many years.

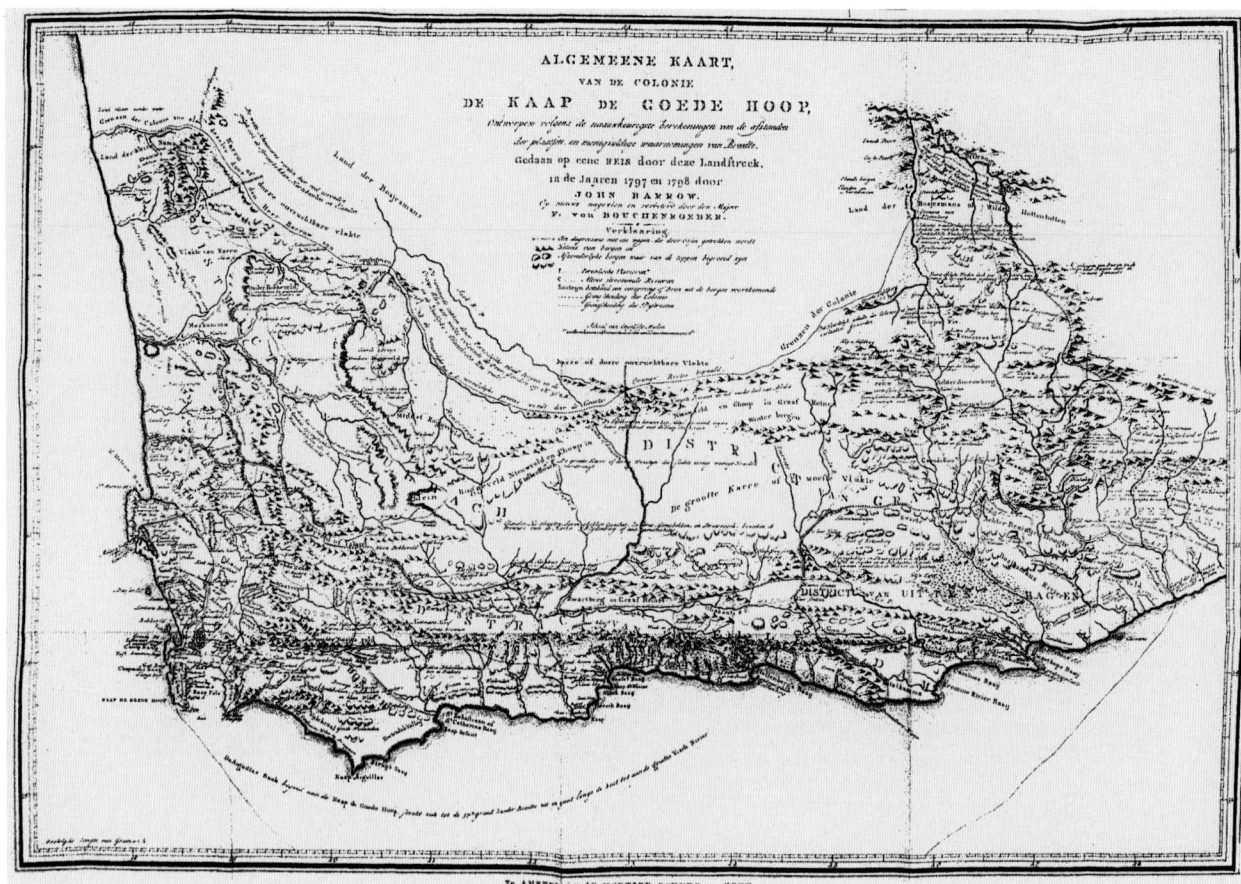

Map 229 BOUCHENROEDER, B.F. von
Algemeene Kaart van de Colonie de Kaap de Goede Hoop ... Gedaan... door John Barrow. (General map of the Colony of the Cape of Good Hope.)
(Amsterdam, Mortier, Covens en Zoon, 1806.)
Map, 46 x 67 cm, coloured outline.
Scale in English miles.
Prime meridian through Greenwich.
In: Bouchenroeder, B.F. von. *Reize in de Binnelanden van Zuid-Afrika dedaan in den jare 1803.*
Amsterdam: Mortier, Covens en Zoon, 1806, fold in front.

The publishers/engravers of this map are the well known firm of Mortier, Covens and Son, of Amsterdam.

The map is of some historical value although it contains a good deal of speculative matter. It is in fact a close copy of the map by Barrow (see Map 228) with a few additions, and includes Barrow's route. The districts of Uitenhage and Tulbach are added and there is also a projected canal from the Berg River to northern Saldanha Bay. The shape of False Bay differs from that of Barrow's. There is also a note saying that the natural northern boundary is defined by the 'Groote Oranje Rivier.'

Baron B.F. von Bouchenroeder, a former Major in the Dutch Services, made a journey into the interior of South Africa in 1803, with the intention of exploring the different bays and harbours along the southeast coast in order to find out how suitable they would be for coast and overseas traffic, especially for the export of products at a profit to Cape Town and the Colony.

256

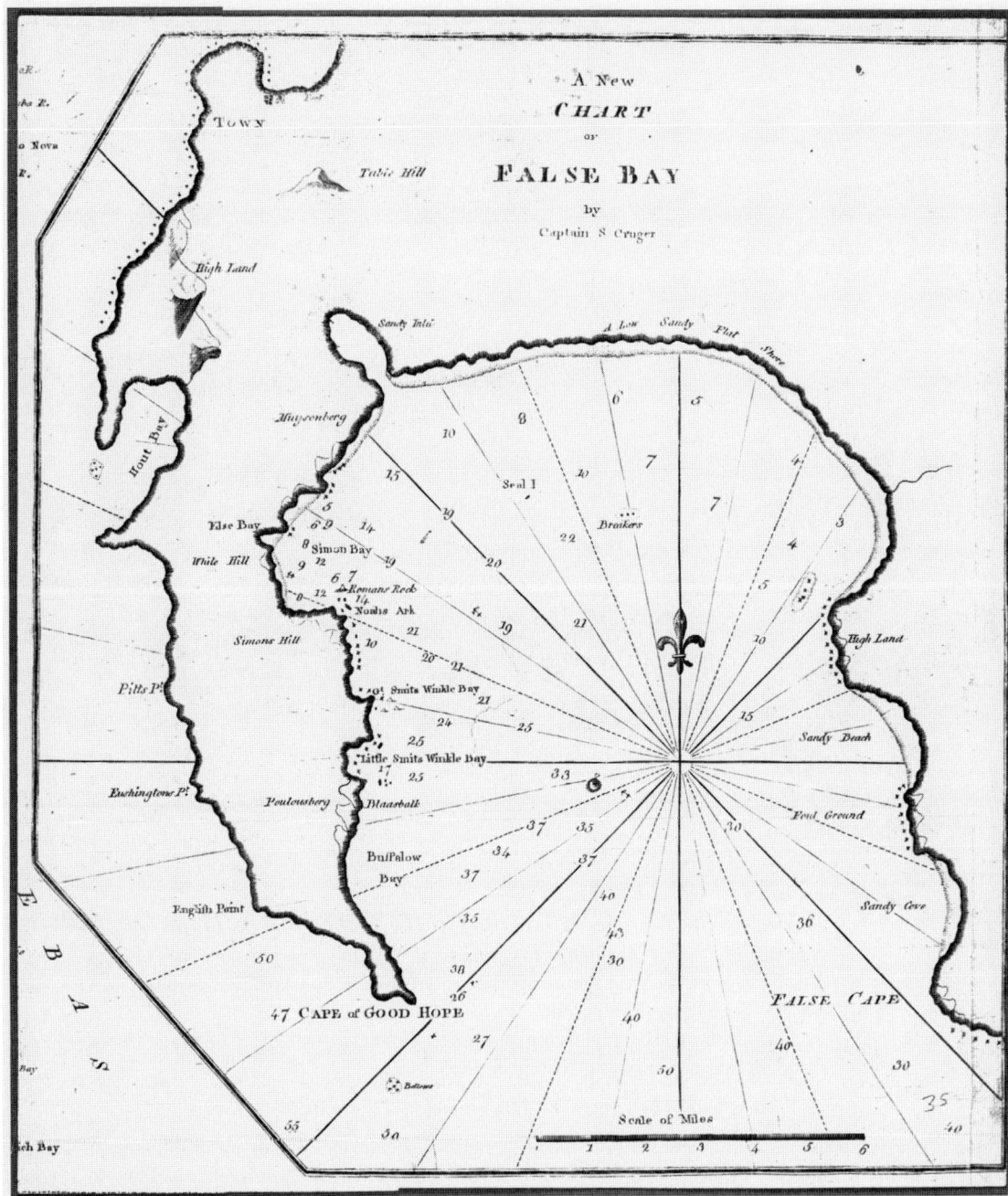

A New
CHART
OF
FALSE BAY
by
Captain S Cruger

Map 230   CRUGER, Capt. S.
A new chart of False Bay. (1808.)
Sea chart, 35 x 30,5 cm, uncoloured.
Scale in miles.

This sea chart of False Bay, with numerous soundings and rhumb lines, by Captain S. Cruger, could not at first be traced in the usual references available, nor from knowledgeable colleagues in England.

After further searching it was traced to the Greenwich Museum Map Library, where it was found to have been part of an inset of a map of the Cape of Good Hope by W. Heather, in the *Marine Atlas or Seamen's Complete Pilot* (of 50 charts) of 1808. The chart catalogued here is identical to the corresponding portion of the inset of the whole of the Cape of Good Hope.

257

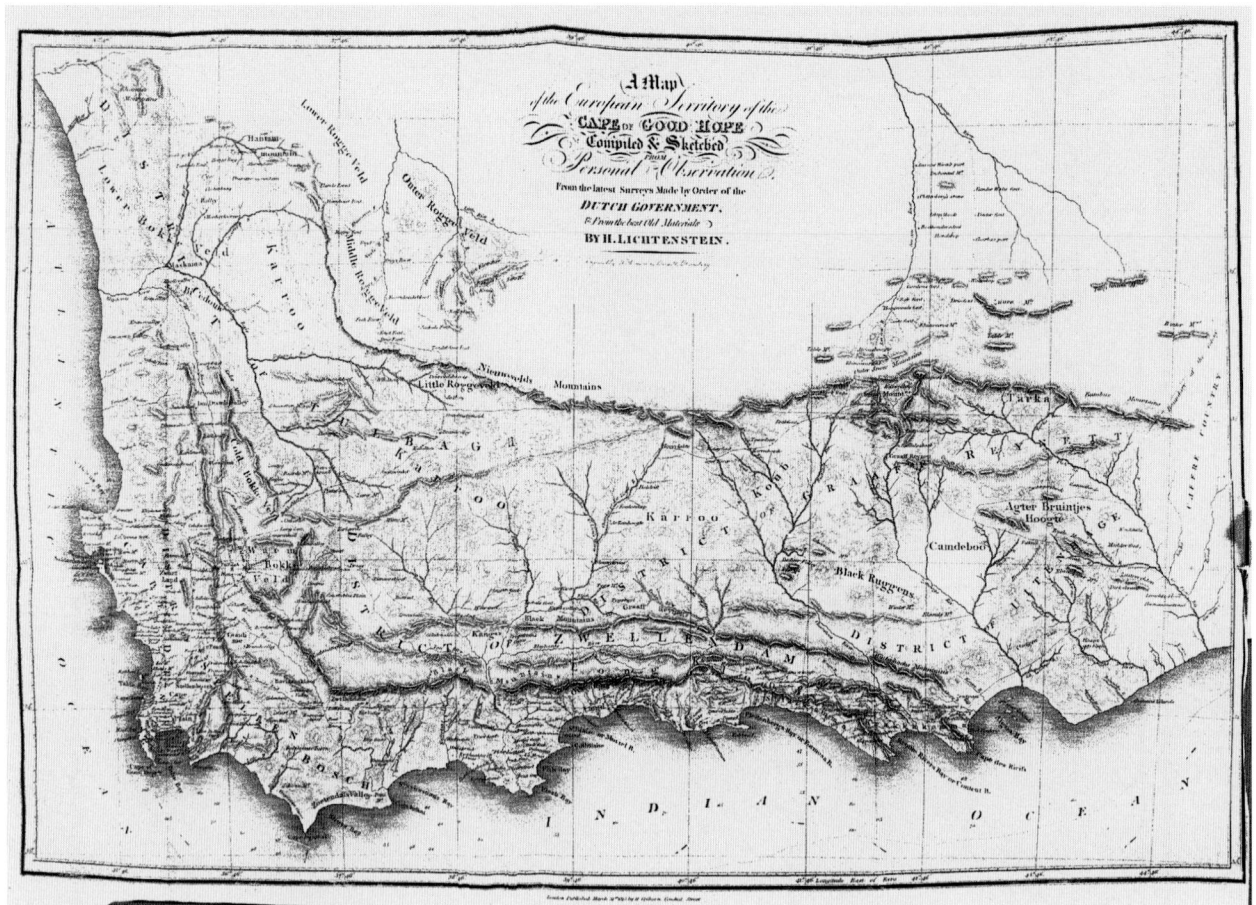

Map 231

LICHTENSTEIN, Henry
A Map of the European Territory of the Cape of Good Hope Compiled and Sketched from Personal Observation. Engraved by M. Thomson. (London, H. Colburn, 1815.)
Map, 34,5 x 51 cm, uncoloured.
In: Lichtenstein, Henry. *Travels in Southern Africa in the years 1803, 1804,1805 and 1806*, vol. 2, tr. by Anne Plumtree. London, H. Colburn, 1815, fold xi.

This map is engraved by M. Thomson of Bloomsbury, London and is similar to Barrow's earlier map, with the district and colonial boundaries and rivers, etc. The coastal bays are still rather exaggerated, especially St. Helena Bay. The country to the east is sparsely recorded. Tribal names adorn the interior. The sources of the Orange River appear mainly in the vicinity of mountain ranges, especially the Wolwekop and Taaybosch ranges. This map appears in the second volume of Lichtenstein's *Travels in Southern Africa in the years 1803, 1804, 1805 and 1806.*

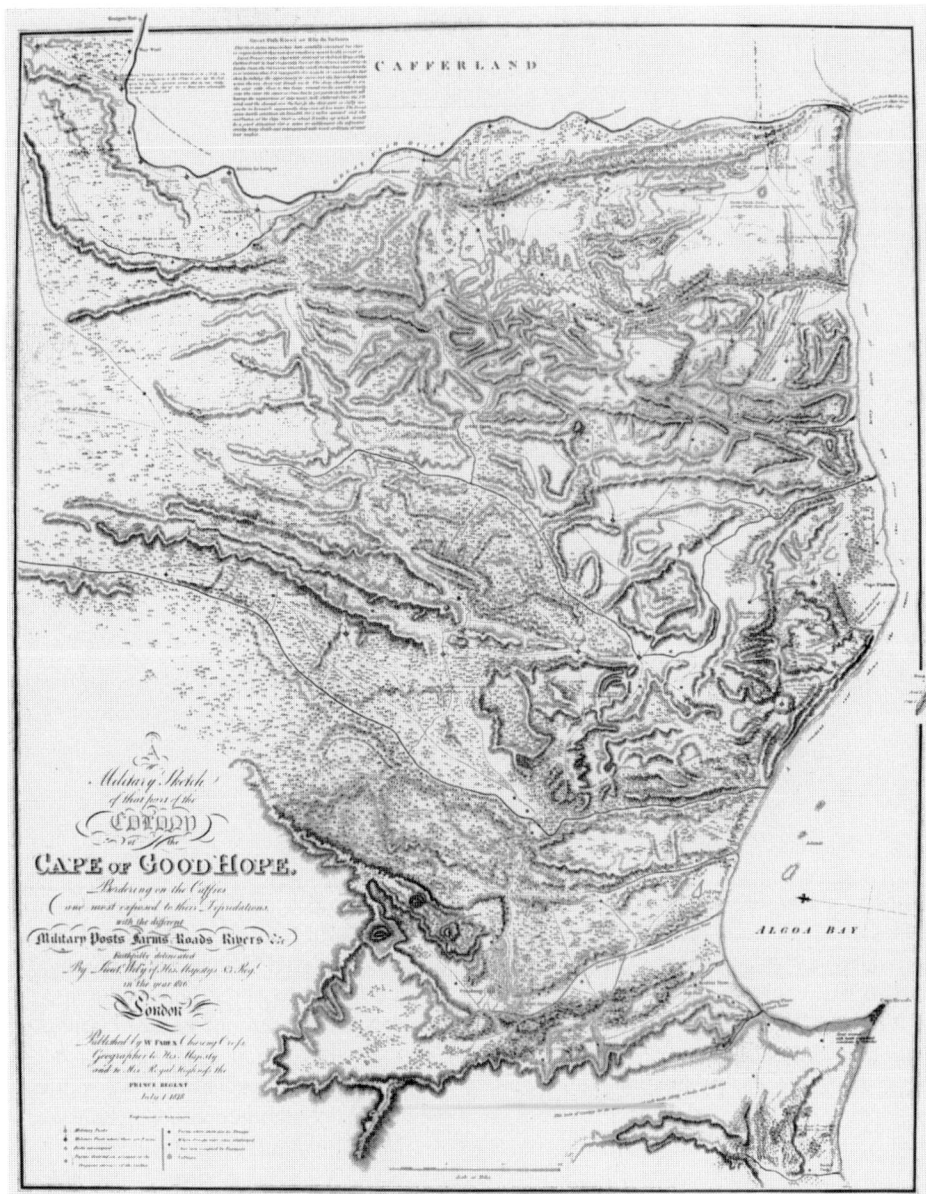

Map 232

FADEN, William (1750-1836)

A Military Sketch of That Part of the Colony of the Cape of Good Hope, Bordering on the Caffres and most exposed to their Depredations with the different Military Posts, Roads, Rivers &c. Drawn by Lieut. Wily, engraved by J.W. Walker. (London, W. Faden, 1818.)
Map, 60 x 48 cm, uncoloured. Scale in miles.
Inset: Note on the Great Fish River.

William Faden was a London publisher, cartographer and geographer to the King and the Prince of Wales. He was succeeded by T. Jefferys and later by J. Wyld the elder. Faden is especially well known for maps of the Cape of Good Hope and military plans of the Cape Colony, and for the accuracy of his rivers.

This so-called military sketch is a greatly detailed map of the Algoa Bay – Cape Recife area and inland, following the course of the Great Fish River, which is also given its Portuguese name, Rio de Infanta, in a descriptive paragraph at the top of the map. A legend refers in detail to military posts, posts with farms, troops' quarters, villages, etc. The mountain ranges are prominently drawn in detail, with a record of forest and bush areas. All these details were delineated by W. Wily of His Majesty's 83rd Regiment in 1816, during the long Kaffir Wars of the eastern Colony. The title on the lower left is an example of fine varying types of printing, making it a most attractive map of that period.

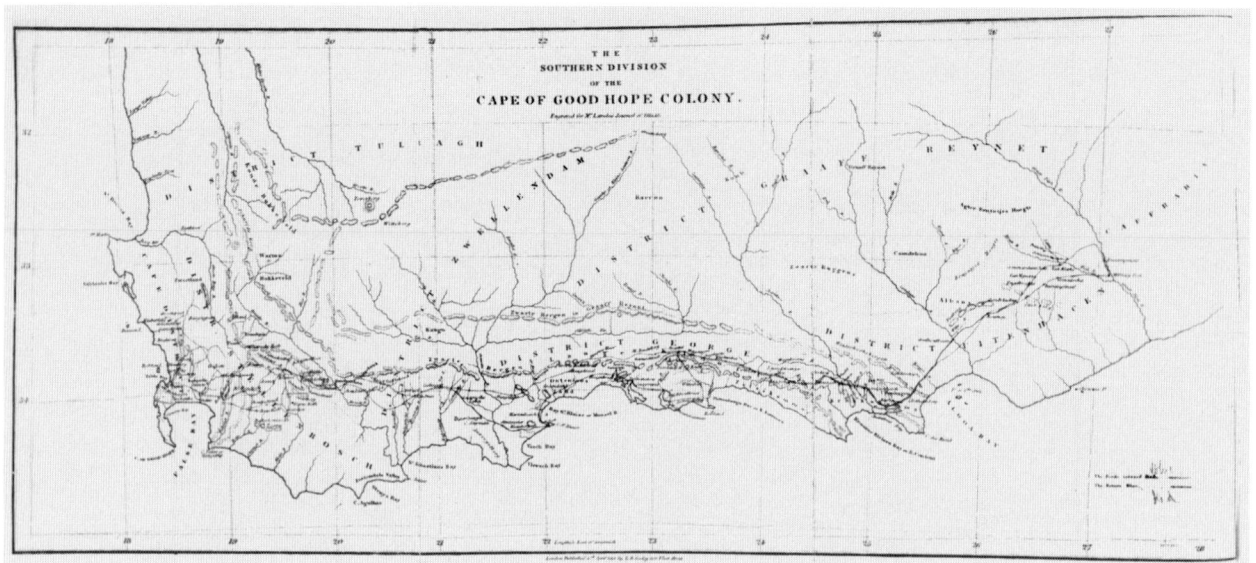

Map 233

LATROBE, C. I.

The Southern Division of the Cape of Good Hope Colony, engraved for Mr. Latrobe's Journal of 1815-1816. (London, L.B. Seeley, 1818.)

Map, 22,5 x 54 cm, uncoloured.
Prime meridian through Greenwich.
In: Latrobe, Rev. C.I. *Journal of a visit to South Africa in 1815 and 1816.* London: L.B. Seeley, 1818, fold in front

The Rev. Latrobe was a Moravian minister, a friend of Burchell, sent to the Cape to select a suitable location for a third mission station. His outward journey took him through Stellenbosch, Cape St. Blaise and the Uitenhage district to the Great Fish River, and he returned via Avantur and Hunyklipkloot, with a diversion to Caledon.

This map is from his *Journal of a Visit to South Africa in 1815 and 1816.* The greatest detail is confined to the area south of the Karre Bergen and the inland ranges. The coastline is becoming more accurate, especially round the Cape, except that Cape Agulhas should be more southerly. For the first time Cape Hangeklip appears correctly spelt. Rivers are indicated but greater detail is confined mostly to Latrobe's route. He also gives details on the farms he visited. Tulbach and George have been added to the districts. Newlands, Wynbergen and Constantia are marked.

260

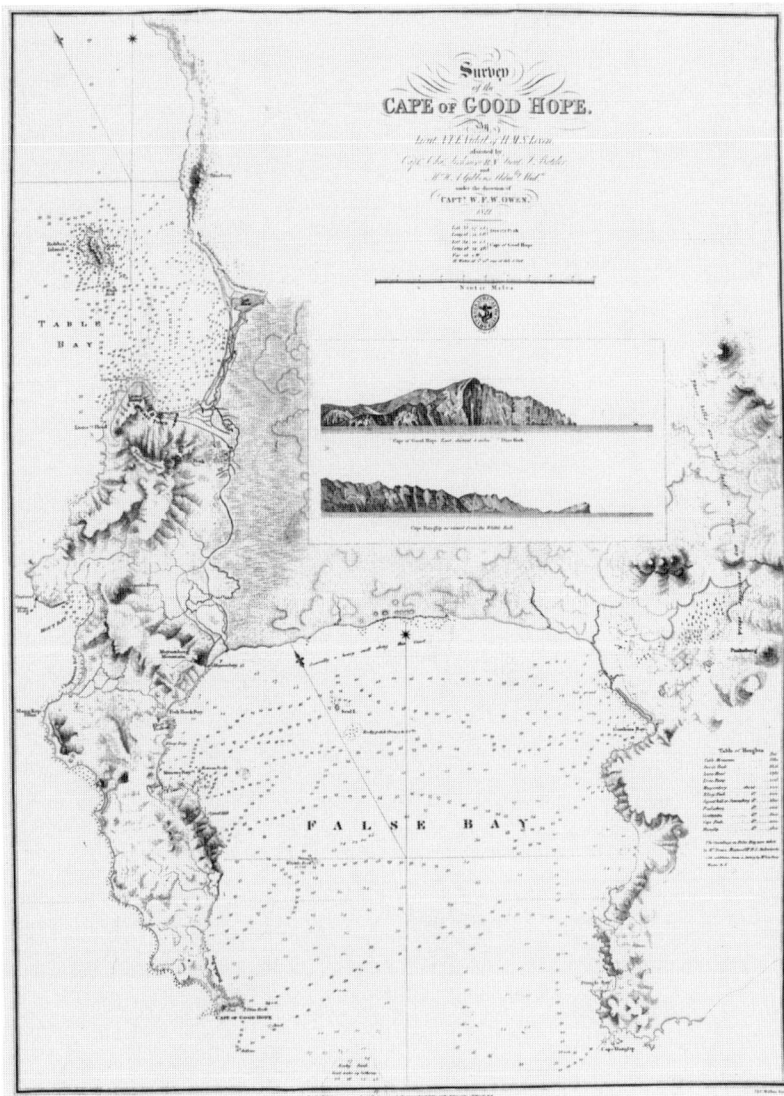

Map 234

OWEN, W.F.W.
Survey of the Cape of Good Hope by Lieut. A.T.E. Vidal of HMS Leven assisted by Capt. Lechmere, Lt. T. Boteler and Mr. M.H.A. Gibbons under the direction of... 1822. Engraved by J. and C. Walker.
(London, Hydrographical Office of Admiralty, 1828.) Sold by R.B. Bate.
Map, 59 x 43 cm, uncoloured.
Scale in nautical miles.
Inset: Dias Rock and Cape Hangklip.
Bottom right: 636

This is an early nineteenth-century map of the Cape of Good Hope published at the Hydrographical Office of the British Admiralty in 1828 under the direction of Capt. W.F.W. Owen assisted by Lieut. Vidal of *HMS Leven*, Capt. Lechmere, Lieut. Boteler and Mr M.H.A. Gibbons. It appears as a map in the two-volume book by Capt. Owen, *Narrative of Voyages to Explore the Shores of Africa, Arabia and Madagascar* (vol.2, 1833).

Table Bay and False Bay are illustrated in great detail, with the surrounding mountain ranges, and there is a table of heights of mountains and peaks. This illustrates the advance in techniques of cartography and survey in the period after 1726 when Gerard van Keulen, the well-known Dutch publisher of sea charts, put out maps for the Dutch East India Company. It was in this period that Amsterdam lost much of its standing in hydrographical activity, due to the sudden decline in the Dutch East India Company, and the hydrographical departments of the British and French authorities began to assume leadership in this sphere.

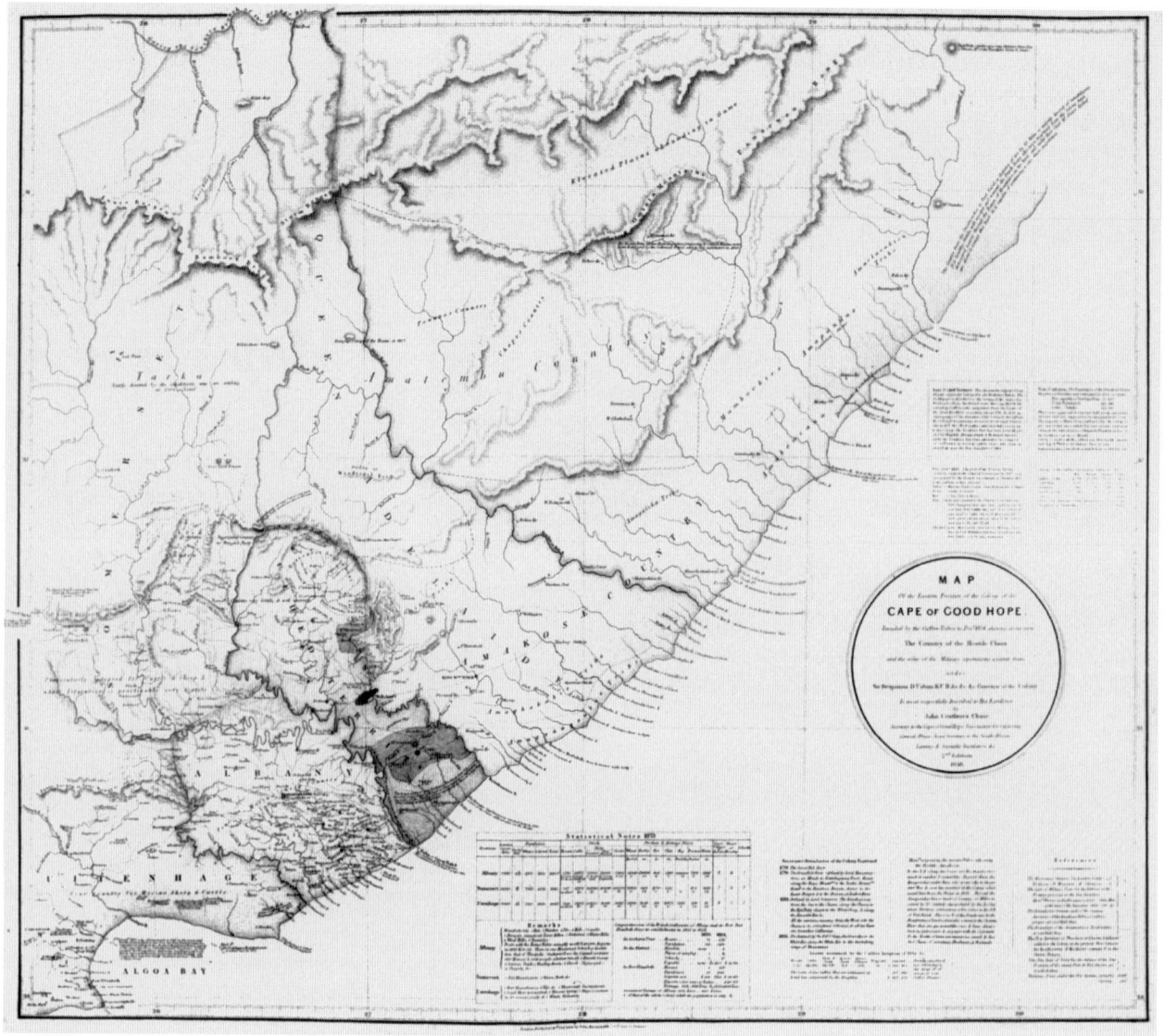

Map 235

CHASE, John Centlivres
Map of the Eastern Frontier of the Colony of the Cape of Good Hope.
(London, John Arrowsmith, 2nd ed. 1838.)
Map, 50,5 x 58 cm, coloured.
Inset: Statistical notes 1833.
Successive boundaries of the Colony eastward.
Dedicated to Sir Benjamin D'Urban.

This coloured map, dated 1838, of the 'Eastern Frontier of the Colony of the Cape of Good Hope invaded by the Caffre tribes in December 1834, as the title describes it, depicts one view of 'the country of the hostile clans and the scene of the military operations against them under Sir Benjamin D'Urban, Governor of the Colony.' It is inscribed to His Excellency the Governor by John Centlivres Chase, who was secretary to the Cape of Good Hope Association for exploring Central Africa and also secretary to the South African Literary and Scientific Institution. It was first published in 1836 by John Arrowsmith of London.

This is regarded as a rare map. In *The Map Collector* (vols. 6 and 17, dealing with maps of the Cape of Good Hope) no mention is made of any of Chase's maps. The whole plate negative of this map is housed in the Royal Geographical Society, London.

The map provides a great deal of historical and geographical information regarding the important part of the Cape Colony which was the scene of nine frontier wars (1779-1878) between the Cape government and the Xhosa tribe. Ample legends are provided describing Caffraria, ceded territory and giving statistical notes on population groups, stocks and prices in 1833. Changing boundaries, showing the progress of the Colony eastward, are illustrated in colour with an 1835 census of the Caffre population.

The details of the map are accurate, demarcating the important districts with remarks on Uitenhage, Albany, Queen Adelaide and a number of tribal territories within the Amatemba and Amakoese countries.

In Algoa Bay mention is made of St. Croix Island, where Bartholomew Diaz administered the Eucharist to his crew when in search of the Cape, and it is also mentioned by the famous Portuguese poet Camoens in his *Lusiad* as the 'Isle of the Holy Cross.'

John Centlivres Chase landed in South Africa as an 1820 Settler and made a journey in 1825 to the relatively unknown northern regions of the country. Apart from his secretarial duties he started to compile a comprehensive account of the various journeys undertaken by travellers in South Africa throughout the previous years. This stimulated his interest in all matters relating to geography and map making and from then on he began to gather, from many sources, every scrap of information relative to the exploration of South Africa.

In addition to his geographical investigation and map making Chase wrote a number of volumes on the Eastern Cape, Natal and the Cape of Good Hope. He was associated with map making of the Eastern Cape, although no record exists of the assistance he provided at that period.

In April 1834 the publishers Baldwin and Cradock published a map of South Africa, under the superintendence of the Society for the Diffusion of Useful Knowledge of the United Kingdom, compiled from the manuscript maps in the Colonial Office of Capt. Owen's survey. When Chase saw a copy of this map he was aghast at finding that in it was incorporated everything that he had included in the sketch he had sent to Viscount Goderich in 1831, and which had not been acknowledged. He complained to the Governor of the Cape, Sir Benjamin D'Urban, who on investigation considered that Chase had been dealt with unjustly. Chase had a similar experience with Andrew Steedman. In his book *Wanderings and Adventures in South Africa*, published in London in 1835, Steedman used the same Society map, without acknowledgement, although he was fully aware of Chase's geographical researches.

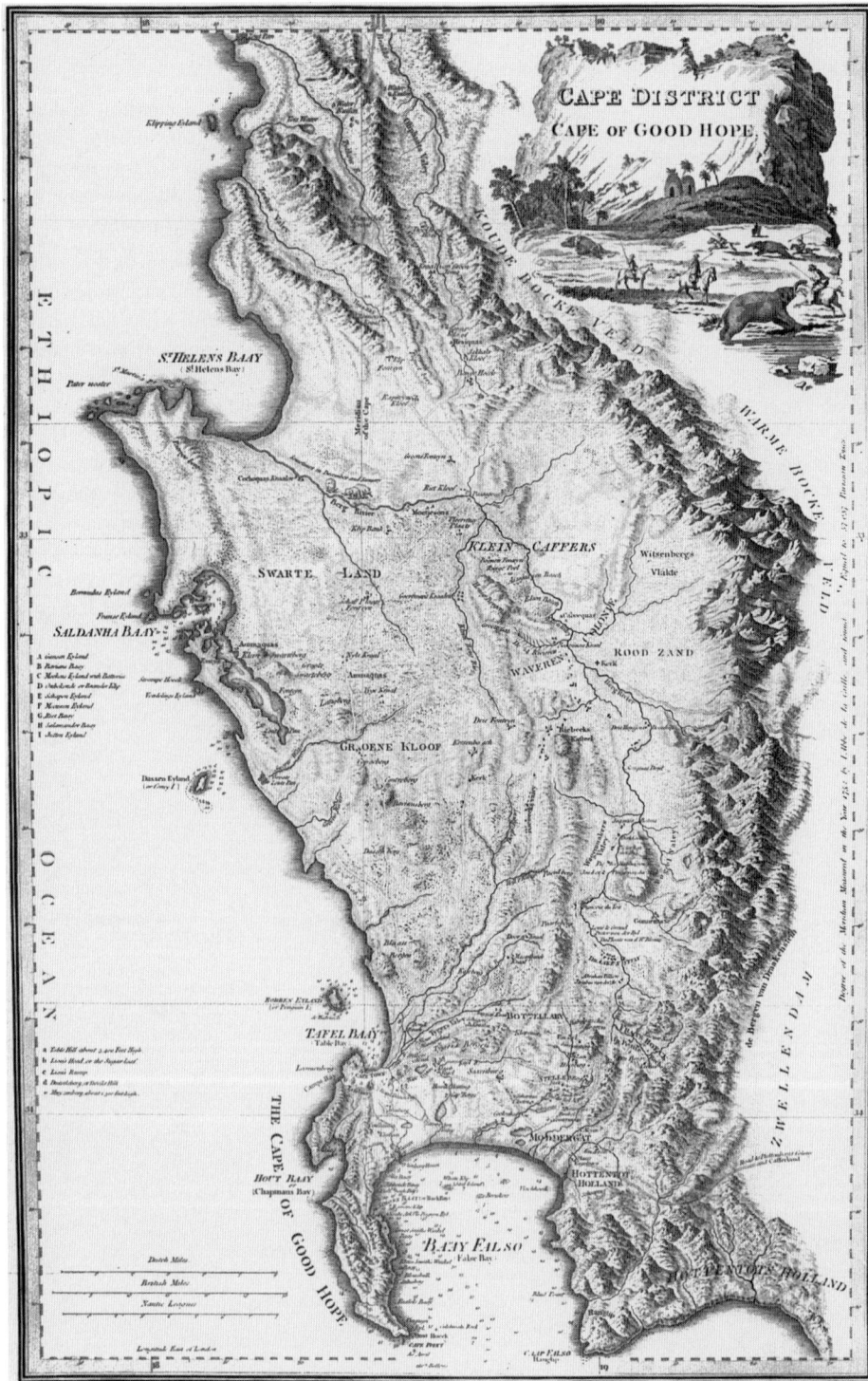

Map 236

DE LA
ROCHETTE, Louis
Stanislas d'Arcy
(1731-1802)
*Cape District, Cape
of Good Hope.*
(London, Ja. Wyld
successor to Mr Faden
geographer to the Queen,
Charing Cross East, 1838.)
Map, 50,5 x 33 cm,
uncoloured.
Scale in Dutch and British
miles, nautical leagues.

This map of the Cape District and Cape of Good Hope was first published by de la Rochette in 1782, with a second edition in 1795 (see Map 225). James Wyld (successor to W. Faden), Geographer to the King and H.R.H. the Duke of York, produced his edition in 1838. The names of de la Rochette and the engraver are omitted from the title and the imprint has changed; the note on Hottentot kraals and the route of the British army have been omitted, 'The Cape' is changed to 'Cape Town' and St. Martin's Point (which seems to be the only addition) appears near St. Helen's Bay.

264

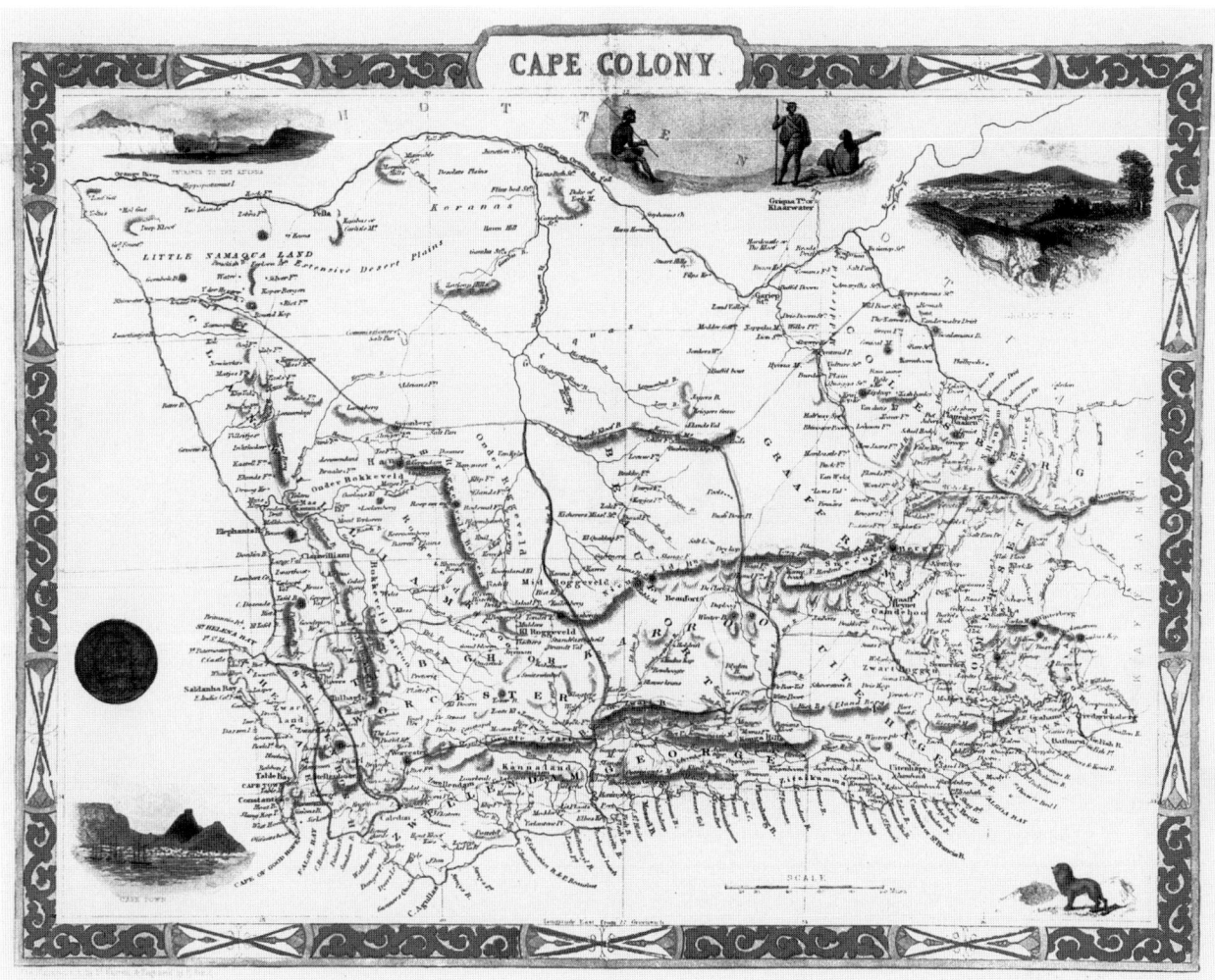

CAPE COLONY

Map 237 TALLIS, John
Cape Colony. Engraved by J. Rapkin. Illustrations by H. Warren and engraved by H. Bond.
(London, J. Tallis and Co. 1851.)
Map, 24 x 32 cm, coloured outline.
Scale in miles.
Prime meridian through Greenwich.
Inset: Cape Town. Entrance to Knysna. Graham's Town.

This map of the Cape Colony appears in the *Illustrated Atlas and Modern History of The World* published by John Tallis and Co., London and New York, in 1851. It was one of the last decorative atlases, with all the maps engraved on steel and all adorned with small vignette views. In this particular map Cape Town, the entrance to Knysna and Grahamstown are faithfully engraved. It has in addition a scene of three natives in traditional dress and another of a lion.

The atlas was originally published in parts, of which the seventh contains the maps of Africa and Arabia with four pages of text. The geographical content is accurate for its time and the Cape Colony is well filled with placenames, rivers and mountains.

265

SEA CHARTS

Sea Charts and Pilot Books

In the early sixteenth century European seamen rarely relied on navigational aids: their voyages were based almost entirely on experience. The incredible feats of these skilled mariners of old who accomplished their hazardous voyages with so few aids still excite our admiration.

The small 'rutters,' or sets of sailing directions, which really served the seamen as memory aids, developed into more detailed guides, with charts and profiles of the coast. They were intended to light the seaman on his way in not easily recognisable waters, moorings and bays of the European sea coast. The titles of these books were apt: *Light of Navigation, Sea Mirror, Sea Torch*, etc.

In 1584, alongside the numerous manuscripts, charts and rutters used on board ship, there first appeared the spectacular pilot book of Lucas Janszoon Waghenaar – *De Spiegel der Zeevaerdt*. It introduced a new form of hydrographical publication, containing an atlas of nautical charts (paskaarten) and a rutter, or set of sailing instructions (leeskaarten), for navigation on the west and northwestern coasts of Europe. It was published as a large folio volume beyond the financial means of the pilot and fisherman – nor were the charts on sale separately. The fact that few could afford *De Spiegel* led Waghenaar to publish his second pilot book in a more acceptable oblong format: the *Thresoor der Zeevaert* (1592), printed at the Plantin Press, Leiden. This *Thresoor* served as a model for future chart books.

The name of Willem Janszoon Blaeu dominated the first decades of the seventeenth century. After Waghenaar's death in 1606, Blaeu produced a new, original and superior pilot book, *The Light of Navigation* (1608), of which at least twenty editions were issued between 1608 and 1637, with Dutch, English and French text.

Stimulated by the rivalry of his competitors in the trade, Blaeu issued a larger version of his pilot book in 1625 – the *Zee-Spiegel*, of which fifteen editions antedated 1652 – five of them with English text. This work was enlarged not only in size (it was altered from the oblong to small folio) but also in scope, as the number of charts increased from 42 to 111. After 1652 there followed a further enlarged *Groote Zeespiegel*, which marked the end of the activities of the Blaeu family as publishers of pilot books.

Jacob Aertsz Colom appeared on the chart scene when he first issued a book of charts in 1632. This work brought the pilot publishing trade a step nearer to its final form, which appeared with van Keulen's pilot book. Colom was the first to introduce the folio pilot book in parts, and in a popular size. His book was a tremendous success and he became the leading chart publisher in the first half of the seventeenth century, pushing Blaeu out of the market and holding his own against the newcomers in the field.

Anthonie Jacobsz, however, appears to have threatened Colom's position with his pilot book – *De Lichtende Colomne ofte Zeespiegel* – a title which clearly meant a challenge to Colom. After Jacobsz's death his productions were continued by his two sons, Jacob and Casparus, who are known by the name of Lootsman, in a series with Dutch, English and French text, from 1643 to 1715.

This brought Pieter Goos and Johannes Janssoon into the field, who also acquired and retailed Jacobsz's work. A fourth publisher, Hendrik Doncker, appeared on the market with a very similar book of charts and started to print charts of coasts outside Europe in 1670. Until then these areas had been rather a neglected field in the industry.

It was not only the knowledge of navigation but also the commercial enterprise of Amsterdam printers and chart-makers that made this city an international centre of marine cartography in the seventeenth century. The Dutch East India Company, with its widespread overseas enterprises, formed a central organisation in Amsterdam and was responsible for a major share of the marine surveys.

The *Zee-Fakkel* (1681) of Johannes van Keulen was the culmination in the development of Dutch pilot books. Nowhere did this development attain such heights as in Amsterdam. Although John Seller had started his *English Pilot* in 1671, English seamen were still mainly dependent on the output of Amsterdam publishers. The same was true with regard to France, and so we find Amsterdam publishers printing pilot books in English and French. Gerard van Keulen, son of the founder of the van Keulen publishing house, gradually enlarged the *Zee-Fakkel* – with the exception of Part Six, which was not allowed to appear in print until 1753. Though the sea-chart business was very much a family affair, none of Gerard van Keulen's successors appear to have brought out any new editions. The House of van Keulen was the largest unofficial hydrographic office in the world from 1678 to 1885, though it had outlived its reputation by the nineteenth century.

In the second half of the eighteenth century the official hydrographic departments of the English and French admiralties were instituted. Their work must be regarded as scientifically superior to the traditional surveys of the Dutch navigators.

Old sea atlases have attracted a great deal of interest in recent years, not only because of their beauty and documentary value but also for what they reveal of the history of the art of navigation. If ever maps should be labelled as functional tools, sea maps or charts have the first claim to the title. Van Keulen's large *Zee-Fakkel* or *Sea Torch* is the most attractive and most informative of all sea atlases ever printed. It contains over 240 double-page charts showing in detail navigable waters, coasts, rivers, estuaries, inlets and harbours known in the first half of the eighteenth century.

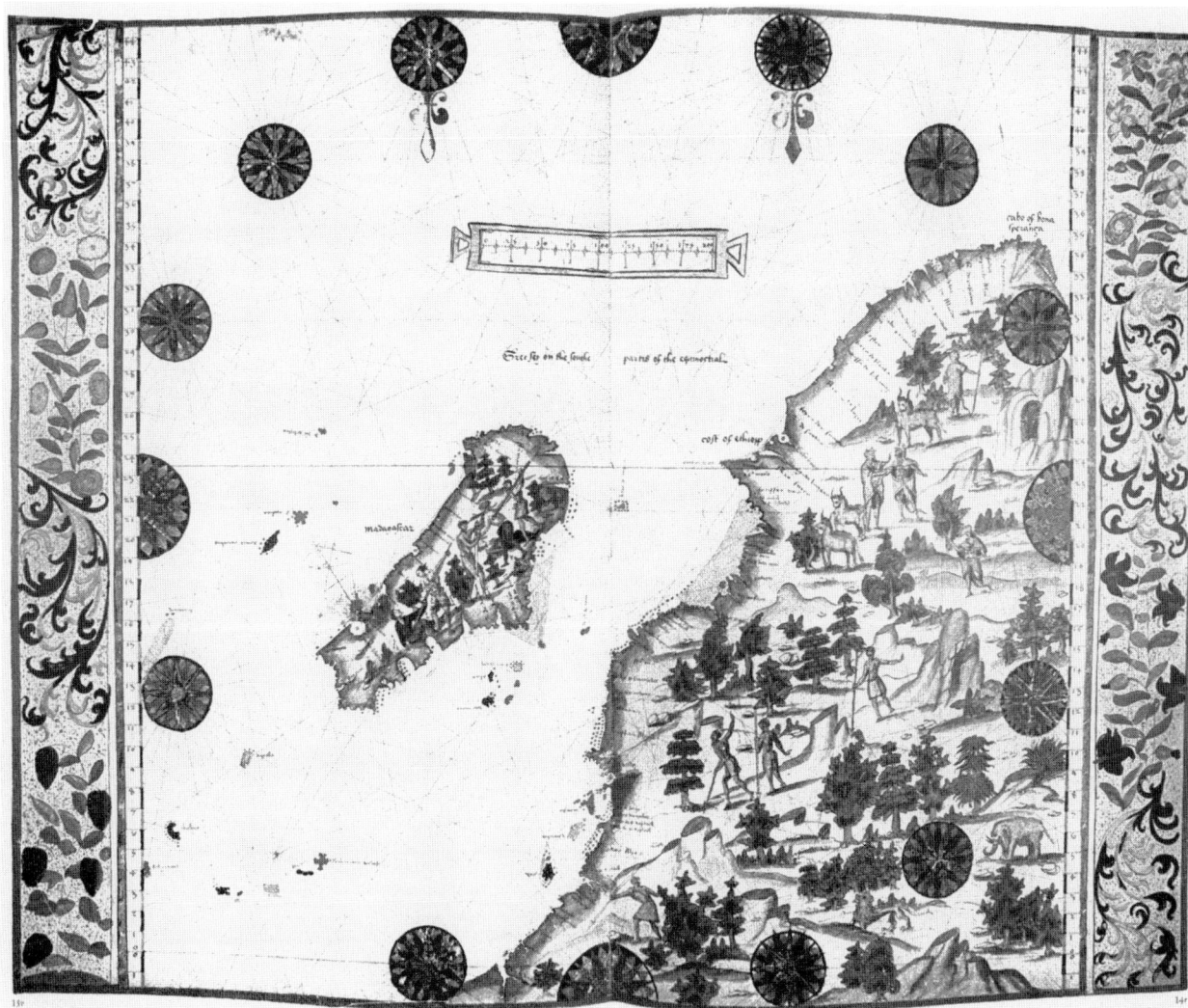

Map 238
ROTZ, Jean (c. 1505-1560)
(West Africa, Southern Africa, South-east Africa and Madagascar.)
Three sea charts, 58 x 74 cm each, coloured.
In: Rotz, Jean. *Boke of Idrography dedicated and presented to Henry VIII in 1542*. Facsimile edition
published by the Oxford University Press for the Roxburghe Club, 1982.

The original manuscript of this famous maritime atlas, *The Boke of Idrography*, was dedicated and presented to Henry VIII by Jean Rotz in 1542, and came to the British Museum as one of the treasures of the Royal Library in 1757. Now in the British Library, to which the library collections of the British Museum were transferred in 1973, this atlas was chosen by Viscount Eccles to be reproduced for presentation to the Roxburghe Club, a well-known group of literary notables and personalities in Great Britain and the United States of America. Three hundred copies of this facsimile atlas were printed by the Oxford University Press in 1982,

and it is considered one of the most finely produced books of the twentieth century.

Rotz claimed noble Scottish ancestry through his father, David Ross, who left Scotland to settle in Dieppe. David Ross's mercantile activities had brought him into contact with the maritime entrepreneur and wealthy ship-owner of Dieppe, Jean Ango, who was the guiding spirit of a host of seamen who set France on the path to maritime expansion. Jean Parmentier, the first Frenchman to pilot a ship to Brazil, was one of Ango's navigators.

269

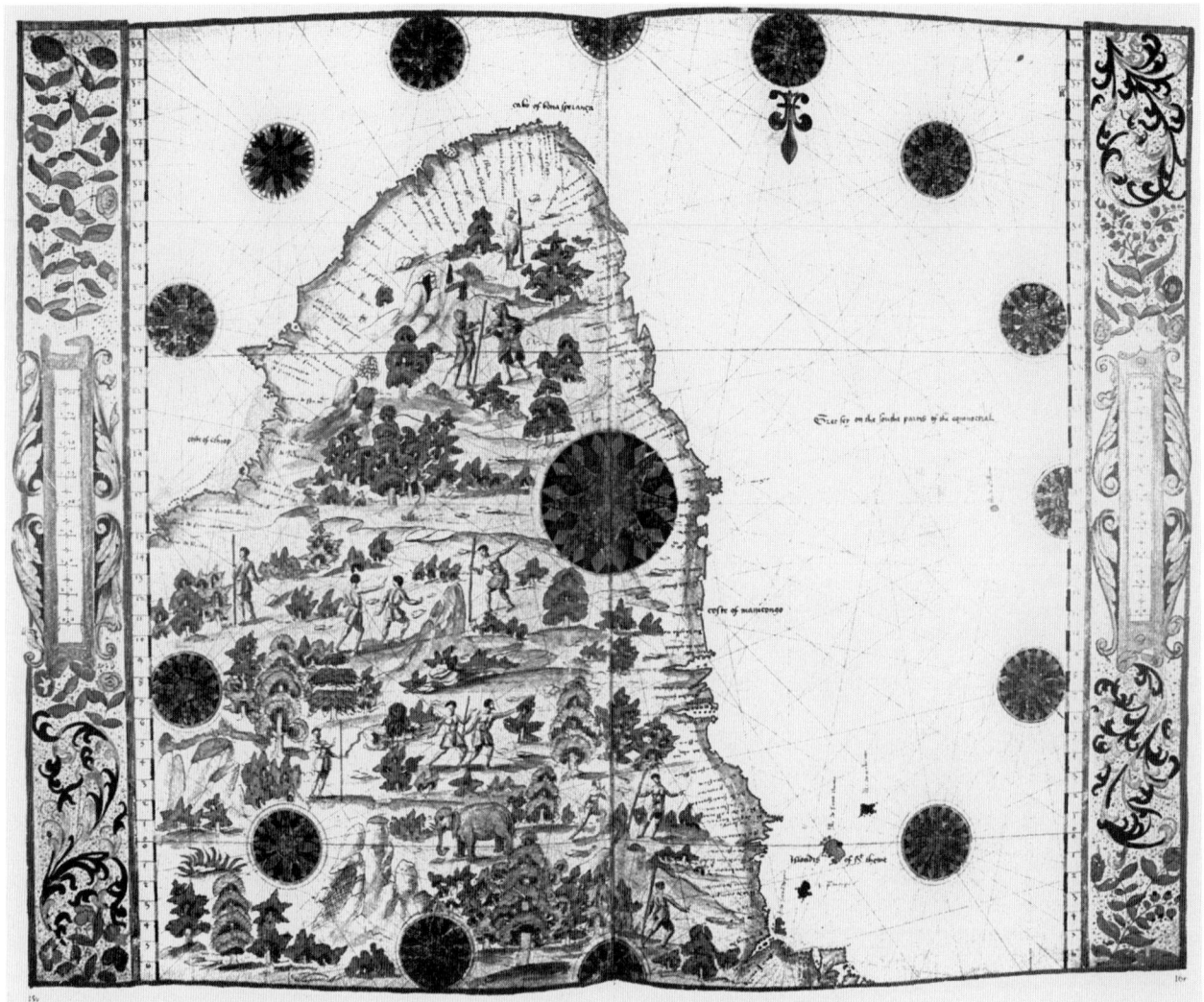

In his youth Jean Rotz, as he was known in France, sailed in Ango's ships. His first recorded voyage was in 1529-1530 – the expedition of Jean Parmentier to Sumatra – and he thus gained professional skills in one of the finest schools of European seamanship.

Parmentier sailed for Sumatra with two vessels, the *Sacré* and the *Pensée*, leaving Dieppe on 3 April 1529. They arrived at Ascension Island on 29 May and named it Ile de France, apparently unaware that it had been discovered twenty-eight years previously by the Portuguese navigator, Juan de Nova. Continuing south, they reckoned that they had rounded the Cape on 23 June and on 24 June they landed at Madagascar, where three Frenchmen were murdered by the inhabitants. In August they were among the Comoro Islands and in September they reached one of the most southerly of the Maldives. Eventually, on 20 October, the ship arrived off the western coast of Sumatra. After various experiences in Sumatra and its environs they prepared to depart on the first day of 1530. On reaching the Cape they lay at anchor for about a month and saw herds of horned cattle with people tending them. The ships obtained provisions at St. Helena and travelled home to Dieppe together.

Rotz appears to have been attracted to the service of Henry VIII by the high salaries which Henry offered to foreign experts to further his policy of building up his navy and strengthening his coastal defences. To win Henry's favour Rotz presented him with a compass of his own invention and a navigating manual in manuscript describing its use. He was rewarded with the post of Royal Hydrographer and an annuity of £40, a higher salary than that of Hans Holbein as Court Painter. Following the desired effect of these gifts Rotz completed and presented the *Boke of Idrography*, in which he could now describe himself as 'Servant to the King's Most Excellent Majesty.'

As Royal Hydrographer from 1542-1547 Rotz probably supervised the charting of the south and east coasts of England, for in 1547 he could claim to be the leading authority on the country's ports and landing places. It is possible that some of the navigating instruments recently retrieved from the wreck of the *Mary Rose*, sunk in 1545 in English waters, were provided by Rotz.

When Henry VIII was known to be dying Rotz became fearful for his future and sought to return to France, offering

270

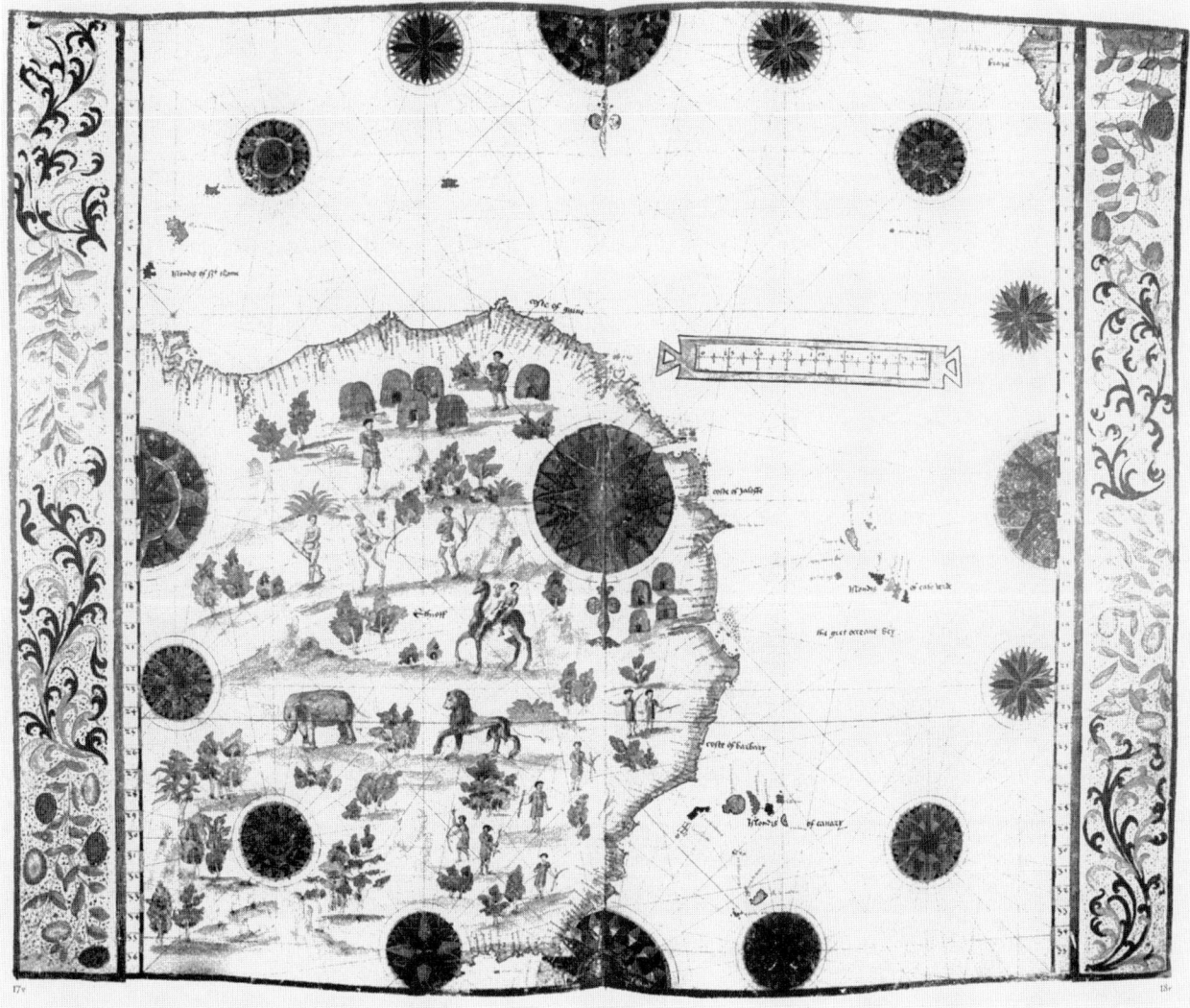

as an inducement to Henri II his knowledge of English and Scottish affairs, and maps marking all their ports. He thus became the leading merchant captain of Dieppe, building ships for the French navy and engaging in the maritime affairs and battles of France until 1560, when his name disappears from the records.

The *Boke* is undoubtedly the earliest major work of Dieppe hydrography still surviving today. It is a manuscript hydrographical atlas made up of sixteen large sheets of vellum, each measuring approximately 59,5 x 77 cm. These sheets are written and painted on one side only, centrally folded to form two folios and individually guarded into the covers. Emblazoned across the title page are the arms of Henry VIII and underneath is written: 'This Boke of Idrography is made by me Johne Rotz, Sarvant to the King's Mooste Excellent Majeste. Gode save his Majeste.' The second folio bears Rotz's dedication to Henry, written in French.

All the sheets except some of those given over entirely to text or bearing the Royal arms and devices have elaborately decorative borders, which all appear upside down in relation to the map. It may be that Rotz was not entirely clear in his own mind which way up his maps were finally to appear. These borders are embellished with designs of blue and gold acanthus interspersed with roses and strawberries, and sometimes with ornamental vases or birds, also upside down. The text is illustrated by diagrams and supplemented with tables. The introductory pages provide instructions to mariners, and the main part of the atlas comprises eleven regional charts covering the world, drawn with the south at the top, on a standard scale of 10° to 123 mm. Helen Wallis states in her article (*The Map Collector*, September 1982) that if the charts were fitted together they would make a large map of the world, 3,965 x 2,135 m (13 x 7 ft). Rotz completed his atlas with a world map in two hemispheres, drawn in 1542 and representing a later state of knowledge than the original charts.

The maps of southern Africa and the Guinea coast included here are highly decorative, with a vertical band of intricate floral design on either side; within each is a

271

longitudinal scale with an attractive floral border. South Africa is depicted as extending from 'Cabo of Bona Spezanca' to 6 or 7 degrees south. There are numerous rhumb lines traversing the seas, with the unusual appearance of five compass roses within the land mass and another ten in the seas surrounding the subcontinent, of which one includes a north-pointing fleur-de-lys.

These charts are characteristically the work of a seaman. Rocks, shoals and sandbanks are minutely marked and islands are distinguished in different colours. Only discovered coast names are shown.

The Rotz maps are much more than hydrographical charts. They display the indigenous flora and fauna, with fascinating scenes of native life. Parts of the coast known to have been frequented by the French are identified by French flags or larger lettering, and Rotz's occasional large lettering in English was probably intended to draw

Henry's attention to major coastal regions and important islands.

In the Guinea map the figures are mostly unbearded males. One type is characterised by a smock to the knees, with bare head and feet, and carrying spears or bows and arrows. The other type wears only a loincloth. A group of huts with a large opening for entrance and apparently constructed in three or four horizontal layers of what appears to be thatch, is illustrated on the Guinea coast. These huts, not repeated elsewhere, do resemble a type of round hut common in parts of Guinea.

The most remarkable figures are those on the two maps of southern Africa. They are dressed in animal skins, complete with the animal's head over their own. These vignettes represent the Hottentots at the Cape of Good Hope, dressed thus as a protection against cold. First sighted in the 1480s by the early Portuguese explorers, the Hottentots were described in a number of manuscript accounts over the next fifty years. The earliest known depiction is the famous series of engravings by Hans Burgkmair, executed to illustrate Balthasar Springer's brief account of his journey to the East in 1505-1506. The one engraving, of a man, woman and child, is reproduced in Walter Hirschenberg's *Monumenta Ethnographica* (Graz, 1962). (See illustration.) The Hottentots in the Rotz maps would seem to be among the very few figures ever shown wearing the head of an animal together with the rest of the skin.

References:

Strangman, Edward. *Early French Callers at the Cape*. 1936.
Schefer, C.H. ed., *Discours de la Navigation de Jean et Raoul Parmentier*. Paris, 1883.
Wallis, Helen. 'The Rotz Atlas, a Royal Presentation,' *The Map Collector*, Sept. 1982.

IN.ALLAGO

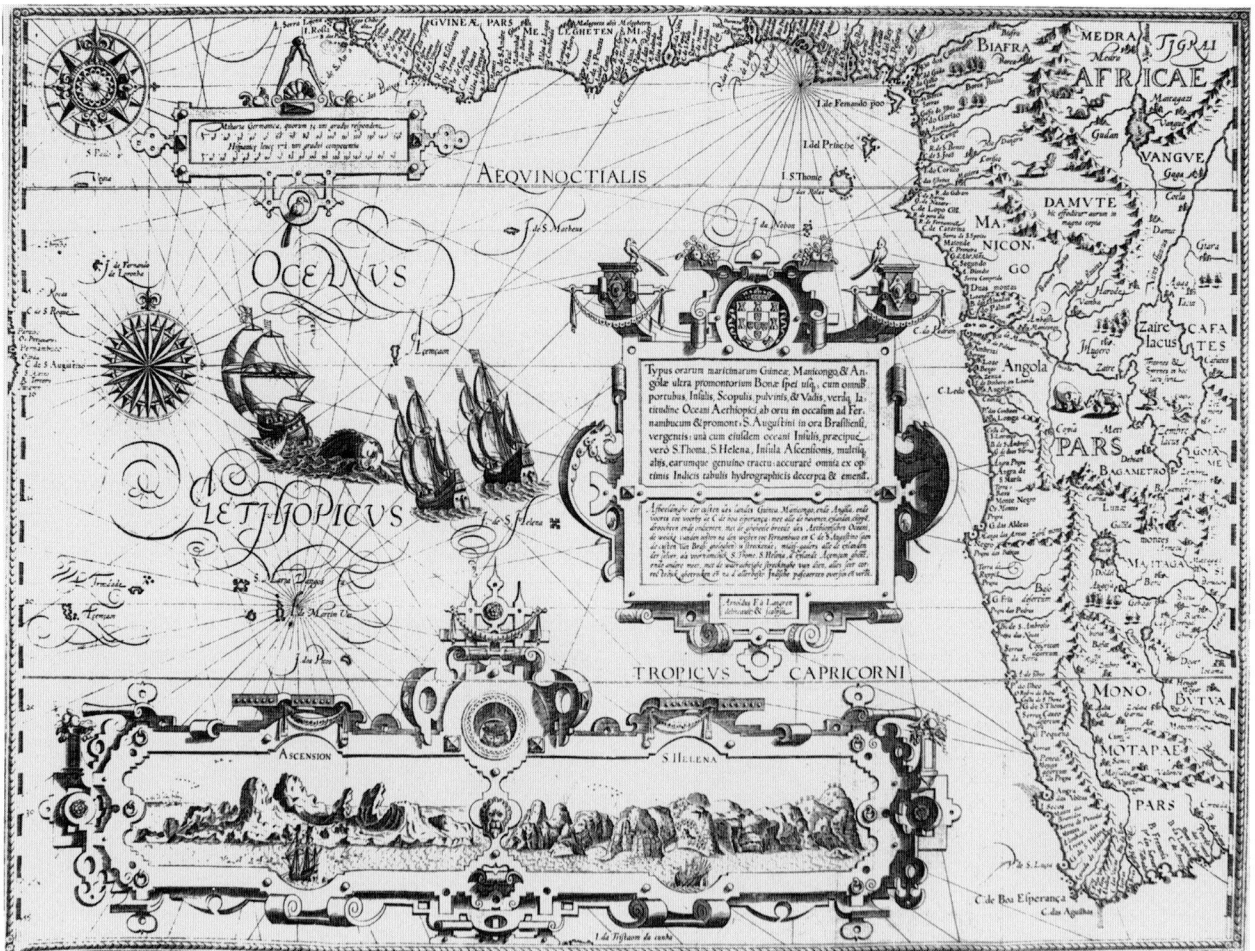

Map 239a

LANGREN, Arnold Florent van (1580-1644)
***Typus orarum maritimarum Guinae, Manicongo & Angolae ultra
promontorium Bonae Spei...*** (Marine map of Guinea, Manicongo and Angola
to the Cape of Good Hope...) Engraved by A. Langren. (Amsterdam, A. Langren, 1596.)
Sea chart, 38 x 53 cm, uncoloured.
Scale in German miles and Spanish leagues.
Inset: Views of Ascension and St. Helena.

This map and Map 239b are companions, together forming a map of Africa and the surrounding seas. The western map is from the Dutch edition, published in 1596, and the eastern one is from the English edition, published in 1598. They are known as the Linschoten maps, and are celebrated for their decorative features, delicate engraving and attractive embellishments.

Jan Huygen van Linschoten (1563-1611), well known as a Dutch voyager, was born in Haarlem. He travelled to the Indies in the service of the Archbishop of Goa and on his return in 1592 he retired to Holland and wrote an account of his travels, the *Itinerario*, which was published in Dutch in 1596 and in Latin in 1599. Subsequent editions of the book appeared in Dutch, in 1605, 1614, 1623, and 1644, and in French in 1610, 1619 and 1638.

Florent van Langren (or Langeren), one of a Dutch family of cartographers and globe-makers, published the first Dutch edition in 1596. The English edition (1598), of the maps alone, was engraved by Robert Beckit (as is noted on the cartouche of Map 239b) and published by John Wolfe in London. The author has a copy of the first Dutch edition of the *Itinerario* with the east coast map, which has the same reference to Langren as appears on the east coast map here; but which, having been bound as a fold in the book, would not have reproduced as clearly as the separate English edition.

The cartouche of the western map is rectangular, surrounded by decorative scrollwork and surmounted by the Portuguese coat of arms with birds, fruit and flowers to its sides. Below is a delicately engraved inset of Ascension Island

274

and, to its right, St. Helena. There are two compass roses of differing design on the left, with a scene of three ships in full sail surrounding a whale. On the mainland are an elephant, a rhinoceros and a lion and an unusual feature is the introduction of sirens in Lake Zaire, with an appropriate legend. The fictitious cities of Vigiti Magna and Monomotapa are marked, and the names of cities and mountain ranges give a good indication of Portuguese nomenclature, which was vague until European settlement of the southern extremity of the continent. The lettering is beautifully engraved and clear, with fine flourishes of the larger letters in the Flemish style.

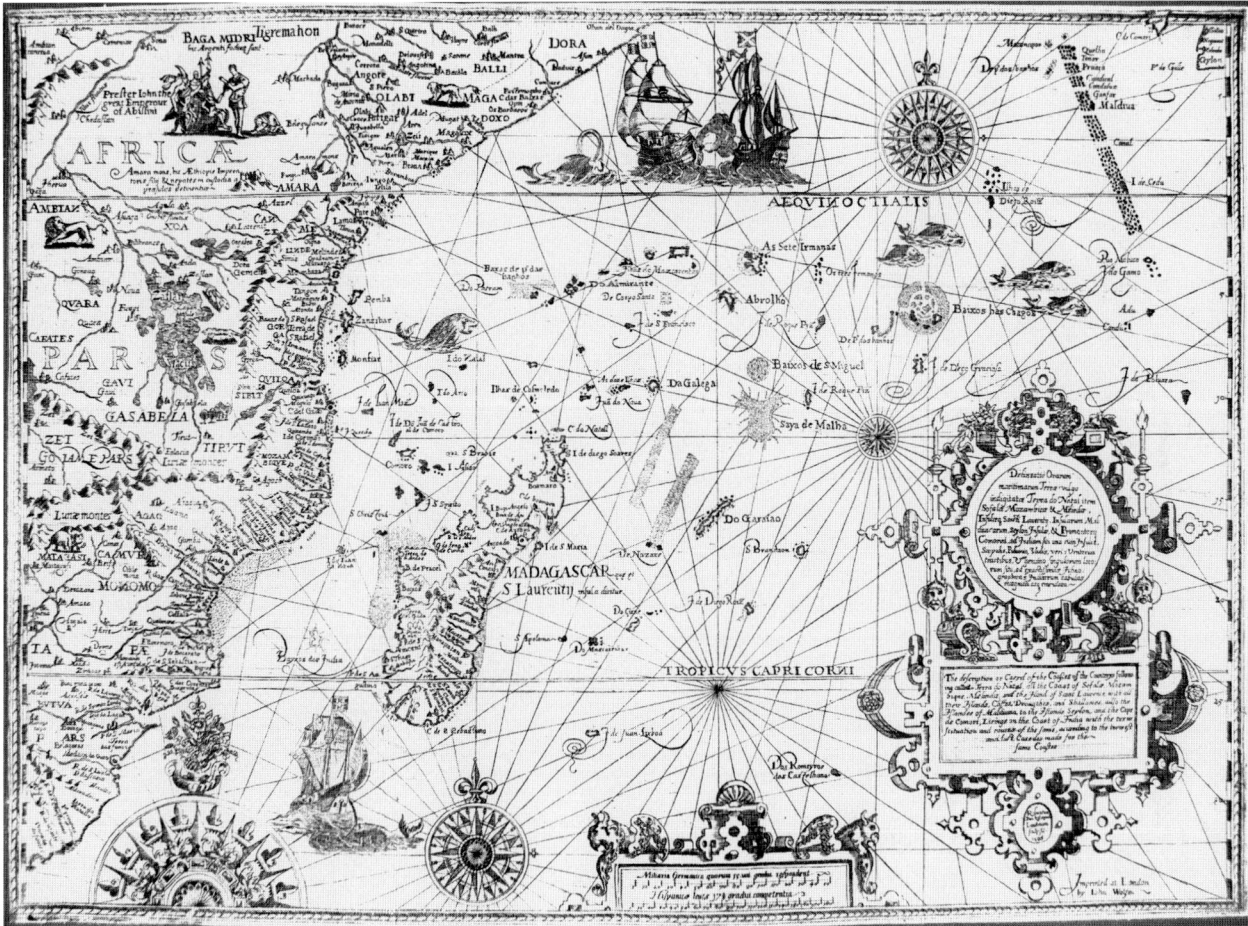

Map 239b
LANGREN, Arnold Florent van (1580-1644)
Delineatio Orarum maritimarum Terrae vulgo indigitatae Terra do Natal item Sofala Mozambicae & Melindae... (Marine map of Natal, Sofala, Mozambique and Melinde...) Engraved by Robert Beckit. (London, John Wolfe, 1598.)
Sea chart, 38 x 53 cm, uncoloured.
Scale in German miles and Spanish leagues.

This map, the companion part to the west of Africa (Map 239a) is the English edition of 1598. It has a decorative cartouche of a circle and a rectangle connected by delicate and attractive scrollwork with, at the bottom, a smaller circle giving the name of the engraver and the date. The scale is to the left of the cartouche at the foot of the map and is decorated with two dragon heads. A wind-rose appears to the top left of the cartouche and there are two fair-sized compass roses at the top and bottom of the map with a further half of one of larger proportions on the lower left border. There are three ships in the sea, with a number of whales and dolphins, and on the mainland are lions and an elephant. 'Prester John the great emperour of Abyssinia' appears with his entourage. The rivers (including the Spirito Santo, reputed by geographers to be the Limpopo of today), the larger lakes, and the mountains (including the Lunae Montes) are typical of the maps of that time.

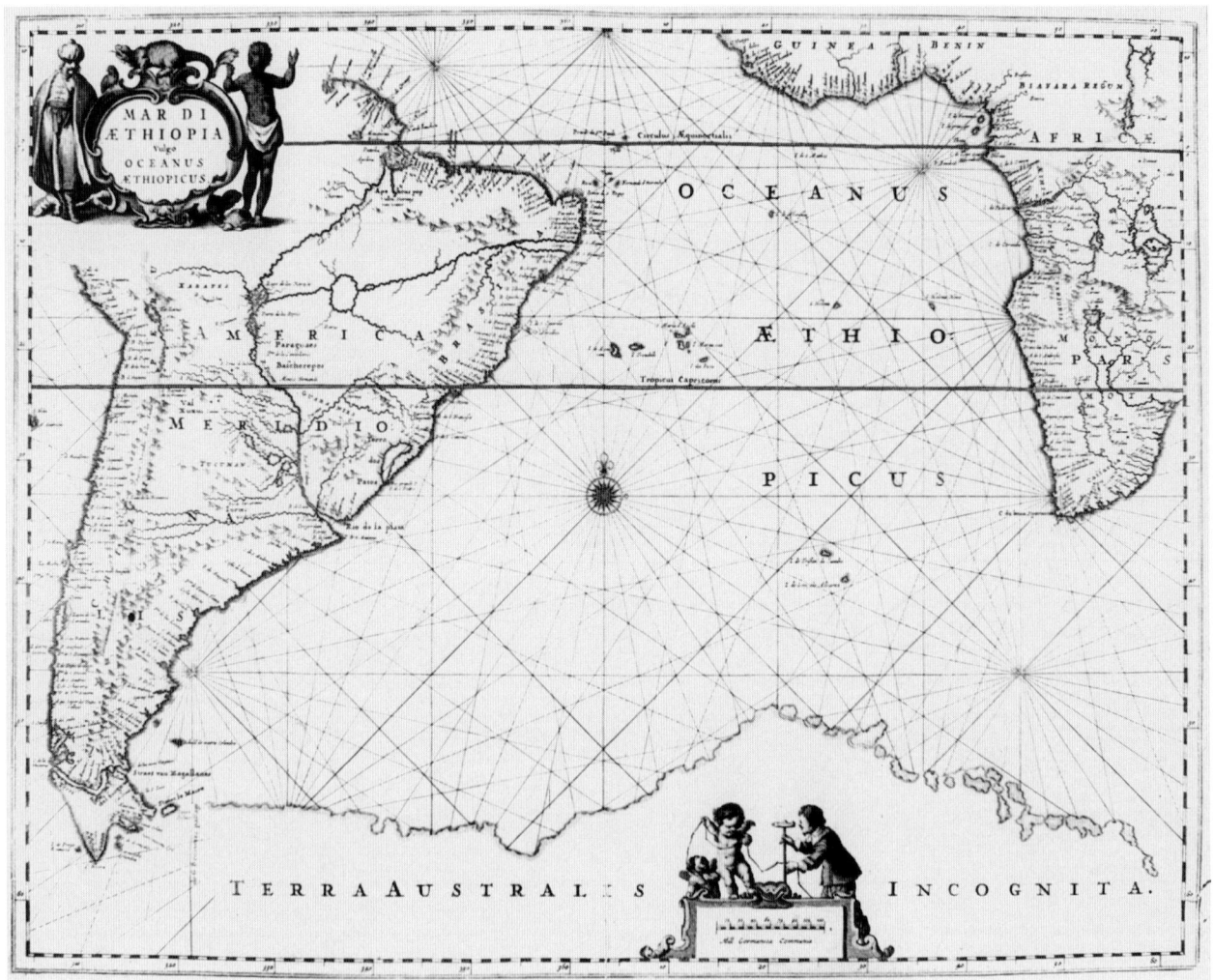

Map 240

JANSSON, Joannes (1588-1664)
Mar di Aethiopia vulgo Oceanus Aethiopicus. (Sea chart of the Atlantic Ocean.)
(Amsterdam, J. Jansson, 1650.)
Sea chart, 43 x 56 cm, slightly coloured.
Scale in German miles.
Latin description on verso.

The sea chart of the Atlantic Ocean featured here first appeared in Jansson's *Atlantis Majoris* and includes almost the whole of South America and the western and southern coastlines of Africa.

The scale appears in a decorative box at the lower border, surmounted by a cherub holding what appears to be a measuring tape, with a man in dress of that period holding a vertical stick with a small cross piece at the top, probably also a measuring device. There is a compass rose in the centre of the ocean between the two continents. The cartouche appears in the top left corner and has a Latin title within an oval shield flanked by two figures. On the left is a man in Moorish costume and on the right, a semi-naked African with a bird perched on his right hand. On the top of the shield is

another bird and a large lizard-like reptile. An elongated land mass along the lower border is labelled *Terra Australis Incognita* – probably one of the earliest representations of Australia.

Jan (John) Jansson married Elizabeth Hondius, daughter of Jodocus I. On the death of Jodocus II, Jansson, together with his other brother-in-law, Hendricus, published a series of atlases. On the death of Hendricus, Jansson continued expanding the atlases, one of which was the *Novus Atlas*. The first sea atlas in the real sense of the word was published by Jansson in 1650 as the fifth volume of his *Atlantis Majoris* – Latin text. Jansson is also noted for his pilot guides. As early as 1620 he had already published the *Licht-der Zeevaert*, which was copied from Blaeu's original work.

276

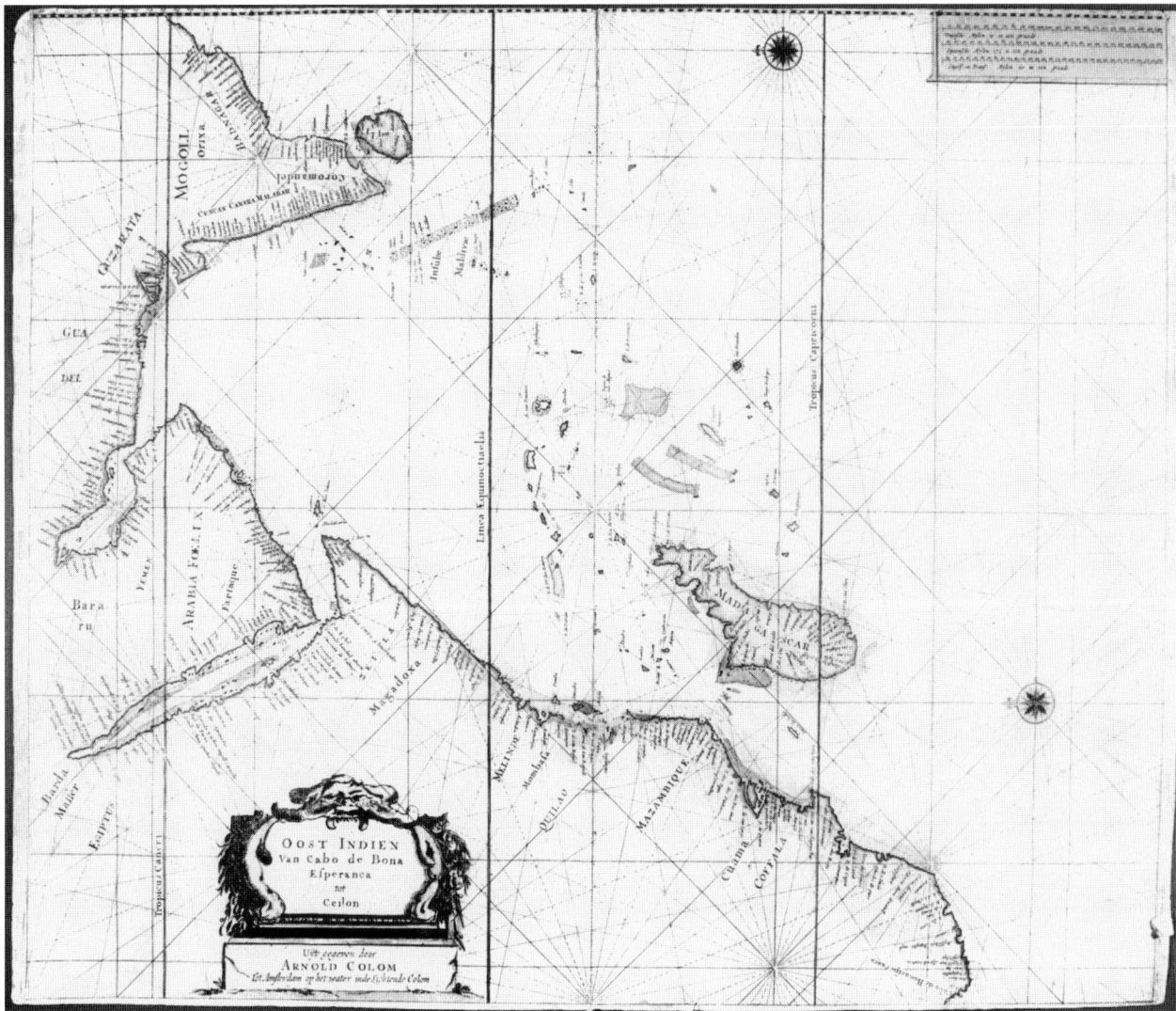

Map 241

COLOM, Arnold (1624-1668)
Oost Indien van Cabo de Bona Esperanca tot Ceilon. (East Indies from
Cape of Good Hope to Ceylon.) (Amsterdam, Arnold Colom, 1654.)
Sea chart, 63 x 54 cm, coloured.
Scale in Dutch, Spanish, English and French miles.

This marine chart from the *Zee Atlas* shows the coast of
Africa from the Cape of Good Hope to just beyond the Red
Sea. It includes a portion of Arabia, the Middle East, India
and Ceylon. The title is depicted on the left of the chart
with a decorative cartouche. The top of the cartouche shows
a large open-mouthed animal with its fangs overlapping the
top edge.

The chart itself has two compass roses in the lower Indian
Ocean, the one to the right of the tip of southern Africa and
the other far to the east in mid-Indian Ocean. The
geographical details of the coastline of all the continents
involved appear to be accurate. Africa includes Caffala,
Cuama, Quiloa, Mombasa, Melindi, Magodoxa and Zeila.
The equator and the Tropic of Capricorn are prominently

featured. Intercrossing rhumb lines radiate in profusion from
many sea harbours, the most important from the Natal coast
and the Comoro group of islands just to the north of
Madagascar.

Arnold Colom was a Dutch chart publisher, engraver and
bookseller. He issued a *Zee Atlas* in 1654 and *Lighting Colom
of the Midland Sea* in 1660. His father, Jacob Aertz Colom,
was an important figure in the history of sea charts. His book
of these in 1632 brought the pilot book publishing trade a
step nearer to its greatest achievement in van Keulen's work.
He was the first to reduce the folio pilot book to a popular
size and its success made him the leading publisher in the
first half of the seventeenth century, putting Blaeu out of
the market.

277

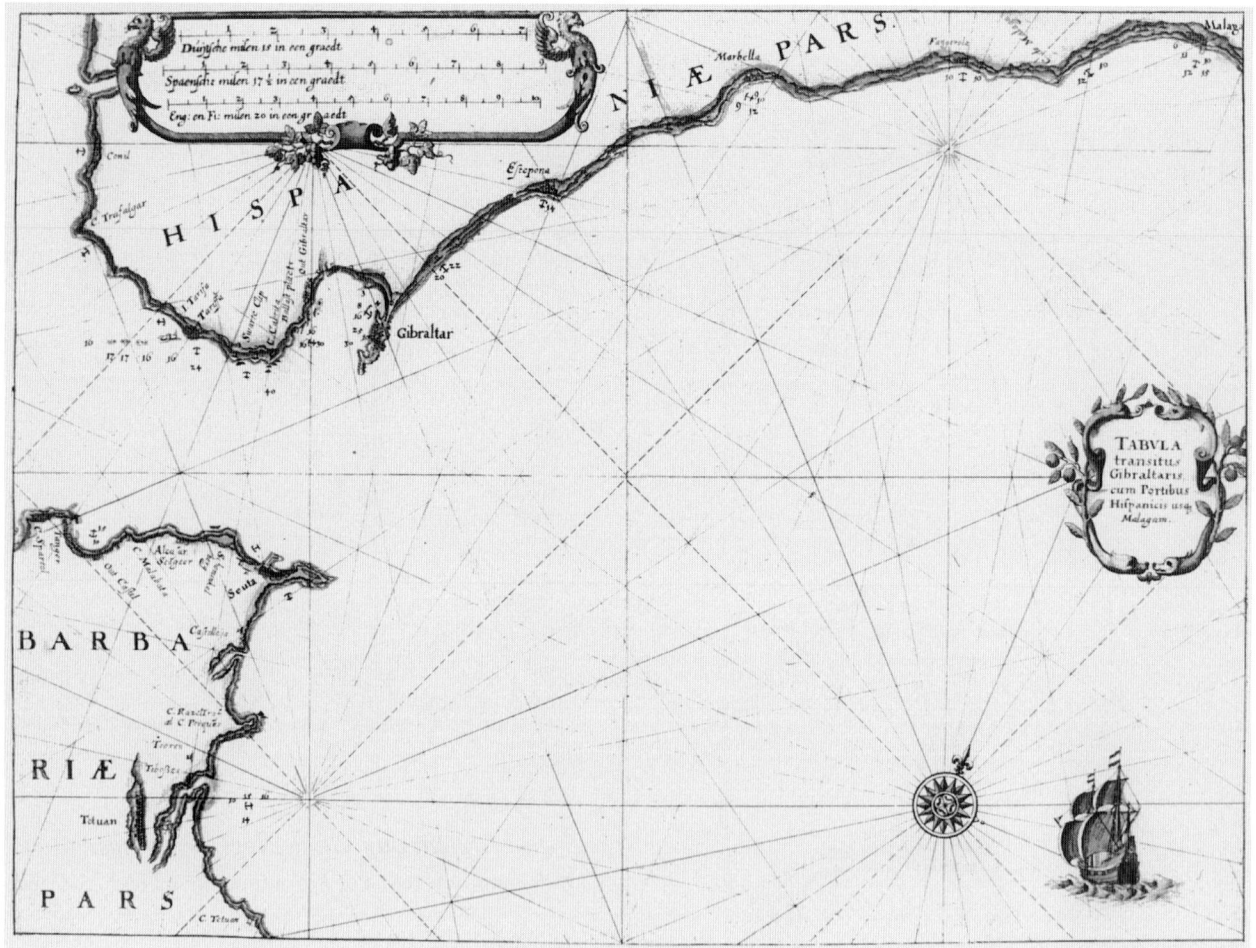

Map 242

MERIAN, Matthäus (1593-1650)
Tabula transitus Gibraltaris cum Portibus Hispanicis usq. Malagam. (Map of the Straits of Gibraltar, Spanish and Barbary coasts.) (Frankfurt, M. Merian.)
Sea chart, 26 x 36 cm, coloured.
Scale in Dutch, Spanish, English and French miles.

This chart is reputed to be by Matthäus Merian, the German topographical engineer and publisher, who reissued a number of Blaeu charts.

This chart is apparently based on Blaeu's 'De Strata van Gibraltar met die Spaansche Cust van daar tot Malaga.' It outlines the southern part of Spain and some of the North African coast, with detailed soundings. The Spanish coast extends as far as Malaga, and the African coast from C. Spartel in the west to C. Tetuan in the east. Cape Trafalgar is shown opposite C. Spartel.

At the right lower corner a sailing ship is illustrated alongside the north-indicating compass rose. Above this is a rectangular cartouche, the border of which is made up of four serpent-like animals, intertwined with a floral design.

278

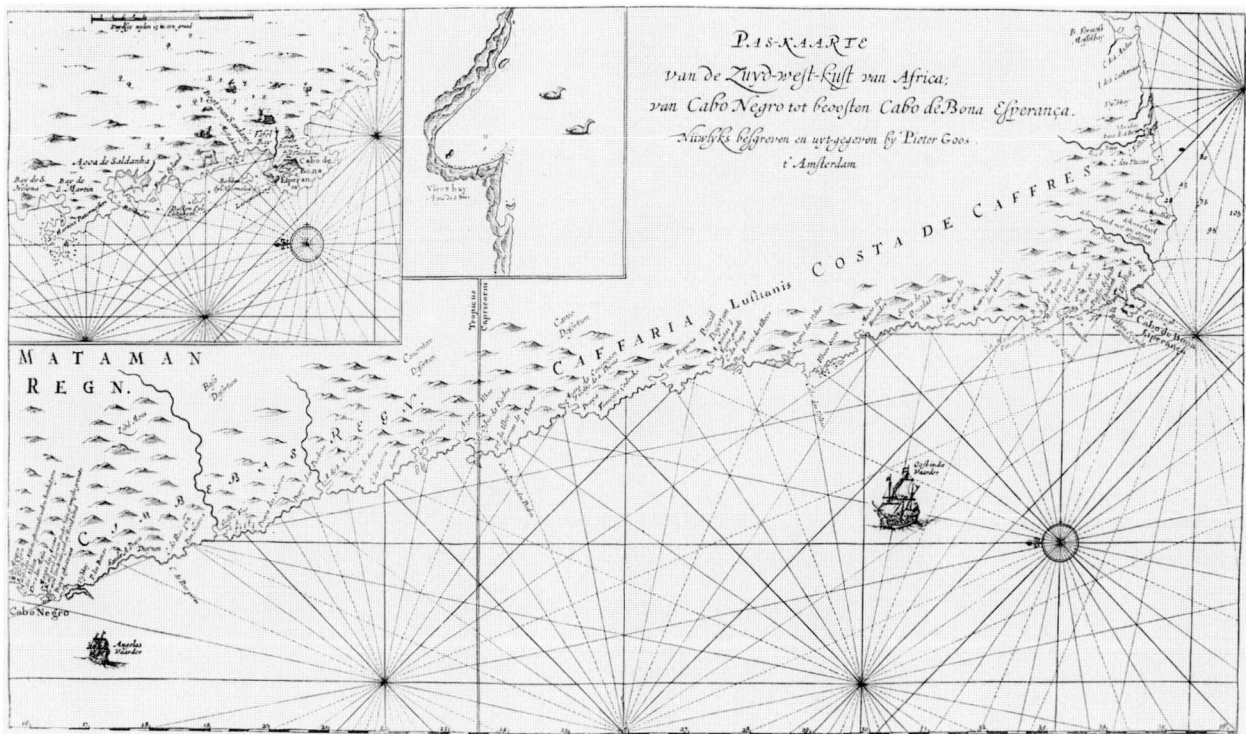

Map 243

GOOS, Pieter (c.1616-1675)

Pas-karte van de Zuyd-west-kust van Africa; van Cabo Negro tot beoosten Cabo de Bona Esperanca. (Map of the south-west coast of Africa from Cape Negro to Cape of Good Hope.) (Amsterdam, P. Goos.)

Sea chart, 29 x 52 cm, uncoloured. Scale in Dutch miles.

Inset: Vlees Bay. Cape of Good Hope.

This sea chart of the west coast of Africa from Cape Negro to the Cape of Good Hope, issued by Pieter Goos, is a fine example of the seventeenth-century marine chart. It is a decisive copperplate engraving with a great amount of detail of the coastline from the Cape of Good Hope in the south to Cape Negro in the north. Further detail is provided in the insets of the Cape of Good Hope and Vlees Bay. There is a compass rose with the north-pointing fleur-de-lys at the lower right, and another in the inset of the Cape. Two sailing ships appear in the north and south, one labelled 'Angolas Vaarder' and the other 'Oost-India Vaarder,' both flying flags.

There is an identical sea chart attributed to Jodocus Hondius, which appears in the *Klare Besgryving van Cabo de Bona Esperanca* of 1652, which is a rare pamphlet showing accurately the knowledge of South Africa possessed by the Europeans in the year that van Riebeeck landed in Table Bay. A facsimile edition was published for the celebration of the tercentenary of van Riebeeck's landing in 1652.

Goos, who was a cartographer, engraver and print-seller, published many sea atlases of a high standard, although he depended on other chart makers for much of his information. He succeeded his father, Abraham, also an engraver and mapseller. Goos's atlases included the *Nieuwe Groot Zee Spiegel*, *De Zee Atlas ofte Watter Wereld* and many others, and were published in Dutch, English and French. After his death his widow continued the business and reissued the *Nieuwe Groot Zee Spiegel* in 1676.

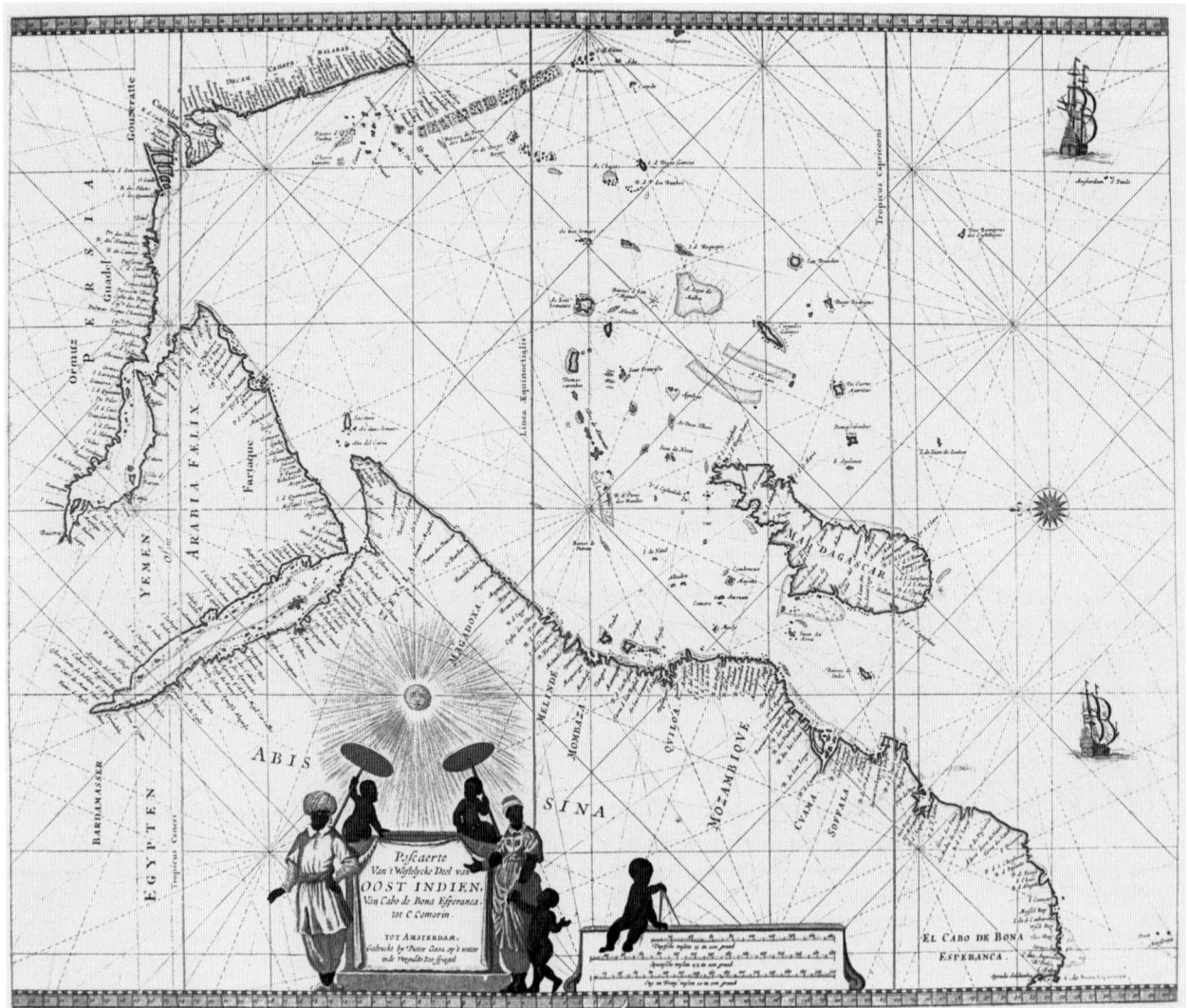

Map 244

GOOS, Pieter (c. 1616-1675)
Pascaerte van 't Westelycke Deel van Oost Indien, van Cabo de Bona
Esperanca tot C. Comorin. (Map of the western part of the East Indies from the
Cape of Good Hope to Cape Comorin.) (Amsterdam, P. Goos, l654.)
Sea chart, 44 x 54 cm, coloured outline.
Scale in Dutch, Spanish, French and English miles.

This sea chart is of the east coast of Africa including the coastline of southern Arabia and the west coast of India. It is an attractive, hand-coloured chart with much detail of the coasts.

The cartouche and title on the lower border depict two Africans in oriental dress standing alongside the drapery on which the title is printed. There are four black children – two seated on top of the masonry holding the drapery, each holding an umbrella protecting them from a strong, bright sun. Another child sits on a stand illustrating the scale,

holding a fine measuring caliper in his right hand. The remaining child is standing holding hands with one of the adults.

There are two sailing ships in the southern sea and a compass rose between them with the usual fleur-de-lys pointing northward.

The title informs us that the map was published in Amsterdam by Pieter Goos in the *Vergulde Zee Spiegel* (Golden Sea Mirror).

280

Sir Robert Dudley's Maps of Africa.

The Dudley maps (245-251) shown here were originally published in his *Arcano del Mare* (Secrets of the Sea), which was the first sea atlas to be issued by an Englishman. It appeared when Dudley was 73 years old, towards the end of a most remarkable career filled with experiences which made him one of the greatest map makers of all time.

Robert Dudley was born in Surrey, England, in 1574, the illegitimate son of the Earl of Leicester. Leicester assumed full responsibility for the child and provided him with a good education, but deprived him of his inheritance to the title. Under the tutorship of Thomas Chaloner, a scientist with a knowledge of shipbuilding, Dudley developed a great love of the sea. On coming down from Oxford, having inherited the property but not the title of his uncle, he began to have ships built and manned. Queen Elizabeth I declined to support his first proposed grandiose maritime adventure because of his tender age. He therefore had to content himself with a modest voyage to the West Indies for which purpose he fitted out a fleet of four vessels at his own expense. Now in his twenty-fourth year, and fortunate in having a master mariner, Abram Kendal, as his pilot, he set off across the Atlantic from the African coast. After gaining a great deal of navigational and maritime experience and skill, Sir Francis Drake gave him command, in the 1580s, of the ship *Cadiz*.

In 1596 he married Alice, daughter of Sir Thomas Leigh. Now resident in London, he began to manage his own estate, and planned commercial ventures by sea. But he began to resent his questioned legitimacy. Soon after the death of Queen Elizabeth he started an action to prove himself born in wedlock, but the court ruled against him.

Dudley reacted vigorously against this decision. He left London in 1605 at the age of thirty, never to return to England. He had to find employment and decided to offer his services to the Grand Duke of Tuscany, Ferdinand I, who was delighted to obtain the services of such a talented naval man. In addition to acting as a naval architect he began to improve the port of Livorno. He was responsible for many new fortifications, and also built a canal and drained marshes.

As he gained increasing favour at the Tuscan Court, the Tuscan minister in London was instructed to use his influence to have Dudley reinstated. Dudley also sought the services of James I and Prince Henry to assist him, but the Prince's premature death put an end to his chances of reinstatement. As a result of these continuing failures Dudley became increasingly obstinate and cantankerous. He was, however, at the height of his fame as an Italian nobleman in his country of adoption.

After the death of his second wife he went into semi-retirement and began to concentrate on his writing. This consisted of two works, the *Dirretto Marittima* and the *Arcano del Mare*, a compendium of all the naval and maritime knowledge of the day. The *Arcano* was originally published in three volumes divided into six parts or books. This, his principal work, was published in Florence in 1646 in small folio size, and contained numerous charts, diagrams and plates of astronomical instruments. The volume was dedicated to Ferdinand I of Tuscany. This monumental sea atlas, the first by an Englishman, was one of the most important cartographical works ever produced and the first to contain a series of charts laid down on the Mercator projection.

This great sea atlas was not the creation of Dudley alone. The Italian craftsman Antonio Francesco Lucini was the master engraver who produced the most brilliant engravings. The clear decisive outline with fine florid calligraphy in the Florentine style which was characteristic of the Lucini engravings is the most impressive lettering in the entire history of map making. The magnitude of the project of engraving these plates is suggested by Lucini himself, who says in the dedicatory epistle that he worked on them for twelve years, in seclusion in an obscure Tuscan village, using no less than 5000 pounds of copper in the process. The beauty of these maps seen in their black and white impression is a perfect example of the art of copperplate engraving.

It is not difficult to imagine that the *Arcano* was widely acclaimed, and it went (posthumously) into its second edition in 1661, Dudley having died in 1649.

Reference:

Ritchie, Neil: 'Sir Robert Dudley: Expatriate in Tuscan Service.' *History Today*, June 1976.

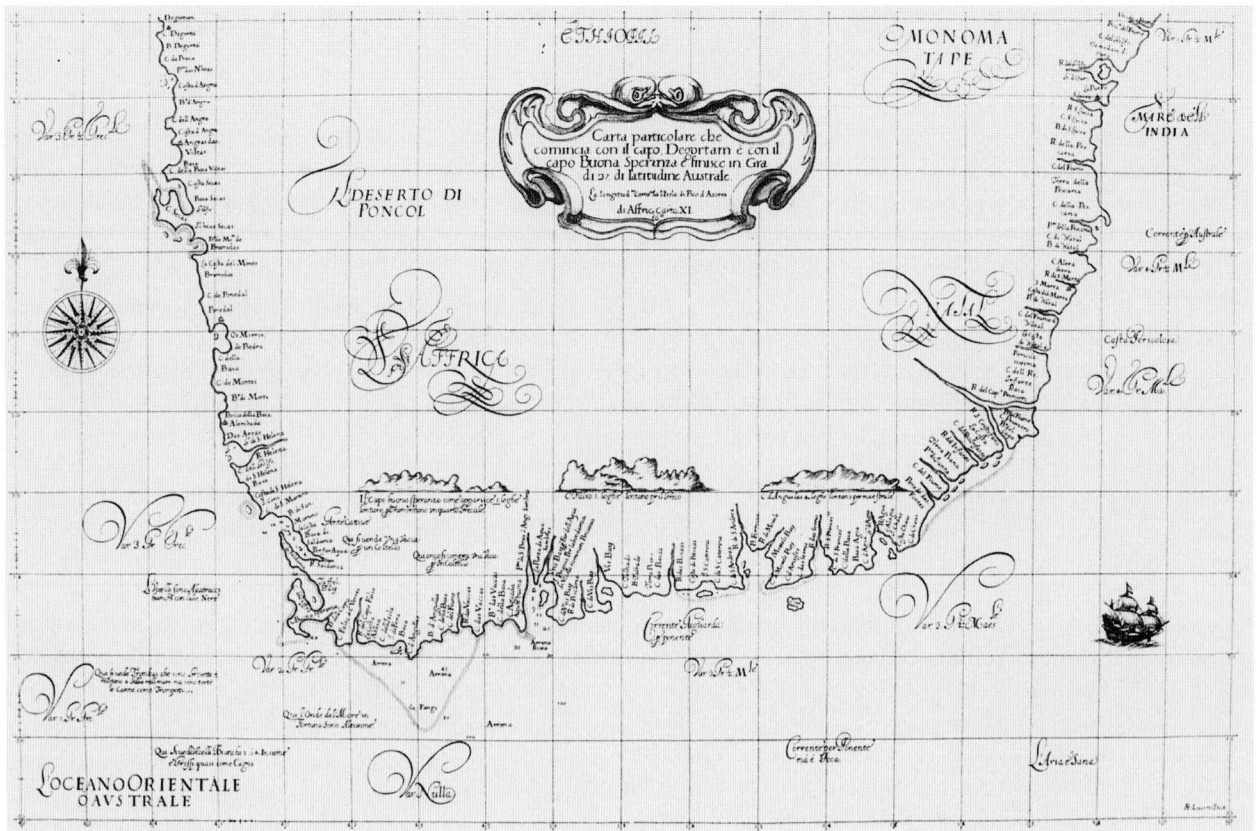

Map 245

DUDLEY, Robert (1573-1649)
Carta particolare che comincia con il Capo Degortam è con il capo Buona Speranza è finisce in Gradi 27 di Latitudine Australe. (Map of South Africa from Cape Degortam round the Cape of Good Hope to lat. 27° S.) Engraved by A. F. Lucini. (Florence, R. Dudley, 1646-7.)
Sea chart, 46 x 73 cm, uncoloured.

This fine sea chart first appeared in Book VI of Dudley's *Arcano del Mare* in 1646-7, and was reissued with the second edition of 1661.

It has the conventional Dudley cartouche scroll in the centre, at the bottom of which appears 'di Affrica Carta XI 16°.' The word 'Affrica' on the left of the map is in florid lettering, as is almost all the lettering on the mainland and in the sea.

An interesting feature in the empty southern part of the continent is the three silhouettes of mountain ranges labelled, from left to right, 'Il Capo Buona Speranza....' 'C. Falso...' and 'C. d'Anguilas.' At the lower right appears a small, neatly engraved sailing vessel. Bays and capes are labelled with the lettering placed inland, leaving the coastline clear. Table and Saldanha bays are marked.

This chart is a more common edition of South Africa than Map 246 (which shows Madagascar), and is found more readily in catalogues of Dudley charts.

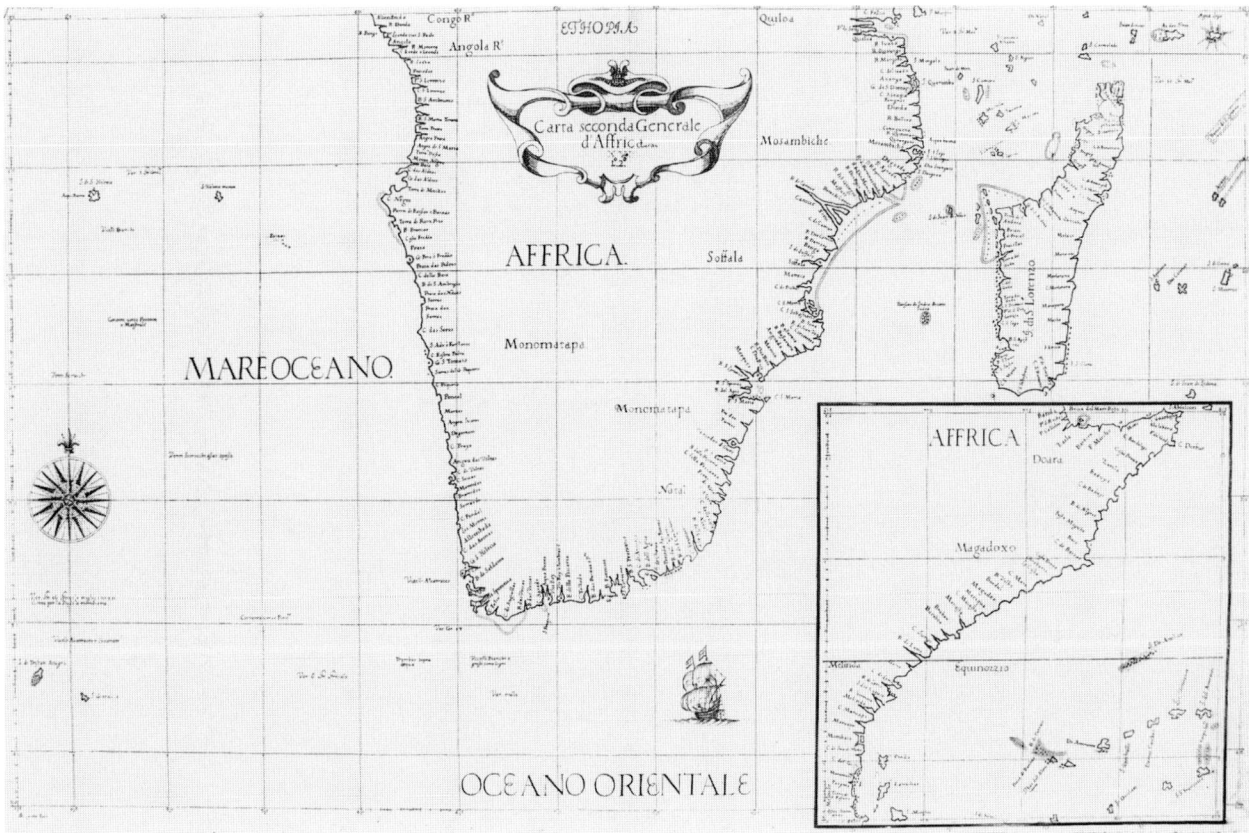

Map 246 DUDLEY, Robert (1573-1649)
Carta Seconda Generale d'Affrica. (Map of southern Africa.) Engraved by A.F. Lucini.
(Florence, R. Dudley, 1646-7.)
Sea chart, 45 x 72 cm, uncoloured.
Inset: Somaliland.

This chart giving Dudley's second version of southern Africa is unusual because until recently it was practically unknown; nor was it publicised in articles on maps of Africa. It had been in the author's possession for many years when this publication prompted him to research its provenance.

In visits to the Map Room of the British Library the previous map of southern Africa (see Map 245) was located in Book VI (the atlas volume) of the *Arcano del Mare*. It seemed likely that the two versions appeared in different editions of 1646, and it was only by chance, by thumbing through all the *Arcano* atlases, that this elongated map was found - not in the atlas volume, but in the text of Book II, opposite p. 39. It transpired that this map appears in both editions (confirmed by the Philips Library of Congress Catalogue) and the Grenville edition, although a smaller folio size, contains the map, folded down to fit the size of the page.

The map illustrated here extends from Ethiopia in the north and involves the whole coast of southern Africa from the Congo River in the west to Quiloa in the east, including

bays and islands. 'Ethiopia' appears in the interior and Monomotapa is described twice – this may refer to the kingdom and the mountains of the same name, the one being more to the west than usually sited in older maps. Sofala and Natal are also featured on the map. The island of Madagascar is sited in its correct relation to Africa but is described as I. di S. Lorenzo, a name commonly used by the early Italian cartographers and publishers. Above the name 'Affrica' in the centre of the map is an attractive oval scroll cartouche with the lettering 'Carta Seconda Generale d'Affrica' followed by the Roman numeral XVIII. Below this in small print is L.°2°. This oval scroll cartouche is used in almost all the Dudley maps from the *Arcano*. A small inset map appears in the lower right corner depicting the east coast of Africa (the Horn) and what today would be Somaliland (Magodoxa.) There are numerous islands noted in the Atlantic and Indian Oceans, with a small well-defined sailing vessel off the southeastern coast. A compass rose with the traditional fleur-de-lys is sited in the lower Atlantic Ocean.

Map 247 DUDLEY, Robert (1573-1649)
Carta particolare che mostra il Capo buona Speranza con il Mare Verso Ponte è con L'isole di Tristan D'Acunha e di Martin Vaz. (Map showing Cape of Good Hope and islands of Tristan da Cunha and Martin Vaz.) Engraved by A.F. Lucini. (Florence, R. Dudley 1646-7.)
Sea chart, 47 x 74 cm, uncoloured.

This chart of the South Atlantic includes the west coast of Africa from the Cape of Good Hope to approximately l9°S (C. Fria). The cartouche at the lower left shows some variation in shape but the calligraphy of the title and ornate scrollwork are in the same style as in the preceding maps.

The calligraphy is particularly florid in this map, and two sailing vessels appear in the Atlantic, with a compass rose to the left. The Cape of Good Hope and Saldanha Bay are accurately noted and, except for two names, all information is printed inland from the coast.

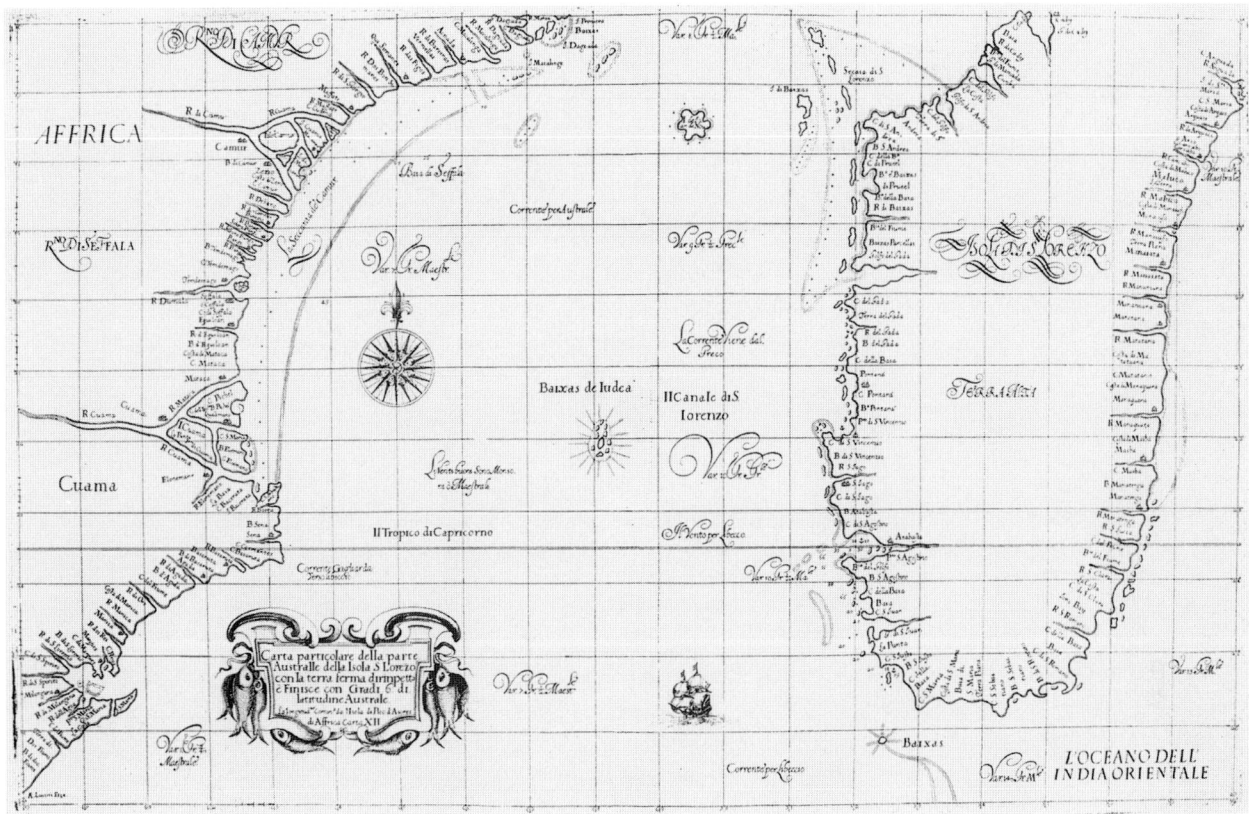

Map 248

DUDLEY, Robert (1573-1649)
Carta particolare della parte Australle della Isola S. Lorenzo con la terra ferma dirinpetto è Finisce con Gradi 6. di Latitudine Australe. (East coast of Africa with the southern part of the Island of Madagascar.) Engraved by A.F. Lucini.
(Florence, R. Dudley 1646-7.)
Sea chart, 47 x 75 cm, uncoloured.

This chart complements Map 245, showing almost all the Isola S. Lorezo (Madagascar), with the coast of Mozambique, where the Portuguese had had a settlement since 1565.

The cartouche with its title appears in the lower left ocean, in a rectangular scroll-like outline with the addition of beautifully engraved fish adorning the sides and lower border. A sailing vessel and compass rose are sited between the mainland of Africa and the Isola S. Lorenzo's eastern coastline. All the well-known Portuguese settlements on the eastern Mozambique coastline are noted, with numerous rivers and bays. As with the other Dudley charts coastal names are sited inland, leaving the sea free of text.

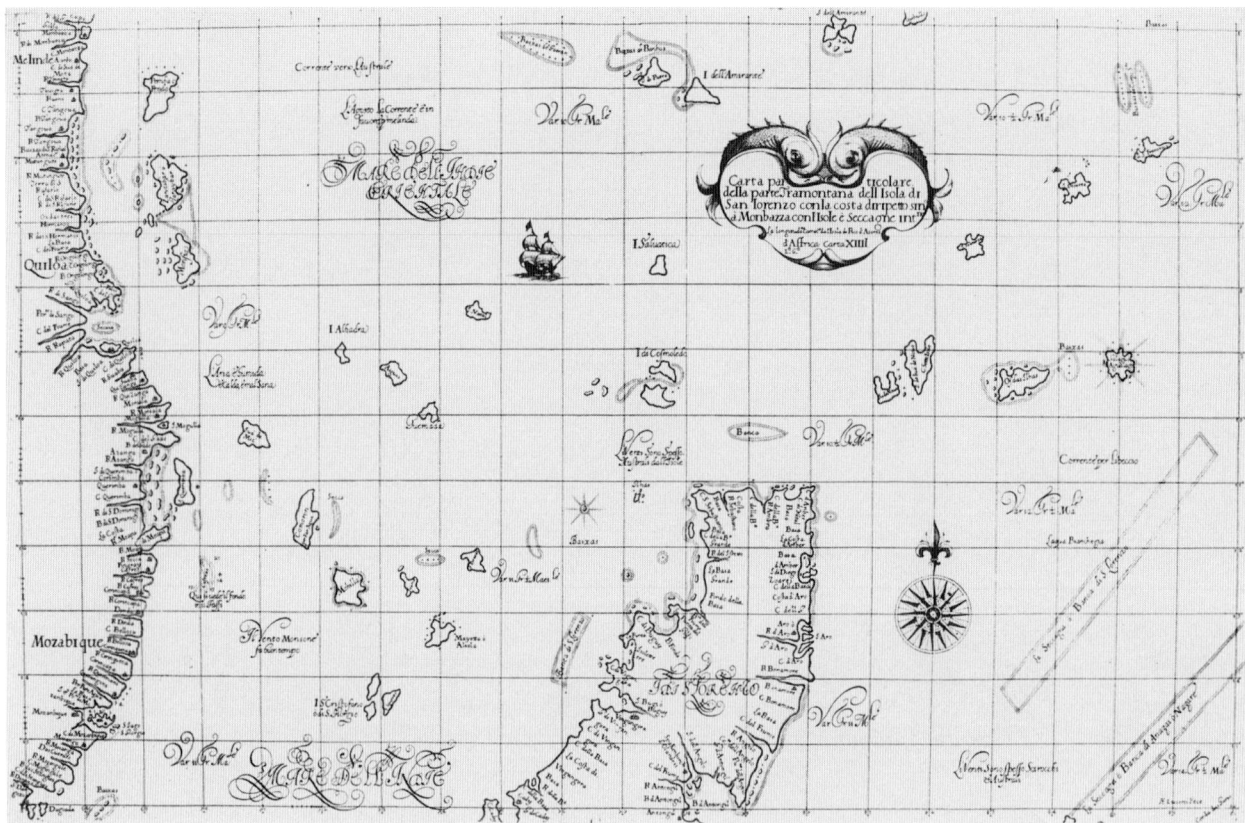

Map 249   DUDLEY, Robert (1573-1649)
Carta particolare della parte Tramontana dell Isola di San Lorenzo con la costa diripetto sin. à Monbazza con l'Isole è Seccagne Int. (East coast of Africa including the northern part of Madagascar.) Engraved by A.F. Lucini. (Florence, R. Dudley 1646-7.)
Sea chart, 45 x 73 cm, uncoloured.

This chart includes the northern aspects of Madagascar and the East African coast from northern Mozambique to Tanzania, Kenya and Zanzibar.

The cartouche is typical of the Lucini style – oval in shape with a scroll effect achieved on the upper border by means of two curved fish. 'Isola di S. Lorenzo' (Madagascar) appears in fine designer's line copperplate engraving. The mainland of the east coast of Africa includes Mozambique, Quiloa and Melinde in the north.

One sailing vessel appears in mid-ocean with a compass rose to the west of upper Madagascar. The numerous islands in these seas are meticulously drawn, each individually named.

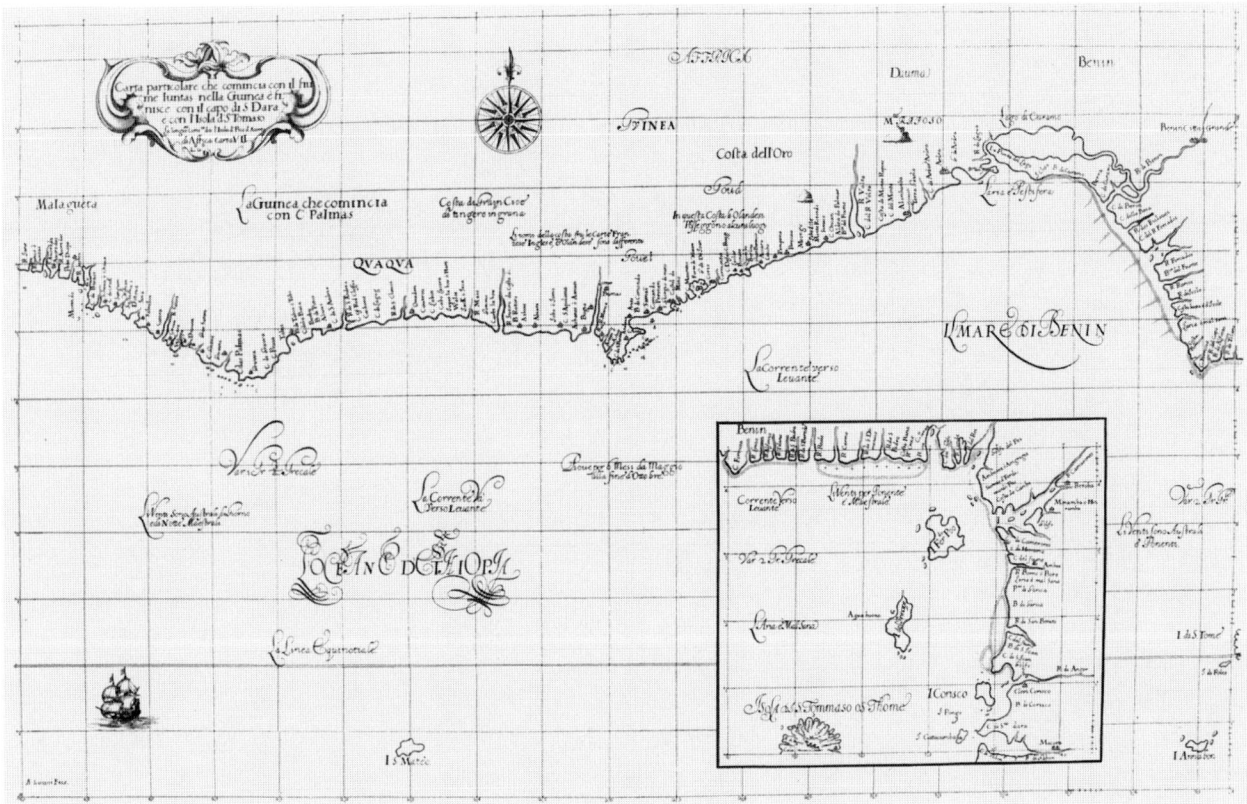

Map 250 DUDLEY, Robert (1573-1649)
Carta particolare che comincia con il fiume Juntas nella Guinea è finisce con il capo di S. Dara è con l'Isola d'S. Thomaso. (West coast of Africa from the river Juntas to Cape St. Dara.) Engraved by A.F. Lucini. (Florence, R. Dudley, 1646-7.)
Sea chart, 46 x 74 cm, uncoloured.
Inset: Island of St. Thomas.

A sea chart created by Dudley and his engraver Lucini of the Gulf of Guinea, the Gold Coast, the Ivory Coast and the Slave Coast. There is a small inset map of the Cameroons and Gabon with the Island of St. Thomas.

On the Guinea coastline mention is made also of the Grain Coast and of the ancient Castel di Mino (Elmina) which was the centre of the gold trade on the Gold Coast.

The oval ornamental scroll cartouche is sited in the upper left corner and a sailing vessel appears at the lower left. There is a compass rose in the upper centre. The fine engraving is of the high standard revealed in all the Dudley charts, confirming that the atlas is undoubtedly one of the world's treasures in the field of cartography, for its beauty and its known accuracy at that date.

Map 251 DUDLEY, Robert (1574-1649)
Carta particolare dell mare Oceano fra l'Ierlandia è l'Isole di Asores. (Chart of the western ocean with the Azores.) Engraved by A.F. Lucini. (Florence, R. Dudley 1646-7.) Sea chart, 46 x 36 cm, uncoloured.

This chart of the Azore islands in the North Atlantic has the usual Dudley oval cartouche containing the Italian inscription. There is a compass rose and a sailing ship between the islands. This chart is one of the half-size series featured in the *Arcano del Mare*.

288

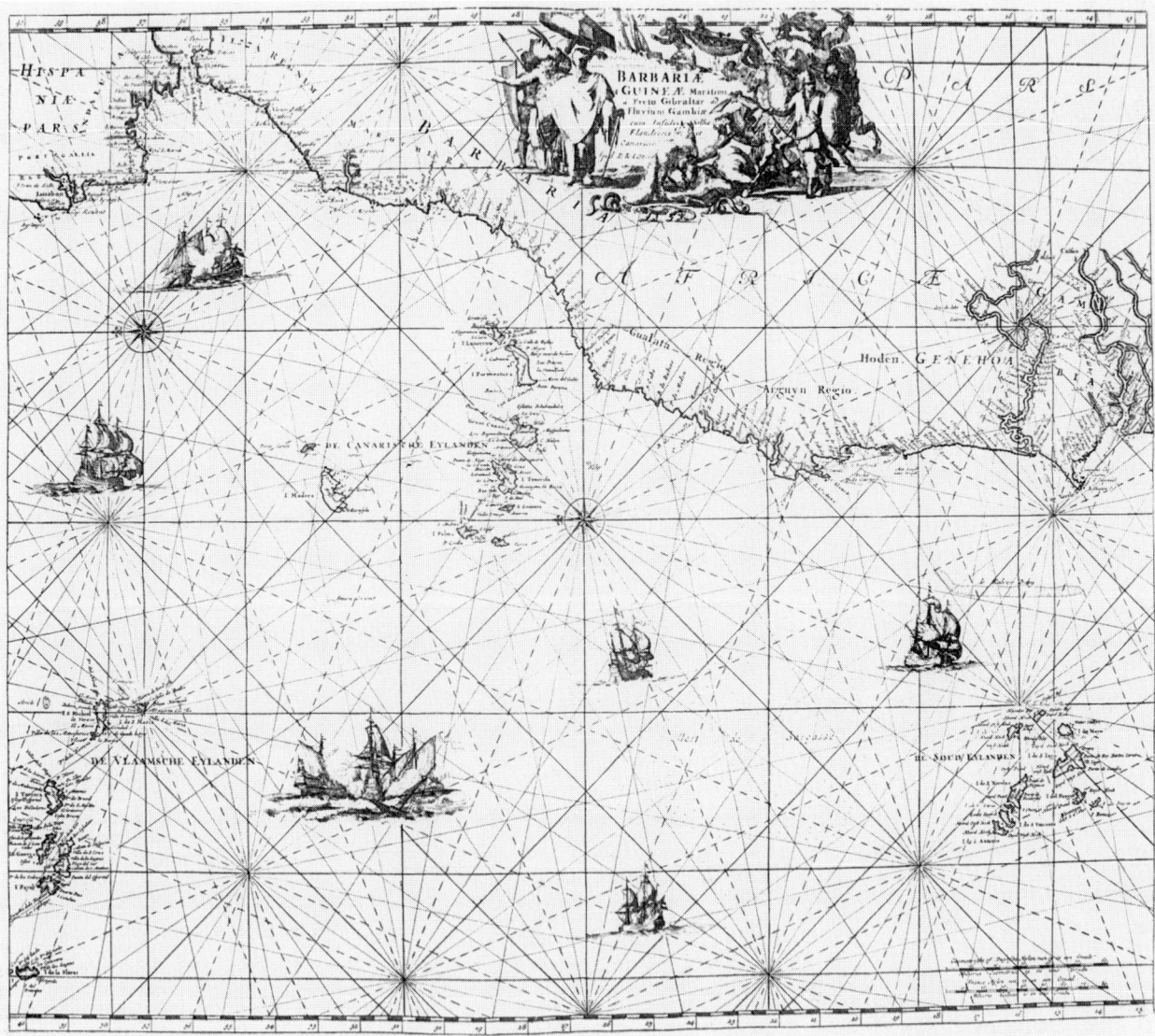

Map 252

OTTENS, Reinier and Josua (fl. 1725-1750)
Barbariae et Guineae Maritimi a Freto Gibralta ad Fluvium Gambiae cum Insulis Salsis Flandricis et Canaricis. (Sea chart of the Barbary and Guinea coasts from the Straits of Gibraltar to Cape Verde.) (Amsterdam, R. and J. Ottens.)
Sea chart, 48 x 56 cm, coloured.
Scale in Dutch and French miles.

This map of the Barbary and Guinea coasts with the southeast coast of Spain and Portugal is a sea chart, with four sailing vessels scattered in the sea, and four others engaged in battle in pairs. Two compass roses are joined by a multitude of radiating rhumb lines. The cartouche depicts an animated scene of men in Moorish clothes and Moorish soldiers on horseback, appearing to be engaged in battle. In the foreground a Moor with a scimitar in his left hand has just completed the decapitation of a white person whose head he holds in his right hand. In front of this scene is a back and side view of a fat-tailed sheep, together with a lizard and snakes. This cartouche, except for the title, is identical to that of the sea chart issued by Johannes van Keulen in

1680 in his *Nieuwe Pascaert van Oost Indien.* (See Map 267.) The use of the fat-tailed sheep in cartouches is described in the author's article 'Fat-tailed Sheep on Maps of Africa' in *The Map Collector* of June 1979. The Canary Islands and those of the North Atlantic feature prominently.

This fine detailed sea chart was issued by the Ottens family, publishers of Amsterdam. The founder of the firm, Joachim, died in 1719 and was succeeded in the business by his widow and his sons, Reinier (who died in 1750) and Josua (who died in 1765). They made plates and maps for Valentyn's *Oost Indies.*

289

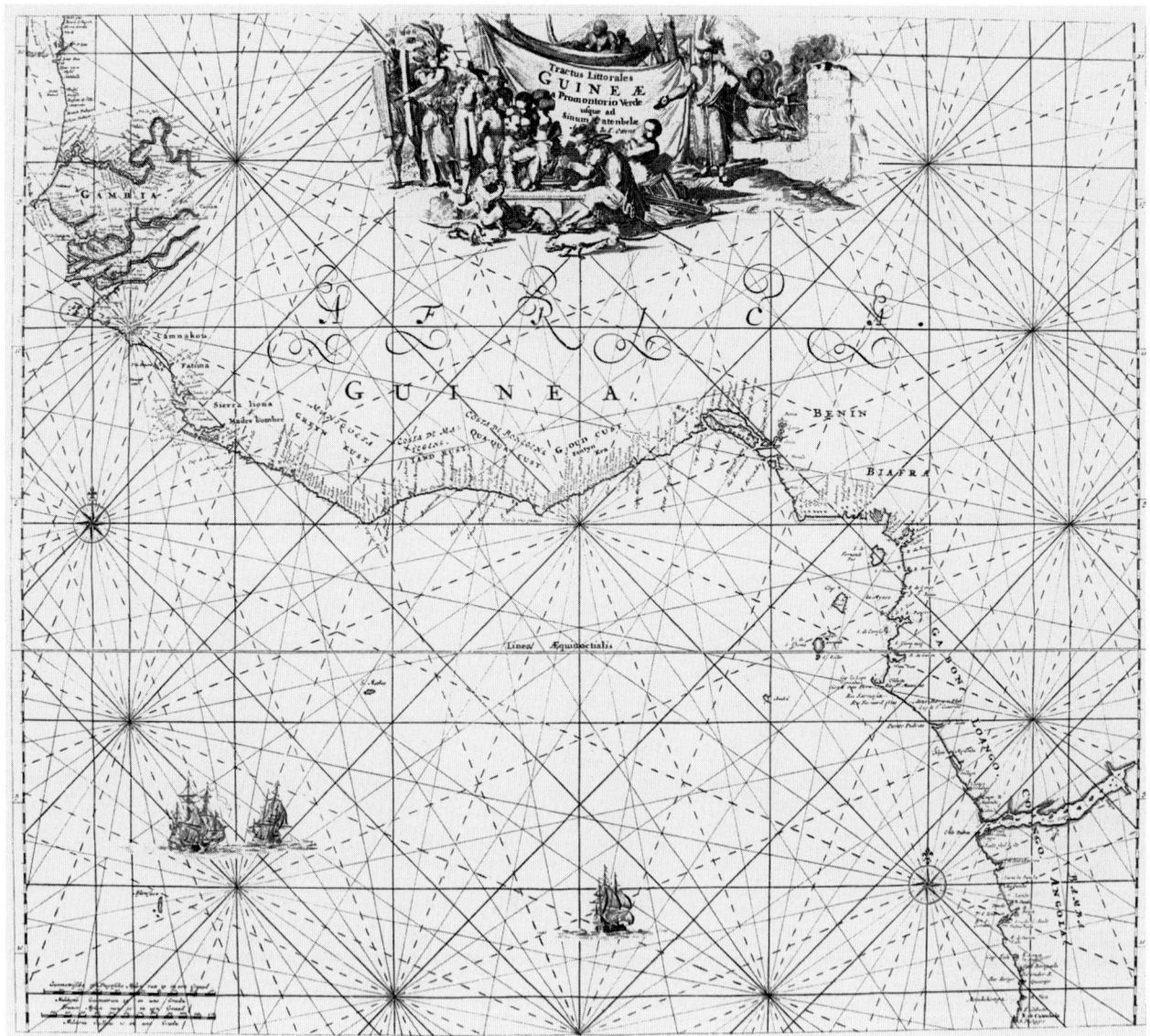

Map 253 OTTENS, Reinier and Josua (fl. 1725-1750)
Tractus Littorales Guineae a Promontorio Verde usque ad Sinum Catenbelae.
(Map of coast of Guinea from Cape Verde to Catenbela Bay.) (Amsterdam, R. and J. Ottens.)
Sea chart, 49 x 56 cm, slightly coloured.
Scale in Dutch and French miles.

This chart of the northwest coast of Africa features mainly Guinea. It is one of a series issued originally by De Wit, and extends from the Congo-Angola region to include the Gold, Ivory and Grain Coasts up to Cape Verde and beyond. There are three ships in the lower seas, and two compass roses with rhumb lines.

The cartouche in the upper centre depicts Europeans, one of whom appears to be holding a scale and is presumably weighing some precious metal. Some black women appear and above the title are an adult and child swinging in a hammock. An undecorated scale appears in the lower left corner.

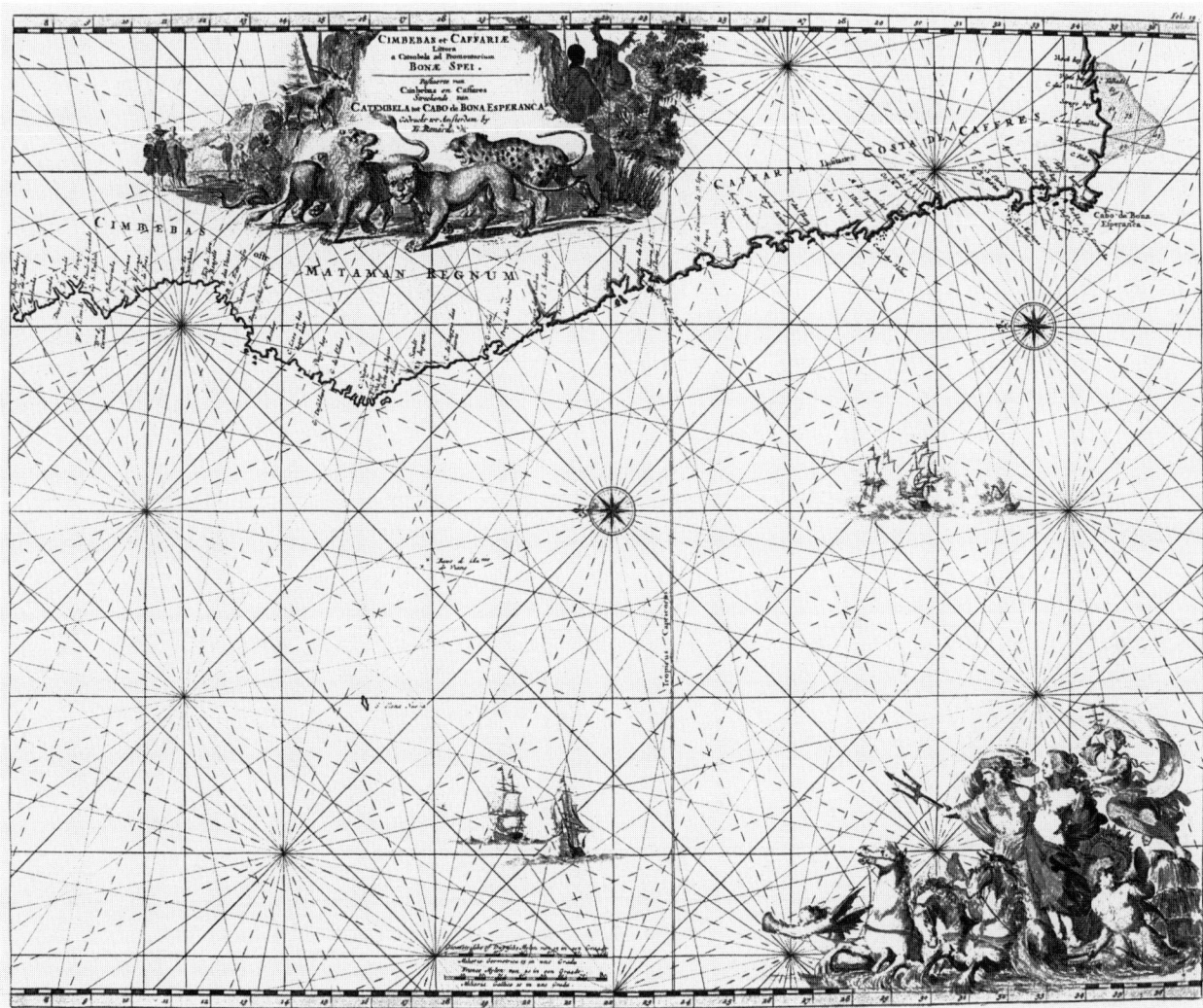

Map 254

RENARD, Louis (fl. early 18th C)
Cimbebas et Caffariae Littora a Catenbela ad Promontorium Bonae Spei.
Pascaerte van Cimbebas en Caffrares Streckende van Catembela tot Cabo de
Bona Esperanca. (Map of Cimbebas and Caffraria from Catembela to Cape of
Good Hope.) (Amsterdam, L. Renard, 1715.)
Sea chart, 43 x 53 cm, coloured.
Scale in Dutch and French miles.
Top right: Fol. 19.

This sea chart is identical to one produced by the Ottens family (see Map 255) except for the cartouche at the lower right, replacing the Cape of Good Hope inset, and the addition of a number of ships.

The title cartouche, in Latin and Dutch, is more or less centred at the top. At the far left appears a group of Europeans apparently conducting trade negotiations. On the cliff which forms the division between this part of the cartouche and a group of three lions and a leopard there is a goat and a little lower down a snake. On the right of the cartouche is another cliff and two partly concealed natives; vegetation surrounds the whole cartouche. In the lower right corner is a highly decorative group consisting of Neptune

and a female figure in a chariot drawn by three fine horses. On either side is a triton blowing his horn.

In the Atlantic there are two groups of ships, two near the centre of the lower part of the map, and three ships to the right engaged in battle. The scale is not decorated. The captions to the scale are in Dutch and Latin, and given in French miles and Dutch geographic miles. The coast is at the top, and the names are sometimes written one way up and sometimes another as appears to be the custom in charts of this style. The coastline shows an unusual number of inlets and the island of S. Elena Nuova is somewhat further south than usual.

291

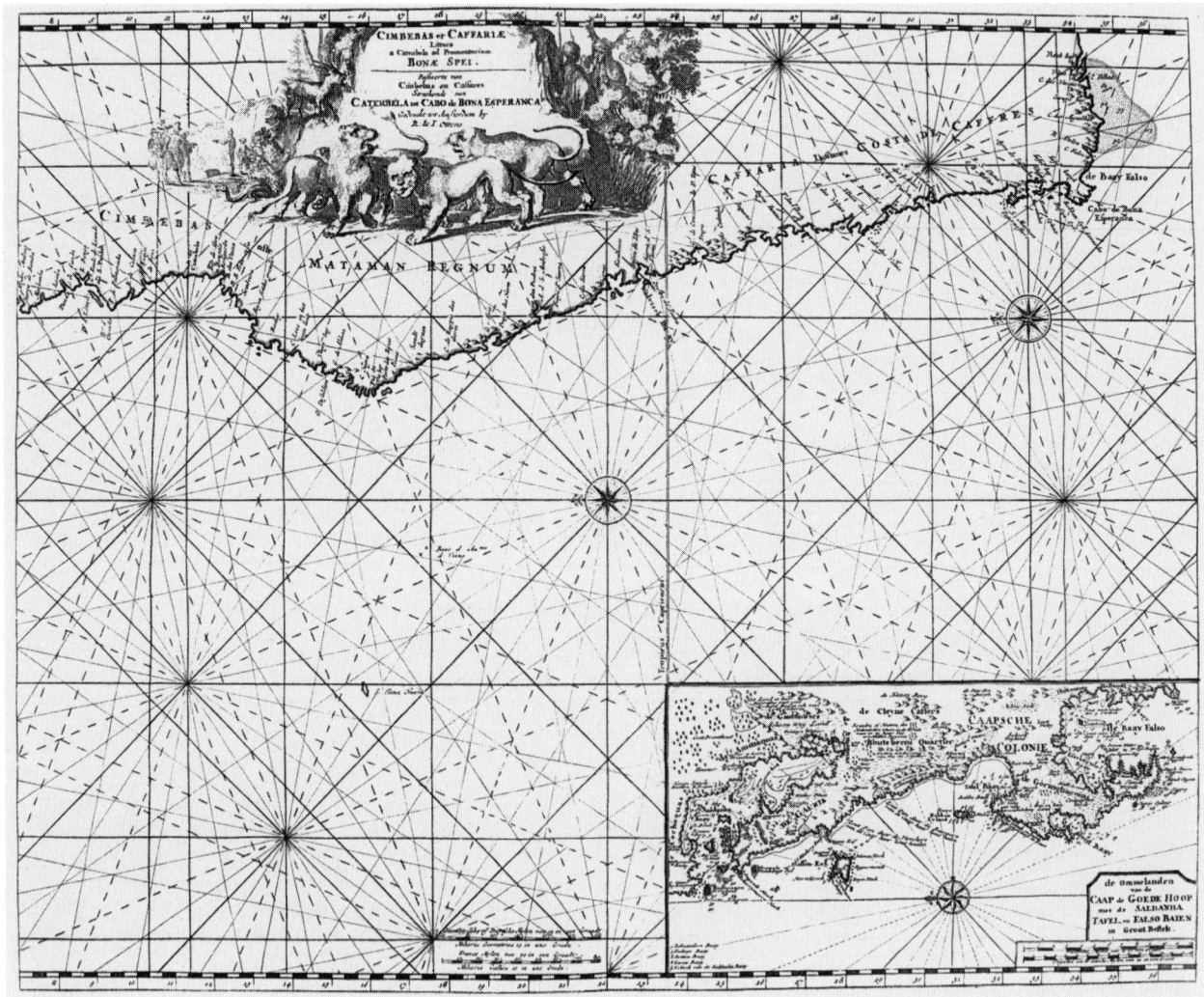

Map 255
OTTENS, Reinier and Josua (fl. 1725-1750)
Cimbebas et Caffariae Littora a Catenbela ad Promontorium Bonae Spei.
Pascaert van Cimbebas en Caffrares Streckende van Catembela tot Cabo de
Bona Esperanca. (Map of Cimbebas and Caffraria from Catembela to Cape of
Good Hope.) (Amsterdam, R. and J. Ottens, 1680-1745.)
Sea chart, 42 x 53 cm, slightly coloured.
Scale in Dutch and French miles.
Inset: *De Ommelanden van de Caap de Goede Hoop met de Saldanha, Tafel*
en Falso Baien in Groot Bestek. (Enlargement of the Cape of Good Hope area
with Saldanha, Table and False Bays.)

This sea chart is identical to that issued by Louis
Renard (see Map 254) except for the imprint, the lack of
ships in the Atlantic and the lower right corner, in which
a detailed inset map of the Cape Colony replaces the
Neptune cartouche.

292

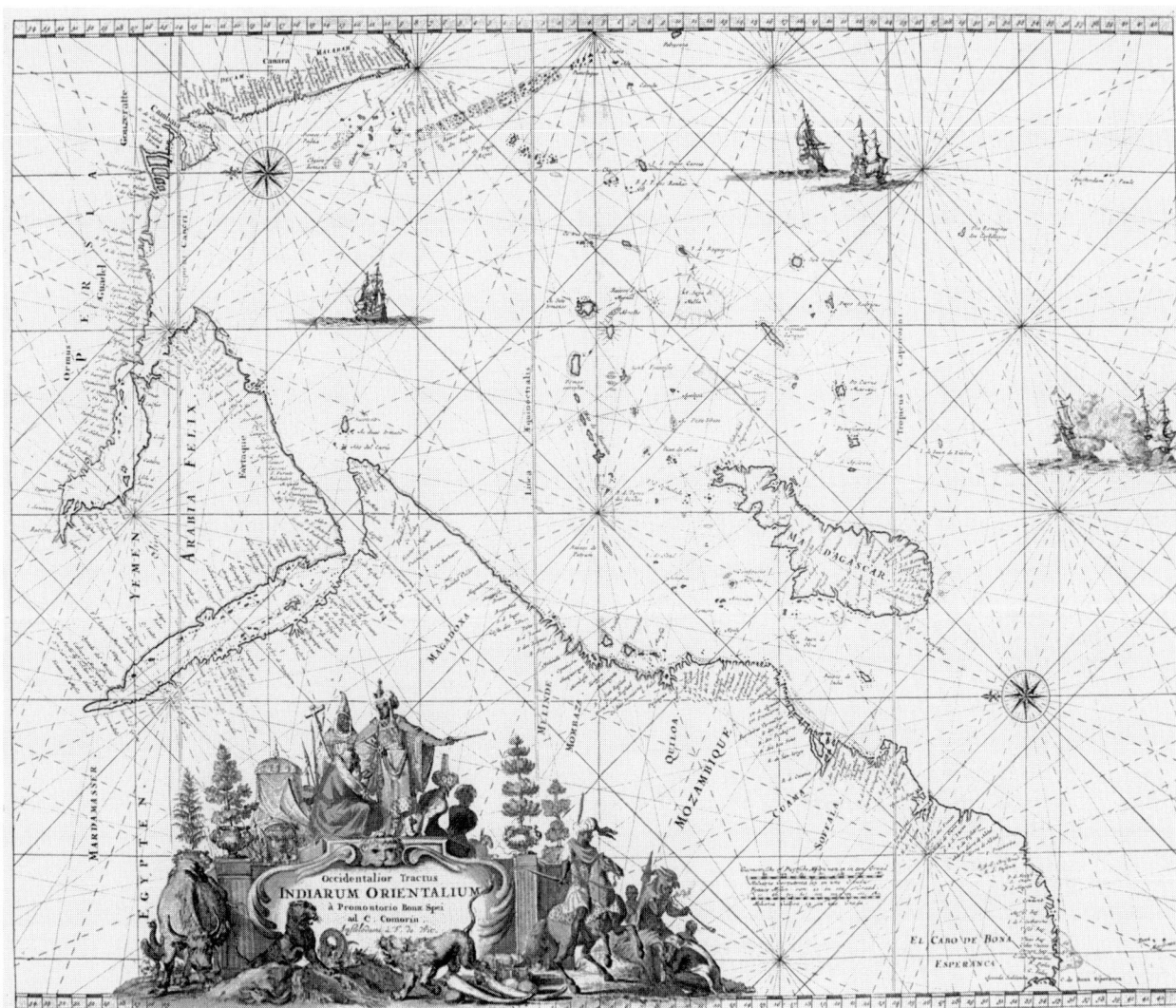

Map 256

WIT, Frederick de (1610-1698)

Occidentalior Tractus Indiarum Orientalium à Promontorio Bonae Spei ad C. Comorin. (Amsterdam, F. de Wit, 1680.)

Sea chart, 44 x 54 cm, coloured outline.
Scale in Dutch and French miles.

The map illustrated here has a most elaborate cartouche at the lower left. There is decorative masonry on which the title is written, surrounded by decorative scroll work. The upper half of a lion's head appears just above the centre of the inscription. A tiger, a snake and a lion appear in the foreground of the platform. To the left is a small ostrich and a cow. On the platform itself there is a group of figures in sumptuous garb. On the right below the platform a horseman, a crouching figure and a woman smoking a long pipe are depicted. On the left in the northern Indian Ocean there is one sailing ship, with two more to the right near the top of the map, and a further group of three ships in the southern sea. This is one of four charts of the coast of Africa, that first appeared in de Wit's *Zee Atlas* of 1675. Both Ottens and Renard appear to have adapted some of the plates for their own use.

The de Wits, father and son, were prominent Dutch map sellers in the late seventeenth century. Frederick the elder was born in 1610 and died in 1698, and his son, also Frederick, died in 1706. When the firm of Blaeu and Jansson terminated in the 1670s the de Wits and the Visschers, became two of the foremost publishing houses in Holland. The de Wits secured many of Blaeu's and Jansson's copperplates, which they then reissued as their own publications without printed text on the back. In 1680 de Wit, the elder, issued a well-known general world atlas, and also one of the Netherland provinces. In 1700 the younger de Wit issued an atlas of town plans, mainly from the Blaeu and Jansson plates, also with no text on the back. An attractive map of Africa, with a figured border, was issued by de Wit the elder in about 1660, but apparently only in small numbers, making the map very scarce.

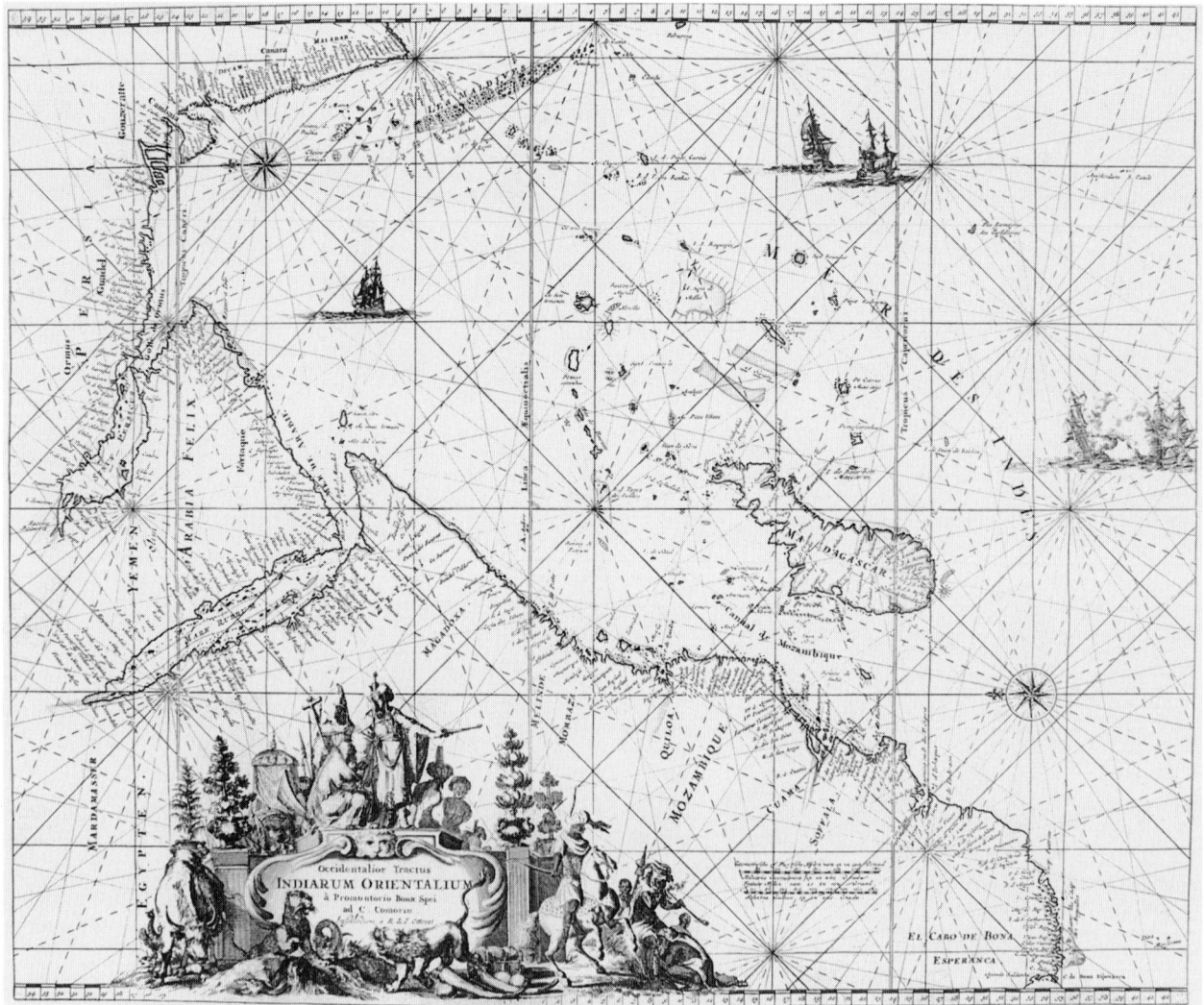

Map 257 OTTENS, Reinier and Josua (fl. 1725-1750)
Occidentalior Tractus Indiarum Orientalium à Promontorio Bonae Spei ad C. Comorin. (The Indian Ocean from Cape of Good Hope to Cape Comorin.)
(Amsterdam, R. and J. Ottens, 1680-1745.)
Sea chart, 44 x 54 cm, slightly coloured.
Scale in Dutch and French miles.

This sea chart is identical to the one issued by de Wit (Map 256) except that the title cartouche bears the names R. & J. Ottens and a number of names have been added, eg.

Les Maldives, Mer des Indies, Canal de Mozambique, Mare Rubrum and Sinus Persicus.

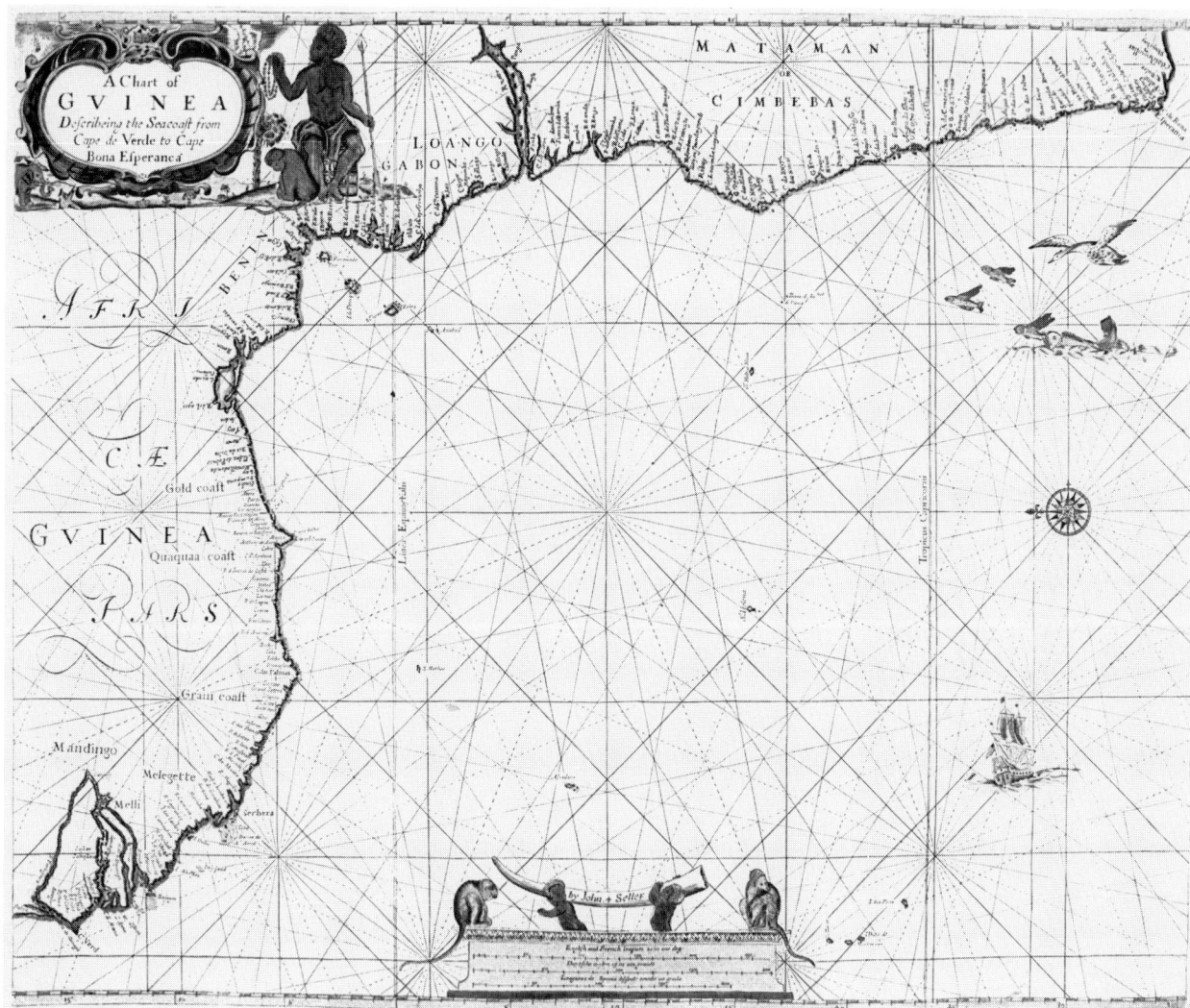

Map 258

SELLER, John (fl. 1664-1697)
A Chart of Guinea Describing the Seacoast from Cape de Verde to Cape Bona Esperanca. (London, J. Seller, 1675.)
Sea chart, 43 x 53 cm, coloured outline.
Scale in English, French, Dutch and Spanish miles.

The title of this sea chart of Guinea from Cape Verde to the Cape of Good Hope appears in a decorative oval cartouche in the upper left corner. The scale in English, French, Dutch and Spanish miles is on the lower border. The chart appears in the third book of the *English Pilot*, and also in the *Atlas Maritimus* (1675).

John Seller was appointed Hydrographer to Charles II and worked mainly in Cornwall and Wapping, London. Wapping became a centre for maritime and nautical publishers and Seller issued numerous charts and maps and books on geography and navigation. Although he was enterprising he never achieved great success. He conceived the idea of the *Sea Waggoner for The Whole World* and the first part of his *English Pilot* appeared in 1671, to be followed by the second and third. Seller's own contribution was limited. Unable to gather sufficient English material and surveys,

he acquired Dutch copperplates from which he erased the titles, replacing them with English names and issuing them under his own name. The finance and distribution of this venture was beyond him and he joined John Thornton, William Fisher and John Colson to promote the sales. Fisher and Thornton expanded the English Pilot. Richard Mount entered the partnership by marrying Fisher's daughter and, a few years later, took Thomas Page into partnership, after which the firm of Mount and Page continued to issue editions of the *English Pilot* up to 1794.

John and Samuel Thornton, both hydrographers to the East India Company, collaborated with John Seller to issue the *Atlas Maritimus*. William Fisher put some capital into a partnership with John Thornton in the production of the *English Pilot*, and in 1703 they issued Book Three.

295

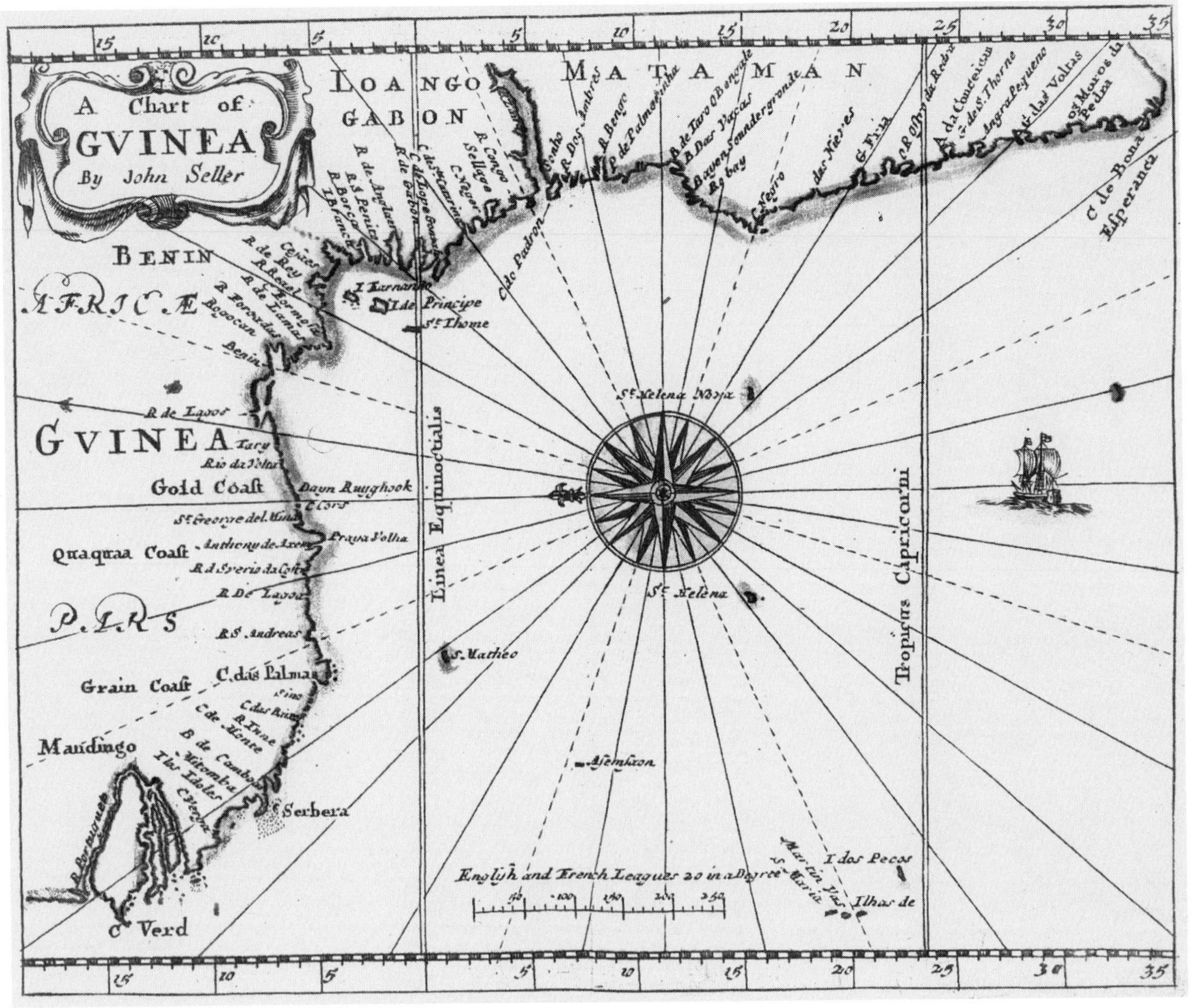

A Chart of
GVINEA
By John Seller

Map 259

SELLER, John (fl. 1664-1697)
A Chart of Guinea. (London, J. Seller, 1679.)
Sea chart, 11 x 14 cm, coloured outline.
Scale in English and French leagues.

This is a miniature sea chart of Guinea, including the whole west coast from C. de Bona Esperanca to Cape Verde, issued by John Seller. This chart is one of a series of the west coast, probably issued in the *Atlas Maritimus* of 1675.

The map itself has a large prominently placed compass rose radiating rhumb lines, with a sailing vessel below it. It names many of the important inlets and bays of the west coast.

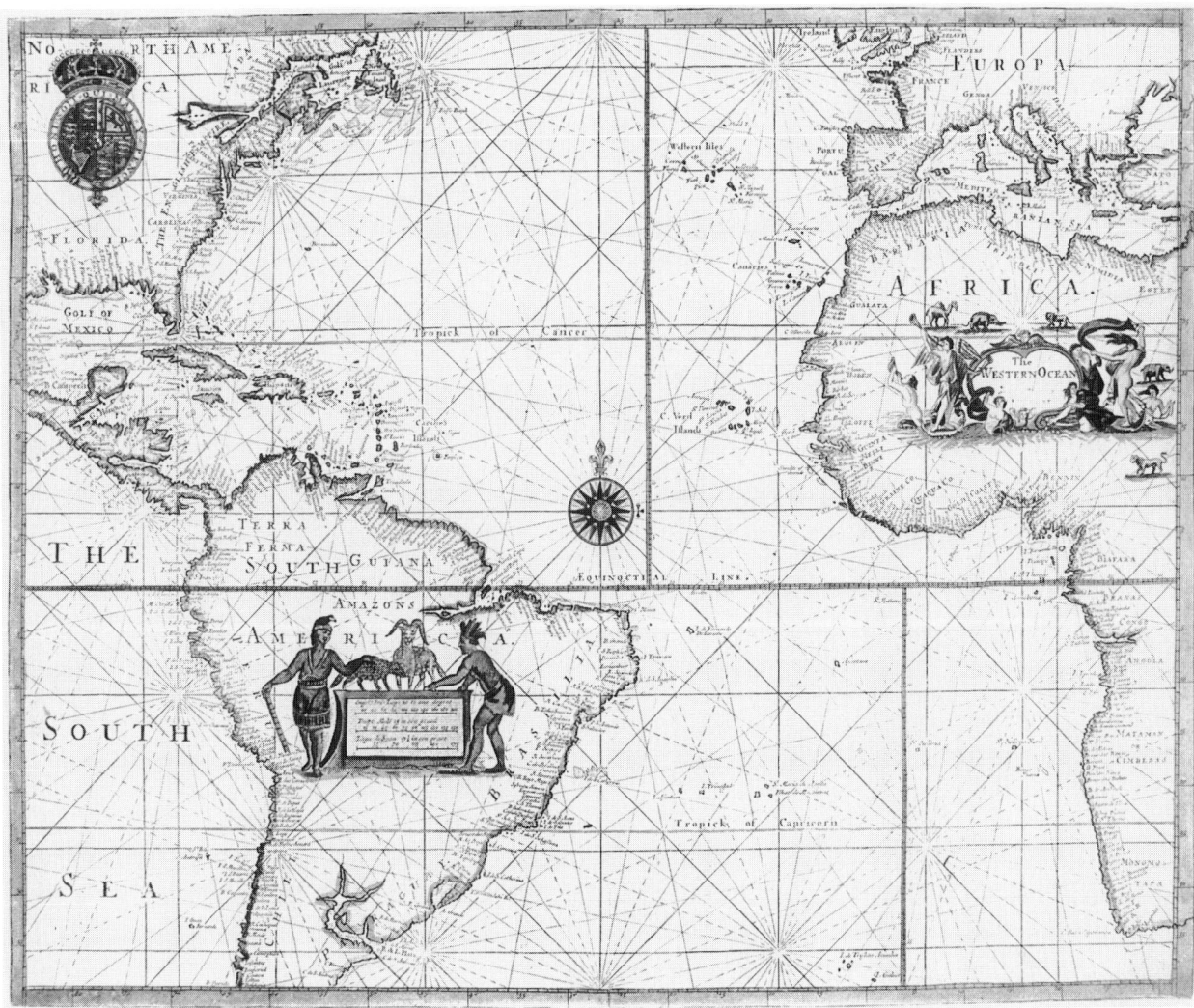

Map 260

SELLER, John (fl. 1664-1697)
The Western Ocean. (London, J. Seller.)
Sea chart, 43 x 53 cm, coloured outline.
Scale in English, French and Spanish leagues, Dutch miles.

This chart includes part of North and South America, western Europe and southern England, together with the whole of the west coast of Africa. The ornate cartouche appears in the upper central continent of Africa. Surrounding the title there are several semi-nude figures waving drapery and blowing long and short horns. Directly below the title it appears as though a name has been rubbed out. Small animals appear around the top to the right of the cartouche.

In the upper left corner of the chart is a coat of arms. In the continent of South America there is another cartouche showing two figures standing on either side of a rectangular block containing the scale in four languages. Standing on top of the masonry are two animals resembling a goat and a spotted leopard. A central compass rose radiates rhumb lines to all four continents.

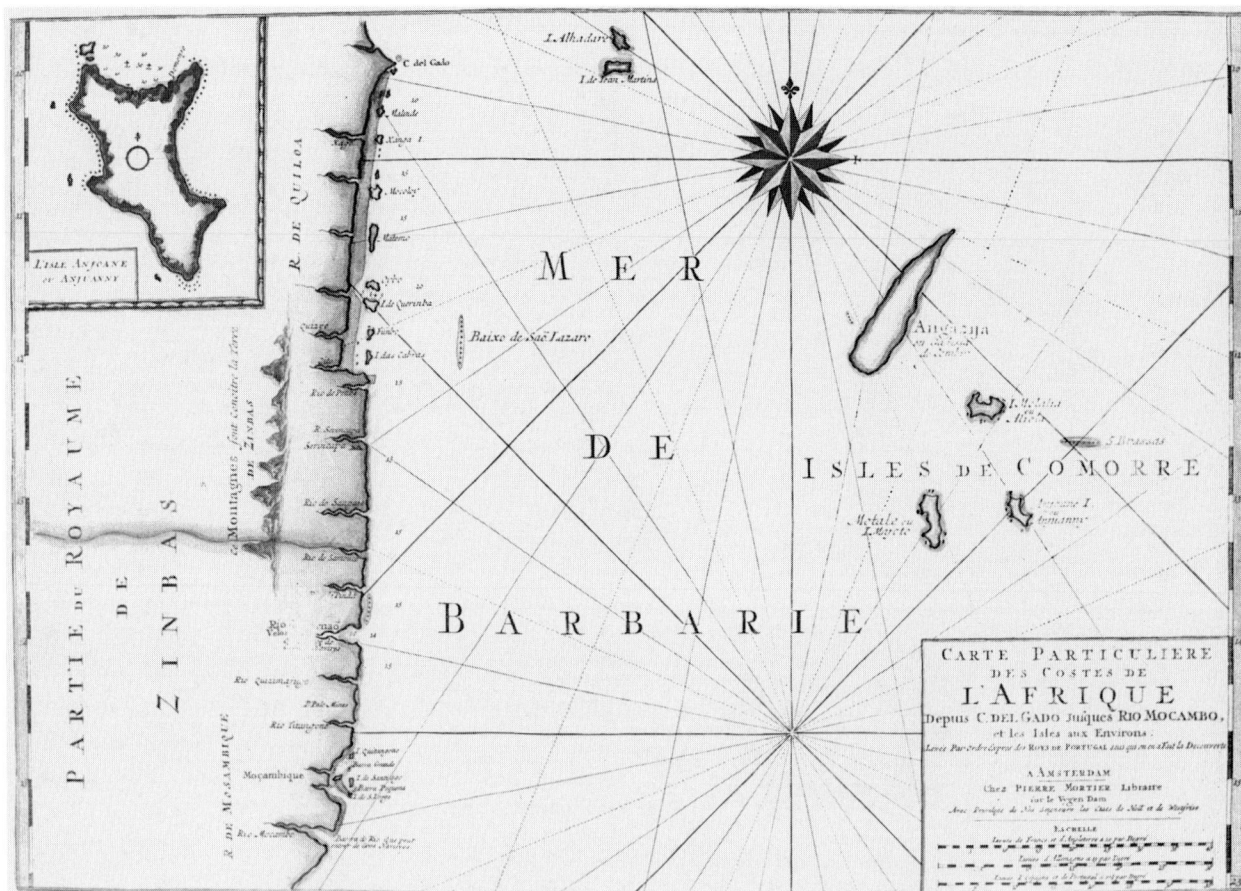

Map 261

MORTIER, Pierre (1661-1711)
Carte particulière des Costes de l'Afrique depuis C. del Gado jusques Rio Mocambo, et les Isles aux Environs. Levée Par Ordre Exprès des Roys de Portugal sous qui on en a Fait la Decouverte. (Specific map of the African coast from Cape del Gado to Rio Mocambo and the adjacent islands.) (Amsterdam, P. Mortier, 1700.)
Sea chart, 40 x 58 cm, coloured.
Scale in French, English, German, Spanish and Portuguese leagues.
Inset: *L'Isle Anjoane ou Anjuanny.*

Pierre Mortier was a publisher and map-seller in the Vygendam, Amsterdam, in 1685, and in 1690 was publishing maps for the French geographers Sanson and Jaillot. He published the *Atlas Suite de Neptune Français*, an atlas for de Fer and for Frederick de Wit, and in 1705 an *Atlas Antiquus*. He was succeeded by his son Corneille who, with Jean Covens, continued the firm as Covens and Mortier.

This chart comes from Mortier's *Atlas Suite de Neptune Français* or the *Atlas Nouveau des Cartes Marines* (1700). The atlas was initiated by Louis XIV, who granted Charles Péne leave to publish an atlas of twenty-nine charts and six pages

of text, which he duly completed in 1693. Mortier and Jaillot subsequently reproduced and added to the plates, and various editions appeared.

The chart shows part of the east coast of Africa, from Rio Mocambo in the south of the Mozambique Channel to Cape del Gado in the north, with an inset of one of the Comoro islands. The Comoro group features prominently in the Indian Ocean (Mer de Barbarie). The sixteen-point compass rose is an attractively multicoloured decoration. The inlets of the many rivers from 23°S to 9°N are featured and the cartouche, title and scale appear in the lower right corner.

298

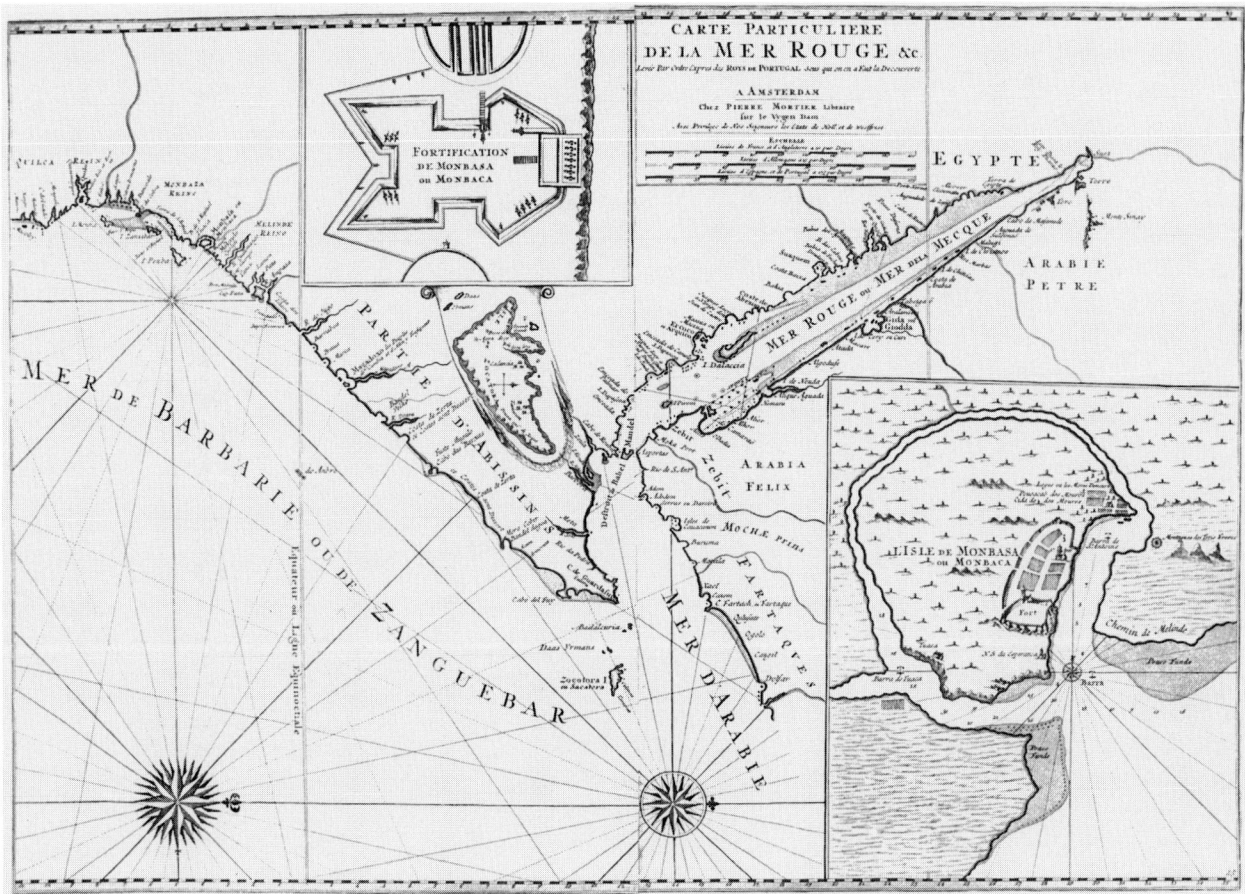

Map 262
MORTIER, Pierre (1661-1711)
Carte Particulière de la Mer Rouge &c. Levée Par Ordre Expres des Roys de Portugal, sous qui on en a Fait la Decouverte. (Specific map of the Red Sea.)
(Amsterdam, P. Mortier, 1700.)
Sea chart, 52,5 x 76 cm, coloured outline.
Scale in French, English, German, Spanish and Portuguese leagues.
Inset: *Fortification de Monbasa ou Monbaca. L'Isle de Monbasa ou Monbaca. Zocatora Island.*

This interesting sea chart was published in the *Atlas Suite de Neptune Français*. It shows the coast of Africa from modern Tanzania to Suez, including southern Arabia, from 11°S to about 35,5°N of the equator.

There is an inset, in the Horn of Africa, of the Island of Zocatora, with details of its coastline. The island and fort of Mombasa are shown in great detail in separate insets at the

top and lower right of the chart. Babel Mandel is prominent in the seaway between the southern Red Sea and the Gulf of Aden, and Zanzibar also appears, with many names along the coast.

This chart has the same features – namely the sixteen-point compass rose and the type used – as all the maps coming from the *Atlas Suite de Neptune Français*.

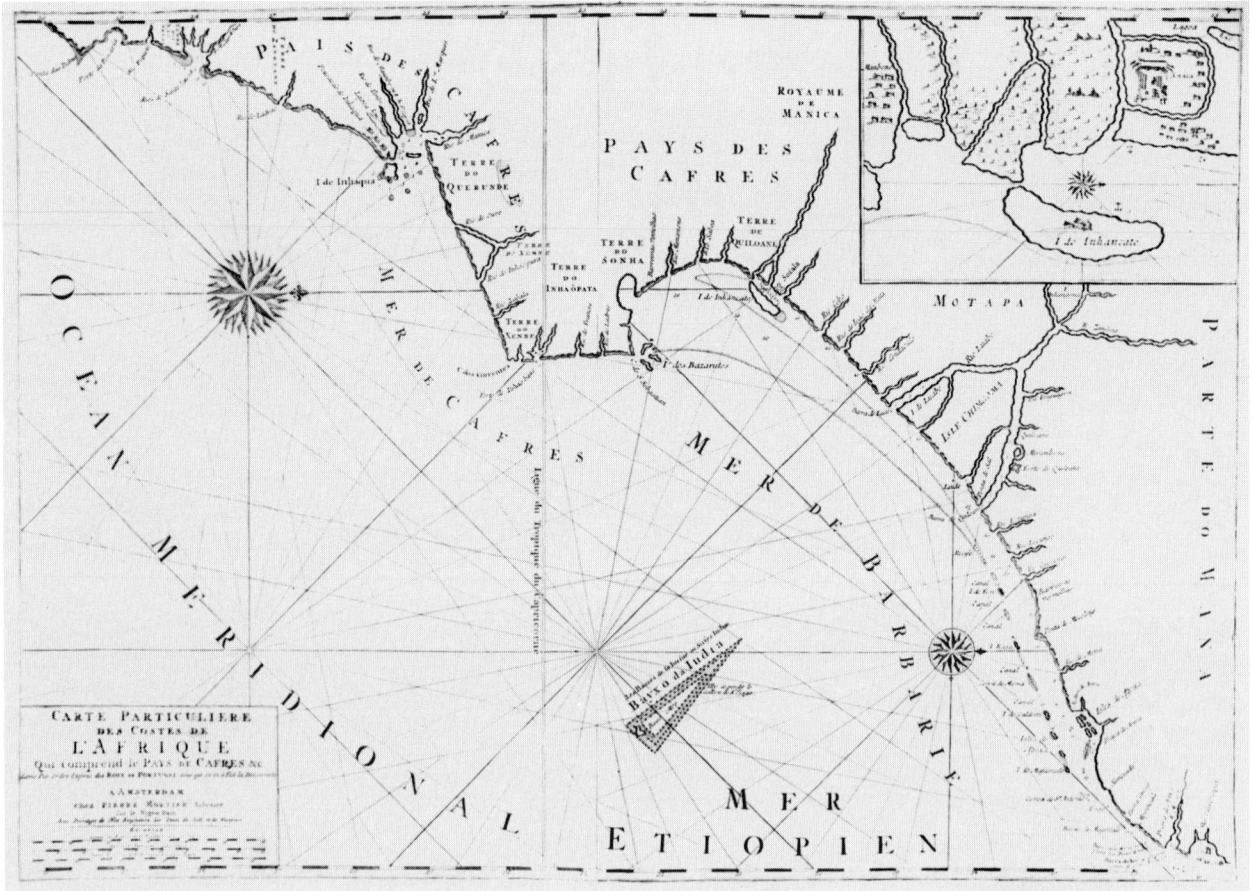

Map 263 MORTIER, Pierre (1661-1711)
Carte Particulière des Costes de L'Afrique qui Comprend le Pays de Cafres &c. Levée par Ordre Exprès des Roys de Portugal sous qui on en a Fait la Decouverte. (Specific map of the African coast including Caffraria.) (Amsterdam. P. Mortier, 1700.)
Sea chart, 57 x 82 cm, uncoloured.
Scale in French, English, German, Spanish and Portuguese leagues.
Inset: *Sofala* and *I. do Inhancato.*

This sea chart extends across the Tropic of Capricorn from approximately latitude 30°S to approximately 15°S, labelled across the centre as Pays de Cafres. It includes a large number of the rivers, bays and settlements known to the Portuguese and named by them during the course of their extensive early discoveries on their way to find a route to the East. These are all clearly marked together with the coastal islands from St. Lucia Bay to Mozambique. The bay of Lourenço Marques (Maputo) is illustrated, though not named as such, but Bazarutes and the island of Inhancato are well-known modern names. In the top right corner there is a separate inset depicting in detail the island of Inhancato

together with the fortress of Sofala and the early settlement there.

This chart has the same features as the other Mortier maps, with the same sixteen-point compass rose. The cartouche is sited in the lower left corner with the title and various scales.

This chart comes, presumably, from the *Atlas Suite de Neptune Français*, but is considerably larger than the previous two Mortier charts, which were perhaps housed in an atlas with single-page maps.

300

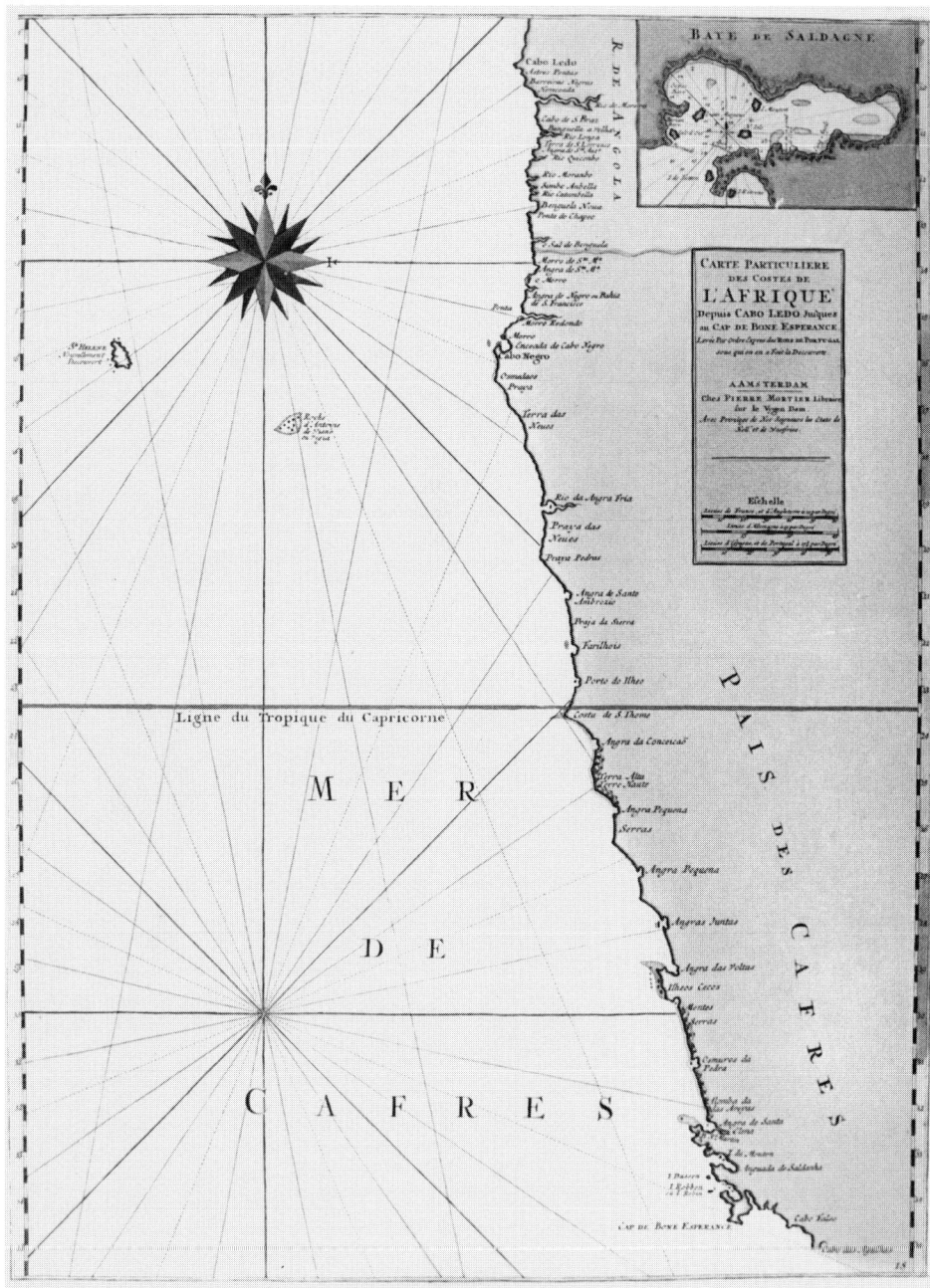

Map 264 MORTIER, Pierre (1661-1711)
Carte Particulière des Costes de l'Afrique depuis Cabo Ledo jusques au Cap de Bone Esperance. Levée par Ordre Expres des Roys de Portugal sous qui on en a fait la Decouvert. (Specific map of the African coast from Cape Ledo to the Cape of Good Hope.) (Amsterdam, P. Mortier, 1700.)
Sea chart, 58 x 43 cm, coloured.
Scale in French, English, German and Spanish and Portuguese leagues.
Inset: *Bay de Saldagne.*

   This sea chart illustrating the west coast of southern Africa from the Cape of Good Hope in the south to Cape Ledo (Angola) in the north (latitudes 35° to 9° south), also appears in the *Atlas Suite de Neptune Français.* In the top right corner is a detailed inset map of Saldanha Bay. Like the other Mortier maps from this atlas the colourful sixteen-point compass rose appears, in the left upper sea. The title cartouche and scales appear in the rectangular block in the upper mainland.

301

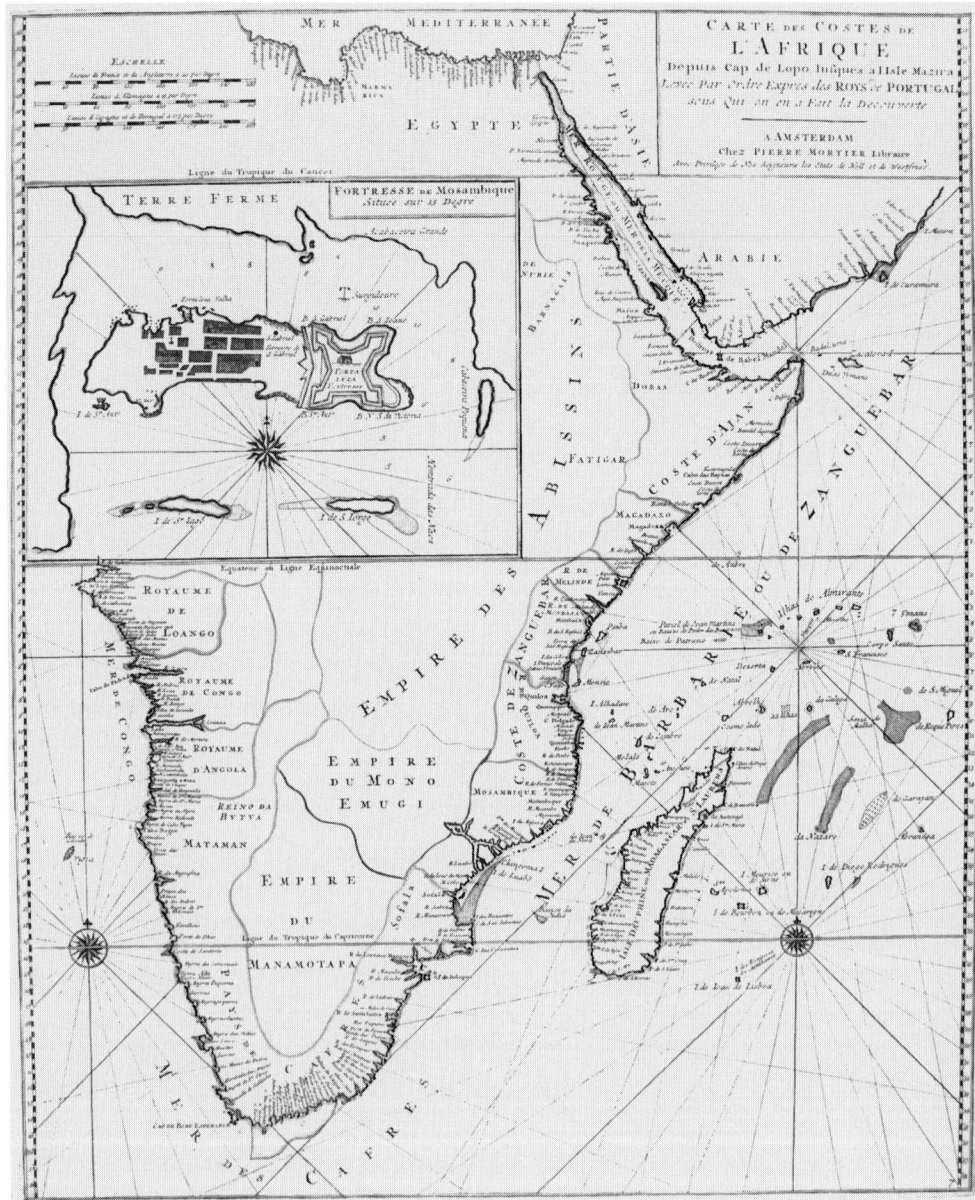

Map 265

MORTIER, Pierre (1661-1711)
Carte des Costes de L'Afrique depuis Cap de Lopo jusques à l'Isle Mazira.
Levée par Ordre Expres des Roys de Portugal sous qui on en a Fait la
Decouverte. (Map of the coast of Africa from Cape Lopo to the island of Mazira.)
(Amsterdam, P. Mortier, 1700.)
Sea chart, 59 x 48 cm, slightly coloured.
Scale in French, English, German, Spanish and Portuguese leagues.
Inset: *Fortresse de Mozambique.*

This sea chart of Africa is confined to the lower half of the west coast, the whole of the east coast and a portion of Arabia. Occupying much of the interior is a large rectangular inset with a detailed map of the fortress of Mozambique and the surrounding small islands.

The major kingdoms of 'Monomotapa,' 'Mono Emugi' and Abyssinia are depicted along almost the whole of the continent extending as far as the Pays de Cafres in the south. No rivers or lakes appear in the interior, and the coastal towns and bays are those of the early Portuguese discoverers. The surrounding islands feature prominently and a compass rose is present in the east and in the west. A rectangular cartouche appears at the top right corner.

302

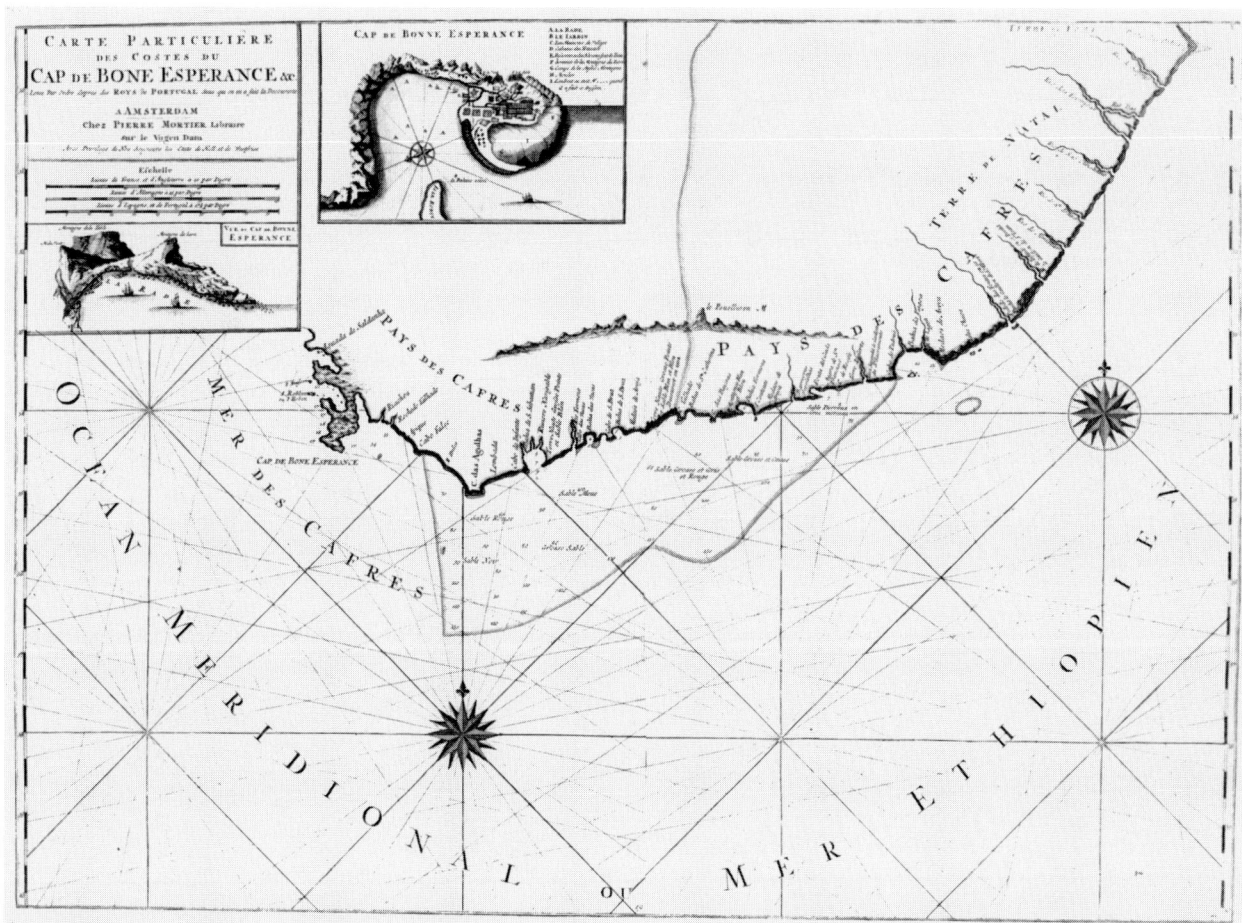

Map 266

MORTIER, Pierre (1661-1711)
Carte Particulière des Costes du Cap de Bone Esperance &c. Levée Par Ordre Expres des Roys de Portugal sous qui on en a Fait la Decouverte. (Map of the coast of the Cape of Good Hope...) (Amsterdam, P. Mortier, 1700.)
Sea chart, 56 x 80 cm, coloured.
Scale in French, English, German, Spanish and Portuguese leagues.
Inset: View of the Cape of Good Hope. Plan of the Cape of Good Hope.

This is a detailed sea chart of the Cape of Good Hope incorporating portions of the Cape Colony and the southern Natal coast. There are two detailed insets of the Cape harbour area. One is of Table Mountain and its companions as seen from the sea, together with coastal buildings, and the other is a plan of the Bay and the Cape Town settlement. The other also illustrates the original garden laid down by the Dutch East India Company to provide fresh vegetables and fruit to the East Indiamen rounding the Cape on their way to the East. The homes of the local inhabitants and those of the Hottentots are pinpointed by a lettered legend together with the fort and the mill. This map is a large folding sea chart from the *Atlas Suite de Neptune Français* by Pierre Mortier, 1700.

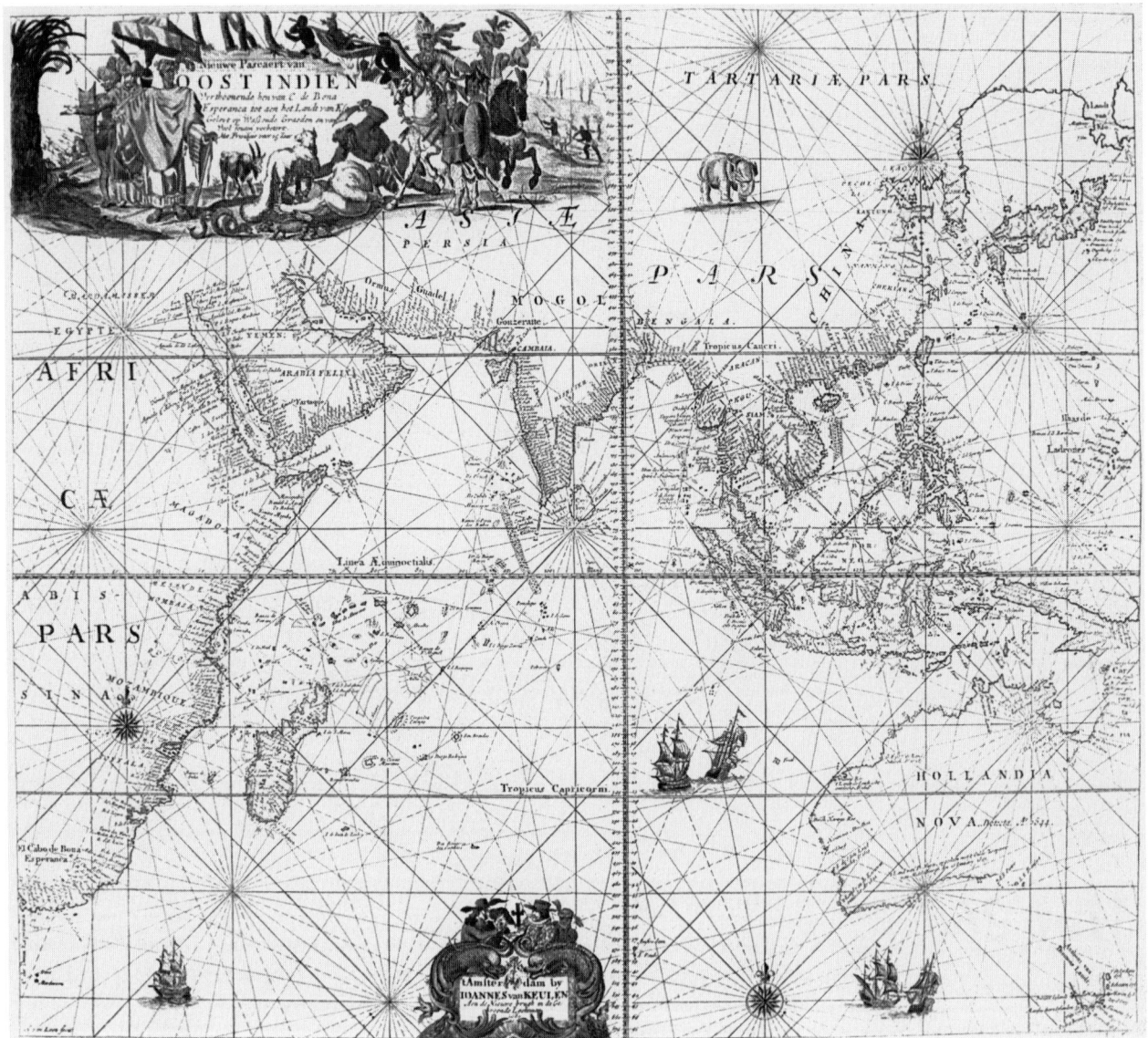

Map 267

KEULEN, Johannes van (1654-1715)
*Nieuwe Pascaert van Oost Indien Verthoonende hen van C. de Bona
Esperanca tot aan het Landt van Eso. Geleyt op Wassende Graeden en van Veel
Fouten Verbetert.* (New map of East Indies from Cape of Good Hope to Japan
and Australia, revised.) Engraved by H. van Loon. (Amsterdam, J. van Keulen, 1680.)
Sea chart, 51 x 58 cm, coloured outline.

This is an extremely decorative sea chart showing the
Cape of Good Hope, the east coast of Africa, Arabia, India,
Japan and the Far East, with the western half of Hollandia
Nova (Australia).

Two cartouches adorn the top and lower margins of the
map. The one at the upper left shows men on horseback
and on foot, all in attractive costumes. In the centre of this
cartouche is a scene of a man holding in his right hand the
decapitated head of what appears to be a white man, and in
the left he holds the sabre with which this act was performed.
Adjacent to this scene is the back view of a fat-tail sheep.

This breed of sheep is not uncommon on cartouches of
antiquarian maps, having been described in biblical times,
and occuring in Arabia, the Middle East and southern Africa.
(See *The Map Collector*, June 1979.)

The second cartouche at the bottom of the map provides
the name and address of the author and is surrounded by
two dolphins and the figures of what appear to be European
traders. Ships adorn the seas. This chart comes from van
Keulen's *Zee Atlas*, appearing in various editions from 1680
onwards.

304

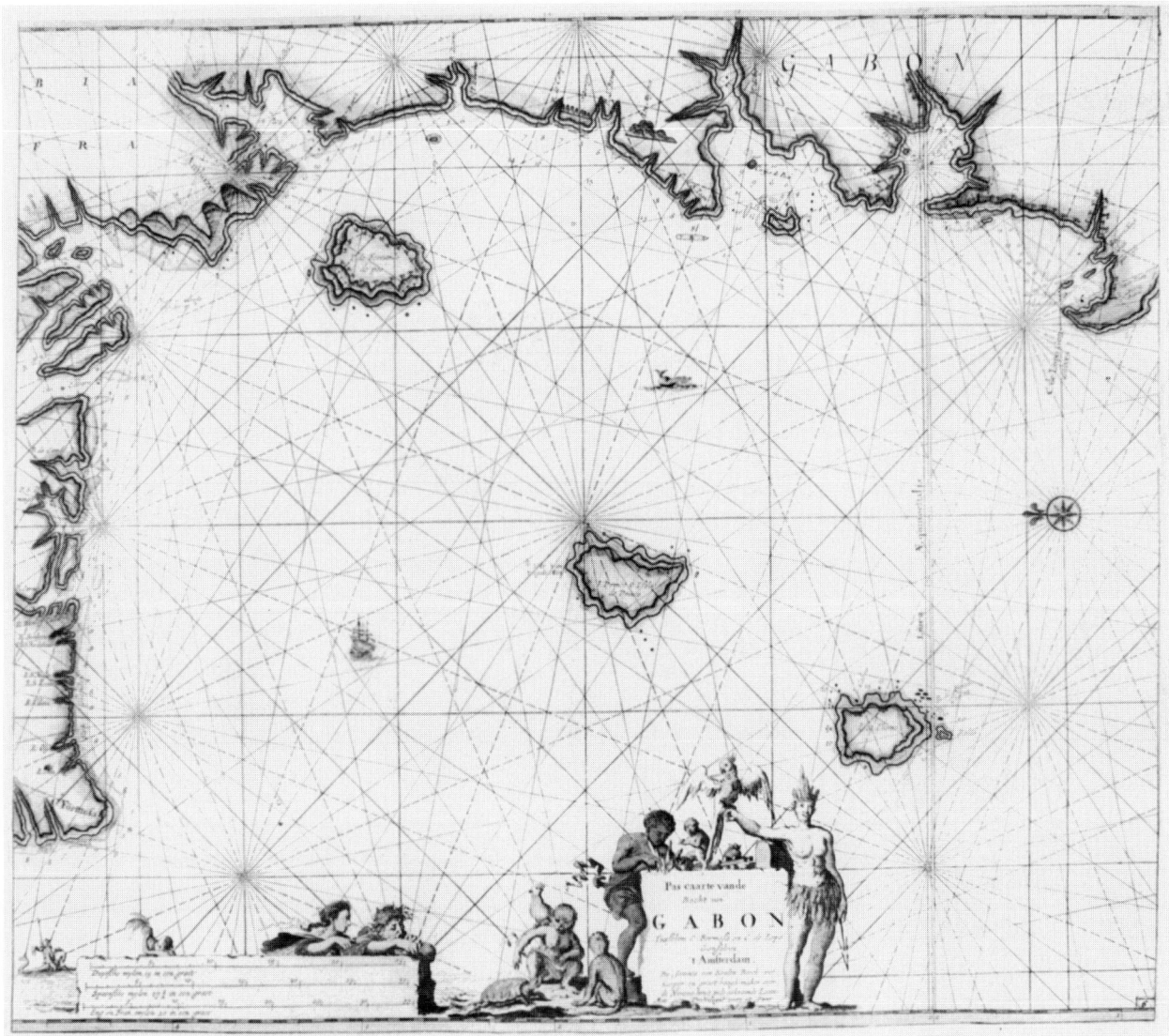

Map 268
KEULEN, Johannes van (1654-1715)
Pascaarte van de Bocht van Gabon tuschen C. Formosa and C. de Lopo. (Map of the coast of Gabon from Cape Formosa to Cape de Lopo.)
(Amsterdam, J. van Keulen, 1680.)
Sea chart, 50,5 x 58,5 cm, coloured.
Scale in Dutch, Spanish, English and French miles.
Lower right: 6.

This sea chart was published by Johannes van Keulen in the *Zee Atlas* of 1680 with thirty-eight maps and text by C.J. Vooght. This was a popular atlas, of which nine editions were published within five years, including French and English versions.

This map depicts the west coast of Gabon from Cape Formosa to Cape de Lopo, with the islands of St. Thomas and Principi. A colourful cartouche is situated at the lower border, showing the title on a rectangular piece of masonry, on each side of which are partially nude natives. One has a parrot perched on her right hand and appears to be wearing

a feathered hat. To the left of the title is an ape-like figure with a parrot on its shoulder, playing with a tortoise. At the lower left border the scale is illustrated on another rectangular stone block with the head and shoulders of a male and female leaning on the masonry. To the left of the scale is a young person appearing to drink from a horn of plenty and further to the left is a naked person wearing what appears to be a top hat, astride a sea horse, holding a trident in his left hand. There is one ship and one sea monster in the sea and to the right, also in the sea, is a compass rose with numerous rhumb lines.

305

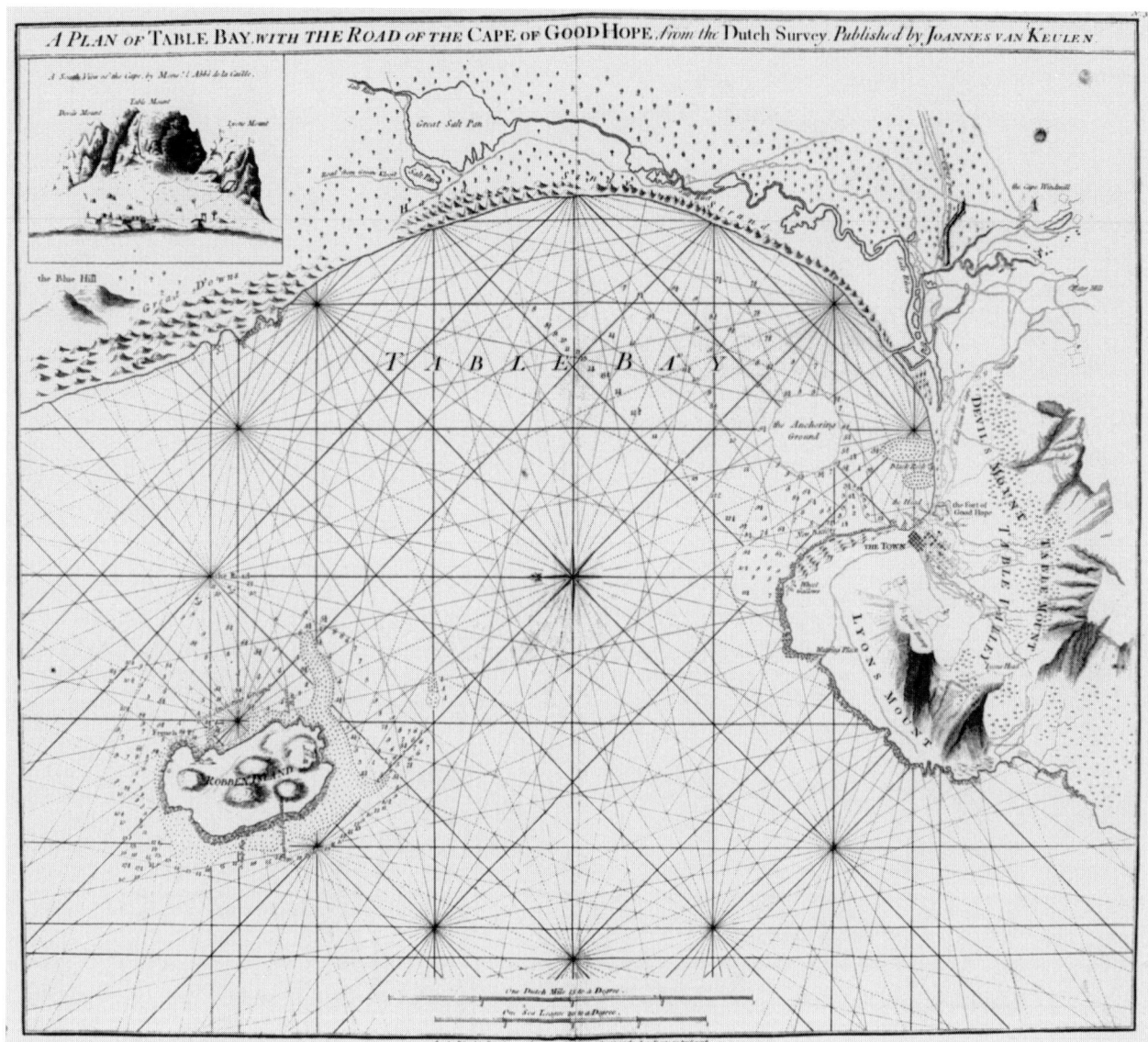

Map 269   KEULEN, Johannes van (1654-1715)
A Plan of Table Bay, with the Road of the Cape of Good Hope, from the Dutch Survey. Published by Johannes van Keulen. (London, Robert Sayer and John Bennett, No. 53 Fleet Street, 1778.)
Sea chart, 48 x 55 cm, uncoloured.
Scale in Dutch miles and sea leagues.
Inset: *A South View of the Cape, by Mons. l'Abbé de la Caille.*
Top right: No. 3

This detailed chart of Table Bay with the road of the Cape of Good Hope provides a wealth of information on the Bay, Robben Island and the topography of Table Mountain. There is a particularly fine inset engraving of Table Mountain from the south, by the Abbé de la Caille, showing sparse settlements at its foot.

A plan of Cape Town, showing the Company's garden, is placed on the harbour. The fort of Good Hope and a gallows, Devil's Mount, Lyons Mount and Lyon's Rump are depicted in attractive detail. The Salt River entering the sea and the road to Drakenstein and Stellenbosch are clearly marked. The anchoring ground in the harbour is surrounded by soundings, as is Robben Island, and the sea has a wealth of rhumb lines.

Nicholas Laurie (1712-1762), a mathematician and astronomer, visited the Cape in 1751 and wrote a volume of detailed description on the area and its people, especially the Hottentots. He was thus able to depict Table Mountain in this fine engraving.

Johannes van Keulen was the founder of a family well known for over two hundred years as nautical chart makers, publishers of nautical text books, books on shipmaking and almanacs. His *Zee Atlas* was sufficiently popular to run to nine editions. This map was published in 1778 by Sayer and Bennett and later, in 1794, by Laurie and Whittle, also of London.

A South View of the Cape, by Monsr l'Abbé de la Caille.

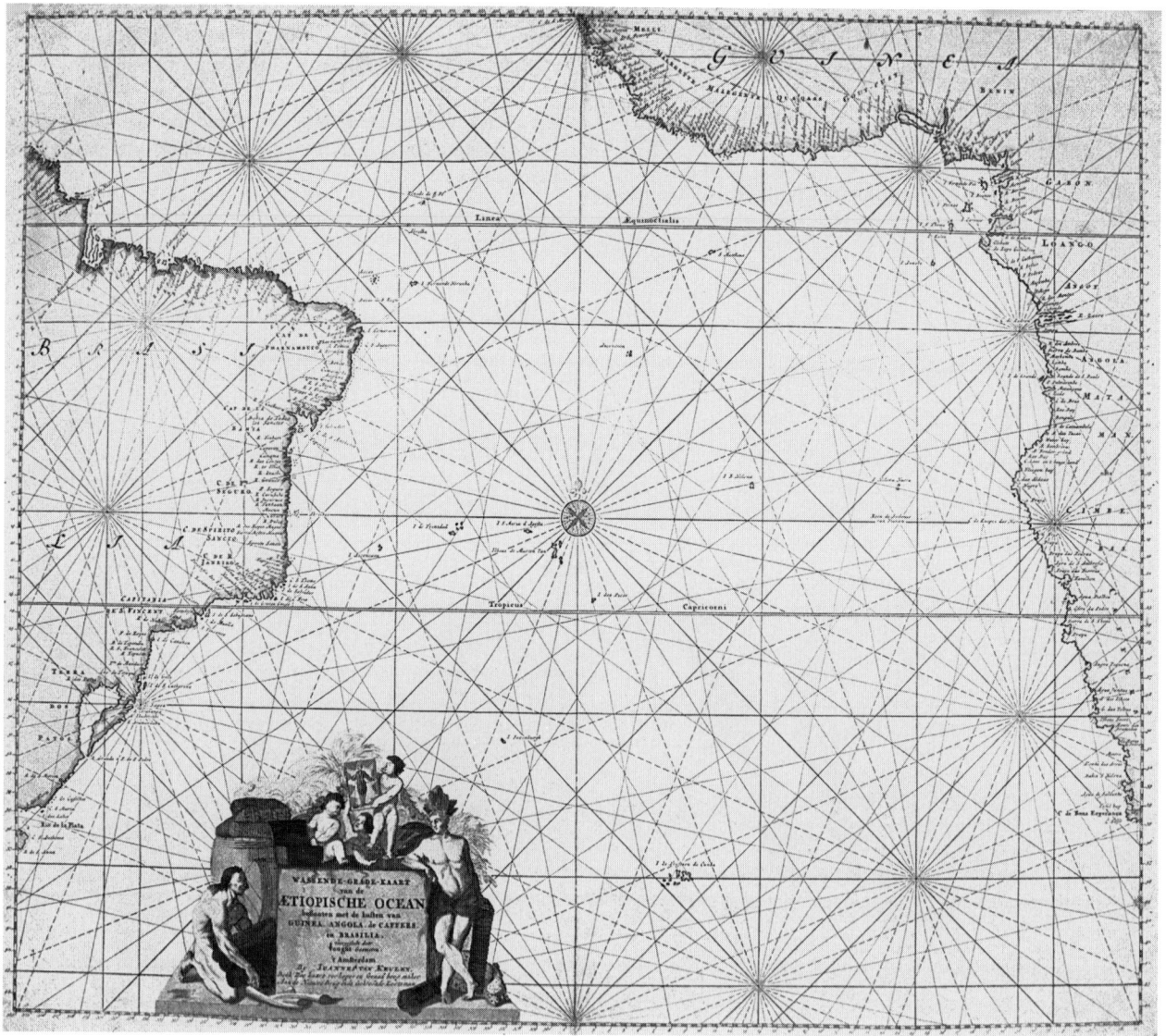

Map 270 KEULEN, Johannes van (1654-1715)
Wassende-Grade-Kaart van de Aetiopische Ocean beslooten met de kusten van Guinea, Angola, de Caffers en Brasilia. (Sea chart of the Atlantic Ocean with the coasts of Guinea, Angola, Caffraria and Brazil.) Engraved by C.J. Vooght. (Amsterdam, J. van Keulen, over de Nicubrug in de Gekroonde Lootsman, 1753?)
Sea chart, 51 x 59 cm, coloured.

This large sea chart of the west coast of Africa from the Cape of Good Hope to the western bulge of Africa, including a portion of Brazil, was issued by the family of van Keulen, probably by the founder Johannes because in the title reference is made to his original address 'over de Nieubrug in de Gekroonde Lootsman.'

The west coast of Africa is accurately depicted with its numerous Portuguese-named harbours. The compass rose is almost centrally placed in the Atlantic Ocean with many rhumb lines between West Africa and what there is of Brazil.

In the ocean many islands are noted, even the two St. Helenas, one marked as 'the new one.'

In the lower left border is a cartouche with the title on a stone, flanked by two figures, the right-hand one having a wreath of coloured feathers around his head. Standing above the title are three children holding a portrait.

Reference is made in the title to C.J. Vooght, who was one of the Dutch personalities assisting the van Keulen family with their great task of chart making.

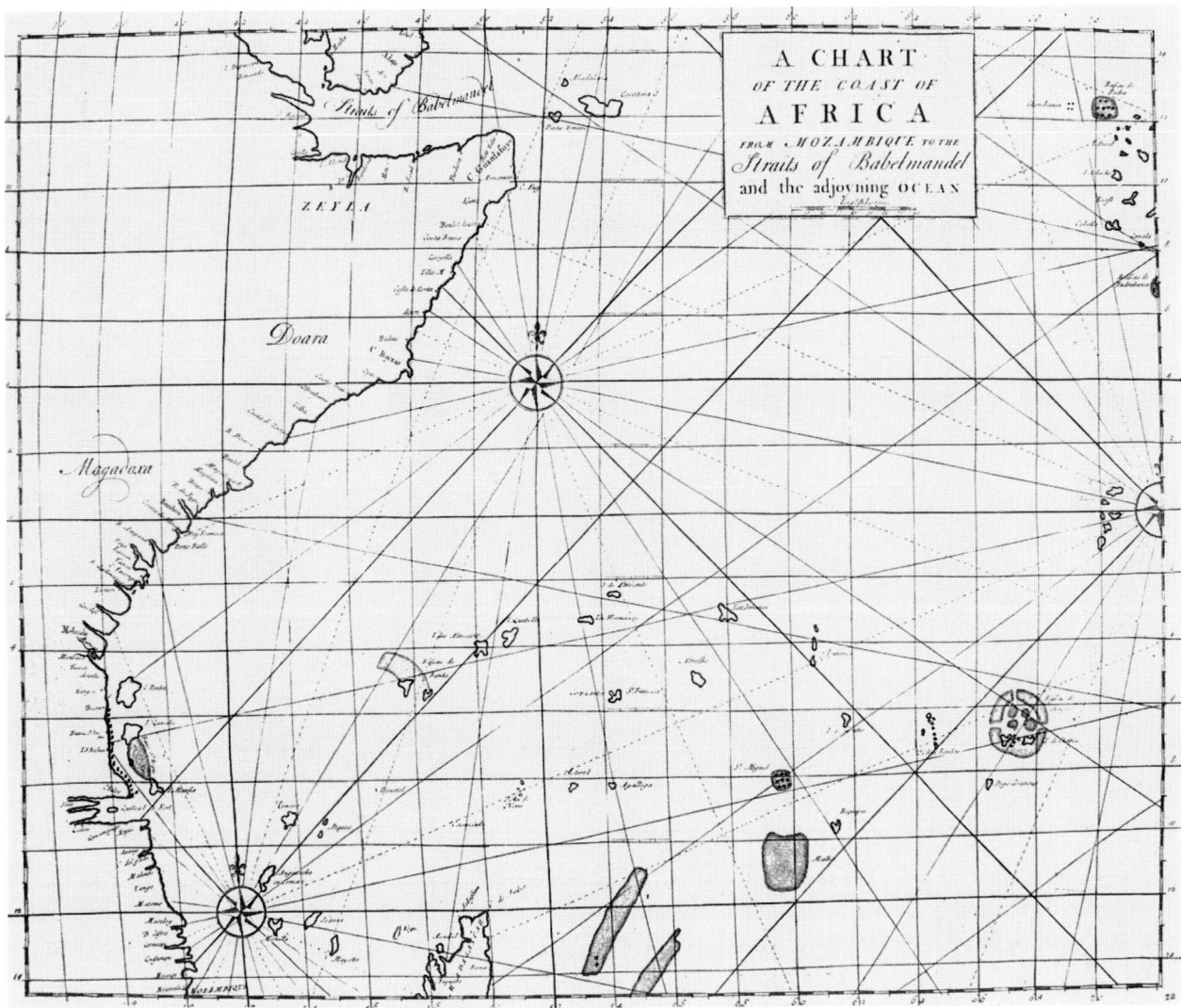

Map 271

KNAPTON, James (fl. 1687-1738) and John (d. 1770)
A chart of the coast of Africa from Mozambique to the Straits of Babelmandel and the adjoyning Ocean. (London, J. Knapton, 1728.)
Sea chart, 48 x 58 cm, uncoloured.
Scale in English leagues.

This sea chart of a portion of eastern Africa from Mozambique to the Straits of Babelmandel, includes the Horn of Africa and numerous islands of the northern Indian Ocean. A simple title is contained in a box at the upper right border. Compass roses appear on the coastline radiating numerous rhumb lines.

This chart appears in the *Atlas Maritimus et Commercialis*, printed in London for James and John Knapton in 1728.

James Knapton printed maps for Dampier's *A New Voyage Round The World* (1705-1717). John Knapton, presumably related, was co-author of the *Atlas Maritimus* and issued maps of North America, Camden's *Britannia* and Moll's *Atlas Manuale* of 1723.

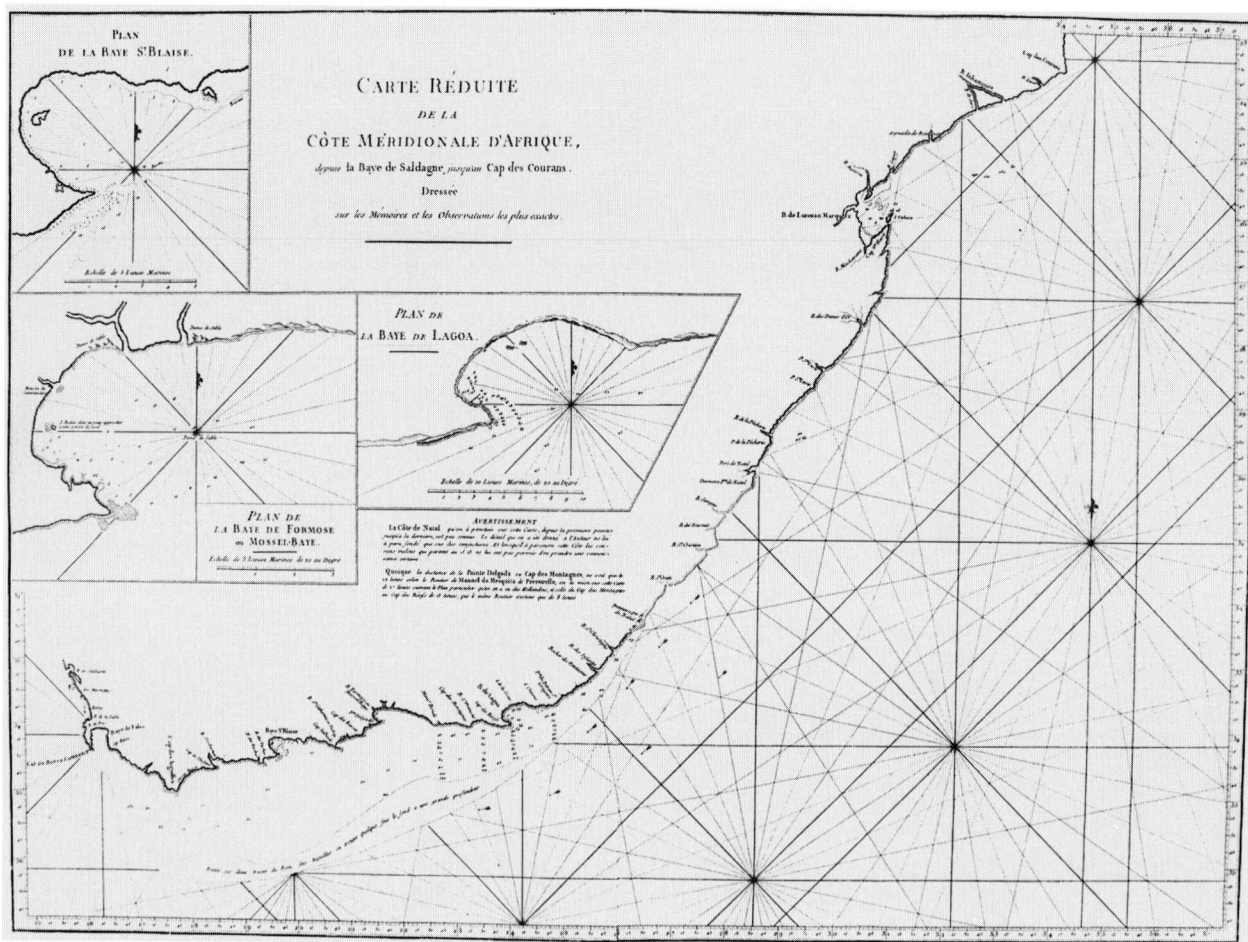

Map 272 MORTIER, Pierre (1661-1711) and JAILLOT, Jean Baptiste (1710-1780)
Carte Réduite de la Côte Méridionale d'Afrique, depuis Ia Baye de Saldagne jusqu'au Cap des Courans... (Map of southern coast of Africa from Saldanha Bay to Cape Courans...) (Amsterdam, 1775.)
Sea chart, 48 x 66,5 cm, uncoloured.
Inset: *Plan de la Baye St. Blaise. Plan de la Baye de Formose ou Mossel-Baye. Plan de la Bay de Lagoa.*
Top right: 11.

A French sea chart of the southeastern coast of Africa from the *Atlas Neptune Oriental* of 1775. It extends from Saldanha Bay to north of Cape Courans, including the coast of Natal and southern Mozambique, with insets of the bays of St Blaise, de Lagoa (Algoa Bay) and de Formosa (Mossel Bay). This is a fine detailed chart, typical of the Mortier and Jaillot series put out under the control of the French Maritime Authentique.

310

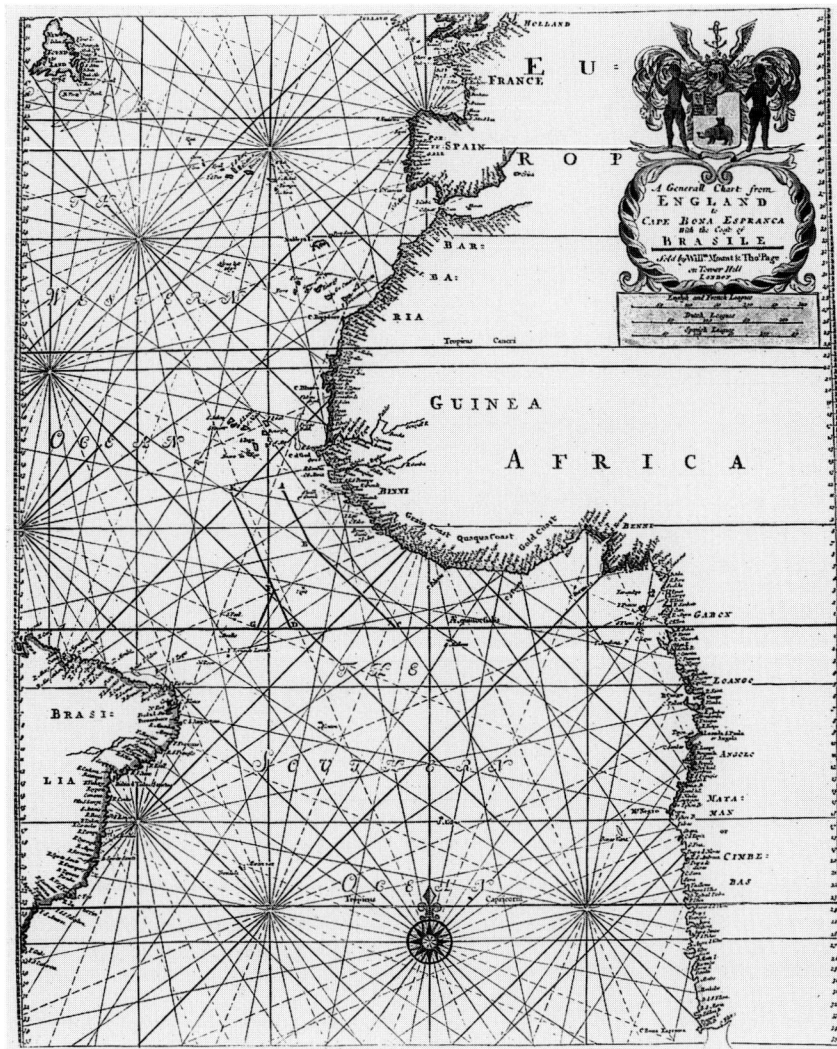

Map 273

MOUNT, William and PAGE, Thomas
A Generall Chart from England to Cape Bona Espranca with the Coast of Brasile. (London, W. Mount and T. Page on Tower Hill.)
Sea chart, 53 x 43,5 cm, slightly coloured.
Scale in English, French, Dutch and Spanish leagues.

This is an attractive sea chart of the whole extent of the west coast of Africa from the Straits of Gibraltar to 'Bona Espranca,' including part of Spain, the west coast of France and Holland and the south of England. There is a great deal of detail on the west coast of Africa relating to bays, capes and adjacent islands. No geographic detail is illustrated on the mainland except for a portion of two rivers entering the ocean, named the R. Senegal and the R. Gambia, lying parallel to each other.

The cartouche is situated at the top right of the map with an oval intertwined design in which the title is placed, surmounted by a striking coat of arms. The heraldic device consists of a square subdivided into four sections, two of which contain flowers similar to the fleur-de-lys, and the remaining two elongated animals resembling reclining lions. Below this is the side view of an elephant with a small tower

strapped to its back. Above is a helmeted and crowned head supporting two outstretched wings. Above the crown stands an anchor with what appears to be a snake twisted around the upright. This is reputed to be the coat of arms of the English Africa Company.

There are numerous rhumb lines directing the mariner to a great variety of destinations, and one compass rose in the South. It is interesting to note that the Atlantic is labelled Southern Ocean whereas further north it is known as the Western Ocean.

The title text refers to this map as 'sold by Will Mount and Thos. Page,' a large firm of publishers of marine charts in London throughout the eighteenth century, who also published editions of the *English Pilot*.

311

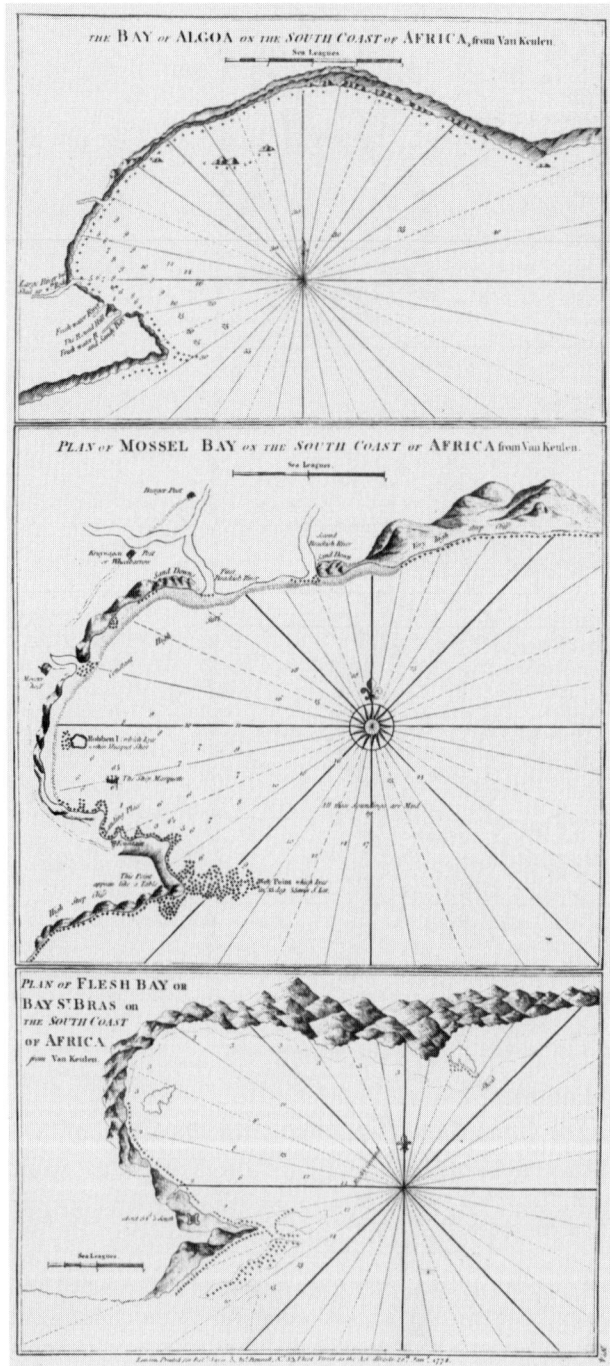

Map 274

KEULEN, Johannes van (1654-1715)
The Bay of Algoa on the South Coast of Africa; Plan of Mossel Bay on the South Coast of Africa; Plan of Flesh Bay or Bay St. Bras on the South Coast of Africa; from van Keulen.
(London, Robert Sayer and John Bennett, No. 53 Fleet Street, 1778.)
Three sea charts, 58 x 26 cm, uncoloured.
Scales in sea leagues.
Bottom right: No. 4.

Here are three sea charts showing three adjoining bays on the southern coast of South Africa, published by Sayer and Bennett and based on the original work of the van Keulen family, famous chart and instrument makers of Holland. The top chart illustrates the Bay of Algoa, the middle one Mossel Bay and the lower one Flesh Bay or Bay St. Bras (today's name). These are all finely executed engravings with soundings, rhumb lines and silhouettes of mountains.

312

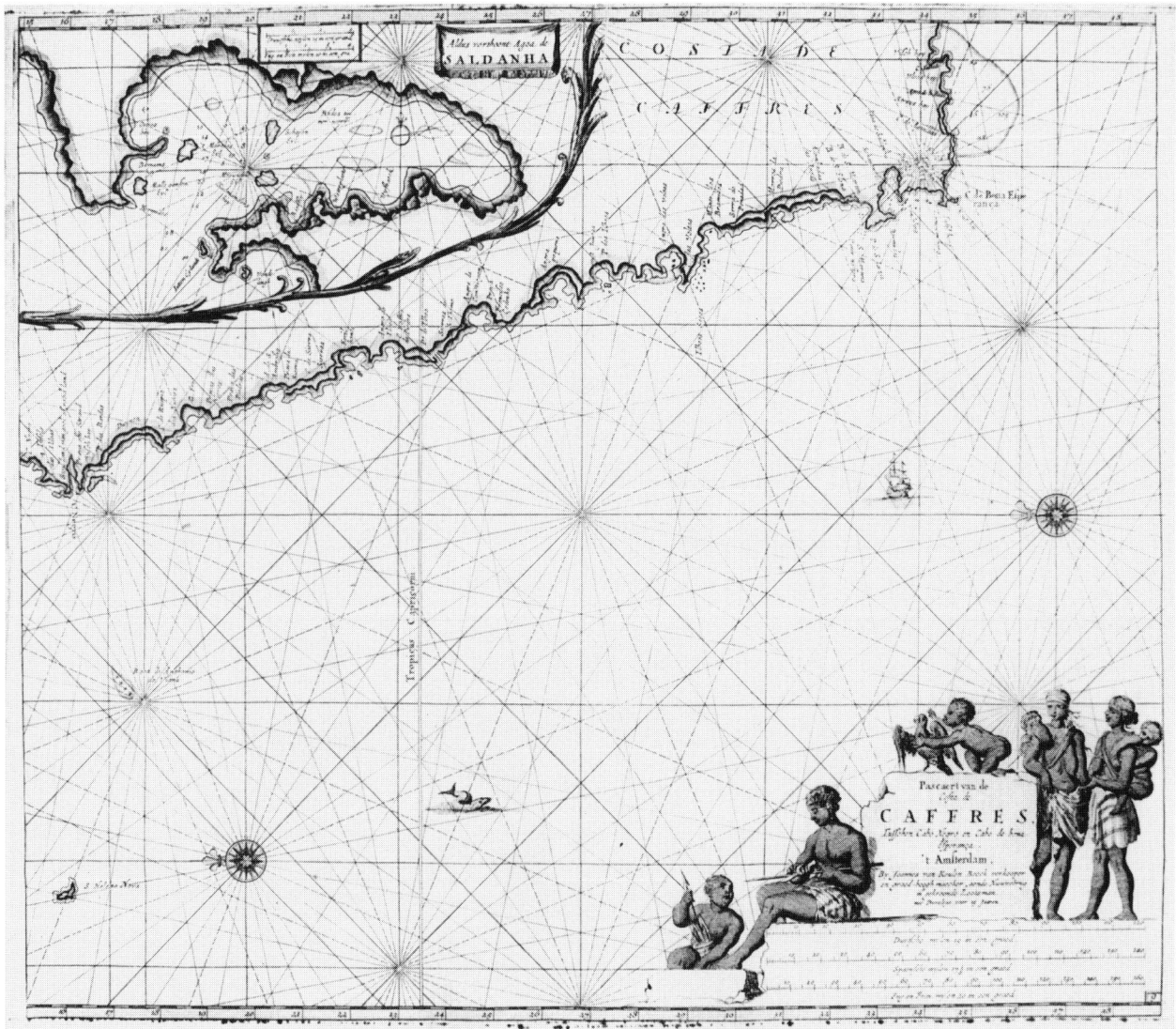

Map 275

KEULEN, Johannes van (1654-1715)

Pascaert van de Costa de Caffres Tusschen Cabo Negro en Cabo de bonae Esperança. (Map of the coast of Kaffraria between Cape Negro and the Cape of Good Hope.) (Amsterdam, J. van Keulen, over de Nieubrug in de Gekroonde Lootsman, 1716.)
Sea chart, 50 x 58 cm, slightly coloured.
Scale in Dutch, Spanish, English and French miles.
Inset: *Aldus verthoont Agoa de Saldanha.*
Bottom right: 9.

On this chart there is an inset of Saldanha Bay demarcated by a decorative branch. Soundings are given off Saldanha Bay; the sea has two compass roses with rhumb lines, one sailing ship and one sea monster

The van Keulens were an extraordinarily conscientious Dutch family of sea chart publishers in addition to being instrument makers and publishers of almanacs and books on geography, navigation, shipbuilding and lighthouses. The founder, Johannes van Keulen, settled in Amsterdam as a book and map seller and his descendants carried on the business for well over two hundred years.

Johannes van Keulen published his first *Zee Atlas* in 1680. His charts were bold, handsomely designed and engraved with large title pieces and scales, decorated with impressive costumed figures in the dress of the region represented. The charts were issued coloured, uncoloured and in special copies heightened with gold.

The *Zee Atlas* was gradually expanded to five volumes, with the help of Johannes's two sons, Gerard and Johannes junior. Gerard continued with the issue of the much larger map, which had to be folded four or more times to fit the atlas. He succeeded the Blaeus as Hydrographer to the Dutch East India Company.

313

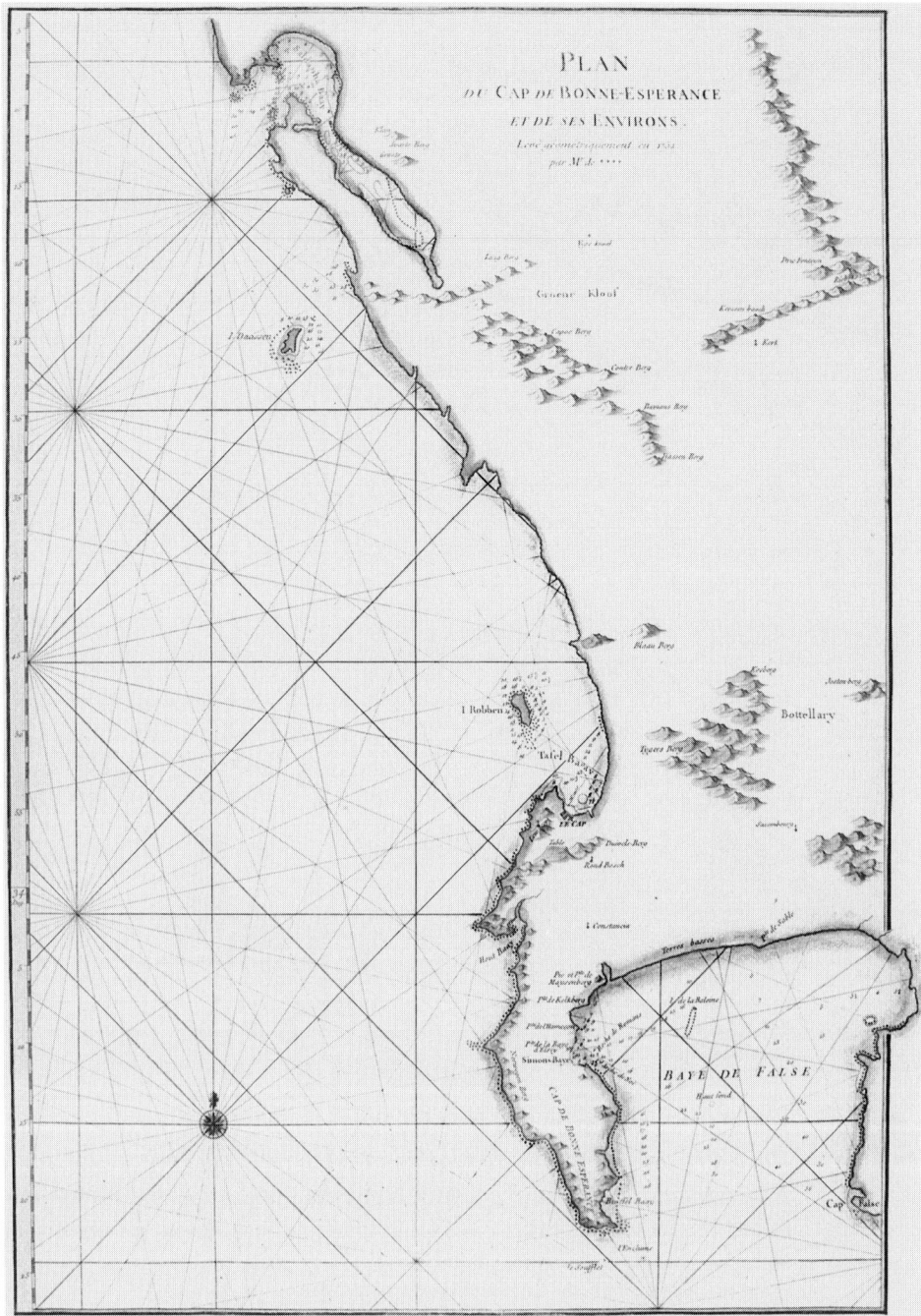

Map 276

APRÈS DE MANNEVILLETTE, Jean Baptiste Nicolas (1707-1780)
Plan du Cap de Bonne-Esperance et de ses Environs. (Map of the Cape of
Good Hope and its surroundings.) (Paris, G.N. Delahaye, 1752.)
Sea chart, 48 x 33 cm, coloured.
Top right: 8.

A detailed sea chart of the Cape of Good Hope from
Saldanha Bay to False Bay. In the interior are noted mountain
ranges and a limited number of place names. Saxembourg
appears as the old name for the present-day Stellenbosch,
although on some maps the two names coexist, suggesting
one to be a farm and the other a town. Saldanha Bay, Dassen

and Robben Islands are accurately depicted. Soundings,
presumably in French fathoms, appear in the sea surrounding
the coastline and are accurately placed. The colour is
probably original. It is on the whole a fine, accurate sea chart
of this area of the Cape for the year of its publication in
1752.

314

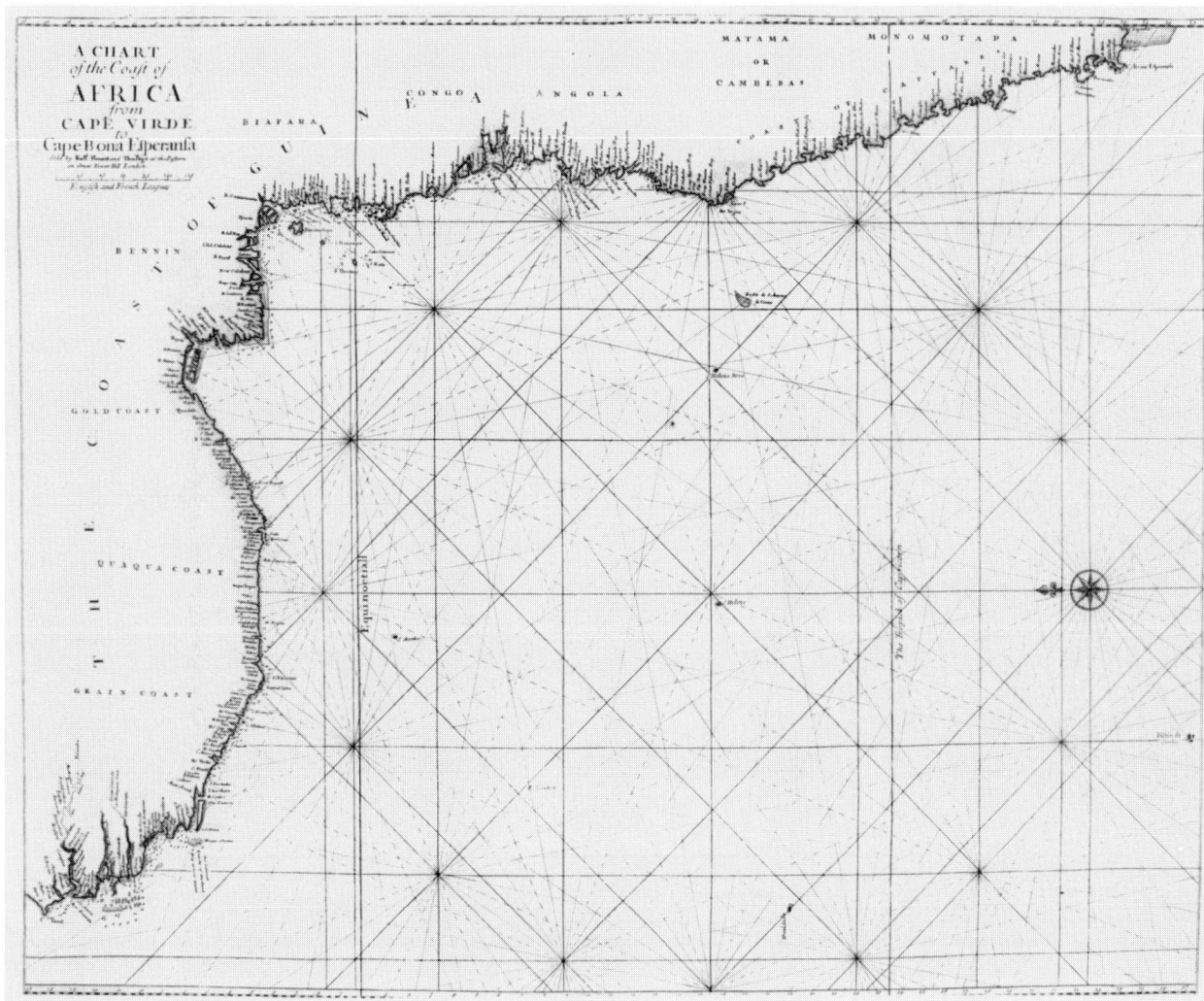

Map 277

MOUNT, Richard and PAGE, Thomas
A Chart of the Coast of Africa from Cape Virde to Cape Bona Esperance.
(London, R. Mount and T. Page, 1708.)
Sea chart, 44 x 55 cm, slightly coloured.
Scale: English and French leagues.

Richard Mount and Thomas Page were in partnership in London as publishers of marine charts in the eighteenth century. Mount, as a publisher and bookseller, had been apprenticed to W. Fisher, a London publisher who joined John Seller and John Thornton to print a series of marine atlases. Mount married into the Fisher family and took over part of Fisher's business on his death in 1691. Mount and Seller joined to publish Seller's *Atlas Maritimus* in 1675 and, with Thornton, the English edition in 1698. Thomas Page in partnership with Mount was succeeded by three further generations, all Thomas Page.

This sea chart of the west coast of Africa from Cape Verde to Cape Bona Esperanca has a simple title. The coastal placenames are numerous and based on Portuguese discoveries and exploration. The compass rose with the usual fleur-de-lys above it appears in the lower western ocean. This chart appears to have come from a later edition of the *Atlas Maritimus*, which was originally published in the third book of the *English Pilot* by John Seller, assisted by Richard Mount.

315

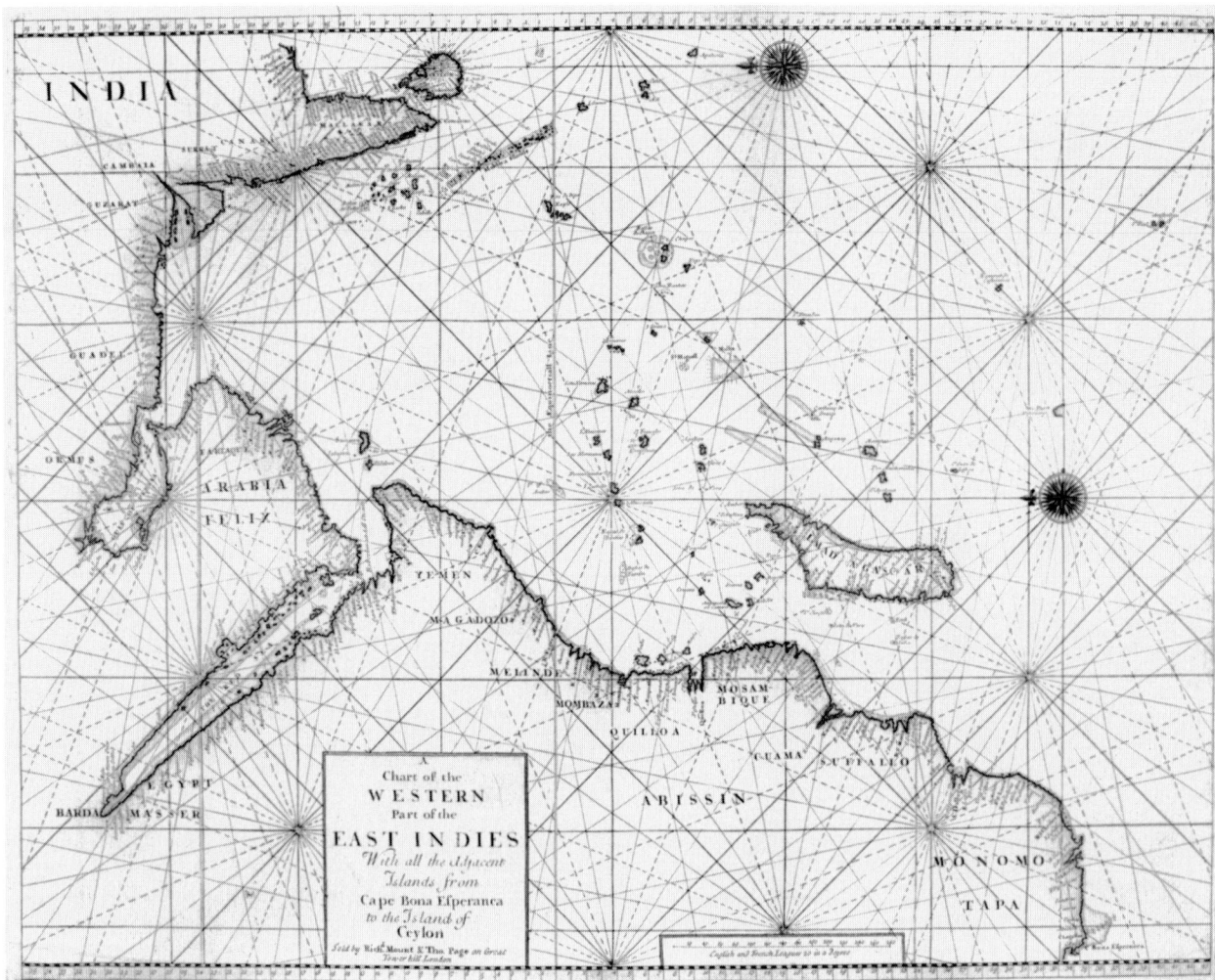

Map 278

MOUNT, Richard and PAGE, Thomas
A chart of the Western part of the East Indies with all the Adjacent Islands from Cape Bona Esperanca to the Island of Ceylon.
(London, R. Mount and T. Page, 1708.)
Sea chart, 44 x 56 cm, slightly coloured.
Scale in English and French leagues.

This sea chart, which covers the area from East Africa to India, was published by Mount and Page, and is a companion chart to the one previously described (see Map 277). It is from a later edition of the *Atlas Maritimus,* published originally by John Seller and re-engraved for this edition, with the original decorative title-piece and scale of miles omitted. Like its companion sheet it has a simple cartouche; two compass roses appear in the Indian Ocean and Portuguese placenames occupy most of the coast.

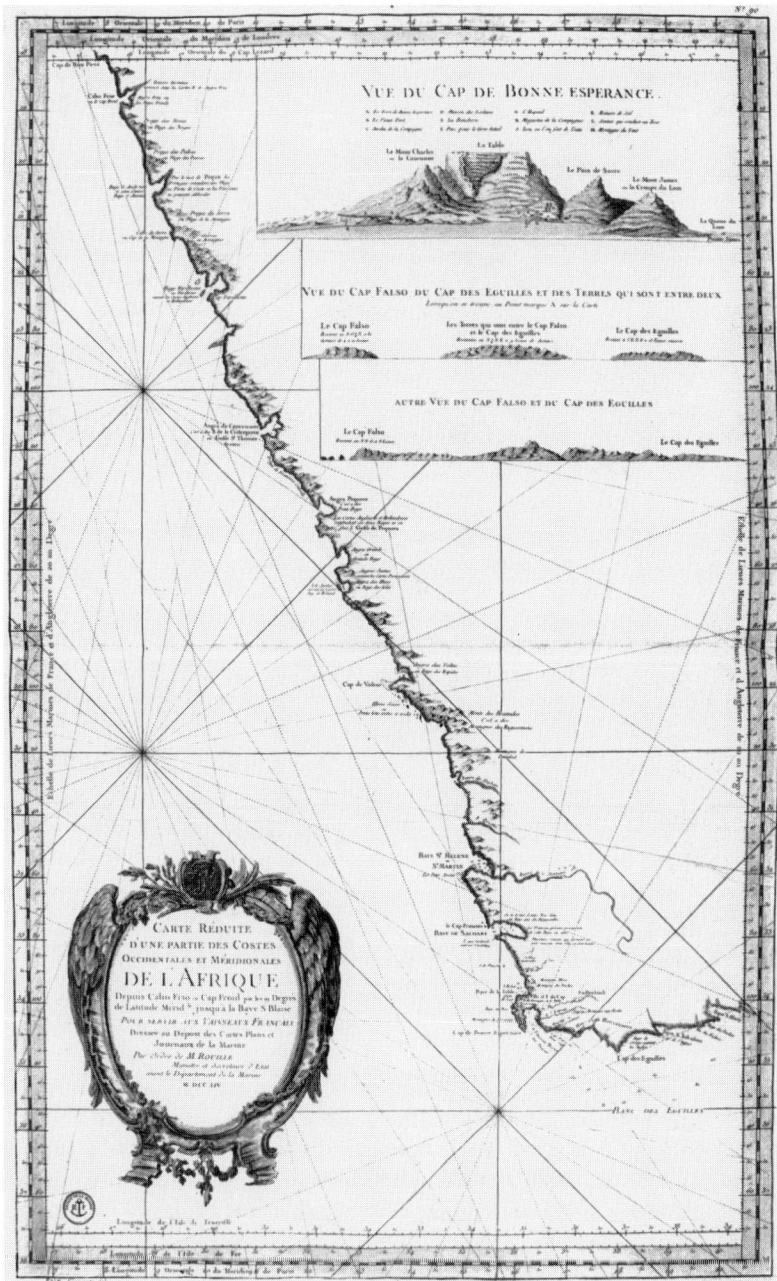

Map 279

BELLIN, Jacques Nicolas (1703-1772)
Carte Réduite d'une Partie des Costes Occidentales et Méridionales de l'Afrique depuis Cabo Frio ou Cap Froid par les 19 degrès de Latitude Merid. jusqu'à La Baye S. Blaise pour Servir aux Vaissaux de La Francais... par ordre de M. Rouille... (Map of southwest coast of Africa from Cap Froid to the Bay of St. Blaise.)
(Paris, J.N. Bellin, 1754.)
Sea chart, 89 x 55 cm, slightly coloured.
Prime meridian through Paris, Ferro Island, Tenerife, Cap Lezard and London.
Inset: View of the Cape of Good Hope.
Views of Cape Agulhas and False Bay headland.
Top right: No. 90.

A large attractive sea chart of the south-west coast of Africa from 19°S latitude to the Cape of Good Hope and beyond to the Bay of St. Blaise.

When King Louis XIV ordered the compilation of a complete survey of the coast of France under the Académie des Sciences the work was so slow that it was eventually put into the hands of Bellin to augment and correct. It was finally published in 1754. Bellin, attached to the Dépôt de la Marine, became hydrographer to the King and a member of the Royal Society of London. He was commissioned to map all the known coasts of the world. This was undertaken with great care and zeal, resulting in the *Hydrographique Français* which appeared in two volumes in 1756-1765. Bellin also compiled the maps for Prévost's twenty-five volume *Histoire Générale des Voyages* of 1747-1780. Volume Five, published in

1748, had a large number of maps of Africa – mainly of the south.

This particular chart has large silhouettes of the coast on the upper right; namely 'Vue du Cap de Bonne Esperance,' 'Vue du Cap Falso, du Cap Des Eguilles et des terres qui sont entre deux,' 'Autre Vue du Cap Falso et du Cap des Eguilles.' Cape Town, Constance (Constantia), Stellenbosch and Hangklip are accurately marked.

There were five variations of this chart, with and without Bellin's name. On this particular variation it is replaced by 'Par Ordre de M. Rouille, Ministère et Secretaire d'Etat'. The arms of France at the top of the cartouche are replaced by a monogram, R(épublic) F(rançais), and the original fleur-de-lys removed from the stamp of the Dépôt de la Marine. At the left lower border appears the price: cinquante sols.

317

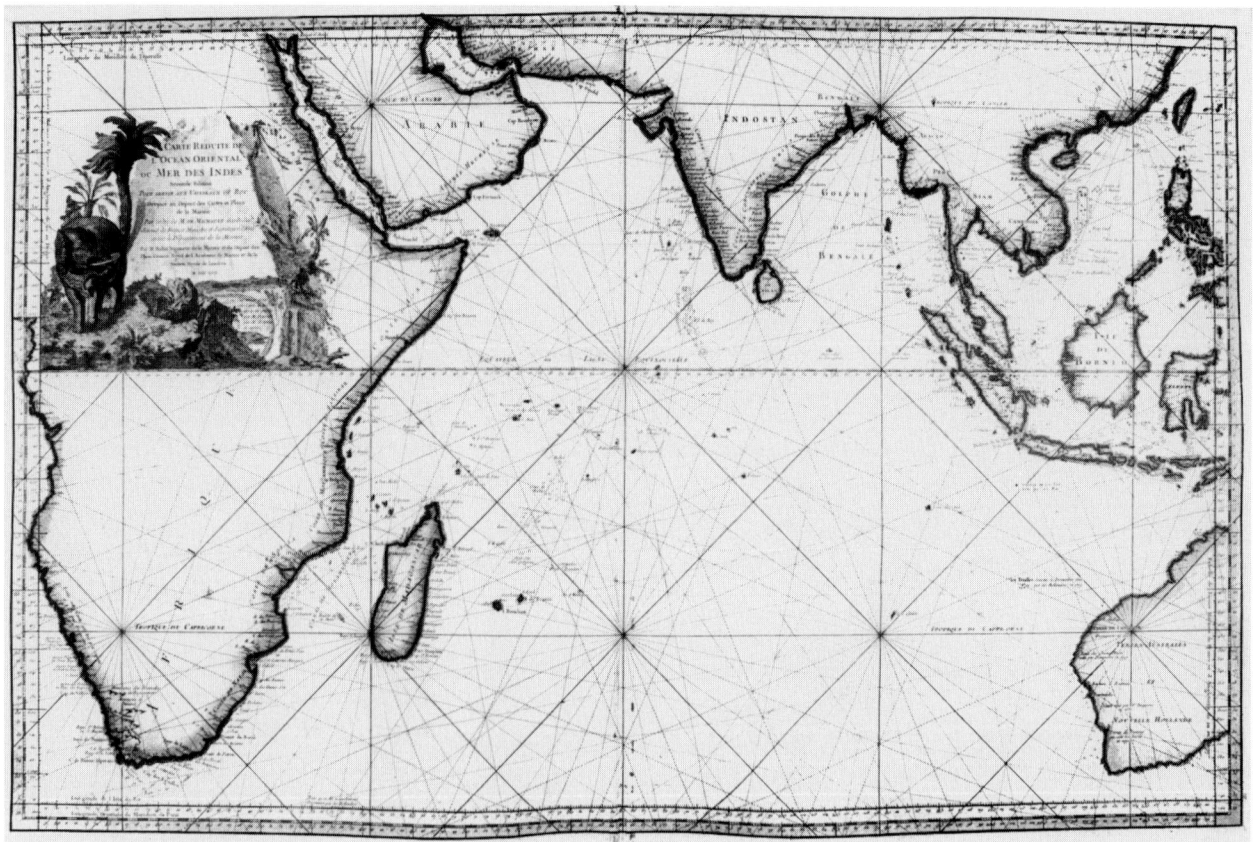

Map 280

BELLIN, Jacques Nicolas (1703-1772)
Carte Réduite de L'Ocean Oriental ou Mer des Indes... Par Ordre de M. de Machault... (Map of the Indian Ocean.)
(2nd edition, Paris, J. Bellin, 1757.)
Sea chart, 55 x 86 cm, coloured.
Prime meridian through Ferro Island, Paris and Tenerife.

A large sea chart, issued by Jacques Nicolas Bellin, covering the area from Africa to the west coast of Australia. A large cartouche with a long title is sited in the left upper corner, including trees, foliage, an elephant and a crocodile.

Except for coastal names the whole of Africa is left blank, with one town, Ville du Cap (Cape Town), noted inland.

The west coast, between Cape Frio and Cape Voltas, is marked 'little known and not frequented.' There are numerous rhumb lines involving all the mainland mentioned, including West Australia.

318

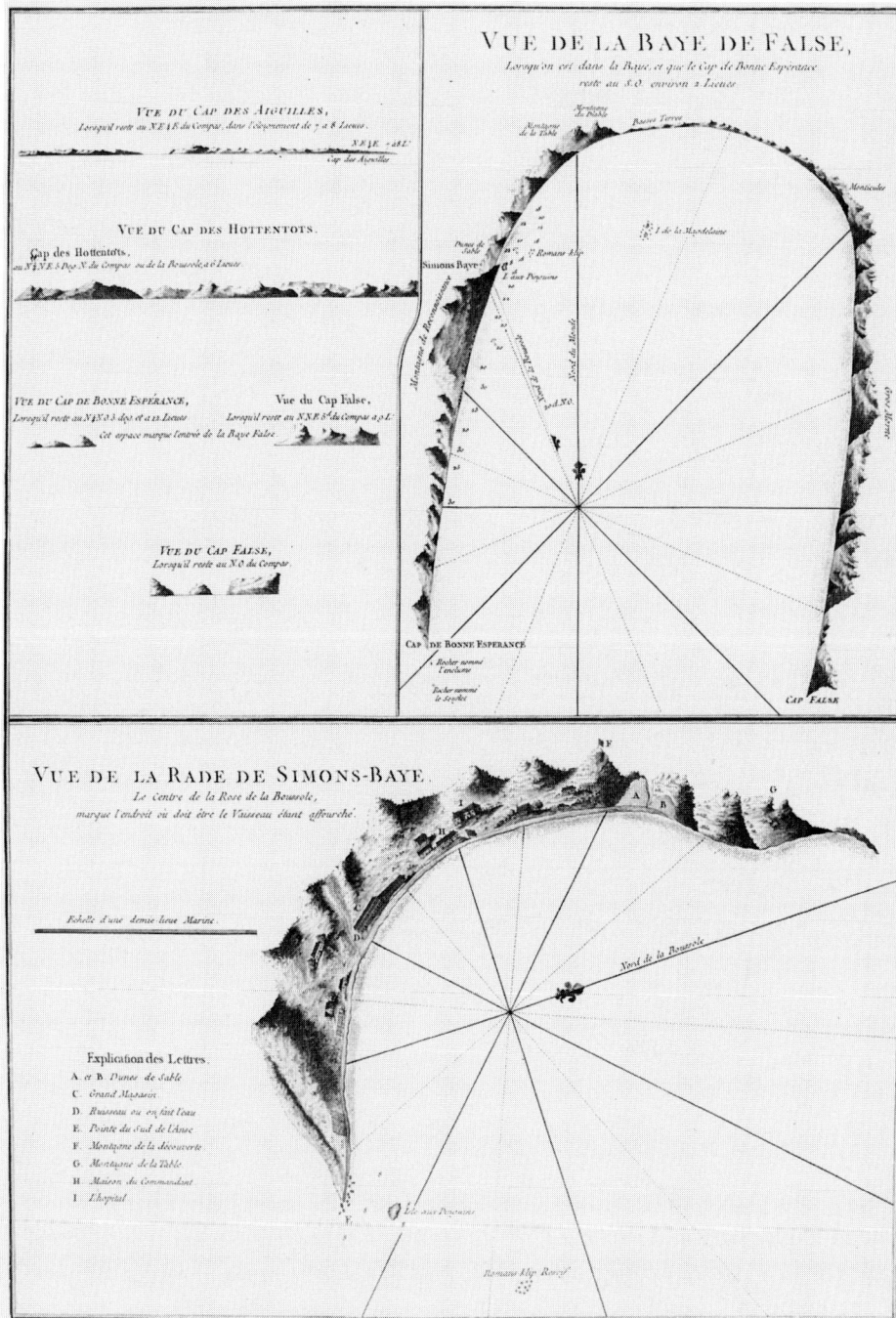

VUE DU CAP DES AIGUILLES,
Lorsqu'il reste au N.E à E du Compas, dans l'éloignement de 7 à 8 Lieues.

N.E à E 182.
Cap des Aiguilles.

VUE DU CAP DES HOTTENTOTS.
Cap des Hottentots.
au N.E à E Deg. N du Compas, ou de la Boussole, à 6 Lieues.

VUE DU CAP DE BONNE ESPÉRANCE,
Lorsqu'il reste au N.E S. 3 deg. et à 12 Lieues.

Vue du Cap False.
Lorsqu'il reste au N.N.E 8°. du Compas à 9 L.
Cet espace marque l'entrée de la Baye False.

VUE DU CAP FALSE,
Lorsqu'il reste au N.O. du Compas.

VUE DE LA BAYE DE FALSE,
Lorsqu'on est dans la Baye, et que le Cap de Bonne Espérance reste au S.O. environ 2 Lieues.

VUE DE LA RADE DE SIMONS-BAYE.
Le Centre de la Rose de la Boussole, marque l'endroit où doit être le Vaisseau étant affourché.

Echelle d'une demie lieue Marine.

Explication des Lettres.
A et B *Dunes de Sable*
C *Grand Magasin*
D *Ruisseau où on fait l'eau*
E *Pointe du Sud de l'Anse*
F *Montagne de la découverte*
G *Montagne de la Table*
H *Maison du Commandant*
I *L'hopital*

Map 281

APRÈS DE MANNEVILLETTE, Jean Baptiste Nicolas (1707-1780)
Vue de la Baye de False. Vue de la Rade de Simons-Baye. (View of False Bay. View of the roadstead in Simon's Bay.)
(Paris, J.B. Après de Mannevillette, 1775.)
5 silhouettes on one sheet,
47 x 33 cm, coloured.
Top right: 8.2d.

This sea chart by Après de Mannevillette gives views of False Bay and Simon's Bay, with finely engraved reliefs (in original colour) of all the important mountains, especially the Table Mountain range and its components, referred to in a key in the Simon's Bay chart. Separate silhouettes of Cape Agulhas, the Cape of Hottentots and False Bay are depicted alongside the False Bay chart. As the title indicates, the chart depicts the Simon's Bay roadstead with its buildings, accurately drawn, such as the commandant's residence, the hospital and the warehouse, together with rivulets and the southernmost tip of the Cape of Good Hope – Cape Point.

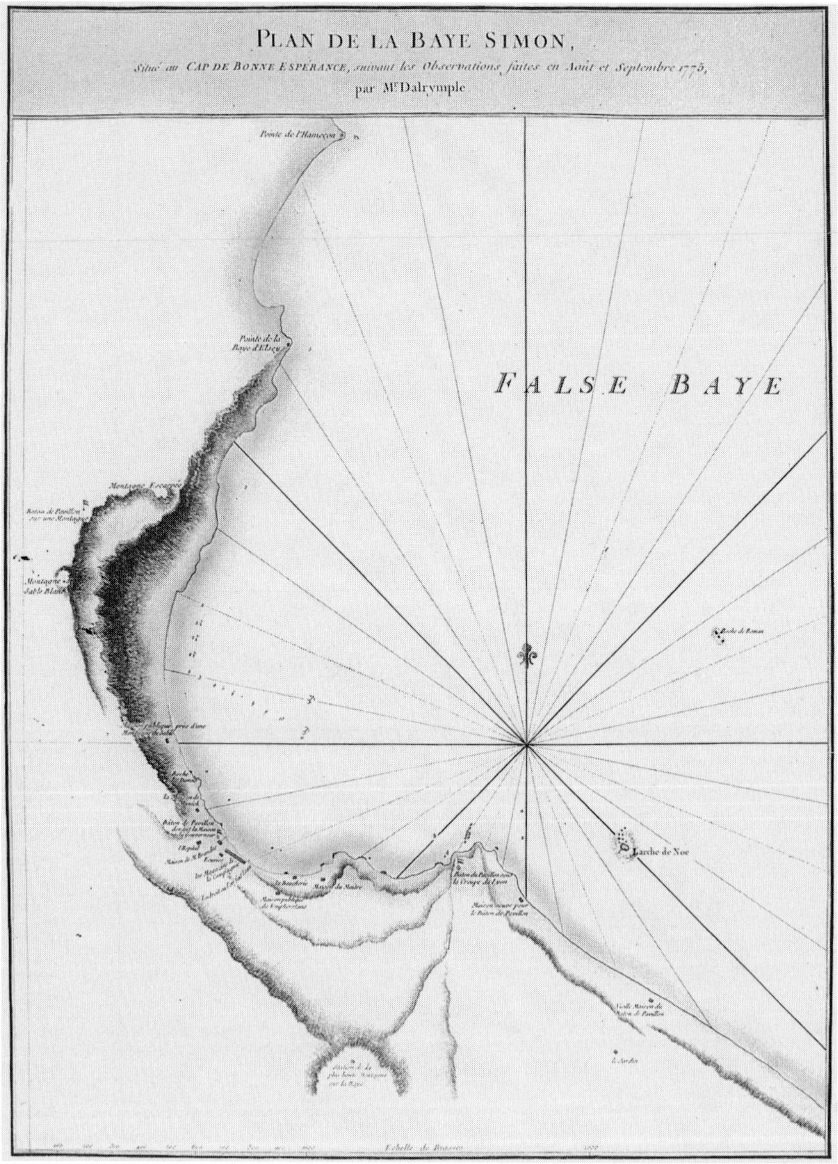

PLAN DE LA BAYE SIMON,
*Situé au CAP DE BONNE ESPERANCE, suivant les Observations faites en Août et Septembre 1775,
par M.* Dalrymple

Map 282 APRÈS DE MANNEVILLETTE, Jean Baptiste Nicolas (1707-1780)
*Plan de la Baye Simon, Situé au Cap de Bonne Esperance, suivant les Obser-
vations faites en Août et Septembre 1775, par Mr. Dalrymple.* (Plan of Simon's
Bay at the Cape of Good Hope.) (Paris, J.B. Après de Mannevillette, 1775.)
Sea chart, 45 x 33 cm, coloured.
Scale in French fathoms.
Top right: 8.d.

This is one of a series of fine sea charts of the Cape of
Good Hope and its bays, also depicting relief views of
mountains and buildings along the coastline, especially that
of False Bay. It is an accurate map of the area at that time.
Like Map 281 it is in original colouring. The soundings are
noted in French fathoms.

Après de Mannevillette was a French sailor and
hydrographer born in Le Havre. He was educated in Paris
and in 1745 produced the *Neptune Oriental*, with a second

folio edition in 1775. He was the first to determine longitude
by measuring the distances of the sun and moon. The
English hydrographer Alexander Dalrymple provided him
with considerable assistance in publishing his sea charts. As
captain of the *Glorieux* he conducted the French astronomer
de la Caille to the Cape of Good Hope in 1750. It was on
this trip that he obtained much of the information he used
in creating his series of fine and accurate sea charts of the
Cape of Good Hope.

320

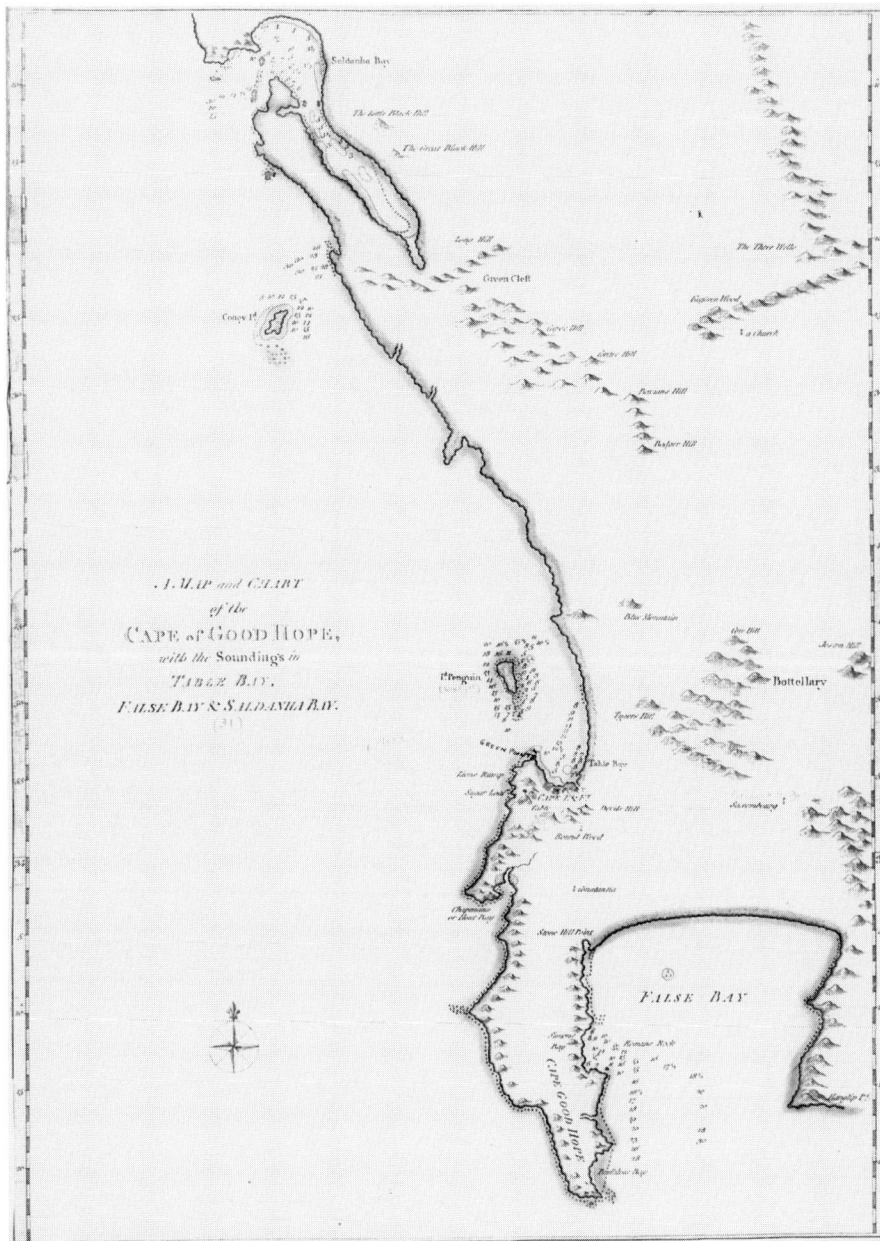

Map 283

BEW, John (fl. 1774-1793)
A Map and Chart of the Cape of Good Hope, with the Soundings in Table Bay, False Bay and Saldanha Bay. Engraved by John Lodge Jnr.
(London, J. Bew, Pater Noster Row, 1781.)
Sea chart, 37,5 x 27,5 cm, slightly coloured.

This late eighteenth-century chart attempts, in common with most others of this period, to abolish all fictitious detail and to record what is really known. The geographic positions of most of the places marked are reasonably correct, although some of the English translations of the original Dutch names have to be looked at twice before recognising them, eg. 'Round Wood' for Rondebosch, 'Cow Hill' for Koeberg, 'Blue Mountain' for Blauberg, 'Green Cleft' for Groenkloof. The name Saxenbourg appears close to the site of Stellenbosch, and may be the original name of the town. A scale and certain place names have been added in by hand on this copy. The only decoration is a simple compass rose with no rhumb lines. Soundings are given for Table Bay, Penguin (Robben) Island, Simon's Bay and part of False Bay – presumably the only places where ships normally approached shore at that period.

This map was published by John Bew in 1781 as the English edition of an identical sea chart of the coast from Saldanha to False Bay, published in 1775 by Après de Mannevillette, the French sailor and hydrographer who produced the *Neptune Oriental* in 1745.

321

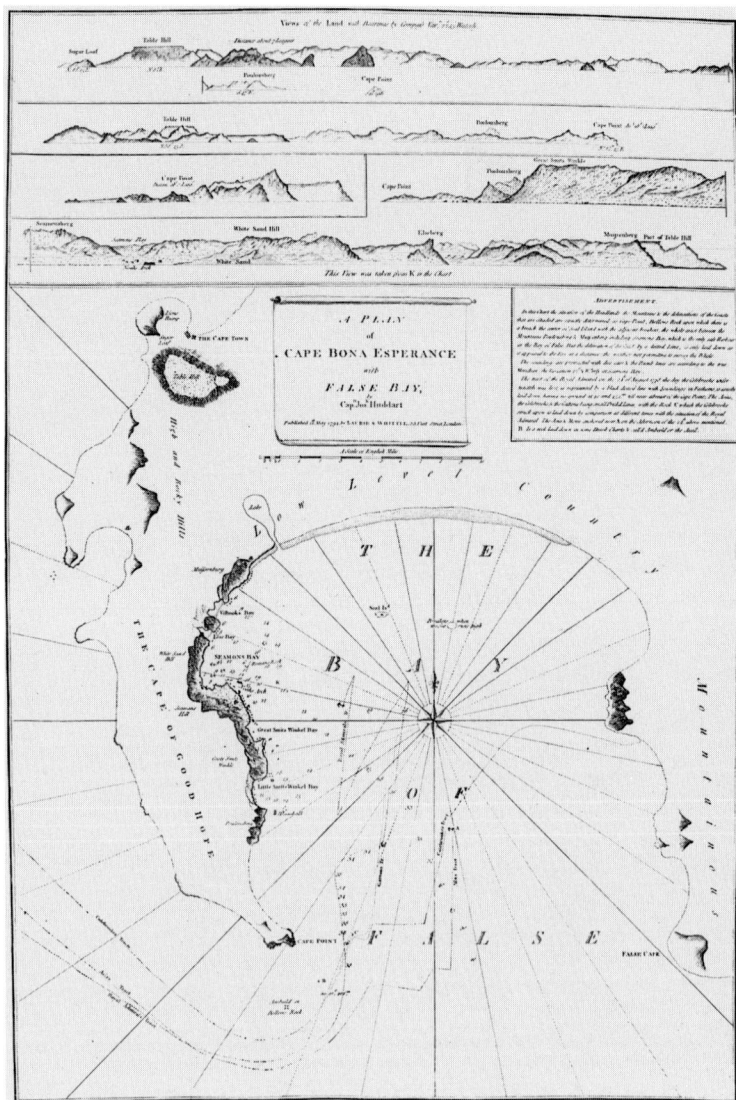

Map 284

HUDDART, Joseph (1741-1816)
A Plan of Cape Bona Esperance with False Bay.
(London, Laurie and Whittle, 53 Fleet Street, 1794.)
Sea chart, 66 x 46 cm, uncoloured.
Scale in English miles.
Inset: *Views of the land with bearings by compass.*

An interesting large-scale plan of the Cape of Good Hope with False Bay prepared in 1778 by Captain Huddart, hydrographer to the East India Company, on his first voyage in command of the *Royal Admiral*, an East Indiaman. The *Colebrook*, another East Indiaman sailing in company, struck a rock off the Cape, which prompted Huddart to compile this chart and endeavour to find the rock. He and his son narrowly escaped drowning when the boat they were working in overturned in the surf.

The chart appeared in some editions of Sayer and Bennett's *Oriental Pilot* and was also published in conjunction with Laurie and Whittle, publishers and engravers originally of Fleet Street, London. Huddart also published many sea charts of the channels of England and Ireland.

An advertisement at the top right of the False Bay chart describes the bay in some detail. All known places in the bay are illustrated, with soundings of the sea. Cape Point appears with the outline of a mountain range behind it and the coastline continues as far as Table Mountain and Cape Town. Seamons (Simon's) Bay shows the large adjacent rocks and Seal Island further offshore.

The cartouche appears in the middle of the map as a rectangular sheet with its title. At the top of the map are four silhouettes of all the mountains depicted in the chart from Table Mountain to Seamensberg. The compass rose, radiating rhumb lines, is situated within False Bay.

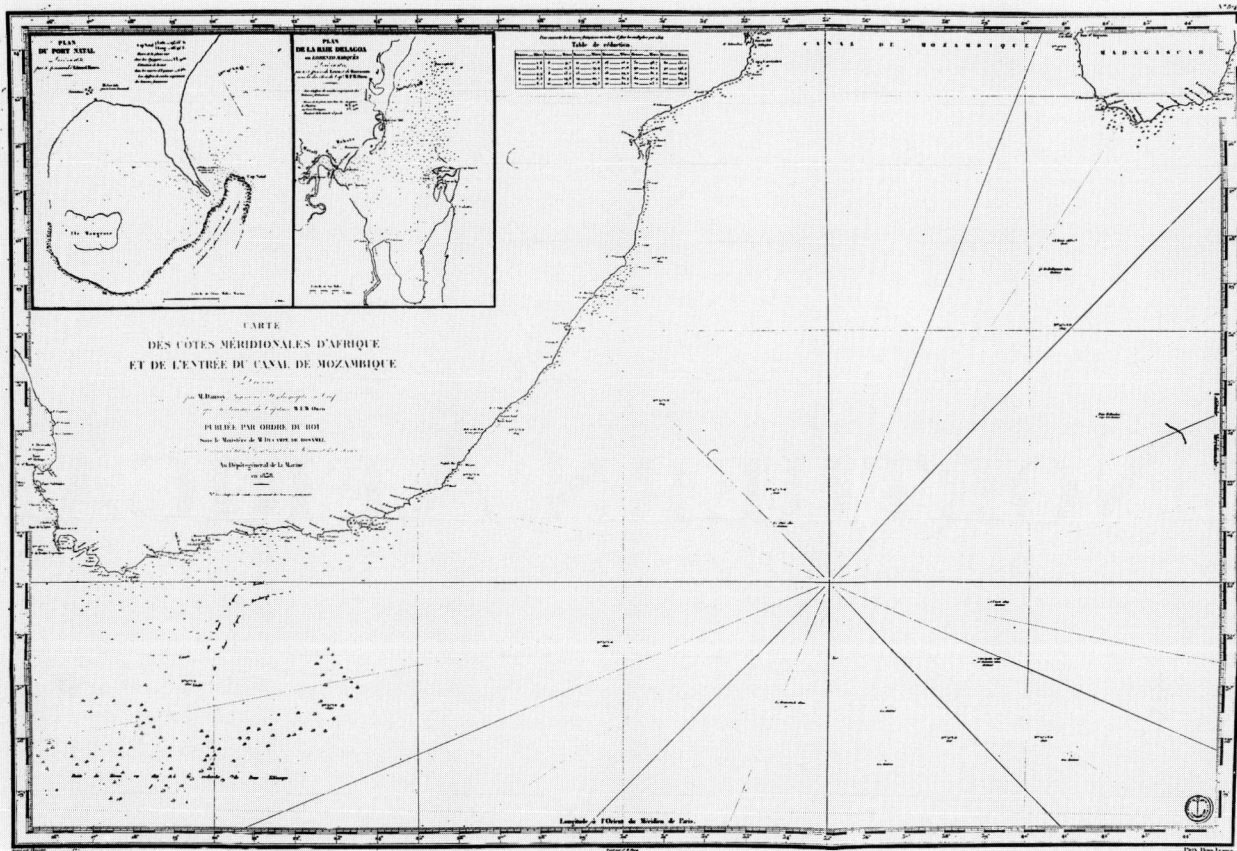

Map 285

DAUSSY, Pierre (1792-1860)
Carte des Côtes Méridionales d'Afrique et de l'Entrée du Canal de Mozambique... d'après les Travaux de Capt. W.F.W. Owen. (South coast of Africa and the Mozambique Channel.) Engraved by Chassant and Hacq.
(Paris, P. Daussy, 1838.)
Sea chart, 59,5 x 90 cm, uncoloured.
Scale in miles.
Prime meridian through Paris.
Inset: *Plan de Port Natal* by Edward Hawes 1831. *Plan de la Baie Delagoa* by W.F.W. Owen
Top right: No. 874.

A large marine map including some of the lower west coast of Africa, the whole of the southern and east coasts up to the southern tip of Madagascar, and the entrance into the Mozambique Channel at 24 1/2°S. It was published in 1838 by Pierre Daussy, a French engraver and astronomer and Directeur d'Hydrographie Dépôt de Marine, and based on the travels of Capt. W.F.W. Owen. It includes two detailed insets: one of Port Natal, with the settlement that was to become D'Urban in 1835, published by Commander Edward Hawes in 1831; and the other based on the travels of Capt.

W.F.W. Owen of the Royal Navy with his ships, the *Leven* and the *Barracouta*, detailing Delagoa Bay and 'Lorenzo Marques' in 1822.

This is a finely engraved map indicating many soundings and the routes of some of the explorers of the southern seas. Chassant, engraver for the Dépôt de la Marine, is noted as such on the map, and Hacq is also mentioned, probably as the script engraver, also for the Dépôt de la Marine. *Prix deux francs* is noted at the bottom right.

323

NORTHERN AFRICA

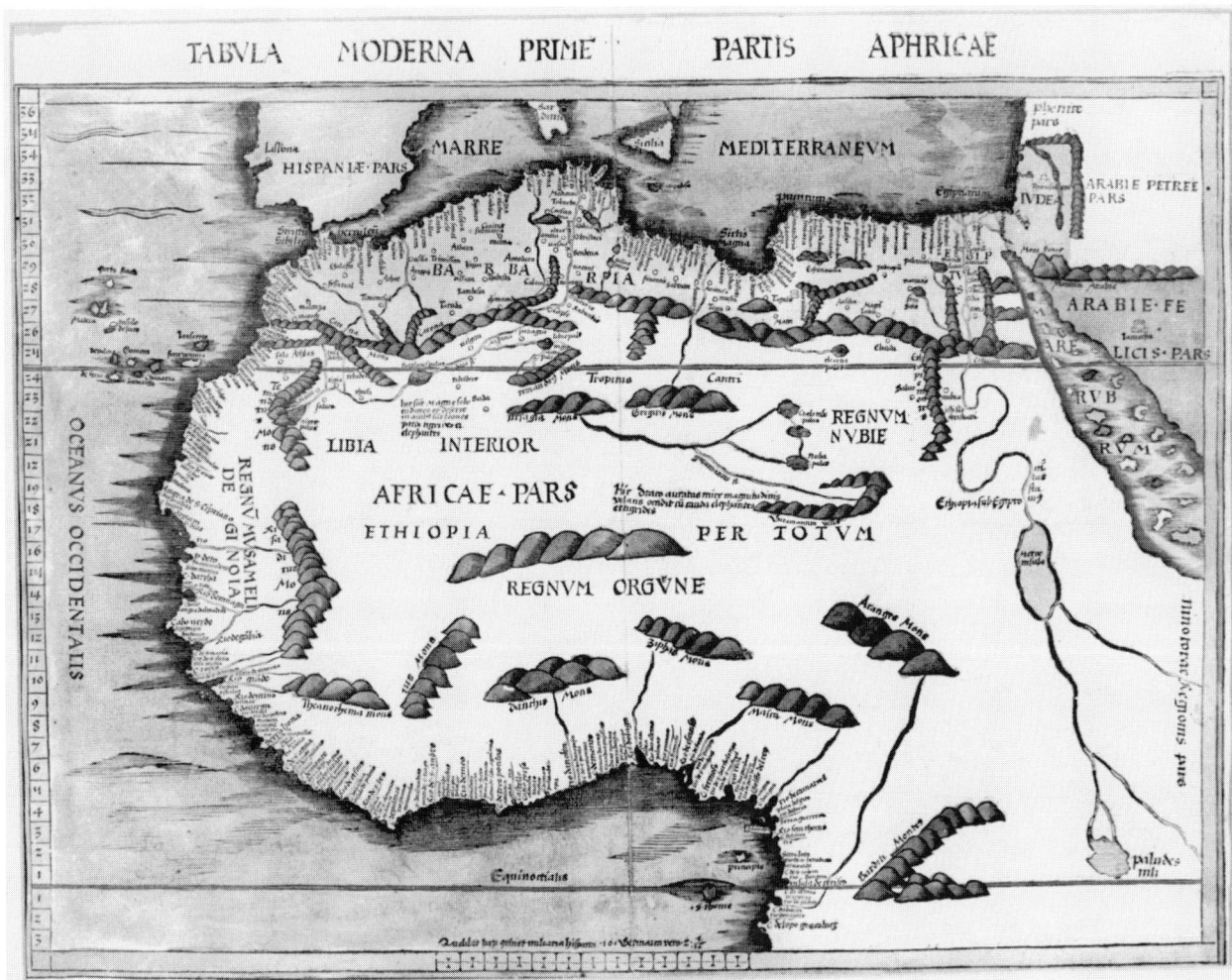

Map 286 WALDSEEMÜLLER, Martin (1470-1518)
Tabula Moderna Prime Partis Aphricae. (Modern map of first part of Africa.)
(Strasbourg, Johannes Schott, 1513.)
Map, 41 x 57 cm, coloured.
Scale in German and Italian miles.

This map of North Africa is from the Strasbourg edition of Waldseemüller's Ptolemy atlas of 1513, published by Johannes Schott. As northern Africa was better known than southern Africa at that period it has more geographical information, especially in the interior. Various kingdoms are described, with a multitude of coastal names, and as with the southern part (see Map 149) these are placed within the coastline. The Mediterranean and Red Seas are prominently featured as well as a number of islands on the northwest coast – Madeira, the Canaries, etc. – which at that time were already well known to the Portuguese in their initial efforts to discover a route to the Indies.

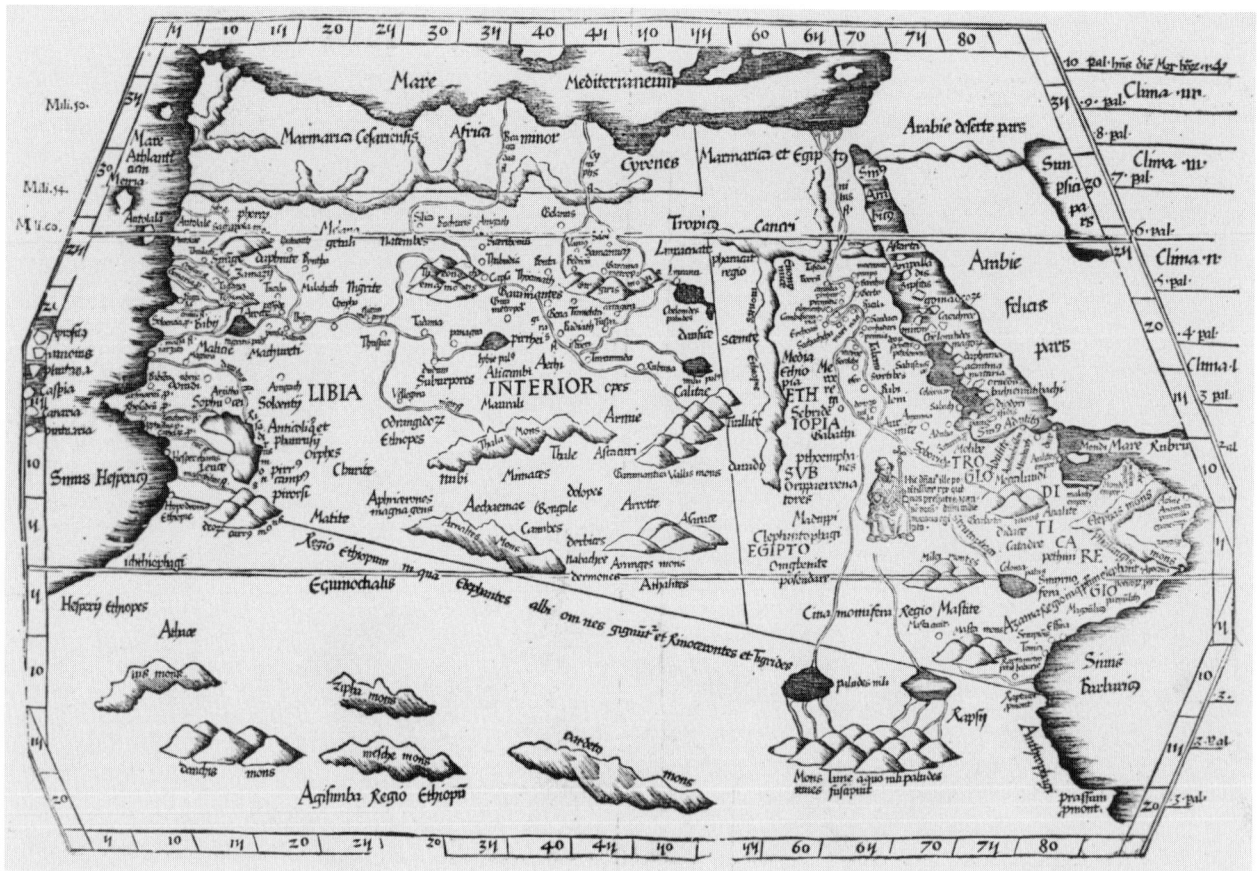

Map 287 PTOLEMY, Claudius (87-150)
(North Africa.) (Lyons, Melchior and Gaspar Trechsel, 1535.)
Map, 29 x 44 cm (shape irregular), uncoloured.

This woodcut of North Africa, with detailed engraved Latin text on the verso, is from a volume the title page of which reads *Clavdii Ptolemaei Alexandrini Geographicae Enarrationis Libri Octo. Ex Bilibaldi Pirckeymheri tralatione, sed ad Graeca & prisca exemplaria a Michaele Villanouano iam primum recogniti... Lvgdvni cx officina Melchioris et Gasparis Trechsel fratrum. M.D.XXXV.*

As seen in the preface printed on the reverse of the titlepage, the text of this 1535 Lyons edition – or rather of its commentaries – was edited by the celebrated martyr to freedom of thought, Michael Servetus or Villanovanus. The edition is therefore designated 'edito prima Servili' (A. E. Nordenskiöld).

The maps were printed from the blocks of Laurent Fries, the famous physician, astrologer and geographer of Metz and Strasbourg. Stevens, in his *Ptolemy Geography*, reports that the woodcut border and ornaments on the verso are said to be the work of Hans Holbein. He also mentions that many copies of the book are said to have been burned by the order of Calvin at the time of the execution of its editor Servetus in 1553. The maps are the same as those in the Ptolemaic editions of 1522 and 1525.

Geographically it presents the characteristic features of the Ptolemaic edition, trapezoid in outline with convoluted mountain ranges. The origin of the Nile conforms to the Ptolemaic tradition and the seated figure below Ethiopia is undoubtedly that of the mythical Prester John.

For further interesting reading and more detailed biography, refer to the article 'Burned for his Beliefs. The story of Michael Servetus' by László Gróff in *The Map Collector*, December 1982, p.8.

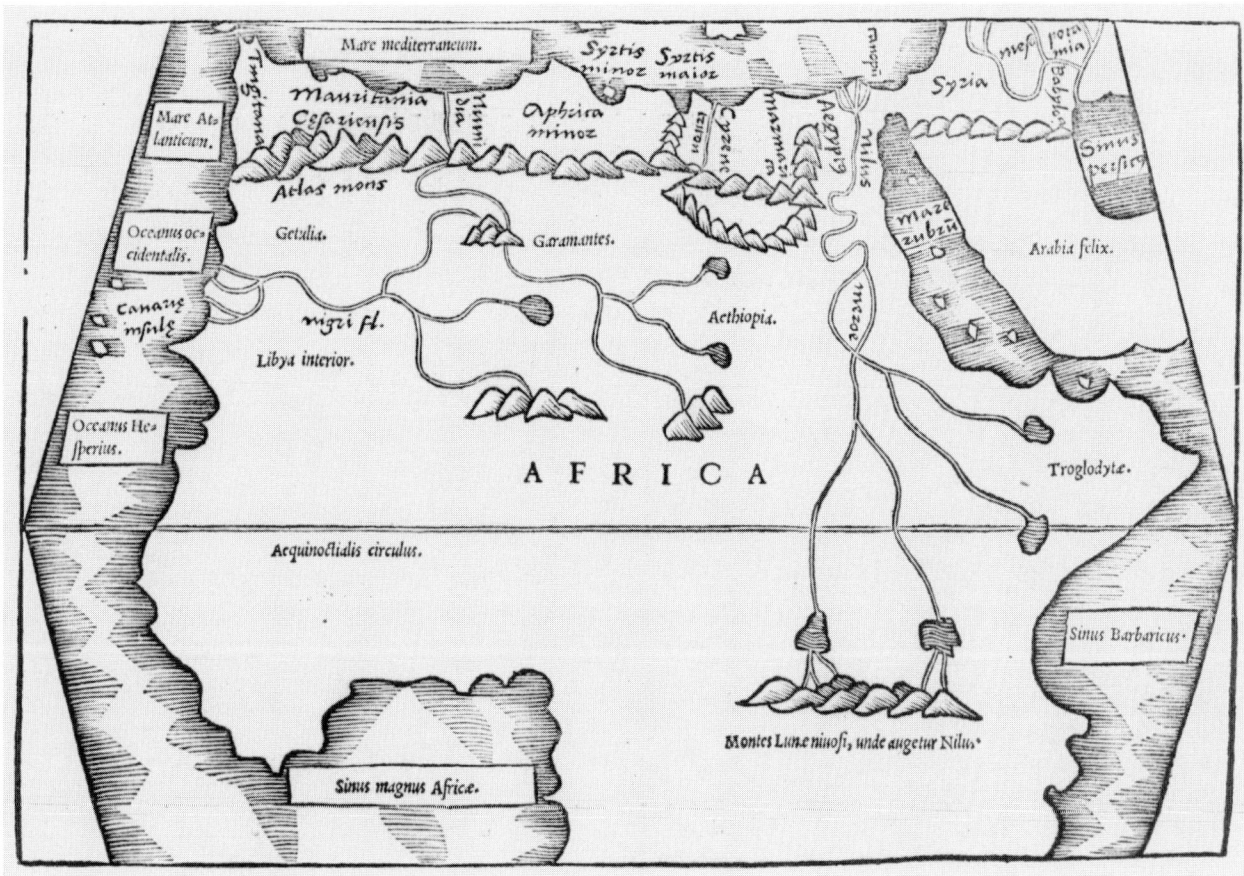

Map 288   MELA, Pomponius (1st century AD)
(North Africa.)
(Basle, 1538)
Map, 16 x 24 cm, uncoloured woodcut.
Top right: 12.
In: Solinus, Caius Julius. *Polyhistor; Rerum Toto Orbe...* Basle, 1538, fold p. 82.

This woodcut map was published in the *Polyhistor* by Solinus and Mela. Pomponius Mela was born in Spain and wrote his *De Orbis Situ* (first published in 1540) during the reign of Emperor Claudius. Solinus worked in the time of the emperors Augustus and Vespasian. His *De Situ Orbis Terrarum* was first published in Venice in 1473 and his *Polyhistor Rerum Toto Orbe* was published in Basle in 1538, with maps and commentaries by Sebastian Münster. (Also see Map 303.)

This map depicts Africa above the equator, with the Mediterranean. The Niger and the Nile are featured separately, each arising from a multiplicity of lakes; the Nile has the Montes Lunae as its origin. Geographically the interior is devoid of information except for rivers and mountain ranges, and a few names appear, among them 'Troglodyte,' on the east coast.

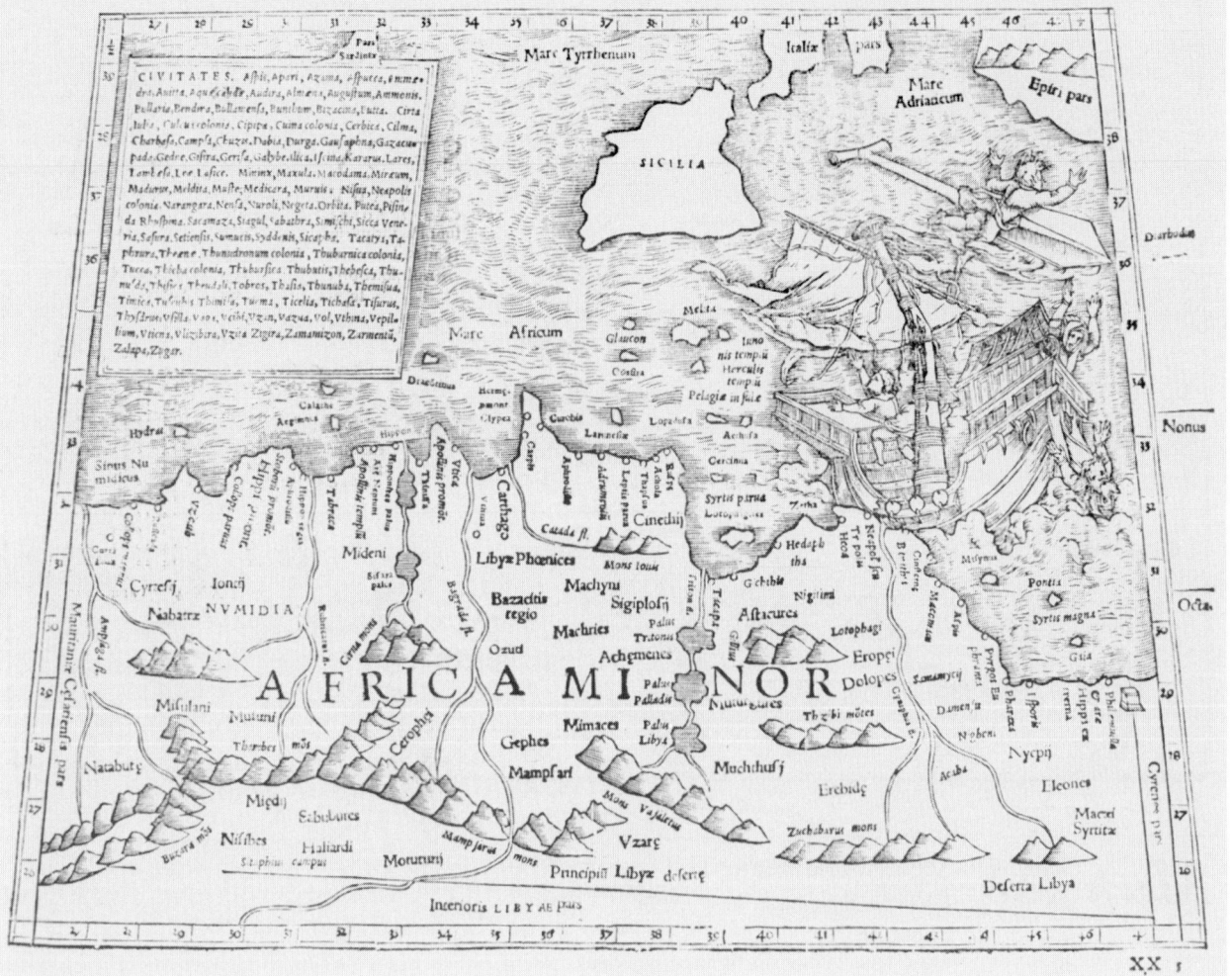

Map 289 MÜNSTER, Sebastian (1489-1552)
Aphricae Tabula II. (Map of North Africa.)
(Basle, S. Münster, 1544.)
Map, 25 x 33 cm, uncoloured.
Top left: 964.
Top right: 965.
Bottom right: XX5.
Text on verso.

This woodcut map of North Africa has a striking illustration of a shipwrecked sailing vessel which dominates the eastern Mediterranean. The size of the ship is an indication of the courage and ingenuity of these early travellers under hazardous conditions.

In the upper left is a legend in Latin containing all the names of towns on the map. The mountain ranges are depicted in the manner typical of the early Ptolemaic fashion, together with the wavy lines depicting the sea. Large rivers appear to be entering the Mediterranean in a most bizarre and inaccurate fashion. It is the typical woodcut illustration that makes these Münster maps fascinating and sought after by map collectors. This map was obviously designed as a double page woodcut illustration, because there are two numbers, 964 and 965, at the top corners. The Latin title in the top border, between two fleurons, would indicate its origin from an unidentified Latin edition, and there is also text in Latin on the verso.

Map 290 MÜNSTER, Sebastian (1489-1552)
Altera Generalis Tab. Secundum Ptol. (Another map according to Ptolemy.)
(Basle, S. Münster, c. 1550.)
Map, 25 x 34 cm, coloured.
Another Latin title on verso.

A map of the northern hemisphere with a Latin title across the top. It also appears in Münster's *Cosmographia* of 1550, though in German and smaller. It is a finely executed woodcut, particularly striking for the surrounding billowing cloud effect and the numerous wind gods blowing their air onto the earth.

Europe, North Africa, Asia, the Middle East and India are included, and an unusual feature is the island of Taprobana below India, which could be Sri Lanka (Ceylon).

Africa is prominent on the left of the map with well-defined, rope-like mountain ranges. The Nile appears with its multiple origins in an unnamed range of mountains and the land to the south is labelled as *terra incognita*.

Reference is made to Ptolemy in another title with text on the verso of the map. Münster paid tribute to the ancients in his reproduction of Ptolemy's *Geographia*, but at the same time kept abreast of contemporary discoveries in subsequent editions.

329

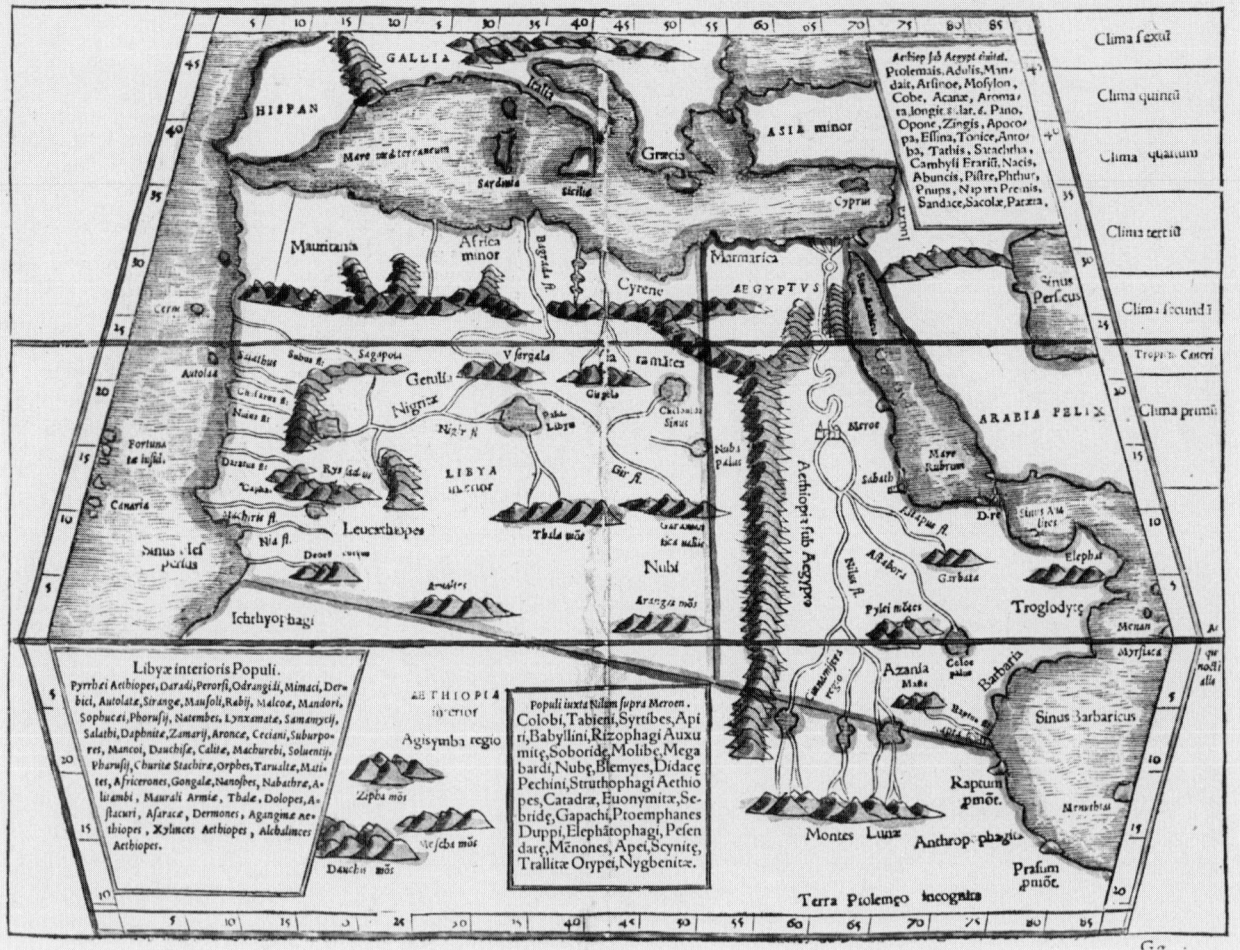

Map 291 PTOLEMY, Claudius (87-150)
Aphricae Tabula IIII. (Map of North Africa.) (Basle, H. Petri, 1571.)
Map, 25 x 34 cm, coloured.
Top right: 519.
Top left: 518.
Title on verso: *Quarta Aphricae Tabula, continent interiorem Libyam*
 et Aethiopiam quae sub Aegypto...
Latin text on verso.

This map shows Africa to approximately 24°S, beyond which it is labelled *terra Ptolemeo incognito*. It is a fine woodcut showing mountain ranges, lakes and rivers. The shape of Africa opposite Spain is sharply angled and the projection is trapezoidal, as were many editions of this period.

There are two legends within the lower section of the map – one on the inhabitants of the interior of Libya and the other on those of the Nile above Meroe. A third legend at the upper right lists the towns of Aethiopia and Egypt.

This map is identified as a Ptolemy edition of 1571, probably by Münster, after Strabo, as Nordenskiöld (*Facsimile Atlas Basileae*, 1571) notes it as 'Strabonus rerum Geographicum libri septem decum... Henripetrini 1571.'

330

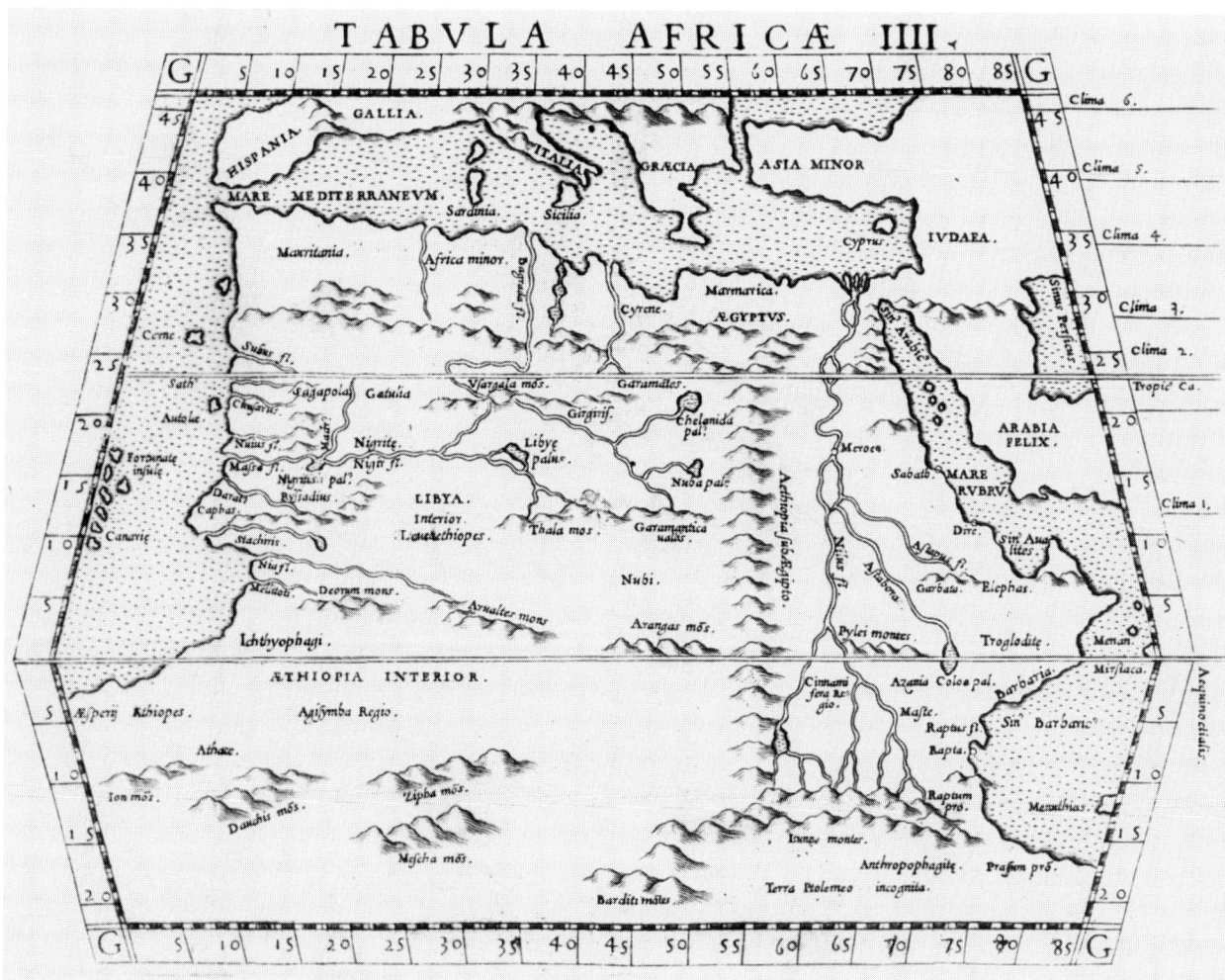

TABVLA AFRICÆ IIII.

Map 292   PTOLEMY, Claudius (87-150)
Tabula Africae IIII. (Map of Africa.) (Venice, G. Ruscelli and V. Valgrisi, 1561.)
Map, 18 x 24 cm, uncoloured.
Italian text on verso.

This Ptolemaic map of North Africa is from the atlas, *La Geograpfia di Claudio Tolomeo Alessandrino, Nouvamente tradatta di Greco in Italiano...* edited by Girolamo Ruscelli and printed by V. Valgrisi, both of Venice, in 1561. The projection is trapezoidal and the technique copperplate engraving, with the sea stippled in the Italian style. The Mediterranean countries are illustrated up to Judaea, with northern Africa as far as approximately 20° S. The large rivers – especially the Nile, arising from a range of mountains, and the Niger, in the west – are featured prominently. The mountain ranges are also typically shown with shadows on their western sides. The verso has text in Italian. This work achieved commercial success and editions were printed through 1599. The book had a series of Ptolemaic maps and their counterparts as 'modern' maps. On the next page we show two such 'modern' maps of North Africa from the last edition of this work published by Heirs of Melchior Sessa, in Venice in 1599. We note that the maps in the 1599 edition, although printed largely from the same plates as earlier editions, often had sea creatures and sailing ships added in the oceans. These maps were rarely (if ever) colored at the time they were made.

331

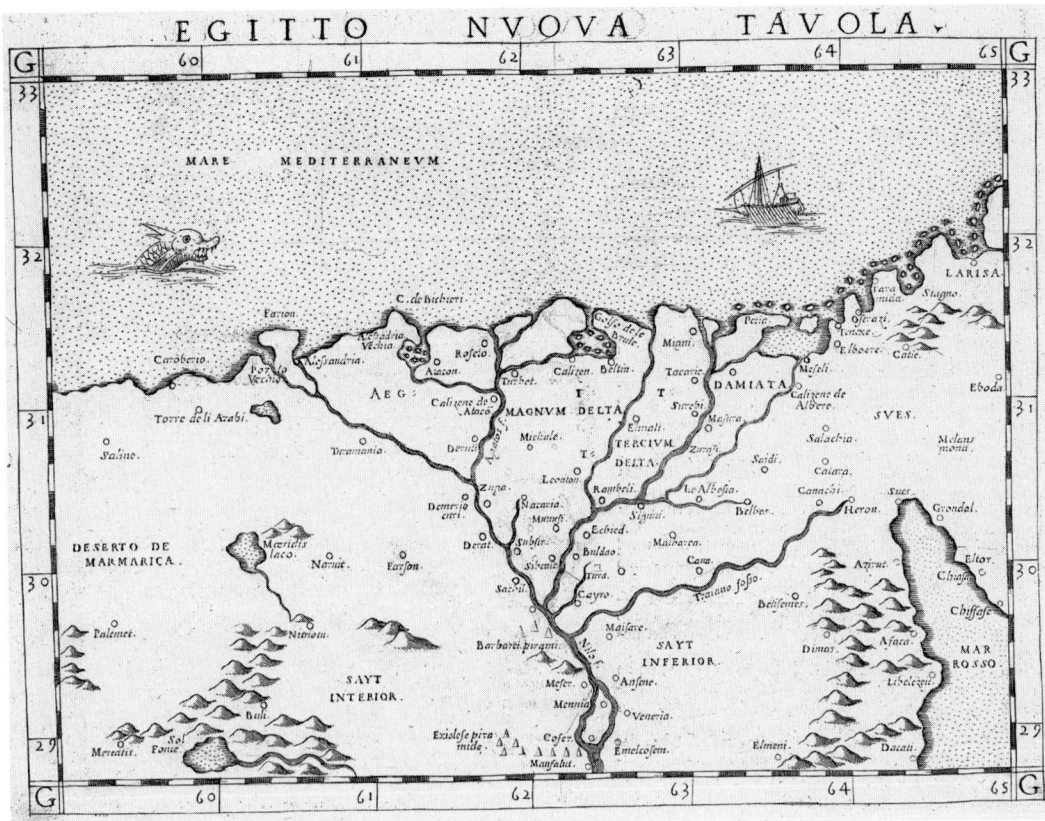

Map 292a

Egypt.
(Venice, G. Ruscelli, 1599.)
Map, 18 x 24 cm, uncoloured.
Italian text on verso.
In: *La Geograpfia di Claudio Tolomeo Alessandrino, Nouvamente tradatta di Greco in Italiano...* Heirs of Melchior Sessa, Venice, 1599.

Map 292b

Marmarica.
(Venice, G. Ruscelli, 1599.)
Map, 18 x 24 cm, uncoloured.
Italian text on verso.
In: *La Geograpfia di Claudio Tolomeo Alessandrino, Nouvamente tradatta di Greco in Italiano...* Heirs of Melchior Sessa, Venice, 1599.

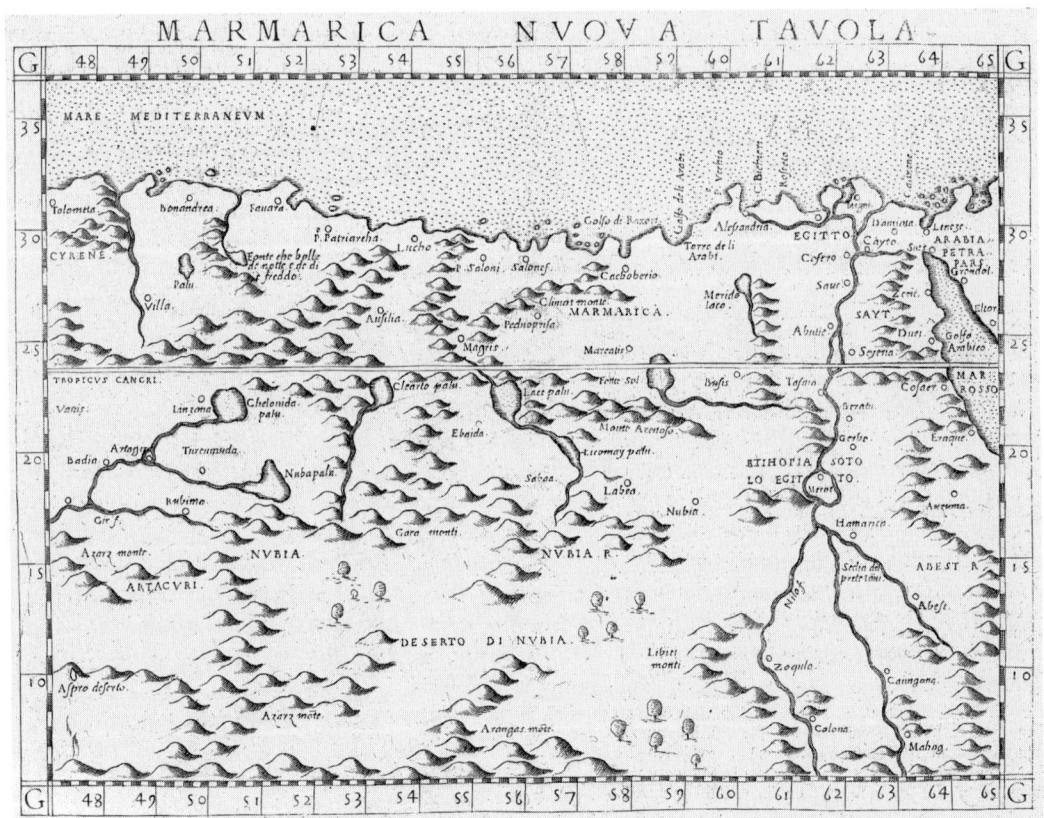

Map 293   PIGAFETTA, Filippo (1533-1603)
The Kingdome of Congo. Engraved by Will Rogers.
(London, John Wolfe, 1598.)
Map, 8,5 x 12 cm, uncoloured.
English text above map.

This map appeared in the English translation of Johan Huygen van Linschoten's *Relatione del Reame di Congo*. It is an illustration at the bottom of the title page of *The Second Booke. The true and perfect description of the whole of Guinea, Manicongo, Angola, Monomotapa...* published by John Wolfe in 1598.

The map is based on information gained by Linschoten, in collaboration with Filippo Pigafetta, from Duarte Lopes's *Travels in Central Africa*.

333

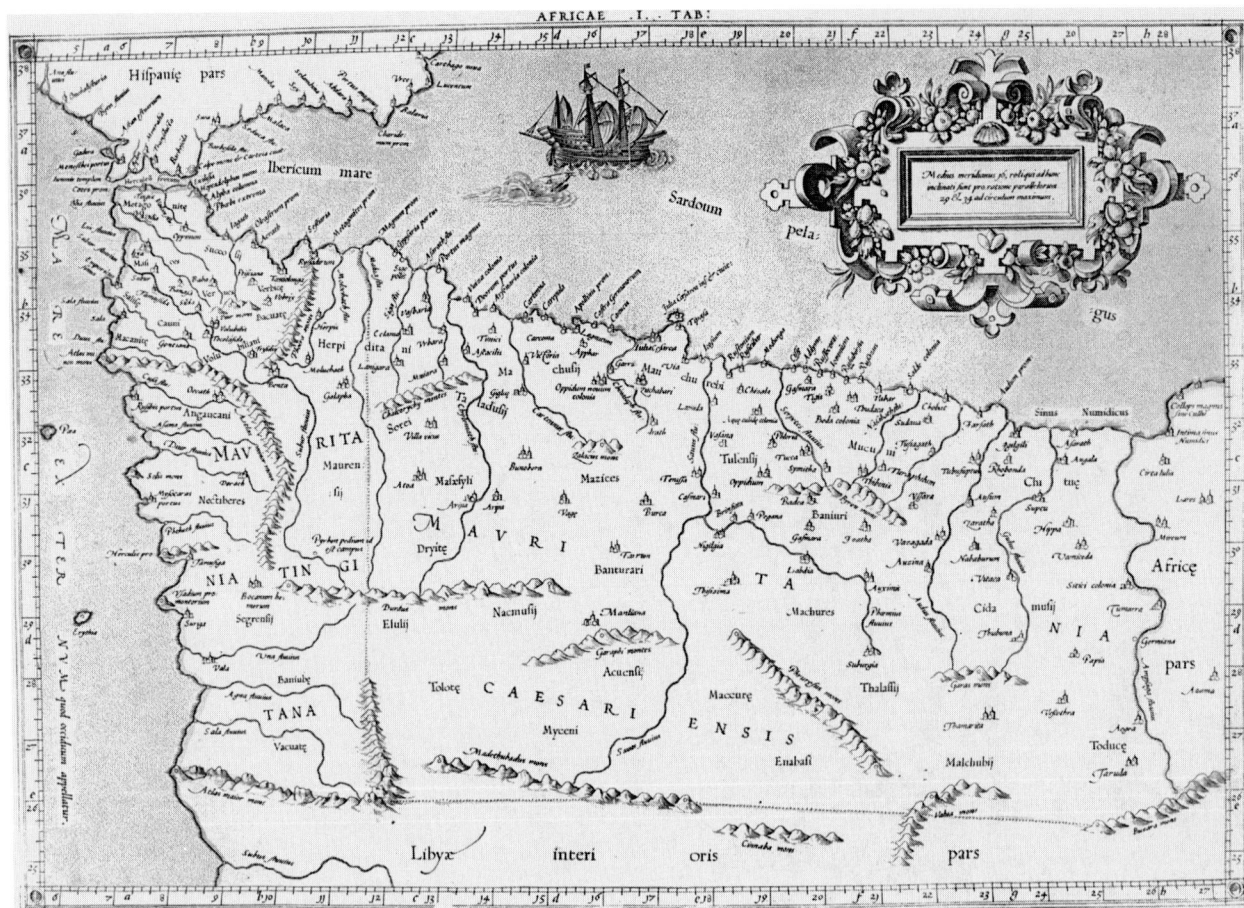

Map 294

PTOLEMY, Claudius (87-150)
Medius meridianus 16, reliqui ad hunc inclinati funt pro ratione parallelorum 29 & 34 ad circulum maximum. (Coast of North-West Africa and part of Spain.)
(Petrus Bertius, 1618.)
Map, 32 x 46 cm, outline coloured.
Title across the top: *Africae I Tab.*

This map and Map 295, marked Africae I Tab and II Tab. respectively, first appeared in the 1584 edition of Ptolemy published by Gerard Mercator in Cologne. They are also found in his 1778 edition. The maps in Mercator's first edition were folded and bear a text on the right hand side of the verso but this particular edition is folded without any text and, according to sources and information supplied by the Map Library of the British Library, was published by Petrus Bertius in 1618.

The cartouche situated at the top right corner is a decorative oval design of fruit and flowers containing the title. The map illustrates a fair amount of detail of the North African coast, with numerous rivers and mountain ranges. The tip of southern Spain appears with the Straits of Gibraltar *(Herculeii fretum)*. In the Mediterranean appears a sailing ship and alongside it a sea monster.

334

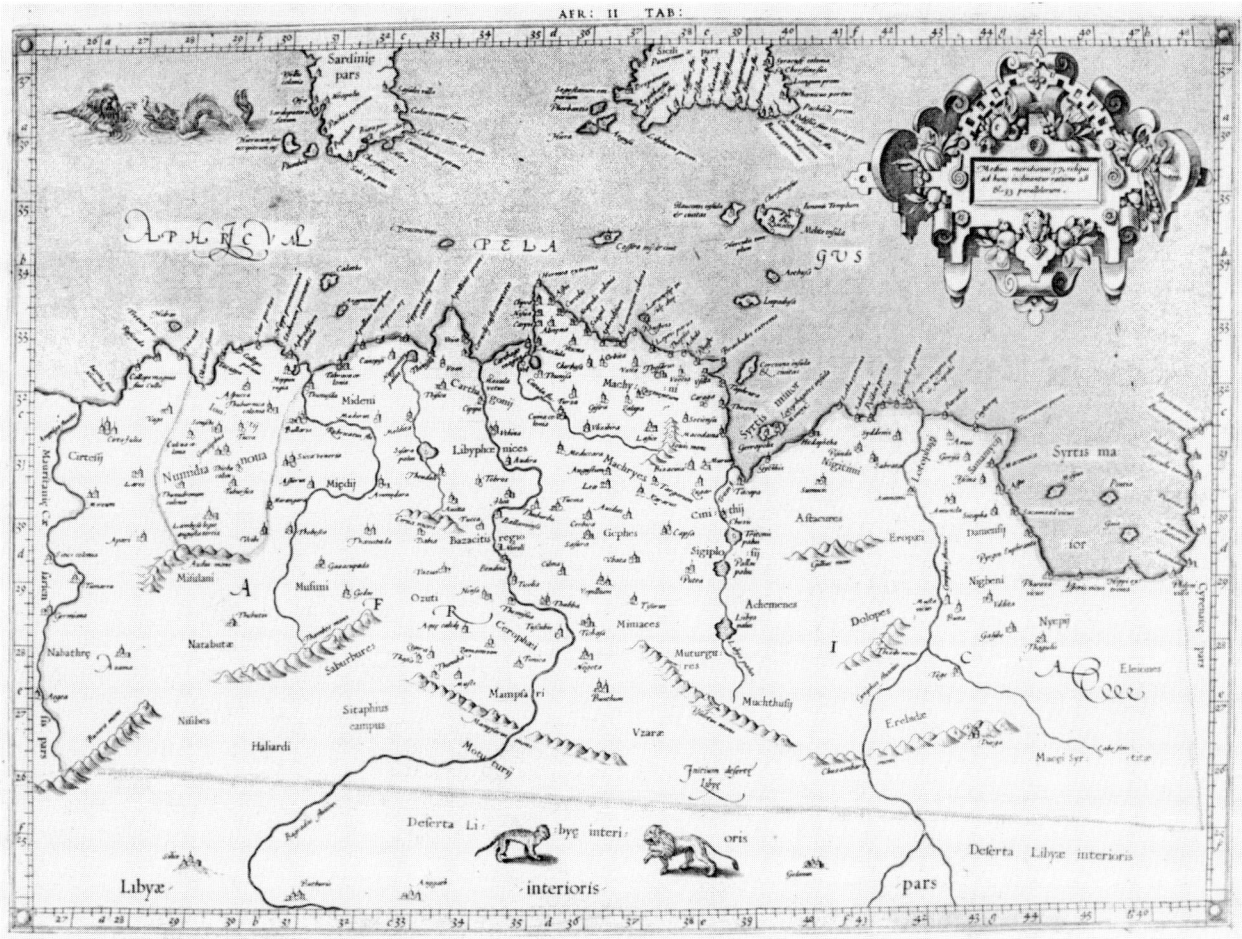

Map 295

PTOLEMY, Claudius (87-150)
Medius meridianus 37, reliqui ad hunc inclinantur ratione 28 & 33 parallellorum. (Mediterranean Sea and north coast of Africa from latitude 37 degrees)
(Petrus Bertius, 1618.)
Map, 33 x 46 cm, outline coloured.
Title across the top: *Africae II Tab.*

This is the *Africae II Tab* of North Africa. This map is more interesting than the previous. It has the same type of cartouche in the upper right and obviously comes from the same atlas. It shows a more westerly position and includes the southern end of Sardinia and Sicily. The sea is stippled and has two sea monsters. At the lower border are a lion and a leopard.

A map of North Africa identical to this one, except for the cartouche, is seen in the *Tabulae Geografficae Orbis Terrarum...* by Jan van Vianen, a Dutch engraver of the Ptolemaic edition of 1704.

Map 296

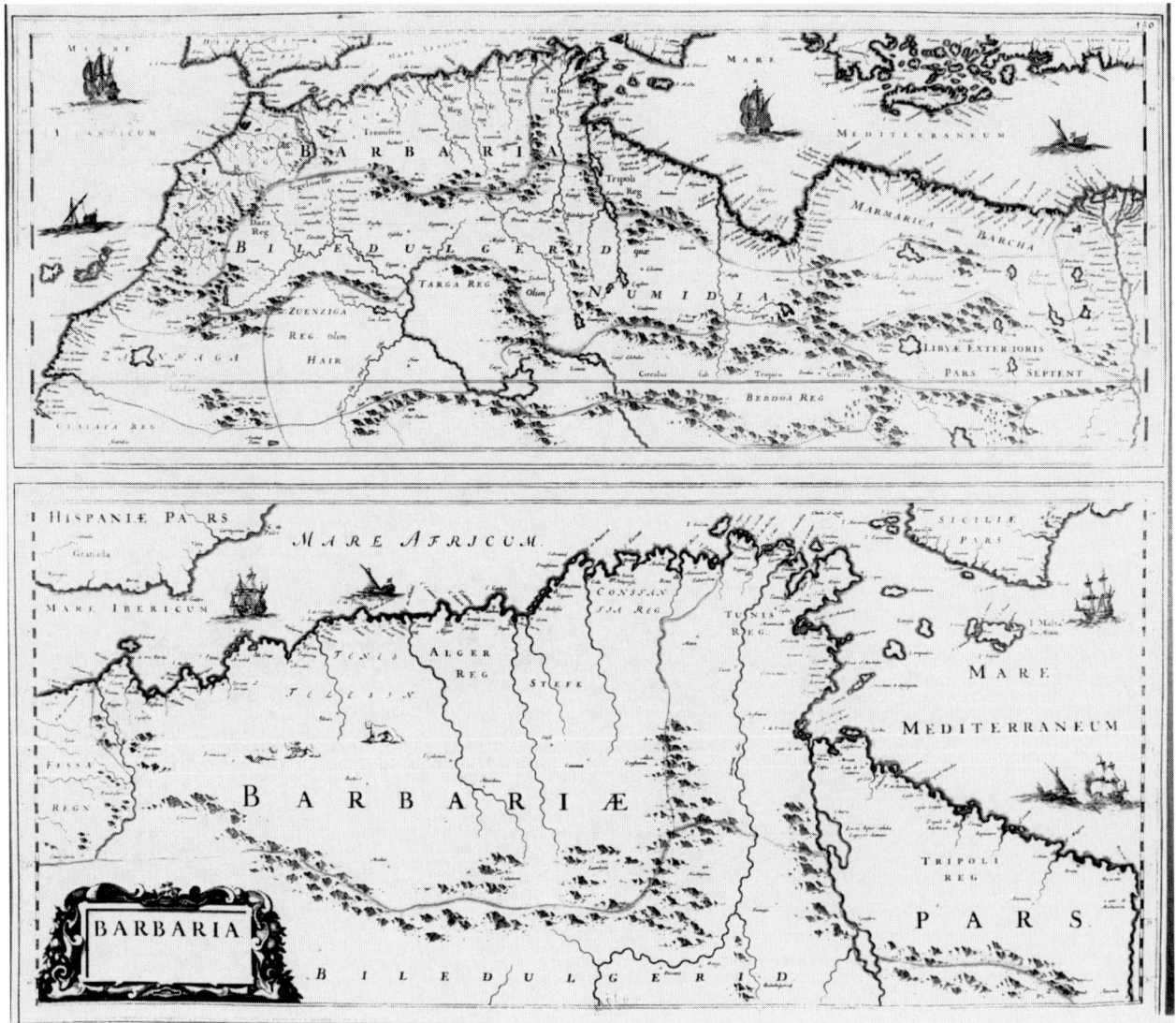

BLAEU, Willem Janszoon (1571-1683)
Barbaria. (Amsterdam, W. J. Blaeu, 1663.)
Two maps, 21 x 57 cm and 26 x 57 cm, coloured outline.
Top right: 150.

This sheet of two parallel maps of North Africa was published in Blaeu's *Le Grand Atlas* of 1663, in the 'Barbaria' section of the volume on Africa. A rectangular cartouche with a floral outline shows only 'Barbaria' as the title.

The upper map of the whole of North Africa, from C. Bajador in the west to the Nile delta in the east, extends south to about 20° latitude. The southern area of Spain is featured with the Straits of Gibraltar.

The lower map shows an enlarged area of Barbaria, from Mare Ibericum to the island of Malta. The features of this map with sailing vessels are typical of the maps of the Blaeu family.

336

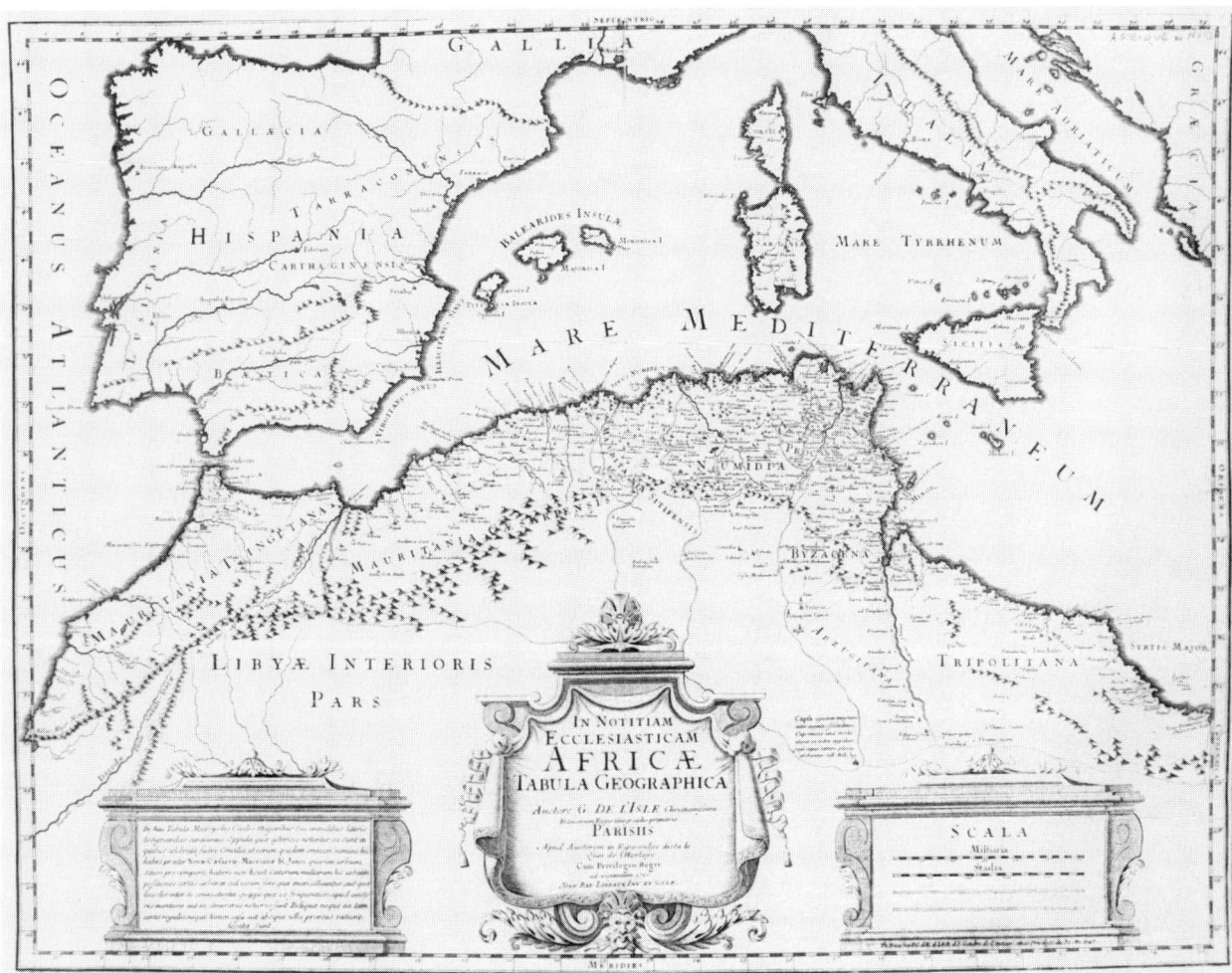

Map 297

L'ISLE, Guillaume de (1675-1726)
In Notitiam Ecclesiasticam Africae Tabula Geographica. (Map of north coast of Africa, Spain and the Mediterranean.) Engraved by Jean Baptist Liebaux.
(Paris, G. de L'Isle, 1745.
Map, 48,5 x 64 cm, coloured outline.
Scale in miles.

This de L'Isle map is of North Africa from western Mauritania to Cyrenica, with the countries bordering the Mediterranean to the north, as far as Greece. It is from the 1745 edition of the 1722 atlas *Carte d'Afrique dressée...*

The cartouche, situated at the lower centre, is in the form of a curtain scroll carrying the title and surrounded by ornamentation at the top and bottom. The scale appears at the right corner in decorative bordered masonry, and on the left is similarly designed rectangular masonry which contains a legend in Latin. Below the scale is the 1745 date, accompanying a series of names and initials. The cartographical centre of North Africa is detailed, with towns, rivers and mountain ranges. The Straits of Gibraltar *(Fretum Herculeum)* are prominently named. The countries and islands north of the African mainland are particularly devoid of detail except for rivers and mountain ranges.

337

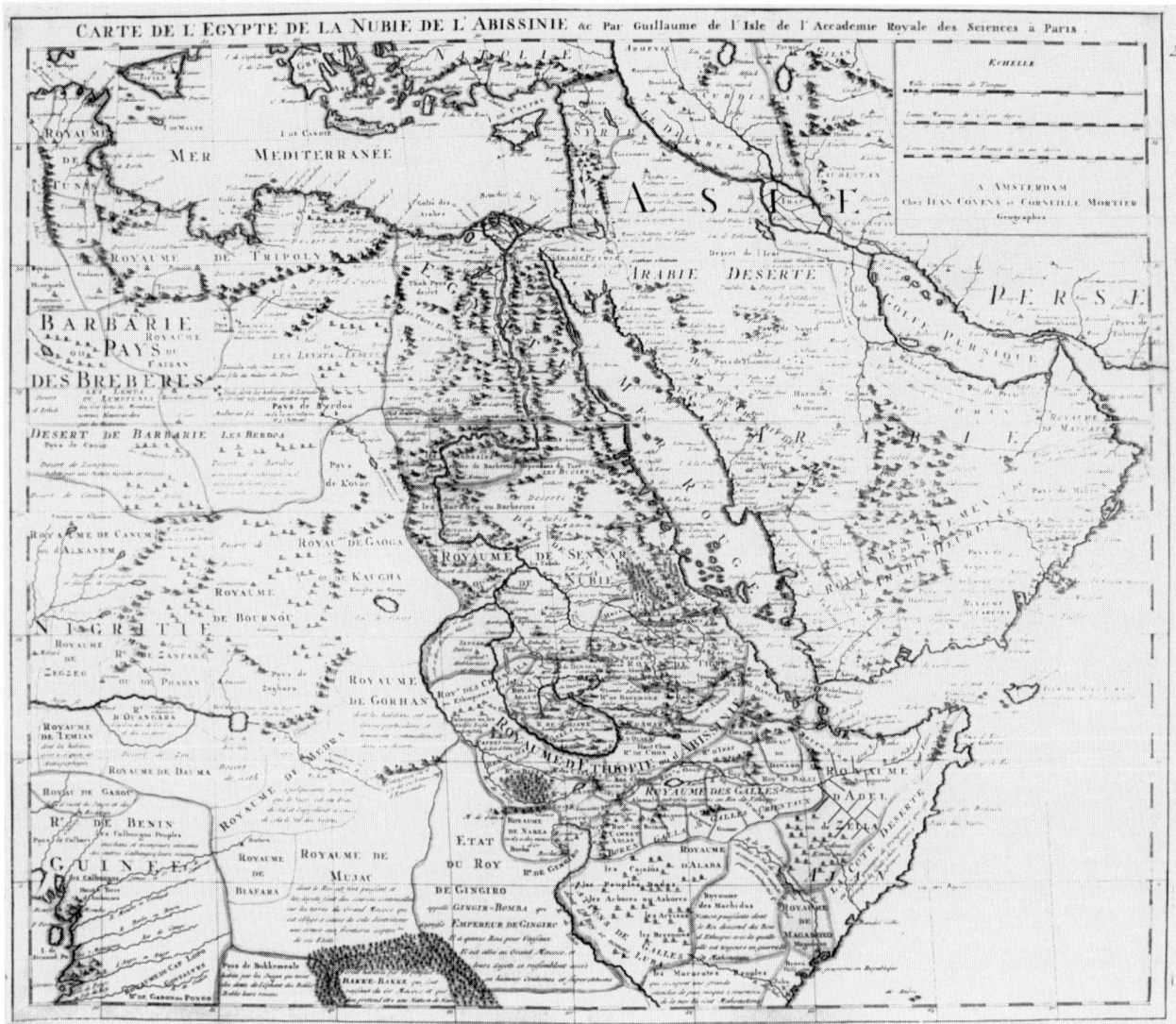

Map 298 L'ISLE (Insulanus), Guillaume de (1675-1726)
Carte de l'Egypte, de Ia Nubie, de l'Abissinie. (Map of Egypt, Nubia and Abyssinia.) (Amsterdam, Jean Covens and Corneille Mortier.)
Map, 49 x 57 cm, coloured.
Scale in Turkish miles, sea leagues, French leagues.
Top right: 22.

This companion map to the previous one of North Africa (see Map 297) depicts Egypt, Nubia and Abyssinia, with the title across the top. This map does, however, have a cartouche, containing the scale, in the upper right corner, stating that the map was published by the geographers Jean Covens and Corneille Mortier of Amsterdam, who issued many atlases after de L'Isle, Sanson and Jaillot. This northern African map provides a great deal of detail above the equator, including the Horn of Africa, the whole of Arabia and a portion of Persia.

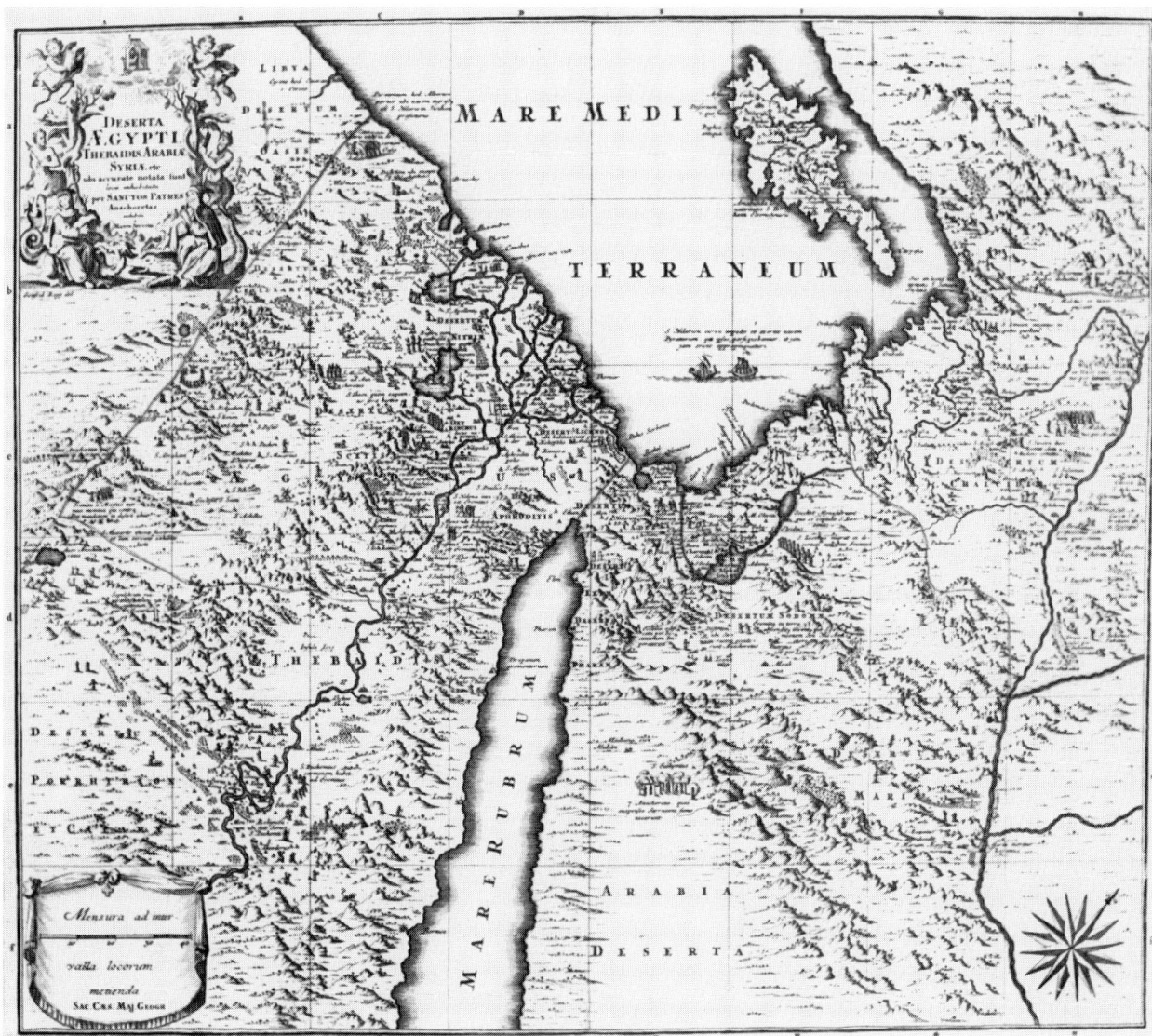

Map 299

SEUTTER, George Matthäus (1678-1757)
Deserta Aegypti, Thebaidis, Arabia, Syriae, etc. (Desert regions of Egypt, Sudan, Arabia, Syria, etc.) Engraved by Gottfrid Rogg. (Augsburg, G.M. Seutter, 1725.)
Map, 49,5 x 57 cm, coloured.
Scale unidentified.

This large map is of Egypt and the surrounding deserts, including Thebaidis, Arabia, Syria, Palestine, Phoenecia, etc. The upper portion of the Red Sea appears to divide the map into two halves, with Egypt to the left and Arabia, Palestine and Syria to the right. There is a plethora of minute details of mountains, animals, people and towns, and events and places of biblical interest are noted.

The Island of Cyprus is prominent in the eastern Mediterranean. Palestine with its ancient cities and places such as Jerusalem, Jericho, Ashkalon and Gaza are illustrated.

There is a decorative cartouche at the top left corner. At the top of the title is a wooden church radiating light; on each side is a symmetrical pattern of cherubs flying, one holding a book, the other an arrow. Below each is a tree entwined by a serpent. On the left is a centaur and on the right a faun. Seated in the foreground on each side is a hooded monk. The one on the right is sitting on a crocodile, and the other on the left is sitting on what appears to be a dragon. In the background are two unusual small animals. At the lower left of the map is a squarish drapery at the top of which is a cross. Inside this is the scale.

The author of this map is the geographer and publisher George Matthäus Seutter the elder, of Vienna and Augsburg. Apprenticed to Homann in 1697, he produced globes and many atlases. The engraver, Gottfrid Rogg, is known for maps of southern Africa, America and Palestine.

339

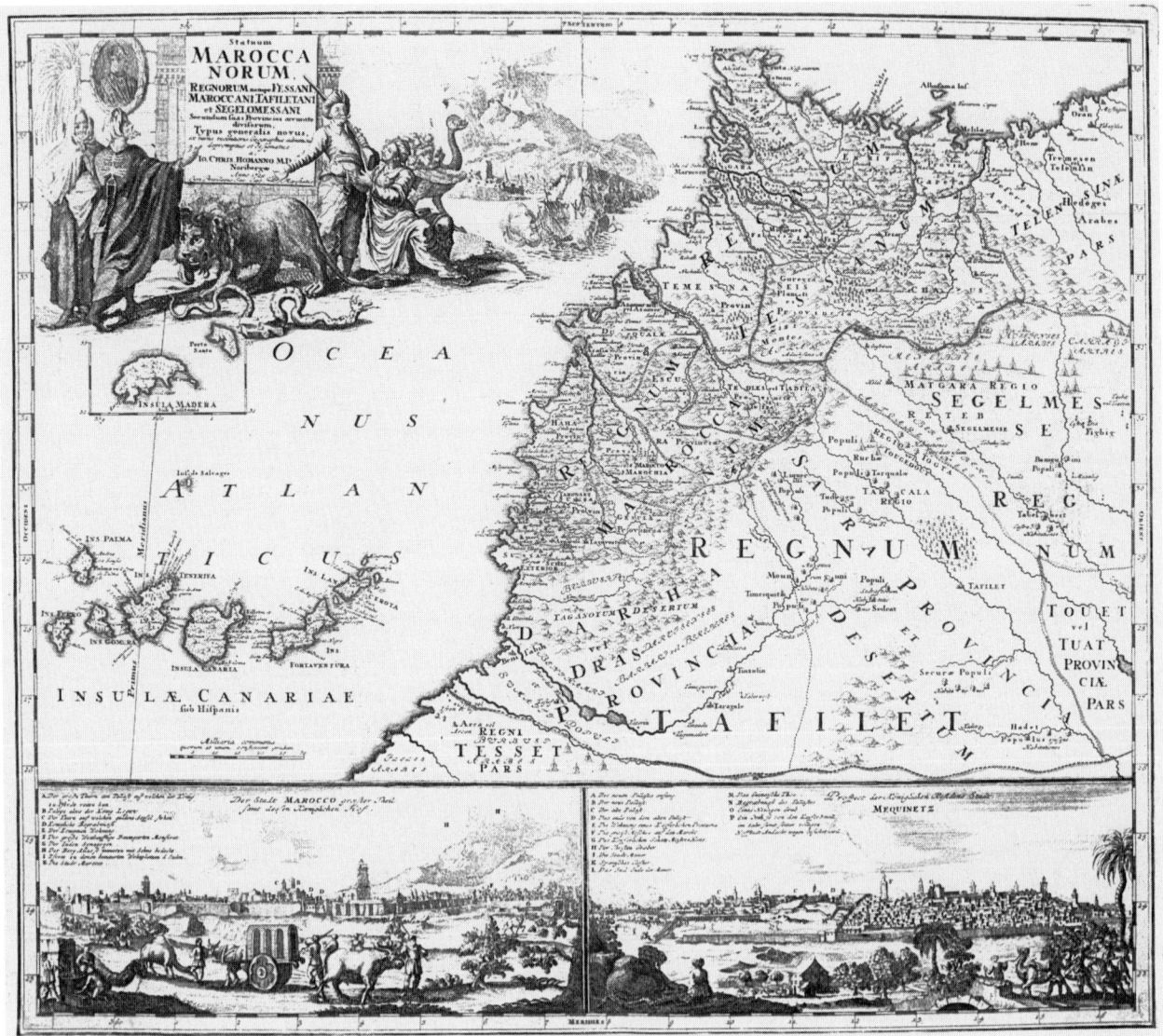

Map 300

HOMANN, Johann Christoph (1701-1730)
Statuum Maroccanorum, Regnorum nempe Fessani, Maroccani, Tafiletani et Segelomessani Secundum suas Provincias accurate divisorum typus Generalis Novus... (Morocco.) (Nürnberg, J.C. Homann, 1728.)
Map, 48 x 55 cm, coloured.
Scale in universal miles.
Prime meridian through Tenerife.
Inset: *Der Stadt Marocco... Prospect der Koniglichen Residens stadt Mequinetz.*

Johann Christoph Homann, the son of Johann Baptist, mapseller and cartographer, issued this fine map of Morocco in 1728. At the top left there is a particularly colourful and striking title cartouche, with a ship in sail in front of a settlement at the base of a mountain. In the foreground an unusual horned serpent is keenly watched by a lion flanked by men in Moorish clothes, with an ostrich on the right to complete a somewhat active picture. Two detailed and colourful insets across the bottom depict the towns of Marocco and Mequinetz. Madeira and the Canary group of islands feature prominently off the mainland.

340

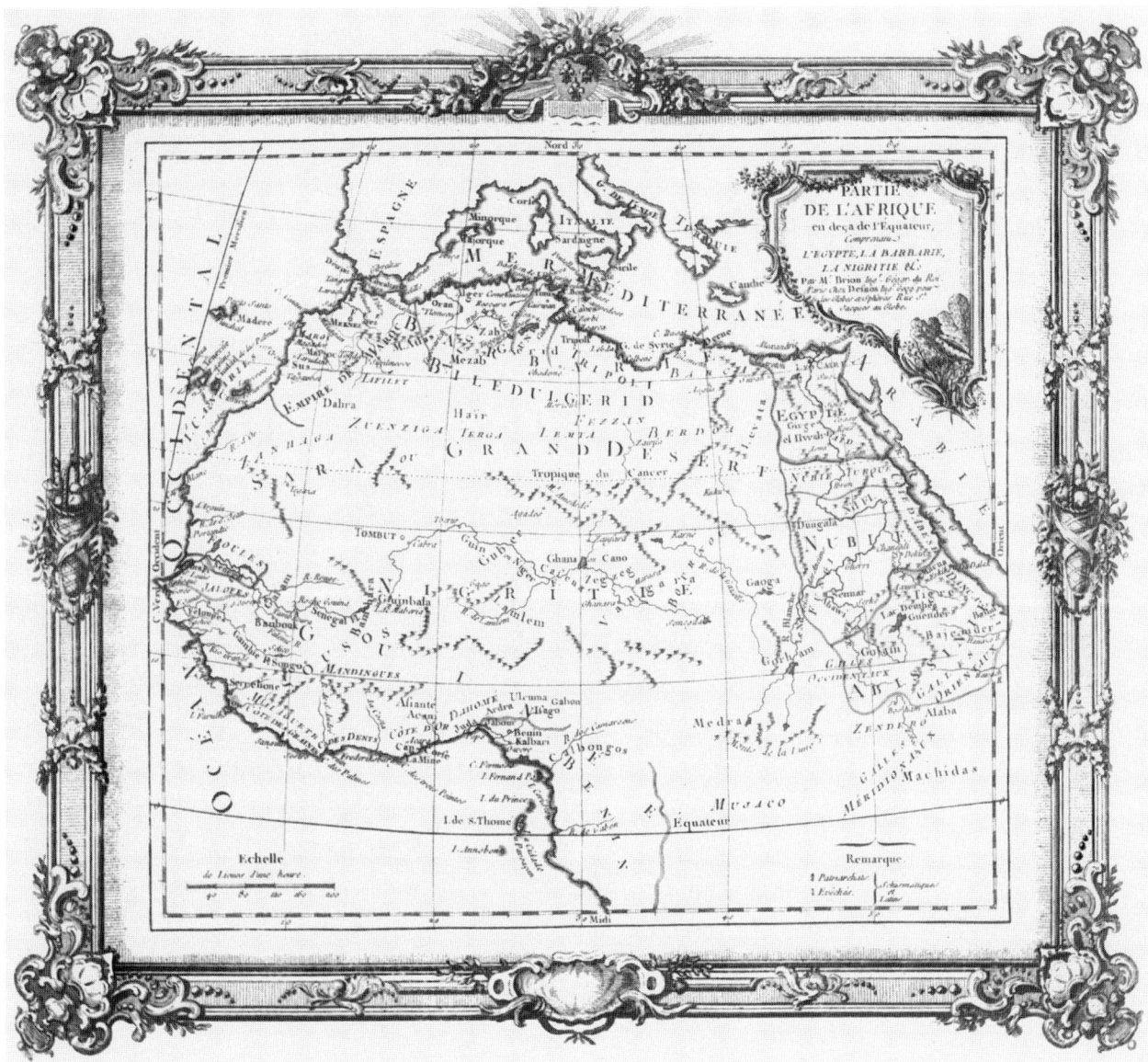

Map 301

BRION DE LA TOUR, Louis (1756-1823)
Partie de l'Afrique en deça de l'Equateur... (Part of Africa north of the equator.)
(Paris, L.C. Desnos.)
Map, 23 x 25,5 cm, partly coloured.
Scale in leagues.
Prime meridian through Tenerife.

The border on this decorative map of North Africa by the French cartographer Brion de la Tour, simulates an ornate frame fairly commonly used by the French geographers and printers of other maps of Africa.

The geographic content is limited, with very few of the great rivers, and with the Nile originating in the Mountains of the Moon, from two small lakes side by side. In the upper right corner a cartouche appears with floral borders within which the title is printed, referring also to Desnos (who appears to have published the other Brion de la Tour maps) as publisher.

341

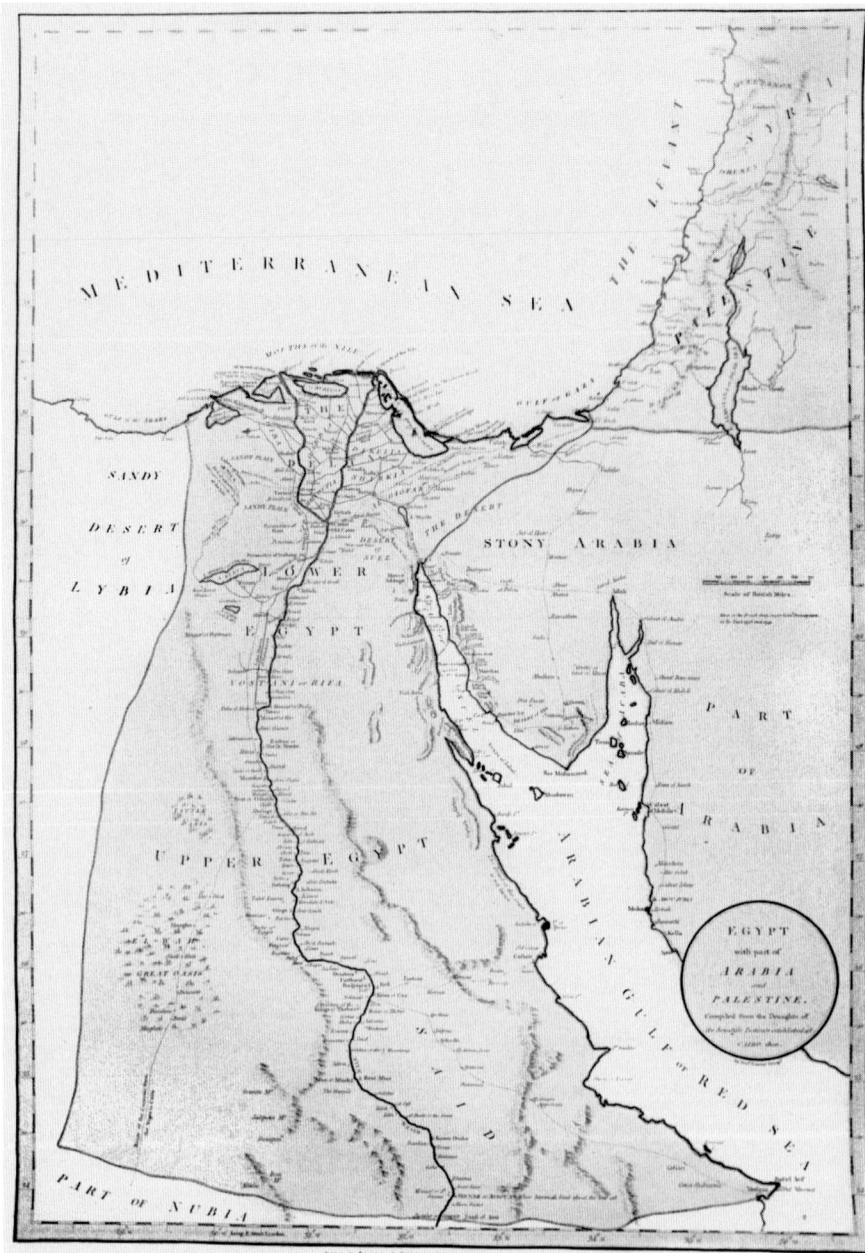

Map 302

ENOUY, Joseph
Egypt with part of Arabia and Palestine. Compiled from the draughts of the Scientific Institute established at Cairo 1800. (London, Laurie and Whittle, 1801.)
Map, 65 x 46 cm, coloured.
Scale in British miles.
Prime meridian through London.

This map of North Africa by Joseph Enouy, published by Laurie and Whittle, was compiled from the drafts of the Scientific Institute which was established at Cairo in 1800. The title is contained in a circular cartouche at the lower right.

The map extends from approximately 24° to 39° north and 28° to 37° east and includes the Nile basin, part of Arabia, the western Mediterranean and the Levant, with Palestine and Syria. The Red Sea (or Arabian Gulf) extends to the horn of Somaliland at Jebbeh Macomo.

Palestine and Syria are geographically well illustrated with the course of the River Jordan with its Dead Sea and

the Sea of Galilee. At the mid-right there is a scale in British miles and below it an explanatory legend using a thin dotted line which refers to the routes of the French army under General Bonaparte in the years 1798 and 1799, namely Bonaparte's route to Acre and to Suez; the route of General Andeofsi in 1800; General Friant's route and the siting of Generals Regnier and Kleber's divisions. On the Mediterranean coast is shown the spot east of Alexandria where Lord Nelson defeated the French fleet on August 1st, 1798.

342

EASTERN AFRICA

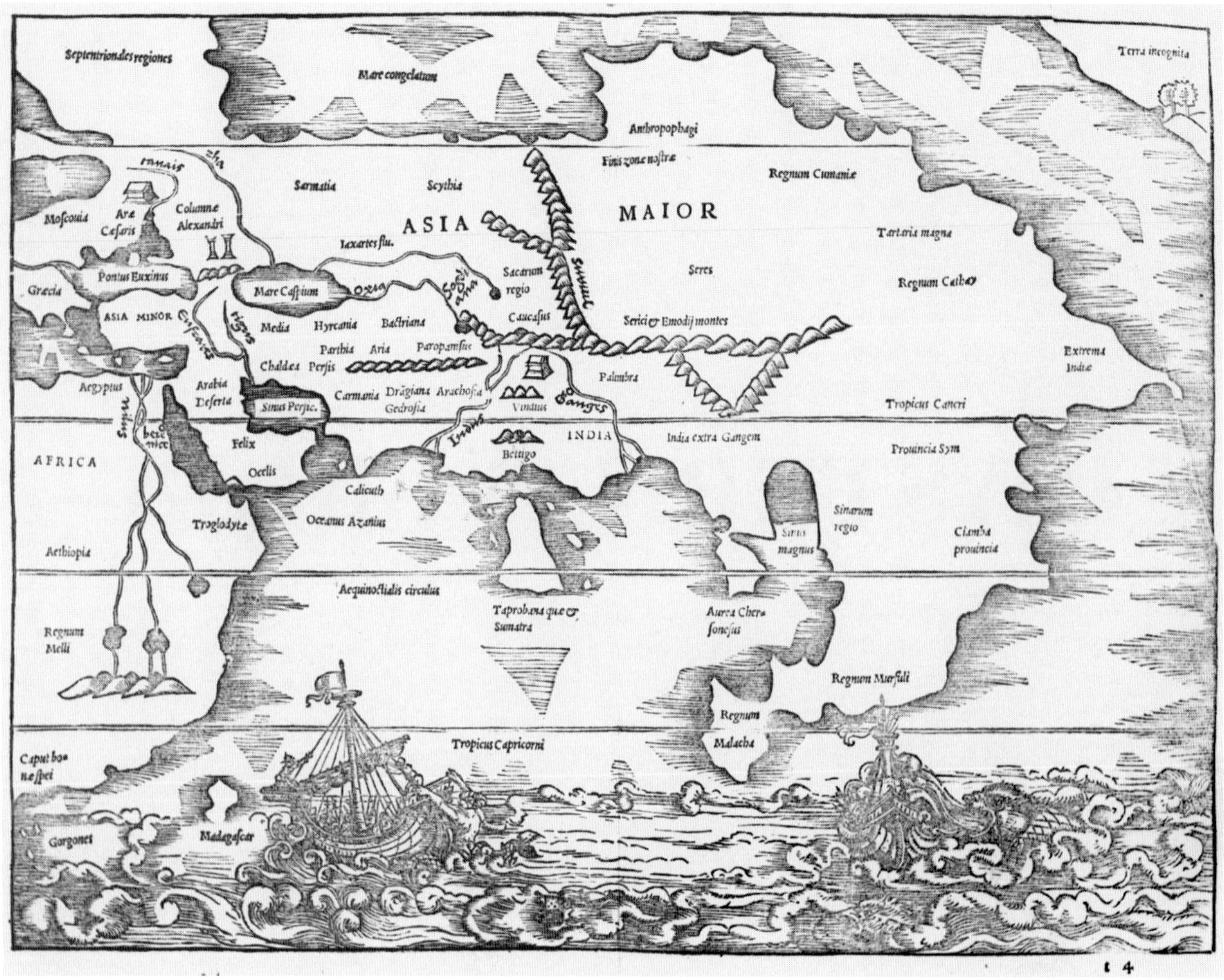

Map 303

MELA, Pomponius (1st century AD)
Asia Maior. (Basle, 1538)
Map, 24,5 x 33 cm, woodcut.
Bottom right: 14.
In: Solinius, Julius. *Polyhistor, Rerum Toto Orbe...* Basle, 1538, fold p.150.

This woodcut map of the eastern hemisphere first appeared in Solinius's *Polyhistor* (1538), which featured a world geography by Pomponius Mela and maps by Münster. Mela, born in Spain, was a geographer under Emperor Claudius in the first century AD. His *Cosmographia* was first printed in 1471 and thereafter many editions appeared under the name of *De Orbis Situ*. Solinius, also a Roman geographer, was responsible for the text. Another woodcut map of Africa from the same source appears in the North Africa section (see Map 288).

This map is folded in the 1538 edition of the *Polyhistor* and lacks a title. It covers an area from the Cape of Good Hope at the lower left to a small portion of North America at the top right, labelled *terra incognita*. This is the earliest known map to feature this part of the North American continent.

The whole length of the eastern half of Africa appears, marked 'Aethiopia' above the equator, 'Regnum Melli' in Central Africa and 'Caput Bonae Spei' to the south. 'Troglodyte' appears in the Horn of Africa, and 'Aegyptus' in the north. The Nile arises from three lakes: two side by side, with their source in a range of mountains, and one further to the northeast.

At the bottom of the map is an animated scene of two galleons sailing in billowy waves, surrounded by sea monsters. A land mass opposite the southern tip of the continent is labelled Madagascar and further to the south is another island, marked 'Gorgones.'

344

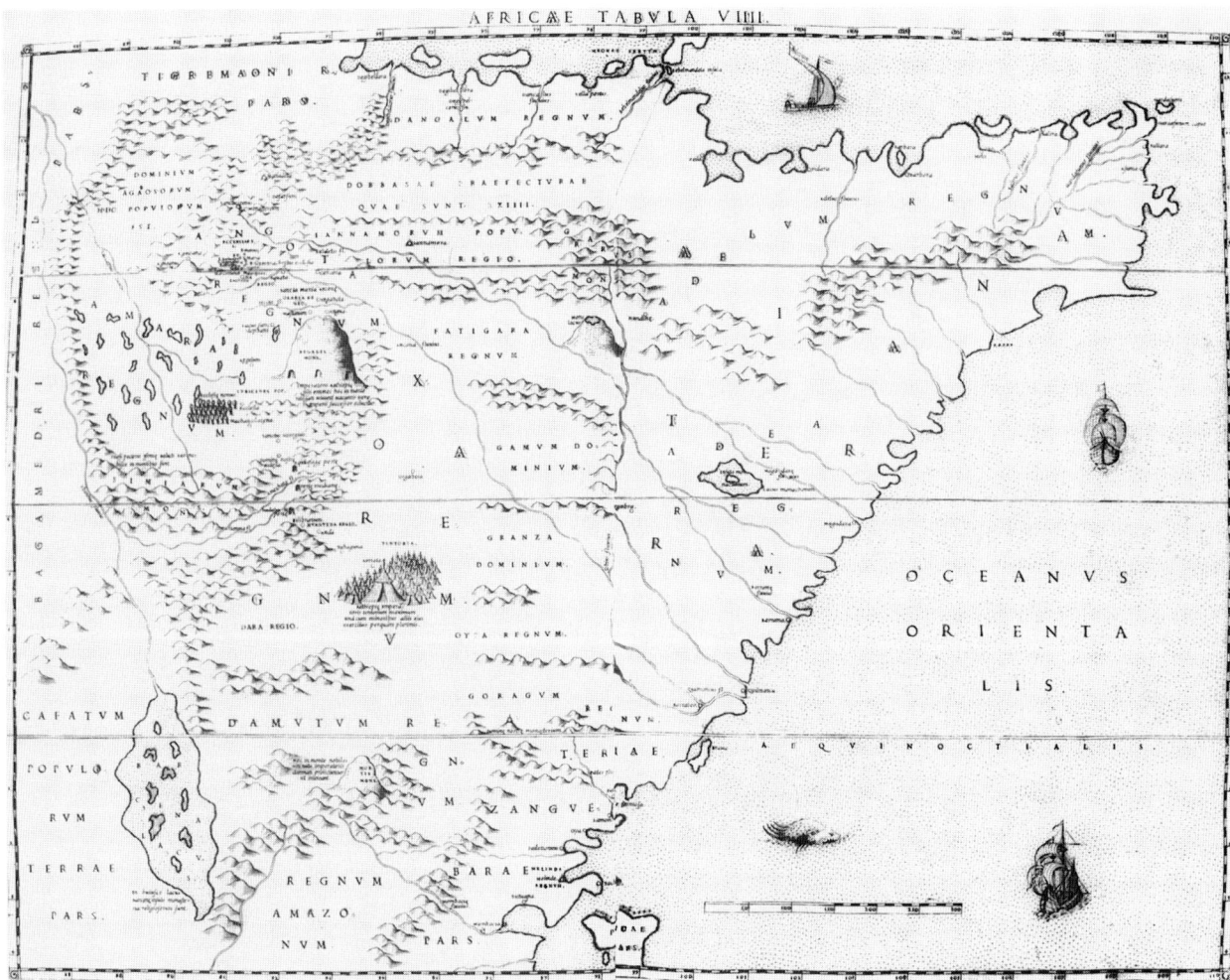

Map 304

SANUTO, Livio (1520-1576)
Africae Tabula VIII. (Venice, L. Sanuto, 1588.)
Map, 38 x 50,5 cm, uncoloured.
Scale in millaria.
In: Sanuto, Livio. *Geographia*. Venice: L. Sanuto, 1588.

'Africae Tabula VIII' is a finely engraved map of Abyssinia, giving a great deal of interesting information. Hills are shaded or hatched on the east slope and towns are represented by a circle with towers added in profile to indicate their importance. The pictorial detail includes the tents of the Emperor of Abyssinia, commonly believed to be the mythical Prester John, and his army (as in medieval maps).

Also shown is Amara, the 'royal mountain' where the sons of the emperor were reputed to live, incarcerated, until the order of succession called them to the throne (as Sanuto reports in his caption underneath the mountain). This mountain appears on Gastaldi's eight-sheet map of Africa (see Map 8) and on many other maps until the late seventeenth century. The legend of Mount Amara was inspired by Francisco Alvares, chaplain to the Portuguese diplomatic expedition to Ethiopia in 1520, and became a

geographical 'fact' after the publication of his journal in 1540. Mount Amara is commonly represented as standing on the equinoctial line. In Gastaldi's map it has a line of entrances reminiscent of a Venetian palace, but in Sanuto's version it has taken an anthill shape, with three entrances. Amara lived on in literary myth for several hundred years: John Milton, in his *Paradise Lost* (1674) writes: '..where Abassin Kings thir issue Guard,/Mount Amara, though this by some suppos'd/ True Paradise under the Ethiop Line/By Nilus head, enclos'd with shining rock,/A whole day's journey high...' (Book IV, 11. 280-84) and in 1759 Samuel Johnson *(Rasselas)* confines his young prince to a palace in a lofty valley in the 'kingdom of Amhara.'

(For further information on Sanuto see Maps 15 and 152.)

Reference: Eversole, Richard. 'An Ethiopian Mountain in Maps and Literature,' *Mapline* 29, March 1983.

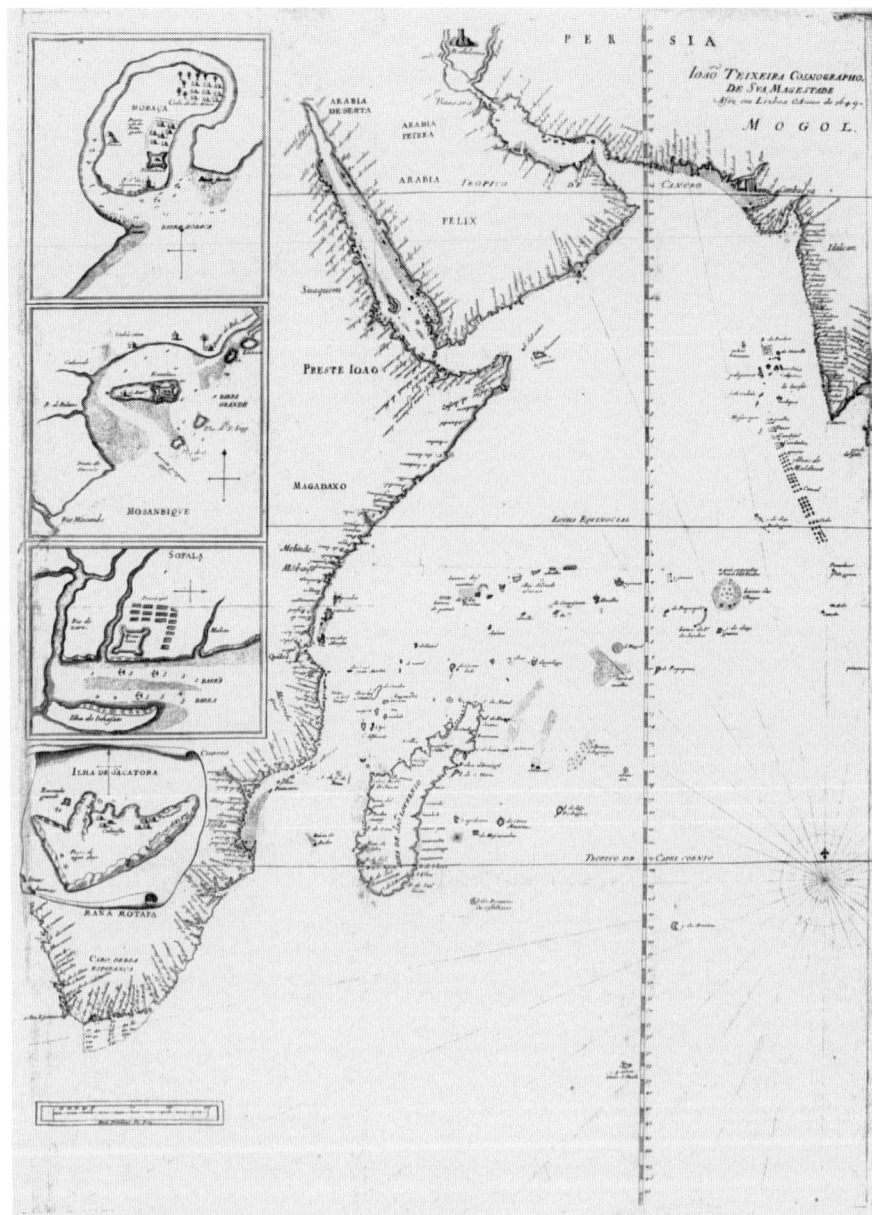

Map 305 TEIXEIRA, Joao (fl. 1620-1650)
Cosmographo de sua Magestade. (East Indies navigational chart or 'carreira'.)
(Lisbon, J. Teixeira, 1649.)
Sea chart, 74 x 51 cm, uncoloured.
Scale unidentified.
Inset: Plans of Mombassa, Mozambique, Sofala and the island of Sacatora.

This Portuguese navigational chart or 'carreira' was published by Joao Teixeira, son of Luis – one of a family famed for their map and chart making. It is extremely important and its excellence and value are indicated by the fact that, more than a century after its compilation, the great cartographer d'Anville (as noted by the geographer Viscount de Santarem) utilised Teixeira substantially for his chart of 1761.

The chart shows the eastern coast of Africa from a little north of the Cape of Good Hope to Ceylon, with Madagascar and the islands of the southwestern Indian Ocean. Detailed insets of Mombasa, Mozambique, Sofala and Sacatora appear to the upper left.

The map is dated 1649 but it probably appeared in Thevenot's *Relations de Divers Voyages Curieux* (1663-1672). The original chart is in the Bibliothèque Nationale, Paris. Thevenot wrote that this 'carreira' is 'after the pattern of those given to the pilots of ships which go from Lisbon to the East Indies...'

346

Map 306

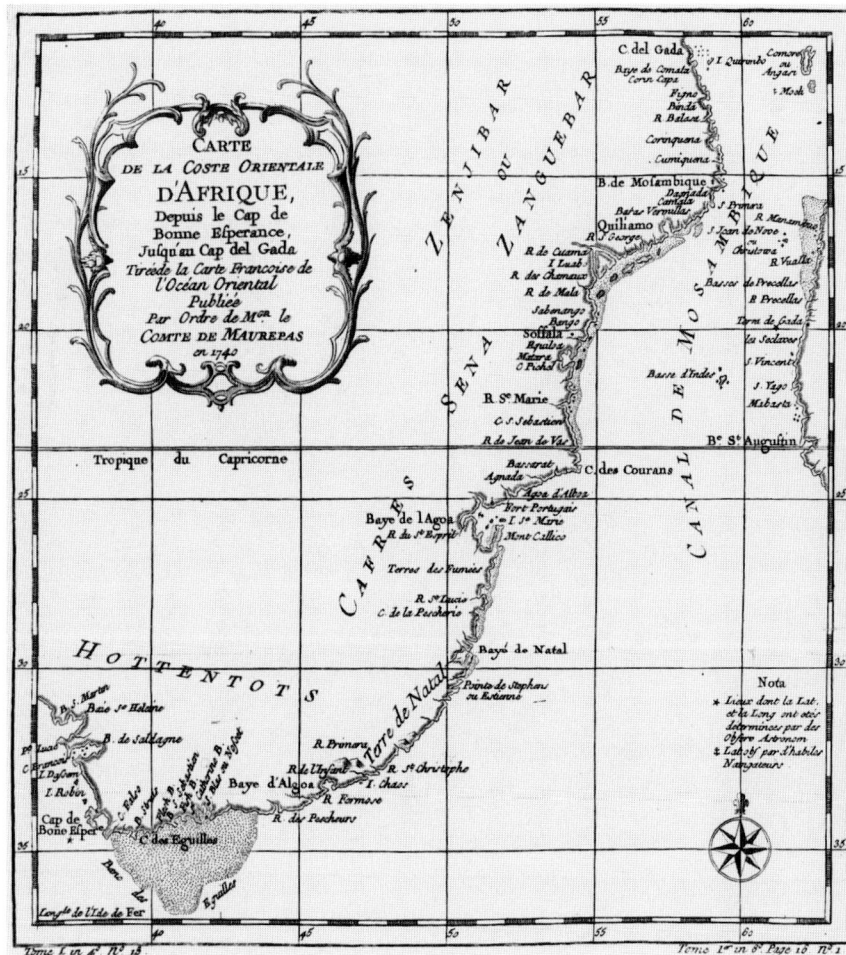

BELLIN, Jacques Nicolas (1703-1772)
Carte de la Coste Orientale d'Afrique, Depuis le Cap de Bonne Esperance, Jusqu'au Cap del Gada... Publieé par ordre de Mgr. le Comte de Maurepas. (East Coast of Africa from the Cape of Good Hope to Cape Delgada.)
(Paris, P. de Hondt, 1740.)
Map, 25 x 23 cm, coloured outline.
Prime meridian through Ferro Island.
Lower left: *Tome I in 4°. No. 18.*
Lower right: *Tome Ier in 8°. Page 16. No. 1.*

This small French map of the southeastern parts of southern Africa appears in the same source as Map 321, namely the *Histoire Générale des voyages* of A.F. Prévost d'Exiles, and in Bellin's atlas. It is slightly narrower than the other map and has a similar rectangular floral cartouche situated in the upper part of the map itself, containing the title, the date and also the fact that it was commissioned by the Comte de Maurepas.

The style of this map is very similar to its western counterpart. It extends from C. del Gada in the north to C. des Eguilles in the south. It appears to have more geographic detail than the west coast map but the place names – mostly early Portuguese strongholds – are almost entirely limited to the coastline, with a few native names within the land mass. There are no geographic features within the map itself except for a number of rivers merely named but not mapped. A small portion of the western coastline of Madagascar appears at the right border with the bay of St. Augustin featured prominently. At the southern tip of Africa the same bays and capes are depicted as are noted in Map 321. At the top right the Comoro Island group is illustrated.

At the lower right, as in Map 321, appears a note that all places marked with an asterisk were determined by astronomical observations or the findings of the navigators. The conventional compass rose appears at the lower right corner.

347

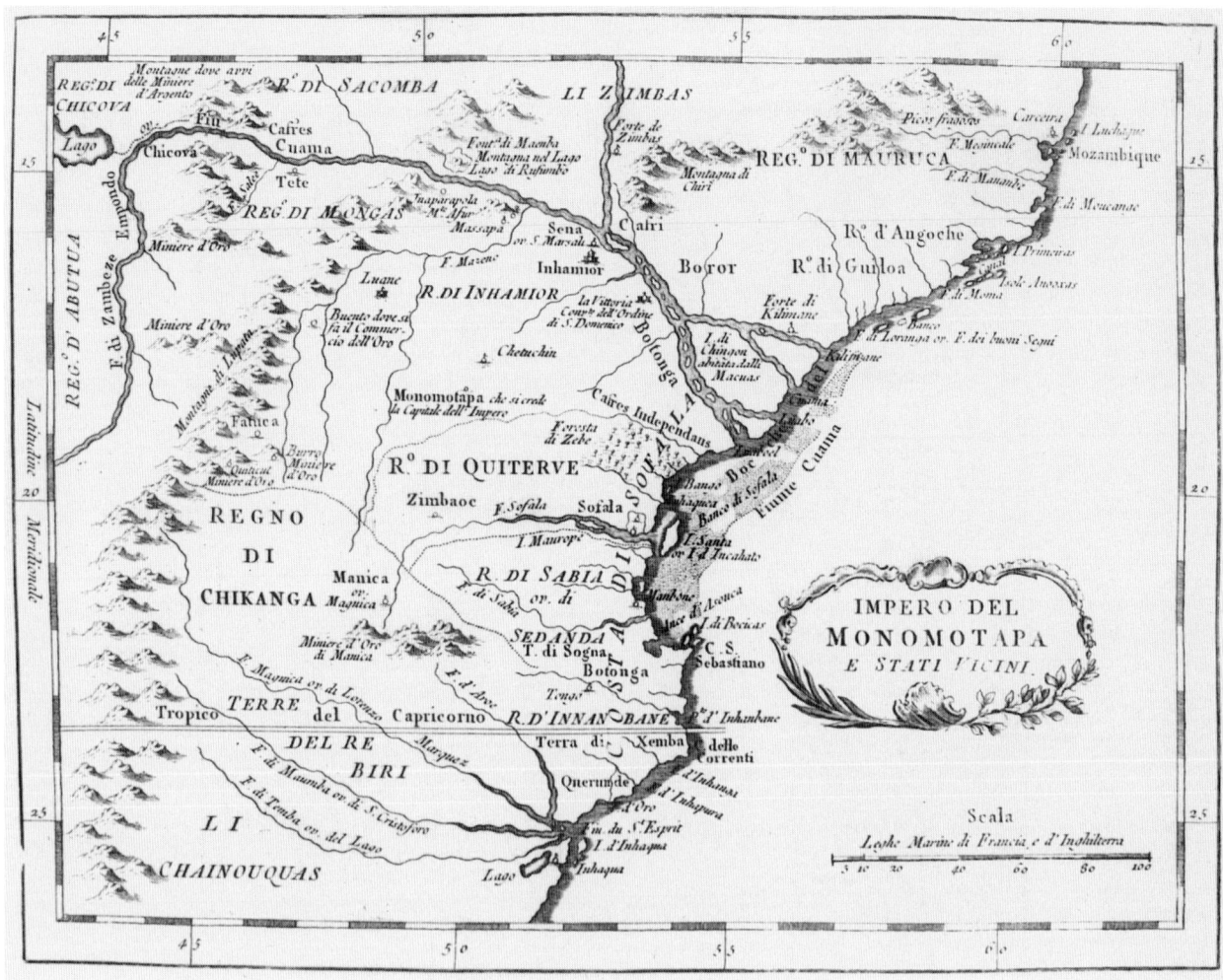

Map 307

BELLIN, Jacques Nicolas (1703-1772)
Impero del Monomotapa e Stati Vicini. (Kingdom of Monomotapa and neighbouring states.) (Venice, 1781.)
Map, 20 x 26 cm, uncoloured.
Scale in French and English leagues.

This small Italian map probably first appeared in an atlas drawn up by Bellin for Prévost's *Histoire Générale des Voyages*, an account of his travels 1738-1777. It depicts a great volume of the geographical detail known of S.E. Africa at the time of its publication in 1781. The coastline features many of the Portuguese settlements and strongholds established during their early search for a sea route to the East. There is abundant evidence of gold and silver having been discovered in the area: in the northeast, near Chicova, a silver mine appears, and below this there is further evidence of at least four gold mines. The Forte de Zimbas in the north could possibly be the Zimbabwe Ruins (although situated further

north than its correct site near the present-day Fort Victoria). The Zambeze/Cuama River appears prominently, with the early Portuguese settlements of Tete and Sena on its banks.

The cartouche, within an oval floral design, referring to the 'Kingdom (or Empire) of Monomotapa and its neighbouring states,' is unusual and important, because the great majority of the many maps featuring Monomotapa (or Benomatapa) within the map itself give no indication of the kingdom's boundaries. This map's title appears to refer to Monomotapa as a definite area within what is today

348

Zimbabwe, and 'Monomotapa, believed to be the capital of the kingdom' appears in the interior.

D.N. Beach and Professor D.P. Abraham, who have done a great deal of vital research over many years in the area of the African continent that is now Zimbabwe, have revealed that Monomotapa was the Mutapa Dynasty, which is known to have existed in the last quarter of the fifteenth century as a well-defined site occupying a triangular area. This area is bounded by the Zambezi River in the north, the Hunyani River and Umvukwe mountain range on the southwest and the Mazoe and Ruenya rivers in the southeast. It thus consisted of a small segment of the southern Zambezian plateau and an arc of the Zambezi valley lowlands.

The Mutapa State was only one of four known major states – Mutapa, Zimbabwe, Torwa and Changamire, each with its own capital. The Mutapa State is of special interest because it is the only one to have escaped being obliterated entirely by new settlements of people for any length of time, despite being closest to the early Portuguese centres encroaching from the eastern coast where their heavily manned and equipped fortresses were established. Their increasing trading development, however, and sporadic search for gold and their introduction of firearms and ammunition to the original inhabitants soon caused the disintegration of the Mutapa State.

The Zimbabwe Ruins – the legendary 'King Solomon's Mines' – are quite erroneously thought to be the site of the origin of the vast quantities of gold in Solomon and Sheba's times (see notes on Ophir, Map 310), and are only one of very many such built-up fortresses and settlements in southern Africa outside the Mutapa State. They are also known and labelled on maps as Zimboaes and Simboaes, appearing in many and varied positions. According to Professor Tom Huffman (Professor of Arcaeology at the University of the Witwatersrand) there were about 120 such Zimbabwean settlements all over this part of Africa.

In many antique maps imaginary towns are depicted and named in various parts of southern Africa. One of these is the town of Monomotapa, usually shown situated on a river. These imaginary towns were regarded as frontier cities of the state of Monomotapa. The first printed map of Africa – the Contarini of 1506 – shows a place marked 'Virgict Magna' on the north bank of the Infante River just north of the Tropic of Capricorn. The Waldseemüller map of 1516 and subsequent maps usually gave the name as Vigiti Magna. Ortelius, in his map of 1570, marked a place, Bavagal, known in other maps as Davagal. Linschoten's well-known map placed Davagal on the Rio do Esperitus Santo, which flowed into the bay of Lourenço Marques, and Vigiti Magna and other cities north of it on the banks of rivers running parallel to the west coast of southern Africa and entering the sea at Cape Agulhas.

Jan van Riebeeck – the first Dutch Commander of the Cape – knew Linschoten's work and attempted to make contact with the people of Monomotapa. He had heard that there was a civilised tribe residing in stone houses twenty or thirty days' travel north of Table Bay, whom he took to be the Monomotapians. In 1659 he encouraged seven free burghers under Christiaan Jansen to proceed in search of them, but lack of fodder prevented this party from going beyond the Great Berg River. Undaunted, van Riebeeck initiated further parties, all of which, however, ended in failure for one reason or another.

Although these expeditions in search of the legendary cities and peoples of Monomotapa were unsuccessful, they did greatly increase Dutch knowledge of southern Africa. The lands of Monomotapa were eventually reached from the south in the middle of the nineteenth century.

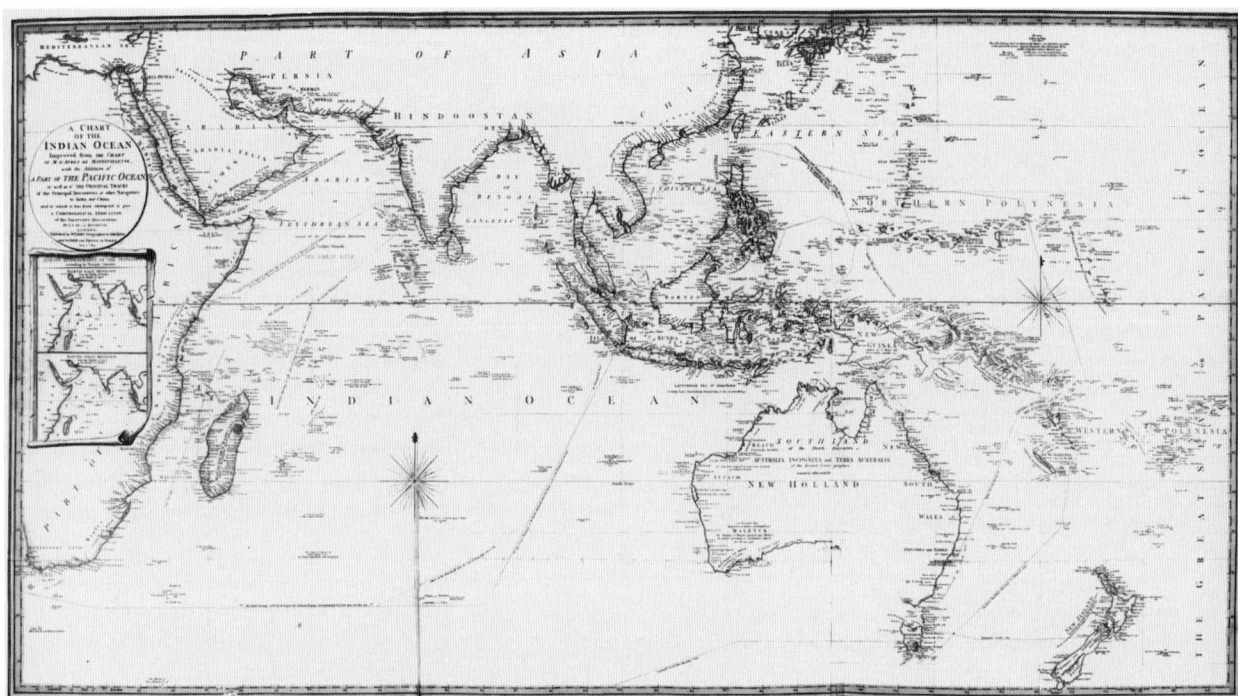

Map 308

DE LA ROCHETTE, Louis Stanislas d'Arcy (1731-1802)
A chart of the Indian Ocean improved from the chart of M. d'Apres de Mannevillette; with the addition of a part of the Pacific Ocean, as well as of the original tracks of the principal discoverers or other navigators to India and China; and in which it has been attempted to give a chronological indication of the successive discoveries. Engraved by J. Hatchett. (London, W. Faden, 1803.)
Map, 60 x 102 cm, coloured outline.
Inset: *Anemo-hydrography of the monsoons.*

A large chart of the whole of the east coast of Africa including the Indian Ocean, Arabia, India, the Far East, New Holland (Australia) and New Zealand, together with all the known islands of this vast area. The title appears in a simple oval cartouche on the upper left-hand side of the map.

De la Rochette is the well-known cartographer and engraver previously described in association with his map of the Cape of Good Hope (Map 236). The cartouche mentions that this map is an improvement on the chart of the well known French cartographer M. Après de Mannevillette, and the equally famous William Faden,

Geographer to the King and H.R.H. the Prince of Wales in 1803, appears as publisher.

Below the title are two insets in a rectangular scroll illustrating the anemo-hydrography of the monsoons, according to Vicomte Grenier, a French hydrographer. The whole coastline affected by the monsoons is depicted in miniature within this box. The upper inset shows the north-east monsoon from the middle of October to April, and the lower inset shows the southwest monsoon from the middle of April to the middle of October.

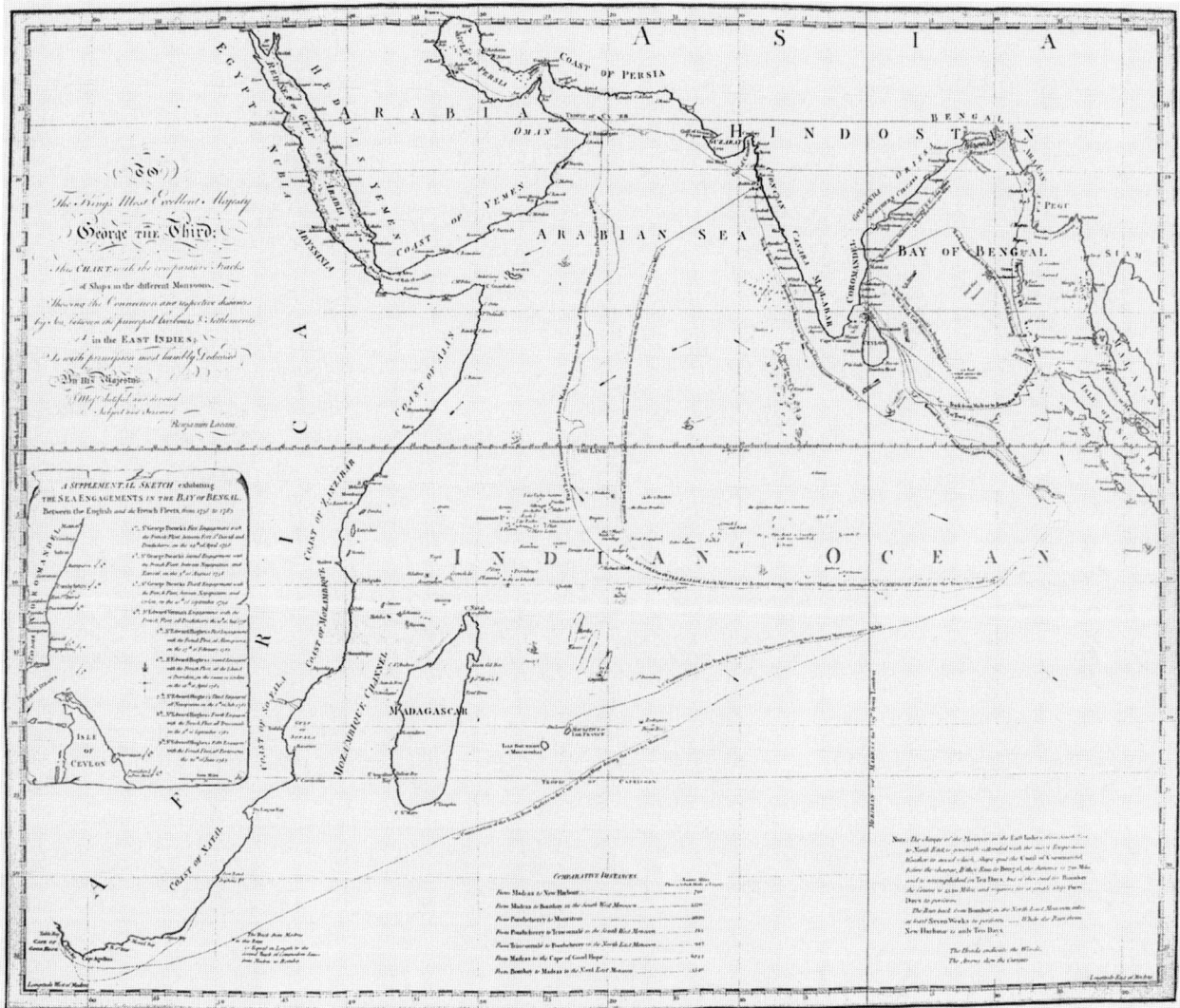

Map 309

LACAM, Benjamin

'To the King's most Excellent Majesty George the Third; This chart, with the comparative Tracks of Ships in the different Monsoons; showing the Connection and respective distances by sea between the principle harbours and settlements in the East Indies; is with permission most humbly dedicated by...'

Engraved by W. Palmer. (London, Laurie and Whittle, 53 Fleet Street, 1808.)

Map, 49 x 59 cm, uncoloured.

Prime meridian through Madras.

Dedicated to George III.

Inset: A supplemental sketch exhibiting the sea engagements in the Bay of Bengal between the English and the French from 1758-1783.

This is an unusual map showing the east coast of Africa from Suez in the north to the Cape of Good Hope in the south. It is intended, as the title in the top left border indicates, to show comparative tracks of the travelling ships in the monsoons, especially in the Arabian Sea and the Bay of Bengal. The coastlines of East Africa, India, Siam and part of Malaya are faithfully depicted.

On the left side of the map below the title is an inset listing in detail the engagements in the Bay of Bengal between the English and French fleets from 1758 to 1783. At the lower right mention is made of the change of the monsoon in the East Indies from southwest to northeast, which is generally attended by tempestuous weather.

Benjamin Lacam was a map publisher who appears to have concentrated on the north part of the Bay of Bengal, Hughley River and the Pilot Bay of Bengal. Mention is made, at the map's lower border, of this being a new edition.

351

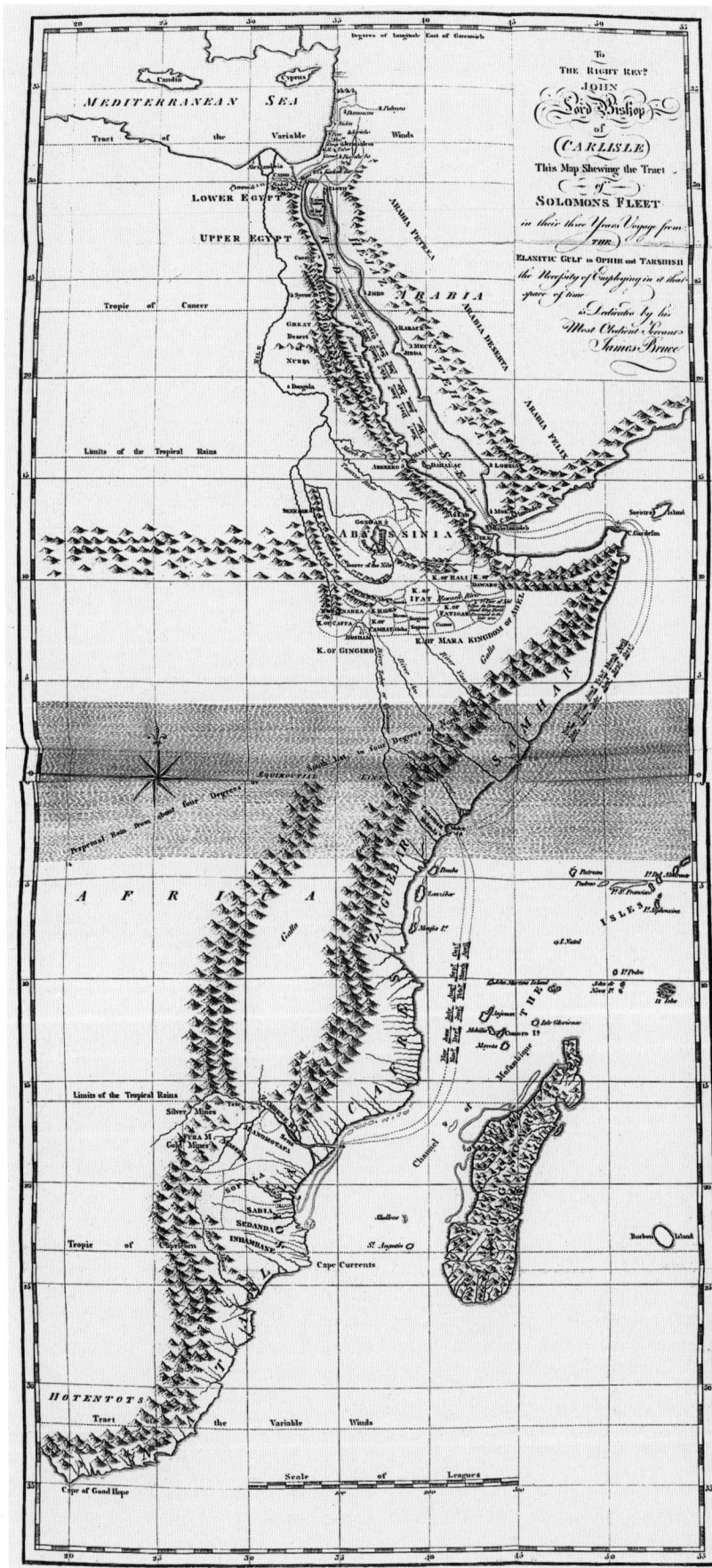

Map 310

BRUCE, James (1736-1794)
To the Right Revd. John Lord Bishop of Carlisle. This Map Shewing the Tract of Solomon's Fleet in their three Years Voyage from the Elanitic Gulf to Ophir and Tarshish...
(Engraved by J. Walker). (Edinburgh and London: G.G.J and J. Robinson, 1790.)
Map, 69 x 31 cm, uncoloured.
Scale in leagues.
Dedicated to the Bishop of Carlisle.
In: Bruce, James. *Travels to Discover the Source of the Nile in the years 1768, 1769, 1770, 1771, 1772 and 1773.* Edinburgh and London: G.G.J and J. Robinson, 1790, vol. 5 fold.

This map appears in volume five (the appendix) of James Bruce's *Travels to Discover the Source of the Nile in the years 1768-1773*, published in 1790.

James Bruce was a Scottish laird – a keen sportsman and amateur scientist, who received a good classical education at Harrow, which no doubt fostered his fondness for antiquities and the East. He visited Spain, which led to his interest in Moorish subjects, and some two-and-a-half years as British Consul in Algiers made him adept in Arabic. His travels took him from Tunis through Tripoli to Syria and then Egypt, whence he journeyed, amid perilous incidents, via the Red Sea and Jeddah, to Massawab and, finally, Gondar, the Abyssinian capital.

There, his knowledge of medicine put him in the good graces of the ruling powers: he was appointed Commander and accompanied the King in the field. The goal of his ambition was reached when he visited Lake Dembea (or Tana), arriving at the village of Geesh, close to the Bubbling Fountains of the Blue Nile. His assertion that he had discovered the source of the Nile was met with hostility, however, as he totally rejected the claims of the Jesuit fathers Pedro Paez and Jerome Lobo, who had compiled an account of their travels towards the eastern headwaters 150 years previously, in which they claimed to have reached the springhead of the Abai or Blue Nile.

The White Nile was known to exist, although its origin was not yet determined, and Bruce deliberately ignored its claim to be a major branch of the river, so determined was he to be regarded as the discoverer of the origin of the Nile. It was seventeen years before his friends could prevail on him to publish an account of his travels, and his remaining years were embittered by the ensuing controversy, though he did witness the rise of the African Association, which was to rehabilitate his credit.

The map includes the whole of the east coast of Africa, from the Cape of Good Hope to the eastern Mediterranean. It shows 'the Tract of Solomon's Fleet in their *Three Years Voyage from the Elanitic Gulf to Ophir and Tarshish*; a voyage which starts at Eloth, at the top of the Red Sea, and is marked by arrows around the horn of Somaliland and along the east coast of Africa to a site marked Yoka, Tarshish and Melinda, and from there to the mouth of the 'Zambeza' River, adjacent to 'Manomatapa.' This is probably the site of Ophir, though it is not marked as such. From here Solomon's fleet return along the same route to Tarshish and Eloth.

An almost continuous range of mountains appears along the length of the coast and, except for some major rivers, there are only a few places, mainly of Portuguese origin, named in the south. The source of the Nile is marked, close to Lake Tana.

This map stimulated the author to do some research into the probable site of the Land of Ophir, twelve references to which appear in the Old Testament: firstly as a Joktanite or southern Arabian tract (Gen. X: 29 *et seq.*), and later as the port of destination of Solomon's fleet. The earliest reference to Ophir in this connection is in I Kings IX: 26 *(et seq.)*, where it is said that Solomon built a navy in Ezion-Geber, near Eloth on the Elanitic Gulf in the Red Sea, manned by an expert crew given to him by Hiram, and sent it to Ophir, whence it brought him 420 talents of gold (18 tons). In a later reference the navy brought back 'great plenty of almug-trees and precious stones,' and from that time on, Ophir was to the Hebrews a land of fabulous riches, according to other references in the Old Testament.

Ophir has been assigned variously to America, South Africa, Arabia, the West Indies, Peru, the coast of India, Ceylon and Spain, only a few of which deserve serious consideration. It also appears on some maps in central East Africa, in the coastal district of Sofala, and there is a long tradition that the Zimbabwe ruins can be traced to the Queen of Sheba and to King Solomon's mines. However, Professor T. Huffman (Professor of Archaeology at the University of the Witwatersrand) states that recent, more accurate carbon-dating proves these ruins to date from around the thirteenth or fourteenth century AD. The word Sofala has no connection with Ophir, but is derived from the Arabic *Safala*, meaning 'low land' *(Heber Shefela)*. There is sufficient evidence to exclude India, as no gold was exported from the area named, and as the Jews became acquainted with India only in the Greco-Persian period. The most probable view is that Ophir is situated in Arabia (Yemen), which is indicated in Gen X: 29. An old tradition, according to the *Jewish Encyclopedia*, is recorded by Eupolemus (c.150 BC), who identifies Ophir with the Island of Uphre in the Red Sea. Both the east and west coasts of Arabia have been considered as Ophir's site. Glaser favours the east coast, the more likely object of a three year voyage, comparing the cuneiform name 'Aper' applied to the northern coast of the Persian Gulf; and the Arab geographer Hamadoni states that gold mines were situated in the northeastern part of Arabia. However, since the references to the three years required for the voyage are not found in the earlier account, there is ample justification for the view that prefers the western coast of Arabia, especially in view of the number of references in ancient authors (confirmed by recent archaeological findings) to rich gold deposits on the southern coast (Yemen), and as Solomon and Hiram would hardly have sent their ships past Yemen to fetch gold from the Persian Gulf, which was so much further away. To the southwest coast of Arabia may be added the adjacent Somaliland coast (both called 'Ptint' by the ancient Egyptians). This theory gains in probability if the authors of I Kings X: 22 meant to imply that the exports were native to Ophir itself; for apes, ivory and ebony are among the products of Somaliland. Added to this view are the writings of Pritchard in his book *Solomon and Sheba*, in which he states that these adjacent portions of land are the only regions known to have grown myrrh and frankincense – products of inestimable value to the ancients for their cosmetic and healing powers.

Tarshish certainly did not appear where Bruce places it, on the East African coast near Moka and Melinda. In ancient and biblical references Tarshish was a distant port from which iron, tin, lead, ivory, monkeys and peacocks were brought to Palestine. According to Gen. X: 4 (*cf* Isa. 23:1), it must be a Mediterranean port since Tarshish is said to be a 'Son of Javan' (Greece). Others, however, identify it with a mining village in southwestern Spain which was, according to Herodotus 4:52, 'beyond the Pillars of Heracles' and, according to Plinius and Strabo, in the Guadalquivir Valley; this is very possible.

References:

Pritchard, James B. *Solomon and Sheba*. Phaidon Press, 1974.

Brown, Robert. *The Story of Africa and its explorers*, vol. 2. Cassell, 1893.

Jewish Encyclopedia, New York and London: 1905.

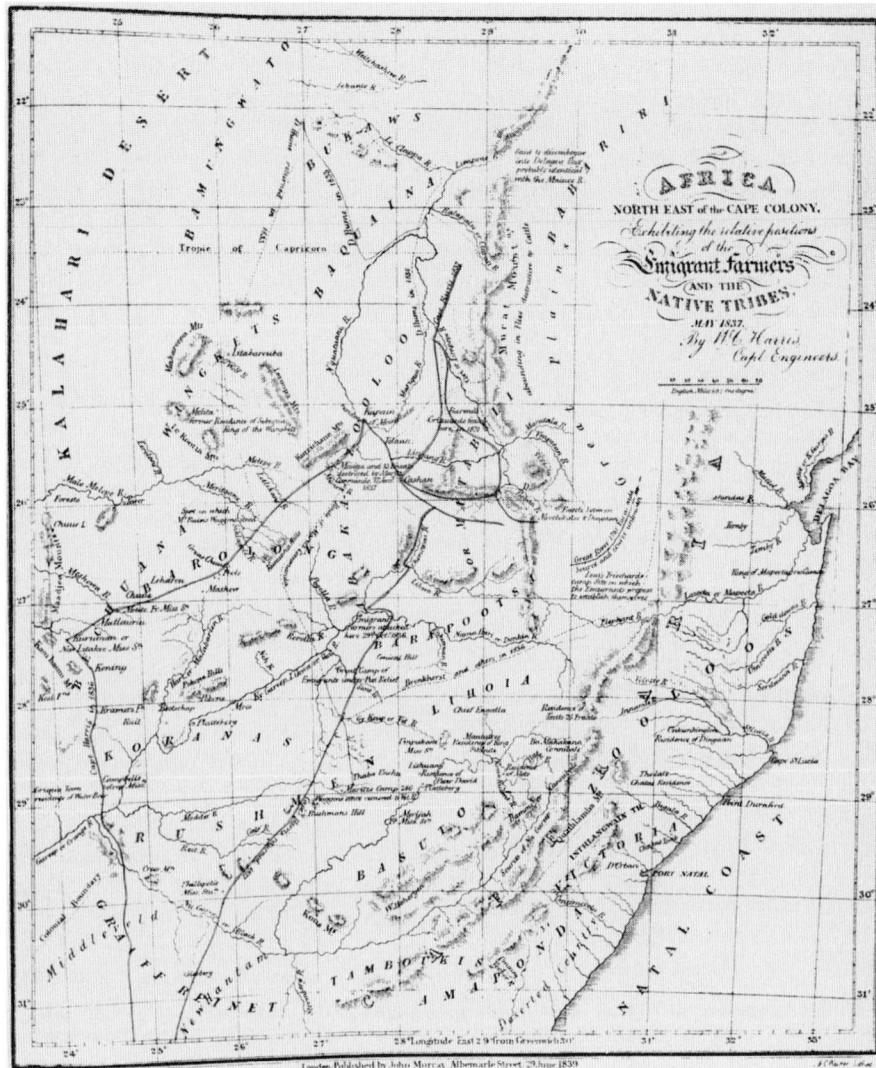

Map 311

HARRIS, William Cornwallis (d. 1848)
Africa North East of the Cape Colony, exhibiting the relative positions of the emigrant farmers and the native tribes, May 1837. J. and C. Walker, lithography. (London, John Murray, 1839.)
Map, 20 x 17 cm, uncoloured.
Scale in English miles.
Prime meridian through Greenwich.
In: Harris, Capt. William Cornwallis. *The Wild Sports of Southern Africa*, 4th ed. London: Pelham Richardson, 1844, fold at back.

This map comes from a travel book by Capt. William Cornwallis Harris, a military man, explorer and hunter. He was already an officer with the Engineers at the age of sixteen and in 1836 he sailed from Bombay for the Cape, arriving there when the Great Trek was the burning question of the day. On board ship be met a Mr William Richardson, who agreed to join him on an expedition. In Cape Town they met Dr Andrew Smith, from whom they received valuable information as to their journey. They resolved to visit New Lattakoo and thence to proceed to Moselekatse's country, and made a start early in September, 1836.

The colonial boundary appears in the lower left of the map, while the district of Victoria appears around Port Natal, between the Tugela River and the Quathlamba Mountains. The routes taken by Capt. Harris via Kuruman, Mosega and Matabele country are shown together with the routes of the Bronkhorst (1836) and Maritz commandoes into Natal.

There is a note on the source of the Limpopo River, which states that it is 'probably identical with Maince River', and Dingaan's kraal, 'Unkunginglove,' appears.

354

Map 312

HULLMANDEL, Charles (1789-1850)
(Zoolu country.) (London, W. Crofts, 1836.)
Map, 20 x 21 cm.
Scale in miles
In: Gardiner, Capt. Allen F. *Narrative of a Journey to the Zoolu country in South Africa undertaken in 1835.* London: W. Crofts, 1836, fold at end.

A map of Natal lithographed by Charles Hullmandel, featuring mainly 'Zoolu country,' which appears in Capt. Allen Gardiner's *Narrative of a Journey to the Zoolu Country in South Africa.*

The coastlines are based on Arrowsmith's map but the interior is filled from personal observation. The map covers the country from 'Unkunginglove,' north of the Tugela River, and includes rivers and villages, with a distinctive mark for the military towns. There are occasional notes on the type of country encountered.

355

WESTERN AFRICA

Map 313 RUSCELLI, Girolamo (c. 1504-1566)
Mauritania nuova tavola. (New map of Mauritania.) (Venice, V. Valgrisi, 1561-2.)
Map, 18 x 24,5 cm, uncoloured.
Italian text on verso.

This map of West Africa appears to come from the same Ptolemy atlas (1561) as Map 151. The title is in Latin and place names are in Latin and Italian, describing many rivers, lakes and mountains. The plates for this atlas were, with only minor changes (principally in the ocean with the addition of ships and sea-creatures) used through the final edition of 1599. For more examples, see Maps 292-292b.

Map 314

RAMUSIO, Giovanni-Battista (1485-1557)
Parte del Africa. (Part of Africa.) (Venice, G. Gastaldi, 1550-1559.)
Map, 27 x 37 cm, woodcut
Top left: 370.
Top right: 371.

This woodcut map is filled with mountains, trees and bushes shown in great detail, and many animals – elephant, camel, monkey and lion. There are natives in the south, many in loincloths, carrying bows and arrows, and one group appears as soldiers in uniform. The southern seas, engraved in horizontal wavy lines, are alive with native canoes and sea monsters, and there are also two European ships. The one on the right has the arms of Portugal and is going from Elmina to the island of St. Thomas, and the vessel on the left bears the fleur-de-lys.

Compiled by the Venetian cartographer, Gastaldi, it appears in the third volume of Ramusio's *Delle Navigationi et Viaggi nel qual si Contiene les Descrittione dele Africa.* Ramusio was born in Treviso in 1485 and died in Padua in 1557. He acted as Secretary to the Council of Ten in Venice for forty-three years and his collection of 'Voyages' was highly esteemed during his own and later times.

The map includes the coast of Guinea, with a scene of natives rendering homage to a cross-legged chieftain. Below this appears a fortified Portuguese fort – Castel de la Mina (modern Elmina), which was founded by the Portuguese in 1482 and became an export centre for gold from the interior (Lamina de Portogal). Today Elmina is part of Ghana (called the Gold Coast until 1957). From 1500 to at least 1530 the Elmina trade was very prosperous, and it remained a post of great importance, flourishing as it did on gold and slavery.

The Senegal (Senega) and Gambia (Gambra) Rivers and the Rio Grande are linked with a west-flowing Niger, the river that the Arab geographer Al-Idrisi called the 'Nile of the Negroes.' At that time the Niger and the Nile were thought to rise from the same source.

At the time of this map French traders were beginning to break Portuguese monopoly of the coast and were doing a brisk trade in Senegal. They travelled to Costa de la Melegeta (Melaquetta) on the Grain Coast (modern Liberia) but they seldom got as far as Elmina and never reached Benin.

358

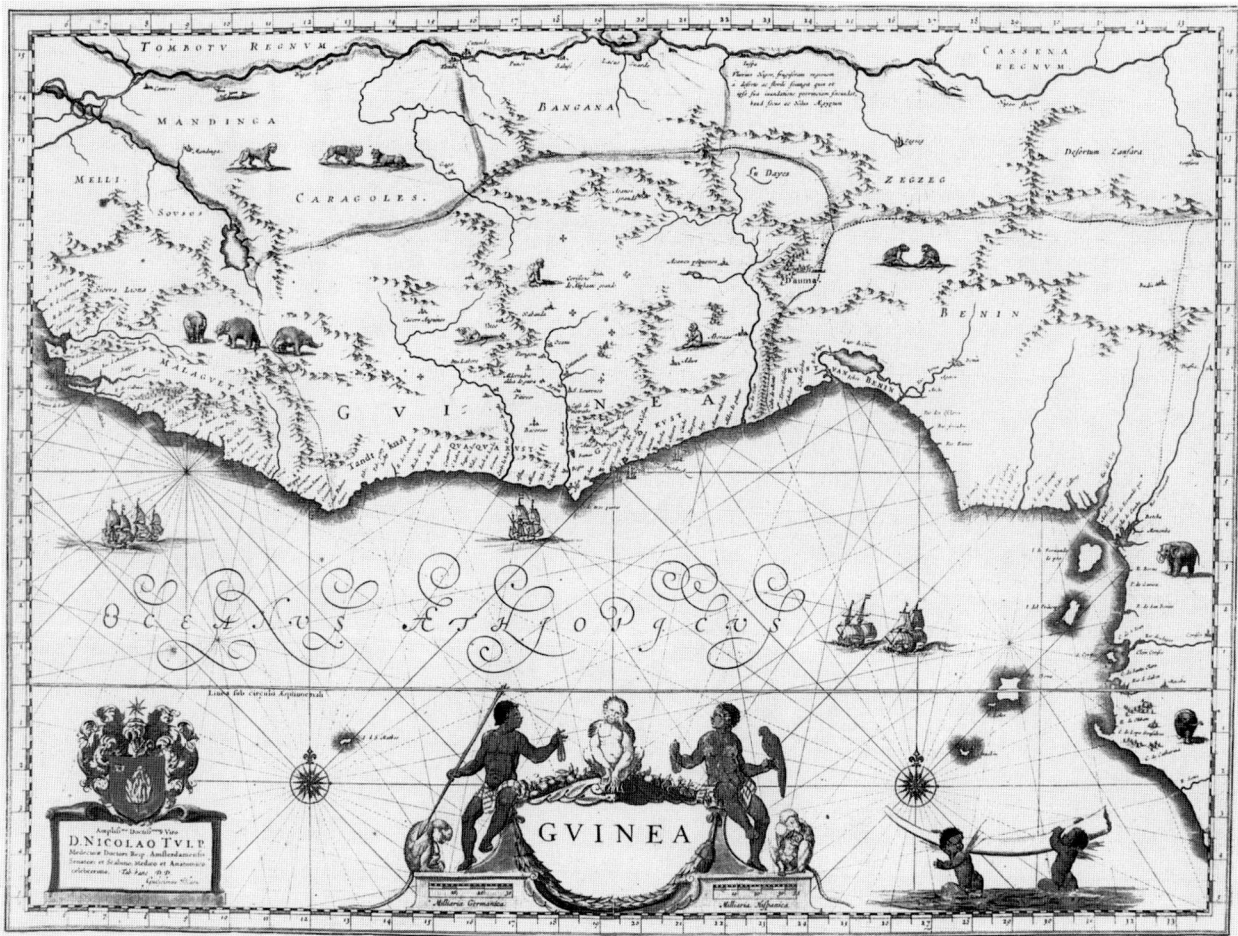

Map 315

BLAEU, Willem Janszoon (1571-1638)
Guinea. (Amsterdam, W.J. Blaeu, 1663.)
Map, 38 x 52 cm, coloured outline.
Scale in German and Spanish miles.
Dedicated to D. Nicolao Tulp.
Description of the country in French on verso.

This map appears in the section on Guinea in vol. 10, *Afrique*, of Blaeu's *Grand Atlas*, published in Amsterdam in 1663.

The cartouche on the lower border is varied and colourful. Two natives sit at the sides of the title, above which is a scroll with flowers and fruit and, in the centre, a very human-looking monkey. Each native holds a string of beads in one hand, and in the other hand one has a spear, and the other a parrot. Sitting at each side of the masonry surround is a monkey and on the lower right border are two children carrying a large elephant tusk on their shoulders.

At the lower left is an armorial cartouche at the base of which is a dedication to D. Nicolao Tulp. Tulp is the Dutch doctor who appears in Rembrandt's famous painting *The Anatomy Lesson*, as the teacher of the group around the body.

The interior is typical of Blaeu's calligraphy, with various animals, rivers, lakes and mountains. Five sailing ships are depicted in the seas along the coastline.

359

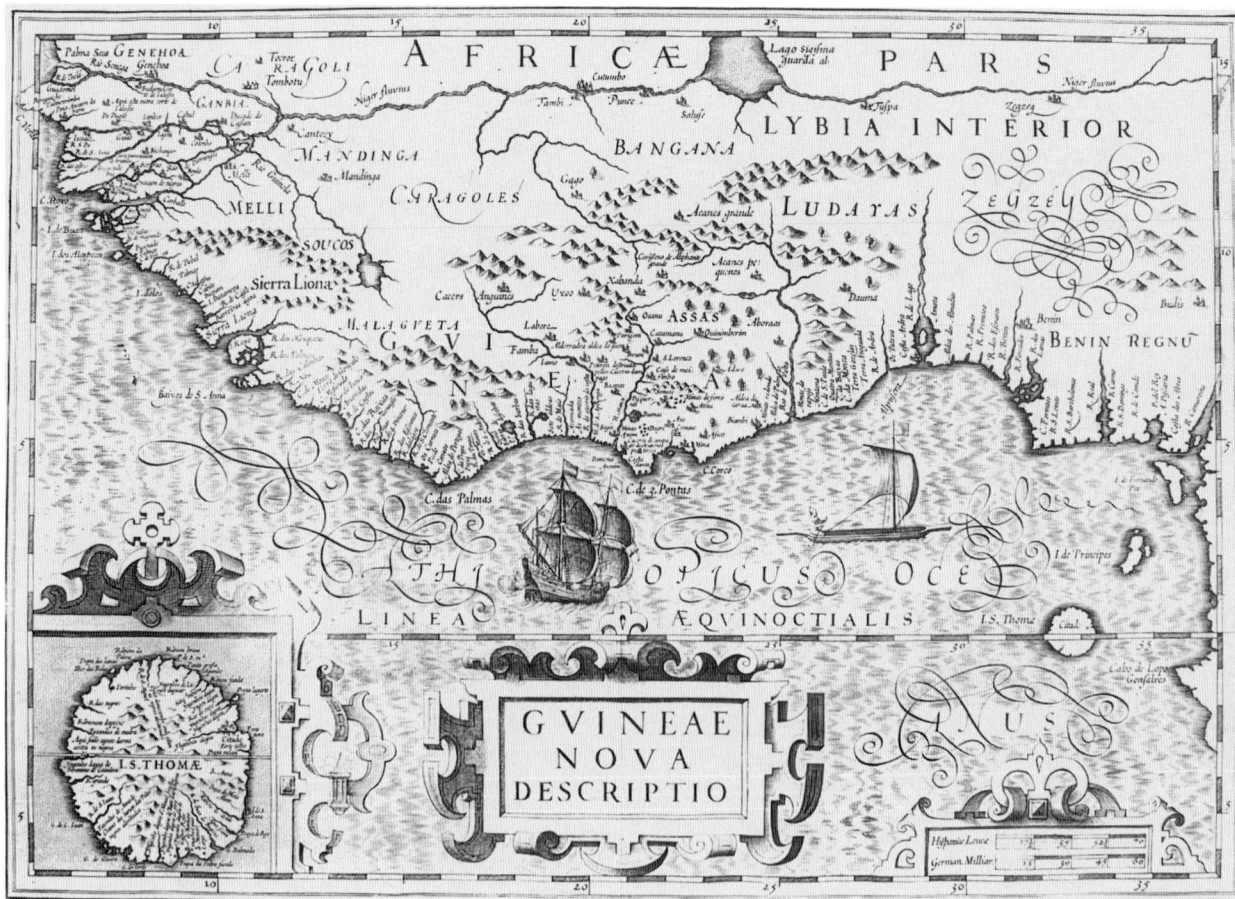

Map 316

HONDIUS, Jodocus (1563-1612)
Guinae nova Descriptio. (New map of Guinea.)
(Amsterdam, J. Hondius, 1606.)
Map, 34 x 49 cm, uncoloured.
Scale in Spanish leagues and German miles.
Inset: Enlargement of the Island of St. Thomas.

The Dutch cartographer Jodocus Hondius added this map to his version of Mercator's *Atlas* in 1606. The florid cartography is typical of his style and, although he was accused of borrowing from other cartographers, his 1606 *Atlas* was very complete, having six maps of Africa against only one in the great posthumous Mercator *Atlas* of 1595.

The map shows the west coast from Senegal (Rio Senego) to Cape Lopez, just below the equator, and includes Guinea, with a portion of the Benin kingdom and of Lybia. There is a detailed inset of the Island of St. Thomae which, together with the I. de Principe, formed a Portuguese province.

The Portuguese role in the exploration and settlement of West Africa – on the decline at this time, due to lack of manpower – is reflected in many seventeenth-century maps like this one.

AFRICÆ PARS

Sep tentrio

GENEHOA

Geneboa

Tamboira

Cutumbo

CARAGOLI

Niger flu

Niger fluvius

zegzeg

Tombu

Bildua

Ancao barro

Talafio

Guiao

Iccrato

Colenho

Tombi

Salufe

Punce

Tulpa

LYBIA INTERIOR

Togra

Tunts

Melli

MANDINGA

Mandinga

BAGANA

Anni gra de

ZEGZEG

MELLI

CARAGOLES

LUDAXAS

BENIN

Soucos

MALAGVE

Cor feno

anes peque

nos

REGNV

Palmar

Palmar

TA

Corf feno

de Eliphare gran de

Benin

Sierra Liona

GVINE

Cacero An gumes

Xabanda

A

Dauma

R. de Caffe

Sierra Liona

Labore

Uxoo

Octanti

ASSAS

Aboraas

Bignam

R. das Palmas

Quinim berim

R. das Palmas

Caractro

R. da Volta

R. de Escravos

Baixo de S. Ana

R. das calinas

as aftru

das pelles Charberin

Aldes

Aldea Acaradaha

R. de Conde

P. del Rey

Pdo faco

ores da imp280

Minas de

Biambi

Afuti

Athiopicus Oce

Bo

gio

Mina

G. Corco

C. do 3 Pentas

J. de Fernando Poo

Praya das Vacas

Ribiera da palma

Citade

Pravalay aria

J. do Principes

Partin ho

J. de S. Nicolai

Singl tribueta

Yngelios dagoa

R. doura

Yngelios dagoa

Forte novo

J. S. Thomæ facchari feracissima

Citade

Citade

Aquinoctialis Linea

Cabo de Lopo Gon falves

I. S. THOMÆ

Forte velho Trava melan

R. Grande

Praya d Aleo fenna

J. de S. Anto

GVINEA

ANVS

Praya de Rey

Dalmada

Milliaria Germanie

Merid.

Occidens

Map 316a

HONDIUS, Jodocus (1563-1612)
Guinea. (Amsterdam, J. Hondius, 1621.)
Map, 14,5 x 18,5 cm, fully coloured.
In: *Atlas Minor.*

This map, from a miniature atlas of 1621 (the Mercator-Hondius *Atlas Minor*) resembles the immediately preceding larger map numbered Map 316. In full wash colour it is very attractive and the quality of engraving is such that even small detail is very legible. The typical Mercator moiré pattern to the sea is shown and there is even space for a sea creature. The Portuguese Island of St. Thomæ is prominent in an inset. Coastal detail is good and we see a portion of the Kingdom of Lybia and also Benin. Miniature maps were an important vehicle for transmitting cartographic knowledge. The *Atlas Minor* went through several editions and was a particularly attractive atlas. There is a title and page number printed by letterpress. Since the map is a copperplate engraving, two impressions were needed: one for the image and one for the type. Registration was not always as good as seen here and many maps in this atlas had the typeset text encroach on the image. The plate shows many tiny imperfections that reveal themselves as dark dots.

Map 317 L'ISLE, Guillaume de (1675-1726)
Carte de la Barbarie, le la Nigritie et de la Guinée (Map of Barbary, Nigeria and Guinea.) (Paris, G. de L'Isle, 1700-1712.)
Map, 49 x 57 cm, coloured.
Scale in sea leagues and French leagues.
Prime meridian through Ferro Island.

A map of North-West Africa from Barbary, Nigeria and Guinea to below the equator. Guillaume de l'Isle, a French publisher and Geographer to the King in 1718, included maps of North Africa in his *Atlas de Geographie* (1700-1712). There is a mass of geographic detail in the interior of these states, with numerous lakes and rivers, major and minor. As so often with de l'Isle's maps, there is no cartouche, but a simple printed title across the top border.

From an early age de l'Isle made a scientific study of geography, which he was determined to reconstitute on a

new basis. His 'Mappemonde,' maps of Africa, Europe and Asia, were based on the most recent astronomic and geographic studies and were so highly regarded that de l'Isle has since been hailed as the principal creator of the modern system of geography. He published more than a hundred maps.

In this map he corrects his previous impression of the Senegal river which, in earlier maps, he joined to the Nile by a hypothetical line. He now records the Niger as an 'arm of the Nile.'

362

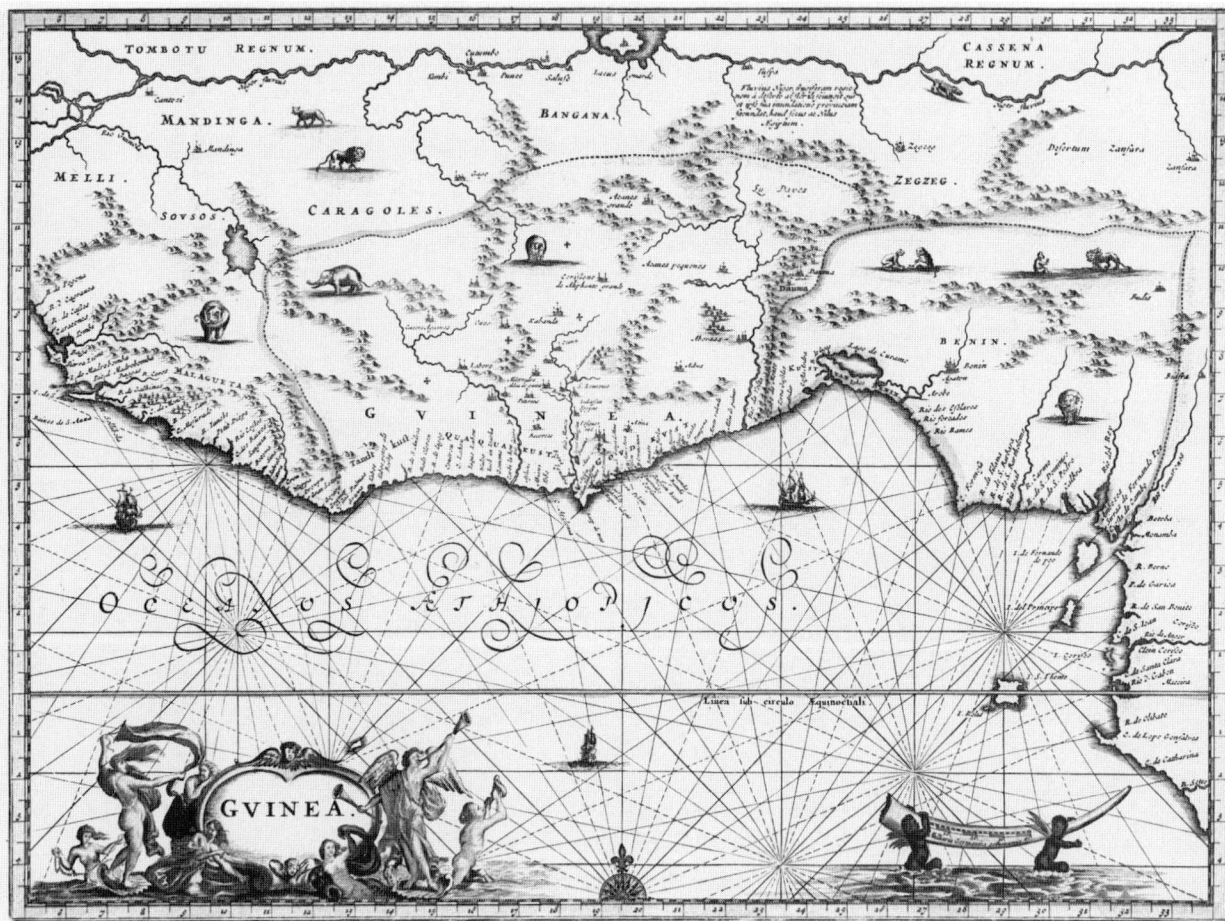

Map 318

OGILBY, John (1600-1676)
Guinea.
(London, J. Ogilby, 1671)
Map, 26 x 36 cm, coloured.
Scale in German miles.

This map of Guinea has a colourful cartouche on the left lower border – an oval frame with the title surrounded by angels and sirens, two of whom are standing blowing horns. There are numerous small animals within the map and a number of towns are represented by buildings with steeples. The calligraphy is flowery, reminiscent of Blaeu. The scale is printed on a curved tusk supported by two cherubs.

John Ogilby, a London geographer, historian and publisher, was born in Edinburgh in 1600. In addition to being a bookseller, printer and translator he was also a dancing teacher and theatre owner. As an equally famous surveyor and cartographer he was appointed Royal Cartographer in 1671. He lost two fortunes during his lifetime, but recovered each time. His *Britannia* was famed for being the first survey of the roads of England and Wales and appeared a year before he died, in 1675. His three maps of Africa appeared in an accurately described and complete *History of Africa*, published in 1670 and based on the Dutch work by Montanus. The three maps appearing in the descriptive volume of Africa are of the continent of Africa, the Guinea coast and southern Africa. The maps appear to be very similar to those of Blaeu, with similar lettering and various small animals placed on the mainland, together with small sailing vessels in the oceans. Nowhere is a cartographer or engraver named but on the map of the whole of Africa Jacob Meursius appears, and it is generally agreed that all were engraved by him.

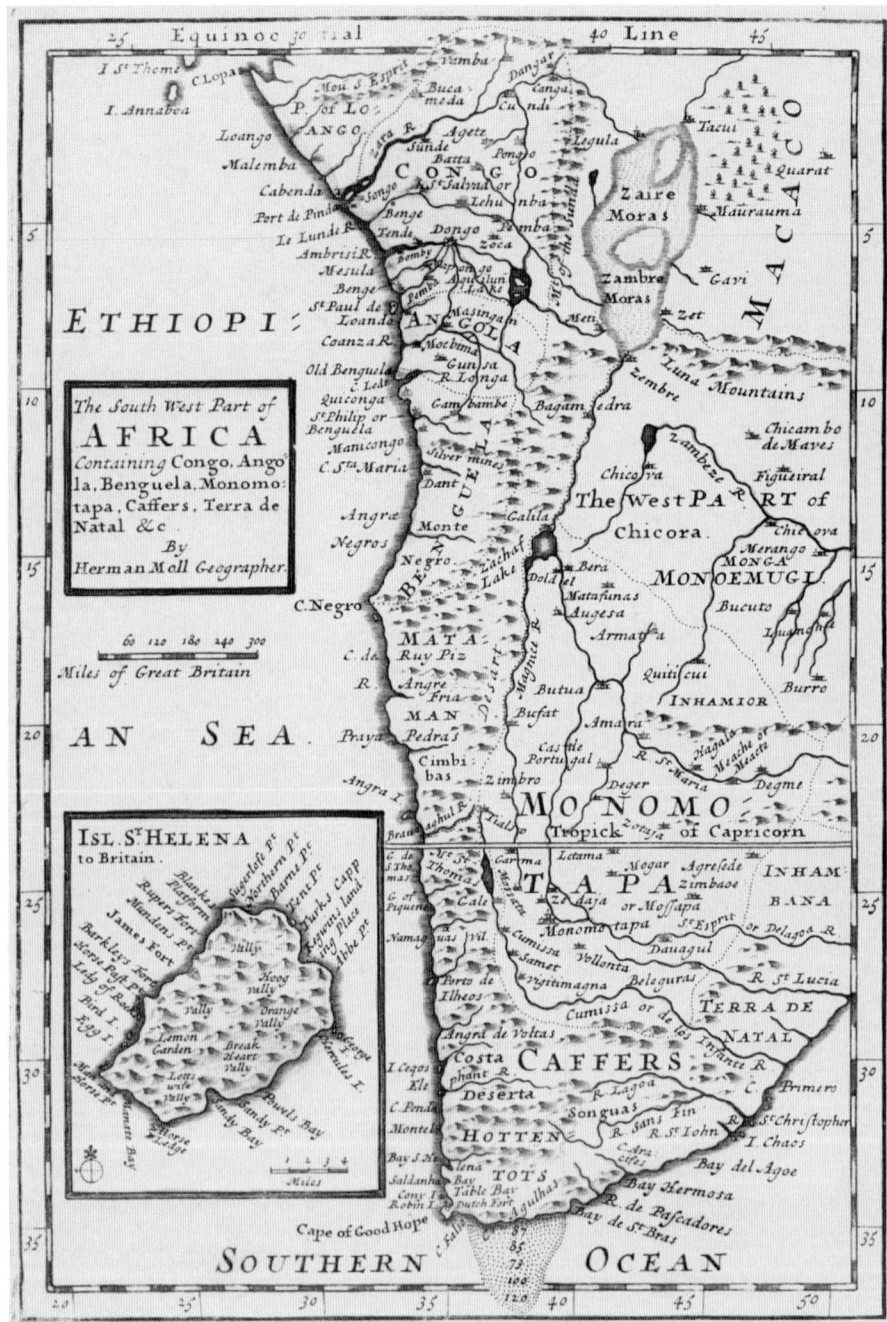

Map 319 MOLL, Herman (d. 1732)
The South West Part of Africa: containing Congo, Angola, Benguela,
Monomotapa, Caffers, Terra de Natal &c. (London, H. Moll. 1714.)
Map, 26 x 18 cm, uncoloured.
Scale in British miles.
Inset: Island of St Helena.

The title of this map of the southwest part of Africa is enclosed in rules at the left, above a scale of 'miles of Great Britain.' Below this is an inset of the Island of St. Helena with its own small north indicator and its own scale. The caption 'Dutch Fort' appears to be the only indication of European settlement in southern Africa. Imaginary cities, rivers and boundaries are marked in the interior and the Zambeze River is noted. This map is found in both Harris's *A Collection of Voyages and Travels*, published in 1745, and in H. Moll's *Atlas Geographus*. The coastline of St. Helena (inset) is abundantly named.

364

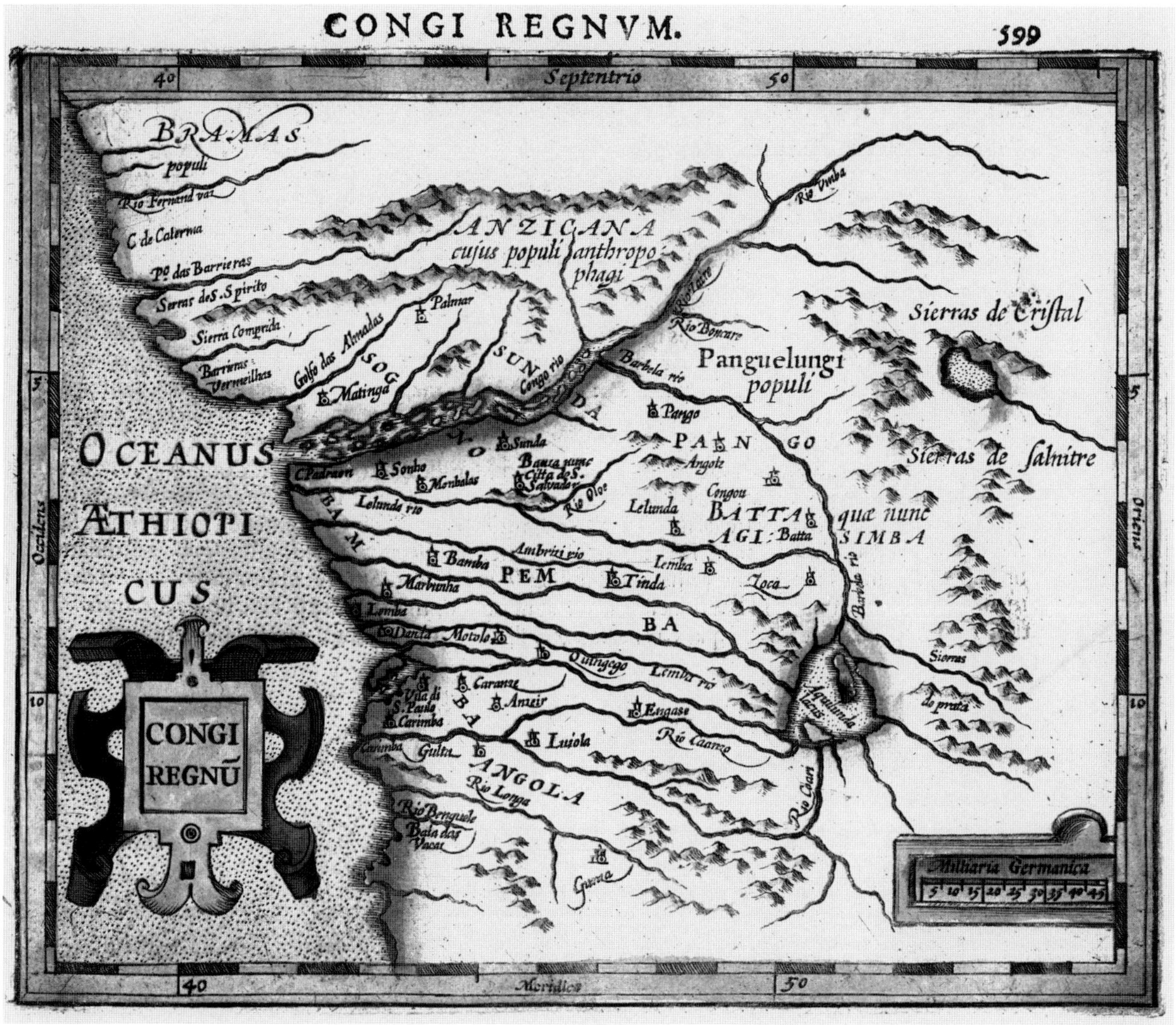

Map 319a   HONDIUS, Jodocus (1563-1612)
Congi Regnum. (Amsterdam, J. Hondius, 1621.)
Map, 14,1 x 17,2 cm, fully coloured
In: *Atlas Minor.*

Another map from the Mercator-Hondius *Atlas Minor.* This map is in partial colour, with the ocean a pale blue and the mountains in brown. Boundaries, such as there were, are outlined in a pale wash. This is an early detailed map of the coastal region and some attempt was made to show interior detail. It is compelling because of the large areas that are not only without detail, but that were not coloured by the colourist. In this map, the blank spaces are truly blank. The map is shown almost full size. Because this map is from a late edition, the plate shows wear in the form of surface imperfections that hold ink and are seen as small dark specks in the image.

365

Map 320

DE LETH, Andries (1662-1731) and Hendrik (1692-1759)
Carte Nouvelle de la Mer du Sud, dressée par ordre des principaux Directeurs et tirée des Memoires les plus récents... (Map of the Southern Oceans.)
(Amsterdam, A. and H. de Leth, 1730.)
Map, 57 x 94 cm, coloured.
Prime meridian through Ferro Island.
Inset: Straits of Gibraltar. Cape of Good Hope. Mexico City. Panama. Bay of Porto Bello. Scenes from the French Islands. Bay of Rio de Janeiro. Vera Cruz and Havana.

This map is attributed to Andries de Leth and his son Hendrik, both Amsterdam engravers and map makers. According to Anna Smith's *Exhibition of Decorative Maps of Africa up to 1800*, the map dates from 1730 and is also recorded by the Library of Congress as appearing in R. and J. Ottens's *Atlas sive geographia compendiose* (Amsterdam, c.1756).

This is a large, elaborate copperplate map, with its title across the top. A Latin inscription, *Nova mona Pacific tabula...*, towards the lower left corner on drapery held by three angels, indicates that this is really a map of the Pacific. It extends from the western half of Africa in the east to the East Indies in the west. On a map of such a large part of the earth not much detail can be recorded but the main divisions of western Africa are noted. It is interesting to see that modern-day California is represented as an island.

There are numerous insets, mostly of places in Central and South America. At the top left an inset shows a plan of Vera Cruz, the Bay de Rio de Janeiro and La Havana, together on one panoply; at the bottom left a map of the Bay of Porto Bello by itself on a panoply and next to it a map of the Panama isthmus, by itself in a box. To the right of this is a note in French on the accuracy and value of this chart. A Dutch title, 'Neeuwe Kaart van die Zuyd Zee' is held by two cherubs, and next to the Latin cartouche is a medallion-like inset made up of four vignettes with captions in French, depicting various unusual scenes such as turning tortoises or turtles onto their backs, and working on a sugar mill. In the centre of the map, surrounded by scrollwork and acanthus leaves, is a group of six Mexican scenes round a plan of La Ville de Mexico. One scene is of a pyramid as it existed in the time of the Mayans of Mexico, at the base of which ritual

366

sacrifice is being conducted by the removal of a live heart from a human chest.

Of chief South African interest is the inset in the lower right corner. It is a view of Table Bay measuring 26,5 x 17,5 cm, in a rococo cartouche. It is a quaint and most attractive picture, making Table Mountain the nearest thing to a sugar loaf. The harbour is full of shipping, and a seal hunt appears to be in progress in the forefront. There is an identification table in the right-hand corner, in which Table Mountain is labelled 'La Montagne de Lions' and the Governor's garden is quoted as being much appreciated by the passengers to India. The legend also states that the Hottentots who live in the surrounding mountains are very savage, eat putrid food and make no profession of Christianity.

A further legend provides a history of the discovery of the Cape and the construction of the Fort. On the decorative frame enclosing the view a fox, two leopards, a lion and a monkey can be seen. On the top of the centre frame is an additional piece of framework enclosing a ground plan of the Castle which is labelled 'Fort au Cap de Bonne Esperance.' The numerous decorations and insets are completed by a map of the straits of Gibraltar in the upper right-hand corner.

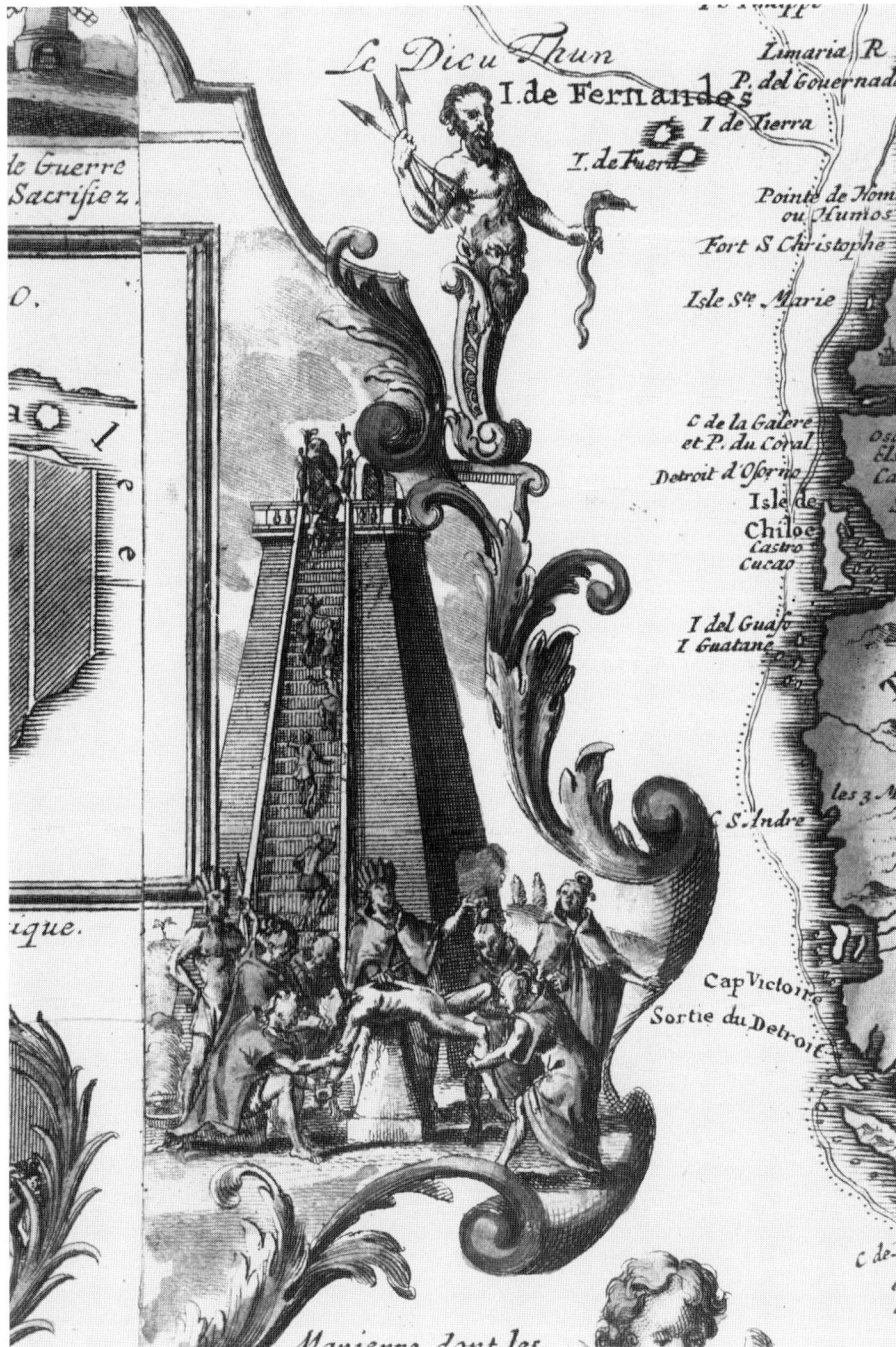

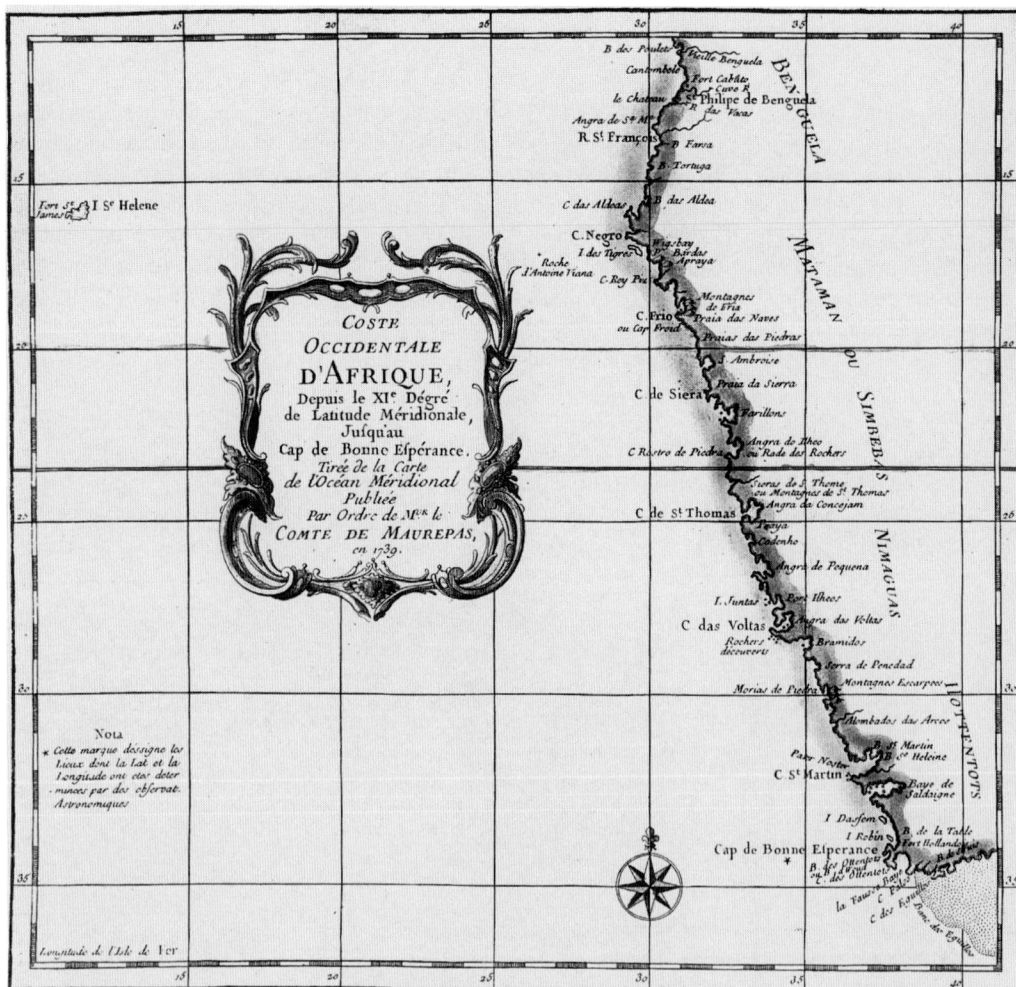

Map 321

BELLIN, Jacques Nicolas (1703-1772)
Coste Occidentale d'Afrique Depuis le XIe Dégré de Latitude Méridionale,
jusqu'au Cap de Bonne Espérance... Publiée par ordre de Mgr. le Comte de
Maurepas. (West coast of Africa from 11° S to the Cape of Good Hope.)
(Paris, J. Bellin, 1739.)
Map, 25 x 26,5 cm, coloured outline.
Top right: *Afrique Occidentale No. 3.*
Bottom right: Tome 3 in 8°. Page 307.
Bottom left: Tome I in 4°. No. 16.

This small French map of the southwestern coast of
Africa first appeared in volume 1 of the twenty-five volume
Histoire Générale des Voyages of Antoine Francois Prévost
d'Exiles. Jacques Nicholas Bellin, the French Royal
Hydrographer, produced the atlas to accompany these
voyages, which took place from 1738 to 1775.

A rectangular floral scroll design cartouche appears
prominently in the Indian Ocean containing the title, the
date and an indication that it was commissioned by the
Comte de Maurepas, a French notable of that time. The
map extends from B. des Poulets, north of C. Negro at 10°
south, to C. des Eguellas in the south. Many Portuguese-
named bays, capes and some rivers are noted along the
coastline. In the south appears C. St. Martin, sited close to

Baye de Saldaigne, B. Ste. Helene, B. de la Table and I.
Roben (Robben Island). Fort Hollandois (The Fort) and the
Cap de Bonne Esperance are also named. Ste. Helene Island
with its Fort St. James is noted in the top left corner. In the
lower seas a compass rose appears with the conventional
fleur-de-lys pointing to the north.

At the left lower border a note appears stating that those
places on the map marked with an asterisk had their
longitude and latitude determined according to astronomical
observations. The Map Library of the British Library has in
its collection two later editions of this map, published by
Didot and de Hondt.

A map from the same source, featuring the southeastern
part of the subcontinent, appears as Map 306.

368

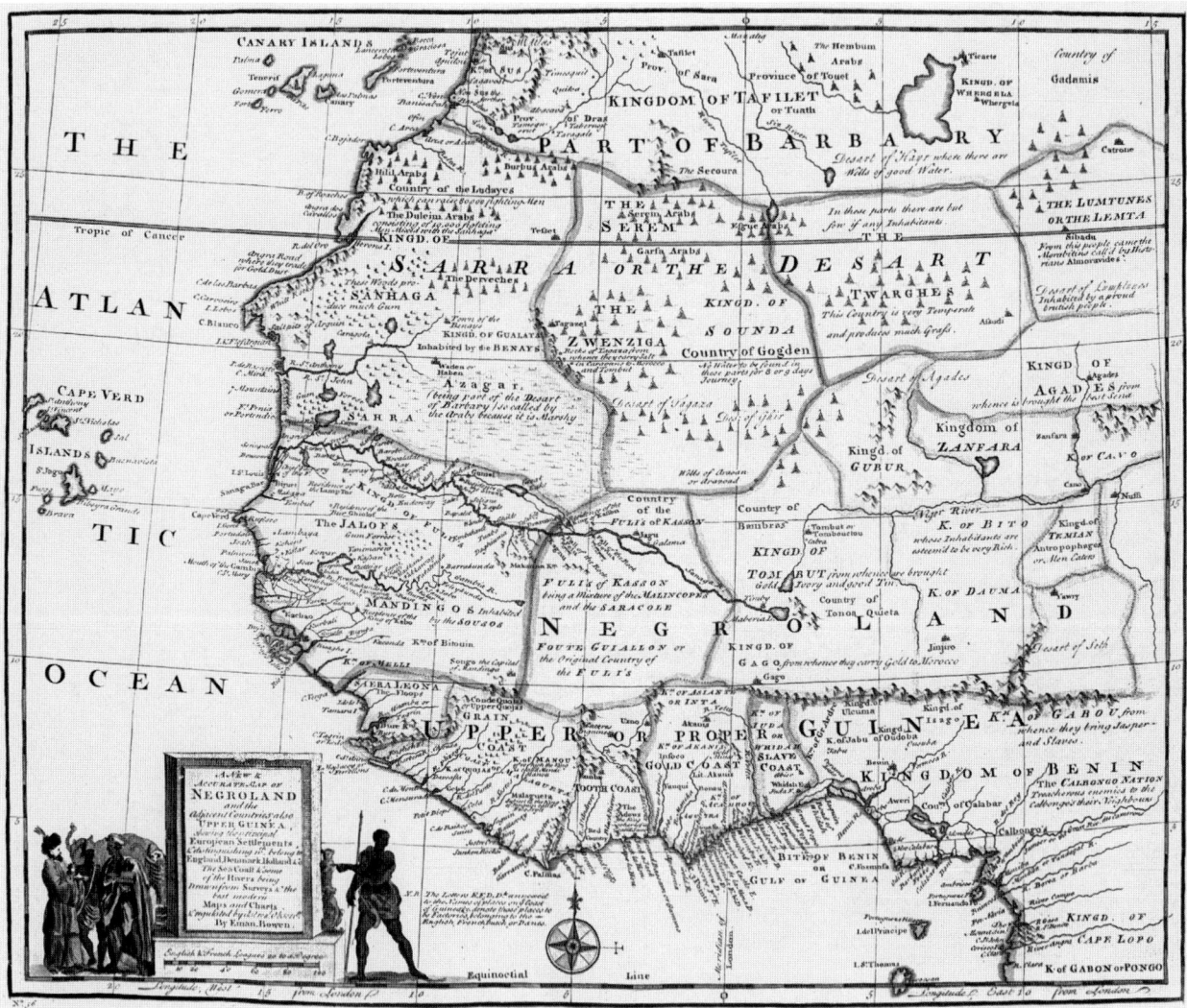

Map 322

BOWEN, Emanuel (fl. 1720-1767)

A New and Accurate Map of Negroland and the Adjacent Countries; also Upper Guinea showing the principal European Settlements, and distinguishing those belonging to England, Denmark and Holland. The Sea Coast and some of the Rivers being drawn from Surveys and the best modern Maps and Charts...

(London, E. Bowen, 1747.)
Map, 24 x 42 cm, coloured.
Scale in English and French leagues.
Prime meridian through London.
Bottom left: No. 56.

A map of 'Negroland' which includes the Gold, 'Tooth,' Grain and Slave Coasts of West Africa, issued by Emanuel Bowen, father of Thomas, who assisted his father as well as producing his own maps. Emanuel produced a large number of maps of Africa and southern Africa. A cartouche in the lower left corner reveals a boxed title sitting on the scale which is in French and English leagues. Flanking the cartouche on the left are costumed Africans standing in front of an elephant, and on the right a semi-nude figure with a long spear in his right hand, pointing to two monkeys holding onto the edge of the title masonry.

The map itself shows the various kingdoms, especially Upper Guinea, and the Kingdom of Benin and further north the Sahara Desert. The Kingdom of Tombut, in the lower centre, is described as the place from whence gold, ivory and good tin are brought. There are numerous legends referring to warring tribes, and various geographical features.

369

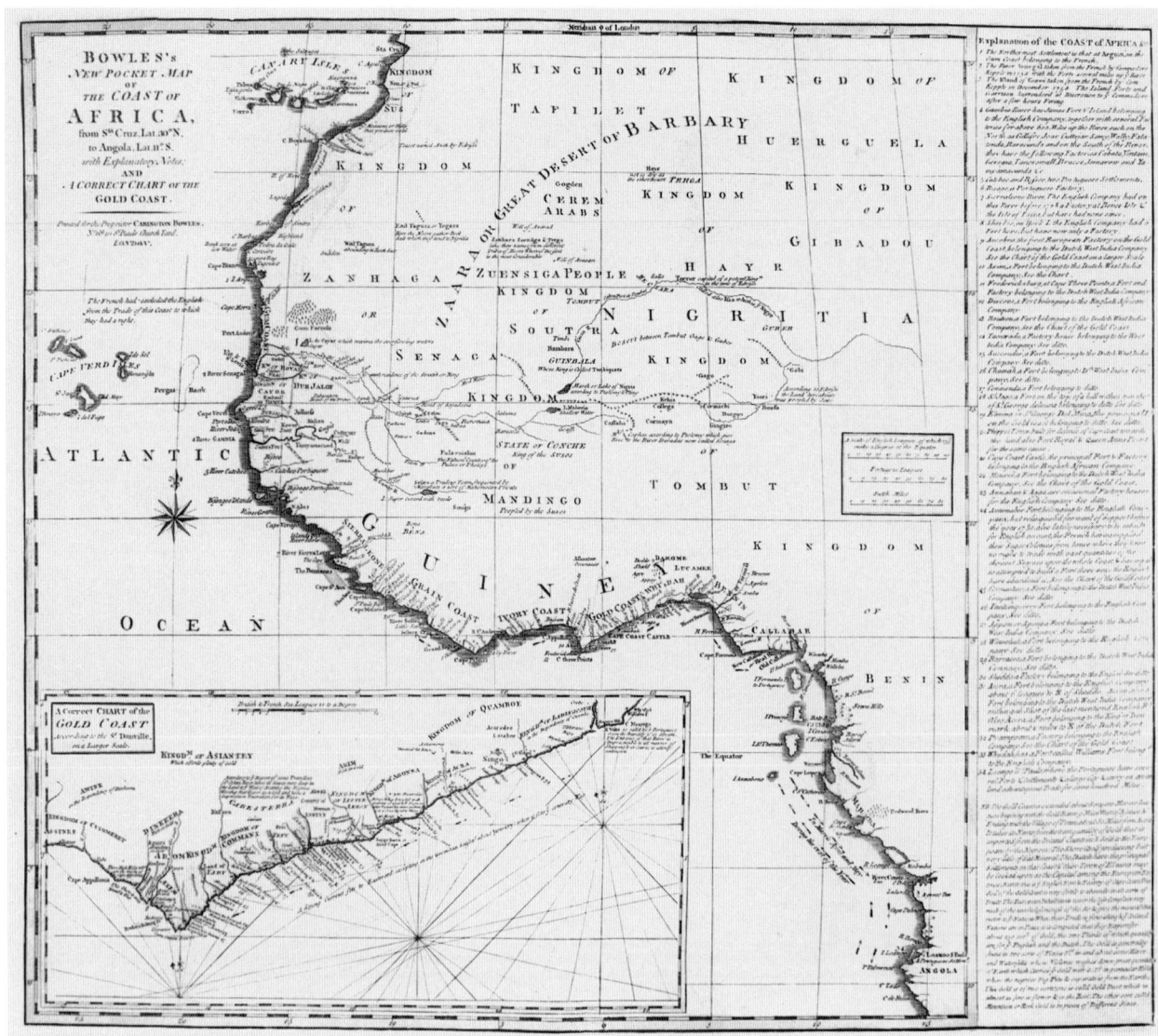

Map 323

BOWLES, Carington (1724-1793)
*Bowles's New Pocket Map of the Coast of Africa, from Sta. Cruz... to Angola...
with explanatory notes and a correct chart of the Gold Coast.*
(London, C. Bowles, 1752.)
Map, 49 x 57 cm, coloured.
Scale in English and Portuguese leagues and Dutch miles.
Inset: *A Correct Chart of the Gold Coast according to the Sr. Danville on a larger scale* (scale in British
and French sea leagues).

The west coast and the Cape Verde Isles on this map are fairly detailed for that period, while the interior is somewhat sparse in geographic detail. A descriptive panel of text appears on the right with historic and topographic details. The inset at the lower left is concerned with 'a correct chart of the Gold Coast' where all the kingdoms are mentioned, with the addition of many rhumb lines.

On the main map Sierra Leone, the Grain Coast, the Ivory Coast and the Gold Coast are noted. The Niger, as with all maps of this period, is incorrectly sited. The

geographers and publishers of that period were unaware of the Niger's south-flowing course into its mouth on the lower west coast of Africa.

Carington Bowles, a printer and publisher of maps in London, put out this map of the west coast of Africa probably in the mid-eighteenth century. He was appointed engraver of maps to George II of England and Louis XV of France and his atlases were mainly of England and America, though he also produced a complete atlas in 1752, in which this map probably appeared.

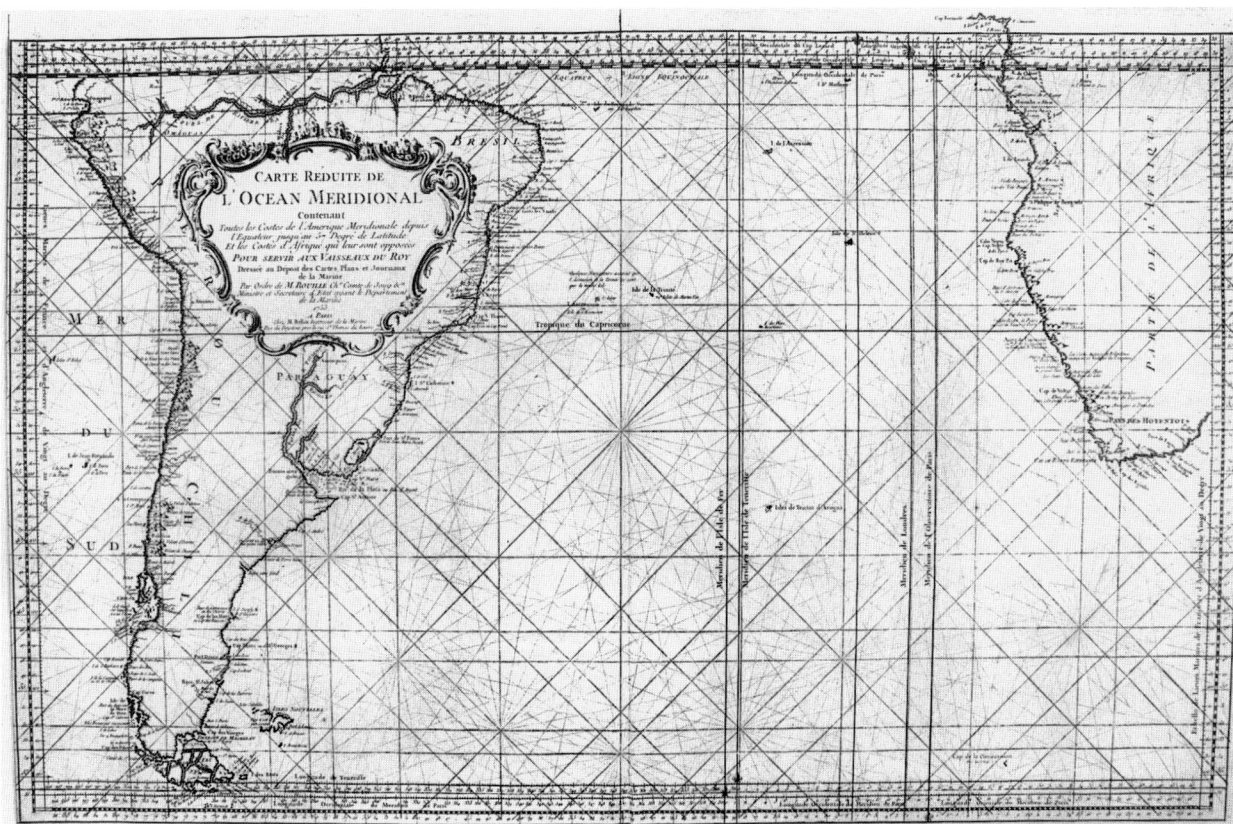

Map 324 BELLIN, Jacques Nicolas (1703-1772)
Carte Reduite de L'Ocean Meridional... (Map of the southern ocean.)
(Paris, J.N. Bellin, 1753.)
Map, 54 x 86 cm, uncoloured.
Scale in French and English sea leagues.
Prime meridian through Paris, Tenerife, Cap Lezard, London, Ferro Island.

This map of the southern oceans is included in this cartobibliography of Africa because it shows a large area of the southwest coast of Africa, including the Cape of Good Hope up to the Eastern Cape Colony.

The cartographer is the Frenchman Jacques Nicolas Bellin, who was attached to the Dépôt de la Marine as a hydrographer to the King. Louis XIV ordered the compilation of a complete survey of the world's coastlines, the *Neptune Français*, originally under the orders of the Governor of the Académie des Sciences. This work was conducted in such a dilatory fashion that it was eventually put into the hands of Bellin to correct and augment and was finally completed in 1753. This map of 1753 is from Bellin's edition of the *Neptune Français*.

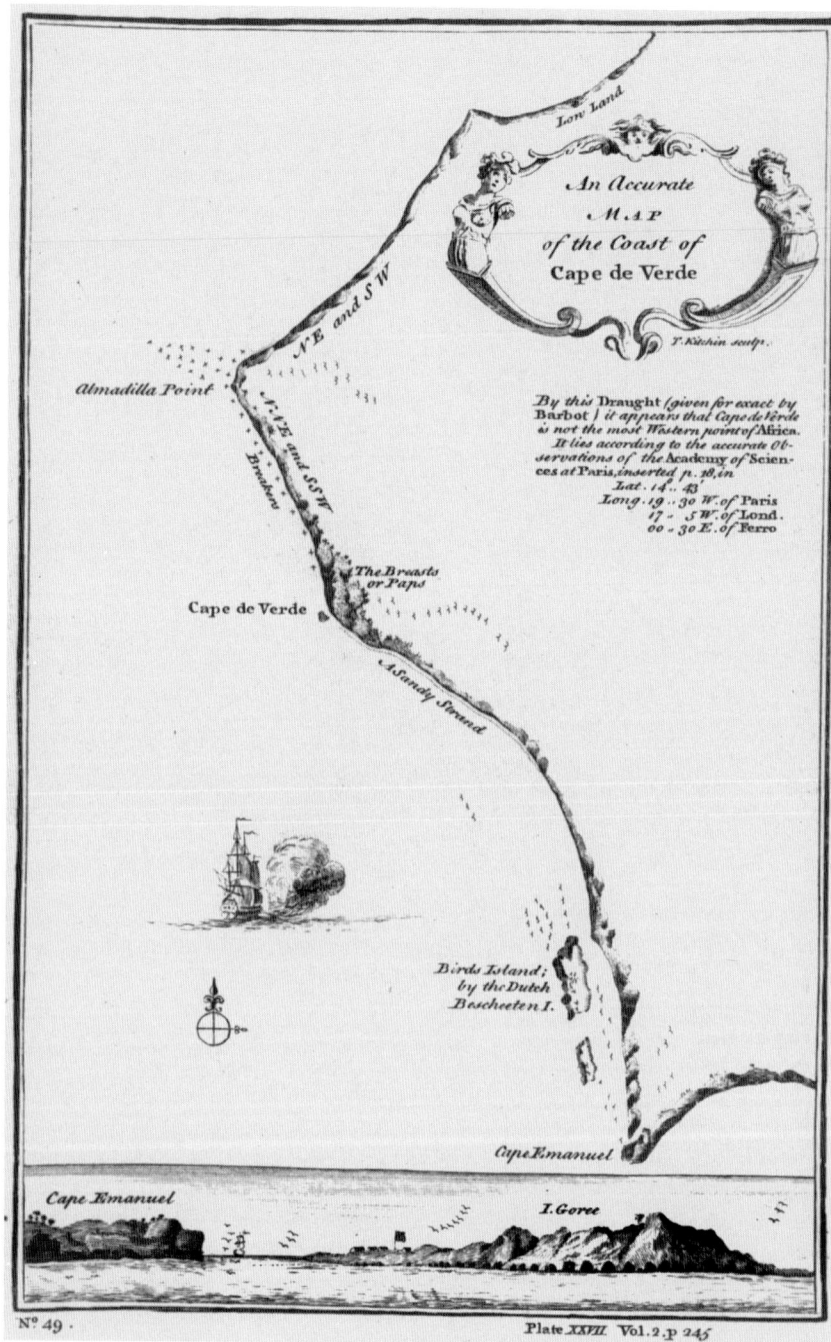

An Accurate MAP of the Coast of Cape de Verde

T. Kitchin sculp.

Low Land

N.E. and S.W.

Almadilla Point

N.N.E. and S.S.W.

By this Draught (given for exact by Barbot) it appears that Cape de Verde is not the most Western point of Africa. It lies according to the accurate Observations of the Academy of Sciences at Paris, inserted p. 16, in
Lat. 14..43
Long. 19..30 W. of Paris
17..5 W. of Lond.
00..30 E. of Ferro

The Breasts or Paps

Cape de Verde

A Sandy Strand

Birds Island; by the Dutch Boscheeten I.

Cape Emanuel

Cape Emanuel

I. Goree

Nº 49.

Plate XXVII. Vol. 2. p. 245

Map 325

KITCHEN, Thomas (1718-1784)
An Accurate Map of the Coast of Cape de Verde.
(London, T. Kitchin, 1765.)
Map, 22 x 14 cm, uncoloured.
Bottom left: No. 49.
Bottom right: Plate XXVII. Vol. 2. p. 245.
Inset: Cape Emanuel and Goree Island.

This is an engraved map of the west coast of Africa, showing clearly that Cape Verde is not the most westerly part of the continent. The map has an oval engraved cartouche with a head at the top and at each side curved figures attached to curved tapering columns similar to those figures on the bow of a ship. In the sea is a small sailing vessel and at the bottom border a relief map showing Cape Emanuel and Goree Island.

Thomas Kitchen (or Kitchin), the engraver of the map, was a publisher and hydrographer to the King in London. His output of engravings was prolific and he worked for important publishers such as Dalrymple, Elphinstone, Jefferys and Willdey.

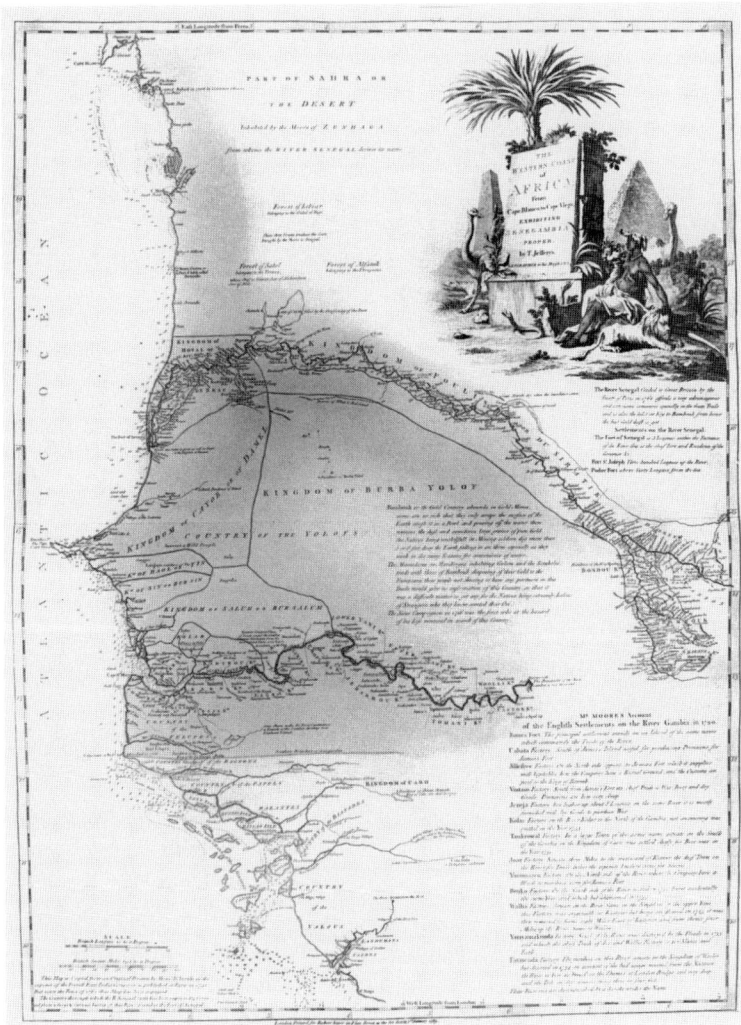

Map 326

JEFFERYS, Thomas (c. 1719-1771)
The Western Coast of Africa from Cape Blanco to Cape Virga, Exhibiting Senegambia Proper.
(London, Robert Sayer, Fleet Street, 1789.)
Map, 71 x 52 cm, coloured.
Scale in British leagues and British statute miles.
Prime meridian through Ferro Island.
Note at bottom left: *This map is copied from an Original Drawn by Mons. d'Anville at the expense of the French East India Company and published at Paris in 1751. But since the Peace of 1762 that Map has been supressed. The Country through which the R. Senegal runs has been improved & corrected from a large & Curious Survey of that River found in the Fort of Senegal.*

This map has an attractive large cartouche in the right upper corner showing a seated woman stroking a recumbent lion with her left hand. On her right is a large title placed on a broad slab of masonry and in her right hand she holds a horn of plenty. On one side of the title is a camel and on the other an ostrich, with snakes in the foreground. In the background is a pyramid. Trees and foliage are plentiful above and to the right of the masonry. Below the main title is a legend describing the River and Fort of Senegal. Below this to the right is 'Mr. Moore's account of the English settlements on the River Gambia in 1730 '

This detailed map of a rather small area of the west coast of Africa was published by Thomas Jefferys, an engraver, geographer and publisher of London, and one of the most prolific and important map publishers of the eighteenth century. He was appointed geographer to Frederick, Prince of Wales, and later to George III. He published large-scale county maps and atlases of London and England, and engraved the maps for Salmon's *Geography* (1749), in which appears a small map of Africa. He also produced important maps of America and the West Indies. Due to some financial difficulties in 1765 Robert Sayer, the London publisher and mapseller, acquired a large part of his interests. Jefferys was later joined by William Faden, also a London map publisher, who succeeded to the business on Jefferys's death in 1771.

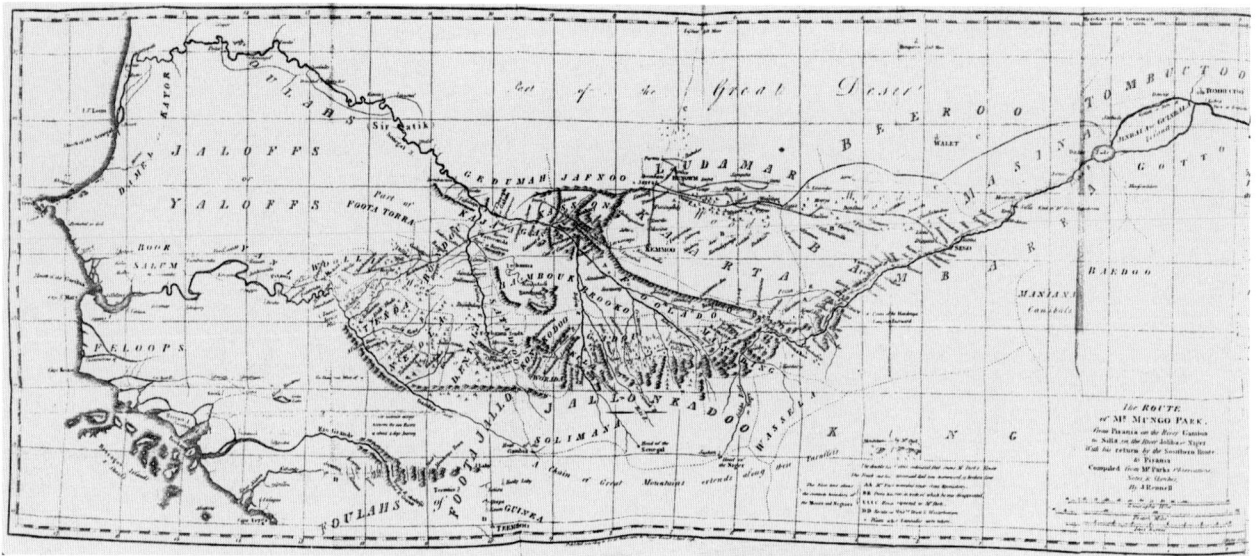

Map 327

RENNELL, James (1742-1830)
The Route of Mr Mungo Park from Pisania on the River Gambia, to Silla on the River Joliba or Niger... Engraved by J. Walker.
(London, J. Rennell, 1798.)
Map, 26,5 x 65 cm, uncoloured.
Scale in geographic miles, English miles and days journey.
Prime meridian through Greenwich.
In: Park, Mungo. *Travels in the Interior Districts of Africa,* 4th ed. London: W. Bulmer, 1800, fold in front.

The closing years of the eighteenth century saw a renewed interest, in Great Britain, in the exploration of inner Africa. The African Association was founded in London in 1788 and it was through its agents that knowledge was gained of the Niger region. Mungo Park was the first to reach that river, having been commissioned to do so in 1795, and was astonished to find that it flowed eastwards, a fact hitherto unknown. This map shows the Niger area south of the 'Great Desert' as far as Guinea, and east from Cape Verde and the mouth of the Senegal to Timbuctoo.

This is an historic map of West Africa, showing the route of Mungo Park, the Scottish surgeon, who travelled from Pisania on the River Gambia to Silla on the River Niger (Joliba) and returned by the southern route in the years 1795, 1796 and 1797. These travels are recorded in his book together with a map showing the routes by James Rennell, who was Surveyor General to the East India Company. Rennell is also known for maps of Hindustan, Bengal and Delhi.

374

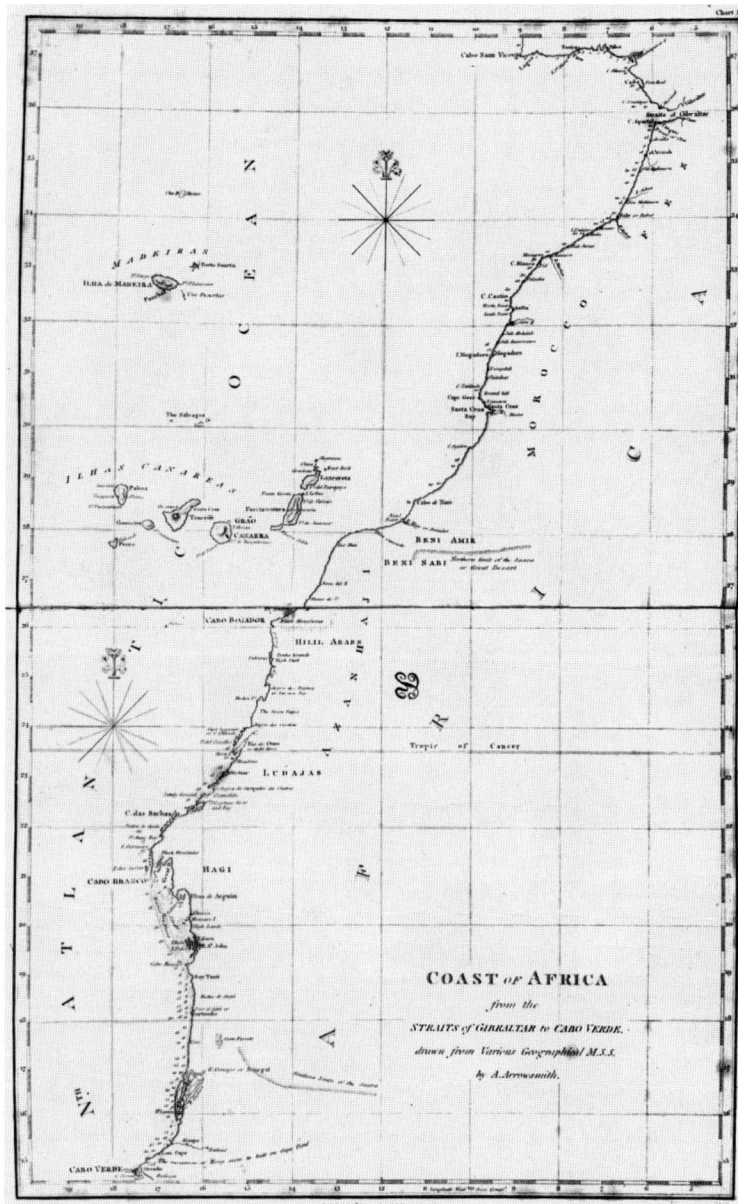

Map 328 ARROWSMITH, Aaron (1750-1833)
Coast of Africa from the Straits of Gibraltar to Cabo Verde... Engraved
by S.J. Neele. (London, Cadell and Davies, 1803.)
Map, 112 x 25,5 cm, uncoloured.
Prime meridian through Greenwich.
Top right: *Chart 1.*
In: Untitled atlas containing maps of Africa. London: Cadell and Davies, 1803

This chart extends from the Straits of Gibraltar to Cabo
Verde and the coastline is detailed with soundings. The
northern and southern limits of the Sahara desert are noted
and it is observed that 'the mountains of Kong seem to butt
on Cape Verd,' but otherwise, details are limited to the
coastline.

The map is from an 1803 atlas, put together to contain a
series of four double-page linen-backed coastal maps of the
western side of Africa, with a single map depicting the

Calabar and Bonny rivers of Nigeria (see Maps 329-331 and
Map 227).

The maps were drawn from a variety of geographical
information by Aaron Arrowsmith, the founder of the well-
known Arrowsmith family, who were British publishers of
approximately 750 maps between them. Aaron held the
position of Hydrographer to the Prince of Wales and, in 1820,
to His Majesty. He was assisted by his sons Aaron II and
Samuel.

375

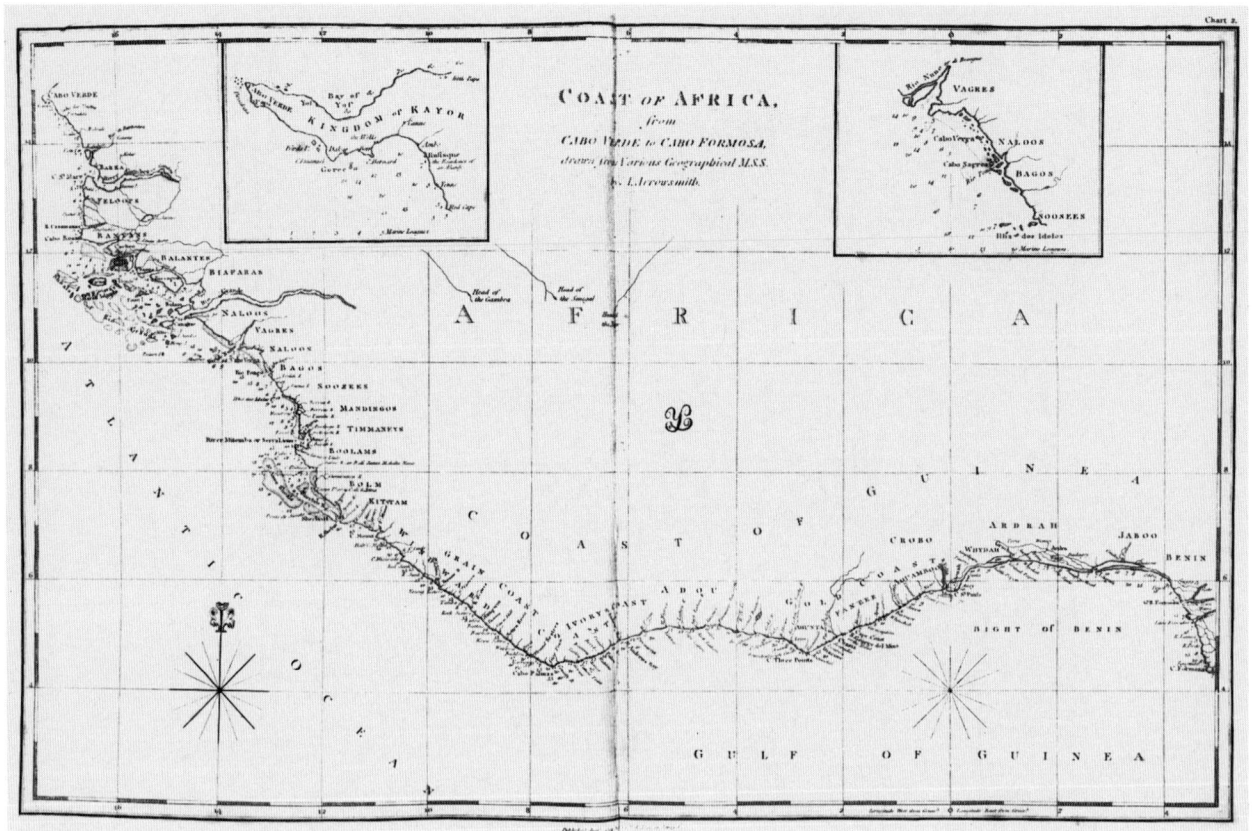

Map 329

ARROWSMITH, Aaron (1750-1833)
Coast of Africa from Cabo Verde to Cabo Formosa...
(London, Cadell and Davies, 1803.)
Map, 25,5 x 40,5 cm, uncoloured.
Prime meridian through Greenwich.
Inset: Kingdom of Kayor (with a scale of marine leagues).
Coastline from Rio Nuno to Illia dos Idolos (with a scale of marine leagues).
Top right: *Chart 2.*
In: Untitled atlas containing maps of Africa. London: Cadell and Davies, 1803.

This chart covers the coast of Africa from Cape Verde to Cape Formosa and shows the coast of Guinea, with the Grain Coast, the Gold Coast and the Ivory Coast. The two insets give enlarged details of Cape Verde itself and of an area slightly to the south whose natives are called Vagres, Naloos, Bagos and Sooses. For details on Arrowsmith and on the atlas from which this map was taken see Map 328.

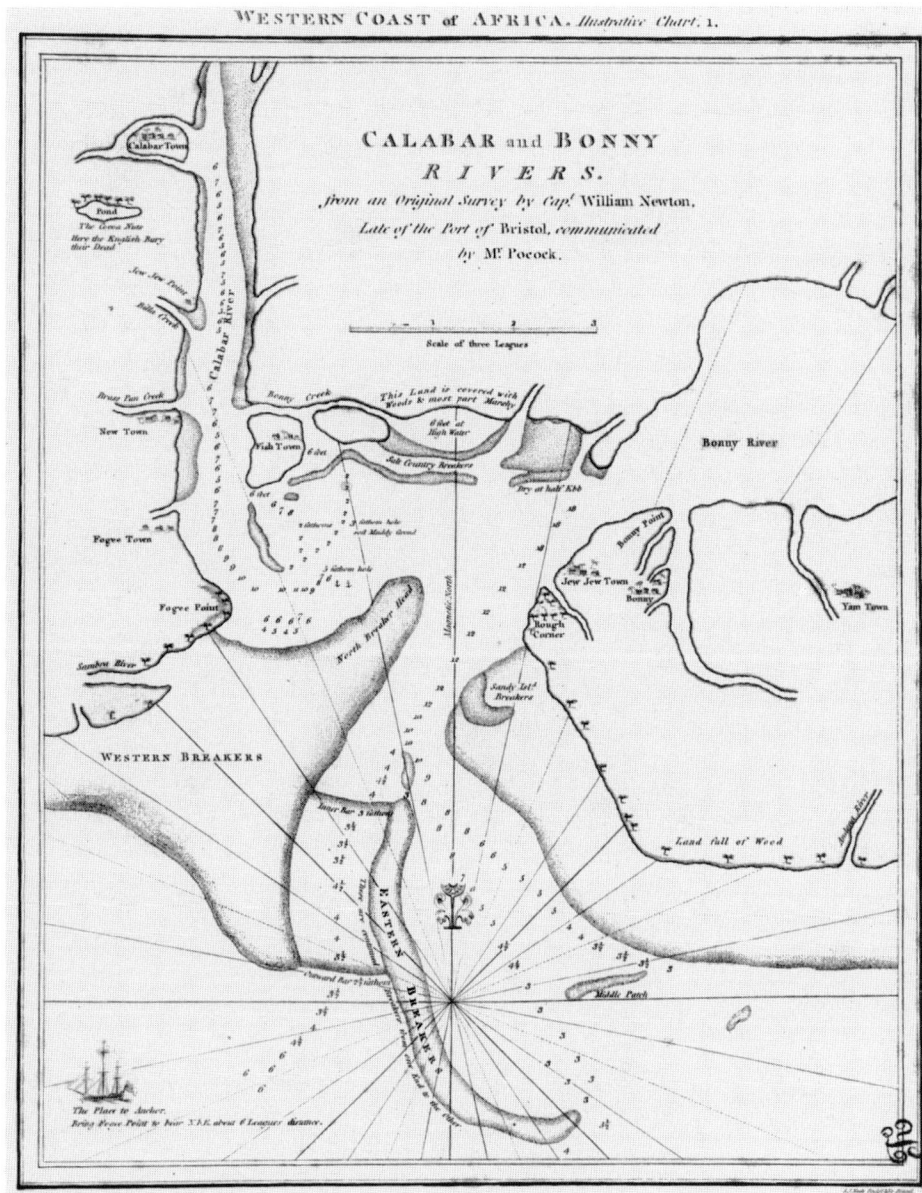

CALABAR and BONNY
RIVERS.
from an Original Survey by Capt. William Newton,
Late of the Port of Bristol, communicated
by Mr Pocock.

Map 330 NEWTON, William
**Calabar and Bonny Rivers. From an Original Survey by Capt. William New-
ton, Late of the Port of Bristol, communicated by Mr Pocock.** Engraved by S.J.
Neele. (London, Cadell and Davies, 1803.)
Map, 26 x 20,5 cm, uncoloured.
Scale in leagues.
Title across the top: *Western Coast of Africa. Illustrative Chart 1.*
In: Untitled atlas containing maps of Africa. London: Cadell and Davies, 1803.

This little chart acts as a guide to the waters of the Calabar
and Bonny estuaries. There is an abundance of soundings
and bars, and breakers and points that become dry on the
ebb tide are clearly marked. At the lower left is a small
engraving of a ship with the note 'The Place to Anchor. Bring
Fogee Point to bear N. bE. about 6 Leagues distance.'

Various points of interest are marked on the land, such as
Rough Corner, Jew Jew Town, Yam Town and the English
burial-place. For details on the atlas in which this map
appears see Map 328.

Map 331

ARROWSMITH, Aaron (1750-1833)
Coast of Africa from Cabo Formosa
to the Cabo de Boa Esperaca...
Engraved by S.J. Neele.
(London, Cadell and Davies, 1803.)
Map, 41 x 25,5 cm, uncoloured.
Prime meridian through Greenwich.
Inset: Cabo de Santa Catherina.
St. Elena Bay, with a scale in one league.
Top right: *Chart 3.*
In: Untitled atlas containing maps of Africa.
London: Cadell and Davies, 1803.

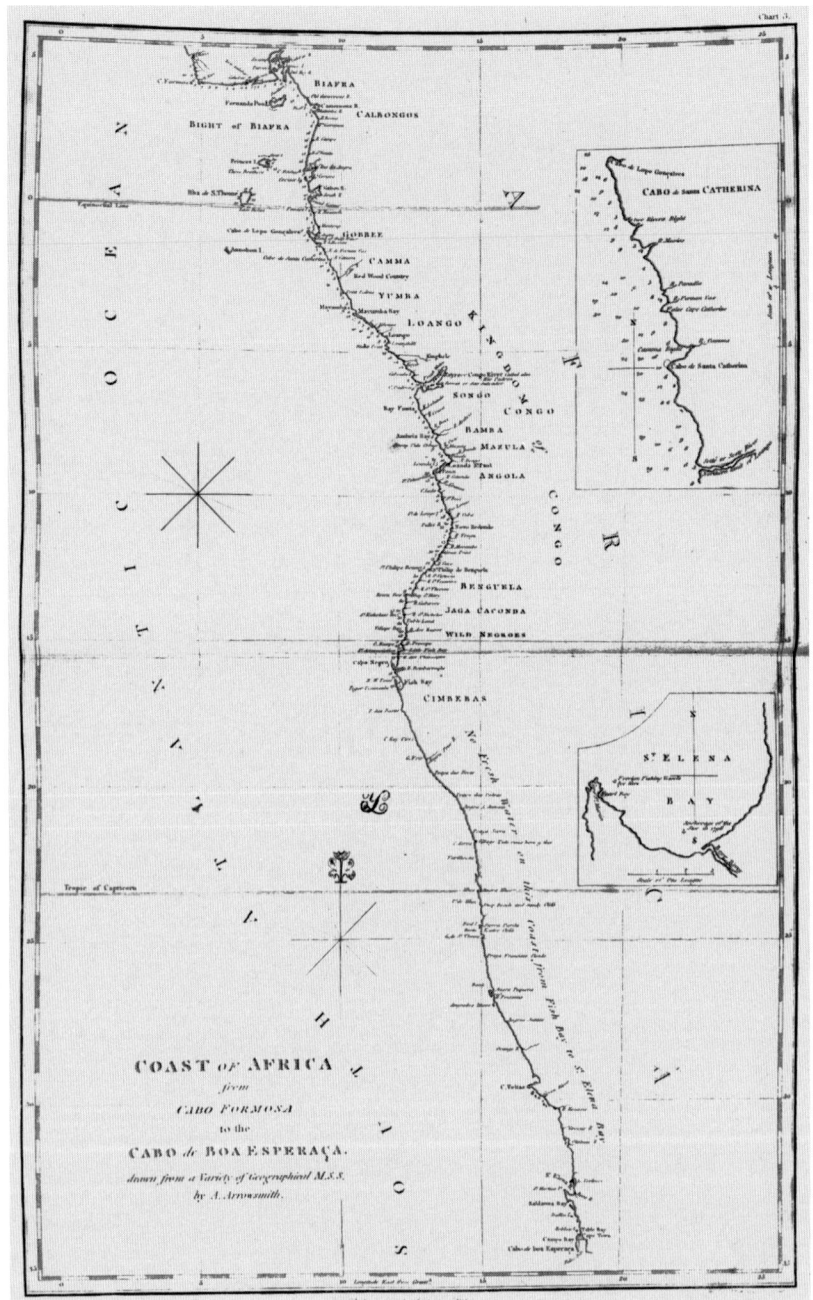

This map extends from about six degrees above the equator down to the Cape of Good Hope. Coastal names (mostly rivers) are quite accurately marked and along the coastline from Cimbebas to the Cape Province appears 'No Fresh Water on this coast from Fish Bay to St. Elena Bay.'

At the Cape Camps Bay, False Bay and Cape Town are named.

For details on Arrowsmith and the atlas in which this map appears see Map 328.

ISLANDS, TOWN PLANS AND PORTS

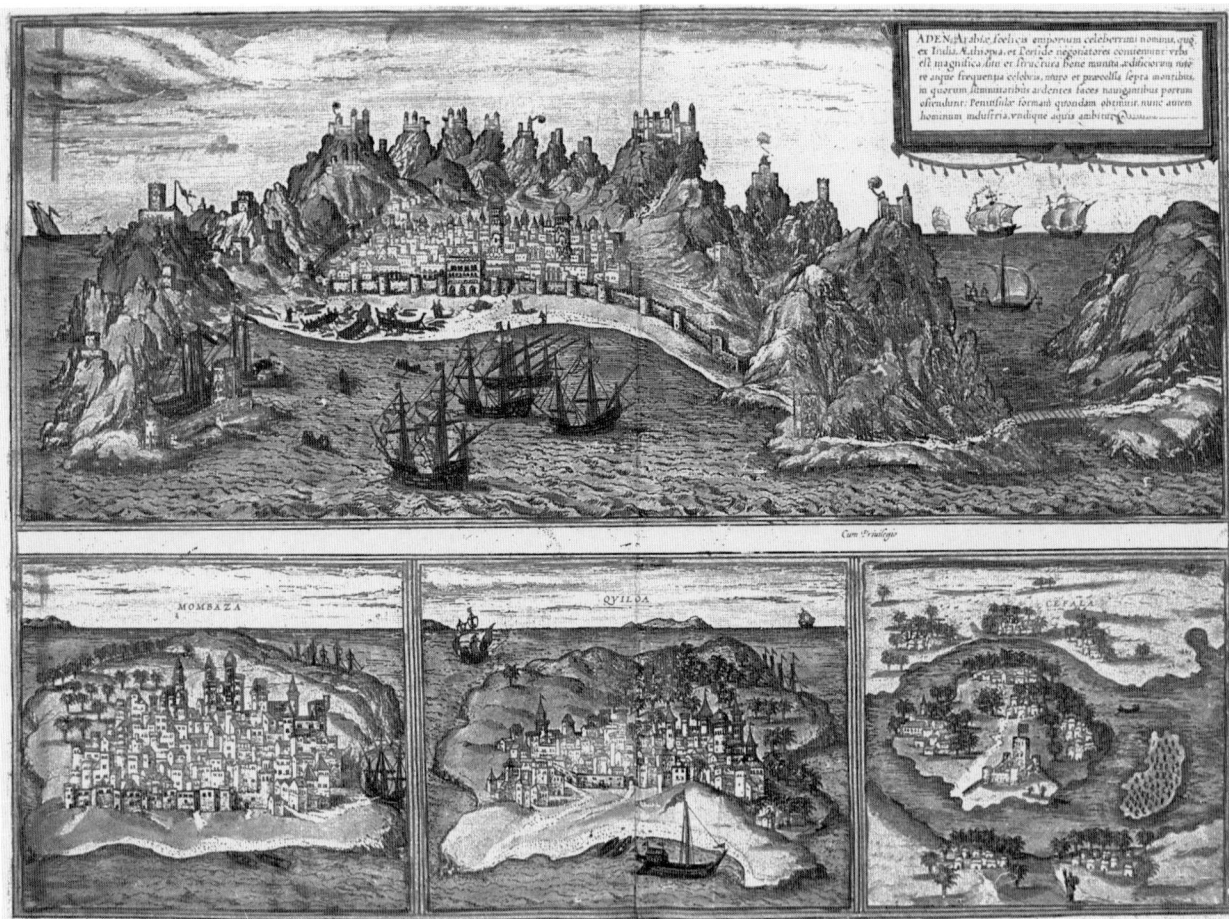

Map 332

BRAUN, George (Joris) and HOGENBERG, Frans (1535-1590)
(Aden, Mombaza, Quiloa,Cefala.) (Colgne, Braun and Hogenberg, 1572)
Four copperplate views, 47 x 32 cm, coloured.

This illustrated map shows the ports and harbours of the east coast of Africa which became well known and were occupied by the Portuguese travellers and invaders. They are Aden, Mombaza, Quiloa and Cefala, and give a vivid impression of their forts, castles, boats and habitable quarters, probably in original hand colouring. The view of Aden has a Latin description in the top right corner.

George Braun and Frans Hogenberg compiled the famed *Civitates Orbis Terrarum* of six volumes in Cologne, 1572-1618. It was the first atlas of towns, places and views embracing the whole world. Much of the appeal of the views lies in their variety, many have contemporary hand colouring. Hogenberg's style of engraving ensured a degree of uniformity, while some of the views he executed came from artists commissioned especially, and others filtered into Cologne through agents and collectors. The series appeared in haphazard order – most views were published according to availability rather than region by region.

The attempts of Braun and Hogenberg to encompass the whole world followed after Volume II was produced in 1575. Little material was available other than the verbal descriptions of merchants and travellers, but in the Hanse merchant Constantin van Lyskirchen of Cologne the editors found a willing agent, who supplied views of the towns of India, Asia, Africa and Persia never portrayed before. Lyskirchen obtained these views from a manuscript produced by an unknown Portuguese illustrator. Apart from these Portuguese views some of the African illustrations were taken from military plans concerned with the expeditions of Emperor Charles V in 1535 and 1541 to Turin and Algeria.

DESCRITTIONE
DELL'ISOLA
DI S. LORENZO.

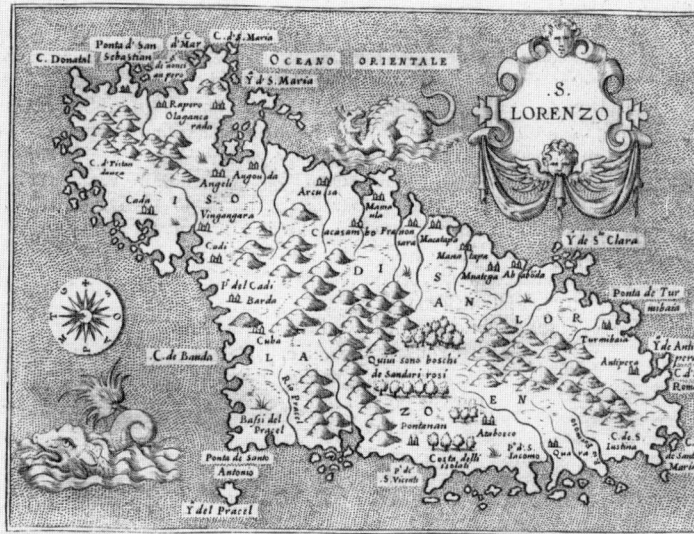

'ISOLA di S.Lorenzo è delle grandi, che nell'India fia-
no state discoperte: & tale che gli Spagnuoli hanno hauu-
to à dire, ch'ell'è maggior del Regno di Castiglia, & di
Portogallo. Come s'ha uoltato il Capo di Buona Speranza per Greco & Garbino; corre quest'Isola da XII. fino
à XXVI. gradi & mezo, uerso l'Antartico, lontana mille
miglia dall'Isola Soccotera, secondo alcuni, & secondo altri Scoira. Andrea Teuet Francese nel suo primo libro,
con molto notabil diuario da gli altri, dice, ch'ella ha settantadue di lunghez-
za, & undici gradi & trenta minuti di larghezza: ilche à settanta miglia nostre
per ciascun grado; risulterebbe gran numero di miglia. Con l'opinione di questo Scrittor Francese si confrontano alcuni Scritti in Spagnuolo, ch'io ho d'un
Piloto Portoghese, che fu in quelle parti: il quale del tutto si conforma nel trat-
tare di quest'Isola co'l detto Francese: & mi son uenute queste Scritture nelle
mani,

Map 333 PORCACCHI, Thomas (1530-1585)
S. Lorenzo. (Madagascar.) (Venice, T. Porcacchi, 1590.)
Map, 22 x 14 cm, uncoloured.
Title across the top: *Descrittione dell'Isola di S. Lorenzo.*
Top left: 178.
Italian text below map.

This miniature map of San Lorenzo, the early Italian name for Madagascar, was issued by Thomas Porcacchi of Castiglione Aretino in the second edition of his atlas, *L'Isole Piu Famose del Mondo* (Venice, 1590), dealing with famous islands of the world. It first appeared in 1572, and various editions appeared up to 1686. The maps were engraved by Girolamo Porro, a publisher and engraver of Venice.

There is a decorative cartouche at the upper right, surrounded by drapes and two faces at the top and bottom. There is a compass rose to the left in the sea, and below this a large unusual sea monster. Another sea monster appears at the top, to the left of the cartouche. The interior of the island is filled with places known at that period. The Italian text appears at the bottom of the map.

VIII.
INSVLA MADAGASCAR,
ALIAS S. LAVRENTII DICTA:
& huius conditio. 2.

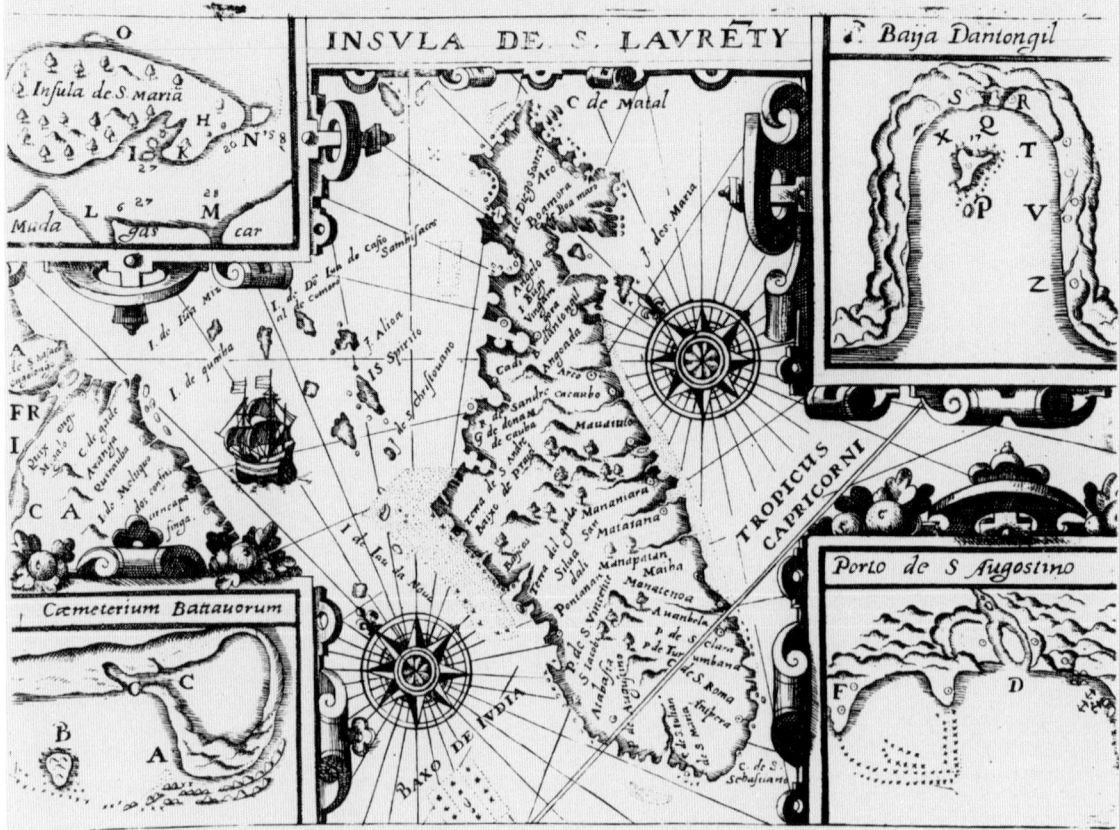

Adagascar Insulæ vnà cum suis breuiis, vadis, & scopulis, sicut & aliæ similes in-
sulæ, hoc se modo habent, quo hic depictæ sunt. A. Locus est, quo anchoras misi-
mus. B. Insula, seu Battauorum cæmiterium : ita dictum, quod è gente nostra ibi
quamplurimi sepulti sunt. C. Lacus aquam dulcem præbens. D. Riuus aquæ dul-
cis. E. Battauorum castrum. F. Angulus anterior de Porto S. Augustin. G. Par-
uæ insulæ, versus Meridiem sparsæ. H. Insula S. Maria. I. Sinus eiusdem insulæ.
K. Insulæ dictæ pagus primarius. L. M. Duo maris brachia. N. Cautes ad faciem
insulæ occidentalem P. Insula parua sinus maioris, vnde aqua dulcis anfertur.
Q. Riuus est. R. Vicus S. Angela. S. Vicus Spakenburg. T. Vicus ad Septentrionem. V. Vicus, ad quem
primo anchoras misimus, X. Pagi s, quo primo negotiati sumus. Y. Z. Pagi duo aly.

C

Map 334 BRY, Théodore de (1528-1598)
Insula Madagascar alias S. Laurentii dicta: & huius Conditio.
(Madagascar.) (London, T. de Bry, 1590-1634.)
Map, 14 x 20 cm, uncoloured.
Latin text below map.

This map of the island of Madagascar is by Theodore de Bry, an engraver, printer, publisher and bookseller who was born in Liege, and worked in England from 1586-1588. This, like many other maps by the de Bry family, is from a collection of early voyages, published between 1590 and 1634. There are two distinct series, the *Grands Voyages*, so called because the pages are taller and wider than the second, which is called the *Petits Voyages*. The first series consists of accounts of North and South America and the second series relates mainly to the East Indies and the Arctic regions.

The inspiration for these series came from Thomas Hariot's *A Briefe and true report of the newly found land in Virginia* (London, 1588), an account of the colony settled by Raleigh. The artist John White had accompanied Raleigh and made a large series of magnificent drawings and watercolours. Either Hariot contacted de Bry or vice versa, but the outcome was the first part of the *Grands Voyages*, consisting of a magnificent volume of text and engravings plus a map, a plate of Adam and Eve and a fine engraving of ancient Picts. All except the Adam plate were engraved after White's drawings. The work was issued in 1590 in English, French, German and Latin.

Théodore de Bry followed this work with Part II (issued in German and Latin only) of the *Grands Voyages*, a volume dealing with the French exploits in Florida. By the time of his death in 1598, de Bry had issued six parts. His widow and two sons, Johan Théodor and Johan Israel, supplied the seventh, eighth and ninth parts, the last in 1600. After an interruption the parts were recommenced in 1619 by another member of the family, Matthäus Merian, who completed the last portion of the fourteen German and thirteen Latin parts by 1630.

The *Petits Voyages* of thirteen parts in German and twelve in Latin were started in 1598 by Théodore and completed in 1628 by Merian. Almost every part was profusely illustrated with engravings on single sheets with text below and the verso blank. A complete set of de Bry consists of fifty-four parts: fourteen parts in German, thirteen in Latin, one in English and one in French in the *Grands Voyages*; thirteen parts in German and twelve in Latin for the *Petits Voyages*. Two other unillustrated parts were also issued as an appendix to each of the two volumes.

This collection is one of the most important of its kind and quite understandably at the same time a nightmare for bibliographers. Credit must be accorded to John G. Garratt who, in his detailed article in *The Map Collector* of December 1979 did more to clarify the situation than any previous bibliographer.

This finely engraved map of Madagascar and its surrounding islands also appears in Pieter Willemsz's and Verhoefs *Voyage to the Moluccas 1607-1609*, first published in Frankfurt-am-Main in 1612 entitled *Indiae Orientalis, part IX*. It has a description in Latin underneath with an alphabetical key to places of particular interest.

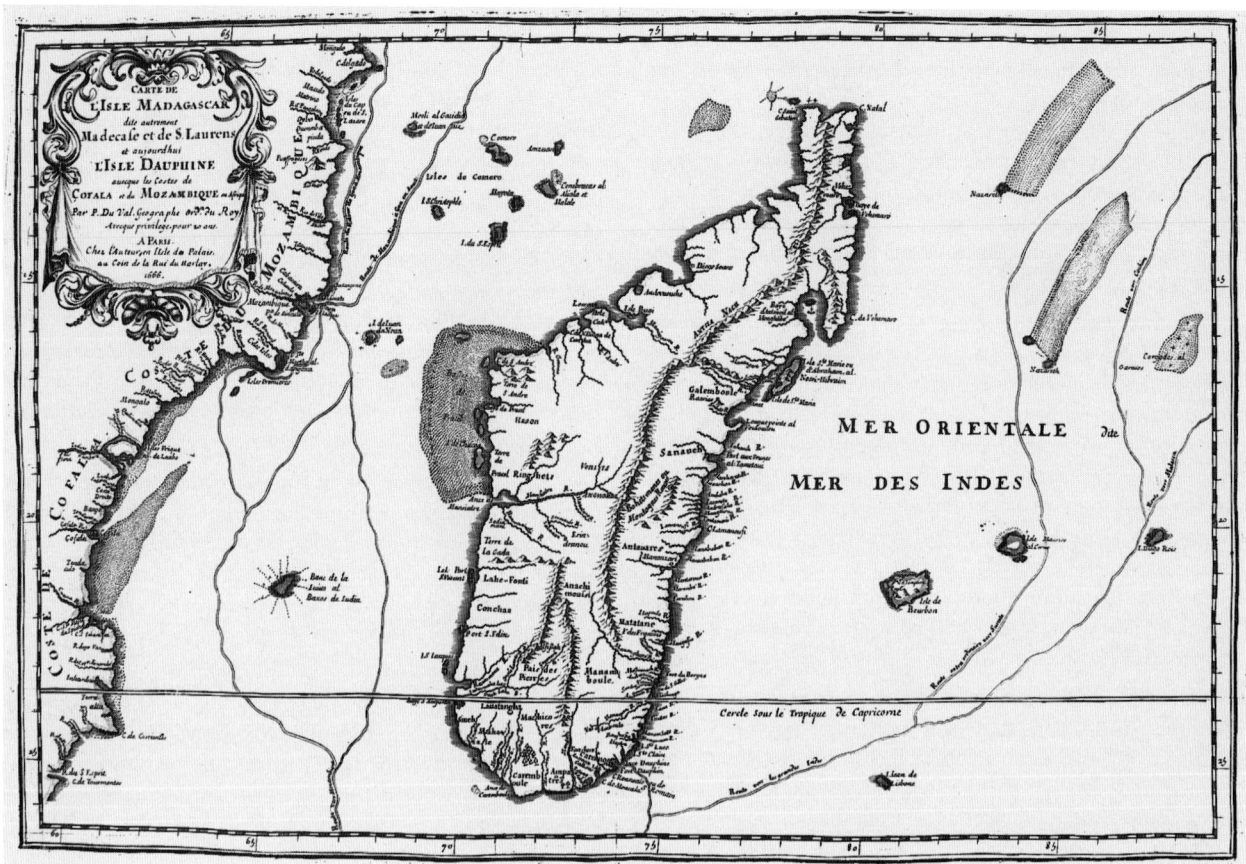

Map 335

DU VAL, Pierre (1618-1683)
Carte de l'Isle Madagscar dite autrement Madecase et de S. Laurens et aujourd'hui L'Isle Dauphine avecque les costes de Cofala et du Mozambique en Afrique. (Map of Madagascar... and the coast of Mozambique.)
(Paris, R du Val, 1666.)
Map, 37 x 56 cm, coloured outline.

A map of Madagascar with the Mozambique coast put out by du Val, the French geographer to the King, and nephew and pupil of Nicolas Sanson. For its period of 1666 it is an accurate map. This map is detailed, showing the surrounding islands, especially the Comoro Group to the northwest and the Isle of Bourbon (now known as Reunion). Although not a sea chart *per se*, it depicts many sea routes from the west coast of Africa to Mozambique and from Fort

Dauphin on the southeast coast of Madagascar to the north, south, and to the east, to India, Surate and Malacca.

Du Val issued many atlases in French and also executed maps for Jansson's *Ancient Atlas*. His wife continued his business after his death, when she reissued his four-sheet map of 'Amerique' in 1684. Their daughter eventually inherited the business.

384

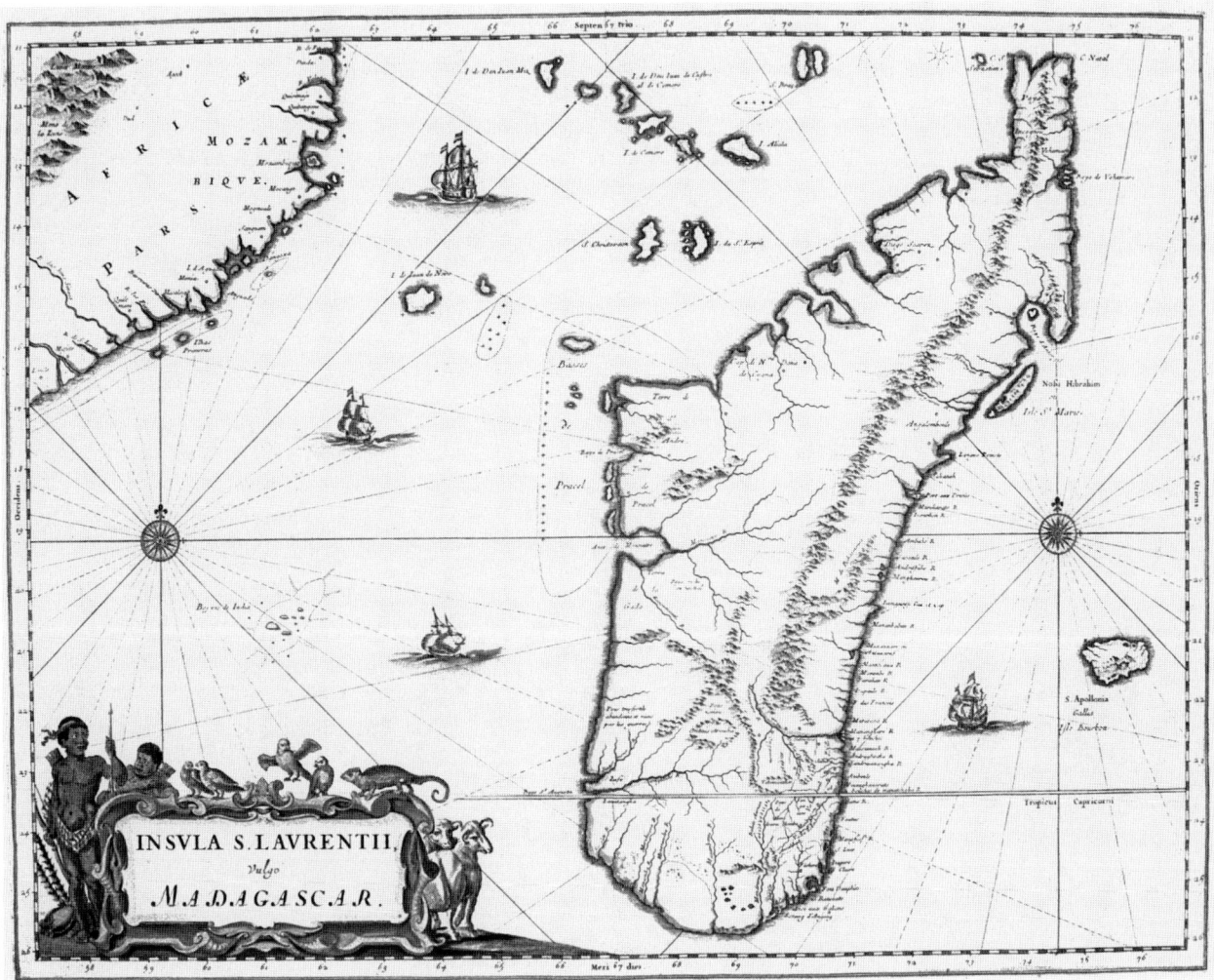

Map 336

BLAEU, John (1596-1673)
Insula S. Lavrentii, vulgo Madagascar. (The Island of Madagascar.) (Amsterdam, J. Blaeu, 1666.)
Map, 42 x 55 cm, slightly coloured.

This map from Blaeu's *Grand Atlas* shows the whole of Madagascar, together with the numerous surrounding islands and part of the Mozambique coastline. Present-day Réunion Island is featured under its original French name of Isle de Bourbon. The title is contained within a large decorative scrollwork cartouche adorned with birds and a chameleon and flanked by natives and sheep. The sea is embellished with two compass roses and four sailing ships.

Map 337   CHATELAIN, Henry Abraham (1684-1743)
Carte de l'Ile de Madagascar dite aujourd'hui l'Ile Daufine. (Map of the
island of Madagascar called Dauphine Island.) (Amsterdam, H.A. Chatelain, 1719.)
Map with eleven engravings and French text, 36 x 43,5 cm, coloured.
Scale in French and German leagues.
Title across the top: *Carte de l'Ile de Madagascar contenent sa description et diverses particularitez
curieuses de ses habitans tant blancs que negres.*
Top right: Tom. VI No. 34 Pag. 142.

This is a map of the Island of Madagascar with a
description in French below on its topography and
inhabitants. The margins are profusely illustrated in colour
with eleven descriptive vignettes of the dress, the events
and the life of the inhabitants. Flanking these vignettes are
eighteen smaller medallions, each one illustrating the facial
features of the various types. At the top of each side column
of illustrations is another scene: on the left, of two seated
Africans with drums and spears, and on the right, two seated
white persons with branches of foliage and fruit.

The map itself is situated at the top centre with a
rectangular title cartouche. Many of the surrounding islands
are wrongly placed. In fact the map of Madagascar itself is
quite inaccurate, revealing a greater artistic and pictorial
quality than its cartographical content.

A colourful map from the seven-volume *Atlas Historique*,
an ambitious work on all aspects of the peoples of the world,
with maps by Henri Abraham Chatelain, the geographer and
publisher of Amsterdam. The text was compiled by
Gueudeville, Garillon and H.P. de Limiers.

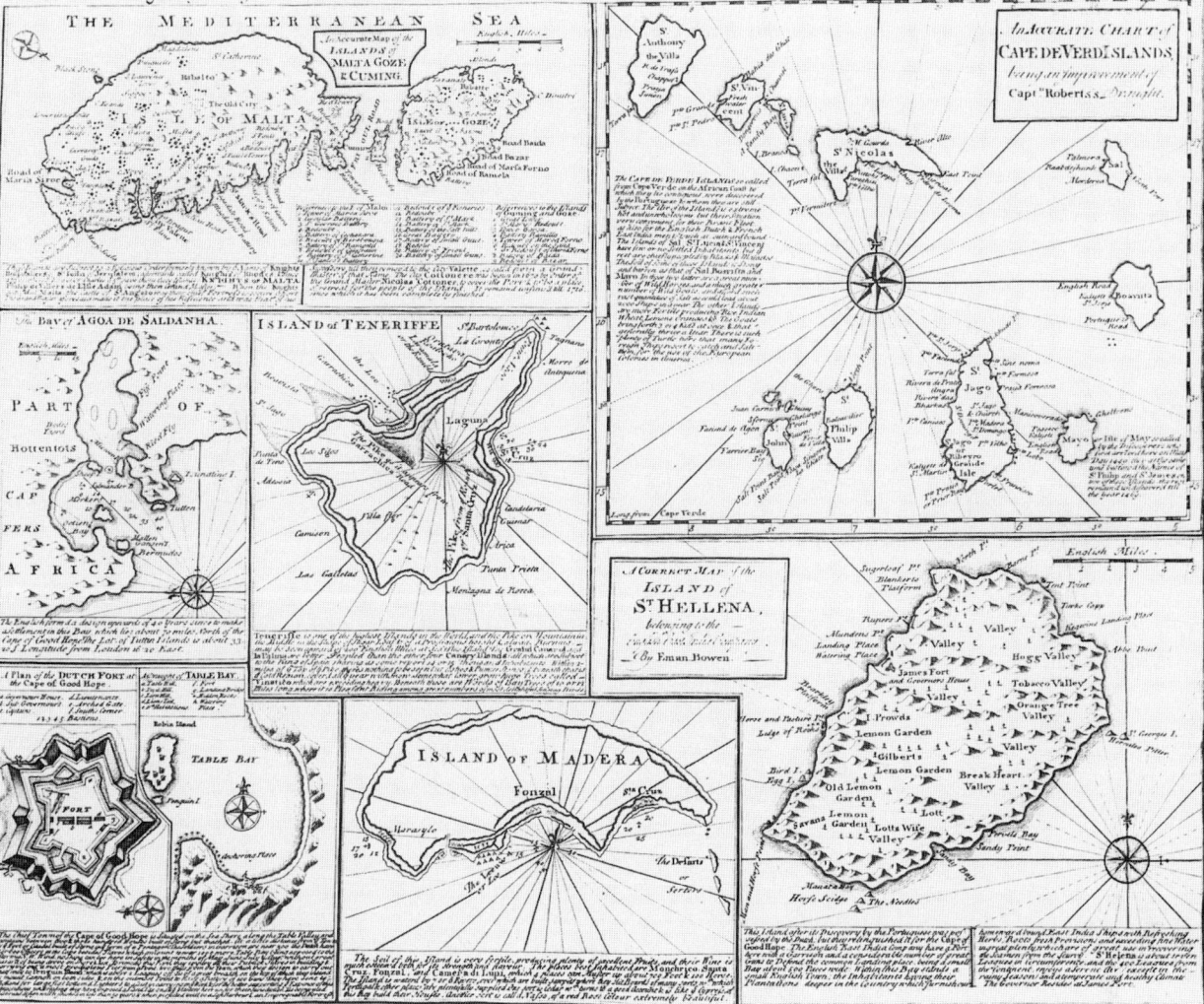

Map 338

BOWEN, Emanuel (c.1720-1767)
Particular Draughts of some of the chief African Islands in the Mediterranean, as also in the Atlantic and Ethiopic Oceans. (London, E. Bowen, 1752.)
Seven maps, 34,5 x 43 cm, uncoloured. Scale in English miles.
Inset: A Plan of the Dutch Fort at the Cape of Good Hope.

Emanuel Bowen was an engraver, printseller and publisher in London. Assisted by his son, Thomas, he produced a wide variety of maps of Africa, especially of the southern subcontinent and the Cape. This map appears in his *Complete System of Geography* and in his *Complete Atlas*, issued in 1752. It is made up of eight separate sections or plans: 'An accurate map of the islands of Malta, Goza and Cuming'; 'An accurate chart of Cape Verde Islands'; 'The Bay of Agoa de Saldanha'; 'Island of Teneriffe'; 'Plan of the Dutch Fort at the Cape of Good Hope'; 'A draught of Table Bay'; 'Island of Madera'; 'A correct map of the Island of St. Hellena, belonging to the English East India Company.'

Each inset has either a fleur-de-lys or compass rose and separate scales appear for the insets of Malta, Bay of Agoa de Saldanha and St. Hellena, each with descriptive matter. Tables of identification accompany Malta and the Dutch Fort.

The plan of the Castle and Table Bay has a descriptive note: *The chief town of the Cape of Good Hope is situated on the sea shore along the Table Valley and contains between 200 and 300 houses built of stone but thatched. At a little distance from ye Town is ye Fort or Citadel built of stone in ye form of a pentagon.* This note also mentions the 300 soldiers and 500 to 600 officers at the Cape, and about 1000 slaves. Mention is also made of severe storms with great hazard on shore: *To remedy this Evil they employ all their Publick Slaves in building and carrying out a most prodigious pier from a point two miles from the Town which they design to carry out halfway to Penguin Island, which is about 2 leagues off. Tis of great breadth, on the top of which they intend to mount cannon on both sides and on the outmost point to have a battery of 100 guns.* Mention is then made of importing flat-bottomed lighters to assist in building, with the help of a Land Tax, *a safe Harbour and an Impregnable Fortress.*

387

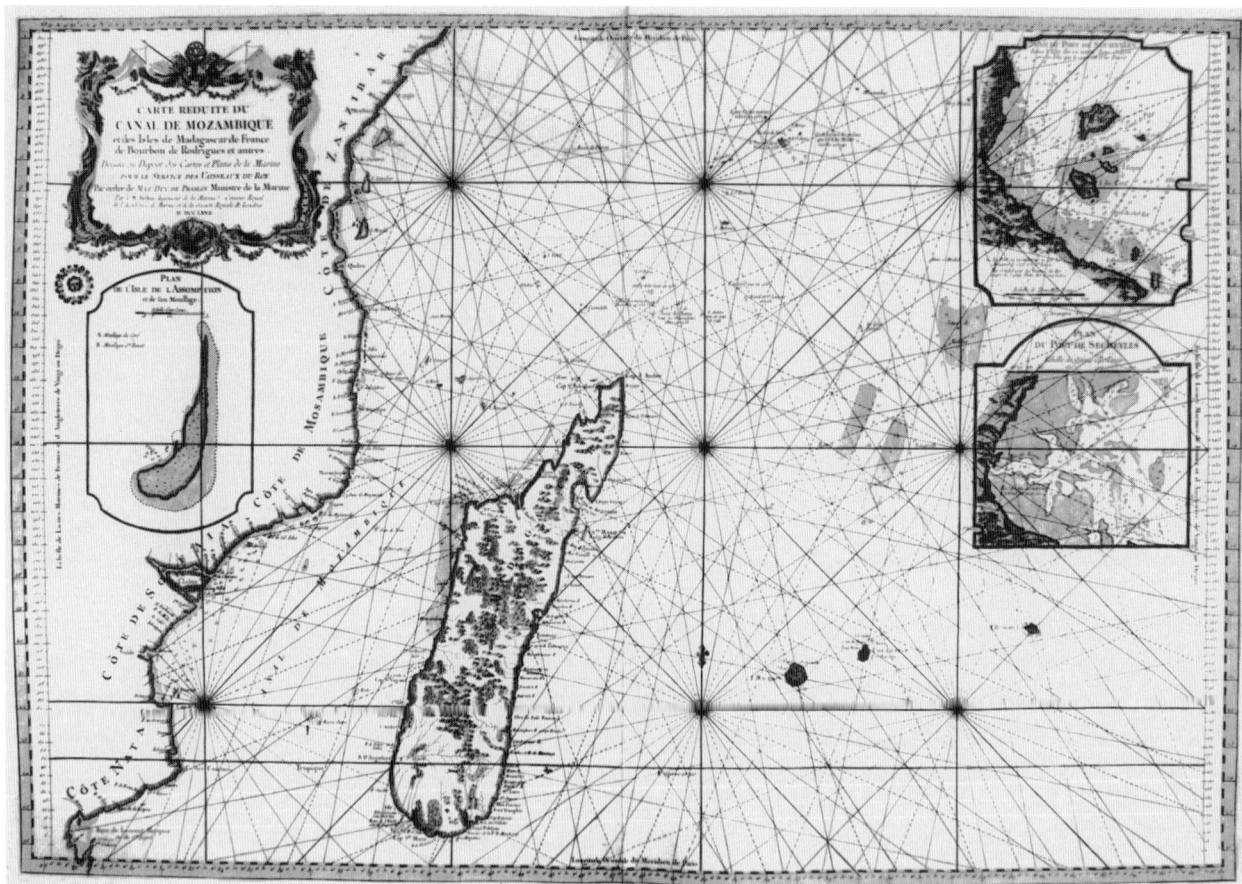

Map 339

BELLIN, Jacques Nicholas (1703-1772)
Carte Réduite du Canal de Mozambique et des Isles de Madagascar, de France, de Bourbon, de Rodrigues et autres. (Sea chart of the Mozambique channel and the Islands of Madagascar, France, Bourbon and Rodrigues.)
(Paris, J.N. Bellin, 1767.)
Sea chart, 59 x 87 cm, coloured.
Scale in French and English sea leagues.
Prime meridian through Paris.
Inset: Island of Assumption.
Two plans of the port of Secheyelles.

A highly decorative sea chart of the Mozambique Channel with the Islands of Madagascar, France, Bourbon (Reunion) and Rodriques, together with two inset plans of the port of Seychelles and another of the Island of Assumption, below the cartouche.

The title is situated within a rococo cartouche at the top left and the map depicts the east coast of Africa from the Baye de Laurent Marques to Zanzibar. The mainland includes the geographic details known to the many explorers who rounded the Cape to the East, with no real knowledge of the interior. Madagascar is profusely illustrated with coastal and interior information.

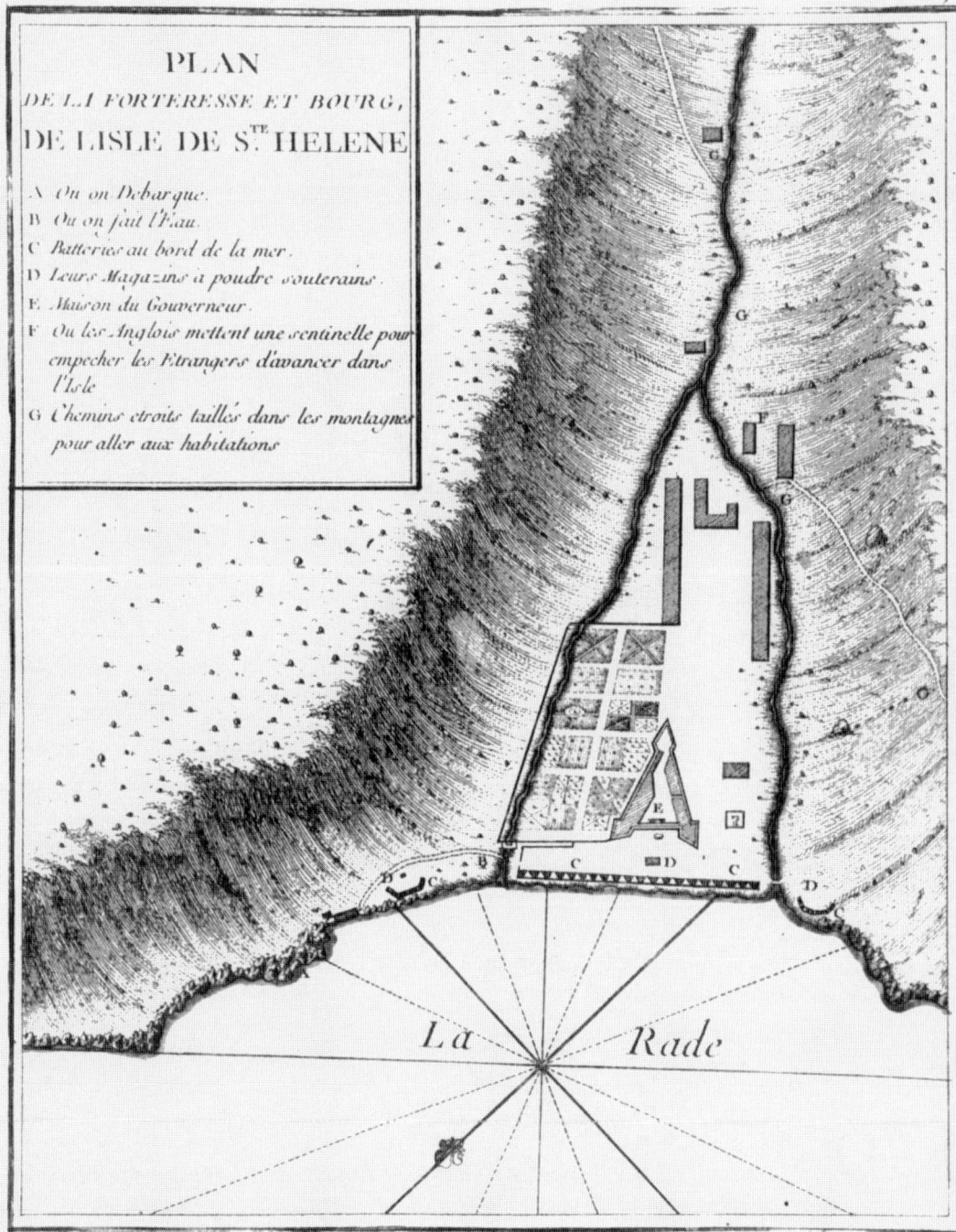

PLAN
DE LA FORTERESSE ET BOURG,
DE L'ISLE DE S.ᵀᴱ HELENE

A   *Ou on Débarque.*
B   *Ou on fait l'Eau.*
C   *Batteries au bord de la mer.*
D   *Leurs Magazins à poudre souterains.*
E   *Maison du Gouverneur.*
F   *Ou les Anglois mettent une sentinelle pour*
    empecher les Etrangers d'avancer dans
    l'Isle
G   *Chemins etroits taillés dans les montagnes*
    pour aller aux habitations

La   Rade

Map 340   ANON
Plan de la Forteresse et Bourg, de l'Isle de Ste. Helene. (Plan of the fort and town on St. Helena.)
Map, 22 x 17 cm, coloured outline.
Top right: Tome 3. No. 124.

This is a French plan of the Island of St. Helena in the Atlantic Ocean off the west coast of Africa showing the town and fort on the island. At the top left appears a legend marking the important sites on the island. The provenance of this French map has been unidentified, although the top right border refers to a volume and page number. The island is important because of its association with Napoleon. In some older maps two St. Helena islands appear and are referred to in cartographical circles as the old and the new.

389

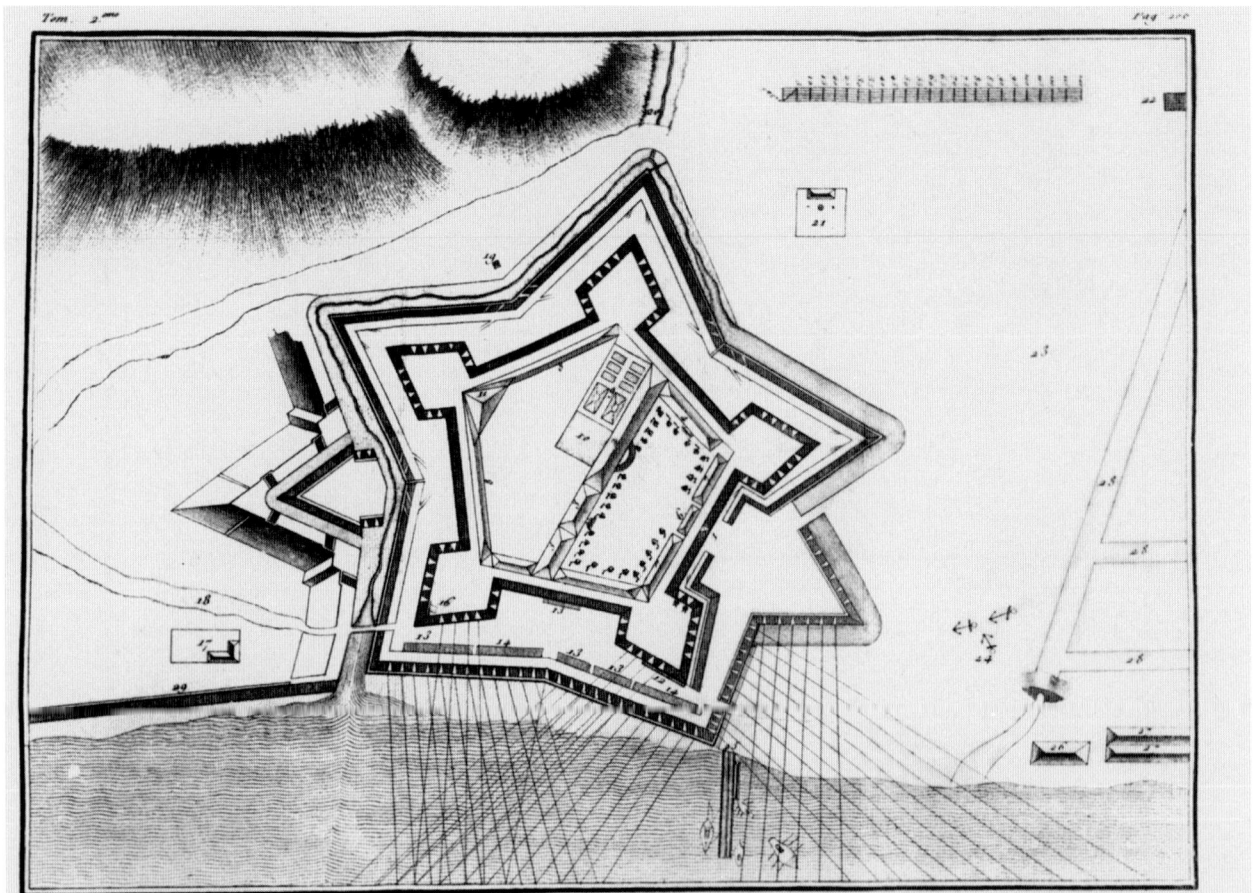

PLAN DE LA CITADELLE DU CAP DE BONNE ESPÉRANCE,
Levé sur les Lieux en 1793, par L. De Grand-Pré.

1. Porte du fort.
2. Maison du gouverneur.
3. Maison du second gouverneur.
4. Logement du commandant militaire.
5. Bureaux du gouvernement.
6. Corps de garde.
7. Logement des officiers militaires.
8. Casernes.
9. Jardin du gouverneur.
10. Reservoir à l'usage de la garnison.

11. Arsenal et poudrerie.
12. Corps de garde.
13. Préposés aux travaux de la campagne, forges &c.
14. Bagne des galériens en temps de guerre.
15. Atelier du charronage d'artillerie.
16. Pavillon de la place.
17. Tuilerie.
18. Grande route la seule qui aboutisse à la ville.
19. Reservoir pour laver.
20. Ravin où coule le ruisseau qui descend de...

— la table et fournit l'eau du fort.
21. Lieu des exécutions criminelles.
22. Commencement de la ville.
23. Esplanade entre le fort et la ville.
24. Lieu où l'on ancre pour le radoub des navires qui peuvent en avoir besoin.
25. Lieu d'où l'on aperçoit les bâtiments.
26. Rochers.
27. Magasins des vivres de la compagnie.
28. Comme qui environnent la place.
29. Commencement de la ligne.

Map 341

DEGRANDPRÉ, L.
Plan de la Citadelle du Cap de Bonne Espérance, Levé sur les Lieux en 1793, par L. De Grand-Pré. (Plan of the castle at the Cape of Good Hope.) (Paris, Dentu, 1801.)
Plan, 20,5 x 28,5 cm, uncoloured.
Top right: Pag. 200.
Top left: Tom 2eme.
In: Degrandpré, L. *Voyage à la Côte Occidentale de l'Afrique...* vol. II. Paris: Dentu, 1801, p. 206.

This detailed plan of the Castle at the Cape appears in Degrandpré's book on his voyage to the west coast of Africa. As an officer of the French navy, he concerned himself – among other matters of interest at the Cape – with a description of military matters. He embellished his description with a view map and this detailed plan of the five-cornered 'Citadelle,' drawn on the spot, with a numbered legend of twenty-nine points of interest in the castle and its environs.

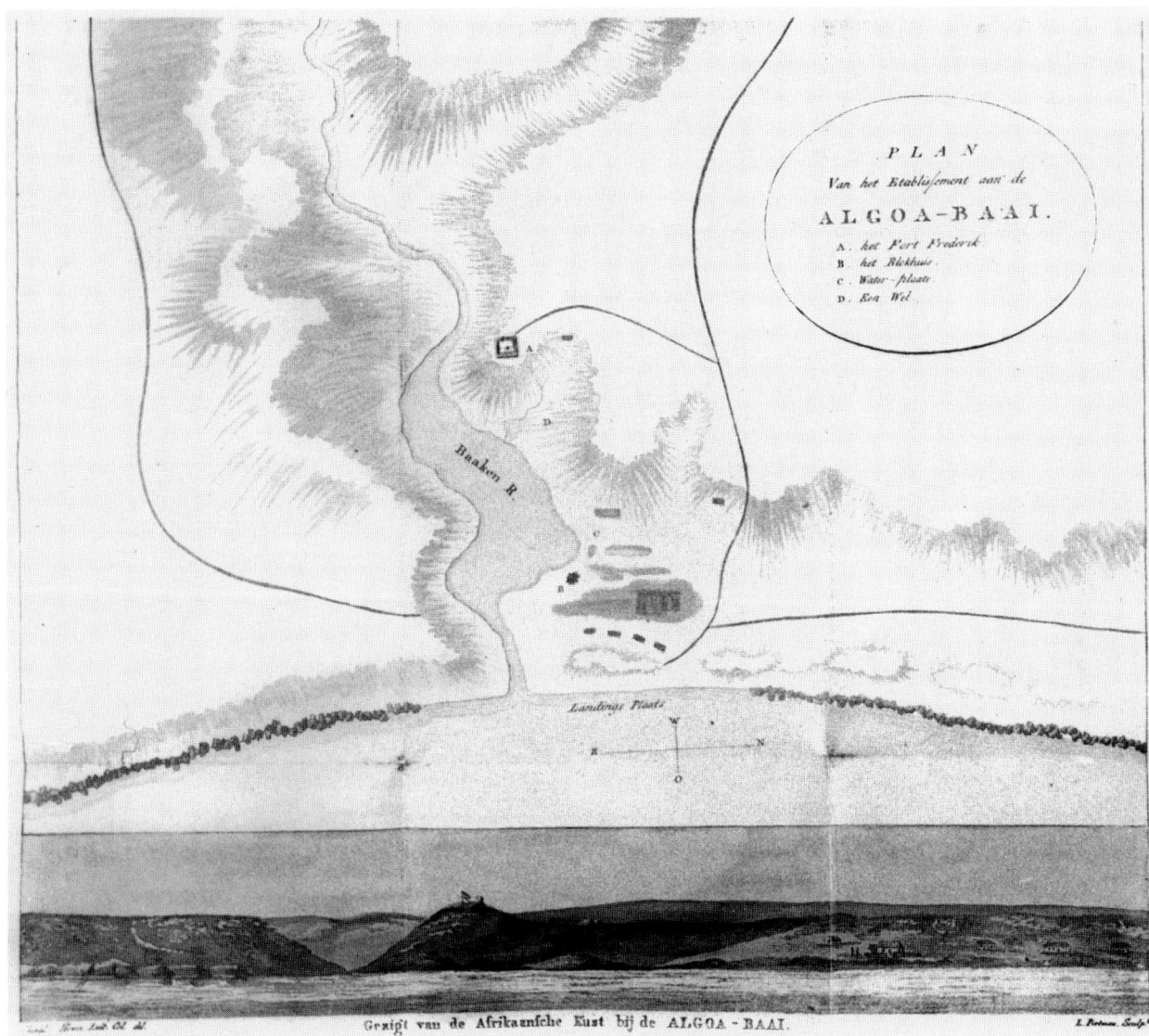

Gezigt van de Afrikaansche Kust bij de ALGOA-BAAI.

Map 342

ALBERTI, Lodewyk
Plan Van het Etablissement aan de Algoa-Baai. (Plan of the fortifications
at Algoa Bay.) Engraved by L. Portman. (Amsterdam, E. Maaskamp, 1810.)
Plan, 25 x 28,5 cm, coloured.
Inset: *Gezigt van de Afrikaansche Kust bij de Algoa Baai.*
In: Alberti, Lodewyk. *De Kaffers aan de Zuidkust van Afrika.* Amsterdam: E. Maaskamp, 1810, fold
in front.

This plan of Algoa Bay comes from the 1810 Dutch
edition of Lodewyk Alberti's *An Account of the Tribal Life and
Customs of the Xhosa* (first published in German in 1807).

The author of this book accompanied General J.W.
Janssen when he proceeded to the Cape as Governor of the
Colony under the Batavian Republic in 1802. He was sent
to Fort Frederick in Algoa Bay, where he acted as landdrost
and concerned himself with the affairs of the tribes in the
area. He gives a fine account of all this in his book, in which
he also describes the vagaries of Chief Gaika. His work was
accompanied by a separately printed album of four prints,

depicting historical events that took place while he was
landdrost – especially the meeting of Gaika and General
Janssen. These prints are commonly referred to as the Alberti
prints and constitute rare items to the Africana collector.

The top of this plan of Algoa Bay is west, with the Baaken
river running down the centre. On the northern bank of the
mouth are some fortifications, while Fort Frederick is up on
the heights above the town. Below this map is an engraving
of the entrance to the river and bay, illustrating also the
adjacent hilly coastline and the landing site.

391

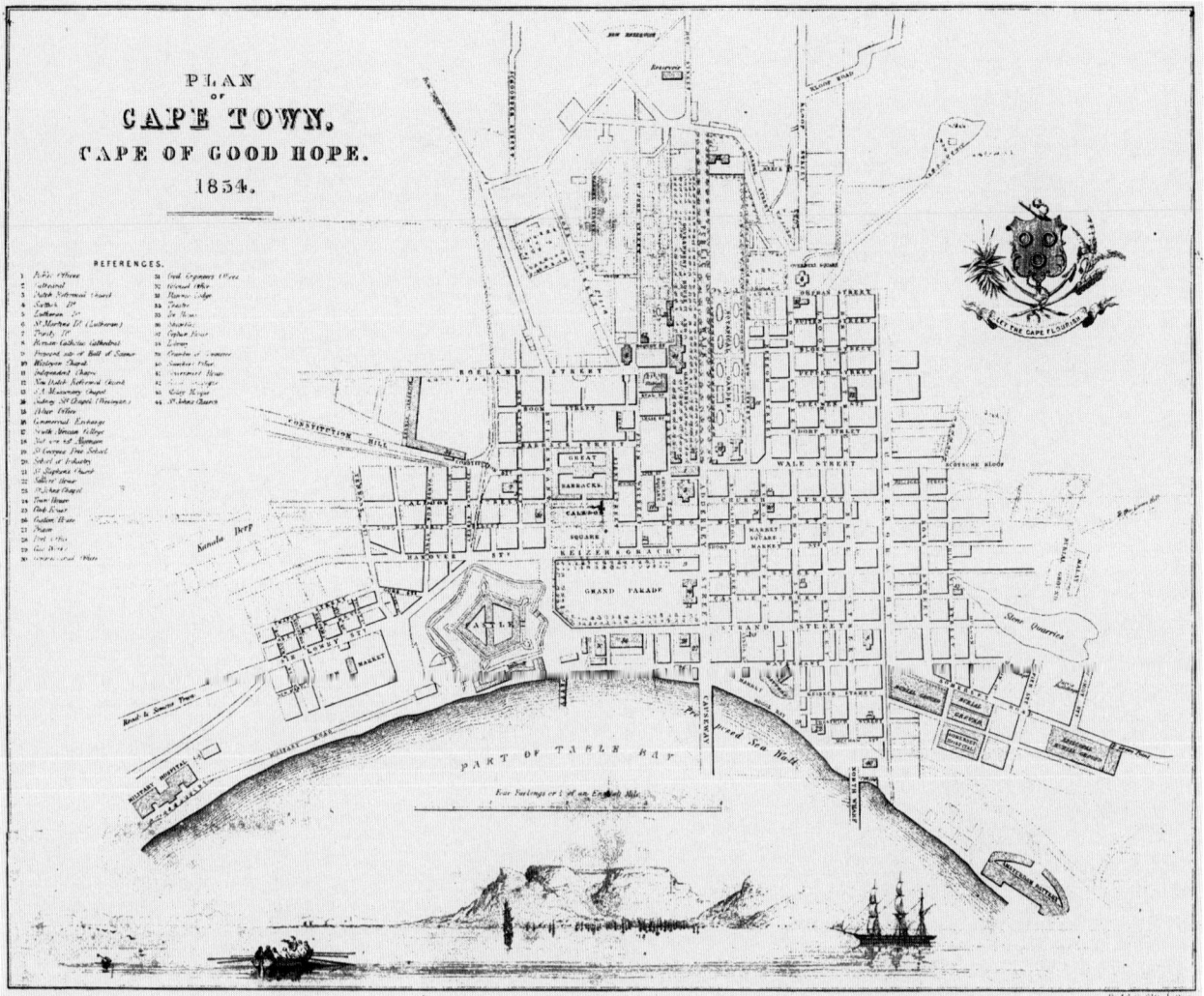

Map 343 DAY, John and Son
Plan of Cape Town, Cape of Good Hope.
(London, Day and Son, 1854.)
Map, 31,5 x 39,5 cm, uncoloured.
Scale in furlongs.

A detailed plan of Cape Town in 1854 with forty-four numbered references to places existing at that time in the town and its surroundings. The coat of arms of Cape Town appears at the top right. This is an extremely fine print, probably a lithograph, printed by Day and Son, lithographers to the Queen. There is a neatly executed drawing at the bottom showing Table Mountain and Table Bay with ships and a rowing boat in the left foreground. The origin or source of this print is unknown, in spite of extensive researches in South Africa and London, but it also appears in a book by A.H. Honikman, *Cape Town – City of Good Hope* (Howard Timmins, 1966).

392

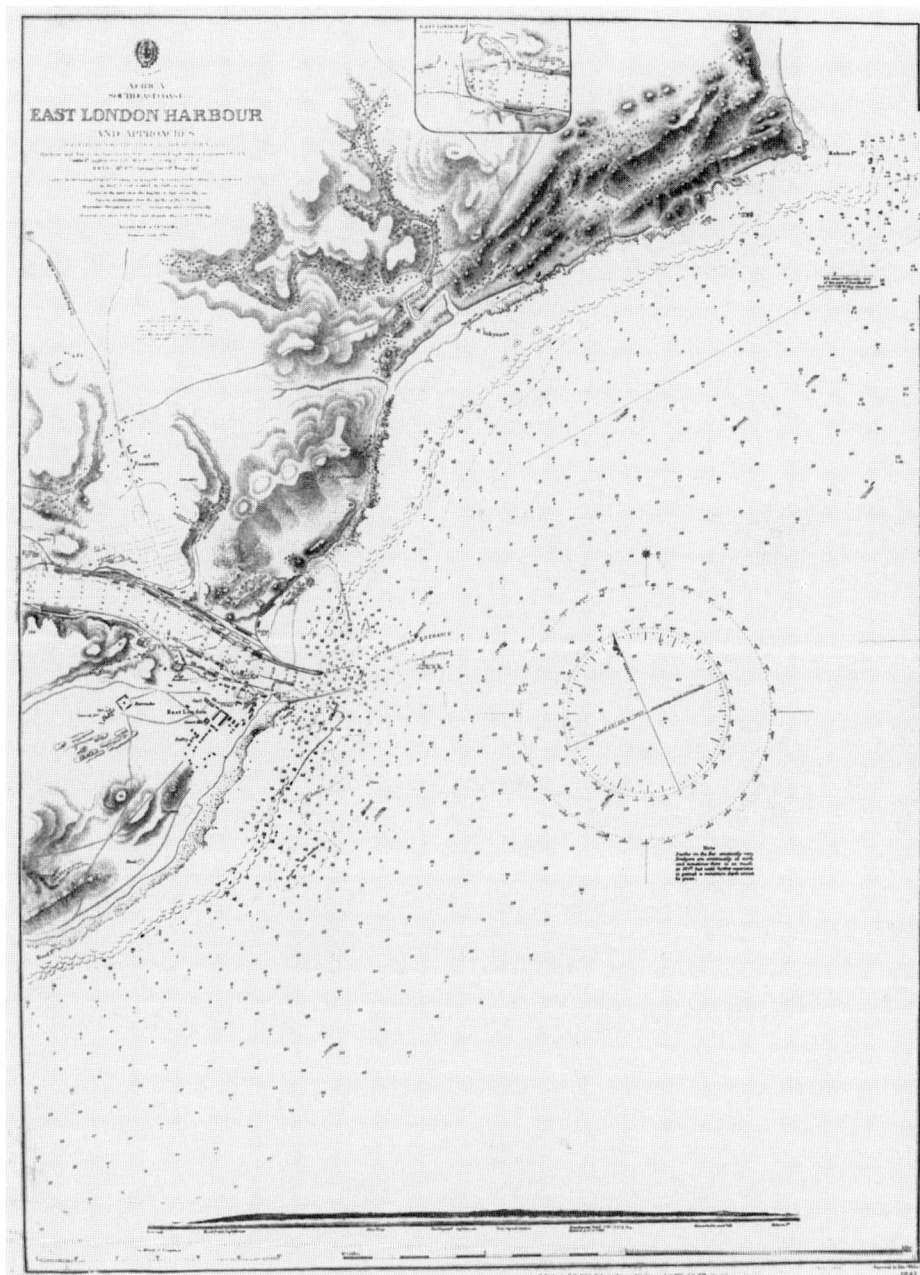

Map 344 ARCHDEACON, Lieut. W. E. and CRABTREE, W. R.
Africa. South-east coast. East London harbour and approaches. Surveyed by... Engraved by Edward Weller, 1843.
(London: Admiralty under the superintendence of Capt. G.H. Richards, 1869)
Map, 65,5 x 48,4 cm, uncoloured.
Natural scale of 1/12 200 soundings in fathoms and of minutes of longitude.
Insets: East London harbour. Elevation of coastline.

A detailed chart of the East London harbour showing a great deal on land and giving numerous soundings in the sea. A relief map at the bottom indicates landmarks around the harbour. Legends on the sea and on land refer to the varying depths of the bar and the use of a time signal with a time ball dropped by electricity from the Cape Town Observatory. This fine chart came from the Hydrographical Office of the British Admiralty, and was first published in August 1869 under the superintendence of Capt. G.H. Richards R.N., T. R. S (later Admiral), with subsequent new editions in June 1889, July 1898, January 1901 and May 1912. The engraver is Edward Weller, a publisher and cartographer of London.

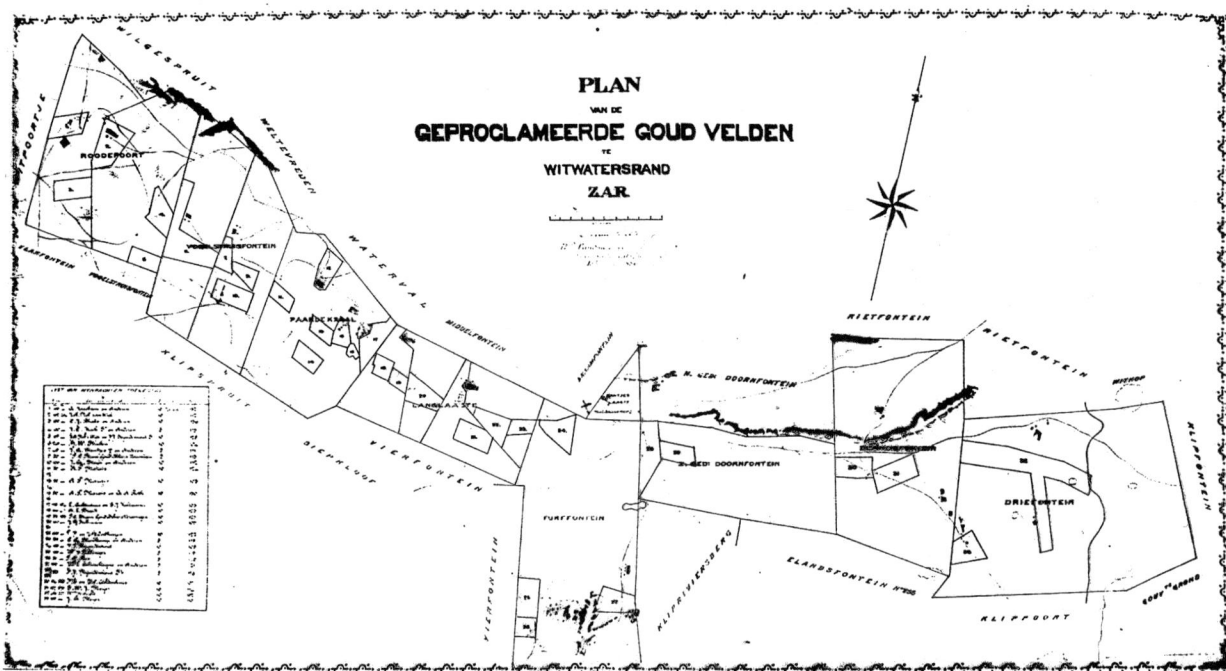

Map 345

RISSIK, Johann F.B. (1857-1925)
Plan van de Geproclameerde Goud Velden te Witwatersrand, Z.A.R. (Plan of the proclaimed gold-fields at the Witwatersrand.)
(Pretoria, Landmeter Generaal's Kantoor, 1886.)
Surveyor's plan, 12 x 23.5 cm, uncoloured.
Inset: *Lyst van Mynpachten Toegestaan.*

This map is from the collection of the Strange Africana Library, Johannesburg. The city of Johannesburg is the centre of the largest and most productive gold fields discovered to date. This discovery was not a sudden happening, but one that took place gradually. On 31 July 1886 the Government of the Zuid-Afrikaansche Republiek (ZAR), the Government of the Transvaal before the Anglo-Boer War of 1899-1902, was petitioned by 13 farm owners to proclaim a gold field. The Surveyor-General Gideon Retief von Weilligh (1859-1932) was unable to provide a compilation diagram of these farms.

On August 3, 1886 a Commission consisting of Christian Johannes Joubert (1834-1911), Head of the ZAR Mines Department, and Johann Friedrich Bernhard Rissik (1857-1925), First Clerk and Inspector of Diagrams in the Surveyor-general's Office was instructed by the Government to report on those farms which should be proclaimed as the Witwatersrand gold fields. At the same time they were asked to select a suitable site for laying out a village, preference to be given to Government ground.

Joubert and Rissik submitted a report on their visit to the Witwatersrand on August 12, 1886, and the Executive Council decided to proclaim nine farms as well as to allow the establishment of the village of Johannesburg on the Government ground known as Randjeslaagte. An official notice announcing the proposed proclamation of public diggings appeared in the *Staatscourant* (the Government gazette) No. 291 of August 18, 1886. The proclamation declaring the ground open to pegging was published in the *Staatscourant* No. 294 of September 8 1886 and was signed by President Stephanus Johannes Paulus Kruger (1825-1904) and State Secretary Willem Eduard Bok (1846-1904).

Even at this stage, no accurate compilation diagram of these nine farms existed, and as Rissik was only admitted to practice as a Land Surveyor on September 15, 1886, this diagram is of great historical importance, as it is believed to be the first compilation diagram emanating from the office of the ZAR Surveyor-General, the first diagram depicting the location of the village of Johannesburg on Government ground Randjeslaagte, the first diagram showing the location of Mynpachts – mining right leases – on the Witwatersrand gold fields, and the first official diagram signed by Rissik on December 15, 1886 in his capacity as Acting Surveyor-General.

To date (April 1996) the original of this diagram has never been located, but it includes a bar scale of 1000 Cape Roods

394

(1000 Cape Roods = 37,782,972 meters or 2,347,725 statute miles). The Cape Rood is a unit of length formerly used in land surveying in South Africa, but with the advent of metrication on January 1, 1970, it was no longer used. It is believed that the original compilation was at 1:29750.

It is interesting to note that the name of Johannesburg, after many years of speculation and research, was finally decided to have been based on the name of Johannes, arising as a component of the forenames of a number of prominent figures in the history of the city of Johannesburg at the time of its founding. These are: Stephanus Johannes Paulus Kruger – President of the ZAR; Christian Johannes Joubert – Head of the ZAR Mines Department; Johannes Petrus Meyer (1842-1919) – Field Cornet for the Ward Kliprivier-district Heidelberg. (Gerry Levin, Consulting Geologist.)

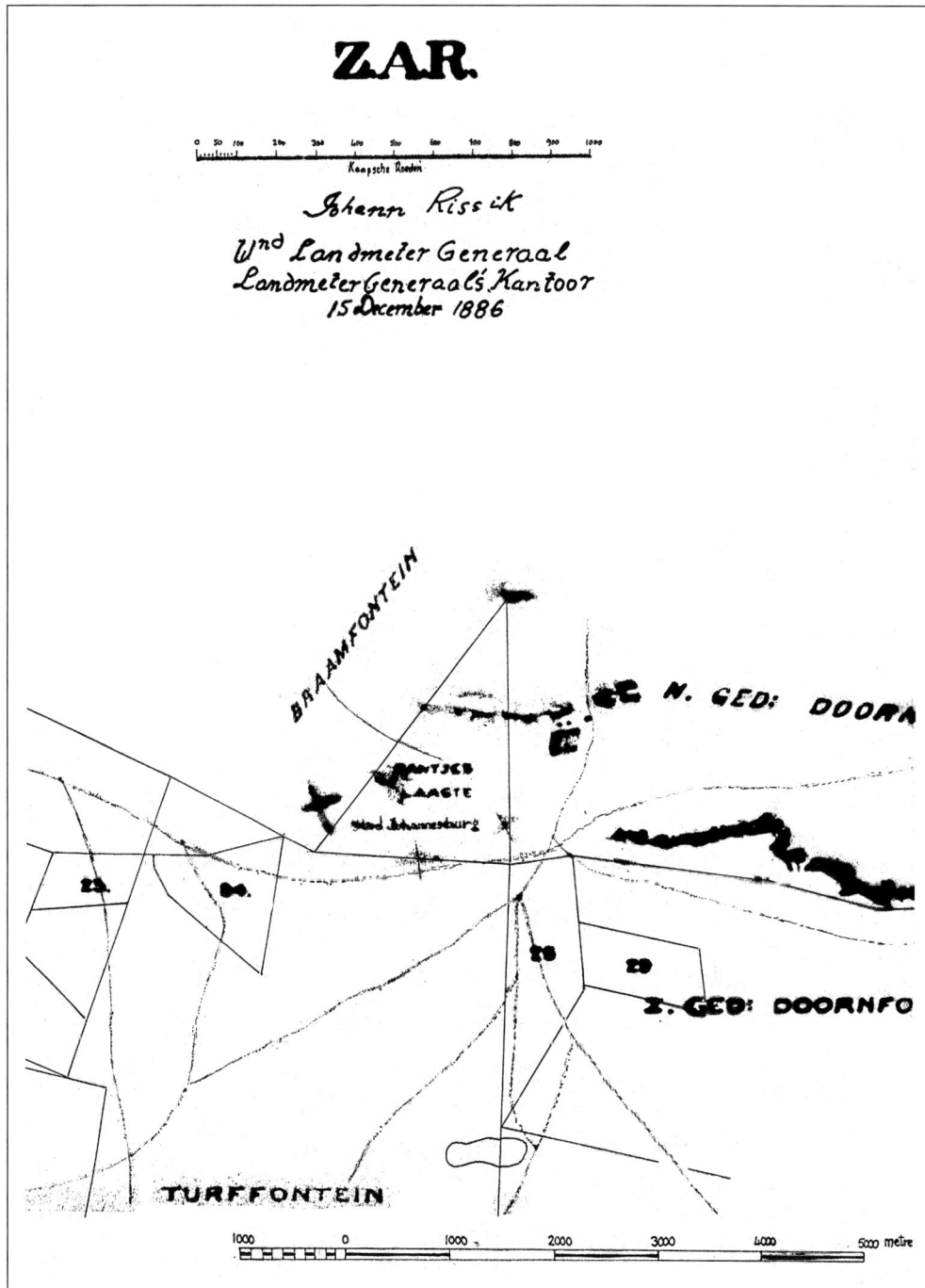

Index

Numbers refer to pages, map page numbers are in bold face. As much as possible, we have tried to include variant spellings.

399

Addendum: Folding maps of Africa

From the nineteenth century onwards, increasingly large scale maps of various parts of Africa were published as shingle sheets, rather than as parts of atlases. As single sheets of paper, they were obviously much more vulnerable to damage unless special provision was made for their storage as happened in the map libraries of national geographical societies or government offices. Their storage was more of a problem to the private individual, whilst their preservation in legible condition was a problem for the increasing number of travellers to Africa. Hence there was a potentially larger market for large scale maps if these problems could be resolved. The response of map publishers throughout the nineteenth century and in the early years of the twentieth century was to cut the map into squares or rectangles and to mount the pieces on a single sheet of linen, leaving narrow gutters between the pieces to facilitate folding. The folded map was then placed between two covers which further protected the map and also allowed it to be stored conveniently on a shelf alongside books. A printed label could be mounted on the top board. Some large format maps were also mounted on linen without dissection; many of these were placed in pockets at the end of travel books.

The large scale maps of parts of Africa from the nineteenth century have not yet found much favour with collectors. They are more difficult to display and are not always as visually attractive as their earlier and smaller counterparts. However, they are often much more important historically than the small scale atlas maps of the period, in that they were often the format in which new geographical information was first published. Indeed, the amount of detail that they contain could only be published at larger scales. Hence, the mounted and folded editions of large sheet maps (which are at least easy to store) are much to be commended to collectors looking for a more challenging field than maps of Africa from nineteenth and early twentieth century atlases. Editions were often small and an enhanced collector interest would help ensure their preservation.

This list does not purport to be complete, nor does it contain much descriptive matter. It has been compiled largely from dealer catalogues which is why, perhaps, it dominated by British maps. Nonetheless, it will give the reader insight into the *type* of folding map produced and is appended here in order to give the interested individual entree into this medium. Dimensions are approximate.

1. *To Captain Carmichael Smyth of the Corps of Royal Engineers — Who obligingly furnished many of the materials — This Chart of the Cape of Good Hope is Inscribed by his obedient and most humble Servant, A. Arrowsmith.* Linen-backed coloured map. 49.25 x 57.5 inches. Scale: approximately 13 miles to the inch. A combination hydrographic chart and map of the Cape area. Covers an area bounded by 29-37 deg. south latitude and 17-28 deg. east longitude. Ocean currents, sea depths and type of ocean bottom are shown. Map gives a good indication of rivers, mountains, settlements and roads. Notations indicate what game is to be found in various areas. 1805, London, A. Arrowsmith.

2. *Africa.* "To the Members of the British Association for Discovering the Interior parts of Africa. November 1, 1802. Additions to 1811." Linen-backed, coloured map. Marbled self-covers. 50.25 x 58.25 inches. A statement on the map indicates "The general opinion in the interior of Africa is that the Niger and the Nile of Egypt are one and the same." Cartouche in lower left shows pyramids, natives and wild animals. 1811. London, A. Arrowsmith.

3. *Cape of Good Hope by J. Arrowsmith.* This map is copied from orig. ms. drawings in the Colonial Office. Linen-backed map coloured in outline. 20.75 x 33.25 inches. Scale: 38 miles to the inch. 15 February, 1834. London, J. Arrowsmith.

4. *North Western Africa by J. A. Arrowsmith.* Linen-backed map hand coloured in outline. Case and label. Matching self cover on map. 22 x 26.5 inches. Scale: 130 miles to the inch. The Mediterranean to the Bight of Biafra. 15 February 1834. London, J Arrowsmith.

5. *South Africa. Compiled from the M.S. Maps in the Colonial Office Capt. Owen's Survey. Drawn by J & C Walker.* Linen-backed coloured map. Marbled self-covers. 13.5 x 16.25 inches. Early S. Africa map. Most of the detail confined to Cape area. 4 insets. April 1, 1834. London, Baldwin & Cradock.

6. *Cape of Good Hope* by J. Arrowsmith. (as above) 15 February, 1840 London J. Arrowsmith.

7. *South Africa.* Linen-backed map, coloured in outline. Boards. No title on cover. Marbled eps. 22 x 33.75 inches. Scale: 40 miles to the inch. An early map of South Africa showing the evolution of the country. The Cape Colony, Orange Free State not yet identified. Durban not yet a recognized town. [nd] ca 1840. London, James Wyld.

8. *A New Map of Africa Exhibiting the Recent Discoveries. Constructed from the Most Recent Travels with Additions to 1852.* Linen-backed map coloured in outline. 24 x 29.75. Case with label. Shows the entire African continent. [nd] ca 1852. London, C F Crutchley.

9. *Map of the Eastern Frontier of the Cape Colony compiled by H. Hall (Draughtsman to the Royal Engineers) From Military and other surveys.* Cover title *Map of the Eastern Frontier of the Cape of Good Hope.* Linen-backed coloured map. Case and label. 38.25 x 39.75 inches. Scale: 8 miles to the inch. A striking map showing great detail. St. Francis Bay along the coast to the Umzimyoobo River and inland to the Mogokare River. 1856. London, Edward Stanford.

402

10. *South Africa. Compiled from all the available official authorities.* Cover title: *South Africa 16 deg S Latitude.* Linen-backed map, coloured in outline. Box with original label. 27.5 x 33 inches. Scale: 69.1 miles to a degree. Shows the division between British Colonies, Boer Republics and Native Tribes. The Cape Colony is further divided into major Divisions. An uncommon map. 1857. Cape Town, JA Crew.

11. *Maps and Plans To Accompany the Records of the Expedition to Abyssinia. Compiled by Major T J Holland & Capt. H M Hozier.* 5 large folding, linen-backed coloured maps in slip case. 1868. London, Topographical Dept War Office.

12. *Map of the Transvaal and Surrounding Territories.* Linen-backed map coloured in outline. Marbled self covers. 23.25 x 24.75 inches. Shows Transvaal, Zulu Land, most of Natal and the Orange Free State. 1878. Pretoria, F Jeppe.

13. *Stanford's Library Map of Africa. This edition of the map first published July 2,1866 with corrections to 1878. Four sheets.* Linen-backed coloured map. Each sheet has marbled self-covers. Each sheet 31.25 x 33.5 inches. Scale: 94.34 miles to the inch. Possessions of all major colonial powers coloured in outline. 1878. London, Edward Stanford.

14. *Africa. Published by the Society for Promoting Christian Knowledge.* Linen-backed, coloured map. Case with label. Marbled self-covers. 31.5 x 27 inches. Scale 230 miles to the inch. January 1, 1879. London, Edward Stanford.

15. *Original Map of Great Namaqualand and Damaraland. Compiled by Th. Hahn P.D. October 1879.* Linen-backed map. Cover. 62.5 x 42.5 inches. Scale: 60 miles to the inch. 13-20 deg. east longitude and 19-30 deg. south latitude. An outstanding early map of this sparsely settled area. Mountains shown very clearly in shaded relief. Elevations of major peaks provided. Shows all roads, settlements and regions occupied by various Hottentot Tribes. 1879. Cape Town, (lithographed at the Surveyor General's Office and printed by Saul Solomon Cape Town bound by R Scott 42 Long Street Cape Town).

16. *Wyld's Military Staff Map of the Theatre of War In Egypt (Large Scale). Wyld's Egyptian War Series No. 8.* Linen-backed coloured map. Board cover and label. 26.25 x 35.25 inches. Scale: 3 miles to the inch. Cairo to the Mediterranean, Port Said and Ismalia. Very interesting to see the sharp contrast between the cultivated areas and the sandy desert. [nd] ca 1880. London, James Wyld.

17. *Juta's Map of South Africa Containing Cape Colony, Natal, South African Republic, Orange Free State, Criqualand, Kaffraria, Basutoland, Zululand, Damaraland, Betshunaland and other Territories. Compiled from the best available Colonial and Imperial Information including Dr. T. Hahn's Damaraland and the Official Cape Colony Map. Cape Town. New and Revised edition. Orig. publ. Nov 17,1884, London, Sampson Low, Marston, Searle & Rivington.* Cover title: *Map of South Africa.* Linen-backed coloured map. Gilt embossed board covers. Marbled eps. 26.75 x 34.5 inches. Scale: 40 miles to the inch. From approx. 20 deg south latitude to tip of southern Africa. An early edition. 1885. Cape Town, JC Juta.

18. *Korti to Khartum, Berber and Suakin.* Cover title: *Large Scale Map of the Nile between Korti, Khartum and Suakin.* Linen-backed map. Board covers and label. 20.5 x 29.75 inches. Scale: 16 miles to the inch. Very detailed map of the area north and east of Khartum. Inset of Khartum. Feb 16, 1885. London, Edward Stanford.

19. *Juta's Map of South Africa From the Cape to the Zambesi. Compiled from the best available Colonial and Imperial information incl. the official Cape Colony Map, Dr. T. Hahn's Damaraland and F.C. Selous' journals & sketches. New and revised edition. Published by J.C. Juta & Co., Cape Town 1889.* Linen-backed coloured map. Case and label. 37 x 49 inches. Scale: 40 miles to the inch. From 16 deg. south latitude to the tip of southern Africa. Tremendous detail. 1889. London, Edward Stanford.

20. *Stanford's Map of the Transvaal Goldfields, British Zululand, The Delagoa Bay Railway and the Routes from Cape Colony & Natal 1889.* Linen-backed coloured map. Board covers and label. Marbled ep. 27 x 37 inches. Scale: 16 miles to the inch. Inset of Witwatersrand Gold Field. Good representation of towns and rivers. Feb. 21, 1889. London, Edward Stanford.

21. *Africa (South Sheet).* Cover title *South Africa.* Linen-backed map. Board cover and label. 22 x 26 inches. Scale: 105 miles to the inch. Africa south of 3 deg. louth latitude. [nd] ca 1890. Printed by George Philip & Son, London and sold by Edward Stanford.

22. *Bacon's Map of South Africa.* Cover title: *Bacon's New Large-Scale Map of South Africa. Showing all the Latest discoveries, new boundaries, railways, roads, & c. Together with Guide and complete index.* Folding paper map. Coloured. 23 x 30.5 inches. Scale: 85 miles to the inch. 3 insets: Port Elizabeth, Durban, Cape Town. Combination map/guide book. Has a 35pp. handbook bound to front cover which gives a complete list of all towns and villages as well as general description and overview of country. [nd] ca 1890. London, GW Bacon & Co.

23. *Bain's Railway Map of South Africa.* Linen-backed map. Printed cover. 14.75 x 20 inches. Scale 62 miles to the inch. 3 insets. Railways, mail routes and steamer routes from South Africa through Rhodesia. [nd] ca 1890. Cape Town, John G Bain

24. *Bartholomew's General Map of Africa.* Linen-backed coloured map. Printed self covers. 28.25 x 31.25 inches. Not dissected. Scale: 190 miles to the inch. 4 insets – Cape Town, Nile Delta, Cairo, The British Isle. Shows the entire continent as well as Madagascar and colonies and protectorates of major colonial powers. [nd] ca 1890. Edinburgh, John Bartholomew & Son..

25. *Map of Rhodesia under the Administration of the British South Africa Company.* Cover title: *Rhodesia.* Linen-backed coloured map. Case and label. 18.25 x 16 inches. Scale: 65 miles to the inch. VG. A small map without a lot of topographical detail. Railway and telegraph lines shown clearly. Districts colour coded to stand out clearly. One inset. [nd] ca 1890. London, Edward Stanford.

26. *Marocco, Algeria & Tunis with Parts of Senegal and the Military Territories of the Western Sudan.* Cover title *London Atlas Map of Marocco, Algeria & Tunis.* Linen-backed coloured map. Board covers and label. 21.25 x 27.75 inches. Scale: 94 1/3 miles to the inch. [nd] ca 1890. London, Edward Stanford.

27. *Northern Zambesia. Compiled for George Cawston.* Cover title *Map of Northern Zambesia.* Linen-backed coloured map. Board covers and label. 39 x 36.5 inches. Scale: 16 miles to the inch. Marbled ep. Lake Tanganyika on the north to Kafue River on the south. Incl. Lake Meru and Lake Bangweolo. Sept. 1, 1890. London, Edward Stanford.

28. *Stanley in Africa 1867 to 1889.* Cover title: *Stanley's Explorations in Africa.* 21.25 x 27.25 inches. Scale: 150 miles to

NORTH WESTERN AFRICA,

BY J. ARROWSMITH.

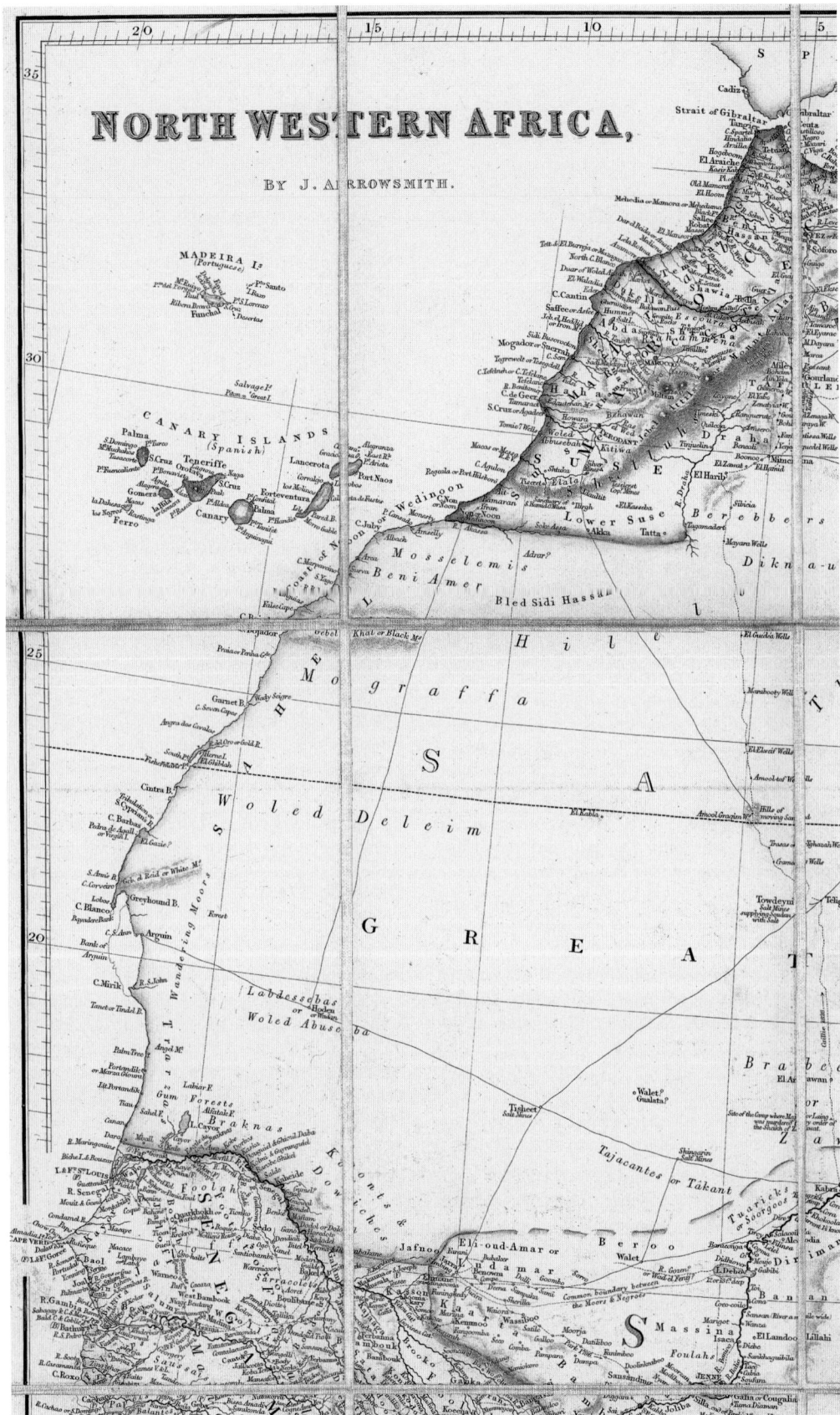

Detail from Nº4, *North Western Africa.*

the inch. Folding coloured paper map. The map shows 7 major African expeditions made by Stanley, each marked in colour and dated. On the verso are printed 5 pages of text giving a complete resume and discoveries of the great explorer. [nd] ca 1890. London, George Philip & Son.

29. *Troye's New Map of the Transvaal Colony. Compiled from all available information. Geological features revised and edited by Troye.* Folding, printed on linen, coloured maps. 9 sheets each 31.25 x 35.75 inches. Sheet 1 scale: 16 miles to the inch, all others scale: 4.69 miles to the inch. Each sheet folds into its own linen cover that has an identifying title. All 9 maps fit into a linen case with a pictorial key map printed on cover. The set gives physical, geographical and geological features of the Transvaal. Sheet 1 is a Railway & Postal Map. All others cover the colony and give Farm Names and their number and are coloured with 13 different colours to indicate different geological formations. A remarkably detailed work. [nd] ca 1890. Johannesburg, Gocott & Sherry.

30. *Carte De L'Algerie.* Linen-backed coloured map. Marbled self-covers. 2 sheets 37 x 57 inches. Scale: 22 km to the inch. Very good detail showing villages, hills, tracks and rivers. French text. [nd] ca 1892. London, Sold by Edward Stanford

31. *A Map of Mashonaland, Matabeliland, Khama's Country & c. The British South Africa Company's Territory. South of the Zambesi 1893.* Linen-backed coloured map. Board covers and label. 36.75 x 53.5 inches. Scale: 16 miles to the inch. Very good detail of river system and trails in a sparsely settled area. 1893. London, Edward Stanford.

32. *A Map of Mashonaland, Matabeliland, Khama's Country & c. The British South Africa Company's Territory. South of the Zambesi 1894.* Linen-backed coloured map. Board covers and label. 36.75 x 53.5 inches. Scale: 16 miles to the inch. Very good detail of river system and trails in a sparsely settled area. 1894. London, Edward Stanford.

33. *Map of Part of British and German East Africa Including the British Protectorate of Uganda.* Cover title: *Map of Southern Portion of East Africa.* Folding, paper map. 27 x 29.5 inches. Scale 25 miles to the inch. 1894. London, Edward Stanford.

34. *Juta's Map of South Africa From the Cape to the Zambesi. Compiled from the best available Colonial and Imperial information incl. the official Cape Colony Map by the Surveyor General-Cape Town, Dr. T. Hahn's Damaraland and F.C. Selous' journals & sketches. New and revised edition. Published by J.C. Juta & Co., Cape Town 1895.* Linen-backed coloured map. Case and label. 38 x 50 inches. Scale: 40 miles to the inch. From 16 deg. south latitude to the tip of southern Africa. 1895. London, Edward Stanford.

35. Cover title: *Map of Africa.* Linen-backed coloured map. Board covers with label. 26.5 x 22 inches. Scale: 260 miles to the inch. Added delineation of colonies and protectorates of major colonizing countries. [nd] ca 1895. London, Edward Stanford.

36. *Map of the Colony of the Cape of Good Hope and Neighbouring Territories.* Linen-backed map on 4 sheets each in separate case. Each sheet 27 x 40 inches. Scale 12.62 miles to the inch. Elevations shown by hachuring. Shows most of South Africa except for the Transvaal. 1895. London, Edward Stanford.

37. *Stanford's Map of British South Africa.* Linen-backed coloured map. Board covers and label. 22.25 x 29 inches. Scale: 94 1/3

miles to the inch. 8 deg. south latitude to tip of southern Africa incl. Madagascar. June 1,1895. London, Edward Stanford.

38. *Stanford's Map of the British Possessions in West Africa 1895.* Cover title: *British Possessions in West Africa.* Linen-backed coloured map. Board covers and label. 20.25 x 29.5 inches. Scale: 94 miles to the inch. Great detail on the "Bulge" of the African coastline. Interesting colour coding show the territory under control of the major colonial nations. 1895. London, Edward Stanford.

39. *Bartholmew's Central and South Africa.* Linen-backed coloured map. 25 x 34 inches. Self-cover. From 12 deg. N latitude to the Cape. 1896. Edinburgh, Bartholomew.

40. *Bartholomew's Special Large Scale Map Of The Sudan With General Map Of North East Africa and Enlarged Plan of Khartum.* Linen-backed coloured map. 29 x 39 inches. Self cover. Scale is 31.5 miles to the inch. 1896. Edinburgh, Bartholomew.

41. *Juta's Map of South Africa From the Cape to the Zambesi. Compiled from the best available Colonial and Imperial information incl. the official Cape Colony Map by the Surveyor General-Cape Town, Dr. T. Hahn's Damaraland and F.C. Selous' journals & sketches. New and revised edition. Published by J.C. Juta & Co., Cape Town 1896.* Linen-backed coloured map. Case and cover label. 37 x 50 inches. Scale: 40 miles to the inch. From 16 deg. south latitude to the tip of southern Africa. 1896. London, Edward Stanford.

42. *Mombasa, Victoria Lake Railway.* Linen-backed map. Marbled self-covers. 26.25 x 30.5 inches. Scale: 4.3 miles to the inch. This is sheet 1 of a 7-sheet series covering the entire route. Mombasa to Tsavo station shown this sheet. Photozincographed at the Ordnance Survey Office Southampton 1893. Reprinted in 1896. London, Edward Stanford.

43. *British South Africa.* Linen-backed coloured map. Board covers; label. 22 x 29 inches. Scale: 94 miles to the inch. From 8 degrees S latitude to the tip of southern Africa, including Madagascar. 1899. London, Edward Stanford.

44. *Crises in South Africa- Stanford's New Map of the Orange Free State, the southern part of the South African Republic, the northern frontiers of the Cape Colony, Natal, Basutoland, and Delagoa Bay in Portuguese East Africa 1899.* Cover title: *Stanford's New Large Scale Map to Illustrate the Crisis in South Africa Oct 1899.* Linen-backed map. Board covers with labels. Each sheet 27 x 39 inches. Scale: 16 miles to the inch. 1899. London, Edward Stanford

45. *Jeppe's Map of the Transvaal or S.A. Republic and Surrounding Territories. Compiled from surveys by Fred. Jeppe, of the Surveyor General's Department and C.F.W. Jeppe of the Mining Department. Pretoria 1899. Lithographed by Wurster, Randegger & Cie. (J. Schlumpf) Winterthur, Switzerland.* Folding, coloured, shaded relief map printed on thin linen. In folder with printed cover depicting map. 6 sheets each 38.75 x 25.5 inches. Scale: 7 miles to the inch. The detail of this map could hardly be exceeded. Gives all the appearance of a modern-day topographical map. In addition to all the territory there are very detailed layout of Johannesburg and Pretoria. [nd] ca 1899. London, Edward Stanford.

46. *Juta's Map of South Africa From the Cape to the Zambesi. Compiled from the best available Colonial and Imperial information incl. the offical Cape Colony Map by the Surveyor General-Cape Town, Dr. T. Hahn's Damaraland and F.C. Selous' journals & sketches. New and revised edition. Published by J.C. Juta & Co., Cape Town.*

Linen-backed coloured map. Case and label. 37 x 50 inches. Scale: 40 miles to the inch. From 16 deg. south latitude to the tip of southern Africa. 1899. London, Edward Stanford.

47. *Stanford's Map of British South Africa.* Linen-backed coloured map. Self-covers. 22 x 29 inches. 1899. London, Edward Stanford.

48. *Stanford's Map of the South African Republic (Transvaal).* Linen-backed coloured map. Board covers and label. 30.5 x 39 inches. Scale: 16 miles to the inch. Special emphasis on railway and telegraph lines in the Transvaal. July 1,1899. London, Edward Stanford.

49. *The Territories of the Nyassa Company. Sketch map compiled from all available sources.* Cover title: *Nyassa (German East Africa).* Linen-backed coloured map. Board covers and label. 29.5 x 42.25 inches. Scale: 12 miles to the inch. Marbled eps. Rovuma River on the North to the to the Lurio River on the south from the coast to Lake Nyassa. Shows proposed route of a railway from the ocean to Lake Nyassa. Aug. 1899. London, Edward Stanford.

50. *Africa.* Linen-backed coloured map. Self cover with index booklet. 24 x 30 inches. Northeast Africa and the Red Sea in great detail. Shows submarine cables. [nd] c. 1900. London, George Philip & Son.

51. *Bacon's Large-Print Map of South Africa.* Cover title: *Bacon's New Large-Print Map of South Africa.* Linen-backed coloured map. Gilt embossed board covers. 21.25 x 28.75 inches. Scale: 53 miles to the inch. 4 insets. [nd] ca 1900. London, GW Bacon & Co.

52. *Bacon's Large-Print Map of South Africa.* Cover title: *Bacon's Large Print Up-to-Date Map of Transvaal, Cape Colony & C.* Folding paper map. Paper covers. 22 x 30 inches. Scale: 53 miles to the inch. 4 insets. [nd] ca 1900. London, GW Bacon & Co.

53. *Bacon's Large-Print Map of South Africa. Cover title: Bacon's New Map of Transvaal.* Linen-backed coloured map. Gilt embossed board covers. 21 x 29 inches. Scale: 53 miles to the inch. Fine. 3 insets: Durban & Port Natal, Laing's NEK & Vicinity, Mafeking to Pretoria. Also incl. 15 pp. of text on "The War in South Africa." January 1900. London, G.W. Bacon & Co.

54. *A Map Of Rhodesia Divided Into Provinces And Districts Under The Administration Of The British South Africa Company 1900. Scale 1:1,000,000* Six sheets each measuring 68.5 x 96.5 cm, each sheet titled at top right: British south Africa, Sheet 1, 2, 3, 4, 5 or 6. Slip case Slip-in case has title plate: Rhodesia British South Africa Company's Territories. Six Sheets. (There is a key on sheet 6. There are at least six editions of this map, published from about 1895 to 1906.) August 1900. London, Published by Edward Stanford, 26 & 27 Cockspur Street, Charing Cross, S.W.

55. *Briton or Boer- Eastern Province Section.* Cover title: *Briton or Boer (South East) Special Map of the Eastern Province, Cape Colony and Native Territories.* Linen-backed coloured map. Paper cover. 19.5 x 24.25 inches. Scale: 20 miles to the inch. September 1900. Cape Town, Wood & Ortlepp.

56. *Miller's New Map of South Africa.* Cover title: *British South Africa.* Linen-backed coloured map. Embossed cloth covers. Marbled ep. 39.5 x 29.5 inches. Scale: 40 miles to the inch. A wealth of information on the country incl. 11 insets of cities and other areas. Sold by Philips with their catalogue on the inside. [nd] ca 1900. Cape Town, T Maskew Miller.

57. *Crises in South Africa- Stanford's New Map of the Orange Free State, the southern part of the South African Republic, the northern frontiers of the Cape Colony, Natal, Basutoland, and Delagoa Bay in Portuguese East Africa 1899.* Cover title: *Stanford's New Large Scale Map to Illustrate the Crisis in South Africa Oct 1899.* With: *Southern Extension of Stanford's New Map of The Orange Free State and the southern part of The South African Republic.* Cover title: *Stanford's New Large Scale Map of the Seat of War in South Africa- Sheet 2.* Folding, linen-backed maps. Board covers with labels. Each sheet 27 x 39 inches. Scale: 16 miles to the inch. Sheet two is interesting: in addition to showing the Eastern Cape Province it also has an inset showing all of British-controlled Southern Africa on a scale of 94 miles to the inch. Sheet one: October 10, 1899. London, Edward Stanford. Sheet two: January 9,1900. London, Edward Stanford.

58. *Crises in South Africa- Stanford's New Map of the Orange Free State, the southern part of the South African Republic, the northern frontiers of the Cape Colony, Natal, Basutoland, and Delagoa Bay in Portuguese East Africa 1900.* Cover title: *Stanford's New Large Scale Map of the Seat of War in South Africa- Sheet 1.* With: *Southern Extension of Stanford's New Map of The Orange Free State and the southern part of The South African Republic.* Cover title: *Stanford's New Large Scale Map of the Seat of War in South Africa- Sheet 2.* Folding, linen-backed maps. Board covers with labels. Each sheet 27 x 39 inches. Scale:16 miles to the inch. Sheet one: April 21,1900. London, Edward Stanford. Sheet two: March 1, 1900. London, Edward Stanford.

59. Cover title: *London Atlas Map of Africa.* Linen-backed coloured map. Board covers and cover label. 26.5 x 22 inches. Scale: 260 miles to the inch. Shows the entire continent as well as Madagascar. Added delineation of colonies and protectorates of major colonizing countries. [nd] ca 1900. London, Edward Stanford.

60. *Map of South Africa. Specially prepared for the Times History of the War in South Africa 1899-1900. Compiled, drawn on stone and printed by W. & A.K. Johnston, Edinburgh.* Cover title: *South Africa.* Linen-backed coloured map. Gilt embossed board covers. 28 x 34.25 inches. Scale: 40 miles to the inch. Good detail. Railroad system highly delineated and mountains shown in shaded relief. Two insets. [nd] ca 1900. London, Sampson Low, Marston.

61. *Map of the British Somali-Coast Protectorate.* Linen-backed map coloured in outline. Board covers and label. 27 x 37 inches. Scale: 16 miles to the inch. Detailed map of a very scarcely populated and hostile area. 1 inset. [nd] ca 1900. London, Edward Stanford.

62. *The Nile Valley including Egypt, Uganda, Abyssinia, British East Africa and Somaliland.* Cover title: *London Atlas Map of the Nile Valley.* Linen-backed coloured map. Board covers and label. 29.25 x 22.5 inches. Scale: 94.34 miles to the inch. Mombassa to the Mediterranean Sea, inland to the Belgian Congo. [nd] ca 1900. London, Edward Stanford.

63. *"The Times" Map of British South Africa, the Transvaal, and Orange Free State.* Cover title: *The Times' War Map of South Africa.* Linen-backed coloured map. Gilt embossed boards. 38 x 28.5 inches. Scale: 40 miles to the inch. Very good detail. Mountains and railways shown. 5 insets. 4 of areas key to the war and one of Wales to give viewer a comparison of South Africa's relative size. Inset of Natal Frontier shown in coloured relief. Table gives distances from ports to places inland and elevation of major towns. [nd] ca 1900. London.

RHODESIA:

BRITISH SOUTH AFRICA COMPANY'S

TERRITORIES.

SHEET No. 1.

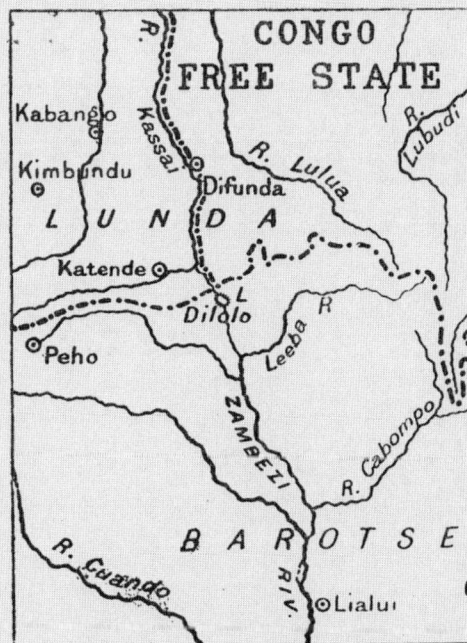

Printed title of Sheet 1 of *Rhodesia: British South Africa Company's Territories.* There were at least 6 editions printed in London between 1895 and 1906 by Edward Stanford.

64. *Natal- Part of Zululand (West) Sheet No. 36. Revised Nov 1901.* Linen-backed map. 19.5 x 24.25 inches. Scale: 2.35 miles to the inch. Very detailed topographical map of an area along the Umvolozi River. 1902. Pretoria, Field Intelligence Dept.

65. *W. & A.K. Johnston's Special Map Of South Africa.* Linen-backed map. 27 x 34 inches. Colour. Self cover with pamphlet in top board. 1902. Edinburgh, W. & A.K. Johnston.

66. *Map Showing the Mombasa-Victoria (Uganda) Railway.* Cover title *Mombasa-Victoria Railway.* Linen-backed map. Board covers and label. 18 x 29 inches. Scale: 25 miles to the inch. All stations along rail route. Good representation of all streams and rivers as well as Lake Nyasa. June, 1903. London, Lithographed at the Intelligence Division of the War Office. Sold by Edward Stanford.

67. *Stanford's Library Map of Africa. New Edition 1904.* Spine title *Map of Africa.* Four sheets. Linen-backed coloured map. Full leather case with fold-over flap. Each sheet with marbled self cover with orig. label. Each sheet 31 x 33.5 inches. Scale: 94.34 miles to the inch. Outstanding map of turn of the century. Large enough to show very good detail. Nov 1,1904. London, Edward Stanford.

68. *Albert Nyanza Sheet 86.* Cover title *Uganda.* From: *Africa 1:1,000,000 Series.* Linen-backed coloured map. Cloth covers and cover label. 22 x 29 inches. Scale: 16 miles to the inch. Area bounded by Lakes Rudolf, Albert Nyanza and Victoria Nyanza. 1905. London, Edward Stanford. Printed for the Topographical Section, at the Ordnance Survey Office.

69. *Kilimanjaro.* Cover title: *Equatorial East Africa.* Linen-backed map. Board covers with label. 21 x 29.5 inches. Scale: 16 miles to the inch. Map shows both Mt. Kenia and Mt. Kilimanjaro. Publ. by the Ordnance Survey Office, Southampton. Part of their *Africa 1: 1,000,000 Series.* Sheets 94 & 95. 1909. London, Edward Stanford.

70. *Northern Nigeria.* Linen-backed coloured map. Board covers. 20 x 29 inches. Scale: 32 miles to the inch. Marbled self covers. Good representation of physical features. Colour shading used to depict elevation. 1909. London, War Office Sold by Sifton Praed & Co. with their orig. label.

71. *A Map of the Peninsula of the Cape of Good Hope and It's Neighbourhood showing the Principal Hard Roads and Homesteads. with corrections/additions from the Government Census Map.* Linen-backed map. Linen covers. 27 x 23 inches. Scale: 4 miles to the inch. Interesting map showing principal roads and railways of the Cape area. Mountains shown in shaded relief. [nd] ca 1910. Cape Town, WB Phillips.

72. *Nigeria.* Linen-backed map with 47 pp. index to all rivers, mountains, villages, provinces and districts. 26 x 33.5 inches. Scale: 32 miles to the inch. Detailed map showing all villages and districts. Geographical Section General Staff War Office 1910. London, Printed for His Majesty's Stationery Office by Harrison and Sons.

73. *The Central & Eastern Provinces of Southern Nigeria 1910. Compiled by officers of Southern Nigeria Regiment and Political Officers in charge of various districts.* Linen-backed, coloured map. Board covers and label. 33.5 x 57.5 inches. Scale: 7.89 inches to the mile. Very large-scale map providing good detail. Administrative districts and Native Tribal areas shown clearly. March 21, 1910. London, Edward Stanford.

74. *Map of Rhodesia under the Administration of the British South Africa Company.* Cover title: *Rhodesia.* Linen-backed coloured map. Cover. Marbled ep. 19 x 17.5 inches. Scale: 80 miles to the inch. VG+. A small map without a lot of topographical detail. Railway and telegraph lines shown clearly. Districts colour coded to stand out clearly. One inset. Feb. 1913. London, George Philip & Son.

75. *Marsabit North A37 & Part of A38. From Africa 1:1,000,000 Series. 25 x 33.5 inches.* Linen-backed coloured topographical map. Marbled self-covers. Northern Frontier District of Kenya. 1915. London, Printed at the War Department Geographical Section of the General Staff.

76. *Map Showing Divisions of Southern Provinces of Nigeria. A. Clennison, Surveyor General, June 1921.* Cover title: *S. Nigeria.* Linen-backed coloured map. Board covers and label. 26.25 x 40 inches. Scale: 15.78 miles to the inch. Rivers, road system and towns but no other physical features. Districts easily identified by colour coding. [nd] ca 1921. London, Edward Stanford.

77. *Philip's Union of South Africa.* Linen-backed coloured map. 21 x 36 inches. Self cover. Shows air mail routes in South Africa. [nd] ca. 1920's. Liverpool, Philip.

78. *The Anglo Egyptian Sudan.* Cover title *Sudan.* Linen-backed coloured map. Case with cover label. 38 x 29.5 inches. Scale: 47.35 miles to the inch. Halfa on the north to Lake Victoria on the south. Good geographical details showing mountains, lakes, villages etc. Printed at the War Office Dec. 1921. London, Edward Stanford.

79. *The Colony and Protectorate of Kenya.* Cover title: *Kenya Colony.* Compiled and drawn by L. Carpenter, Kenya 1924-25. Linen-backed coloured map. Board covers with label. 31.25 x 42.75 inches. Scale: 15 miles to the inch. VG+. The map is intended to be used for economic purposes; a lot of detail has been omitted. Provinces and Districts shown very clearly. 2 insets. [nd] ca 1925. London, Edward Stanford.

80. *Road Map of the Gold Coast-Southern Sheet.* Linen-backed coloured map. Cover. 29.5 x 41.5 inches. Scale: 8 miles to the inch. 1926. Accra.

81. *Ordnance Survey Maps from Africa 1:2,000,000 Series.* Cover title: *Rhodesia Africa-Mozambique* (1929), *Rhodesia* (May 1919), *Transvaal* (July 1922 small corrections Sept 1924), *Upper Congo* (May 1919), *Zanzibar* (July 1927). Linen-backed, coloured maps. All with orig. labels, marbled self covers. Orig. case with orig. label. Each map 22.5 x 30.25 inches. Scale: 31.56 miles to the inch. Compiled at the Geographical Section General Staff, Printed at the War Office. London, Edward Stanford.

82. *Road Map of the Colony and Protectorate of Nigeria.* Linen-backed coloured map (no covers). Shaded relief. 26.75 x 32.5 inches. Scale: 27.62 miles to the inch. Good detail but workmanship poor. [nd] ca 1953. Lagos Survey Dept.

83. *Road Map of Ghana. (In two sheets).* Linen-backed map, coloured in outline. Pictorial cover. Each sheet 32 x 41.5 inches. Scale: 7.89 miles to the inch. A very detailed road map, easy to follow symbols. Sheet for Northern Section- 4th ed. Sept. 1961. Sheet for Southern Section- 5th ed. 1965. Accra Survey of Ghana.

Colophon

We used the Adobe Caslon family of typefaces for most of the
text of this book. Times Roman was used for the index.

The black and white illustrations are, for the most part, from
Dr. Norwich's original photographs. They were digitized by
G.B. Manasek, Inc. with Jared Manasek as technical consultant.

The color illustrations are from film provided by Mrs. Norwich,
with the exception of plates F and G, courtesy
Thomas, Ahngsana and Sainatee Suárez,
and plate E, from G.B. Manasek, Inc.

Proofreading by Alan Berolzheimer.

Dust jacket by Carrie Fradkin of *C Design*, Norwich, VT, USA.

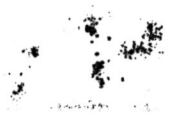